Navy
Basic Military Requirements

Following the Model for Military Education

Navy
Basic Military Requirements

Following the Model for Military Education

Navy Basic Military Requirements: Following the Model for Military Education

PharmaLogika, Inc.
PO Box 461
Willow Springs, NC 27592

www.pharmalogika.com

The user is put on notice that while we have endeavored to reproduce the text provided by the government faithfully, mistakes may exist and that as such the company disclaims any responsibility for anomalies that may be found in the content of the book and that it must not be considered as an official version of the Work.

Other documents that complement these regulations and guidelines are available from the relative branches of government and the documents contained herein may be updated from time to time. It is the responsibility of the user to ensure that the information is still valid and accurate.

Author / Editor: Mindy J. Allport-Settle

Published by PharmaLogika, Inc.

Printed in the United States of America. First Printing.

ISBN 0-9821476-3-5
ISBN-13 978-0-9821476-3-4

Contents

Overview and Orientation

About this Book

The United States Navy Basic Military Requirements have long been the standard used as the pinnacle of training achievement. This manual has been continuously tested and updated to successfully educate every mamber of the United States Navy since its inception. The needs of the instructor, the student, and the Navy are perfectly balanced. This is the model all educators should follow when developing training programs.

This book provides an example of how training materials should be designed for any adult education training program, but especially for any industry that is subject to government regulation.

Included Documents and Features

Basic Military Requirements (NAVEDTRA 14325)

The Role of Education in the Military

Military education and training is a process which intends to establish and improve the capabilities of military personnel in their respective roles.

Military education can be voluntary or compulsory duty. Before any person gets authorization to operate technical equipment or be on the battle field, they must take a medical and often a physical test. If passed, they may begin primary training.

The primary training is recruit training. Recruit training attempts to teach the basic information and training in techniques necessary to be an effective service member.

To achieve this, service members are drilled physically, technically and psychologically. The drill instructor has the task of making the service members fit for military use.

After finishing basic training, many service members undergo advanced training more in line with their chosen or assigned specialties. This range from navy training to studies of explosives. In advanced training, military technology and equipment is often taught.

Many large countries have several military academies, one for each branch of the service, that offer college degrees in a variety of subjects, similar to other colleges. However, academy graduates usually rank as officers, and as such have many options besides civilian work in their major subject. Higher ranking officers also have further educational opportunities.

Personnel Resocialization

While regulated industry does not seek to create soldiers and certainly does not want to employ all of the standard military training techniques, it does need to make certain personnel are trained to follow directions precisely and respect the authority of the regulatory agencies.

Resocialization is an important aspect of inducting a civilian into a military. Resocialization is a sociological concept dealing with the process of mentally and emotionally "re-training" a person so they can operate in an environment other than what they are accustomed to. Successful resocialization into a total institution involves changes to an individual's personality.

Key examples include the process of resocializing new recruits into the military so that they can operate as soldiers – or, in other words, as members of a cohesive unit. Another example is the reverse process, in which those who have become accustomed to such roles return to society after military discharge.

Resocialization from the life of a combat soldier to a civilian member of society is often difficult because of what that soldier saw and did in his military experience. In the transition from civilian to soldier, the individual is trained to solely follow the command of his superiors. In some cases commands would go against certain natural aversions (such as killing) of the individual based on one's moral and ethical principles.

A leading expert in military training methods, Grossman(2001) gives four types of training techniques used; brutalization, classical conditioning, operant conditioning and role modeling.[1] According to Grossman (2001), these techniques were meant to break down barriers to embrace a new set of norms and way of life (brutalization), condition them to pair killing with something more enjoyable and pleasurable (Classical Conditioning), repeat the stimulus-response reaction to develop a reflex (Operant Conditioning), and finally the use of a role model of a superior to provide action by example.

While leaders effectively train their soldiers to accomplish the goal of battle preparedness, these techniques increase psychological trauma experienced in veterans post-combat.[2] It is because of the evident psychological problems in post-combat situations (i.e. Post Traumatic Stress Disorder) that pose a threat to public safety because of the conditioning of the individual who might be made unstable because of his actions. Resocialization following such intense training and conditioning should be further researched and developed to better aide those discharged from the military service.

---

[1] Grossman, D. (2001) Trained to Kill. *Professorenforum-Journal*,2(2).

[2] Kilner, P. (2002, March). Military Leaders' Obligation to Justify Killing in War. *Military Review*, 32(2).

Basic Military Requirements

NAVEDTRA 14325

NONRESIDENT
TRAINING
COURSE

February 2002

Basic Military Requirements

NAVEDTRA 14325

Although the words “he,” “him,” and “his” are used sparingly in this course to enhance communication, they are not intended to be gender driven or to affront or discriminate against anyone.

PREFACE

By enrolling in this self-study course, you have demonstrated a desire to improve yourself and the Navy. Remember, however, this self-study course is only one part of the total Navy training program. Practical experience, schools, selected reading, and your desire to succeed are also necessary to successfully round out a fully meaningful training program.

COURSE OVERVIEW: *Basic Military Requirements*, NAVEDTRA 14325, is a self-study training manual (TRAMAN)/nonresidsent training course (NRTC) that covers the basic knowledges required of the men and women of the U.S. Navy and Naval Reserve. This TRAMAN/NRTC provides subject matter that directly relates to the naval standards for the apprenticeship (E-2/E-3) rates. The naval standards are found in the *Manual of Navy Enlisted Manpower and Personnel Classification and Occupational Standards* (Volume 1), NAVPERS 18068F.

THE COURSE: This self-study course is organized into subject matter areas, each containing learning objectives to help you determine what you should learn along with text and illustrations to help you understand the information. The subject matter reflects day-to-day requirements and experiences of personnel in the rating or skill area. It also reflects guidance provided by Enlisted Community Managers (ECMs) and other senior personnel, technical references, instructions, etc., and either the occupational or naval standards, which are listed in the *Manual of Navy Enlisted Manpower Personnel Classifications and Occupational Standards*, NAVPERS 18068.

THE QUESTIONS: The questions that appear in this course are designed to help you understand the material in the text.

VALUE: In completing this course, you will improve your military and professional knowledge. Importantly, it can also help you study for the Navy-wide advancement in rate examination. If you are studying and discover a reference in the text to another publication for further information, look it up.

2002 Revision
Only Errata Incorporated

Published by
NAVAL EDUCATION AND TRAINING
PROFESSIONAL DEVELOPMENT
AND TECHNOLOGY CENTER

NAVSUP Logistics Tracking Number
0504-LP-101-1377

Sailor's Creed

"I am a United States Sailor.

I will support and defend the Constitution of the United States of America and I will obey the orders of those appointed over me.

I represent the fighting spirit of the Navy and those who have gone before me to defend freedom and democracy around the world.

I proudly serve my country's Navy combat team with honor, courage and commitment.

I am committed to excellence and the fair treatment of all."

CONTENTS

INSTRUCTIONS FOR TAKING THE COURSE

ASSIGNMENTS

The text pages that you are to study are listed at the beginning of each assignment. Study these pages carefully before attempting to answer the questions. Pay close attention to tables and illustrations and read the learning objectives. The learning objectives state what you should be able to do after studying the material. Answering the questions correctly helps you accomplish the objectives.

SELECTING YOUR ANSWERS

Read each question carefully, then select the BEST answer. You may refer freely to the text. The answers must be the result of your own work and decisions. You are prohibited from referring to or copying the answers of others and from giving answers to anyone else taking the course.

SUBMITTING YOUR ASSIGNMENTS

To have your assignments graded, you must be enrolled in the course with the Nonresident Training Course Administration Branch at the Naval Education and Training Professional Development and Technology Center (NETPDTC). Following enrollment, there are two ways of having your assignments graded: (1) use the Internet to submit your assignments as you complete them, or (2) send all the assignments at one time by mail to NETPDTC.

Grading on the Internet: Advantages to Internet grading are:

- you may submit your answers as soon as you complete an assignment, and
- you get your results faster; usually by the next working day (approximately 24 hours).

In addition to receiving grade results for each assignment, you will receive course completion confirmation once you have completed all the assignments. To submit your assignment answers via the Internet, go to:

https://courses.cnet.navy.mil

Grading by Mail: When you submit answer sheets by mail, send all of your assignments at one time. Do NOT submit individual answer sheets for grading. Mail all of your assignments in an envelope, which you either provide yourself or obtain from your nearest Educational Services Officer (ESO). Submit answer sheets to:

COMMANDING OFFICER
NETPDTC N331
6490 SAUFLEY FIELD ROAD
PENSACOLA FL 32559-5000

Answer Sheets: All courses include one "scannable" answer sheet for each assignment. These answer sheets are preprinted with your SSN, name, assignment number, and course number. Explanations for completing the answer sheets are on the answer sheet.

Do not use answer sheet reproductions: Use only the original answer sheets that we provide—reproductions will not work with our scanning equipment and cannot be processed.

Follow the instructions for marking your answers on the answer sheet. Be sure that blocks 1, 2, and 3 are filled in correctly. This information is necessary for your course to be properly processed and for you to receive credit for your work.

COMPLETION TIME

Courses must be completed within 12 months from the date of enrollment. This includes time required to resubmit failed assignments.

PASS/FAIL ASSIGNMENT PROCEDURES

If your overall course score is 3.2 or higher, you will pass the course and will not be required to resubmit assignments. Once your assignments have been graded you will receive course completion confirmation.

If you receive less than a 3.2 on any assignment and your overall course score is below 3.2, you will be given the opportunity to resubmit failed assignments. **You may resubmit failed assignments only once.** Internet students will receive notification when they have failed an assignment--they may then resubmit failed assignments on the web site. Internet students may view and print results for failed assignments from the web site. Students who submit by mail will receive a failing result letter and a new answer sheet for resubmission of each failed assignment.

COMPLETION CONFIRMATION

After successfully completing this course, you will receive a letter of completion.

ERRATA

Errata are used to correct minor errors or delete obsolete information in a course. Errata may also be used to provide instructions to the student. If a course has an errata, it will be included as the first page(s) after the front cover. Errata for all courses can be accessed and viewed/downloaded at:

https://www.advancement.cnet.navy.mil

STUDENT FEEDBACK QUESTIONS

We value your suggestions, questions, and criticisms on our courses. If you would like to communicate with us regarding this course, we encourage you, if possible, to use e-mail. If you write or fax, please use a copy of the Student Comment form that follows this page.

For subject matter questions:

E-mail: n314products@cnet.navy.mil
Phone: Comm: (850) 452-1001, Ext. 1826
DSN: 922-1001, Ext. 1826
FAX: (850) 452-1370
(Do not fax answer sheets.)
Address: COMMANDING OFFICER
NETPDTC N314
6490 SAUFLEY FIELD ROAD
PENSACOLA FL 32509-5237

For enrollment, shipping, grading, or completion letter questions

E-mail: fleetservices@cnet.navy.mil
Phone: Toll Free: 877-264-8583
Comm: (850) 452-1511/1181/1859
DSN: 922-1511/1181/1859
FAX: (850) 452-1370
(Do not fax answer sheets.)
Address: COMMANDING OFFICER
NETPDTC N331
6490 SAUFLEY FIELD ROAD
PENSACOLA FL 32559-5000

NAVAL RESERVE RETIREMENT CREDIT

If you are a member of the Naval Reserve, you may earn retirement points for successfully completing this course, if authorized under current directives governing retirement of Naval Reserve personnel. For Naval Reserve retirement, this course is evaluated at 15 points. Unit 1 – 12 points upon satisfactory completion of assignments 1 through 12. Unit 2 – 3 points upon satisfactory completion of assignments 13 through 15. (Refer to *Administrative Procedures for Naval Reservists on Inactive Duty,* BUPERSINST 1001.39, for more information about retirement points.)

Student Comments

Course Title: *Basic Military Requirements*

NAVEDTRA: 14325 **Date:** ____________________

We need some information about you:

Rate/Rank and Name: ____________________ SSN: ____________ Command/Unit ____________________

Street Address: ____________________ City: ____________ State/FPO: ________ Zip ________

Your comments, suggestions, etc.:

Privacy Act Statement: Under authority of Title 5, USC 301, information regarding your military status is requested in processing your comments and in preparing a reply. This information will not be divulged without written authorization to anyone other than those within DOD for official use in determining performance.

NETPDTC 1550/41 (Rev 4-0)

CHAPTER 1

POLICIES AND PROGRAMS

When you cease to make a contribution, you begin to die.

—Eleanor Roosevelt

What is a *policy* or a *program*? A policy is an overall plan that contains general goals and broad guidelines. Policy usually establishes the *end* to be attained, not the *means* of getting there. A program is a plan or system under which action may be taken towards a goal. Programs provide the *means* to reach the *end* (goal) stated by the policy. In other words, policies tell where to go; programs tell how to get there. This chapter covers some of the policies and programs of the U. S. Navy.

ENVIRONMENTAL POLLUTION CONTROL

Learning Objectives: When you finish this chapter, you will be able to—

- Identify the need for environmental pollution control to include the sources and effects (natural, historical, archeological, physical, and biological) of pollution.
- Recognize the Navy programs for pollution control to include the Clean Air Act, Clean Water Act, and Noise Prevention Ashore.

The Navy's ability to accomplish its mission requires daily operations in land, sea, and air environments. The Navy is committed to operating ships and shore facilities in a manner compatible with the environment. National defense and environmental protection are and must be compatible goals. Therefore, an important part of the Navy's mission is to prevent pollution, protect the environment, and conserve natural, historic, and cultural resources. To accomplish this mission element, personnel must be aware of the environmental and natural resources laws and regulations that have been established by federal, state, and local governments. The Navy chain of command must provide leadership and a personal commitment to ensure that all Navy personnel develop and exhibit an environmental protection ethic.

Environmental pollution is the altering of the natural environment in an adverse way. Pollution can result from the presence of chemical, physical, or biological agents in the air, water, or soil. Some of the worst effects of pollution are economic loss (agricultural and industrial), fewer recreational opportunities, and the marring of the earth's natural beauty. Pollutants affect human health and comfort, fish and wildlife, plant life, water resources, physical structures, and equipment. In other words, environmental pollution results from any substance added to our water, air, or land that makes it less fit for use by plants, animals, or human beings.

SOURCES OF POLLUTION

Nature contributes to pollution by eroding the soil causing silt to build up in streams and by volcanic eruptions that pollute the atmosphere. However, **people** cause most pollution problems in the world. The main sources of pollutants are agricultural, industrial, municipal, and transportation operations.

Agricultural pollutants include insecticides, herbicides, pesticides, natural and chemical fertilizers, drainage from animal feedlots, salts from field irrigation, and silts from uncontrolled soil erosion.

Industrial operations produce a wide variety of pollutants. Industrial pollutants include acids from mines and factories, thermal discharges from power plants, and radioactive wastes from mining and processing certain ores. Industries create pollutants by producing food, chemicals, metals, petroleum products, and poisons, as well as countless other by-products of our country's technology.

The primary **municipal** pollutants are raw or inadequately treated sewage. Other municipal pollutants include refuse, storm-water overflows, and salts used on streets in wintertime.

Transportation pollutants include emissions from aircraft, trains, waterborne vessels, and cars and trucks.

Motor vehicles create most of our air pollutants through their release of unburned fuel vapors (hydrocarbons). Oil becomes a pollutant when ships spill it during refueling operations or as a result of collision or grounding. Several million gallons of oil may ruin and pollute miles of coastline as a result of such accidents.

EFFECTS OF POLLUTION

Pollution produces physical and biological effects that vary from mildly irritating to lethal. The more serious of the two are the biological effects.

Physical Effects

The physical effects of pollution are those that we can see, but they include effects other than actual physical damage.

Oil spills. One obvious physical effect of pollution is the result of oil spills that are caused by ship collisions or other accidents. Oil spilled into the seas coats everything it touches. It fouls boat hulls, pier pilings, and shore structures; spoils the beauty of nature by killing fish and birds; and makes beaches unusable. In addition to the physical effects, oil spills require costly cleanup operations.

Air pollutants. Air pollutants damage a wide variety of materials. Burning oil and coal produce sulfur oxides, which cause steel to erode two to four times faster than normal. When combined with other pollutants (soot, smoke, lead, asbestos, and so on), sulfur oxide particulates cause corrosion to occur at an even faster rate. By themselves, particulates damage and soil materials, structures, and equipment. Air pollutants speed the erosion of statues and buildings, which in some instances, destroys works of art.

Biological Effects

The most serious result of pollution is its harmful biological effects on human health and on the food chain of animals, birds, and marine life. Pollution can destroy vegetation that provides food and shelter. It can seriously disrupt the balance of nature, and, in extreme cases, can cause the death of humans.

Pesticides, which include herbicides and insecticides, can damage crops; kill vegetation; and poison birds, animals, and fish. Most pesticides are nonselective; they kill or damage life forms other than those intended. For example, pesticides used in an effort to control or destroy undesirable vegetation and insects often destroy birds and small animals. Some life forms develop immunity to pesticides used to destroy them. When that happens, we develop more potent chemicals and the cycle repeats itself.

The widespread use of pollutants, such as oil, chemicals, and fertilizers, pollutes our waterways. The biological effect of water pollution is its danger to our water supplies; we require water to survive. Water pollutants are also dangerous to all forms of marine life. Oil is an especially harmful pollutant. It kills surface-swimming animals and sea birds and, once it settles on the bottom, harms shellfish and other types of marine life.

The primary pollution concern of Navy personnel involves pollution produced by shipboard wastes. In addition to oil, shipboard wastes include sanitary wastes, galley and laundry waters, chemicals, solid wastes, and combustion by-products of oil- and gasoline-driven engines. Pollutants produced by ships are similar to those generated by municipal and industrial operations.

NAVY PROGRAMS FOR POLLUTION CONTROL

Based on an Executive Order, all government agencies must provide leadership in the protection and enhancement of the quality of our air and water resources. They also must comply with all environmental laws and regulations. Accordingly, the Secretary of the Navy, the Chief of Naval Operations, and other authorities have issued several pollution control instructions. Those instructions cover the abatement (lowering) of air, water, and noise pollution. In addition, we have a program to preserve our natural, cultural, and historic resources.

Clean Air Act

Under the Clean Air Act, each state has the primary responsibility for assuring air quality. All naval activities must meet both federal and state standards for preventing air pollution.

The Navy has begun taking steps to help meet the nation's goal of reducing air pollution. One step has

Student Notes:

been the conversion of its power plants ashore and afloat to use more efficient pollution control systems for stack emissions.

When required by state or local regulations, activities have set up a program for monitoring and analyzing the exhaust from each vehicle. Vehicles that do not meet emission standards must have corrective maintenance before being returned to service.

Clean Water Act

The Clean Water Act, as amended, represents a clear goal to restore and maintain the chemical, physical, and biological integrity of the nation's waters.

To help reach that goal, the Navy uses municipal treatment plants where possible for its wastewater and sewage. (**NOTE:** Operators of municipal treatment plants must meet the operator-certification requirements of the state in which the Navy facility is located.) Whenever use of a municipal facility is impractical, the Navy provides, installs, and operates its own wastewater treatment and disposal systems. (**NOTE:** The Navy complies with water pollution control standards that govern the types of pollutants that can be introduced into a treatment facility.)

The Navy has installed marine sanitation devices (MSDs) aboard most of its ships. Some of those devices treat sewage to a level acceptable for overboard discharge. Others (collection, holding and transfer [CHT] systems) retain sewage on board for later discharge ashore or in waters in which discharge is allowed. No untreated or inadequately treated sewage may be discharged into navigable waters of the United States. In foreign waters, Navy vessels comply with the applicable Status of Forces Agreement (SOFA) in operating MSDs. If no SOFA exists, vessels operate MSDs based on the sewage-discharge standards observed by the host country's military forces.

Some of the guidelines followed by the Navy are shown below:

- Navy vessels may not discharge hazardous substances in harmful quantities into or upon navigable waters of the United States.
- They also may not discharge those substances upon adjoining shorelines or into or upon waters of the contiguous zone (12 nautical miles [nm] from shore).
- Since Navy vessels use many substances they cannot dispose of at sea outside the contiguous zone, they must store them in containers for shore disposal.
- Vessels may not discharge unpulped trash at sea within 25 nautical miles (nm) from the U. S. coastline and pulped trash within 12 nm of the U.S. coastline.
- They must make every effort to package all trash for negative buoyancy before overboard discharge.
- Submarines may discharge negatively buoyant compacted trash not less than 12 nm from the U.S. coastline only if the water depth is greater than 1,000 fathoms.
- Vessels may not discharge any trash within 25 nm of any foreign coastline.
- Overboard discharge of plastic waste material is prohibited.

Noise Prevention Ashore

The Noise Prevention Program directs federal facilities, including naval shore stations, to comply with all substantive or procedural requirements that apply to environmental noise reduction. Therefore, whenever feasible, the Navy procures (buys) low-noise emission products and provides soundproofing of Navy-owned/operated schools and hospitals affected by noisy operations. It also locates housing and other developments away from major noise sources and cooperates with and in support of neighborhood self-help programs. Aviation facilities consider remote siting, sound suppression equipment, and sound barriers when developing new systems. To the extent possible, the Navy limits the use of noisy tools, machinery, and equipment to normal working hours.

Afloat, the design of new ship systems and equipment is reducing noise emissions. The

Student Notes:

government doesn't prescribe retrofit (modifications) for existing noise sources. The government also exempts military aircraft, combat equipment, and weapon systems from new noise design standards.

NATURAL, HISTORIC, AND ARCHEOLOGICAL RESOURCES PROTECTION

In keeping with federal programs, the Navy has established programs, suitable to its military mission, for the preservation of natural, cultural, and historic resources. The Navy programs consist of land management, forest management, fish and wildlife management, outdoor recreation, and general support for all land under naval jurisdiction.

Land management includes soil and water conservation, land restoration, noxious weed and poisonous plant control, agricultural lands leasing, range management, landscaping, and ground maintenance.

Forest management includes the production and sale of forest products for multiple-use and sustained-yield principles

Fish and wildlife management includes marine mammal protection, migratory fish protection, game and nongame species management, and animal damage control.

Outdoor recreation involves the protection of nature through programs such as migratory bird management, endangered species protection, and preservation of the earth's natural beauty. Recreation programs include the management and use of off-road vehicles, national recreation trails, hiking trails, outdoor recreation areas, wild and scenic rivers, and wilderness areas.

General support programs involve people and nature. Those involving people include youth programs, public participation, and cooperative agreements with public agencies. Those involving nature include resources inventory, conservation awards, coastal area management, wetland protection, flood plain management, natural area preservation, and animal disease eradication.

REVIEW 1 QUESTIONS

Q1. List the four main sources of pollutants.

a.

b.

c.

d.

Q2. What are two primary effects of pollution?

Q3. What is the primary pollution concern of the Navy?

Q4. To help reduce air pollution, what steps has the Navy taken to improve their power plants ashore and afloat?

Q5. Name the agreement that the Navy follows to dispose of treated sewage in foreign waters.

Q6. How many nautical miles from the U.S. coastline can vessels discharge unpulped trash?

ENERGY CONSERVATION PROGRAM

Learning Objective: When you finish this chapter, you will be able to—

- Recognize the policy for and identify the need for the Navy's energy conservation program.

Most of the Navy's ships and all of its aircraft use petroleum-based fuel. In addition, the Navy's vehicles

Student Notes:

used for ground transportation, such as cars, trucks, and buses, also use petroleum. That gives you an idea of how much the Navy depends on petroleum—without it, the Navy couldn't move. Navy bases also use petroleum fuel for heating and running electric plants.

We get most of our petroleum in the form of crude oil from the Middle East countries. We cannot rely on oil from that area forever because political unrest exists there. Besides, only so much oil exists in the earth. Therefore, we must conserve as much of our petroleum resources as possible.

The Navy's policy on energy conservation is that it will make all possible efforts to improve the way it uses energy resources. The Navy will not compromise readiness, effectiveness, or safety in its energy conservation efforts. One of the Navy's energy conservation goals is to ensure it provides the fleet with enough fuel to sustain peacetime and combat operations. Another objective is to reduce energy costs and dependency on unreliable energy sources while conserving petroleum.

You can help to conserve our energy resources by reporting wasteful practices to your supervisor or leading petty officer (LPO).

REVIEW 2 QUESTION

Q1. What is the Navy's policy on energy conservation?

NAVY SPONSOR PROGRAM

Learning Objective: When you finish this chapter, you will be able to—

- Recognize the purpose of and identify the responsibilities of the Navy Sponsor Program.

The Chief of Naval Operations set up the Navy Sponsor Program to ease the relocation of naval personnel and their families when transferred on permanent change of station (PCS) orders. This program has helped thousands of Navy families. Navy policy regarding assignment of sponsors is as follows:

- Assignment of a sponsor is mandatory in the case of PCS orders to or from any activity.
- In all other cases, individuals should request assignment of a sponsor through the use of the sponsor request form, NAVPERS 1330/2, entitled Navy Sponsor Notification.

NOTE

If you are being transferred on PCS orders, desire a sponsor, and have not heard from your new command, you can request assignment of a sponsor using the sponsor request form, NAVPERS 1330/2.

- School commands have developed home port and/or country information packages or packets to all personnel being transferred on PCS orders, especially first-term members.
- If you are assigned as a sponsor for a fellow naval member who is scheduled to transfer to your location, fully accept the responsibility of sponsorship. You can help make the difference between a good move and a bad one for the person being transferred as well as for that person's family. Using the following checklist (table 1-1) may help you be a better sponsor.

OVERSEAS DUTY SUPPORT PROGRAM

Learning Objective: When you finish this chapter, you will be able to—

- Identify the Overseas Duty Support Program to include sources of information available for single Sailors and Navy families.

The Overseas Duty Support Program (ODSP) provides information and support to help personnel who are guests in foreign lands. Whether you find yourself in a foreign country as a result of a PCS or a deployment, the ODSP will provide you with information about the country you are visiting.

Student Notes:

Table 1-1.—Sponsor Checklist

SPONSOR CHECKLIST
1. Write a "welcome aboard" letter to your new shipmate. Include the following: a. Introduce yourself and give a warm welcome aboard. b. Include a copy of the rental and for sale ads from the local paper in any correspondence with the new member. c. Provide the member information on how he/she can contact you (your home address and telephone number). d. Ask him/her family members (if they will accompany the member, their mode of transportation, ages, and so on).
2. Ask him/her to keep you posted on travel and arrival plans. You are required to keep the sponsor program coordinator and pay/personnel administrative support system (PASS) officer informed of any changes the member may make.
3. Ensure follow-up letter or phone calls to answer any questions the new member may have.
4. Ensure transportation is available from place of arrival to the command if the member requests it.
5. Before the member's arrival, check on the housing availability. Inform the member if housing will be available upon reporting or if he/she may need to make arrangements for temporary lodging. Help the member with arrangements if necessary. (Make sure the new member checks in with the housing referral office before renting or buying a house. That is a must.)
6. Escort the individual through the check-in process.
7. Help the member locate the personal property office to check on household goods and private auto shipments. That may be done before the member reports in for duty.
8. Provide a tour of the base and local area if the member desires.

When visiting a foreign country, always remember you are a guest in that country. As a guest, you should respect the local customs and make every attempt to avoid getting into trouble. If possible, try to learn and use the local language, especially if you are stationed in that country. Be careful about taking pictures or recording anything without the permission of the people around you. Most police (or even passersby) will forbid you to take pictures of sights, such as government buildings, no matter how illogical that may seem. Some people will not want you to take their pictures for religious reasons. Some may feel you are making fun of them. They may think you will show the pictures to your friends back home as an example of how "backward" or "primitive" they are.

Dress comfortably, but be careful of your appearance. A woman in shorts is a "no-no" in many countries, and even a woman in slacks can upset people in some rural or conservative areas. A man in shorts can give the impression of insensitivity.

Unless you are similar in appearance to the people in the country you are visiting, they will likely stare, shout, giggle, and point at you, especially in out-of-the-way places. The people of most countries will usually shower attention upon you good naturedly. If you smile and accept the attention in the same way, your hosts will make you feel welcome. If you resent it and get angry, your hosts will be confused and displeased.

Student Notes:

Even though you look different than the local people, if you can manage a few words in the local language, you will reap many benefits. You will see taxi fares miraculously drop, room service drastically improve, and art objects sell for less. The people will receive you more warmly and genuinely than if you had spoken English.

Knowing some basics about the country you are visiting and its culture goes a long way towards helping you have a good time. Before the trip, read all you can about each country you will visit. Look at various publications that contain information about the countries or regions you will be touring. *Africa Report*, for example, is a magazine that offers much insight about the culture of that continent. If your library doesn't have a specific magazine, you can order it from the publisher. You will find the addresses of magazine publishers in your local library.

You may also get information on other countries by visiting the Family Service Center or logging on to the SITES home page on the Internet. The SITES home page gives you up-to-date information of the country you will be visiting.The address for SITES is www.dmdc.osd/mil/sites.

When you visit the towns, villages, and cities of other countries, you'll discover what so many of us have found, the unexpected hospitality and warmth of the people.

MILITARY CASH AWARDS PROGRAM

Learning Objective: When you finish this chapter, you will be able to—

- Recognize the opportunities gained through the Navy Cash Awards Program.

The Military Cash Awards Program (MILCAP) is a special incentive awards program. It is designed to find new ideas to effectively increase performance within the Department of the Navy. The program has been responsible for important savings.

MILCAP provides monetary recognition awards of up to $25,000. It awards personnel for beneficial suggestions, inventions, and scientific achievements that increase efficiency, economy, or productivity or effect other improvements in operations. All active-duty military personnel are eligible to participate in MILCAP.

A beneficial suggestion is the proposal of an idea or a method of doing a task better, faster, cheaper, or safer. An individual or a group can submit a beneficial suggestion. To qualify for the MILCAP, the suggester must show a specific need for improvement and give a workable solution. The suggestion should also be beyond the suggester's normal job capability. Suggestions should do one or more of the following:

- Improve services to the fleet
- Increase productivity
- Conserve energy, manpower, materials, time, or space
- Reduce costs without loss of quality or efficiency

Perhaps you have an idea worthy of a cash award. Submit your suggestion in writing, either on a suggestion form or in a letter format, to your local MILCAP administrator. OPNAVINST 1650.8 contains additional information about the MILCAP.

REVIEW 3 QUESTIONS

Q1. Delete

Q2. Delete

Q3. What program provides you with support and information about foreign lands?

Student Notes:

Q4. Besides ODSP, what other source can you use to get information about other countries?

Q5. What program does the Navy have that rewards you monetarily for beneficial suggestions, inventions, and scientific achievements?

HEALTH AND PHYSICAL READINESS PROGRAM

Learning Objective: When you finish this chapter, you will be able to—

- Recognize the purpose of the Health and Physical Readiness Program.
- Identify the benefits of the Health and Physical Readiness Program to the individual.

People in the Navy and in the civilian community share a common problem—excessive body fat. It usually results from sitting all day at a desk job, eating too much, and getting too little exercise. Excessive body fat spoils our health, longevity, stamina, and military appearance. Maintaining good health and physical readiness helps to keep us combat ready, make us personally effective, and give us high morale.

The Navy's Health and Physical Readiness Program promotes active health and fitness at the command level. The program includes semiannual testing of all personnel to make sure they meet to certain standards. It provides educational programs that help personnel who don't meet the Navy's fitness or body fat standards. It also helps personnel who want to change long-established bad health habits to improve their fitness.

NAVY POLICY ON PREGNANCY AND DEPENDENT CARE

Learning Objectives: When you finish this chapter, you will be able to—

- Identify the servicewoman's responsibilities regarding pregnancy.
- Identify the Navy's responsibility for obstetrical care.
- Recall the Navy policy for assignment of pregnant servicewomen.
- Recall the Navy policy for post delivery convalescent leave.
- Recall the Navy policy for separating pregnant servicewomen.
- Recall the Navy family care policy.
- Recognize the consequences of failing to maintain an up-to-date family care plan.

This section covers your and the Navy's responsibilities on pregnancy and family care.For information about hygiene, you should refer to chapter 14 of this TRAMAN.

PREGNANCY

Getting pregnant can negatively impact your goals. It can also affect the mission of the Navy. By knowing the Navy's policy on pregnancy, you can make informed, personal choices **and** meet your commitment to the Navy.

The Sailor's Responsibilities

Servicewomen are responsible for—

- Planning the pregnancy to meet family **and** military obligations
- Confirming the pregnancy at a military medical treatment facility
- Notifying the commanding officer or officer in charge of the pregnancy
- Performing military duties while pregnant
- Complying with work- and task-related safety and health recommendations

Student Notes:

The Navy's Responsibilities

When a servicewoman is pregnant, the Navy has specific responsibilities for her care. Navy policy governs the assignment of pregnant servicewomen.

RESPONSIBILITY.—The Navy provides obstetrical care. When a pregnant servicewoman remains at her duty station, the military treatment facility provides care if—

- The facility has obstetrical-gynecological capabilities.
- The servicewoman lives in the facility's in-patient area.

POLICY.—The Navy policy for assigning pregnant servicewomen covers several areas, including overseas, CONUS, shipboard, aviation squadron, and military school assignment.

Overseas.—Based on medical considerations, no servicewoman may be assigned overseas or travel overseas after the beginning of the 28th week of pregnancy.

Continental United States (CONUS).—Pregnant servicewomen may be assigned within CONUS as follows:

- Without restriction, provided the servicewoman doesn't fly after the 28th week of pregnancy.
- Will not be transferred to a deploying unit during the period from the 20th week of pregnancy through 4 months after the expected date of delivery.

Shipboard.—The following policies govern the assignment of pregnant servicewomen:

- The CO, in consultation (talking to) with the health care provider and occupational health professional, decides if the Sailor may safely continue her shipboard assigned duties.
- Pregnant servicewomen won't remain aboard ship if the time for medical evacuation to a treatment facility is more than 6 hours.
- Service women can't remain on board a deployed unit beyond the 20th week of pregnancy.

Aviation Squadron.—Assignment of pregnant servicewomen to aviation squadrons is governed by the following:

- Pregnancy disqualifies designated flight status.
- Air controllers may work up to the 28th week of pregnancy; however, they are normally restricted from tower duties after their 27th week.
- Servicewomen who become pregnant while assigned to an aviation squadron due for deployment should be reassigned to a squadron not scheduled for deployment from their 20th week of pregnancy through the recuperative period.

Military schools.—The following rules apply to pregnant servicewomen at military schools:

- A pregnant servicewoman isn't assigned to a school if the projected delivery date or recuperative period will occur during the course of instruction.
- If a servicewoman becomes pregnant during training, the school's CO determines if the Sailor can complete the training based on the projected delivery and recuperation dates.

POST-DELIVERY CONVALESCENT LEAVE.—Normally, the CO grants 6 weeks (42 days) convalescent leave after the servicewoman has delivered the baby.

SEPARATION FROM THE NAVY.—The Navy can discharge servicewomen from the Navy without maternity benefits under the following conditions:

- The servicewoman was pregnant before entry into recruit training.
- The pregnancy is certified during recruit training.

Student Notes:

- The pregnancy occurred during initial training (must be determined by the permanent duty station).

A pregnant servicewoman may request separation from the Navy before the 20th week of pregnancy (normally, such a request is not approved). Under law, the military departments, CHAMPUS, or the Veteran's Administration has the authority to pay civilian maternity care expenses for former servicewomen who separate from the Navy while pregnant.

FAMILY CARE

All single service members and dual military couples having custody of children under 19 or other dependents must have a formalized plan for family care. If you are eligible for family care, you must complete the Family Care Plan Certificate, NAVPERS 1740/6 and Family Care Plan Arrangements, NAVPERS 1740/7. These forms certify that your family members will be cared for during your absence. These forms also identify the logistical, relocation, and financial arrangements that you've made.

Custodian(s) you designate (name) **must** have the following documents:

1. A power of attorney that authorizes medical care and person(s) action in *loco parentis* (as the parent)
2. Identification cards for all eligible dependents

If the person you designate as the custodian doesn't live in the local area, you will also need to make sure of the following:

- A nonmilitary escort for family members that need help; for example, infants, children, or elderly disabled adults
- Financial support to transport the family or caregiver to a designated location

If you don't maintain an up-to-date family plan, you can be separated from the Navy. In fact, the CO can separate members who are unable or refuse to maintain an updated family care plan, who do not remain available for worldwide assignment, or who are unable to perform their professional or military duties.

PUBLIC AFFAIRS AND COMMUNITY RELATIONS PROGRAM

Learning Objective: When you finish this chapter, you will be able to—

- Recall the value of the public affairs and community relations programs.

Public affairs works on the principle that the public has the right to be fully informed about matters of national defense. In the Department of the Navy, the **mission of public affairs** is to inform the public and members of the naval service about the following:

- The Navy as an instrument of national policy and security
- Navy operations and programs
- The responsibilities and activities of naval personnel as U. S. citizens

An **objective of public affairs** is to better the general public's understanding of the following:

- The nature of sea power and its role in preserving the security of the United States
- The reasons underlying the need for an efficient and effective modern Navy
- The contributions of the Navy in scientific research and in community assistance
- The service naval members provide to their country
- The career advantages of naval service

The Navy is a part of the community in which its facilities or personnel are located. The attitude the civilian community has towards Navy personnel affects their morale and effectiveness. Therefore, all Navy personnel are responsible for maintaining good community relations. They can help to do that by taking an active part in civilian activities and organizations. In addition, each command develops a Community Relations Program to ensure Navy personnel and the civilian community live in harmony.

For more information about the Public Affairs and Community Relations Program, see SECNAVINST 5720.44, chapter 2.

Student Notes:

REVIEW 4 QUESTIONS

Q1. What Navy program provides educational programs for people who don't meet the Navy's fitness or body fat standards and who also need to help change long-established bad health habits?

Q2. With regard to pregnant servicewomen, what is the Navy responsibility?

Q3. List the responsibilities of pregnant servicewomen.

a.

b.

c.

d.

e.

Q4. Service members must complete what forms for family care?

a.

b.

Q5. List the areas that the Department of the Navy public affairs office informs the public and service members about.

a.

b.

c.

INTEGRITY AND EFFICIENCY PROGRAM

Learning Objective: When you finish this chapter, you will be able to—

- Recall key provisions of the Navy's Integrity and Efficiency Program to include fraud, waste, and abuse.

The Integrity and Efficiency (I & E) Program carries out the Department of the Navy's policy to detect, deter, and eliminate *fraud*, *waste*, *abuse*, and *mismanagement*. The terms *fraud*, *waste*, *abuse*, and *mismanagement* are defined as follows:

Fraud. Fraud is intentional misleading or deceitful conduct that deprives the government of its resources or rights.

Waste. Waste is the extravagant, careless, or needless expenditure of government resources.

Abuse. Abuse is the intentional wrongful or improper use of government resources.

Mismanagement. Mismanagement is to manage incompetently or dishonestly.

You report fraud, waste, abuse, and mismanagement through any of the following means:

- Chain of command
- Navy hotline
- Naval Criminal Investigative Service (NCIS)
- Congressional communication (writing your congressmen)

STANDARDS OF CONDUCT AND PROFESSIONAL ETHICS

Learning Objectives: When you finish this chapter, you will be able to—

- Recall key points of the Standards of Conduct.
- Identify the need for professional ethics.

The Department of the Navy's ability to maintain public confidence in its integrity is essential to the performance of its mission. To help maintain that integrity, all naval personnel must comply with the following standards of conduct (table 1-2).

Student Notes:

Table 1-2. Standards of Conduct

1. Avoid any action, whether or not specifically prohibited, that might result in or reasonably be expected to create the appearance of the following:
    a. Using public office for private gain
    b. Giving preferential treatment to any person or entity
    c. Impeding government efficiency or economy
    d. Losing complete independence or impartiality
    e. Making government decisions outside official channels
    f. Adversely affecting the confidence of the public in the integrity of the government
2. Do not engage in any activity or acquire or retain any financial interest that results in a conflict between your private interest and the public interest of the United States related to your duties.
3. Do not engage in any activity that might result in or reasonably be expected to create the appearance of a conflict of interest.
4. Do not accept gratuities (gifts) from defense contractors.
5. Do not use your official position to influence any person to provide any private benefit.
6. Do not use your rank, title, or position for commercial purposes.
7. Avoid outside employment or activity that is incompatible with your duties or may bring discredit to the Navy.
8. Never take or use government property or services for other than officially approved purposes.
9. Do not give gifts to your superiors or accept them from your subordinates.
10. Do not conduct official business with persons whose participation in the transaction would be in violation of law.
11. Seek ways to promote efficiency and economy in government operation and public confidence in its integrity.
12. For more information consult SECNAVINST 5370.2.

THE ROLE OF THE INSPECTOR GENERAL

Learning Objective: When you finish this chapter, you will be able to—

- Identify the purpose of the inspector general.

The mission of the naval inspector general (IG) is to "inquire into and report" on any matter that affects the discipline or military efficiency of the DoN. One way the IG can fulfill its mission is by providing a method of receiving and investigating reports of fraud, waste, mismanagement, and related improprieties

Student Notes:

(wrong actions). Generally, the hotline is used when the chain of command cannot or will not take appropriate action.

ALCOHOL AND DRUG POLICIES

Learning Objective: When you finish this chapter, you will be able to—

- Identify the consequences of alcohol and drug abuse.

Our Navy is the most professional, highly trained, and capable force in the world and our Sailors function in a highly complex, technological environment requiring 100 percent of their mental and physical abilities. Alcohol and drug abuse and the incidents it causes impair our readiness and reduce the quality of life of our Navy team.

All DoN bases/installations will strictly conform to the drinking age limitations of the state or country in which they are located, and under no circumstances will drinking be permitted below the age of 18. The perception that alcohol is central to our tradition is wrong. Everyone from the new recruit to admiral must recognize the effect alcohol abuse can have on them, on others (including their families) and on their careers.

Each Sailor is ultimately responsible and will be held accountable for their own actions. Personal responsibility means no drinking and driving, no drinking to the extent that it impairs judgment (resulting in irresponsible behavior or incidents), no public drunkenness, and absolute compliance with the local laws for purchase, possession, and use of alcoholic beverages.

We have a personal responsibility not to abuse alcohol. We also have a responsibility to our shipmates. Shipmates take care of shipmates. All of us must be aware of the warning signs of alcohol abuse and must take positive steps to ensure shipmates stay on the right course. Don't let a shipmate drive after drinking. Shipmates who are drunk need to be watched and not left alone.

Drug use is incompatible with the Navy's high standards of performance, military discipline, and readiness. The Navy has a "zero tolerance" policy, which means using illegal drugs or abusing prescription drugs will result in being discharged from the Navy.

PROVISIONS OF THE PRIVACY ACT

Learning Objective: When you finish this chapter, you will be able to—

- Recall key provisions of the Privacy Act and recognize its purpose.

The Privacy Act primarily protects the rights of personal privacy of people about whom records are maintained by agencies of the federal government. In other words, the Privacy Act protects your privacy about your records the federal government maintains. It isn't legal for an agency of the federal government to maintain records on people without announcing the fact in the Federal Register.

If your duty requires you to keep personal information about others, keep only that information that is needed to do what law requires. Remember, keep personal information private! Don't disclose information about a person to any unauthorized person. If you make an unauthorized disclosure, you may be fined up to $5,000. If you maintain records on your fellow Navy member, you have an obligation to protect this information from unauthorized disclosure.

Also, it's your right to look at any record the Department of the Navy keeps on you. You have the right to copy it and to request to have it corrected if you think it is wrong.

REVIEW 5 QUESTIONS

Q1. What is the Navy's policy on drug abuse?

Q2. What is the purpose of the Integrity and Efficiency Program?

Student Notes:

Q3. List the ways to report fraud, waste, and abuse.

a.

b.

c.

d.

Q4. For the Navy to maintain public confidence in its integrity, naval personnel should comply with ________________________________.

EQUAL OPPORTUNITY IN THE NAVY

Learning Objectives: When you finish this chapter, you will be able to—

- Identify the policies of the Command Managed Equal Opportunity (CMEO) Program.
- Recognize the purpose of equal opportunity and human rights.
- Recall the need for equal opportunity in the following areas: performance evaluations, duty assignments, training and advancement, justice, service and recreational facilities, housing, and equal opportunity off base.
- Recognize the consequences of discrimination.

According to *Navy Regulations*, Article 1164, "Equal opportunity shall be afforded to all on the basis of individual effort, performance, conduct, diligence, potential, capabilities, and talents without discrimination as to race, color, religion, creed, sex or national origin. Naval personnel shall demonstrate a strong personal commitment to stand on these principles and carry them out."

The **key word** in the title of this section is **equal**. Real democracy cannot exist if our society doesn't have equal opportunities for all of its people. Equality can't be legislated; however, the law can serve to make sure that everyone receives equal treatment.

The Navy is made up of people, and people in the Navy come from all parts of the United States and from other countries. They bring with them views common to their racial, social, religious, and economic backgrounds. For many years, the official policy of the Department of the Navy has been one of equal treatment and opportunity for all personnel, regardless of race, color, creed, sex, or national origin. The Navy carries out that policy through the Command Managed Equal Opportunity (CMEO) Program. The Chief of Naval Operations (CNO) provides the guidance and policy for that program.

COMMAND MANAGED EQUAL OPPORTUNITY (CMEO) PROGRAM

The Command Managed Equal Opportunity (CMEO) Program is the Navy's equal opportunity program. The Navy requires all commands to set up a CMEO program. CMEO allows the commanding officer (CO) to create and maintain a positive equal opportunity (EO) environment. CMEO also allows commands to identify and resolve command-level equal opportunity problems and concerns. Commands can closely monitor EO issues, and at the same time, remain flexible enough to use only the manpower required for any given situation.

Commanding officers must make equal opportunity a reality in their commands. They must ensure only merit, ability, performance, and potential affect a person's promotion, training, duty assignment, or other personnel action.

PERFORMANCE EVALUATION IN EQUAL OPPORTUNITY

The Navy evaluates every Navy member's support of the equal opportunity program. Personnel in paygrades E-1 through E-9 receive numeric grades, based on a number scale, on overall annual performance evaluations of specific traits. Your performance evaluation will reflect your attitude toward and your conduct in support of the Navy's equal opportunity program.

Student Notes:

PROCEDURES CONCERNING INSENSITIVE PRACTICES

When people enter the Navy, they have their own feelings, attitudes, prejudices, and ideas based on their individual personal backgrounds. An insensitive practice is behavior that is prejudicial to another person because of that person's race, religion, creed, color, sex, or national origin. To ensure teamwork and to fulfill the Navy's mission, individuals **must** put aside their personal feelings, attitudes, prejudices, and ideas about other people and how they act around others.

Commanding officers take proper action to correct insensitive practices. If a person takes part in insensitive practices, that person receives counseling about their responsibilities with regard to equal treatment. If such counseling is not effective or if further action is warranted, personnel may receive administrative or disciplinary action or both.

DUTY ASSIGNMENTS

The unfair assignment of general administrative and support duties (food service, compartment cleaning, and work details) outside the normal requirements of a rating frequently lowers morale. It also weakens the efficiency and overall effectiveness of a command.

Based on Navy policy, supervisors should assign work not included in a specific rating on a fair, rotational basis. They should make such assignments without regard to race, creed, color, sex, age, or national origin. Although supervisors may consider the seniority of personnel in detailing such duties, they must make positive efforts to ensure fair treatment.

Assignment to duty on ships or stations should also comply with the Navy's equal opportunity goals. The repeal of the combat exclusion law potentially opens all classes of surface ships to women. The Naval Construction Force, or Seabees, has also received women in sea duty construction battalions. All construction battalions are now open to women, opening more than 4,000 seagoing billets to women.

The expanded opportunity for women in the Navy ensures a more equitable rotation between sea and shore duty for all Sailors and provides career paths for women that are consistent with those of their male counterparts.

PROFESSIONAL TRAINING AND ADVANCEMENT

The Navy expects everyone who enters the naval service to increase his/her knowledge and skills. Your command will provide the necessary training so you can develop a skill and properly prepare yourself for advancement. Although advancement is an individual effort, the command has the responsibility to provide you with an equal opportunity for training and advancement. How far you advance depends primarily on your own initiative, capabilities, and qualifications.

The Department of the Navy sets the requirements for advancement for paygrades E-1 through E-9. To advance to E-4 through E-7, Sailors must pass an advancement-in-rate exam. However, just meeting all the requirements does not guarantee advancement. Only the most qualified will be advanced, and they will be advanced **only if vacancies exist** for that paygrade. Improving yourself, your skills, and your education increases your chance for advancement.

MILITARY JUSTICE

To assure equal justice and treatment, your command will continuously review charges, dismissed cases, issued warnings, and all nonjudicial punishment procedures. Such reviews detect racial, religious, ethnic, cultural, or sexual bias affecting either the accusation or the punishment phase of military justice.

SERVICE AND RECREATIONAL FACILITIES

Service and recreational facilities must meet the needs of all segments of the Navy community. Commands must pay special attention to the possibility of discriminatory practices in the operation of exchanges, commissaries, service clubs, and recreational facilities. Segregation, lack of tolerance of cultural preferences, or discriminatory practices in command facilities are inconsistent with equal opportunity.

Navy exchange facilities provide a variety of products. These products include items purchased by minority and female personnel and dependents, such as special categories of cosmetics, books, magazines, and

Student Notes:

records. Barber and beauty shops employ personnel trained and qualified to serve all Navy members and their dependents.

HOUSING REFERRAL OFFICE

The Department of Defense (DoD) has established housing referral offices at locations with large concentrations of military families. DoD has also established housing referral offices at locations where housing discrimination may exist because of race, color, creed, or national origin.

The Navy requires that all personnel offices include the following statement in orders for detachment and TAD of more than 30 days in any one place:

"You are directed to report to the appropriate Housing Referral Office prior to negotiating any agreement for off-base housing."

The HRO provides information about government housing and the type, cost, and availability of private housing. The HRO also maintains a list of facilities banned as housing for military personnel because of discriminatory practices.

Department of the Navy (DoN) policy supports the Federal Fair Housing legislation through its efforts to ensure equal opportunity for available housing. It makes every effort to eliminate off-base housing discrimination toward DoD personnel because of sex, race, color, religion, or national origin. It tries to ensure DoD personnel who meet ordinary standards of character and financial responsibility can obtain off-base housing as easily as any other person.

EQUAL OPPORTUNITY OFF BASE

Discrimination in the civilian community has an adverse effect on the welfare and morale of military personnel and their dependents. Consequently, discrimination off base is harmful to the military effectiveness of a command. Dealing with discrimination in the civilian community is more difficult than within a command. However, the Navy deals affirmatively with such problems to ensure equal treatment for service members and their dependents in nearby communities.

Commands can take a number of actions to promote equal opportunity for its members in the civilian community. At some installations, problems of mutual concern to the base and the community are discussed on an informal but regular basis. The Navy has established command-community relations committees as another approach. Whatever approach they use, commands make every effort to eliminate off-base discrimination for military personnel and their dependents.

Military personnel moving into or changing their place of residence may not enter into rental, purchase, or lease arrangements with facilities under restrictive sanctions. However, such sanctions do not apply to personnel who may be residing in the facility at the time the sanction is imposed. Personnel who intentionally, and contrary to instructions, take residence in restricted facilities are subject to disciplinary action and loss of basic allowance for quarters (BAQ).

Equal opportunity also applies to public facilities, such as schools, parks, playgrounds, libraries, and hotels. Any person who receives discriminatory treatment in such places can seek relief through military or civilian channels. Normally, you should go through military channels (your chain of command). That gives your command the opportunity to try to get the facility to comply with the law.

The First Amendment of the Constitution guarantees you the rights of freedom of speech and assembly. That means you can attend civil rights demonstrations and similar gatherings. However, service members have several conditions attached to their participation in such demonstrations.

- Navy personnel cannot take part in civil rights demonstrations while wearing their uniform or during duty hours.
- Navy personnel cannot take part in a civil rights demonstration held on a military reservation or in a foreign country.
- Navy personnel cannot take part in demonstrations that violate law and order or that could reasonably be expected to result in violence.

Student Notes:

DISCRIMINATION COMPLAINT PROCEDURES

Apart from their individual merits, legitimate complaints can provide valuable information about the existence of discriminatory treatment within a command. Positive action in cases where complaints are found to be valid lends credibility to your command's stated commitment to ensure equal treatment and justice.

No matter how badly they are treated, most people are reluctant to complain about this treatment or to express grievances to their seniors. People feel that if they complain, they will suffer. *Navy Regulations* and the *U.S. Navy Equal Opportunity Manual* guarantee personnel the right to file a complaint of discriminatory treatment without danger of reprisal from the command. This means that if you file a complaint about the way someone treated you, you can expect to be treated fairly and not be discriminated against by that person or anyone else. However, personnel should gather all pertinent facts before forwarding a complaint.

Special Request Chit

When you cannot resolve a complaint among the persons involved or with the help of a supervisor (your LPO or division CPO), submit a special request chit as your first course of action. Attach a written complaint to the special request chit and submit it through the chain of command within a timely manner of the incident. You do not have to follow any special format in writing a discrimination complaint. However, you should fully explain the complaint and include all facts.

Captain's Mast Request

As a Navy member, you have the right to speak with the commanding officer to voice a complaint or get help in resolving a problem. To do this, you request a captain's mast. But, you should request a captain's mast only if your first course of action fails. (Remember, your first course of action is to let the chain of command correct the problem.) You can also request a captain's mast if you feel that the problem you have is so important that immediate action is required. A captain's mast request requires each person in the chain of command to forward the request—whether or not the request is approved.

FALSE DISCRIMINATION COMPLAINTS

Filing false discrimination complaints is just as serious as discrimination itself. The Navy doesn't tolerate the filing of false discrimination complaints for any reason. The Navy takes the same administrative and disciplinary actions to those who file false complaints as it does to those who discriminate.

HAZING

Learning Objectives: When you finish this chapter, you will be able to—

- Define hazing.
- Identify the Navy's policy on hazing.

Military customs and traditions are a part of the Navy and Marine Corps. Ceremonies, initiations, and rites of passage are leadership tools that instill *esprit de corps* and build respect for the accomplishments of other Sailors and Marines. Most ceremonies honor the bravery of our military men and women and commemorate (memorialize) significant events. The accomplishments of Sailors and Marines and significant events are the basis of the Navy's Core Values—Honor, Courage, and Commitment. Graduations, chiefs' initiations, and crossing-the-line ceremonies are used to celebrate and recognize the achievements of individual Sailors, Marines, or entire units. However, hazing behavior (behavior that is degrading, embarrassing, or causes injuries) is illegal.

DEFINITION OF HAZING

Hazing is defined as any conduct whereby a military member or members, regardless of service or rank, without proper authority causes another military member or members, regardless of service or rank, to suffer or be exposed to any activity which is cruel, abusive, humiliating, oppressive, demeaning, or harmful. Soliciting or coercing another to perpetrate any such activity is also considered hazing. Hazing need not involve physical contact among or between military members; it can be verbal or psychological in nature. Actual or implied consent to acts of hazing does not eliminate the culpability of the perpetrator.

Student Notes:

Hazing can include, but is not limited to, the following types of activities:

- Playing abusive or ridiculous tricks
- Threatening or offering violence or bodily harm to another
- Striking
- Branding
- Taping
- Tattooing
- Shaving
- Greasing
- Painting
- Requiring excessive physical exercise beyond what is required to meet standards
- "Pinning," "tacking on," "blood wings"
- Forcing or requiring the consumption of food, alcohol, drugs, or any other substance

DEPARTMENT OF THE NAVY (DoN) POLICY ON HAZING

The DoN's policy on hazing is as follows:

- Hazing is **prohibited** and will not be tolerated.
- No service member in the DoN may engage in hazing or agree to be hazed.
- No commander or supervisor may, by act, word, deed, or omission condone (agree to) or ignore hazing if they know or reasonably should have known, that hazing may or did occur.
- It is the responsibility of every Sailor and Marine to make sure that hazing does not occur. Every service member is responsible to make the appropriate authorities aware of hazing violations.
- Commanders or individuals in supervisory positions are responsible for making sure that all ceremonies and initiations conducted within their organizations or commands comply with this policy.
- Supervisory personnel must make sure that service members participating in command-authorized ceremonies, initiations, and other activities are treated with dignity and respect during these events.
- Reprisal actions against any victim or witness of hazing incidents are strictly prohibited.

REVIEW 6 QUESTIONS

Q1. All naval personnel should be treated equally and be given equal opportunities. What program does the Navy use to achieve this purpose?

Q2. Who provides guidance and policy for the CMEO Program?

Q3. Telling a sexist or racial joke would be an example of what type of behavior?

Q4. To avoid discriminating practices in the operation of exchanges, commissaries, service clubs, and recreational facilities, what do command facilities provide?

Q5. Where is the best place for you or your family to start to resolve an off-base discrimination complaint?

Student Notes:

Q6. Service members can attend civil rights demonstrations and similar gatherings except for what conditions?

a.

b.

c.

Q7. You have submitted a special request chit and feel that the chain of command has not resolved your discrimination complaint. What should be your next course of action?

Q8. What is hazing?

Q9. What is the Navy's policy on hazing?

SEXUAL HARASSMENT

Learning Objectives: When you finish this chapter, you will be able to—

- Recall the definition of sexual harassment.
- Identify the policies that pertain to sexual harassment.

All military and civilian personnel in the Department of the Navy have responsibilities. One of those responsibilities is to maintain high standards of honesty, integrity, and conduct to assure proper performance of business and to maintain public trust. Sexual harassment violates those standards, especially equal opportunity. Both men and women can be victims or harassers.

Sexual harassment is a form of sex discrimination. Sexual harassment is unwelcome sexual advances, requests for sexual favors, and other verbal or physical conduct that is sexual nature. Sexual harassment occurs when—

- Submission to or rejection of such conduct is made, either explicitly or implicitly, a term or condition of a person's job, pay, or career, or
- Submission to or rejection of such conduct by a person is used as a basis for career or employment decisions affecting that person, or
- Such conduct interferes with an individual's performance or creates an intimidating, hostile, or offensive environment.

This means that if a supervisor or someone in a command position makes sexual advances and tells you that if you don't go along you could lose your job, not be promoted, or make it impossible to do your job because of that persons actions, you are being sexually harassed.

Any supervisor or person in a command position who uses sexual behavior to control or affect your career, pay, or job is sexually harassing you. Also, any person who makes unwelcome verbal comments, gestures, or physical contact of a sexual nature is sexually harassing you.

Basically, sexual harassment means bothering someone in a sexual way. For a person's behavior to be considered sexual harassment, it must meet three criteria:

1. Be unwelcome
2. Be sexual in nature
3. Occur in or impact on the work environment

UNWELCOME BEHAVIOR

Unwelcome behavior is behavior that a person doesn't ask for and considers undesirable or offensive. Not everyone has the same perception (idea) of what is *undesirable or offensive.* What's okay for some people isn't okay for others.

So, whose perception should be used, the person who is giving the unwelcome behavior or the person

Student Notes:

receiving (recipient) the unwelcome behavior? The person receiving the behavior is being affected; therefore, it's the recipient's perception that counts. If the recipient is a reasonable person and not overly sensitive, behavior which the recipient finds unwelcome should be stopped. From the view of the recipient, this is a **reasonable personal standard** and is really no more than using common sense.

BEHAVIOR WHICH IS SEXUAL IN NATURE

Behavior that is sexual in nature is fairly easy to determine. For example, if someone tells sexually explicit jokes, displays sexually suggestive pictures, and talks about sex, that person's behavior is sexual in nature.

Some people consider other behaviors, such as touching, to be sexual in some cases but not in others. Not all touching is sexual in nature. However, touching certain parts of the body or done suggestively is sexual in nature. Again, using common sense is normally enough to tell whether a certain behavior is sexual in nature.

OCCUR IN OR IMPACT ON THE WORK ENVIRONMENT

For sexual harassment to occur, unwelcome sexual behavior must occur in or impact on the work environment.

Quid Pro Quo (This for That)

When someone is offered or denied something that is work-connected in return for submitting to or rejecting unwelcome sexual behavior, that person is being subjected to a type of sexual harassment known as *quid pro quo* ("this for that").

A person isn't promoted because he/she didn't submit to unwelcome sexual behavior. This is an example of *quid pro quo* sexual harassment. Other examples include the loss of a job, a demotion, or a bad performance eval.

Basically, if a work-connected decision is made because a person is being subjected to or has rejected unwelcome sexual behavior, sexual harassment has occurred. Normally, this is from a senior to a junior, because the senior person can offer something.

Hostile Environment

If unwelcome sexual behavior of one or more persons in a workplace interferes with another person's work performance, sexual harassment has occurred. Now, suppose the behavior makes the workplace offensive, intimidating, or abusive to another person, whether or not work performance is affected. This type of sexual harassment is called *hostile environment*. The following are examples of a person's behavior that could create a hostile environment:

- Use of sexually explicit or sexually offensive language.
- Display sexually oriented posters or calendars of nude or partially clad individuals.
- Touch someone in a suggestive manner (that is, intentionally brushing against or pinching a person).
- Give someone unwelcome letters, cards, or gifts of a personal nature that have sexual overtones.
- Give unwanted or uninvited pressure for dates.

Some types of unwelcome sexual behavior don't have to create a hostile environment to be sexual harassment. If a person fondles or gropes another person in the workplace, the behavior is considered sexual harassment. This behavior is considered sexual harassment even if it only happened once. Other, less obvious behaviors can become sexual harassment if they are repeated.

RANGE OF BEHAVIORS

There is a wide range of behaviors, from leering to rape, that can be unwelcome, sexual, and work-connected. These behaviors can constitute sexual harassment. Some behaviors may be unwelcome and work-connected, but not sexual (for example, performance counseling). This behavior is not sexual harassment. To make it easier to understand, it is helpful to think of the entire range of possible behavior in terms of a traffic light. The traffic light has three colors—red,

Student Notes:

yellow, and green. Behavior may be divided into three zones.

- Red on the traffic light means **stop**; behavior in the red zone means **don't do it**—it's sexual harassment.
- Yellow on the traffic light means **use caution**; behavior zone may be sexual harassment.
- Green on the traffic light means **go**. Behavior in the green zone means **it's acceptable**—it's not sexual harassment.

Just as with a traffic light, if in the yellow zone long enough, the light will turn red. If yellow zone behavior is repeated enough, especially after having been told it is unwelcome, it becomes red zone behavior—sexual harassment. The following show these three types of behavior, but they are certainly not all-inclusive:

NOTE

Remember that the above examples are only guides. Individuals who believe they are being sexually harassed base their belief on their perceptions. Also, each incident is judged on all the facts in that particular case, and that individual's judgment may vary on the same facts. Therefore, use caution. Any time sexual behavior is introduced into the work environment or among coworkers, the individuals involved are on notice that the behavior may constitute sexual harassment.

REPORTING AN INCIDENT INVOLVING SEXUAL HARASSMENT

Individuals who believe they have been sexually harassed have ways to seek resolution and redress (remedy). Check with your CMEO officer for the avenues available to you. All reported incidents of sexual harassment are investigated and resolved at the lowest appropriate level. All incidents are resolved promptly and with sensitivity. Confidentiality will be maintained to every extent possible.

If you believe that you have been sexually harassed, talk about your concerns or objections directly with the person who is behaving in a harassing way. If you are subjected to or observe objectionable behavior, you should promptly notify the chain of command if one of the following conditions exists:

- The objectionable behavior does not stop,
- The situation is not resolved,

Red zone. These behaviors are always considered sexual harassment. They include sexual favors in return for employment rewards, threats if sexual favors aren't provided, sexually explicit pictures (including calendars or posters) or remarks, using status to request dates, or obscene letters or comments. The most severe forms of sexual harassment constitute criminal conduct; that is, sexual assault (ranging from forcefully grabbing to fondling, forced kissing, or rape).

Yellow zone. Many people would find these behaviors unacceptable, and they could be sexual harassment. These behaviors include violating personal "space," whistling, questions about personal life, lewd or sexually suggestive comments, suggestive posters or calendars, off-color jokes, leering, staring, repeated requests for dates, foul language, unwanted letters or poems, sexually suggestive touching, or sitting or gesturing sexually.

Green zone. These behaviors are not sexual harassment. They include performance counseling, touching that couldn't reasonably be perceived in a sexual way (such as touching someone on the elbow), counseling on military appearance, social interaction, showing concern, encouragement, a polite compliment, or friendly conversation.

Student Notes:

- Addressing the objectionable behavior directly with the person concerned is not reasonable under the circumstances, or

- The behavior is clearly criminal in nature.

If the person whose behavior is objectionable is a direct superior in the chain of command or if the chain of command condones (tolerates) the conduct or ignores a report, the person subjected to or who has observed the objectionable behavior should promptly communicate the incident through other available means. When possible, always use your chain of command; but, if your boss allows the conduct or ignores a report of objectionable behavior, use other means to report it.

REVIEW 7 QUESTIONS

Q1. Improper conduct is considered sexual harassment when—

a.

b.

c.

Q2. List the criteria for a person's behavior to be termed sexual harassment.

a.

b.

c.

Q3. Whose perceptions count when there is unwelcome behavior?

Q4. Your LCPO offers you high marks on your evals in exchange for sexual favors. What type of sexual harassment is this?

Q5. Displaying sexual posters or using sexual explicit language creates what type of environment?

Q6. How has the Navy made it easier to understand the wide range of good to bad behavior?

a.

b.

c.

FRATERNIZATION

Learning Objectives: When you finish this chapter, you will be able to—

- Recognize the fraternization policy.

- Identify the consequences of not observing this policy.

Article 1165 (Fraternization Prohibited) of the *U.S. Navy Regulations* states: "No person in the Navy is to enter a personal relationship that is unduly familiar, does not respect differences in rank, and is prejudicial to good order and discipline." Some relationships between Navy members violate naval traditions. Unduly familiar relationships can exist between officers, officer and enlisted persons, or between enlisted persons. If there is an unduly familiar relationship between a supervisor and a subordinate, the relationship isn't right and harms good order and discipline. Breaking the rules on fraternization among Regular and Reserve personnel may result in administrative or punitive action.

Student Notes:

FAMILY OMBUDSMAN PROGRAM

Learning Objectives: When you finish this chapter, you will be able to—

- Recall the provisions of the Family Ombudsman Program.
- Identify sources of information available to Navy families.

The command ombudsman is a volunteer who is a liaison between the command and families. This person undergoes a regular application and interview process by the command team and is appointed by the commanding officer. Once the CO appoints an ombudsman, he/she attends an intense 21-hour training course that provides the basic tools for performing the duties of the position. In addition, ombudsmen undergo continuous training offered through local assemblies and family service centers.

Since 1973, the focus of the program has shifted away from the grievance-processing role. Today, the enlarged scope of the ombudsman job is assisting COs in their responsibilities for the morale and welfare of the families of the command. In general, any functions that promote these goals may be included in the ombudsman program. It is **always** the CO who determines the content and priorities of that program. The basic roles and functions of an ombudsman are as follows:

- Serve as the primary link/liaison and communicator of information between command families and the command.
- Communicate regularly with command families in ways approved and supported by the command; for example, command newsletter, command-sponsored telephone "Careline," and a phone tree for emergency information as directed by the CO.
- Provide information and outreach to command family members. Interact and cooperate with organizations and military departments, family service centers, chaplain's office, medical treatment facilities, Navy-Marine Corps Relief Society, American Red Cross, Navy Wifeline Association, legal assistance offices, and so forth.
- Refer individuals in need of professional assistance to appropriate resources; possibly provide support to individuals and refer them for counseling.
- Act as an advocate for command family members; help access the appropriate level of chain of command for intervention and for the forwarding of appropriate requests/grievances while exercising confidentiality.
- Participate in indoctrination and orientation programs.
- Assist in welcome programs and act as a family coordinator as part of the command Sponsor Program.
- Represent the command on committees, boards, and working groups in the military or civilian communities concerned with services and support to command families.
- Participate in activities that would promote the morale, health, and welfare or command families.

The ombudsman is trained to help with information and referral, however, they are not a taxi service or babysitter. The Privacy Act and strict rules of confidentiality bind the command ombudsman; they are not a rumor mill.

REENLISTMENT QUALITY CONTROL PROGRAM

Learning Objective: When you finish this chapter, you will be able to—

- Recall the incentives for reenlistment, education, and special duty.

Student Notes:

The Reenlistment Quality Control Program sets standards you should meet to be eligible to reenlist. The purpose of the Reenlistment Quality Control Program is as follows:

- To provide a personnel management program to control rating manning, reduce advancement stagnation, and meet end strength requirements
- To issue reenlistment criteria for the Reenlistment Quality Control Program
- To establish terms of years an individual may reenlist based upon Career Reenlistment Objective (CREO) group of his/her rating and years of service at the time of reenlistment
- To establish standardized professional growth points or high-year tenure (HYT) by paygrade
- To establish procedures by which personnel may request consideration for reenlistment/extension beyond established professional growth points

All first-term Sailors in paygrades E-1 through E-6 requesting reenlistment must be approved through the Enlisted Navy Career Options for Reenlistment (ENCORE) Program. E-1/E-2 cannot reenlist unless involved in special programs, such as the 2YO program, and approval is granted through ENCORE. Personnel serving in paygrade E-3 are eligible to reenlist/extend provided they have met the professional growth criteria and have been approved through ENCORE.

The Reenlistment Quality Control Program uses reenlistment codes to indicate whether you have met professional growth criteria. Reenlistment codes reflect the quality control category and the status of personnel who separate from the Navy. If you don't reenlist at your end of active obligated service (EAOS), your Certificate of Release/Discharge from Active Duty (DD214) indicates your reenlistment code. If you decide to reenlist later on, the reenlistment code indicates whether or not you're qualified to reenlist. The reenlistment codes are as follows:

- RE-R1—Recommended for Preferred Reenlistment
- RE-1—Eligible for Reenlistment
- RE-R3—Eligible for Probationary Reenlistment
- RE-4—Not Eligible for Reenlistment.

STATE AND NATIONAL VOTING PROCEDURES

Learning Objective: When you finish this chapter, you will be able to—

- Identify the procedures for state and national voting to include absentee ballots.

Democracy depends on its members access to vote. Department of the Navy policy is to ensure its members, their spouses, and their dependents may register and vote in all elections within their home districts.

The Chief of Naval Personnel directs and supervises the Navy's voting program. The voting program makes sure all ships and stations receive voting information pamphlets, posters, and materials. All eligible personnel receive in-hand delivery of the Federal Post Card Application for Absentee Ballot (FPCA), SF Form 76, and revised 1987, for all federal elections. They receive the ballots well in advance of the November election. The recommended delivery time for overseas areas is 15 August and for stateside personnel 15 September.

REVIEW 8 QUESTIONS

Q1. Describe why the Navy has a fraternization policy.

Q2. What is the function of the ombudsman?

a.

b.

Q3. How does the ombudsman communicate with Navy families?

Student Notes:

Q4. List some the agencies or organizations ombudsman works with.

a.

b.

c.

d.

e.

f.

g.

Q5. What must a first-term Sailor have before they can reenlist or extend?

Q6. When leaving the Navy, what determines if you will be able to reenlist at a later date?

Q7. When away from their home district, how can a Sailor and his family register to vote?

SUMMARY

Since the Navy is such a large, diverse, and complex organization, it requires numerous programs to help its members resolve a wide variety of problems. For example, the Command Managed Equal Opportunity (CMEO) Program makes sure Navy members have the same basic rights all other citizens in our society enjoy.

Department of the Navy policies govern our day-to-day operations by requiring us to perform to certain standards. Those policies give us a general goal and the guidelines to achieve that goal. Those policies and the programs that support them ensure Navy personnel know how to do their jobs and where to find help to resolve their problems.

REVIEW 1 ANSWERS

A1. The four main sources of pollutants are—

a. **Agricultural runoff**

b. **Industrial**

c. **Municipal**

d. **Transportation operations**

A2. Pollution affects both the **physical and biological world**.

A3. The primary pollution concern of the Navy is the **pollution produced by shipboard waste**.

A4. The Navy has been **converting their power plants to use more efficient pollution control systems for stack emissions**.

A5. The agreement the Navy follows to operate marine sanitation devices (MSDs) and to dispose of treated sewage in foreign waters is the **Status of Forces Agreement (SOFA)**.

A6. Vessels must be **at least 25 nautical miles** from the U.S. coastline before they can discharge unpulped trash.

REVIEW 2 ANSWER

A1. The Navy makes every possible effort to **improve the way it uses energy resources, without compromising readiness, effectiveness, or safety**.

REVIEW 3 ANSWERS

A1. **Delete**

A2. **Delete**

A3. The **Overseas Duty Support Program (ODSP)** provides you with support and information about foreign lands.

A4. Besides the ODSP, you can get information about other countries through the **Overseas Transfer Information Service (OTIS)**.

A5. The **Military Cash Awards Program (MILCAP)** is used to reward individuals for their beneficial suggestions, inventions, and scientific achievements.

REVIEW 4 ANSWERS

A1. If Navy service members don't meet the fitness or body fat standards and need help to change health habits, they take part in the **Health and Physical Readiness Program**.

A2. The Navy responsibility for pregnant servicewomen is **providing obstetrical care at a medical facility if it has obstetrical/gynecological facilities and if the servicewoman lives in the treatment facility area**.

A3. List the responsibilities of pregnant servicewomen.

a. **Plan pregnancy to meet family and military obligation**

b. **Confirm pregnancy with military medical facility**

c. **Notify CO or OIC of pregnancy**

d. **Perform military duties while pregnant**

e. **Comply with work- and task-related safety and health recommendations**

A4. Service members must complete what forms for family care?

a. **Family Care Plan Certificate, NAVPERS 1740/6**

b. **Family Care Plan Arrangements, NAVPERS 1740/7**

A5. The Department of the Navy public affairs informs the public and service members about the following subjects:

a. **The Navy as an instrument of national policy and security**

b. **Navy operations and programs**

c. **The responsibilities and activities of naval personnel as U. S. citizens**

REVIEW 5 ANSWERS

A1. The Navy's policy on drug abuse **is zero tolerance**.

A2. The purpose of the Integrity and Efficiency Program **is to detect, deter, and eliminate fraud, waste, and abuse**.

A3. You can report fraud, waste, and abuse situations by the following means:

a. **Chain of command**

b. **Navy hotline**

c. **Naval Criminal Investigative Service (NCIS)**

d. **Congressional communication**

A4. For the public to feel confident about the Navy's integrity, naval personnel should comply with the **Standards of Conduct and Professional Ethics**.

REVIEW 6 ANSWERS

A1. The Navy uses the **Command Managed Equal Opportunity (CMEO) Program** to ensure that all naval personnel are treated equally and are given equal opportunities.

A2. Guidance and policy for the CMEO Program is provided by the **Chief of Naval Operations**.

A3. Telling an sexist or racial joke is an **insensitive practice.**

A4. Command facilities provide **a variety of products and services in command facilities** to avoid discrimination in the operation of exchanges, commissaries, service clubs, and recreational facilities.

A5. The best place for you or your family to start to resolve an off-base discrimination complaint is **your chain of command**.

A6. Navy personnel can't take part in civil rights demonstrations in the following situations:

a. **While wearing their uniform or during duty hours.**

b. **When held on a military reservation or in a foreign country.**

c. **When law and order are violated or when they could reasonably be expected to result in violence.**

A7. You have submitted a special request chit and feel that the chain of command hasn't resolved your discrimination complaint. **Your next course of action is to request captain's mast.**

A8. Hazing is **any conduct whereby a military member or members, regardless of service or rank, without proper authority causes another military member or members, regardless of service or rank, to suffer or be exposed to any activity which is cruel, abusive, humiliating, oppressive, demeaning, or harmful.**

A9. According to Navy policy, **hazing is prohibited**.

REVIEW 7 ANSWERS

A1. Improper conduct is considered sexual harassment when—

a. **Submission to or rejection of such conduct is made either explicitly or implicitly a term or condition of a person's job.**

b. **Pay, or career; submission to or rejection of such conduct by a person is used as a basis for career or employment decisions affecting that person.**

c. **Such conduct interferes with an individual's performance or creates an intimidating, hostile, or offensive environment.**

A2. A person's behavior to be termed sexual harassment if is—

a. **Unwelcome**

b. **Sexual in nature**

c. **Occurs in or impacts on the work environment**

A3. The **recipient's perceptions** count when there is unwelcome behavior.

A4. If your LCPO offers you high marks on your evals in exchange for sexual favors, it is ***quid pro quo*** sexual harassment.

A5. Displaying posters or using sexually explicit language creates a **hostile** environment.

A6. To make the range of good to bad behavior easier to understand, the Navy has compared behavior ranges to the traffic light.

   a. **Red light—Sexual harassment behavior**

   b. **Yellow light—Many people find behavior unacceptable**

   c. **Green light—Acceptable behavior, not sexual harassment**

REVIEW 8 ANSWERS

A1. The function of the ombudsman is to **promote good order and discipline**.

A2. The ombudsman—

   a. **Acts as a liaison between Navy families and the command, and**

   b. **Keeps the families informed about command policies.**

A3. The ombudsman communicates with Navy families through **command newsletters, command-sponsored telephone "Careline," or phone trees**.

A4. The ombudsman works through the following agencies/organizations:

   a. **Navy-Marine Corps Relief Society**

   b. **American Red Cross**

   c. **Family Service Centers**

   d. **Chaplain's office**

   e. **Navy Wifeline Association**

   f. **Medical treatment facilities**

   g. **Legal assistance offices**

A5. Before a first-term Sailor can reenlist or extend, he/she must have **ENCORE** approval.

A6. If you leave the Navy, your **reenlistment code** tells whether you can reenlist.

A7. When away from their home district, a Sailor and his/her family can register to vote by using a **Federal Post Card Application for Absentee Ballot (FPCA)**.

CHAPTER 2

MILITARY CONDUCT AND JUSTICE

I pledge allegiance to the flag of the United States of America, and to the Republic for which it stands, one nation under God, indivisible, with liberty and justice for all.

Because the United States Navy is a military service and since you are a member of the U.S. Navy, you are expected to be *military* in the best sense of the term. You are expected to know the traditions of the Navy, its customs, and its language. You should understand the organization and mission of the Navy and the "why" behind the Navy's discipline and its drills. In the front of this manual is the Navy Creed; if you haven't read it, read it now. You will see the importance of your responsibilities and duties to your Country and to the Navy.

In this chapter, you will learn about military conduct, which includes the conduct expected of you if you should become a prisoner of war. You will also learn about the military police, the purpose of discipline and punishment, and the three sources that set forth the basic disciplinary laws for the U.S. Navy. Those sources are *U.S. Navy Regulations, Standard Organization and Regulations of the U.S. Navy,* and the *Uniform Code of Military Justice (UCMJ).*

PERSONAL CONDUCT

Learning Objective: When you finish this chapter, you will be able to—

- Identify the personal characteristics of a good Sailor.

Every Sailor in the Navy should set an example of high personal and military ideals. Every Sailor should always set a good example for other, perhaps younger, nonrated personnel. Remember, a good Sailor always does the following:

- Acts in a military and seamanlike manner.
- Puts the good of the ship and the Navy before personal likes and dislikes.
- Obeys the rules of military courtesy and etiquette as well as the rules of military law.
- Demonstrates loyalty, self-control, honesty, and truthfulness.
- Knows what to do in an emergency and how to do it with the least waste of time and with minimum confusion

As a Sailor, you represent the Navy. People form their opinions of the Navy based on your appearance and actions. Always wear your uniform with pride. Conduct yourself in a manner that will reflect credit on you and the Navy. In effect, you conduct the business of public relations for the Navy. The way in which you sell the Navy to civilians and the way you sell yourself to your superiors and shipmates determines their opinion of you and of the Navy.

A good Sailor is morally responsible. That means you know what's right and what's wrong, and you try to do what is right. As a morally responsible person, you perform all assigned duties as correctly and timely as humanly possible without worrying about personal gain or inconveniences.

To succeed in any line of work, you must be devoted to duty and be able to take orders. Shipboard life is so exacting that a team of members must do many tasks; one person alone cannot do them. In battle or in solving a battle problem, all personnel must work as a team; and it doesn't make any difference whether the team consists of a few or many members.

The Navy isn't the place for the immature self-seeker who puts forth his/her best efforts only when some personal advantage is to be gained. Also, the Navy doesn't have room for the resentful, hardheaded, self-important person who can't take an order. Rules and regulations serve as guides for daily living and, if followed by all, make life more pleasant and easier for all hands.

THE CODE OF CONDUCT

Learning Objective: When you finish this chapter, you will be able to—

- Recognize the responsibilities stated in articles I through VI of the Code of Conduct for members of the Armed Forces of the United States.

Because of the conduct of a few Americans during the Korean conflict, President Dwight D. Eisenhower prescribed a Code of Conduct for members of the armed forces in 1955. That code provides American military personnel with a standard of conduct should they be captured by an enemy. It provides a framework of ideals and ethical standards that will help personnel resist the physical, mental, and moral onslaughts of their captor.

Many Americans have been prisoners of war (POWs), and they all agree that life as a POW is hard. A few POWs were unprepared or lacked the ability to maintain their faith and loyalty under extreme pressure. The enemy broke their will, and they gave information and/or acted in a way that hurt their country and their fellow prisoners.

If you ever become a POW, don't make up stories; your interrogator will eventually catch on and could resort to harsher methods to try to gain information. A simpler, "I don't know," is a better answer. Your captors will use many methods to gain information. They will try to get prisoners to collaborate by torturing them or by trying to turn prisoners against each other. Although forbidden by the Geneva Convention, history has shown that some captors have resorted to physical and mental forms of torture to get the information they want. Maintain your faith in your God, your country, and your fellow prisoners.

Remember the first sentence of the first article of the Code of Conduct, "I am an American, fighting in the forces which guard my country..." If you live up to that principle, you don't ever have to worry about an investigation concerning your behavior. You won't live the rest of your life knowing that something you said harmed your fellow prisoners, comrades in arms, or your country and its allies.

In 1988, President Ronald Reagan issued Executive Order 12633, amending the Code of Conduct to use gender-neutral language. First expressed in written form in 1955, the Code is based on time-honored concepts and tradition that date back to the days of the American Revolution. The six articles of the Code of Conduct are as follows:

ARTICLE I

I am an American, fighting in the forces which guard my country and our way of life. I am prepared to give my life in their defense.

ARTICLE II

I will never surrender of my own free will. If in command I will never surrender the members of my command while they still have the means to resist.

ARTICLE III

If I am captured I will continue to resist by all means available. I will make every effort to escape and aid others to escape. I will accept neither parole nor special favors from the enemy.

ARTICLE IV

If I become a prisoner of war, I will keep faith with my fellow prisoners. I will give no information or take part in any action which might be harmful to my comrades. If I am senior, I will take command. If not, I will obey the lawful orders of those appointed over me and will back them up in every way.

ARTICLE V

When questioned, should I become a prisoner of war, I am required to give name, rank, service number and date of birth. I will evade answering further questions to the utmost of my ability. I will make no oral or written statements disloyal to my country and its allies or harmful to their cause.

ARTICLE VI

I will never forget that I am an American, fighting for freedom, responsible for my actions, and dedicated

Student Notes:

to the principles which made my country free. I will trust in my God and in the United States of America.

REVIEW 1 QUESTIONS

Q1. List the three sources that contain the basic disciplinary laws for the U.S. Navy.

a.

b.

c.

Q2. What is one of the most important characteristics of a good Sailor?

Q3. For you to succeed in your work in the Navy, you should possess what quality?

Q4. Why was the Code of Conduct established?

Q5. What total number of articles are there in the Code of Conduct?

Q6. When questioned under article V, what is the only information you are allowed to give?

MILITARY POLICE

Learning Objectives: When you finish this chapter, you will be able to—

- Recognize the authority of the military police/shore patrol.
- Identify the functions of the military police/shore patrol.

All branches of the armed forces assign personnel to duties as military police. In the Air Force, they are called *security police*; in the Army and Marine Corps, they are called *military police* (*MP*); and in the Navy, they are called *shore patrol* (*SP*). The shore patrol consists of officers and petty officers assigned to assist military personnel ashore. They are identified by armbands bearing the letters *SP.*

In areas where units of different armed services are located, the military police may be combined to form one unit instead of a separate unit for each service. This single unit is known as an *Armed Forces Police Detachment* (*AFPD*), and all members are identified by brassards (armbands) with the letters *AFPD*. The primary duties of AFPD are to assist military personnel ashore, maintain good order and discipline among military personnel, and report conditions or practices that appear prejudicial to the welfare of military personnel. They have authority to stop, question, apprehend, or take into custody any member of the armed forces.

When asked to do so by the military police, you must show your ID card, leave authorization, and the like. You must obey any orders given you by the military police.

Some reminders are listed here for you to observe in any dealings with the military police or shore patrol. (For the remainder of our discussion, *patrol* is used to include all armed forces police.)

1. Obey the orders of the patrol.
2. Don't become argumentative if the patrol is questioning you. The patrol will rightfully question you if you are out of uniform, appear drunk, or act in a suspicious manner.
3. Never interfere with the members of the patrol in the performance of their duty. If you are in a place where a fight has begun or is about to develop and the patrol orders you to leave, do so without protest.

Don't feel that the patrol is trying to harass you. You won't have to prove who you are or verify that

Student Notes:

you're entitled to be ashore every time you see a patrol. The patrol will stop you only when you appear to be in, or to be headed for, some kind of trouble or if you arouse their suspicions in some other manner. The military patrol can be a real friend in time of need. The patrol's orders are to be courteous, fair, and reasonable in all dealings with members of the armed forces and with civilians.

Whenever you are away from your ship or station and need advice, directions, or help of any kind, call on the nearest military patrolman or patrol headquarters.

Aboard ships and stations, masters-at-arms (MAAs) and police petty officers have functions similar to those of the shore patrol. The master-at-arms force, headed by the chief master-at-arms (CMAA), works directly for the executive officer. The master-at-arms force enforces Navy and ship regulations, musters restricted personnel, holds reveille, and performs other duties as are required for the maintenance of good order and discipline.

The duties of police petty officers (PPOs) are about the same as those of MAAs, but are on a divisional instead of a shipwide basis. Although PPOs stand their regular watches and perform their normal duties within their divisions, they are assigned additional duties such as making reveille and taps, ensuring compartments are cleaned, and maintaining order. At times they assist the MAA force in searching the ship, and providing bunks for new personnel or passengers.

REVIEW 2 QUESTIONS

Q1. In the Navy, the military police are known as the—

Q2. Shore patrol personnel are identified by—

Q3. When military police from different branches of the armed forces combine to form one unit, they are known as—

Q4. List the primary duties of the shore patrol.

a.

b.

c.

PURPOSE OF DISCIPLINE

Learning Objective: When you finish this chapter, you will be able to—

- Recall the purpose of good order and military discipline.

The word *discipline* comes from a Latin word meaning "to teach." However, discipline involves a certain type of teaching. Discipline is not peculiar to military organizations. Discipline is the training that develops self-control, character, and efficiency, or is the result of such training. **Discipline is a character builder, not a destroyer of individuality**.

The Navy's discipline consists of training its Sailors to behave in certain ways under certain circumstances. It gets them to work as a unit with maximum efficiency. To encourage Sailors to work as a unit, the Navy uses a system of motivation and correction through reward and punishment. Studious Navy men and women, when recommended by their commanding officers, are rewarded by timely promotions; lazy or careless individuals suffer a self-inflicted punishment by missing out on those promotions. Fines, restriction, confinement, demotion, and other forms of disciplinary action punish Sailors who get into trouble because they are negligent or indifferent.

The signs of discipline are shown in smart salutes, proper wearing of the uniform, prompt and correct action in any emergency, and in battle efficiency that brings victory in wars (fig. 2-1). Discipline, obviously, is indispensable to a military organization. Without it almost any effort would be defeated by lack of organization.

The purpose of discipline in the military services is to bring about an efficient military organization—a body of human beings trained and controlled for

Student Notes:

Figure 2-1.—The results of discipline are shown in prompt and correct action in an emergency, and especially in battle efficiency.

concerted action for the attainment of a common goal. Each individual understands how to fit into the organization as a whole. The members understand one another through the sharing of common knowledge. They are bound together by a unity of will and interest expressed by their willingness to follow and obey their leader. A group so organized is effective, not only for the specific purpose intended, but also for an emergency. Thus, a gun crew may be readily converted into a repair party for carrying out any essential job within its capabilities; a company of midshipmen may be turned into a fire-fighting organization. A well-disciplined naval unit responds automatically to an emergency and is not subject to panic.

PUNISHMENT

Learning Objective: When you finish this chapter, you will be able to—

- Recall the Navy's concept of punishment.

Based on the Navy's concept, punishment is not personal, vindictive, or inflicted as revenge for misconduct. The Navy realizes punishment cannot right the wrong resulting from an act of dereliction (failure). The value of punishment is the object lesson the punishment teaches the wrongdoer and others—the offense must not be repeated. That concept is referred to as the deterrent theory of punishment.

To accomplish its purpose, punishment must be consistent and just and must be recognized as such by the recipients and their shipmates. Punishment should neither be of such a nature that it lowers self-esteem, nor should it be so severe that it is out of proportion to the offense. Recipients of Navy punishment should keep two facts in mind:

1. Personnel are punished only as a result of their misbehavior, and
2. They will not be punished again if they learn to conform to Navy standards of conduct.

The administration of punishment is not personal; therefore, those who administer it should be shown no malice (hate). They are carrying out their duties as required by *Navy Regulations*.

REVIEW 3 QUESTIONS

Q1. What method does the Navy use to help Sailors work as a unit with maximum efficiency?

Q2. What is the purpose of discipline in the military?

Student Notes:

Q3. What theory of punishment does the Navy use?

Q4. What two things should a recipient of Navy punishment remember?

a.

b.

REGULATIONS THAT GOVERN THE U.S. NAVY

Learning Objectives: When you finish this chapter, you will be able to—

- Recall various parts of the *Navy Regulations* and *Uniform Code of Military Justice* (*UCMJ*) articles.
- Identify types of courts-martial, purpose of the report of offense, and the procedures for redress of grievance.

Figure 2-2 shows the three official sources that set forth the basic disciplinary laws for the Navy. These sources are the *Uniform Code of Military Justice (UCMJ)* (contained in the *Manual for Courts-Martial, 1995 Edition*), *United States Navy Regulations* (commonly called *Navy Regs*), and the *Standard Organization and Regulations of the U.S. Navy* .

Figure 2-2.—Three official sources for basic disciplinary laws.

You probably have heard the saying: "Ignorance of the law is no excuse." This is a true saying. If it weren't, personnel could excuse their misconduct merely by saying they didn't know there was a law against it. When you entered the Navy, you agreed to live by the Navy's laws and regulations. However, you do need time to learn all the rules you must obey. You should make every effort to learn them as soon as possible to avoid embarrassing situations.

The *U.S. Navy Regulations* you must learn about are discussed first. Then the *Standard Organization and Regulations of the U.S. Navy* and the *UCMJ* are discussed.

U.S. NAVY REGULATIONS

The articles published in *United States Navy Regulations* describe the principal parts of the Department of the Navy. They also describe the duties, authority, and responsibilities of some of the offices within the Department of the Navy, such as the Secretary of the Navy, the Chief of Naval Operations, and the commanding officer. You will find the regulations concerning the honors and ceremonies given to civilian and military officials of the United States and foreign governments in the *Navy Regs.*

U.S. Navy Regulations describe the rights and responsibilities of all Navy members. As you become more familiar with the regulations that govern the Navy, you can see that they are written to protect you and to provide guidance affecting your day-to-day routine.

The Chief of Naval Operations is responsible for ensuring the *U.S. Navy Regulations* conform to the current needs of the Department of the Navy. *U.S. Navy Regulations* and changes to it are issued by the Secretary of the Navy after being approved by the President of the United States.

Summaries and Excerpts from *Navy Regulations*

This section lists articles (with a condensation of their text, if appropriate) from *United States Navy Regulations, 1990,* that all personnel in the Navy should know. This listing serves only as a starting place for you to learn about Navy regulations. **You are responsible for learning and obeying all regulations**. These

Student Notes:

regulations are not punitive articles, but laws under which the Navy operates. Many exist for your own protection. Failure to obey any regulation subjects the offender to charges under article 92, *UCMJ* (Failure to obey an order or a regulation).

The first two digits of the article number indicate the chapter of *Navy Regs* from which the article is taken. If the article is self-explanatory, no further explanation is given; the article will be shown in block quotation exactly as stated in *Navy Regs*. Articles that are lengthy and, in some cases, difficult to interpret, have been paraphrased (rewritten) to give you a brief overview of what the article contains. Remember that in *Navy Regs*, the words *he, his*, or *him* refers to both men and women Navy Sailors.

0818. Publishing and Posting Orders and Regulations

> 1. In accordance with Article 137 of the *Uniform Code of Military Justice*, the articles specifically enumerated therein shall be carefully explained to each enlisted person:
>
> a) At the time of entrance on active duty or within six days thereafter;
>
> b) Again, after completion of six months active duty; and
>
> c) Again, upon the occasion of each reenlistment.
>
> 2. A text of the articles specifically enumerated in Article 137 of the *Uniform Code of Military Justice* shall be posted in a conspicuous place or places, readily accessible to all personnel of the command.
>
> 3. Instructions concerning the *Uniform Code of Military Justice* and appropriate articles of *Navy Regulations* shall be included in the training and educational program of the command.
>
> 4. Such general orders, orders from higher authority, and other matters which the commanding officer considers of interest to the personnel or profitable for them to know shall be published to the command as soon as practicable. Such matters shall also be posted, in whole or in part, in a conspicuous place or places readily accessible to personnel of the command.
>
> 5. Upon the request of any person on active duty in the armed services, the following publications shall be made available for that person's personal examination:
>
> a) A complete text of the *Uniform Code of Military Justice*;
>
> b) *Manual for Courts-Martial*;
>
> c) *Navy Regulations*;
>
> d) *Manual of the Judge Advocate General*;
>
> e) *Marine Corps Manual* (for Marine Corps personnel); and
>
> f) *Naval Military Personnel Manual* (for Navy personnel) or *Marine Corps Personnel Manual* (for Marine Corps personnel).

0917. Dealings With Foreigners

When in foreign ports, officers and enlisted personnel should respect local laws, customs, ceremonies, and regulations; display courtesy and moderation; and cultivate a feeling of good will and mutual respect.

1001. Officers of the Naval Service

> Officers of the United States naval service shall be known as officers in the line, officers in the staff corps, chief warrant officers and warrant officers.

1002. Precedence of Officers

This article discusses the precedence of officers.

1003. Relative Rank and Precedence of Officers of Different Services

This article explains the relative rank of grades of officers of the Army, Air Force, Marine Corps, Navy, Coast Guard, and members of the National Oceanic and

Student Notes:

Atmospheric Administration and Public Health Service serving with the military.

1010. Manner of Addressing Officers

This article describes the proper manner of addressing officers orally and in writing.

1020. Exercise of Authority

> All persons in the naval service on active service, and those on the retired list with pay, and transferred members of the Fleet Reserve and the Fleet Marine Corps Reserve, are at all times subject to naval authority. While on active service they may, if not on leave of absence…, on the sick list, taken into custody, under arrest, suspended from duty, in confinement or otherwise incapable of discharging their duties, exercise authority over all persons who are subordinate to them.

1021. Authority Over Subordinates

This article gives officers the authority necessary to perform their duties.

1022. Delegation of Authority

Although you may delegate authority, that does not relieve you of being responsible. You must make sure the delegated authority is properly exercised and orders and instructions are properly executed.

1023. Abuse of Authority

Persons in authority are forbidden to injure their subordinates by tyrannical or capricious conduct, or by abusive language.

1024. Contradictory and Conflicting Orders

> If an enlisted person in the naval service receives an order, which annuls, suspends or modifies one received from another superior, he or she shall immediately represent the facts to the superior from whom the last order was received. If, after such representation, the superior from whom the last order was received should insist upon the execution of that order, it shall be obeyed. The person receiving and executing such order shall report the circumstances as soon as practicable to the superior from whom the original order was received.

1025. Authority of an Officer in Command

An officer in command, either of the line or of a staff corps, has authority over all officers and other persons attached to the command, whatever their rank and whether they are of the line or of a staff corps.

1033. Authority in a Boat

This article provides the senior line officer eligible for command at sea the authority over all persons embarked in a boat. It also delegates to the officer the responsibility for the safety and management of the boat.

1034. Authority and Responsibility of a Senior Officer Under Certain Circumstances

This article gives the senior person present, whether an officer or an enlisted person, the authority to take necessary action during a riot, a quarrel between naval service members, or circumstances not covered by these regulations.

1037. Authority of Warrant Officers, Noncommissioned Officers and Petty Officers

> Chief warrant officers, warrant officers, noncommissioned officers and petty officers shall have, under their superiors, all necessary authority for the proper performance of their duties, and they shall be obeyed accordingly.

1038. Authority of a Sentry

> A sentry, within the limits stated in his or her orders, has authority over all persons on his or her post.

1052. Orders to Active Service

You may not be ordered to active service without permission of the Chief of Naval Personnel.

Student Notes:

1064. Detail of Enlisted Persons for Certain Duties

Petty officers will not be detailed to perform mess duties, except when nonrated persons are unavailable.

1101. Demand for Court Martial

> Except as otherwise provided in the *Uniform Code of Military Justice*, no person in the naval service may demand a court martial either on him- or herself or on any other person in the naval service.

1102. Limitations on Certain Punishments

> Instruments of restraint, such as handcuffs, chains, irons and straitjackets, shall not be applied as punishment. Other instruments of restraint may not be used except for safe custody and no longer than is strictly necessary....
>
> The punishments of extra duties and hard labor without confinement are not performed on Sunday, although Sunday counts in the computation of the period for which such punishments are imposed.
>
> Guard duty shall not be inflicted as punishment.

1104. Treatment and Release of Prisoners

Persons in confinement must not be subjected to cruel or unusual treatment. They must be visited at least once every 4 hours to check on their condition and to care for their needs. In the event of an emergency, they may be removed to a safe area or released within the limits of the command. No greater force than that required to restrain or to confine an offender should be used to take into custody a person under the influence of alcohol, marijuana, narcotic substances, or other controlled substances.

1105. Places of Confinement

Prisoners must be confined only in brigs or other facilities designated as naval places of confinement by the Secretary of the Navy. In case of necessary temporary confinement, the senior officer present may authorize confinement in spaces that provide sufficient security, safety for both prisoner and guards, and adequate living conditions.

Persons under the influence of alcohol or other drugs should not be confined in any place or manner that may be dangerous to them in their condition.

1110. Standards of Conduct

> All Department of the Navy personnel are expected to conduct themselves in accordance with the highest standards of personal and professional integrity and ethics. At a minimum, all personnel shall comply with directives issued by the Secretary of Defense and the Secretary of the Navy regarding the Standards of Conduct and Government Ethics.

1111. Pecuniary Dealings with Enlisted Persons

No officer should have any dealings involving money with enlisted persons except as may be required in the performance of the officer's duties or as involved in the sale of personal property. An officer may be designated by superior authority to accept deposits from enlisted personnel for the purpose of safeguarding those funds under emergency or operational situations.

1112. Lending Money and Engaging in a Trade or Business

Naval personnel must not lend money to another member of the armed services at an interest rate, for the period of the loan, that exceeds 18 percent simple interest per year. Personnel may not act as a salesperson or an agent or engage in a business on board without permission of the commanding officer.

1113. Endorsement of Commercial Product or Process

> Except as necessary during contract administration to determine specification or other compliance, no person in the Department of the Navy, in his or her official capacity, shall endorse or express an opinion of approval or disapproval of any commercial product or process.

Student Notes:

1115. Report of Fraud

Any suspicions of fraud, collusion, or improper conduct in matters concerning supplies and repairs should be reported to the proper authority.

1121. Disclosure, Publication and Security of Official Information

Naval personnel may not make speeches or write anything that might disclose information of interest to foreign countries or that would aid persons with claims against the United States. If naval personnel publish articles on Navy, political, or international subjects, they must state the views are theirs and not those of the Navy. When such articles are accepted for publication, personnel must forward a complete copy of each article to the Secretary of the Navy.

1122. Adverse Matter in Officer Fitness Reports and Enlisted Performance Evaluation Reports

Information of an adverse nature should not be entered in the record of a person of the naval service unless the member was first afforded an opportunity to submit a written statement regarding the matter. Certain medical and dental entries are excepted.

1125. Inspection of the Record of a Person in the Naval Service

The record of a person in the naval service which is maintained by the Chief of Naval Personnel or the Commandant of the Marine Corps shall be available for inspection by the person or a duly authorized agent, designated as such in writing by the person.

1126. Correction of Naval Records

Any military record in the Department of the Navy may be corrected by the Secretary of the Navy, acting through the Board for Correction of Naval Records, when the Secretary considers that such action should be taken in order to correct an error or to remove an injustice.

Applications for corrections may be made only after exhaustion of all other administrative remedies afforded by law or regulation.

1127. Control of Official Records

No person, without proper authority, shall withdraw official records or correspondence from the files, or destroy them, or withhold them from those persons authorized to have access to them.

1129. Records of Fitness

Records of fitness reflect each officer and enlisted person's fitness for service and fitness for the performance of duties. These records are used to determine promotions and duty assignments.

1132. Compliance with Lawful Orders

All persons in the naval service are required to obey readily and strictly, and to execute promptly, the lawful orders of their superiors.

1133. Language Reflecting on a Superior

No person in the naval service shall use language that may tend to diminish the confidence in or respect due to his or her superior officer.

1134. Exchange of Duty

An assigned duty may not be changed with another person (such as trading watches) without permission from proper authority.

1135. Relations with Foreign Nations

Naval service members must conform to international law and precedents set by the United States in its relations with foreign nations.

1136. Foreign Religious Institutions

Navy personnel visiting foreign nations must respect that country's religious institutions and customs.

Student Notes:

1137. Obligation to Report Offenses

> Persons in the naval service shall report as soon as possible to superior authority all offenses under the Uniform Code of Military Justice which come under their observation, except when such persons are themselves already criminally involved in such offenses at the time such offenses first come under their observation.

You should report all offenses under the UCMJ that you see to the proper authority unless reporting the offense would incriminate yourself. This is known as *self-incrimination.*

1138. Responsibilities Concerning Marijuana, Narcotics and Other Controlled Substances

Personnel may not bring on board any naval activity, or have in their possession at any time, marijuana, narcotics, or any controlled substances.

1140. Capture by an Enemy

A person in the naval service who is captured by the enemy is required to give name, grade or rate, service number, and date of birth. That person will make no statement disloyal to, critical of, or harmful to the United States or its allies.

1142. Unavoidable Separation from a Command

Persons who become separated from their ship, station, or unit by shipwreck, disaster, or other unavoidable happening, should proceed to the nearest U.S. military activity as soon as possible.

1143. Report of a Communicable Disease

Personnel should report any suspicions of communicable disease to their medical representative.

1144. Immunization

Personnel must take the immunizations prescribed for them as scheduled.

1145. Service Examinations

No persons in the Navy, without proper authority, should have or attempt to have in their possession, any examination papers, any part or copy thereof, or any examination answer sheets. They also must not obtain, sell, publish, give, purchase, receive, or reproduce any of these examination products.

1150. Redress of Wrong Committed by a Superior

A person who believes a superior has wronged him/her or is guilty of misconduct should submit a complaint to his/her commanding officer.

1151. Direct Communication with the Commanding Officer

> The right of any person in the naval service to communicate with the commanding officer in a proper manner, and at a proper time and place, shall not be denied or restricted.

1152. Suggestions for Improvement

Any person in the Navy may submit suggestions or constructive criticism about efficiency or economical methods of administration or management within the Department of the Navy. Suggestions or criticism should be submitted to the Secretary of the Navy through the chain of command.

1154. Communications to the Congress

Personnel may not, in their official capacity, apply to Congress for congressional action of any kind or provide information requested by Congress. The only exception to this regulation is such communication as authorized by the Secretary of the Navy or as provided by law.

1155. Dealings with Members of Congress

All persons may write to their congressmen in a personal or private capacity on any subject as long as they do not violate security regulations or the law.

Student Notes:

1156. Forwarding Individual Requests

Requests from persons in the naval service shall be acted upon promptly. When addressed to higher authority, requests shall be forwarded without delay. The reason should be stated when a request is not approved or recommended.

1157. Leave and Liberty

Leave and liberty will be granted to the maximum extent practicable.

1159. Possession of Weapons

Personnel may not have any weapons or explosives in their possession without proper authority.

1160. Possession of Government Property

Personnel shall not possess, without permission, any property of the United States except what is needed in the performance of their duty.

1162. Alcoholic Beverages

The personal possession of any alcoholic beverages aboard any ship is prohibited. The transportation aboard ship of alcoholic beverages for personal use ashore is authorized subject to the discretion of and under regulations established by the commanding officer.

1164. Equal Opportunity and Treatment

All persons in the Department of the Navy regardless of their race, color, religion, sex, or national origin, consistent with requirements for physical capabilities, will be afforded equal opportunity and treatment.

1165. Fraternization Prohibited

No person in the Navy is to enter a personal relationship that is unduly familiar, does not respect differences in rank, and is prejudicial to good order and discipline.

1166. Sexual Harassment

Do not make offensive verbal comments, gestures, or physical contact in the work environment. Do not use implicit or explicit sexual behavior to control other personnel.

1167. Supremacist Activities

> No person in the Naval service shall participate in any organization that espouses supremacist causes; attempts to create illegal discrimination based on race, creed, color, sex, religion, or national origin; advocates the use of force or violence against the government of the United States or the government of any state, territory, district, or possession thereof, or the government of any subdivision therein; or otherwise engages in efforts to deprive individuals of their civil rights.

STANDARD ORGANIZATION AND REGULATIONS OF THE U.S. NAVY

The *Standard Organization and Regulations of the U.S. Navy*, OPNAVINST 3120.32, provides regulations and guidance governing the conduct of all members of the Navy. This instruction specifies duties and responsibilities of personnel within a unit organization—from the commanding officer down to the messenger of the watch.

Naval personnel who fail to comply with regulations may be awarded punishment based on the *Uniform Code of Military Justice (UCMJ)*. Many regulations are printed on large posters and posted in conspicuous locations aboard naval units.

Excerpts from the *Standard Organization and Regulations of the U.S. Navy*

This section contains some of the articles contained in chapter 5, "Regulations," of the OPNAVINST 3120.32. Self-explanatory articles are shown in block quotation exactly as stated in the OPNAVINST 3120.32. Sections that are lengthy or difficult to interpret are paraphrased to briefly explain the contents of the regulation.

Student Notes:

510.5 Armed Forces Identification Cards and Leave Papers

No person without proper authority shall:

a. Have in his/her possession more than one properly validated Armed Forces identification card.

b. Depart on liberty without his/her own properly validated identification card; or, in the case of leave, without his/her own properly validated leave papers and identification card.

c. Have in his/her possession a false or unauthorized identification card; or a mutilated, erased, altered, or not properly validated identification card; or an identification card bearing false or inaccurate information concerning a name, grade, service number, or date of birth.

d. Return from leave without depositing his/her leave papers with the proper authority. Any person returning without an identification card shall report the loss to the OOD in person.

510.14 Customs

Upon arrival of a naval unit in United States territory after visiting a foreign port, it is subject to customs and other inspections by Federal authorities.

a. On such occasions, customs declarations will be distributed to all hands in sufficient time to be filled out and returned before arrival in port.

b. It shall be the duty of all personnel to accurately complete customs declarations prior to arrival in port.

c. No person, without permission from the commanding officer, shall bring on board any article, animal, or any other thing, the introduction of which into U.S. territory is forbidden or restricted under current regulations.

510.16 Divine Services

Accessible and appropriate space shall be provided for divine services. No person shall conduct himself/herself in a manner that would interfere with properly authorized divine services.

510.18 Emergency Equipment

No person shall use emergency equipment for any purpose other than that for which it is intended. Emergency equipment includes items such as battle lanterns, emergency first aid boxes, shoring, wrenches, life rings, equipment in life rafts and boats, portable fire pumps, fire hoses, and fuel for emergency machinery.

510.21 Government Property

No person shall:

a. Conceal or fail to report to proper authority the loss, removal, destruction, or damage of government property entrusted to his/her care or custody.

b. Remove without proper authority from its regular place of stowage or location, for any purpose whatever, any article of government property, including hull and damage control fittings, first aid equipment, life saving and emergency equipment, and stores and foodstuffs.

c. Have in his/her possession any article of government property except as may be necessary for the performance of his/her duty or as may be authorized by proper authority.

510.27. Intoxicated Persons

a. The officer of the deck or the command duty officer shall ensure that the medical officer or a qualified representative shall promptly examine all persons who return on board in an intoxicated condition, or found on board intoxicated.

Student Notes:

b. When restraint is imposed on an individual, it should be in such a manner as to accomplish the desired degree of restraint with a minimum of force. Attachment of an individual to a fixed or immovable object should only be authorized when all else fails, and then a continuous guard should be posted with specific instructions to care for the welfare of the person under restraint in the event of an emergency.

510.34 Motor Vehicles

a. No person shall operate a Government-owned motor vehicle assigned to a naval unit unless specifically designated to do so by the commanding officer, and then only for official unit business.

b. Military personnel operating Government-owned motor vehicles shall comply with all post, station, local, state, and federal directives. U.S. Government operator's permit is not required for vehicles under one ton.

c. All persons operating Government-owned motor vehicles assigned to a naval unit shall obtain the permission of the OOD before driving away from the unit and shall report to the OOD upon return.

510.35 Working Stocks of Narcotics

All narcotics and other controlled substances authorized for medical purposes shall be in the custody of the medical or dental officer. No one shall have access to this material except as prescribed by these officers or the commanding officer.

a. The medical and dental officers shall supervise in person all receipts and issues of narcotics and other controlled substances in their custody and shall keep proper records of all transactions to ensure strict accountability and detect losses promptly.

b. With the exception of medical and dental officers, no person shall prescribe or administer any narcotics or other controlled substances, either to oneself or to another person, except to aid the injured during action or emergencies. The medical and dental officers may authorize certain hospital corpsmen and dental technicians to administer narcotics and controlled drugs to patients in sick bay per the medical and dental officer's prescription.

c. In units to which no medical officer is attached, all narcotics and dangerous drugs shall be in the custody of the controlled substances custodian, except small quantities of necessary narcotics and dangerous drugs that may be issued to the leading petty officer in the medical department. The narcotics and dangerous drugs shall be kept in a three-combination safe or, if this is not possible, under lock and key. All transactions between the bulk custodian and medical department representative shall be receipted for. Issues from the working stock in the sick bay shall be covered by prescription. Narcotics and dangerous drugs shall be inventoried monthly by a special inventory board....

510.44 Photographic Equipment

No person shall:

a. Possess or introduce on board a naval unit any camera or other photographic equipment capable of exposing a photographic plate or film without permission of the commanding officer or authorized representative.

b. Make photographs of a naval unit or its equipment, or of objects from the unit, without permission of the commanding officer, and then only of the objects for which permission was specifically given.

c. While on watch or duty as a sentry or member of a patrol, knowingly permit the introduction of any camera or photographic equipment on board a naval unit unless such equipment is authorized by the commanding officer or authorized representative.

Student Notes:

UNIFORM CODE OF MILITARY JUSTICE

Until 1951, the various branches of our armed forces operated under different military codes. The Army's *Articles of War* guided the administration of discipline and legal processes of the Army and Air Force. The Navy was guided by the *Articles for the Government of the Navy* ("Rocks and Shoals"), and the Coast Guard, by the *Disciplinary Laws of the Coast Guard*. Not surprisingly, an act considered an offense in the eyes of the Navy might not have been judged so in the Army. Even if an act was a breach of discipline in all branches of the armed forces, the type of trial and severity of punishment awarded varied.

A standardized code of military justice was recognized as a logical and necessary unification measure. Secretary of Defense, James Forrestal, appointed an interservice committee to study the measure. After an intensive study, the committee drafted what is now known as the *Uniform Code of Military Justice (UCMJ)*. The *UCMJ* was passed by Congress on 5 May 1950, signed into law by the President, and became effective 31 May 1951.

The *Manual for Courts-Martial, United States, 1951 (MCM)*, consolidated and standardized military legal procedures. Effective 31 May 1951, the same date as the original *UCMJ*, the *MCM* became the new standard of military justice. Case decisions of the Court of Military Appeals and changes in courts-martial procedures have made necessary several changes to the original manual. The current edition is the *Manual for Courts-Martial, 1995 Edition*.

Congress and the Navy have taken steps to ensure you will know the disciplinary laws and regulations most likely to affect your daily life. Article 137 of the *UCMJ* states that certain articles of the Code must be explained carefully to every enlisted person at certain intervals. They must be explained—

- at the time the person enters on active duty,
- after 6 months of active duty, and
- when the person reenlists.

In general, these articles concern the following topics:

Article	Subject
2	Persons subject to the Code
3	Jurisdiction to try certain persons even though they have been separated from the service
7-14	Apprehension and restraint
15	Nonjudicial punishment (captain's mast)
25	Membership of courts-martial
27	Detail of trial and defense counsel
31	Compulsory self-incrimination prohibited
37	Unlawful influence on the court
38	Duties of counsel
55	Certain cruel and unusual punishments prohibited
77-134	Punitive articles
137	Articles that must be explained
138	Complaints of wrongs
139	Payment for injury or loss of property

Navy Regulations supplements article 137 of the *UCMJ* by requiring each command to post the text of those articles in the preceding list in a conspicuous place. *Navy Regs* also requires each command to include these and other appropriate articles of *Navy Regulations* in the command's training and education program. Copies of the complete *UCMJ* (140 articles), *Navy Regulations*, and other general orders are available to any person who wants to read them.

Excerpts from the *Uniform Code of Military Justice*

The purpose of this section is not to make you an expert on the *Uniform Code of Military Justice (UCMJ)* but to give you an overview of each of the articles

Student Notes:

prescribed by article 137. Those articles that are self-explanatory are shown in block quotation as stated in the *UCMJ*; no further explanation is given. Some of the more lengthy articles have been shortened to present only portions of these articles. Articles that are lengthy and, in some cases, difficult to interpret are paraphrased (rewritten) to give you a brief overview of what the article contains.

NOTE

In this section of the chapter, the words "he," "his," and "him" do not indicate gender and are used for economy of communication.

Art. 2. Persons Subject to this Code

> The following persons are subject to this code:
>
> (1) Members of a regular component of the armed forces, including those awaiting discharge after expiration of their terms of enlistment; volunteers from the time of their muster or acceptance into the armed forces; inductees from the time of their actual induction into the armed forces; and other persons lawfully called or ordered into, or to duty in or for training in, the armed forces, from the dates when they are required by the terms of the call or order to obey it.

This article includes all persons on active duty, certain retired persons, prisoners, and prisoners of war.

You should specifically note the following provisions of article 2:

- Any person serving a sentence imposed by a court-martial remains subject to the *UCMJ*. Thus a prisoner who is serving a court-martial sentence may be tried for a crime committed while a prisoner. This applies even though the prisoner's term of enlistment has expired at the time of commission of the crime.

- A reservist on inactive-duty training is subject to the *UCMJ* when (a) the training is authorized by written orders; (b) the orders are voluntarily accepted by the reservist; and (c) the orders specify that the reservist is subject to the *UCMJ*.

- A reservist ordered into the active military service is subject to the *UCMJ* beginning on the date specified in the orders for the reservist to report for active duty.

- The United States Supreme Court has held unconstitutional the exercise of court-martial jurisdiction over civilians in time of peace.

Art. 3. Jurisdiction to Try Certain Personnel

Article 3 states that a person may be tried by court-martial, even after leaving the service, for offenses committed while subject to the *UCMJ*.

Art. 7. Apprehension

> (a) Apprehension is the taking of a person into custody.
>
> (b) Any person authorized under regulations governing the armed forces to apprehend persons subject to this code or to trial thereunder may do so upon reasonable belief that an offense has been committed and that the person apprehended committed it.
>
> (c) Commissioned officers, warrant officers, petty officers, and noncommissioned officers have authority to quell quarrels, frays, and disorders among persons subject to this code and to apprehend persons subject to this code who take part therein.

Enlisted persons performing police duties should not apprehend an officer except on specific orders of a commissioned officer. The exception is when such apprehension is necessary to prevent disgrace to the service, the commission of a serious offense, or the escape of one who has committed a serious offense. In such cases, the apprehending individual immediately notifies the officer to whom he or she is responsible or an officer of the security police, military police, or shore patrol.

An apprehension is effected by clearly notifying the offender that he/she is thereby taken into custody. The order may be oral or written.

Student Notes:

A clear distinction exists between the authority to apprehend and the authority to arrest or confine (article 9). Any person empowered to apprehend an offender, however, is authorized to secure the custody of an alleged offender until proper authority may be notified.

Art. 8. Apprehension of Deserters

Any civil officer having authority to apprehend offenders under the laws of the United States or of a State, Territory, Commonwealth, or possession, or the District of Columbia may summarily apprehend a deserter from the armed forces and deliver him into the custody of those forces.

When a military service sends out a description of a deserter, with a request for the deserter's apprehension, the notice gives civil officers the authority to apprehend the person.

Art. 9. Imposition of Restraint

(a) Arrest is the restraint of a person by an order not imposed as a punishment for an offense, directing him to remain within certain specified limits. Confinement is the physical restraint of a person.

(b) An enlisted member may be ordered into arrest or confinement by any commissioned officer by an order, oral or written, delivered in person or through other persons subject to this code. A commanding officer may authorize warrant officers, petty officers, or noncommissioned officers to order enlisted members of his command or subject to his authority into arrest or confinement.

(c) A commissioned officer, a warrant officer, or a civilian subject to this code or to trial thereunder may be ordered into arrest or confinement only by a commanding officer to whose authority he is subject, by an order, oral or written, delivered in person or by another commissioned officer. The authority to order such persons into arrest or confinement may not be delegated.

(d) No person may be ordered into arrest or confinement except for probable cause.

(e) Nothing in this article limits the authority of persons authorized to apprehend offenders to secure the custody of an alleged offender until proper authority may be notified.

Art. 10. Restraint of Persons Charged with Offenses

Any person subject to this chapter charged with an offense under this chapter shall be ordered into arrest or confinement, as circumstances may require; but when charged only with an offense normally tried by a summary court-martial, he shall not ordinarily be placed in confinement. When any person subject to this chapter is placed in arrest or confinement prior to trial, immediate steps shall be taken to inform him of the specific wrong of which he is accused and to try him or to dismiss the charges and release him.

As the words *normally* and *ordinarily* imply, the provisions of this article may not apply in exceptional cases. Whether to confine, arrest, or restrict a person in lieu of arrest is with the discretion of the officer having the power to do so. What this article says, in effect, is that in most instances confinement is not necessary for persons accused of minor offenses.

Art. 11. Reports and Receiving of Prisoners

(a) No provost marshal, commander of a guard, or master-at-arms may refuse to receive or keep any prisoner committed to his charge by a commissioned officer of the armed forces, when the committing officer furnishes a statement, signed by him, of the offense charged against the prisoner.

(b) Every commander of the guard or master-at-arms to whose charge a prisoner is committed shall, within twenty-four hours after that commitment or as soon as he is relieved from guard, report to the commanding officer the name of the prisoner, the offense charged against him, and the name of the person who ordered or authorized the commitment.

Student Notes:

An arrest is imposed by notifying the person to be arrested that the person is under arrest and informing the person of the limits of the arrest. The order of arrest may be oral or written. A person to be confined is placed under guard and taken to the place of confinement.

Art. 12. Confinement with Enemy Prisoners Prohibited

> No member of the armed forces may be placed in confinement in immediate association with enemy prisoners or other foreign nationals not members of the armed forces.

Members of the armed forces may be confined in the same jails, prisons, or other confinement facilities, however, so long as they are separated from the other categories mentioned.

Art. 13. Punishment Prohibited Before Trial

> No person, while being held for trial, may be subjected to punishment or penalty other than arrest or confinement upon the charges pending against him, nor shall the arrest or confinement imposed upon him be any more rigorous than the circumstances required to ensure his presence, but he may be subjected to minor punishment during that period for infractions of discipline.

The minor punishment permitted under article 13 includes that authorized for violations of discipline set forth by the place in which the person is confined. The article does not prevent a person from being required to do ordinary cleaning or policing or from taking part in routine training and duties not involving the bearing of arms.

Art. 14. Delivery of Offenders to Civil Authorities

> (a) Under such regulations as the Secretary concerned may prescribe, a member of the armed forces accused of an offense against civil authority may be delivered, upon request, to the civil authority for trial.

> (b) When delivery under this article is made to any civil authority of a person undergoing sentence of a court-martial, the delivery, if followed by conviction in a civil tribunal, interrupts the execution of the sentence of the court-martial, and the offender after having answered to the civil authorities for this offense shall, upon the request of competent military authority, be returned to military custody for the completion of his sentence.

Art. 15. Commanding Officer's Nonjudicial Punishment

Article 15 explains commanding officers' nonjudicial punishment. For some offenses, commanders may offer an article 15 instead of court-martial. If accepted, the commander may impose punishment as permitted by regulations (usually at captain's mast). Receiving an article 15 is not a conviction, and it does not give a person a criminal record. This article will be explained in greater detail later in this chapter under "Nonjudicial Punishment."

Art. 25. Who May Serve on Courts-Martial

Any commissioned officer, including commissioned warrant officers, on active duty with the armed forces is eligible to serve on a court-martial. Any warrant officer on active duty with the armed forces is eligible to serve on a general court-martial (GCM) and special court-martial (SPCM) for the trial of any person, other than a commissioned officer. Any enlisted person on active duty with the armed forces who is not a member of the same unit as the accused is eligible to serve on general and special courts-martial for the trial of enlisted persons. However, enlisted personnel may serve as a member of a court-martial only if, before the assembling of such court, the accused has personally requested in writing that enlisted personnel serve as members of the court.

Art. 27. Detail of Trial Counsel and Defense Counsel

Each general and special court-martial must have a trial counsel and a defense counsel, with such assistants as the convening authority deems necessary. The terms *counsel, trial counsel*, and *defense counsel* should be

Student Notes:

interpreted to mean the detailed counsel named in the convening order. The term *individual counsel* refers to the military counsel selected by the accused or the civilian counsel provided by the accused at his/her own expense.

The trial counsel and defense counsel detailed for a general court-martial must have equivalent legal qualifications. Each must be a judge advocate of the Army, Navy, Air Force, or Marine Corps who is a graduate of an accredited law school or is a member of the bar of a federal court or of the highest court of a state. Each must be certified as competent to perform such duties by the Judge Advocate General of the armed forces of which he/she is a member. A civilian counsel must be a member of the bar of a federal court or of the highest court of a state.

In a special court-martial, the accused must be afforded the opportunity to be represented by counsel qualified under article 27, *UCMJ*, unless such counsel cannot be obtained because of the geographical location or pressing military requirements. If qualified defense counsel cannot be obtained, the court may be convened and the trial held. The convening authority makes a written statement that states why qualified counsel cannot be obtained. The following conditions must be met:

- If the detailed trial counsel or any assistant trial counsel is qualified to act as counsel before a general court-martial, the defense counsel must be a qualified person; and
- If the detailed trial counsel or any assistant trial counsel is a judge advocate or a member of the bar of a federal court or the highest court of a state, the defense counsel detailed by the convening authority must be one of the foregoing.

Art. 31. Compulsory Self-Incrimination Prohibited

Article 31 explains your rights not to provide evidence against yourself (self-incrimination), a right given to all citizens under the Fifth Amendment to the U.S. Constitution. The following statements explain your rights against self-incrimination:

- You cannot be forced to answer questions or give evidence that may help to prove your guilt.
- You must be told the nature of the offense of which you are accused; that you do not have to make any statement; and that if you do, it can be used against you.
- You cannot be forced to make a statement or give evidence in a trial that is not related to the case or that may degrade you.
- No statement obtained from you by threats or trickery can be used against you in a court-martial trial.

Art. 37. Unlawfully Influencing Action of Court

> (a) No authority convening a general, special, or summary court-martial, nor any other commanding officer, may censure, reprimand, or admonish the court or any member, military judge, or counsel thereof, with respect to the findings or sentence adjudged by the court, or with respect to any other exercise of its or his functions in the conduct of the proceedings. No person subject to this chapter may attempt to coerce or, by any unauthorized means, influence the action of a court-martial or any other military tribunal or any member thereof, in reaching the findings or sentence in any case, or the action of any convening, approving, or reviewing authority with respect to his judicial acts.

Article 37 is designed to ensure that every court, its members, and its officers are completely free to fulfill their functions without fear of reprisal.

Art. 38. Duties of Trial Counsel and Defense Counsel

The trial counsel prosecutes in the name of the United States and, under the direction of the court, prepares the record of proceedings. The duties of the trial counsel might be compared to those of a civil district attorney. The prosecution must prove beyond a reasonable doubt the guilt of the accused for each offense charged. Of course, such burden of proof is

Student Notes:

relieved by a plea of guilty. The many duties of the trial counsel vary widely beginning at the time of assignment to the trial. The duties change throughout the preparation for trial, the trial itself, and the preparation and disposition of the record of trial.

All accused persons have the right to be represented before special and general courts-martial by defense counsel. This counsel may be a civilian or military lawyer selected by the accused or the convening authority may appoint a defense counsel. If a civilian counsel is selected, the accused must pay the counsel's expenses. If the accused prefers to select a civilian counsel, the detailed counsel and assistant counsel act as associate counsel if the accused so desires; otherwise, they may be excused.

Art. 55. Cruel and Unusual Punishments Prohibited

Article 55 prohibits any cruel or unusual punishment. In particular, courts-martial are forbidden to award sentences that include flogging, branding, marking, or tattooing the body. The use of irons is also prohibited except for the purpose of safe custody.

Punitive Articles of the *UCMJ*

The punitive articles of the *UCMJ* are those numbered 77 through 134. They are the laws of Congress telling you what you must do and must not do, under pain of punishment.

What about civil laws? Can you be given military punishment for nonmilitary offenses? Yes, you can. For example, the only *UCMJ* regulations against drunkenness are for drunken driving and being drunk on duty. Many civilian communities, though, have laws against public intoxication. If you are found guilty in civil court and spend time in jail for public intoxication, the Navy can try you for being absent without leave (*UCMJ*, article 86) and for bringing discredit upon the Navy (*UCMJ*, article 134).

If you willfully refuse to pay just debts, you will be warned to pay them by your commanding officer. Continued failure to pay your debts can lead to an undesirable type of discharge. The Navy has no use for people who don't exhibit integrity and honesty. On the other hand, if unscrupulous dealers are gouging you, see your legal officer for assistance.

The punitive articles that follow are those that you are required to know. If you have any questions about their meaning, ask your division officer for guidance.

Art. 77. Principals

The mere fact that a person is at the scene of a crime does not make the person a principal. To be a principal of a crime, the person must be guilty of intent to aid or encourage the person(s) who committed the crime.

A person who witnesses a crime can be a principal. Evidence must show the witness had a duty to interfere and the witness's noninterference was intended to operate and did operate to encourage or protect the perpetrator.

A person may be a principal even though not at the scene of the crime if he/she commanded, advised, or obtained another person to commit an offense.

Art. 78. Accessory After the Fact

> Any person subject to this chapter who, knowing that an offense punishable by this chapter has been committed, receives, comforts, or assists the offender in order to hinder or prevent his apprehension, trial, or punishment shall be punished as a court-martial may direct.

A person who voluntarily gives an escaped prisoner provisions that permit him/her to avoid pursuers becomes an accessory after the fact to the prisoner's escape. Provisions include transportation, clothing, money, or any other necessities.

Art. 79. Conviction of Lesser Included Offense

> An accused may be found guilty of an offense necessarily included in the offense charged or of an attempt to commit either the offense charged or an offense necessarily included therein.

Student Notes:

Examples of what generally are held to be lesser-included offenses contained in a principal offense include the following:

ART	PRINCIPAL OFFENSE	ART	LESSER INCLUDED OFFENSE
94	Mutiny	92	Failure to obey order or regulation
94	Sedition	116	Riot or breach of peace
95	Resistance, breach of arrest, and escape	134	General article
118	Murder	119	Manslaughter
122	Robbery	121	Larceny and wrongful appropriation
124	Maiming	128	Assault

Art. 80. Attempts

(a) An act, done with specific intent to commit an offense under this chapter, amounting to more than mere preparation and tending, even though failing, to effect its commission, is an attempt to commit that offense.

(b) Any person subject to this chapter who attempts to commit any offense punishable by this chapter shall be punished as a court-martial may direct, unless otherwise specifically prescribed.

(c) Any person subject to this chapter may be convicted of an attempt to commit an offense although it appears on the trial that the offense was consummated.

An accused may be guilty of an attempt even though the crime turns out to be impossible to commit because of an outside intervening circumstance. For example, a pickpocket who puts a hand in the pocket of another person with the intent to steal a billfold is guilty of an attempt to commit larceny, even though the pocket is empty.

Art. 81. Conspiracy

Conspiracy is defined as an agreement between two or more persons to commit a crime. Conspiracy refers to such a plan by a group whose intent usually is to commit a crime of a bold nature, such as overthrowing a government.

The agreement in a conspiracy need not be formal. The agreement need only be a common understanding in the minds of the parties to accomplish the objective of the conspiracy.

Art. 82. Solicitation

(a) Any person subject to this chapter who solicits or advises another or others to desert in violation of...(article 85) or mutiny in violation of...(article 94) shall, if the offense solicited or advised is attempted or committed, be punished with the punishment provided for the commission of the offense, but, if the offense solicited or advised is not committed or attempted, he shall be punished as a court-martial may direct.

(b) Any person subject to this chapter who solicits or advises another or others to commit an act of misbehavior before the enemy in violation of...(article 99) or sedition in violation of...(article 94) shall, if the offense solicited or advised is committed, be punished with the punishment provided for the commission of the offense, but, if the offense solicited or advised is not committed, he shall be punished as a court-martial may direct.

Solicitation may be accomplished by other means than by word of mouth or by writing. Any act or conduct that may reasonably be considered as a serious request or advice to commit one of the offenses named in the article may constitute solicitation. The accused may act through other persons in committing this offense.

Student Notes:

Art. 83. Fraudulent Enlistment, Appointment, or Separation

Any person who:

(1) procures his own enlistment or appointment in the armed forces by knowingly false representation or deliberate concealment as to his qualifications for that enlistment or appointment and receives pay or allowances thereunder; or

(2) procures his own separation from the armed forces by knowingly false representation or deliberate concealment as to his eligibility for that separation;

shall be punished as a court-martial may direct.

An essential element of the offense of fraudulent enlistment or appointment is that the accused shall have received pay or allowances while under that enlistment or appointment. Acceptance of food, clothing, shelter, or transportation from the government constitutes receipt of allowances.

After apprehension, an accused charged with having fraudulently obtained separation from a branch of the armed forces is subject to the *UCMJ*. The accused is subject to the *UCMJ* while in the custody of the armed forces and while awaiting trial for the fraudulent separation.

Art. 84. Unlawful Enlistment, Appointment, or Separation

Any person subject to this chapter who effects an enlistment or appointment in or a separation from the armed forces of any person who is known to him to be ineligible for that enlistment, appointment, or separation because it is prohibited by law, regulation, or order shall be punished as a court-martial may direct.

Art. 85. Desertion

Members of the armed forces who, without permission, leave their place of duty or organization with the intent to remain away permanently are guilty of desertion.

The status of an absentee changes to that of a deserter after 30 days of absence, or sooner if the intent to desert is apparent. For example, suppose a Navy member goes ashore without permission, taking all personal belongings and announcing to shipmates that he/she is leaving the service for good. That person could be immediately declared a deserter.

After an individual is declared a deserter, notification is forwarded to the next of kin, the deserter's hometown police, and various other law enforcement agencies, including the FBI. Deserters are nearly always caught and identified because of nationwide fingerprinting and identification practices. Furthermore, expenses incurred in the return of the deserter to military control are chargeable to the returned absentee.

The effects of desertion can be many; some can be severe. If tried and convicted of desertion, the deserter is almost certainly imprisoned; in time of war, the deserter may be executed. A person whose conviction of desertion in time of war results in a dishonorable discharge can never hold any office of trust or profit in the United States government.

Art. 86. Absence without Leave

Any member of the armed forces who, without authority—

(1) fails to go to his appointed place of duty at the time prescribed;

(2) goes from that place; or

(3) absents himself or remains absent from his unit, organization, or place of duty at which he is required to be at the time prescribed;

shall be punished as a court-martial may direct.

This article covers every case not provided for in the other punitive articles in which an armed forces member, through that member's own fault, is not in a required location at a specified time. As opposed to desertion, whether or not the member intended to remain away makes no difference. The intent is expressed by the member's absence.

Student Notes:

Make sure you avoid the bad habit of taking the last bus, train, or plane when returning from leave. Always allow time for unexpected delays.

Art. 87. Missing Movement

> Any person subject to this chapter who through neglect or design misses the movement of a ship, aircraft, or unit with which he is required in the course of duty to move shall be punished as a court-martial may direct.

Provisions of article 87 should be self-explanatory. However, note that the violator, to be found guilty, need not have known the exact hour or even the exact date of the scheduled movement. If a person had knowledge of only the approximate date, the court may convict the absentee on the charge of missing movement. Missing movement is a serious offense in the Navy. It leaves the ship shorthanded and requires someone else to do the absentee's work and stand the absentee's watches.

Art. 88. Contempt Toward Officials

> Any commissioned officer who uses contemptuous words against the President, the Vice President, Congress, the Secretary of Defense, the Secretary of a military department, the Secretary of the Treasury, or the Governor or legislature of any State, Territory, Commonwealth, or possession in which he is on duty or present shall be punished as a court-martial may direct.

Art. 89. Disrespect Toward Superior Commissioned Officer

> Any person subject to this chapter who behaves with disrespect toward his superior commissioned officer shall be punished as a court-martial may direct.

A superior commissioned officer is a commissioned officer that is superior in rank or command. Disrespect includes insulting words, insolence, impertinence, undue familiarity or other rudeness, and failing to salute.

Art. 90. Assaulting or Willfully Disobeying Superior Commissioned Officer

> Any person subject to this chapter who
>
> (1) strikes his superior commissioned officer or lifts up any weapon or offers any violence against him while he is in the execution of his office; or
>
> (2) willfully disobeys a lawful command of his superior commissioned officer;
>
> shall be punished, if the offense is committed in time of war, by death or such other punishment as a court-martial may direct, and if the offense is committed at any other time, by such punishment, other than death, as a court-martial may direct.

An officer is in the "execution of his office" when performing any act the officer is required or authorized to do. Note that the article is not confined to striking a superior commissioned officer; it takes in brandishing a weapon or waving a fist at that officer.

Willful disobedience, as used here, means intentional defiance of a lawful order. You must presume that any order given by an officer is legal. If you disobey because you think otherwise, you do so at your own risk. It is better to do your questioning after you have carried out the order.

Art. 91. Insubordinate Conduct Toward Warrant Officer, Noncommissioned Officer, or Petty Officer

> Any warrant officer or enlisted member who
>
> (1) strikes or assaults a warrant officer, noncommissioned officer, or petty officer, while that officer is in execution of his office;
>
> (2) willfully disobeys the lawful order of a warrant officer, noncommissioned officer, or petty officer; or
>
> (3) treats with contempt or is disrespectful in language or deportment toward

Student Notes:

a warrant officer, noncommissioned officer, or petty officer, while that officer is in the execution of his office;

shall be punished as a court-martial may direct.

This article has the same general objectives with respect to warrant officers, noncommissioned officers, and petty officers as articles 89 and 90 have with respect to commissioned officers. Namely, it ensures obedience to their lawful orders and protects them from violence, insult, or disrespect.

Art. 92. Failure to Obey Order or Regulation

Any person subject to this chapter who

(1) violates or fails to obey any lawful general order or regulation;

(2) having knowledge of any other lawful order issued by a member of the armed forces, which it is his duty to obey, fails to obey the order; or

(3) is derelict in the performance of his duties;

shall be punished as a court-martial may direct.

A general order or regulation is one that applies generally to an armed force. The President or the Secretary of Defense, the Secretary of Transportation, or the Secretary of a military department may issue it. An officer having general court-martial jurisdiction, a general or flag officer in command, or a commander superior to one of these may also issue it.

Disobedience of "any other lawful order" requires that the person must have had a duty to obey the order and must have had knowledge of the order. An accused may be charged with disobedience of the lawful order of one not a superior, provided the accused had a duty to obey such order. Examples are lawful orders of a sentinel or of members of the armed forces police.

Dereliction in the performance of duties occurs when a person willfully or negligently fails to perform them or performs them in a culpably inefficient manner. To be guilty of inefficiency, an accused must have had the ability and opportunity to perform the assigned duties efficiently, but performed them inefficiently nevertheless.

Art. 93. Cruelty and Maltreatment

Any person subject to this chapter who is guilty of cruelty toward, or oppression or maltreatment of, any person subject to his orders shall be punished as a court-martial may direct.

The cruelty, oppression, or maltreatment must be real, although not necessarily physical. To assault and to subject to improper punishment are examples of this offense. The assignment of necessary or proper duties and the requirement for their correct performance will not constitute this offense even though such duties may be arduous and/or hazardous.

Art. 94. Mutiny or Sedition

(a) Any person subject to this chapter who—

(1) with intent to usurp or override lawful military authority, refuses, in concert with any other person, to obey orders or otherwise do his duty or creates any violence or disturbance is guilty of mutiny;

(2) with intent to cause the overthrow or destruction of lawful civil authority, creates, in concert with any other person, revolt, violence, or disturbance against that authority is guilty of sedition;

(3) fails to do his utmost to prevent and suppress a mutiny or sedition being committed in his presence, or fails to take all reasonable means to inform his superior commissioned officer or commanding officer of a mutiny or sedition which he knows or has reason to believe is taking place, is guilty of a failure to suppress or report a mutiny or sedition.

(b) A person who is found guilty of attempted mutiny, mutiny, sedition, or failure to suppress or report a mutiny or sedition shall be punished by death or such other punishment as a court-martial may direct.

Student Notes:

Art. 95. Resistance, Breach of Arrest, and Escape

Any person subject to this chapter who resists apprehension or breaks arrest or who escapes from custody or confinement shall be punished as a court-martial may direct.

Art. 96. Releasing Prisoner without Proper Authority

Any person subject to this chapter who, without proper authority, releases any prisoner committed to his charge, or who through neglect or design suffers any such prisoner to escape, shall be punished as a court-martial may direct, whether or not the prisoner was committed in strict compliance with law.

Art. 97. Unlawful Detention

Any person subject to this chapter who, except as provided by law, apprehends, arrests, or confines any person shall be punished as a court-martial may direct.

Art. 98. Noncompliance with Procedural Rules

Any person subject to this chapter who—

(1) is responsible for unnecessary delay in the deposition of any case of a person accused of an offense under this chapter; or

(2) knowingly and intentionally fails to enforce or comply with any provision of this chapter regulating the proceedings before, during, or after trial of an accused;

shall be punished as a court-martial may direct.

Art. 99. Misbehavior before the Enemy

Any person subject to this chapter who before or in the presence of the enemy—

(1) runs away;

(2) shamefully abandons, surrenders, or delivers up any command, unit, place, or military property, which it is his duty to defend;

(3) through disobedience, neglect, or intentional misconduct endangers the safety of any such command, unit, place, or military property;

(4) casts away his arms or ammunition;

(5) is guilty of cowardly conduct;

(6) quits his place of duty to plunder or pillage;

(7) causes false alarms in any command, unit, or place under control of the armed forces;

(8) willfully fails to do his utmost to encounter, engage, capture, or destroy any enemy troops, combatants, vessels, aircraft, or any other thing, which it is his duty so to encounter, engage, capture, or destroy; or

(9) does not afford all practicable relief and assistance to any troops, combatants, vessels, or aircraft of the armed forces belonging to the United States or their allies when engaged in battle;

shall be punished by death or such other punishment as a court-martial may direct.

Art. 100. Subordinate Compelling Surrender

Any person subject to this chapter who compels or attempts to compel the commander of any place, vessel, aircraft, or military property, or of any body of members of the armed forces, to give it up to an enemy or to abandon it, or who strikes the colors or flag to any enemy without proper authority, shall be punished by death or such other punishment as a court-martial may direct.

Although these offenses are similar to mutiny, they do not require concert of action. The compulsion to surrender must be by acts, rather than words. To "strike the colors or flag" is to surrender. The offense is committed by anyone subject to the *UCMJ* who assumes the authority to surrender a military force or position when that person is not authorized to do so either by competent authority or by the necessities of battle.

Student Notes:

Art. 101. Improper Use of Countersign

Any person subject to this chapter who in time of war discloses the parole or countersign to any person not entitled to receive it or who gives to another who is entitled to receive and use the parole or countersign a different parole or countersign from that which, to his knowledge, he was authorized and required to give, shall be punished by death or such other punishment as a court-martial may direct.

A *countersign* is a word designated by the principal headquarters of a command to aid guards and sentinels in their scrutiny of persons who apply to pass the lines. It consists of a secret challenge and a password. A *parole* is a word used as a check on the countersign; it is imparted only to those who are entitled to inspect guards and to commanders of guards.

Art. 102. Forcing a Safeguard

Any person subject to this chapter forces a safeguard shall suffer death or such other punishment as a court-martial may direct.

A *safeguard* is a detachment, guard, or detail posted by a commander. It protects persons, places, or property of the enemy or of a neutral affected by the relationship of the opposing forces in their prosecution of war or during a state of conflict. The term also includes a written order left by a commander with an enemy subject or posted upon enemy property for the protection of the individual or property concerned. The effect of a safeguard is a pledge of honor by a nation that its armed force will respect the person or property concerned.

Art. 103. Captured or Abandoned Property

(a) All persons subject to this chapter will secure all public property taken from the enemy for the service of the United States, and shall give notice and turn over to the proper authority without delay all captured or abandoned property in their possession, custody, or control.

(b) Any person subject to this chapter who—

(1) fails to carry out the duties prescribed in subsection (a);

(2)buys, sells, trades, or in any way deals in or disposes of captured or abandoned property, whereby he receives or expects any profit, benefit, or advantage to himself or another directly or indirectly connected with himself; or

(3) engages in looting or pillaging;

shall be punished as a court-martial may direct.

Immediately upon its capture from the enemy, public property becomes the property of the United States. Persons subject to military law have an immediate duty to take those steps within their power and functions to secure such property to the service of the United States. They then have the duty to protect that property from destruction or loss.

Art. 104. Aiding the Enemy

Any person who—

(1) aids, or attempts to aid, the enemy with arms, ammunition, supplies, money or other things; or

(2) without proper authority, knowingly harbors or protects or gives intelligence to, or communicates or corresponds with or holds any intercourse with the enemy, either directly or indirectly;

shall suffer death or such other punishment as a court-martial or military commission may direct.

This article applies to all persons whether or not they are otherwise subject to military law. Enemy denotes citizens as well as members of military organizations. All the citizens of hostile nations, as well as their government, are our enemies.

Student Notes:

Art. 105. Misconduct as Prisoner

Any person subject to this chapter who, while in the hands of the enemy in time of war—

(1) for the purpose of securing favorable treatment by his captors acts without proper authority in a manner contrary to law, custom, or regulation, to the detriment of others of whatever nationality held by the enemy as civilian or military prisoners; or

(2) while in a position of authority over such persons maltreats them without justifiable cause; shall be punished as a court-martial may direct.

Art. 106. Spies

Any person who in time of war is found lurking as a spy or acting as a spy in or about any place, vessel, or aircraft, within the control or jurisdiction of any of the armed forces, or in or about any shipyard, any manufacturing or industrial plant, or any other place or institution engaged in work in aid of the prosecution of the war by the United States, or elsewhere, shall be tried by a general court-martial or by a military commission and on conviction shall be punished by death.

The words *any person* brings within the jurisdiction of courts-martial and military commissions all persons of whatever nationality or military or civilian who commit the offense of spying.

Art. 106a. Espionage

Any person subject to this chapter who, with intent or reason to believe that it is to be used to the injury of the United States or to the advantage of a foreign nation, communicates, delivers, or transmits, or attempts to communicate, deliver, or transmit, to any entity described ... either directly or indirectly, any thing described ... shall be punished as a court-martial may direct, except that if the accused is found guilty of an offense that directly concerns nuclear weaponry, military spacecraft or satellites, early warning systems, or other means of defense or retaliation against large scale attack, war plans, communications intelligence or cryptographic information, or any other major weapons system or major element of defense strategy, the accused shall be punished by death or such other punishment as a court-martial may direct.

The term *entity* can be any of the following:

- A foreign government
- A faction or party or military force within a foreign country...
- A representative, officer, agent, ... of such government, faction, party, or force

Art. 107. False Statements

Any person subject to this chapter who, with intent to deceive, signs any false record, return, regulation, order, or other official document, knowing it to be false, or makes any other false official statement knowing it to be false, shall be punished as a court-martial may direct.

Several articles of the *UCMJ* provide for the punishment of untruths: articles 83 and 84 (fraudulent and unlawful enlistment, appointment, or separation), article 107 (false statements), article 131 (perjury), and article 132 (fraud). You can see how highly *truth* is regarded in the military service.

A statement, whether oral or in writing, is official when it is made pursuant to regulations. A statement is also official when made in response to a request or question from one's commanding officer or a person acting under the commanding officer's authority. Official statements thus include all those made in the line of duty.

Student Notes:

Art. 108. Military Property of the United States—Loss, Damage, Destruction, or Wrongful Disposition

Any person subject to this chapter who, without proper authority—

(1) sells or otherwise disposes of;

(2) willfully or through neglect damages, destroys, or loses; or

(3) willfully or through neglect suffers to be lost, damaged, sold, or wrongfully disposed of;

any military property of the United States, shall be punished as a court-martial may direct.

Whether the property in question was issued to the accused, whether it was issued to someone other than the accused, or whether it was issued at all is immaterial.

Willful means intentional. *Neglect* means inattention to duty or failure to take action that, under the circumstances, should have been taken to prevent the loss, destruction, or damage of any military property.

Art. 109. Property Other Than Military Property of United States—Waste, Spoilage, or Destruction

Any person subject to this chapter who willfully or recklessly wastes, spoils, or otherwise willfully and wrongfully destroys or damages any property other than military property of the United States shall be punished as a court-martial may direct.

Wastes and *spoils* refer to wrongful acts of voluntary destruction, such as burning down buildings, burning piers, tearing down fences, or cutting down trees. To be destroyed, property need be only sufficiently damaged to be useless for the purpose for which it was intended. *Damage* consists of any physical injury to the property. The property must be other than military property of the United States and must belong to one other than the accused.

Art. 110. Improper Hazarding of Vessel

(a) Any person subject to this chapter who willfully and wrongfully hazards or suffers to be hazarded any vessel of the armed forces shall suffer death or such punishment as a court-martial may direct.

(b) Any person subject to this chapter who negligently hazards or suffers to be hazarded any vessel of the armed forces shall be punished as a court-martial may direct.

The word *suffers* means to allow or permit. A person suffers a ship to be hazarded who, although not in direct control of the vessel, knows a danger to be imminent but takes no steps to prevent it. For example, a plotting officer of a ship under way inadvertently fails to report observation of a radar target on a collision course with, and dangerously close to, the ship. The officer has negligently suffered the ship to be hazarded.

Art. 111. Drunken or Reckless Driving

Any person subject to this chapter who operates any vehicle while drunk, or in a reckless or wanton manner, or while impaired by a substance described in… (article 112a(b)), shall be punished as a court-martial may direct.

Operating a vehicle includes not only driving or guiding it while in motion, it also includes starting it or manipulating its controls to cause the vehicle to move. The term *vehicle* applies to all types of land transportation, whether motor-driven or passenger carrying. Drunken or reckless operation of water or air transportation may be charged as a violation of article 134. For the meaning of *drunk(en)*, see the remarks following article 112.

Art. 112. Drunk on Duty

Any person subject to this chapter, other than a sentinel or lookout, who is found drunk on duty, shall be punished as a court-martial may direct.

The term *on duty* in article 112 refers to routine or detailed duties on board a ship or station. The term does not cover periods of leave or liberty (which come under

Student Notes:

a different article), but does include duties of a standby nature. A person whose mental or physical abilities are impaired by either liquor or drugs may be considered drunk.

Art. 112a. Wrongful Use, Possession, etc., of Controlled Substances

(a) Any person subject to this chapter who wrongfully uses, possesses, manufactures, distributes, imports into the customs territory of the United States, exports from the United States, or introduces into an installation, vessel, vehicle, or aircraft used by or under the control of the armed forces a substance described in subsection (b) shall be punished as a court-martial may direct.

(b) The substances referred to in subsection (a) are the following:

(1) Opium, heroin, cocaine, amphetamine, lysergic acid diethylamide [LSD], methamphetamine, phencyclidine, barbituric acid, and marijuana and any compound or derivative of any such substance.

(2) Any substance not specified in clause (1) that is listed on a schedule of controlled substances prescribed by the President for the purposes of this article.

(3) Any other substance not specified in clause (1) or contained on a list prescribed by the President under clause (2) that is listed in schedules I through V of section 202 of the Controlled Substances Act (21 U.S.C. 812).

Don't do drugs! Just say NO!

Art. 113. Misbehavior of Sentinel

Any sentinel or lookout who is found drunk or sleeping upon his post, or leaves it before being regularly relieved, shall be punished, if the offense is committed in time of war, by death or such other punishment as a court-martial may direct, but if the offense is at any other time, by such punishment other than death as a court-martial may direct.

Some actual or imaginary line does not limit a post, nor is it confined to those times when you may be on watch as a sentry. This article covers all periods when you are standing a watch of any kind, such as guarding stores or prisoners or acting as a bow lookout. It also covers periods when you are performing any other duty that requires you to remain alert at all times.

A sentinel on post who is found asleep or drunk is guilty of a serious offense. In time of war, the offense may be punishable by death. For persons in the armed forces, drunkenness is prejudicial to good order and discipline whenever and wherever it appears. Being drunk in public, whether a person is in uniform or civilian clothes, may bring discredit upon the service, while being drunk on station is a breach of military discipline. But being drunk while on duty as a sentinel or lookout in time of war endangers every person in the command.

Art. 114. Dueling

Any person subject to this chapter who fights or promotes, or is concerned in or connives at fighting a duel, or who, having knowledge of a challenge sent or about to be sent, fails to report the fact promptly to the proper authority, shall be punished as a court-martial may direct.

Art. 115. Malingering

Any person subject to this chapter who for the purpose of avoiding work, duty, or service—

(1) feigns illness, physical disablement, mental lapse or derangement; or

(2) intentionally inflicts self-injury;

shall be punished as a court-martial may direct.

Malingering is an offense defined as any act to avoid duty by *feigning* (pretending) to be ill or physically/mentally disabled.

Student Notes:

Art. 116. Riot or Breach of Peace

Any person subject to this chapter who causes or participates in any riot or breach of the peace shall be punished as a court-martial may direct.

The term *riot* is used when a disturbance is caused by a group of three or more persons and is engaged against anyone who may oppose them.

Breach of the peace is an unlawful disturbance by violent or turbulent means that disturbs the peace of the community. Engaging in a fight and using abusive words in public are examples of breach of the peace. As used in this article, *community* includes any military installation or ship, as well as a civilian community.

Art. 117. Provoking Speeches or Gestures

Any person subject to this chapter who uses provoking or reproachful words or gestures towards any other person subject to this chapter shall be punished as a court-martial may direct.

Art. 118. Murder

Any person subject to this chapter who, without justification or excuse, unlawfully kills a human being, when he—

(1) has a premeditated design to kill;

(2) intends to kill or inflict great bodily harm;

(3) is engaged in an act that is inherently dangerous to another and evinces a wanton disregard of human life; or

(4) is engaged in the perpetration or attempted perpetration of burglary, sodomy, rape, robbery, or aggravated arson;

is guilty of murder, and shall suffer such punishment as a court-martial may direct, except that if found guilty under clause (1) or (4), he shall suffer death or imprisonment for life as a court-martial may direct.

Art. 119. Manslaughter

(a) Any person subject to this chapter who, with an intent to kill or inflict great bodily harm, unlawfully kills a human being in the heat of sudden passion caused by adequate provocation is guilty of voluntary manslaughter and shall be punished as a court-martial may direct.

(b) Any person subject to this chapter who, without an intent to kill or inflict great bodily harm, unlawfully kills a human being—

(1) by culpable negligence; or

(2) while perpetrating or attempting to perpetrate an offense, other than those named in clause (4) of... article 118, directly affecting the person;

is guilty of involuntary manslaughter and shall be punished as a court-martial may direct.

Manslaughter is the unlawful killing of another. There are two basic types of manslaughter: *voluntary* and *involuntary*.

Voluntary manslaughter is the unlawful killing of another when there is intent to kill or inflict great bodily harm, but the act is committed in the heat of sudden passion caused by adequate provocation.

Involuntary manslaughter is the unlawful killing of another committed without intent to kill or inflict great bodily harm.

Art. 120. Rape and Carnal Knowledge

(a) Any person subject to this chapter who commits an act of sexual intercourse, by force and without consent, is guilty of rape and shall be punished by death or such other punishment as a court-martial may direct.

(b) Any person subject to this chapter who, under circumstances not amounting to rape, commits an act of sexual intercourse with a female not his wife who has not attained the age of sixteen years, is guilty of carnal knowledge and shall be punished as a court-martial may direct.

(c) Penetration, however slight, is sufficient to complete either of these offenses.

Student Notes:

Art. 121. Larceny and Wrongful Appropriation

(a) Any person subject to this chapter who wrongfully takes, obtains, or withholds, by any means, from the possession of the owner or of any other person any money, personal property, or article of value of any kind—

(1) with intent permanently to deprive or defraud another person of the use and benefit of property or to appropriate it to his own use or the use of any person other than the owner, steals that property and is guilty of larceny; or

(2) with intent temporarily to deprive or defraud another person of the use and benefit of property or to appropriate to his own use or the use of any person other than the owner, is guilty of wrongful appropriation.

(b) Any person found guilty of larceny or wrongful appropriate shall be punished as a court-martial may direct.

Art. 122. Robbery

Any person subject to this chapter who with intent to steal takes anything of value from the person or in the presence of another, against his will, by means of force or violence or fear of immediate or future injury to his person or property or to the person or property of a relative or member of his family or of anyone in his company at the time of the robbery, is guilty of robbery and shall be punished as a court-martial may direct.

Art. 123. Forgery

Any person subject to this chapter who, with intent to defraud—

(1) falsely makes or alters any signature, to, or any part of, any writing which would, if genuine, apparently impose a legal liability on another or change his legal right or liability to his prejudice; or

(2) utters, offers, issues, or transfers such a writing, known by him to be so made or altered;

is guilty of forgery and shall be punished as a court-martial may direct.

A forgery may be committed by a person signing his/her own name to an instrument. For example, presume a check payable to the order of a certain person comes into the hands of another person of the same name. The receiver commits forgery if, knowing the check to be another person's, he/she endorses it with his/her own name with the intent to defraud.

Some of the instruments most frequently subject to forgery are checks, orders for delivery of money or goods, military orders directing travel, and receipts. A writing may be falsely "made" by materially altering an existing writing; by filling in or signing the blanks in a paper, such as a blank check; or by signing an instrument already written.

Art. 123a. Making, Drawing, or Uttering Check, Draft, or Order Without Sufficient Funds

Any person subject to this chapter who—

(1) for the procurement of any article or thing of value, with intent to defraud; or

(2) for the payment of any past due obligation, or for any other purpose, with intent to deceive;

makes, draws, utters, or delivers any check, draft, or order for the payment of money upon any bank or other depository, knowing at the time that the maker or drawer has not or will not have sufficient funds in, or credit with, the bank or other depository for the payment of that check, draft, or order in full presentment, shall be punished as a court-martial may direct....

This article provides specific statutory authority for the prosecution of bad-check offenses. In the absence of evidence indicating otherwise, bad faith might be shown by the maker's or drawer's failure to effect redemption within the 5-day period provided for in the

Student Notes:

article. The offense of wrongfully and dishonorably failing to maintain sufficient funds for payment of checks upon presentment is a violation.

Art. 124. Maiming

> Any person subject to this chapter who, with intent to injure, disfigure, or disable, inflicts upon the person of another an injury which
>
> (1) seriously disfigures his person by a mutilation thereof;
>
> (2) destroys or disables any member or organ of his body; or
>
> (3) seriously diminishes his physical vigor by the injury of any member or organ;
>
> is guilty of maiming and shall be punished as a court-martial may direct.

Maiming includes putting out a person's eye; cutting off a person's hand, foot, or finger; or knocking out a person's front teeth, as these injuries destroy or disable those members or organs. Maiming also includes cutting off a person's ear or scaring a person's face, as these injuries seriously disfigure the person. Injuring an internal organ so as to seriously diminish the physical vigor of a person is also considered maiming.

Art. 125. Sodomy

> (a) Any person subject to this chapter who engages in unnatural carnal copulation with another person of the same or opposite sex or with an animal is guilty of sodomy. Penetration, however slight, is sufficient to complete the offense.
>
> (b) Any person found guilty of sodomy shall be punished as a court-martial may direct.

Art. 126. Arson

> (a) Any person subject to this chapter who willfully and maliciously burns or sets on fire an inhabited dwelling, or any other structure, movable or immovable, wherein to the knowledge of the offender there is at the time a human being, is guilty of aggravated arson and shall be punished as court-martial may direct.
>
> (b) Any person subject to this chapter who willfully and maliciously burns or sets fire to the property of another, except as provided in subsection (a), is guilty of simple arson and shall be punished as a court-martial may direct.

In aggravated arson, danger to human life is the essential element; in simple arson, it is injury to the property of another. In either case, the fact that no one is injured is immaterial.

Art. 127. Extortion

> Any person subject to this chapter who communicates threats to another person with the intention thereby to obtain anything of value or any acquaintance, advantage, or immunity is guilty of extortion and shall be punished as a court-martial may direct.

A threat may be communicated by word of mouth or in writing, the essential element of the offense being the knowledge of the threat to the victim. An acquittance is, in general terms, a release or discharge from an obligation.

Art. 128. Assault

> (a) Any person subject to this chapter who attempts or offers with unlawful force or violence to do bodily harm to another person, whether or not the attempt or offer is consummated, is guilty of assault and shall be punished as a court-martial may direct.
>
> (b) Any person subject to this chapter who—
>
> (1) commits an assault with a dangerous weapon or other means of force likely to produce death or grievous bodily harm; or
>
> (2) commits an assault and intentionally inflicts grievous bodily harm with or without a weapon;
>
> is guilty of aggravated assault and shall be punished as a court-martial may direct.

Section (a) describes the offense of simple assault. Swinging your fist, pointing a gun at a person, or raising

Student Notes:

a club over someone's head, even though no harm is actually done, is each an act of simple assault. When the threat is consummated and force is applied to the victim, the offense becomes assault and battery.

Section (b) describes aggravated assault, of which there are two types. The first is assault with a dangerous weapon and other means of force likely to kill or grievously harm the victim (like shoving a person over the fantail). The second type takes place when an assailant intentionally inflicts severe bodily harm, with or without a weapon. If, after you have knocked an individual down, you repeatedly kick him/her so as to break the person's ribs, you have committed aggravated assault.

Art. 129. Burglary

> Any person subject to this chapter who, with intent to commit an offense punishable under... articles 118 - 128, breaks and enters, in the nighttime, the dwelling house of another, is guilty of burglary and shall be punished as a court-martial may direct.

The house must be a dwelling place at the time of the breaking and entry, but the residents do not have to actually be in it. A simple act such as opening a closed door or window or some other similar fixture or cutting out the glass of a window or the netting of a screen constitutes breaking. Entry gained through a trick, false pretense, impersonation, intimidation, or collusion also constitutes breaking. For the intruder to succeed in carrying out the intent for which the house was broken into is not an essential element.

Art. 130. Housebreaking

> Any person subject to this chapter who unlawfully enters the building or structure of another with intent to commit a criminal offense therein is guilty of housebreaking and shall be punished as a court-martial may direct.

The initial entering must amount to trespassing; this article is not violated if the accused entered the building or structure lawfully, even though the person had the intent to commit an offense therein. This offense is broader than burglary in that the place entered need not be a dwelling house; also, the place need not be occupied. A breaking is not essential. The entry may be either in the nighttime or in the daytime. The criminal intent is not limited to those offenses punishable under articles 118 through 128.

Art. 131. Perjury

> Any person subject to this chapter who in a judicial proceeding or in a course of justice willfully and corruptly—
>
> (1) upon a lawful oath or in any form allowed by law to be substituted for an oath, gives any false testimony material to the issue or matter of inquiry; or
>
> (2) in any declaration, certificate, verification, or statement under penalty or perjury as permitted under section 1746 of title 28, United States Code, subscribes any false statement material to the issue or matter of inquiry;
>
> is guilty of perjury and shall be punished as a court-martial may direct.

Art. 132. Frauds against the United States

> Any person subject to this chapter—
>
> (1) who, knowing it to be false or fraudulent—
>
> (A) makes any claim against the United States or any officer thereof; or
>
> (B) presents to any person in the civil or military service thereof, for approval or payment, any claim against the United States or any officer thereof;
>
> (2) who, for the purpose of obtaining the approval, allowance, or payment of any claim against the United States or any officer thereof—
>
> (A) makes or uses any writing or other paper knowing it to contain any false or fraudulent statements;
>
> (B) makes any oath to any fact or to any writing or other paper knowing the oath to be false; or

Student Notes:

(C) forges or counterfeits any signature upon any writing or other paper, or uses any such signature knowing it to be forged or counterfeited;

(3) who, having charge, possession, custody, or control of any money, or other property of the United States, furnished or intended for the armed forces thereof, knowingly delivers to any person having authority to receive it, any amount thereof less than that for which he receives a certificate or receipt; or

(4) who, being authorized to make or deliver any paper certifying the receipt of any property of the United States furnished or intended for the armed forces thereof, makes or delivers to any person such writing without having full knowledge of the truth of the statements therein contained and with intent to defraud the United States;

shall, upon conviction, be punished as a court-martial may direct.

This article deals with frauds against the United States. It pertains to making false claims against the government to obtain money or property.

It also pertains to the offense of making a writing or other paper known to contain a false statement for the purpose of obtaining the approval, allowance, or payment of a claim. The offense is complete when the writing or paper is made for that purpose, whether or not the use of either one has been attempted and whether or not the claim has been presented.

Art. 133. Conduct Unbecoming an Officer and a Gentleman

Any commissioned officer, cadet, or midshipman who is convicted of conduct unbecoming an officer and a gentleman shall be punished as a court-martial may direct.

Art. 134. General Article

Though not specifically mentioned in this chapter, all disorders and neglects to the prejudice of good order and discipline in the armed forces, all conduct of a nature to bring discredit upon the armed forces, and crimes and offenses not capital, of which persons subject to this chapter may be guilty, shall be taken cognizance of by a general, special or summary court-martial, according to the nature and degree of the offense, and shall be punished at the discretion of that court.

Article 134 makes punishable acts or omissions not specifically mentioned in other articles. Those acts include wearing an improper uniform, abusive use of a military vehicle, the careless discharge of a firearm, or impersonating an officer. They also include offenses involving official passes, permits, and certificates; and the wrongful possession of a habit-forming narcotic drug.

Discredit means to injure the reputation of; that is, to bring the service into disrepute. Examples include acts in violation of state or foreign laws, failure to pay one's debts, adultery, bigamy, and indecent acts.

Crimes and offenses not capital include those acts or omissions, not punishable by another article, denounced as crimes or offenses by enactment of Congress or under authority of Congress and made triable in the federal civil courts. Some of these offenses are punishable wherever committed; others are punishable only if committed within the geographical boundaries of the areas in which they are applicable.

Art. 137. Articles to be explained

Articles 2, 3, 7 through 15, 25, 27, 31, 37, 38, 55, 77 through 134 and 137 through 139 of this chapter shall be carefully explained to each enlisted member at the time of his entrance on active duty, or within six days thereafter. They shall be explained again after he has completed six months of active duty, and again at the time when he reenlists. A complete text of the *Uniform Code of Military Justice* and of the regulations prescribed by the President thereunder shall be made available to any person on active duty upon his request, for his personal examination.

Student Notes:

Art. 138. Complaints of wrongs

Any member of the armed forces who believes himself wronged by his commanding officer, and who, upon due application to that commanding officer, is refused redress, may complain to any superior commissioned officer, who shall forward the complaint to the officer exercising general court-martial jurisdiction over the officer against whom it is made. The officer exercising general court-martial jurisdiction shall examine into the complaint and take proper measures for redressing the wrong complained of; and he shall, as soon as possible, send to the Secretary concerned a true statement of that complaint, with the proceedings had thereon.

This article provides for redress of wrongs inflicted by a commanding officer on subordinates, and it prescribes the procedure to be followed by subordinates to apply for such redress.

Art. 139. Redress of injuries to property

(a) Whenever complaint is made to any commanding officer that willful damage has been done to the property of any person or that his property has been wrongfully taken by members of the armed forces, he may, under such regulations as the Secretary concerned may prescribe, convene a board to investigate the complaint. The board shall consist of from one to three commissioned officers and, for the purpose of that investigation, it has power to summon witnesses and examine them upon oath, to receive depositions or other documentary evidence, and to assess the damages sustained against the responsible parties. The assessment of damages made by the board is subject to the approval of the commanding officer, and in the amount approved by him shall be charged against the pay of the offenders. The order of the commanding officer directing charges herein authorized is conclusive on any disbursing officer for the payment by him to the injured parties of the damages as assessed and approved.

(b) If the offenders cannot be ascertained, but the organization or detachment to which they belong is known, charges totaling the amount of damages assessed and approved may be made in such proportion as may be considered just upon the individual members thereof who are shown to have been present at the scene at the time the damages complained of were inflicted, as determined by the approved findings of the board.

Nonjudicial Punishment

If you break a rule or are negligent, careless, or unmilitary in your conduct, an officer or petty officer may put you on report. Being *put on report* means you may appear before the commanding officer at a specified time for nonjudicial punishment (*UCMJ*, art. 15); that is, you may appear at captain's mast.

Art. 15. Commanding Officer's nonjudicial punishment

(a) Under such regulations as the President may prescribe, and under such additional regulations as may be prescribed by the Secretary concerned, limitations may be placed on the powers granted by this article with respect to the kind and amount of punishment authorized, the categories of commanding officers and warrant officers exercising command authorized to exercise those powers, the applicability of this article to an accused who demands trial by court-martial, and the kinds of courts-martial to which the case may be referred upon such a demand. However, except in the case of a member attached to or embarked in a vessel, punishment may not be imposed upon any member of the armed forces under this article if the member has, before the imposition of such punishment, demanded trial by court-martial in lieu of such punishment. Under similar regulations, rules may be prescribed with respect to the suspension of punishments authorized hereunder. If authorized by regulations of the Secretary concerned, a commanding officer exercising general court-martial jurisdiction or an officer of general or flag rank in command may delegate his powers under this article to a principal assistant.

Student Notes:

(b) Subject to subsection (a) any commanding officer may, in addition to or in lieu of admonition or reprimand, impose one or more of the following disciplinary punishments for minor offenses without the intervention of a court-martial…

Commanding officer's nonjudicial punishment is often referred to as captain's mast. Captain's mast gets its name from the old sailing days when the setting for this form of naval justice was the weather deck near the ship's mainmast.

Cases are heard and punishments given at captain's mast. Anyone who is not attached to or embarked in a vessel may, however, demand trial by court-martial in lieu of punishment at mast, before such punishment is imposed.

The punishments permitted at captain's mast depend upon the rank of the officer holding mast. Figure 2-3 shows the punishment that may be awarded.

A commanding officer that decides an offense deserves a punishment more severe than he/she is authorized to award at mast may order a court-martial.

The following paragraphs explain some of the punishments that may be given at captain's mast.

RESTRICTION.—Restriction is the requirement to remain within certain specified limits (ship, station, etc.). Although required to muster at certain times, the restricted person usually continues to perform his/her regular duties.

CORRECTIONAL CUSTODY.—Correctional custody is the physical restraint (confinement) of a person during duty or nonduty hours, or both. The person may be required to perform extra duties or hard labor. A typical example is an individual who is free to carry out regular duties during the day but is confined in a confinement facility at night.

CONFINEMENT ON BREAD AND WATER OR DIMINISHED RATIONS.—Confinement on bread and water or diminished rations may be imposed only on enlisted persons E-3 and below aboard ship.

EXTRA DUTY.—Extra duty is the assignment of any duty (except guard duty) to be performed after the person's regular working hours. Extra duty is not to exceed 2 hours daily or to be performed on holidays.

ENLISTED		
	Any officer commanding, LCDR and above	**Commanding officers below LCDR; OICs, any grade**
Admonition or reprimand	Yes	Yes
Confinement on B&W or diminished rations	3 consecutive days (aboard ship) E-3 and below	3 consecutive days (aboard ship) E-3 and below
Correctional custody	30 consecutive days E-3 and below	7 consecutive days
Forfeiture of pay	1/ 2 of 1 month pay per month for 2 months	7 days' pay
Reduction in grade	To next inferior grade	To next inferior grade
Extra duty	45 consecutive days	14 consecutive days
Restriction	60 consecutive days	14 consecutive days

Figure 2-3.—One or more of the maximum punishment authorized by article 15, *UCMJ*.

Student Notes:

Petty officers may not be assigned extra duties that would demean their grade or position.

FORFEITURE OF PAY.—Forfeiture of pay is a permanent loss of a specified amount or a temporary withholding of a certain amount of pay. The detention period must be specified. The money detained is normally returned at the end of the detention period, but it can be detained for a period of 1 year.

APPEALS.—If persons consider their punishment under article 15 to be unjust or out of proportion to the offense, they may appeal to the next superior authority in the chain of command. The appeal must be made within a reasonable time (generally 5 days) and promptly forwarded. If the superior authority upholds the appeal, all rights, privileges, and property are restored.

PROTECTION AGAINST SELF-INCRIMINATION.—Under article 31 of the *UCMJ*, compulsory self-incrimination is prohibited. The accused must be informed of the nature of the charges against him/her. The accused must also be advised that he/she does not have to make any statement regarding the offense of which accused, but that any statement made may be used as evidence against him/her in a trial by court-martial. No statement obtained through the use of coercion, unlawful influence, or unlawful inducement may be used as evidence against the accused.

MERITORIOUS AND REQUEST MASTS.—Not all masts are for disciplinary purposes. A meritorious mast may be held by the commanding officer to give awards or commendations to those persons who have earned them.

Article 1107 of *Navy Regs* grants the right for any person to communicate with the commanding officer. You can't just walk up to the captain, however, and start talking. For the purpose of hearing valid requests or complaints from crew members, the CO sets certain times aside. This practice is called *request mast*. The person having a request or grievance should first try to resolve the problem through the chain of command. Failing that, the person may request mast.

COURTS-MARTIAL

Based on article 16 of the *UCMJ*, courts-martial are of three types: summary, special, and general. The captain decides the type of court-martial to award based on the nature, time, and place of the offense.

Summary Courts-Martial

A summary court-martial consists of one commissioned officer. If the commanding officer is the only officer with the command, that officer acts as the summary court officer. A summary court can award any sentence that may be given at mast. It can also award the additional punishments of confinement for 1 month and hard labor without confinement for 45 days. Any person awarded confinement at a summary court-martial will then be held, as appropriate.

Special Courts-Martial

A special court-martial consists of not less than three members; or a military judge and not less than three members; or only a military judge. When a military judge (a qualified lawyer) is detailed to the court, the accused has the right to know the identity of the military judge. The accused also has the right to consult with the defense counsel and to request that the court consist of only the military judge. The request must be in writing, submitted before the court is assembled, and approved by the military judge. A special court-martial may award the same punishment as a summary court, or it may award a more severe punishment. For example, it can award a bad-conduct discharge, confinement for 6 months, loss of two-thirds pay per month for 6 months, and hard labor without confinement for 3 months.

General Court-Martial

A general court-martial consists of a military judge and not less than five members; or only a military judge. Under the conditions described for a special court, the accused may request that the court consist of only a military judge. A general court-martial can award any punishment not forbidden by the *UCMJ*, including death when specifically authorized for the offense.

All accused persons have the right to be represented before special and general courts-martial by defense counsel. This counsel may be a civilian or a military lawyer selected by the accused or a defense counsel appointed by the convening authority. If a civilian counsel is selected, the accused must pay the counsel's expenses.

Student Notes:

REVIEW 4 QUESTIONS

Q1. What chapter of the *Navy Regs* deals with your responsibility and authority while carrying out orders?

Q2. What chapter of the *Navy Regs* deals with rights and responsibilities?

Q3. Who is responsible for ensuring the *Navy Regs* conforms to the current needs of the Department of the Navy?

Q4. New *Navy Regs* and changes to it are issued by whom and approved by whom?

Q5. The instructions found in the OPNAVINST 3120.32 provide guidance and regulations for—

Q6. What was the purpose for developing and signing into law the *Uniform Code of Military Justice*?

Q7. When was the *UCMJ* signed into law?

Q8. Article 137 of the *UCMJ* states that certain articles of the Code must be explained carefully to every enlisted person at what minimum interval?

Q9. List the three types of court-martials.

a.

b.

c.

SUMMARY

Discipline is training that develops self-control, character or orderliness, and efficiency. *Justice* is impartiality—fairness. *Conduct* is the way one acts—behavior. We all deal with discipline, justice, and conduct in our day-to-day dealings as members of the U.S. Navy. We have certain standards of behavior, both on and off duty, by which we must abide. Our justice system sets those standards of behavior; therefore, it should not intimidate us.

We also have standards of conduct by which we must abide if we are taken prisoner. These standards are fundamental to our safety and to our fellow prisoners.

Discipline or conduct could sometimes make the difference between saving or losing a unit. Without discipline, ships would not have the efficient fire or repair parties that have kept many of them afloat after major damage. Imagine the panic that would take place if Sailors didn't have the discipline, self-control, and efficiency they have been taught.

Our justice system has its own checks and balances. For example, if a person does something wrong, the commanding officer is restricted as to the type of punishment he/she may award. A person also has a right to appeal punishment awarded—a right all people enjoy. A person also has the right to communicate with the commanding officer.

Our justice system protects us. Because of naval regulations and standards of conduct, we know what the Navy expects of us. Our conduct, both on and off duty, should reflect our pride in the Navy and in our unit.

Student Notes:

REVIEW 1 ANSWERS

A1. The three sources that contain the basic disciplinary laws for the U.S. Navy are—

a. ***U.S. Navy Regulations***

b. ***Standard Organization and Regulations of the U.S. Navy***

c. ***Uniform Code of Military Justice (UCMJ)***

A2. One of the most important characteristics of a good Sailor is a **sense of moral responsibility**.

A3. To succeed in your work in the Navy, you should value and take part in **teamwork.**

A4. The Code of Conduct was established to **govern situations where Sailors were POW's.**

A5. There are **six articles in the Code of Conduct**.

A6. Under article V, the only information you are allowed to give is your **name, rank, service number, and date of birth**.

REVIEW 2 ANSWERS

A1. In the Navy, the military police are known as the **shore patrol.**

A2. Shore patrol personnel are identified by **armbands bearing the letters *SP*.**

A3. When military police are from different armed forces and combined to form one unit, they are known as an **Armed Forces Police Detachment (AFPD)**.

A4. The primary duties of the shore patrol are to—

a. **Render assistance to military personnel ashore**

b. **Maintain good order and discipline among military personnel**

c. **Report conditions or practices that appear prejudicial to the welfare of military personnel**

REVIEW 3 ANSWERS

A1. **Motivation and correction through reward and punishment** are used to help Sailors work as a unit.

A2. The purpose of discipline in the military is **to bring about an efficient military organization**.

A3. **Deterrent** theory of punishment is used by the Navy.

A4. The two things a recipient of Navy punishment should remember are—

a. **Punishment is a result of their behavior.**

b. **They won't be punished again if they learn to conform to Navy's standard of conduct.**

REVIEW 4 ANSWERS

A1. **Chapter 10 of the *Navy Regs*** deals with your responsibility and authority while carrying out orders.

A2. **Chapter 11 of the *Navy Regs*** deals with your rights and responsibilities.

A3. **Chief of Naval Operations** is responsible for making sure that the *Navy Regs* conform to the current needs of the Department of the Navy.

A4. **Secretary of the Navy issues and the President approves** new *Navy Regs* and changes to *Navy Regs*.

A5. The instructions found in the OPNAVINST 3120.32 provide guidance and regulations for **the duties and responsibilities for all personnel within a unit or organization.**

A6. The purpose for developing and signing into law the *Uniform Code of Military Justice* was **to standardize legal procedure and discipline throughout all branches of service.**

A7. The *UCMJ* was signed into law on **31 May 1951**.

A8. Article 137 of the *UCMJ* states that certain articles of the Code must be explained carefully to every enlisted person **at the time of entrance or no later than 6 days later, 6 months on active duty, and every reenlistment**.

A9. The three types of courts-martial are—

a. **Summary**

b. **Special**

c. **General**

Student Notes:

CHAPTER 3

WATCH STANDING

Thank God I have done my duty.

—Admiral Horatio Nelson

In this chapter, you will learn about the basic shipboard watch organization. You will learn about a typical watch, quarter, and station bill; the terms used during watches; and some typical watches, both ashore and afloat. You will also learn about procedures for reporting bearings and using binoculars.

WATCH STANDING

Learning Objectives: When you finish this chapter, you will be able to—

- Recognize the responsibilities of personnel for the Watch, Quarter, and Station Bill.
- Identify types of watches, general orders of a sentry, procedures to follow when relieving an armed watch, and when a weapon may be fired.
- Recognize the duties of lookouts.

During a ship's entire commissioned life, it will always have Sailors on watch. There are probably more than a hundred different types of watches, depending on the ship or station.

Whatever type of watch, the watch stander must devote full attention to it. The ship's organization and the watches manned by its personnel keep the ship running smoothly 24 hours a day. Watches vary, of course, depending on both the type of ship and whether the ship is under way or in-port. Even when the ship is moored in-port and receiving hotel services (utilities, such as steam, water, and electricity) from the pier or another ship, it's necessary to maintain a watch for communications, security, and safety.

During your time in the Navy, you will be required to stand many watches. Some watches will be of a security nature, such as a pier sentry or roving patrol; others will be operational, such as a telephone talker and/or status board operator. Whatever the type of watch, you must devote your full attention to it. Inattention or negligence on your part can result in serious consequences for the ship and your fellow shipmates.

DECK LOG

Probably the most important log you will maintain is the ship's deck log. The basic requirements for maintaining the deck log are contained in the *U.S. Navy Regulations and Standard Organization* and *Regulations of the U.S. Navy.* The ship's deck log is a complete daily record, by watches, of every event of importance or interest about the crew and the operation and safety of the ship.

A ship's deck log has both historical importance and legal standing. At times, it may be used in naval, admiralty, and civil courts. In an incident involving the ship, the log may be the only available evidence on which to base a legal decision. At sea, the ship's deck log is kept by the quartermaster of the watch (QOOW). In-port, chronological entries are made, but these entries are made by the petty officer of the watch (POOW).

Entries in the ship's deck log are handwritten using a black, ball-point pen. Entries must be neat and legible. Use only standard Navy phraseology. Because the log may be used as evidence in legal proceedings, do not erase an entry. If you make a mistake, draw a single line through the original entry (so that it remains legible), insert the correct entry, and place your initials in the margin. The log is signed at the end of each watch by the OOD. The name of the officer of the deck must also be printed beneath the signature. Facsimile signature is not authorized.

The following are entries that are always recorded:

- Convening of courts-martial or fact-finding bodies
- Inspections held, including administrative, material, personnel, lower deck, and magazine inspections
- Injuries, accidents, and casualties
- Official visits
- Salutes fired and flags displayed

- Arrivals and departures of the commanding officer and executive officer and, if on board, flag officers and civil officials
- Drills held
- Observance of sunrise and sunset
- Reports made to the OOD; for example, fuel and water, chronometer, magazine temperatures, and so forth
- Equipment casualties

WATCH, QUARTER, AND STATION BILL

For any ship to carry out its assigned missions and tasks, it must have an administrative organization. In the organization, every person is assigned one or more tasks. Personnel are trained so they can do their jobs.

The ship's organized plan for action is contained in the battle bill. The battle bill is based on the organization manual and other publications and directives. The battle bill contains lists of stations that must be manned during battle and at other specified times. Using the organization manual and the battle bill as references, each division officer and division chief assigns qualified personnel in the division to the stations and enters their names on the watch, quarter, and station (WQS) bill.

The WQS bill displays in one place your duties for each emergency and watch condition. It also shows your administrative and operational duties.

Contents of the Watch, Quarter, and Station Bill (WQS)

The WQS bill lists, by billet number and rate, divisional stations to be manned for various situations. The billet number consists of either four numbers or a letter and three numerals. The first number (or letter) indicates the person's division; the second number indicates the section; the last two numbers show the person's seniority in the section. Figure 3-1 shows the assignments for personnel in the first section of the first division.

Look at figure 3-1. The first column shows the billet number. The second column shows your name. Your bunk and locker numbers are usually the same. There are three columns under rate: the first column shows the

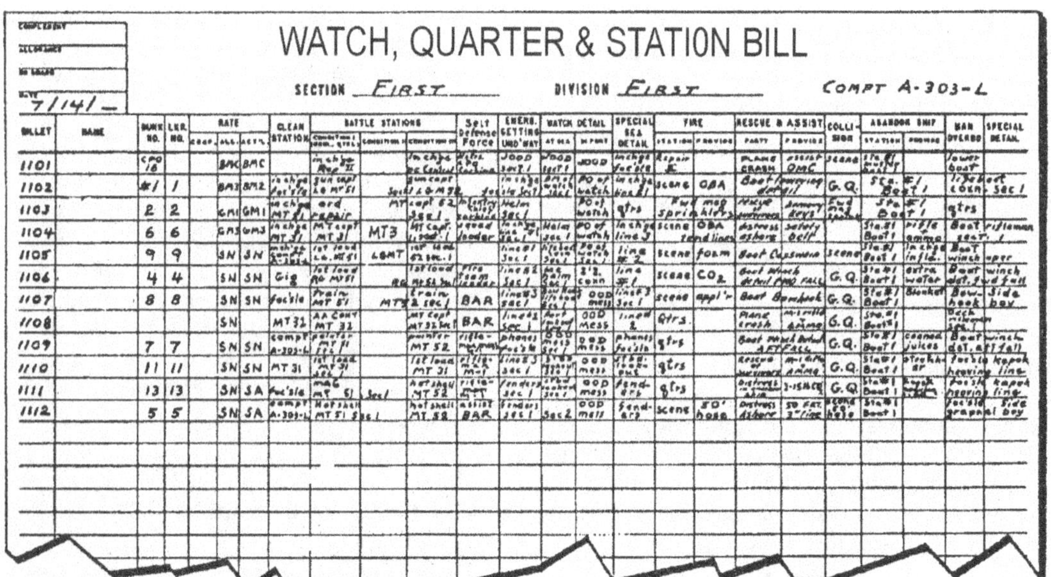

Figure 3-1.—Watch, quarter, and station bill.

Student Notes:

wartime complement, the second the peacetime allowance (usually less than for wartime), and the third lists the rates actually on board.

Often, cleaning stations are omitted, since they are posted in a separate cleaning bill.

There are three columns under the BATTLE STATIONS—Condition I, Condition II, and Condition III. **Condition I** is general quarters. Under Condition I all battle stations are manned, and usually surface or air action is imminent (about to take place). Condition I is sometimes modified to let a few persons at a time rest on station or to let designated personnel draw rations for delivery to battle stations (condition IE). **Condition II** is a special watch used by gunfire support ships for situations such as extended periods of shore bombardment. **Condition III** is the normal wartime cruising watch. Normally, when cruising under Condition III, the ship's company stands watch on a basis of 4 hours on, 8 hours off; about one-third of the ship's armament is manned in the event of a surprise attack.

Assignments to the SELF-DEFENSE FORCE vary according to ship type. The purpose of the self-defense force is to provide a capability for reacting to emergency security situations aboard ship and at pierside to protect the ship, its sensitive equipment, and its personnel.

The next column, EMERGENCY GETTING UNDER WAY, is for use in-port when most of the crew is ashore and the ship must get under way before personnel can be recalled.

There are two columns under WATCH DETAIL. The **left column** is for normal peacetime cruising, or **Condition IV** The number of watch sections depends on the type of ship and the number of personnel aboard. The **right column** lists the type of watch personnel will stand in-port (Condition V). The time of the watch is posted on a separate in-port watch list.

The SPECIAL SEA DETAIL is manned whenever the ship leaves and enters port. Because of the critical nature of mooring or anchoring, getting under way, and maneuvering in restricted waters, only the most experienced persons are assigned to these details. You can expect to be assigned to a station, however, so that you can learn what to do and how to do it.

The remaining columns of the WQS bill (except the last one) show assignments to the ship's emergency bills. Always be prepared to man your emergency station and know where to get the equipment you may be required to provide.

For a MAN-OVERBOARD situation, you go to quarters or some other designated place of muster if you are not assigned a specific detail. The final column is for assignments to such miscellaneous details as mess cooking, MAA duty, and side boys.

Responsibilities

It is your responsibility to check the WQS bill daily. You should check for any changes made in your assignments and to refresh your memory for assignments to seldom-used details (such as to a prize crew). When abandon ship drill is held, for instance, you should not have to take time to find out what your station is and where it is located. A shipmate's life may depend on you to be where you're assigned to be.

WATCHES

Most of the watches in the Navy are of 4 hours duration. Time off between watches depends on the number of sections and the number of personnel in each station. Normally, watches start on the even hours, such as 0400, 0800, or 1200. However, you should arrive at your station **at least 15 minutes ahead of time** to receive any pertinent information from the person you are relieving. Regardless of the type of watch you stand, observe proper military bearing. Proper grooming standards and uniform appearance is a must. Stand your watch in strict adherence to the eleven general orders of the sentry (covered later in this chapter). Know the chain of command as it relates to watch standing. If there is an emergency, it's important to know who and when to call.

Military Time

The Navy uses the 24-hour system of keeping time. The day starts at midnight. Four numbers are used to indicate the time—the first two digits indicate hours and the last two show the minutes. Midnight is expressed two ways—0000 to indicate the start of the day, and

Student Notes:

2400 to indicate the end of the day. Each succeeding hour, starting at midnight, is increased by 100 (0000, 0100, 0200, and so on) until 2400 is reached, then a new day starts.

Time is spoken in hundreds. For example, 0100 (1:00 a.m.) is spoken "zero one hundred"; 2000 (8 p.m.) is pronounced "twenty hundred"; 2315 (11:15 p.m.) is spoken "twenty three fifteen."

Converting time on a 12-hour clock to Navy time is an easy matter. The hours from midnight to noon aren't any problem; from noon to midnight, simply add 12 hours to the time indicated.

The ship's bell may also indicate time. The bell is struck once for each half hour, with a maximum of eight bells. At 0830, for instance, one bell is sounded; at 0900, or two bells; and so on until eight bells are struck at 1200. The use of this system is usually restricted to the hours between reveille and taps.

Watch Terms

Two methods are used in identifying watches. One method uses a descriptive name that identifies the type of watch—for example, pier sentry. The other method also uses a name, but it identifies the time of the watch. Standard watch times and their names are given in the following listing:

TIME	TYPE OF WATCH
0000 to 0400	Midwatch
0400 to 0800	Morning watch
0800 to 1200	Forenoon watch
1200 to 1600	Afternoon watch
1600 to 1800	First dog watch
1800 to 2000	Second dog watch
2000 to 2400	Evening watch

The dog watches permit rotation of the watches; otherwise, personnel would stand the same watch every day. (Usually the 1600-2000 watch is dogged only at sea.) Normally, Sailors having the midwatch are permitted to sleep 1 hour past reveille (late sleepers).

Often, particularly in foreign ports when extra security precautions are required, the ship's crew is placed in two sections—port and starboard. In such instances, one of the normal three sections (usually the third) is split between the first and second sections. The resulting first section becomes the starboard section; the second section, the port section. If the crew is divided into four sections, the odd-numbered sections make up the starboard section; the even-numbered ones, the port section. The actual watches, though, conform to the times described.

Types Of Watches

Hundreds of different types of watches are required throughout the Navy, both under way and in-port. Most of them are of a professional nature; that is, they are peculiar to a specific rating or rating group. In this chapter, you will learn about the watches that are more or less of a security nature and that most Sailors stand at one time or another.

Security watches are stood to prevent sabotage, protect property from damage or theft, prevent access to restricted areas by unauthorized persons, or protect personnel. Security watches include sentry duty, barracks watches, fire watches, and watches stood under way.

WATCH OFFICERS

Some key assignments for officers in the watch organization include the command duty officer (CDO), officer of the deck (OOD), junior officer of the deck (JOOD), and junior officer of the watch (JOOW). (**NOTE**: Senior petty officers can qualify for some of the officer's assignments.)

Command Duty Officer (CDO)

Although an official watch stander, the command duty officer (CDO) may be on duty for a period of several watches. The CDO is eligible for command at sea and is designated and empowered by the captain to advise, supervise, and direct the officer of the deck (OOD) in matters concerning the general operation and safety of the ship or station.

Student Notes:

Officer of The Deck (OOD)

The officer of the deck (OOD) is in charge of the ship and is responsible to the commanding officer (CO) for the safe and proper operation of the ship or station. That includes navigation, ship handling, communications, routine tests and inspections, reports, supervision of the watch, and carrying out the plan of the day (POD).

Junior Officer of the Deck (JOOD)

The junior officer of the deck (JOOD) is the principal assistant to the OOD. Anyone making routine reports to the OOD normally makes them through the JOOD or the JOOW.

Junior Officer of the Watch (JOOW)

The junior officer of the watch (JOOW), when assigned, is in training for qualification as the OOD.

Combat Information Center (CIC) Watch Officer

The combat information center (CIC) watch officer supervises the operation of the CIC, which reports, tracks, and evaluates air, surface, and submarine contacts.

WATCH STANDERS

Senior enlisted watch standers also have a number of important assignments. While there are scores of other enlisted watch assignments, those described in the following sections are the most important and the most responsible. The majority of other enlisted watch standers report to, or through, watch officers.

Quartermaster of the Watch (QMOW)

The quartermaster of the watch (QMOW) is an enlisted assistant to the OOD while under way (and in-port on certain classes of ships). The QMOW assists the OOD in navigational matters and maintains the ship's deck log. Additional duties include reporting and recording weather changes and executing required ship's navigational lighting changes. The QMOW, who must be a qualified helmsman, supervises the helmsman if senior to the BMOW.

Boatswain's Mate of the Watch (BMOW)

The boatswain's mate of the watch (BMOW) is an enlisted assistant to the OOD during under way watches. The BMOW must see that all deck watch stations are manned with qualified personnel and all watch standers in previous watch sections are relieved. Although the section leader and the division petty officer have the duty of instructing the personnel they send on watch, the BMOW must verify that every person in the watch has been properly instructed and trained. A BMOW must be a qualified helmsman and supervises the helmsman if senior to the QMOW.

Lookouts, Sky and Surface

The lookout watch mans assigned lookout stations and performs duties as prescribed in the ship's lookout instructions. Lookouts should be rotated at least hourly. They are under the direct supervision of the OOD. Lookouts are trained in their duties by the CIC officer. The posting and training of lookouts will, as a minimum, conform to the requirements of the *International Regulations for Preventing Collisions at Sea.*

Messenger of the Watch (MOW)

The messenger of the watch stands the watch on the bridge (under way) and the quarterdeck (in-port). The MOW delivers messages, answers telephones, and carries out such duties as the OOD may direct. Messengers need to be familiar with various departments of the ship and ship's company. The underway messenger is normally assigned from the weapons/deck department.

Fog Lookouts

Fog lookouts are required during fog or reduced visibility. The watch is stood in those locations where approaching ships can best be seen or heard (normally in the bows). The fog lookouts stand a vigilant watch to detect, either by hearing fog signals or actually sighting, approaching ships or craft. Posting and training of fog lookouts will, as a minimum, meet the requirements of the International Regulations for Preventing Collisions at Sea. This watch will consist of two personnel—one phone talker and one lookout. The addition of the phone

Student Notes:

talker allows the fog lookout to work without his or her hearing being impaired by wearing sound-powered phones. As with other lookouts, the fog lookouts are in contact with the OOD through the bridge phone talker.

Helmsman

The helmsman is a qualified steersman who steers courses prescribed by the conning officer. The helmsman alternates with other members of the deck watch as directed by the BMOW and as approved by the conning officer (who is generally the OOD or the JOOD). The helmsman is normally assigned from the weapons/deck department.

Lee Helmsman

The lee helmsman who stands watch at the engine order telegraph on the bridge rings up the conning officer's orders to the engine room, making sure all bells are correctly answered. The lee helmsman alternates with other members of the deck watch as directed by the BMOW and as approved by the conning officer. The lee helmsman is normally assigned from the weapons/deck department.

After Steering Watch

This watch, stationed in after steering, is set when positive steering control must be maintained, such as during general quarters, under way replenishment, and sea and anchor detail. During these evolutions, a qualified member from the navigation department is assigned as the after steersman along with a member from the engineering department. This watch is responsible for lining up and operating the steering engines according to orders received from the conning officer. During normal under way steaming, this watch is not usually manned, except on board the larger types of vessels (such as CVNs or LHAs).

Gangway Watch

When required, the gangway watch is posted at the foot of the brow or gangway to perform such duties as directed by the OOD. These duties normally include security of the brow and ceremonial duties.

Security Watches and Patrols

Security watches and patrols, in addition to those described elsewhere in this chapter, may be assigned at the discretion of the CO. Security watches and patrols are established to increase the physical security of the ship. Sailors assigned to security watches and patrols will be trained and qualified by the department head responsible for the areas to which specific watches and patrols are assigned. Duties of security watches and patrols include but are not limited to the following:

- Maintaining continuous patrols above decks and below decks
- Checking classified stowage, including spaces containing classified equipment
- Being alert for evidence of sabotage, thievery, and fire hazards
- Checking security of weapons magazines
- Obtaining periodic sounding of designated tanks and spaces
- Periodically inspecting damage control closures

Sounding and Security Patrol

The watch of the sounding and security patrol is regular and continuous. It is essential that only well-indoctrinated, experienced personnel are assigned this watch. The patrol follows an irregular route while conducting a continuous inspection of all spaces (except those on which a watch is posted or those spaces designated as limited or exclusion areas) to detect and prevent fire hazards, fire, flooding, theft, sabotage, or other irregularities affecting the physical security of the ship. Soundings and results of the inspection will be reported to the OOD and are logged in the ship's deck log. Any unusual conditions are reported to the OOD immediately.

Fire Watch

The purpose of a shipboard fire watch is to immediately extinguish fires caused by welding or burning operations. (Burning means cutting through

Student Notes:

metal with an oxyacetylene torch.) Often two persons are assigned to this duty—one is stationed at the scene, the other in the space behind the one in which the cutting or welding is being done. Heat generated by welding or burning can pass through a bulkhead or deck and ignite material on the other side.

When assigned a fire watch, you will be given a portable fire extinguisher and eye protectors, such as goggles. If you use the fire extinguisher or if the seal is broken, be sure you inform the person who issued it so that the bottle can be weighed to determine if it needs refilling.

The fire watch may become boring, but you must remain alert at all times. For example, when undergoing a shipyard overhaul, the ship's fire mains may be inoperative. The shipyard fire department then assumes responsibility for fighting shipboard fires. If you are goofing off on your watch or are absent from your station, a fire could gain considerable headway before arrival of the fire department, resulting in extensive (and unnecessary) damage to the ship and possible personnel casualties.

Barracks Security Watch

A security watch is maintained in all barracks for protection against fire, for the safety of personnel and material, and for carrying out routines. A security watch stander is responsible for knowing and carrying out the provisions of the fire bill, emergency bill, barracks regulations, and the like. The barracks security watch stander is responsible for maintaining prescribed standards of order and discipline.

If you are standing the security watch and an officer approaches, you salute and sound off with your name and rate. The formula for sounding off may vary from place to place, but it goes something like this: "Smith, Seaman, security watch, Barracks K, Sir/Ma'am."

A barracks security watch is usually a roving one and, depending on the type of barracks, it may cover two or more wings and/or decks. You must be alert to spot any fires that may be started by personnel smoking in their bunks (which is against regulations). If you see anyone smoking in a bunk, have that person put out the cigarette. Persons returning from liberty after taps must be prevented from disturbing the sleep of other personnel. Watch standers have the additional responsibility of making sure that unauthorized personnel don't enter the barracks. Usually, you must report periodically to the duty officer in the barracks office or, in some instances, to the OOD by telephone. Normally, you report that all is secure; however, you must report all disturbances and any unusual circumstances, such as illness or mishap. You may also have to enforce taps and hold reveille.

In the event of a fire, your duties are to do the following:

1. Report the fire. (Know the fire department number and the locations of the fire alarms.)
2. Spread the alarm—pass the word. Ensure all personnel except fire parties are clear of the area.
3. If time permits, close doors and windows to confine the fire and prevent drafts. Do not endanger yourself or others in this effort.
4. Fight the fire if possible, using the proper equipment at hand to extinguish the fire, pending the arrival of the fire department.

SENTRIES

Sentries are required at a number of locations, such as at gates to military bases, aboard ship, along a fence, or in a hangar. Although our discussion concerns a sentry walking a post, the same provisions apply to other types of sentry watches.

Sentries are governed by two types of orders—**general** and **special**. General orders (which do not change) cover situations of a routine nature common to most sentry posts. Special orders cover a certain time or situation peculiar to a particular post and are issued in addition to the general orders. Special orders may be written or verbal. Usually, written orders are issued by the CO and remain in effect until canceled or changed with subsequent (new) orders from the CO. These instructions are called standing orders. Verbal orders may be issued by any responsible officer or petty officer. Normally, they remain in effect for a limited period of time.

Student Notes:

Eleven General Orders of a Sentry

Normally, general orders for a sentry call for reporting to the petty officer of the guard. However, at any given station, you will make your reports to the petty officer of the watch, officer of the day, officer of the deck, or to the person designated as your immediate supervisor of the watch. Before standing watch, you need to know your chain of command for the watch.

There are 11 general orders for a sentry. They are reproduced here in **bold** letters, along with a brief explanation of each. You should memorize them word for word. You don't need to memorize the explanations, but you should understand the meaning of each order.

1. **To take charge of this post and all government property in view**. The number of the post, type of sentry duty, and limits of your post are part of your special orders. Within the limits of your post, you have authority over all persons, and it's your duty to challenge and, if necessary, detain all persons acting in a suspicious manner. You should apprehend all persons involved in disorder or discovered committing a crime. All persons detained or apprehended are turned over to the petty officer of the guard. You should fire your weapon only as a last resort. Smoking in a prohibited area, for example, is hardly a shooting offense. There are times, however, when firing at another person may be justified, but only after all means of defense or crime prevention have failed. In general, such times are as follows:

   a. To protect your own or another's life

   b. To prevent the escape of a person known to have committed a serious crime, such as armed robbery, rape, or murder

   c. To prevent sabotage, espionage, arson, and other crimes against the government

   d. If you must fire your weapon, try to wound instead of kill the person you're aiming at

2. **To walk my post in a military manner, keeping always alert and observing everything that takes place within my sight or hearing**. Keep turning your head as you walk your post, observing everything ahead and to the sides. If you hear a strange noise, investigate it.

You cannot expect to stand all your watches in fair weather. When the weather is bad, you will be issued appropriate clothing. Do not stand under a tree to keep out of the rain or stay behind a building to get out of a cold wind; during times of bad weather and darkness, you must be particularly alert.

3. **To report all violations of orders I am instructed to enforce**. If a person is acting from thoughtlessness, you need only remind the offender of the regulation being broken. For example, if you see a person starting to light a cigarette in a no smoking zone or a visitor blundering into a restricted area, you need only tell the person the regulation in effect. If the person is willfully violating a regulation, however, like trying to jump the fence or stealing Navy property, you must stop the person and place the offender under apprehension; then call for the petty officer of the guard. If the person tries to escape, give the order to halt. If the person does not obey, fire into the air; if the person does not stop, fire at the fleeing party's legs, subject to the limitations given under general order 1. If the offender escapes, report the matter as quickly as you can to the petty officer of the guard. In every instance, try to remember what the offender looked like so that you may identify the person. Do not leave your post to chase the offender unless immediate action is essential.

By firing your weapon and shouting, you can alert other sentries to intercept the offender. Do not fire at an offender if anyone else is around who could be hit by your shot. It is better to let the wrongdoer escape for the time being than to shoot an innocent person.

4. **To repeat all calls from posts more distant from the guardhouse (quarterdeck) than my own**. Suppose your post is number 3. To call the petty officer of the guard for any purpose other than relief, fire, or disorder, you call, "Petty officer of the guard (or corporal of the guard), post number 3." Sentry number 2 will repeat your call, giving your number, and so will sentry number 1. Thus the petty officer will know immediately which post to go to. Similarly, if sentry number 4 calls out, repeat the call, giving his or her number.

5. **To quit my post only when properly relieved**. If you aren't relieved on time, don't abandon your post, but call the petty officer of the guard for instructions. If you require a relief because of sickness or other reason,

Student Notes:

call, "Petty officer of the guard, post number ____, relief."

6. **To receive, obey, and pass on to the sentry who relieves me all orders from the commanding officer, officer of the day, and officers and petty officers of the guard only**. During your tour of duty, you are subject to the orders of the CO, XO, OOD, and the officers and petty officers of the guard only. Other officers and petty officers have no authority to take or inspect your weapon, to tell you how to stand your watch, or to order you to leave your post. Such other officers, however, still have the authority to investigate your conduct and to report it. Thus an enemy agent cannot dress up in an officer's uniform and order you from your post. You obey orders only from officers whom you know are authorized to give you commands related to your sentry duty. However, a passing naval officer who believes you are standing a poor watch may ask your name and post and report any observations to your superiors.

7. **To talk to no one except in the line of duty**. When you challenge or talk with a person, take the position of port arms. Answer questions briefly but courteously. Normally, if you maintain silence and military bearing, visitors will not try to engage you in long conversations. If, however, visitors or other naval personnel show a desire to pass the time of day with you, you must say politely to them "Excuse me, I am on duty and cannot talk with you further. Please move on." If they refuse to move on or show signs of becoming disorderly, you should call for the petty officer of the guard. Remember, if your superiors see you chatting while on duty, they will hold you responsible-not your visitor.

8. **To give the alarm in case of fire or disorder**. In case of fire, you immediately call, "Fire, post number ____" and sound whatever alarm is available. When you are sure your alarm has been heard by the other sentries or by the petty officer of the guard see what you can do to put out the fire. (If you can do so safely and without leaving your post, do so; otherwise, remain where you can direct apparatus to the fire.)

Remember that the fire may be a trick to lure you away from your post. You must remain vigilant (alert), even amid the confusion and excitement that accompanies a fire.

What we have said about fire applies also for disorder. In the event of a disorder, call the guard immediately; then try to quiet the trouble. If you approach the disorder first, you might be overcome and then could not give the alarm. Sometimes you can stop a disorder before it becomes too serious by calling to the persons involved, "I have reported you to the guard, who will be along immediately. Come to order now; further trouble will make matters worse for you." The persons concerned might realize you are right and follow your orders. If they do, maintain watch over them but do not approach too closely. Keep your weapon at port arms.

9. **To call the petty officer of the guard in any case not covered by instructions**. When you do not know what to do, call the petty officer of the guard.

10. **To salute all officers and all colors and standards not cased**. As used here, colors and standards both refer to the national ensign. The ensign is called the national colors (or just colors) when it is flying from a staff or pike carried by an individual or displayed in a fixed location, as from a flagpole. When mounted on a vehicle, the ensign is called the national standard. (Colors and standards are cased when they are furled and placed in a protective covering.) For sentries, the rules for saluting are the same as those described in chapter 9 of this manual with the following modifications:

a. If you are walking your post or patrolling while armed with a rifle, you halt and salute by presenting arms; when at sling arms, you render the hand salute.

b. If you're in a sentry box, you stand at attention in the doorway when an officer approaches; if you're armed with a rifle, you present arms. If otherwise armed, render the hand salute. If you're on duty in front of a building or passageway entrance where there is heavy traffic of officers, you may render the rifle salute at order arms. If you're in conversation with an officer, you don't interrupt the conversation to salute another officer. If the officer with you salutes a senior, however, then you also salute.

c. During the time of challenging, you don't salute an officer until the officer has advanced and has been duly recognized. You don't salute if to do so will interfere with the proper execution of your specific duties.

Student Notes:

11. **To be especially watchful at night and during the time for challenging, to challenge all persons on or near my post, and to allow no one to pass without proper authority**. When you see a person approaching your post, take the position of port arms and call, "Halt! Who is there?" The challenge must be made at a distance sufficient to prevent your being rushed by the person being challenged. If the person answers "Friend" or "Petty officer of the guard" or gives another reply indicating a friendly nature, call, "Advance (friend, and so on) to be recognized."

If you challenge a party of persons, after receiving a reply indicating the party is friendly, you call, "Advance one person to be recognized." When you have identified the one, you have the person bring up the rest of the party and identify each individual.

You must positively identify all persons challenged before permitting them to pass. If you can't identify them to your satisfaction, detain them and call the petty officer of the guard.

Never let more than one person advance at a time. If two persons approach at the same time, have them halt; then advance the senior and pass that person (if properly identified) before advancing the other person.

If the people are in a vehicle, you halt the vehicle and inspect the driver's or the passengers' credentials, as appropriate. (Normally, inspecting the driver of a military vehicle is sufficient; but for a commercial truck or taxi, you should check the passengers too.) If you believe there's something suspicious about the vehicle or its occupants, direct one of the occupants to get out and approach you for recognition. If you aren't satisfied beyond a reasonable doubt that the people are authorized to pass, detain the person or party and call the petty officer of the guard.

When challenging, advancing, and passing persons and patrols, always stand where you can get a good look at them in such a way that you are protected from a surprise attack.

Relieving an Armed Watch

Two methods are used for relieving armed sentries. One way (usually used ashore) is for the Petty Officer of the Watch (POOW) to fall in the reliefs and march them to their posts. Normally, each person in the relieving detail is armed with a rifle. At each post, the petty officer halts the ranks, and both the sentry being relieved and the reporting sentry come to port arms while the person being relieved passes any special orders or other information the relief should know.

In the other method (usually used aboard ship), each relieving sentry goes alone to the post. This sentry normally is unarmed and will relieve the sentry of the rifle or pistol as well as the post. The relief reports to the sentry, "I am ready to relieve you." The sentry executes inspection arms and port arms and repeats the orders; the relief says, "I relieve you." The relieving procedure is completed when the sentry being relieved passes the rifle to the relief and says, "I stand relieved."

NOTE

Refer to the ship's Standard Operating Procedures (SOP) for relieving an armed watch.

When standing an armed watch with a pistol, you must strictly observe the following additional precautions:

1. Keep the pistol in its holster except when the watch is relieved or circumstances require you to use it. Never engage in horseplay with the pistol—it is a deadly weapon and must always be treated as such.

2. Do not surrender the pistol to any unauthorized person.

3. The pistol normally is carried loaded aboard ship with one round in the chamber. Two loaded clips (magazines) are in the pouches attached to the pistol belt. Leave the clips in their pouches.

4. When being relieved, a safe area for unloading a pistol must be established. In a safe area, remove the magazine from the pistol. With the weapon pointed in a safe direction (i.e., barrel full of sand), carefully jack the slide to the rear and remove the round from the chamber. Check the chamber, ensuring no rounds are present. Release the slide and let the hammer go home (weapons terminology for returning the hammer to the uncocked position). Dry fire the weapon and then engage the safety.

Student Notes:

CIRCUMSTANCES UNDER WHICH A WEAPON MAY BE FIRED

Only the CO can authorize the use of deadly force. (The term *deadly force* is defined as that force which, if used, has the potential to cause death or serious bodily harm.) The pistol or rifle should be used only as a last resort and then only under the following conditions:

1. To protect your life or the life of another person where no other means of defense will be effective in the particular situation
2. When no other effective means is available to prevent the commission of or to prevent the escape of a person known to have committed robbery, murder, rape, arson, or kidnapping
3. To prevent acts of sabotage, espionage, or other crimes against the government after failure of all other available means of preventing such crime

LOOKOUTS

You may wonder why visual lookouts are needed today when U.S. Navy radar and sonar are the best. Well, there are some objects radar can't detect, and water conditions may severely limit the sonar detection range. For example, you might be able to see a submarine's periscope that's beyond sonar detection range and whose radar indication is lost in the surrounding sea return echoes.

Lookouts are important members of the ship's operating team. As mentioned above, there are some objects radar can't detect. Smoke, flares, swimmers, torpedo wakes, debris, low-flying aircraft, and life rafts are either impossible or very difficult to detect. Sometimes, radar also indicates the presence of objects that actually are not there. A lookout may be able to verify the validity of a radar contact report and identify the objects detected. During conditions of electronic silence, lookouts are the only means of detection.

The number of lookout stations varies according to the type of ship and whether it is peacetime or wartime. Naturally, large ships have more personnel available than do small ships; therefore, they can man more lookout stations. More lookouts are required in wartime than in peacetime. When enough personnel are available in peacetime, and always in wartime, three basic lookout searches are established.

1. **Surface lookouts**, who search from the ship to the horizon
2. **Low sky lookouts**, who search from the horizon to 5 degrees above it
3. **High sky lookouts**, who search from the horizon to the zenith (directly overhead)

Additionally, several persons may be assigned to each search, each person being responsible for a specified sector. Adjacent sectors have about 10 degrees overlap so that no area will be overlooked.

The normal peacetime lookout organization has three persons in each watch section.

- Two persons are located on the bridge or atop the pilothouse (for destroyer-type ships)—one searches to port, the other to starboard. Their sectors extend from just abaft the beam forward to dead ahead.

- The third person is stationed aft and is called the *after lookout/life buoy watch*. This sector extends from the starboard beam aft and around to the port beam. In addition to reporting all objects behind the ship, you would have the responsibility for promptly throwing overboard a life buoy if you see a person fall over the side, hear the cry "Man overboard," or hear cries for help coming from the water. If you are the first to see the accident, you call, "Man overboard, (port/starboard) side." You also relay reports made by others.

When you are on lookout watch, always report everything you see or hear. Trash in the water may seem unimportant to you, but it indicates a vessel has passed that way. In wartime, such a disclosure could lead to the sinking of the vessel. Discolored water may mean you are entering a shoal area. The OOD will never reprimand you for reporting objects but will reprimand you for not reporting them. There is no excuse for letting the OOD see something before you do.

Student Notes:

REVIEW 1 QUESTIONS

Q1. List the three main reasons for a ship to maintain a watch.

a.

b.

c.

Q2. To find the ship's organized plan for action, you would look in the—

Q3. What person assigns qualified personnel to stations and enters their names on the Watch, Quarter, and Station Bill?

Q4. Write the condition on the right that matches the battle station situation on the left.

SITUATION	CONDITION
a. A special watch used by gunfire support	
b. The normal wartime cruising watch	
c. All battle stations manned	

Q5. You are relieving a watch. How many minutes ahead of time should you arrive at your station?

a. 15

b. 20

c. 25

d. 30

Q6. The ship's bell is usually restricted to what hours?

Q7. How many bells are sounded at 0700?

Q8. List the type of the watches for the times listed on the left.

TIME	TYPE
a. 0000 to 0400	
b. 0400 to 0800	
c. 0800 to 1200	
d. 1200 to 1600	
e. 1600 to 1800	
f. 1800 to 2000	
g. 2000 to 2400	

Q9. What is the purpose of the dog watch?

Q10. What type of watch is stood by most Sailors?

Q11. The watch system is divided into what two parts?

a.

b.

Q12. List four purposes of a security watch.

a.

b.

c.

d.

Student Notes:

Q13. What person is responsible for maintaining the ship's deck log while under way?

Q14. The fog lookout normally stands—

Q15. List the conditions under which you would normally stand an after steering watch.

a.

b.

c.

Q16. The sound and security watch reports directly to the (a) ___________, and the results of their inspections are logged in (b) _______________.

Q17. There are 11 general orders and these orders don't change. General orders cover what situation(s)?

Q18. List the precautions that must be strictly adhered to while standing an armed watch with a pistol.

a.

b.

c.

d.

e.

Q19. List the conditions under which deadly force may be used.

a.

b.

c.

BEARING

Learning Objectives: When you finish this chapter, you will be able to—

- Recognize the procedures to use when reporting bearings, to include scanning procedures and reports.
- Identify the procedures to follow when using binoculars to include night vision.

The direction of an object from a ship is called the *bearing*. Bearings are measured in degrees clockwise around a circle from 000° to 360°. There are three types of bearings.

1. Relative bearings use the ship's bow as a reference point.
2. True bearings use true north (the geographic north pole) as the reference point.
3. Magnetic bearings use the magnetic north pole as the reference point.

Sometimes, all three types of bearings coincide, but such situations are rare and of a temporary nature. Lookouts report objects (contacts) in degrees of relative bearing.

Figure 3-2 shows the relative bearings around a ship. An object dead ahead bears 000°, while an object abeam to starboard bears 090°, and so on. Study this figure, practice pointing to various objects. Compare your estimates of their bearings to what the objects actually bear. With practice, you should be able to report a contact within 5° to 10° of its actual bearing.

Student Notes:

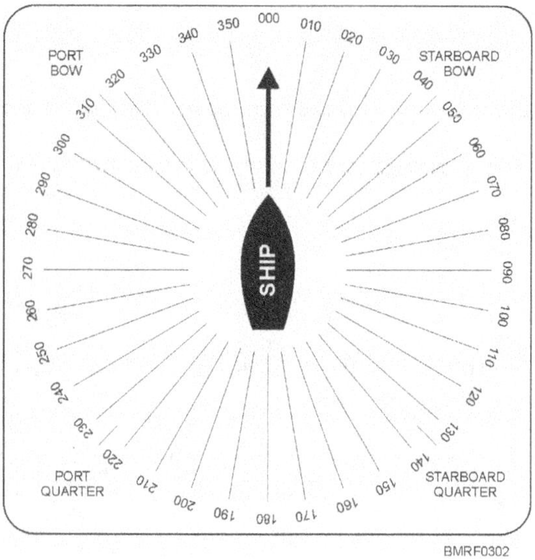

Figure 3-2.—Relative bearings.

To prevent confusion, the Navy uses a standard system for pronouncing numerals. The following list shows how numerals (numbers) are spoken:

NUMERAL	PRONOUNCED
0	Zero
1	Wun
2	Too
3	Tree
4	Fo-wer
5	Fife
6	Six
7	Seven
8	Ate
9	Niner

Bearings are always reported in three digits and spoken digit by digit, except that objects dead ahead or astern (000° or 180°), on either beam (090° or 270°), or on either bow (045°or 315°) or quarter (135° or 225°) may be reported as such. For example, a ship bearing 090° may be reported as being "abeam to starboard."

Do not become excited when you report contacts or other sightings. Failing to use the proper terminology can result in the OOD wasting time trying to find the object. Take a few seconds to think about how you are going to report the sighting. Taking that few seconds could mean the difference between the entire bridge looking on the wrong side of the ship for a sighting that is actually on the other side. Note that the word *relative* was not included. It is understood that lookouts report only in relative bearing.

REPORTING TARGET ANGLE

Target angle is the relative bearing of your ship from another ship. You may wonder why you would care what your ship bears from another ship. The OOD uses target angles as an aid in determining the course of actions when another ship is sighted. (Target angles are useful during gunnery and antisubmarine operations.)

Look at figure 3-3. You are the starboard lookout and you detect a ship on your starboard bow heading at a right angle across your course. You report to the OOD, "Bridge, starboard lookout, ship broad on the starboard bow (or zero, fo-wer, fife), target angle tree wun fife." Assuming that your course is 000°, the OOD knows the other ship's course is approximately 270° and, depending upon the speed of the two ships, the possibility of collision exists. Your target angle report has alerted the OOD that a change of course or speed or both may be required. A change in target angle can mean that the contact has changed course, which is not always immediately apparent to the radar plotters in CIC.

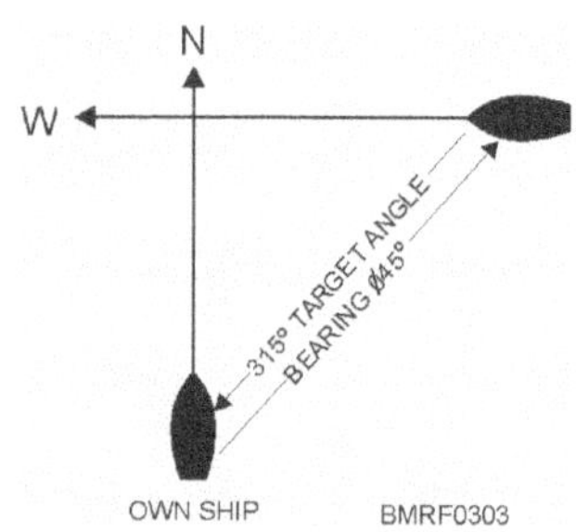

Figure 3-3.—Target angle.

Student Notes:

REPORTING POSITION ANGLE

An object located in the sky is reported by its bearing and position angle. The position angle of an aircraft is its height (in degrees) above the horizon as seen from the ship. The horizon is 0° and directly overhead is 90°. A position angle can never be more than 90°, as shown in figure 3-4. Position angles are reported in one or two digits and spoken as a whole—not digit by digit.

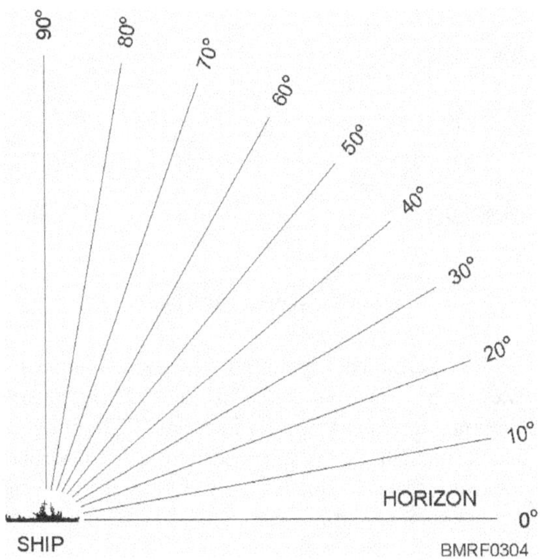

Figure 3-4.—Position angle.

Position angles should be reported on all aircraft. Look at figure 3-5. As the aircraft approaches the ship, the position angle increases. Whenever the position angle changes significantly, all stations should be informed. To help you more accurately determine an aircraft's position angle, you can use the aids shown in figure 3-6. The width of the thumb between the horizon and the aircraft is approximately 2°; the width of the closed fist, approximately 8°; and the open hand, approximately 15° (at arm's length).

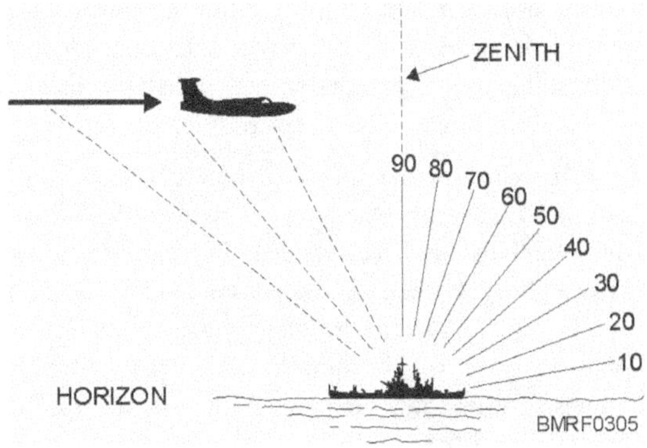

Figure 3-5.—Position angle.

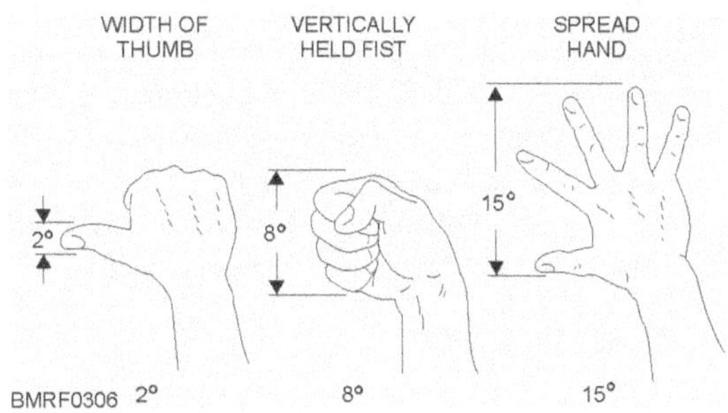

Figure 3-6.—Position angle aids.

REPORTING RANGES

Most of the time, if you give reasonably good bearings and position angles when reporting contacts, the OOD will have little difficulty in locating them. But suppose you sight a submarine periscope, a person, or some other object low in the water. In these instances, you can save valuable time by reporting the object's approximate distance. Ranges are always reported in yards. Estimating distances over water is difficult for the inexperienced lookout. Distances can be very deceptive.

Student Notes:

A ship that looks like it's 1/2 mile away may actually be twice, or more than twice, that distance from you. Sometimes objects that seem to be half the distance to the horizon may actually be considerably closer.

Knowing your height above the water helps you to estimate ranges. For example, at a height of 50 feet, the distance to the horizon is about 16,000 yards (8 miles); at a height of 100 feet, the distance is about 23,000 yards (11 1/2 miles). Practice estimating distances to known objects. Until you become proficient at estimating ranges, use phrases, such as "close aboard," "on the horizon," and "hull down."

Ranges are reported in yards and spoken digit by digit, except that multiples of hundreds and thousands are spoken as such.

USE OF BINOCULARS

Using binoculars for searching isn't always better than using the naked eye. Several factors govern when and how binoculars should be used. For example, in fog binoculars should not be used. At night, they should be used quite often. Another factor is their field of view, which is about 7°. Depending on the type of search, such a narrow field may hamper proper scanning techniques.

Adjusting Binoculars

Three adjustments are required to obtain proper focus and to gain maximum benefit from the light-gathering quality of binoculars—two adjustments for focus and one for the proper distance between lenses.

To properly focus your binoculars, you should do the following:

1. Set both eyepieces to the +4 mark. Place the binoculars firmly against the eyebrows and locate a small, well-defined object about 1/2 mile away.
2. Cover one lens. (Do not touch the glass.)
3. Slowly turn the other eyepiece until a sharp image is obtained, then back off as far as possible without losing the sharpness. (Keep both eyes open; closing one will give an incorrect focus.)
4. Note the reading on the scale; then repeat the previous procedures two or three times to obtain the exact setting. Follow the same procedure for the opposite eye.

The final adjustment is to establish the interpupillary distance (IPD), which is the distance between your eyes. Move the barrels up and down until you see a single circle (fig. 3-7). Then note the reading on the IPD vernier between the barrels. An incorrect IPD setting will strain the eyes and waste part of the binoculars' light-gathering ability.

Figure 3-7.—Proper IPD setting.

You won't have your own personal binoculars. They are passed from watch to watch. Therefore, it's important for you know your focus and IPD settings so that the binoculars may be properly adjusted at night or when there are no objects on which to focus in the daytime. For nighttime use, the focus setting is one mark less than for daytime.

Daytime use of binoculars depends upon the type of search being conducted. Surface lookouts should use them to scan across their sector—they should then use the naked eye on return sweeps. Sky lookouts should use them only to identify a contact detected with the naked eye.

The binoculars should be used more frequently at night than during daylight, but searches should still be made with the naked eye. You often can see objects, particularly moving ones, out of the corner of your eye. These objects might not be detected with the binoculars because of their narrow field of view.

Binoculars should never be used in fog, rain, snow, or thick haze.

Student Notes:

Care of Binoculars

Binoculars are fairly delicate instruments; they cannot stand much knocking about. Therefore, keep them on a short strap when wearing them to prevent their banging against solid objects. **Always** keep the strap around your neck. **Never** hold binoculars over the side of the ship without the strap being around your neck. Many pairs of binoculars have been lost over the side in this manner. Keep the lenses dry; otherwise, you will not be able to see properly. Don't let them become overheated; the cement around the lenses might melt. Above all, keep them clean. You must be careful, however, not to damage the lenses when cleaning them. First, blow off loose dust; then breathe on the lenses (except in freezing weather) and gently clean them with lens paper. Rags, plain paper, handkerchiefs, or your sleeve or shirttail should not be used, as they might scratch the lens. You can usually get a supply of lens paper from the QMOW.

NIGHT VISION

Have you ever walked from a lighted theater lobby into the darkened theater? You would almost be blind for a few minutes. As your eyes become accustomed to the weak light, your vision gradually improves. The same situation exists when you go on night watch directly from a lighted compartment. After 10 minutes, you can see fairly well. After 30 minutes, you reach your best night vision. This improvement of vision in dim light is called *dark adaptation.*

Specially designed red goggles are provided for you to use before you go on night lookout duty. These goggles prepare your eyes for darkness without affecting your ability to play games, write letters, or read before going on watch. You should wear them without interruption for at least half an hour before going on watch. Even then, it will still take you at least 5 minutes more in darkness to develop your best night vision.

After your eyes are dark adapted, you must learn to use your *night eyes*. In the daytime, you should look directly at an object to see it best. In the dark, you need to look above, below, or to one side of an object to see it. This is called *off-center vision*. At night, it's also easier to locate a moving object than one standing still. Because most objects on or in the water have a relatively slow speed, we move our eyes instead, and the effect is nearly as good. Therefore, while scanning at night, lookouts move their eyes in slow sweeps across the area instead of stopping the eyes to search a section at a time.

Your ship may be equipped with night vision equipment. Before standing watch, be sure you are trained in operating the night vision equipment assigned to your ship.

SCANNING PROCEDURES

A well-trained lookout will see much more than a "green" hand would see. In good weather, lookouts can easily spot planes with the naked eye at 15 miles. With binoculars and in unusually clear weather, lookouts have detected planes at 50 miles. At night, skilled lookouts will detect objects that the untrained lookout would never suspect were there.

The lookout's technique of eye search is called *scanning*, which is a step-by-step method of looking. It is the only efficient and sure way of doing the job. Scanning does not come naturally. You must learn to scan through practice. In the daytime, your eyes must stop on an object to see it. Try moving your eyes around the room or across the water rapidly. Note that as long as your eyes are in motion, you see almost nothing. Allow your eyes to move in short steps from object to object. Now you can really see what is there.

Figure 3-8 shows how you should search along the horizon. (You also must cover the surface between your ship and the horizon.) Search your sector in 5° steps, pausing between steps for approximately 5 seconds to scan the field of view. At the end of your sector, lower the glasses and rest your eyes for a few seconds; then search back across the sector with the naked eye.

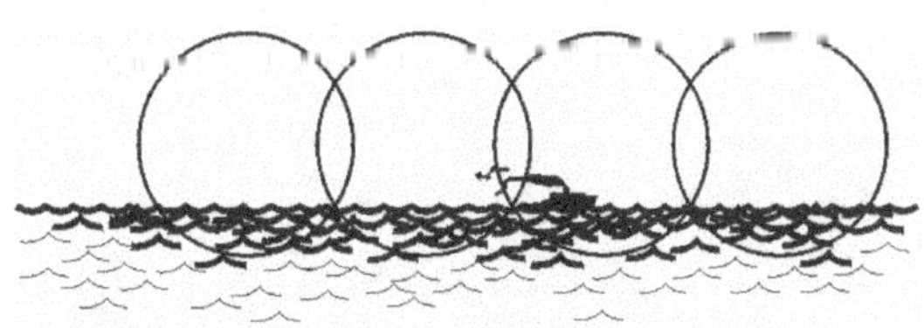

BMRF0308

Figure 3-8.—Scanning using the step-by-step method.

Lookouts also search from the horizon to the zenith (overhead), using binoculars only to identify a contact. Move your eyes in quick steps (about 5°) across your

Student Notes:

sector just above the horizon. Then, shift your gaze upward about 10°, and search back to the starting point. Repeat this process until the zenith is reached; then rest your eyes for a few seconds before starting over.

When searching at night, keep your eyes moving. Try to adhere to (stay with) the sector scan (and upward shift) even though the horizon may not be visible. If you spot a target (or even think you have), don't stare at it. Instead, look slightly to either side.

REPORTS

Every object sighted should be reported, no matter how insignificant it may seem to you. The initial report consists of two basic parts—what you see and its bearing (direction) from the ship. Aircraft sighting reports also include altitude (position angle). Report the contact as soon as you see it, then follow with an amplifying report. Include the object's identity (destroyer, periscope, log, and so on) and direction of travel (closing, crossing, and so on). Refer to the ships SOP on reporting procedures.

REVIEW 2 QUESTIONS

Q1. Describe how bearings are measured?

Q2. List the three different types of bearings.

a.

b.

c.

Q3. Explain the difference between reporting bearing angles and position angles.

Q4. You are using your hand as an aid to determine the position angle of an aircraft. What is the approximate width of a closed fist (in degrees)?

Q5. If the binocular IPD is adjusted properly, what will you see when viewing through them?

a. One circle

b. Two separate circles

c. Two circles

Q6. List the three adjustments that must be made when using binoculars.

a.

b.

c.

Q7. The improvement of vision in dim light is known as—

Q8. How many minutes will it take for you to reach your best night vision?

Q9. Explain the difference between the method used for a day lookout and a night lookout.

SUMMARY

In this chapter, you learned about the basic fundamentals of the watch organization and some of the procedures associated with standing a proper watch. We also covered the importance of communications in relation to watch-standing duties. Having well-trained and competent watch standers would be useless without

Student Notes:

a means of relaying information. You also learned how bearings are reported.

Every person in the Navy has, at one time or another, been assigned some type of watch. Your safety, and that of your shipmates, depends on how well you execute these duties. Just one moment of inattention could mean the difference between a shipmate that has fallen overboard being recovered or lost. A minute of "slacking off" as a fog lookout may be the difference in reaching home port safely or being involved in a collision at sea. No watch is more or less important than others. Every watch on board your ship or station is interdependent. The safety of all crew members depends upon each watch stander carrying out his or her assigned duties in a proper military fashion and according to the eleven orders of the sentry.

REVIEW 1 ANSWERS

A1. The three main reasons for a ship to stand watch are—

a. **Communications**

b. **Security**

c. **Safety**

A2. The ship's organized plan for action is located in the **battle bill**.

A3. The **division officer** and **division chief** are responsible for assigning qualified personnel to stations and entering their names on the Watch, Quarter, and Station Bill.

A4. The condition on the right matches the battle station situation on the left.

SITUATION	CONDITION
a. A special watch used by gunfire support	**Condition II**
b. The normal wartime cruising watch	**Condition III**
c. All battle stations manned	**Condition I**

A5 When relieving a watch, you should arrive at your station **15 minutes** ahead of time.

A6. The ship's bell is usually restricted to the hours between **reveille and taps**.

A7. **6 bells** are sounded at 0700.

A8. The type of the watches for the times listed.

TIME	TYPE
a. 0000 to 0400	**Midwatch**
b 0400 to 0800	**Morning watch**
c. 0800 to 1200	**Forenoon watch**
d. 1200 to 1600	**Afternoon watch**
e. 1600 to 1800	**First dog watch**
f. 1800 to 2000	**Second dog watch**
g. 2000 to 2400	**Evening watch**

A9. The purpose of the dog watch is to **rotate watches**.

A10. Most Sailors stand **security watches**.

A11. The watch system is divided into (a) **underway** and (b) **in-port watches**.

A12. Security watches—

a. **prevent sabotage**

b. **protect property from theft**

c. **prevent access to restricted areas**

d. **protect personnel**

A13. The **QMOW** maintains the ship's deck log while under way.

A14. The fog lookout normally stands **in the bow where approaching ships can better be seen and heard**.

A15. Normally, an after steering watch is stood under the following conditions:

a. **General quarters**

b. **Under way replenishment**

c. **Sea and anchor detail**

A16. The sound and security watch reports directly to the **(a) OOD**, and the results of their inspections are logged in **(b) ship's deck log**.

A17. General orders cover situations of a **routine nature common to most sentry posts**.

A18. The precautions to be strictly adhered to while standing an armed watch with a pistol include—

a. **keep the pistol in the holster.**

b. **don't engage in horseplay with the pistol.**

c. **don't surrender the pistol to any unauthorized person.**

d **leave two loaded magazine clips in their pouch and remember there is one round loaded in the chamber.**

e. **when relieved, unload the pistol in a safe designated area. Remove the round from the chamber and check the chamber clear. Release the slide and let the hammer go home. Dry fire the pistol then engage the safety.**

A19. Deadly force can be used—

a. **to protect your life or the life of another person where no other means of defense will be effective**

b **when no other means is available to prevent the commission of or to prevent the escape of a person known to have committed robbery, murder, rape, arson, or kidnapping**

c. **to prevent acts of sabotage, espionage, or other crimes against the government after failure of all other available means of preventing such crime**

REVIEW 2 ANSWERS

A1. Bearings are measured **in degrees, clockwise around a circle from 000° to 360°.**

A2. The three different types of bearings are—

a. **Relative**

b. **True**

c. **Magnetic**

A3. Bearings are reported in **three digits, spoken digit by digit**; positions are reported in **one or two digits and spoken whole.**

A4. When using your hand as an aid to determine the position angle of an aircraft, your closed fist is **approximately 8°.**

A5. If the binocular IPD is adjusted properly, you will see **one circle** when viewing through them.

A6. T he three adjustments that must be made when using binoculars are **to adjust each eyepiece and set the IPD.**

A7. The improvement of vision in dim light is known as **dark adaptation.**

A8. It will take **30 minutes** for you to reach your best night vision.

A9. The different methods used for a day lookout and a night lookout are the day lookout **moves his/her eyes in 5 steps, pausing at each step**; the night lookout **keeps moving his/her eyes.**

CHAPTER 4

COMMUNICATIONS

A glance at a globe is all it takes to appreciate the meaning of control of the sea in the nuclear age.

—Admiral Arleigh Burke

Communications are of vital importance to a shipboard organization and are sometimes referred to as *the voice of command.* Without proper communication among the different parts of the ship, the whole organization could break down and fail in its mission.

Communications, as discussed in this chapter, are grouped into two basic categories—interior and exterior. Interior communications are concerned only with the exchange of information between individuals, divisions, and departments aboard a single ship or station. Exterior communications deal with conveying information between two or more ships, stations, or commands.

One of the most important communications systems used aboard ship is the sound-powered telephone. Sometime in your Navy career, you **will** "man" a sound-powered telephone set. You must become familiar with the proper usage and care of the equipment. In addition, you must learn the correct procedures used with the sound-powered telephone system, including the use of the phonetic alphabet.

THE PHONETIC ALPHABET

Learning Objective: When you finish this chapter, you will be able to—

- Identify the phonetic alphabet as applied to communications.

It is easy to confuse the sounds of certain letters, such as bee and dee, cee and zee. To avoid confusion, the Navy requires that phonetic equivalents of letters be spoken instead of the letters themselves.

The Navy has had a phonetic alphabet for many years. From time to time, it's been changed in attempts to use words that would instantly bring to mind the letter represented by the word. The phonetic alphabet (table 4-1) was adopted by the armed forces of the various NATO nations as a means of overcoming many language difficulties. Each word is accented on the

Table 4-1.—Phonetic Alphabet

LETTER	EQUIVALENT	SPOKEN
A	ALFA	AL fah
B	BRAVO	BRAH voh
C	CHARLIE	CHAR lee
D	DELTA	DELL ta
E	ECHO	ECK oh
F	FOXTROT	FOKS trot
G	GOLF	GOLF
H	HOTEL	hoh TELL
I	INDIA	In dee ah
J	JULIETT	JEW lee ett
K	KILO	KEY loh
L	LIMA	LEE mah
M	MIKE	Mike
N	NOVEMBER	no VEM ber
O	OSCAR	OSS cah
P	PAPA	pah PAH
Q	QUEBEC	kay BECk
R	ROMEO	ROW me oh
S	SIERRA	see AIR rah
T	TANGO	TANG go
U	UNIFORM	YOU nee form
V	VICTOR	VIK tah
W	WHISKEY	WISS key
X	XRAY	ECKS ray
Y	YANKEE	YANG key
Z	ZULU	ZOO loo

capitalized syllable. You should memorize the phonetic alphabet and use it along with correctly pronounced numbers, as described earlier in chapter 3, for all telephone and lookout reports.

REVIEW 1 QUESTION

Q1. You are manning the sound-powered telephone in a repair locker. DC central calls and wants the serial number of the P-100 pump (23DBCX14) in your repair locker. How should you say this number over the phone?

SOUND-POWERED TELEPHONES

Learning Objectives: When you finish this chapter, you will be able to—

- Recognize the components of the sound-powered telephone.
- Identify the procedures to follow when using sound-powered telephones.

Sound-powered phones are just what the name implies—phones that operate on your voice power and require no batteries or external electrical power source.

When you speak into the mouthpiece, the sound waves of your voice cause a diaphragm to vibrate. The vibrations are transferred from the diaphragm through a drive rod to an armature centered in a wire coil. The coil is located in a magnetic field supplied by two permanent magnets. Movement of the armature in the magnetic field causes a current to be induced into the coil. The current then is transmitted to a receiver (the earpiece) where the process is reversed, and the person at the other end of the circuit hears the same sounds you transmitted.

The mouthpiece and earpiece, though shaped differently, function in the same manner and thus can be used interchangeably. You can talk into an earpiece and hear through a mouthpiece. This feature is important to remember not only in the event of a breakdown of one or the other pieces but also because undesired noises can be fed into the system through an earpiece turned away from your head.

THE HEADSET

Figure 4-1 shows a headset type of sound-powered telephone. The mouthpiece is suspended from a yoke that is attached to a metal breastplate. The earpieces are connected by an adjustable band. The mouthpiece and earpiece are connected by wire from a junction box on the breastplate. The plug cord is also connected into this junction.

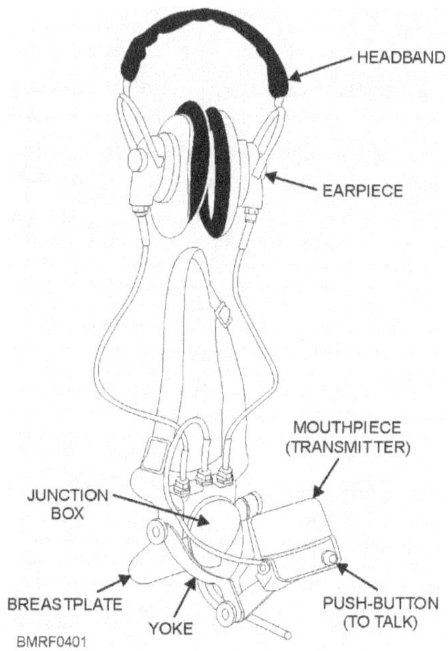

Figure 4-1.—Headset.

The headset is delicate and can be easily damaged. When you pick up the set to put it on, hold the entire unit in your left hand. You will find the headset is hung over the transmitter's supporting yoke and the lead wires are coiled.

To put the gear on—

1. Unhook the right side of the neck strap from the breastplate, put the strap around your neck, and then fasten it to the breastplate again.
2. Take off the coil of lead wires; then put the earpieces on and adjust the headband so that the center of the earpiece is directly over the opening of the ear.

Student Notes:

3. Insert the plug into the jack box and screw the collar on firmly.

Adjust the mouthpiece to bring it directly in front of your mouth when you stand erect. When you speak into the transmitter, it should be about 1/2 to 1 inch from your mouth. In making this adjustment, remember that the fine wire that goes to the transmitter can be broken easily. Be sure there aren't any sharp bends in it, and don't allow it to get caught between the transmitter and the yoke.

When you are wearing the headset, always keep some slack in the lead cord and be sure it is flat on deck. If you have the cord stretched taut (tight), someone may trip over it and damage the wires, receive an injury, or injure you. Don't allow objects to roll over or rest on the cord.

After plugging in the phones, test them with someone on the circuit. If the phones aren't in order, report that fact to the person in charge of your station and don a spare set; do not attempt to repair the set yourself.

If you are on lookout and should be listening as well as searching, cover only one ear with an earpiece so that you can hear outside noises as well as telephone communications. Keep the unused earpiece flat against the side of your head so that noises will not enter the circuit.

Never secure the phones until you have permission to do so. When permission is given, make up the phones for stowage according to the following instructions:

1. Remove the plug from the jack box (fig. 4-2) by holding the plug in one hand and unscrewing the collar with the other. When the collar is loose, grasp the plug and pull it out. Don't pull on the lead to remove the plug; that will weaken and eventually break the connection. When the plug is out, lay it carefully on the deck. Immediately screw the cover on the jack box, as dust and dirt will soon cause a short circuit in a jack box left uncovered. (**NOTE**: If you see an uncovered jack box, cover it, even though you were not responsible for the carelessness.)

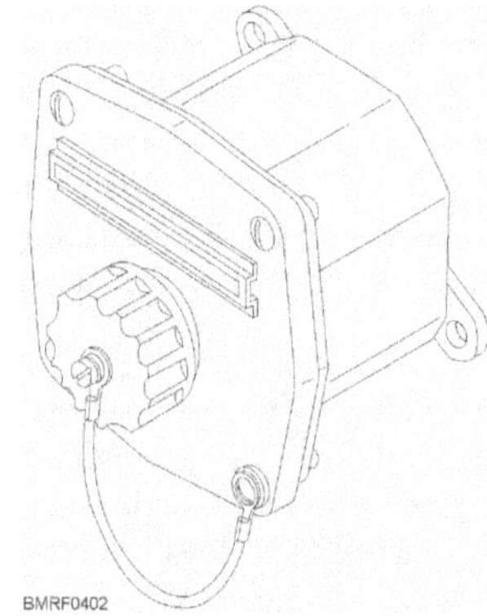

Figure 4-2.—Sound-powered telephone jack box.

2. Remove the headset and hang it over the transmitter yoke, as shown in figure 4-3.

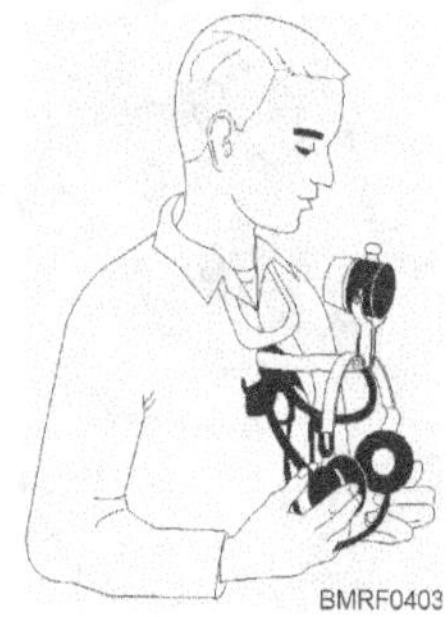

Figure 4-3.—Hanging the headset on the yoke.

3. Coil the lead cord, starting from the end at the phone. Coil the lead in a clockwise direction, holding the loops in one hand, as shown in figure 4-4. The loops should be 8 to 10 inches across, depending on the size of the space where the phones are stowed. When you are coiling the lead, be careful not to bang the plug against the bulkhead or deck.

Figure 4-4.—Coiling the lead cord.

Student Notes:

4. When the lead is coiled, remove the headset from the transmitter yoke and put the headband in the same hand with the coil. Use this same hand to hold the transmitter while you unhook one end of the neck strap from the breastplate. Fold the transmitter yoke flat, being careful not to put a sharp bend in the transmitter cord.

5. Wrap the neck strap around the coil and the headband two or three times and snap the end back on the breastplate; then fold the mouthpiece up against the junction box. You now have a neat, compact package for stowage, as shown in figure 4-5.

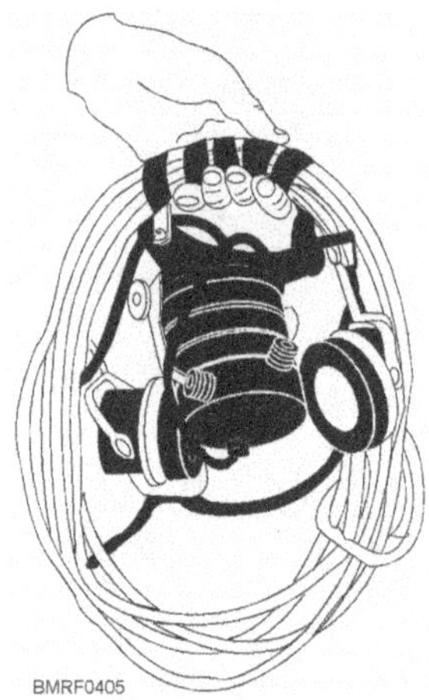

Figure 4-5.—Coiling completed.

6. Put the phone into the box or hang them on the hook provided. Be careful not to crowd or jam the leads.

Headset phones should always be unplugged when they are not in use. If they are left plugged in, the earpieces will pick up noise and carry it into the circuit. Never place the phones on the deck. Not only is it possible that someone may step on them, but decks are good conductors of noise, which can be picked up by the phones.

THE HANDSET

The handset telephone shown in figure 4-6 is held in one hand with the receiver over one ear and the transmitter in front of the mouth. A button, located on the bar connecting the transmitter and the receiver, is pushed down for talking. (The button must also be depressed [pushed down] to listen.) (**NOTE**: If the button is held down at other times, all of the noise at the talker's station will go throughout the circuit and make it difficult for other talkers on the line to understand each other.)

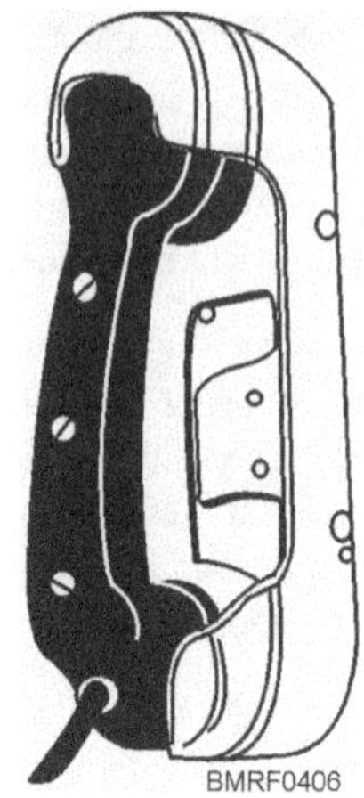

Figure 4-6.—Handset.

When not in use, the handset telephone is held on a bracket on a bulkhead with a lever or spring attachment that keeps it from being jarred loose. When you replace the handset in its bracket, be sure it is secured so that it cannot fall to the deck and be damaged.

REVIEW 2 QUESTIONS

Q1. True or False. The mouthpiece and earpiece of a sound-powered phone are interchangeable.

Q2. Describe the reason why you pick the headset phones up as a whole unit.

Q3. You are finished using the headset sound-powered phone. You should then unplug the headset for what reason?

Student Notes:

Q4. When using a handset sound-powered phone, what action should you take to talk or listen through the phone?

SOUND-POWERED CIRCUITS

Learning Objective: When you finish this chapter, you will be able to—

- Recognize the function of the primary, auxiliary, and supplementary systems of a sound-powered circuit.

Sound-powered telephone circuits aboard ship fall into three categories—primary, auxiliary, and supplementary systems.

The **primary system** includes all circuits necessary for controlling armament, engineering, damage control, maneuvering, and surveillance functions during battle. These circuits are designated JA through JZ.

The **auxiliary system** duplicates many of the primary circuits for the purpose of maintaining vital communications in the event of damage to the primary system. Auxiliary circuits are separated as much as possible from primary circuits. Circuit designations are the same as the primary system, preceded by the letter *X* (XJA, X1JV, and so on).

The **supplementary system**, X1J through X61J, consists of several short, direct circuits, such as from the bridge to the quarterdeck or from the quarterdeck to the wardroom. Circuits in the primary and auxiliary systems can be tied together at various switchboards or individual stations may be cut out of the circuits, but the supplementary system does not have these provisions. Because circuits in the supplementary system usually are not manned, most circuits contain a buzzer system so that one station can alert another station that communications between the two are desired.

Circuit designations are characterized by a letter and number code. The 21JS4 primary battle circuit, for example, is identified as follows: numerals 21 indicate the specific purpose of the circuit; the letter *J* denotes sound power; the letter *S* means general purpose (radar, sonar, and ECM information); and the numeral 4 indicates a particular station in the circuit. The same circuit in the auxiliary system is X21JS4. All auxiliary and supplementary circuit designations are preceded by the letter *X*, but supplementary circuits are easily identified as such because they have no letter after the letter *J*.

The following are some typical shipboard sound-powered circuits:

JA	Captain's battle circuit
JC	Weapons control
JL	Lookouts
21JS	Surface search radar
22JS	Air search radar
61JS	Sonar information
1JV	Maneuvering and docking
2JZ	Damage control
X8J	Replenishment at sea

If you are on a lookout watch, your reports will go over the JL circuit to the bridge and the ship's CIC. On small ships, the JL circuit sometimes is crossed with another circuit, such as the 1JV, to reduce manning requirements. The bridge talker then has the lookout, CIC, engineering, and after steering (emergency) stations on the same circuit.

TELEPHONE TALKERS

Learning Objective: When you finish this chapter, you will be able to—

- Identify the responsibilities of telephone talkers to include telephone talking procedures.

As you have learned, you'll probably stand some form of watch aboard ship as a telephone talker. A ship at sea requires many talkers even during a peacetime cruising watch. In addition to the lookouts, there are talkers on the bridge, in firerooms, and in engine rooms, to mention only a few of the many spaces. To be a good sound-powered telephone talker, you must learn proper telephone procedures.

Student Notes:

Sound-powered telephone talkers are essential to the operation of a ship at sea because the ship must have a reliable interior communicating system. Imagine the difficulties the captain would have without means of communication with the engine room, with gunnery stations during battle, or with all the other spaces that help run the ship.

GENERAL TELEPHONE TALKING PROCEDURES

Here are some tips on how to be a telephone talker.

- Because all the power for the phones is generated by your voice, you must speak loudly and clearly if your message is to get through. However, do not shout unnecessarily. Do not run your words together; make every part of your message stand out clearly. Repeat all messages word for word to the intended receiving station; if you try to paraphrase a message, its meaning may be changed.

- Never have gum or food in your mouth while you are using the phones. Talk from the front of your mouth, never from the corners. Remember, you must project your voice to every station on the circuit.

- You gain nothing by talking too rapidly; a message spoken slowly, so that it is understood the first time, is better than a message spoken so rapidly that it must be repeated.

- During an emergency, remember that it is doubly important to get the message through. By talking slowly, some of your own excitement will subside. If you are calm and sure of yourself, you will influence other talkers on the circuit to behave in the same way.

- Nearly everyone has a manner of speech that reveals to others what part of the country they are from. On occasion, you may have found it difficult to understand the speech of a person from a different part of the country. With this thought in mind, try to speak without local accents.

CIRCUIT DISCIPLINE

The sound-powered system resembles a party line—everyone can talk and listen at once. For that reason, strict circuit discipline must be maintained. Otherwise, the circuit will become clogged with private conversation just when someone is trying to transmit an important message.

The rules for circuit discipline are as follows:

1. Transmit only official messages.
2. Keep the button in the OFF position except when actually transmitting.
3. Use only standard words and phrases.

Don't use slang or profanity on the phones. Use correct nautical terms. If naval terminology is new and unfamiliar to you, make it your business to learn the correct terms.

You, as a phone talker, are a very important link in the interior communication chain; that chain is no stronger than its weakest link. Unauthorized talking means there are at least two weak links in the chain. Be efficient. If someone else on your circuit persists in useless talking, remind the person that the line must be kept clear at all times.

Circuit discipline also means you must never show impatience, anger, or excitement. You must talk slowly, clearly, and precisely. Circuit discipline means self-discipline.

STANDARD TELEPHONE TALKING PROCEDURES

Most messages are divided into the following parts:

1. Name of the station called
2. Name of the station calling
3. The message

You call the station for which you have a message, identify yourself, and send the message without waiting for the receiving station to answer.

When a message is received, it must be acknowledged (receipted for) as soon as it is understood. You acknowledge a message by identifying your station and saying "Aye."

Student Notes:

NOTE

"Aye" is not used as an answer to a question; instead, "Affirmative," "Negative," or other appropriate reply is given.

When a message is received, it must be repeated back word for word. An example would be "Catapult center deck, primary; raise the starboard jet blast deflector." The response would be "Primary, catapult center deck; raise the starboard jet blast deflector, aye." The catapult center deck operator would then wait for a few seconds for the primary operator to confirm that the order was understood. The catapult center deck operator would then raise the jet blast deflector.

Communications on the sound-powered phone system is phrased in the declarative (statement) instead of the interrogative (question). For example, the questions "What is the status of the jet blast deflector?" or "When will the jet blast deflector be repaired?" would be rephrased to "Report the status of the jet blast deflector" and "Report the estimated repair time of the jet blast deflector."

Slang expressions or locally devised codes should not be used. The use of abbreviations should be avoided. Some abbreviations may be easily misunderstood, such as SSTG, SSDG, and SFMG.

When a subordinate station requests permission to carry out an action, do not say, "Permission granted." Another station might think you are giving it permission to carry out some other action. Respond to a request with a direct order. For example, when permission is requested to change phone talkers, the proper response, if approved, would be "Change phone talkers."

If you "belay an order," immediately order what action is needed. For example, when the throttleman is given an order to "Close the throttle" and that order is belayed, then you tell the throttleman what you want him or her to do, such as "Return throttle to original position" or "Open throttle to ________."

Never receipt for a message unless you are sure you understand it. If you do not understand, tell the sender, "Say again." If the message is long and you need only a part of it to be repeated back, you can say, "Say again all after ..." or "Say again all before ..."

When you are leaving the circuit for any reason, you must obtain permission from the controlling station. You may be leaving the circuit to change headphones because of a faulty set, to be relieved by someone else, or to secure. In any case, when leaving the circuit, **request permission**.

When a circuit is in use and a station has a more important message to transmit (to report a fire, for example), the talker says, "Silence on the line." Whenever you hear that command, you must immediately stop talking so that the message can be transmitted.

EXAMPLES OF TELEPHONE TALKER PROCEDURES

The following examples of sound-powered telephone transmissions are representative of the types of messages sent over the phones. Study them until you are sure you have the procedures correct; only practice can make you into a reliable talker.

Circuit Test

To find out if telephone stations are manned and ready, the talker at control says, "All stations, control; phone check."

Each talker then acknowledges in assigned order. On a gun circuit it would go like this:

> Each station responds in order, but does not wait more than a few seconds for the station immediately preceding to acknowledge. If you are on gun 3, and gun 2 does not respond in a few seconds, you acknowledge and let gun 2 come in at the end. A circuit test is not complete until every person has answered and faults in equipment have been checked

Sending

In sending a message, first call the station you want, and then identify your own station; finally, state the message:

> "Foc'sle, bridge; prepare to anchor in fife minutes."
>
> "Fantail, bridge; slack off stern line."

Student Notes:

Receiving

When receiving a message, first repeat back the message, identify yourself, and then acknowledge the message.

"Prepare to anchor in fife minutes; foc'sle, aye."

"Slack off stern line; fantail, aye."

Both Sending and Receiving

The following are examples of sending and receiving a message:

"Fantail, bridge; report the status of slacking off the stern line."

"Report status of slacking off the stern line; fantail, aye; stern line is slack"

"Main engine control, bridge; report which boilers are on the line."

"Report which boilers are on the line; main engine control, aye; wait."

"Bridge, main engine control; boilers too, tree, and fo-wer on the line."

"Boilers too, tree, and fo-wer on the line; bridge, aye."

Repeats

When a message is not clear to the listener at the receiving end, the receiver should say, "Say again." For example, damage control central wants repair two to send a submersible pump to repair three. The central talker says, "Repair too, central; send one submersible pump to repair tree."

Repair two does not understand this message, so the talker there says, "Central, repair too; say again."

Central repeats the message and repair two acknowledges by saying, "Send one submersible pump to repair tree; repair too, aye."

Spelling

Difficult words are spelled by using the phonetic alphabet preceded by the prowords (procedural words) "I spell." Pronounce the word before and after spelling it. For example:

"Foc'sle-I spell—FOXTROT OSCAR ROMEO ECHO CHARLIE ALFA SIERRA TANGO LIMA ECHO, Foc'sle."

Temporarily Leaving the Circuit

When a phone talker is relieved by another talker, the phone talker must request permission to change phone talkers. If a talker is exchanging a faulty set of phones for a good set, the phone talker must request permission to change phones.

"Bridge, after steering; request permission to change phone talkers."

"Bridge, combat; request permission to change phones."

Once the talker has been given permission to go off the circuit and the talker rejoins the circuit, the report given is,

"Bridge, combat; back on the line."

Securing

Before securing the phones, you must always get permission.

Fantail asks, "Bridge, fantail; request permission to secure."

Bridge says, "Request permission to secure; bridge, aye; wait."

The bridge talker gets permission from the OOD for the person on the fantail to secure, then says,

"Fantail, bridge; secure."

Fantail replies, "Fantail, aye; going off the line."

Student Notes:

REVIEW 3 QUESTIONS

Q1. List the three categories of sound-powered phone circuits.

a.

b.

c.

Q2. An XJZ circuit is what type of circuit?

Q3. List four tips that you should use to be a good phone talker.

a.

b.

c.

d.

Q4. Sound-powered phone circuits are like a party line; therefore, some phone talker disciplines must be followed. List four types of good discipline.

a.

b.

c.

d.

Q5. A sound-powered-phone circuit has to be cleared to transmit an important message. What should the sender say over the circuit?

Q6. When you receive a message, what is the proper response?

DIAL TELEPHONES

Learning Objectives: When you finish this chapter, you will be able to—

- Recognize the purpose of dial telephones.
- Identify the procedures to follow when using dial telephones.

At home, ashore, and at sea, the telephone is a part of everyone's life. It is an important and essential instrument in every Navy office, and you must know how to use it properly. By observing proper techniques, you will give and receive information correctly and quickly. Remember, the success of a telephone conversation depends almost entirely upon your ability to express yourself in words; whereas, when speaking to a person directly, your facial expressions, gestures, and the like, help get your point across.

TYPES OF DIAL TELEPHONES

Different types of dial telephones currently in use are shown in figure 4-7. The desk set is used in staterooms, cabins, offices, and similar areas. A bulkhead-mounted telephone can be used in any station except those on weather decks. It is a nonwatertight unit that should not be exposed to the weather. A bulkhead-mounted telephone, is a splashproof unit that may be installed on weather decks and other areas exposed to moisture. All the phones in figure 4-7 are type 'G' telephones, general use.

USE OF THE DIAL TELEPHONES

Good telephone technique starts with answering your telephone as promptly as possible. Don't let it ring several times while you finish what you are doing. After lifting the receiver, you should speak immediately to the person calling. Identify yourself when answering the telephone; usually the person making the call will tell you who is calling. This procedure puts the conversation on a business-like basis and eliminates that hazy feeling one has when unsure of the identity of the person on the other end.

Student Notes:

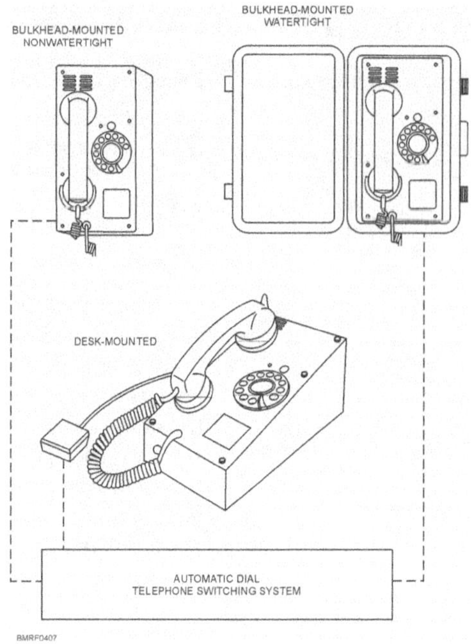

Figure 4-7.—Telephones.

Don't go on talking to someone in the office as you answer the telephone. You never know who your caller may be, and information inadvertently given out in this way could be harmful to national security. In addition, it is discourteous to make the caller wait while you finish your office conversation.

When you answer the phone for someone who is absent from the office, give some facts to the person making the call. Do not merely say, "He is not in right now." Rather, tell the caller when you expect the person to return, or volunteer to help if you can. If you have no information concerning the whereabouts of the person called, ask the caller if you may take a message.

Always make sure you have a pencil and pad beside the telephone for taking messages. This practice eliminates needless rummaging about while the other person is holding the line open. Also, it is worth remembering that the message will mean little to the person for whom it is intended unless you leave the following information:

1. Name of the caller
2. The message
3. Time and date of the message
4. Your name

Sometimes, you may have to leave the telephone to obtain additional information for a call. When this delay is necessary, you should make it known to the caller. If it takes more time to obtain the required information than you anticipated, give the caller an occasional progress report, such as "I'm sorry I did not find it there. If you do not mind waiting, I will look elsewhere."

When making a telephone call, there are certain rules you should observe.

1. Be sure that the number you dial is the correct one. When you dial wrong numbers, you waste other people's time as well as your own.
2. When making a call to another office, identify yourself immediately.
3. If you make the call for another person or an officer, so inform the person at the other end of the line. This courtesy eliminates the need for the other party to question you in this regard.

Student Notes:

If you make a call and are informed that the person called is not in, ask the person answering the telephone to take a message, if appropriate. You should make sure that the person to whom you are speaking understands the message, knows how to spell your name or the name of the person for whom you are making the call, and has your correct telephone number.

The tonal quality of your voice may or may not be subject to improvement. But by speaking correctly and distinctly and by speaking clearly and unhurriedly, you should have little difficulty in making yourself understood. Do not shout; it probably will not help and is likely to hinder.

Some people become nervous when speaking over the telephone. They take a deep breath, start at the beginning of their notes, and rush through to the end, all in the same breath. Naturally, the person at the other end of the line cannot absorb so much information so quickly, with the result that the whole conversation is unintelligible. Do not race through a conversation. The person on the other end is just as anxious to hear your information as you are to give it, so avoid the need (and the waste of time) of having to repeat your message.

REVIEW 4 QUESTION

Q1. You are taking a telephone message. List the four elements that you should include when taking a message.

a.

b.

c.

d.

INTEGRATED VOICE COMMUNICATIONS SYSTEM (IVCS)

Learning Objective: When you finish this chapter, you will be able to—

- Recognize the purpose of an integrated voice communications system (IVCS).

The IVCS is an integrated communications system that solves some of the shortcomings of older systems installed on older ships. IVCS combines the features of sound-powered telephones, dial telephones, and intercommunications units into one system. The IVCS also can interface with other shipboard communications systems. The system consists of terminals (user access devices), accessories, and two computer-controlled Interior Communications Switching Centers (ICSCs).

NOTE

Whenever IVCS are installed, sound-powered telephone circuits are designated as secondary communications circuits.

TERMINAL DEVICES

Two types of terminal devices (network terminal and dial terminal) are used with the IVCS. The type of terminal and the way it is connected into the system determines the type of service that is provided to you the user.

Network Terminal

The network terminal (fig. 4-8) provides service comparable to that provided by sound-powered telephone systems. By depressing one of the five numbered push buttons, you're connected to any one of four networks. Each network circuit is also connected to one of the ICSCs. The network circuits are manned for certain shipboard operations, similar to sound-powered telephones.

Student Notes:

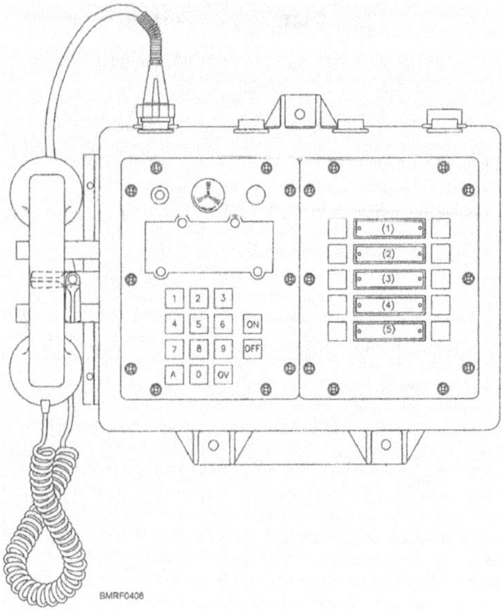

Figure 4-8.—Network terminal.

Dial Terminal

The dial terminal provides services that can be most easily compared to that provided by a dial telephone system. The dial telephones terminals (fig. 4-9) are connected to ICSCs. They are used similar to a commercial dial telephone with push-button dialing.

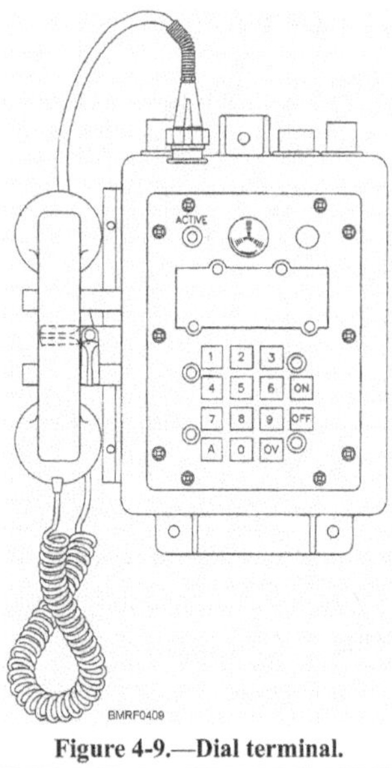

Figure 4-9.—Dial terminal.

Terminal Accessories

There are several types of accessories designed for use with the dial and network terminals. These accessories include headsets, handsets, spray-tight enclosures that permit the installation of the terminals in exposed areas, and loud speaker units. The loudspeaker units (fig. 4-10) are designed for use with either the dial or network terminals. Both units are equipped with press-to-talk switches. Additionally, by depressing the hands-free push switch on the unit, the operator can communicate without using the press-to-talk switch. This permits you to communicate without a handset or headset.

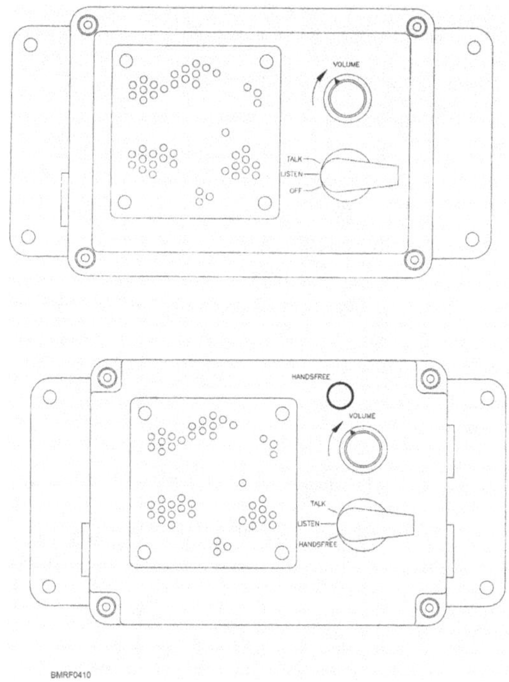

Figure 4-10.—Loudspeaker units.

INTERIOR COMMUNICATIONS SWITCHING CENTER (ICSC)

The ICSCs are the heart of the IVCS. They perform the switching actions necessary to connect the calling party to the called party, similar to the automatic switchboards of a dial telephone system. Figure 4-11 shows the relationship between ICSC and the IVCS.

Student Notes:

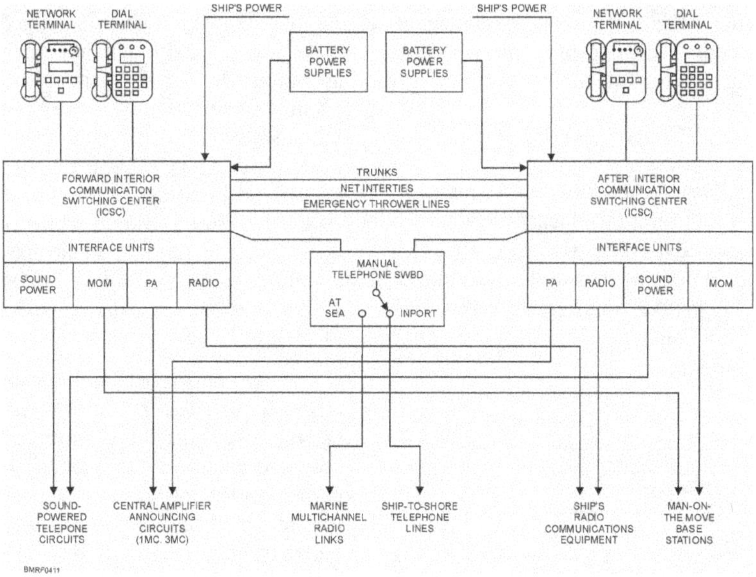

Figure 4-11.—IVCS block diagram.

REVIEW 5 QUESTIONS

Q1. List the terminal devices used with IVCS.

a.

b.

Q2. What is the purpose of the ICSC within the IVCS?

COMMUNICATIONS SECURITY

Learning Objective: When you finish this chapter, you will be able to—

- Identify basic communications security procedures.

Communications security is defined as the protective measures taken to deny unauthorized persons information derived from telecommunications of the United States government that are related to national security and to ensure the authenticity of each telecommunication.

Classified information may not be discussed in telephone conversations except as may be authorized over approved secure communications circuits. When in doubt about the classification of information necessary to answer a question asked in a telephone conversation, you should say nothing. When answering a telephone on a nonsecure communications circuit, you should inform the caller that the telephone is nonsecure. For example: "Quarterdeck, USS Never Sail messenger of the watch speaking, sir this is a nonsecure telephone."

ANNOUNCING AND INTERCOMMUNICATION SYSTEMS

Learning Objective: When you finish this chapter, you will be able to—

- Recognize the purpose and use of the announcing and intercommunication systems.

The general purpose of shipboard announcing and intercom systems, circuits 1MC through 59MC, is to transmit orders and information between stations within the ship by amplified voice communication by either a central amplifier system or an intercommunication system. A central amplifier system is used to broadcast

Student Notes:

orders or information simultaneously to a number of stations. An intercom system is used for two-way transmission of orders or information.

GENERAL ANNOUNCING SYSTEM

The basic MC circuit is the 1MC shown in figure 4-12. This is the general announcing system, over which word can be passed to every space in the ship. The ship's alarm system is tied into it as well. Transmitters are located on the bridge, quarterdeck, and DC central/central control station; additional transmitters may be located at other points.

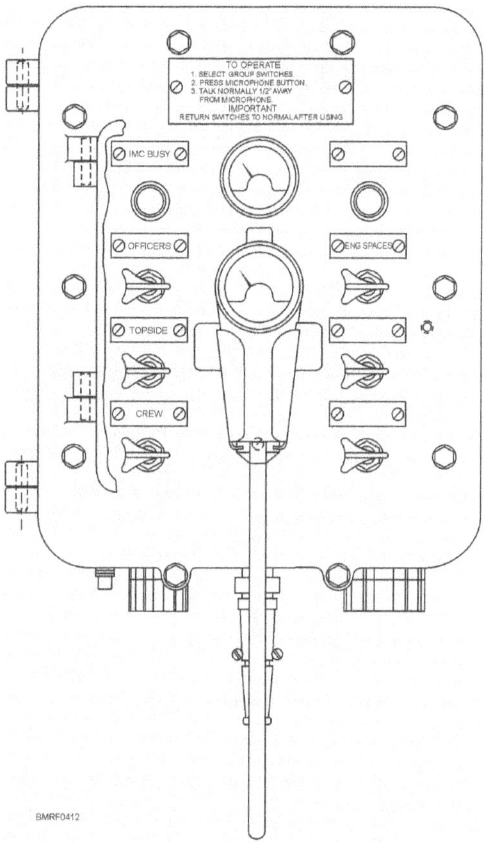

Figure 4-12.—Loudspeaker transmitter

The OOD is in charge of the 1MC. No call may be passed over it unless it is authorized by the OOD, the executive officer, or the captain, except for a possible emergency call by the damage control officer.

Normally, the 1MC is equipped with switches that make it possible for certain spaces to be cut off from announcements of no concern to them. The captain's cabin, for instance, should not be blasted with calls for individuals to lay down to the spud locker. The BMOW is responsible for passing the word over the 1MC. If the BMOW is absent and you are required to pass the word yourself, be sure you know which circuits should be left open. Some parts of the ship have independent MC circuits of their own, such as the engineers' announcing system (2MC) and the hangar deck announcing system (3MC).

The bullhorn (6MC) is the announcing system from one point to another. It can be used to communicate between two ships. It is a convenient means for passing orders to boats and tugs alongside or to line-handling parties beyond the range of the speaking trumpet. If the transmitter switch is located on the 1MC control panel, you must be careful to avoid accidentally cutting in the bullhorn when you are passing a routine word.

The 1MC, 2MC, 3MC, and 6MC are all one-way systems. A partial list of loudspeaker systems is shown in table 4-2.

INTERCOMS

MC circuits, such as the 21MC (commonly known as "squawk boxes"), differ from the preceding systems in that they provide two-way communications. Each unit has a number of selector switches. To talk to one or more stations, you only need to position the proper switches and operate the PRESS-TO-TALK switch. A red signal light mounted above each selector switch shows whether the station is busy. If it is busy, the light flashes; if it burns with a steady light, you know that the station is ready to receive. Typical IC circuits are as follows:

4MC	DC
19MC	Aviation ready room
20MC	CIC
21MC	Captain's command
22MC	Radio central
24MC	Flag officer
26AMC	Machinery control

Student Notes:

Table 4-2.—Shipboard Announcing Systems

CIRCUIT	SYSTEM	CIRCUIT	SYSTEM
1MC	General	35MC	Launcher captains'
2MC	Propulsion plant	39MC	Cargo handling
3MC	Aviators	40MC	Flag administration
4MC	Damage control	42MC	CIC coordinating
5MC	Flight deck	43MC	Unassigned
6MC	Intership	44MC	Instrumentation space
7MC	Submarine control	45MC	Research operations
8MC	Troop administration and control	46MC	Aviation ordnance and missile handling
9MC	Underwater troop communication	47MC	Torpedo control
18MC	Bridge	49MC	Unassigned
19MC	Aviation Control	50MC	Integrated operational intelligence center
21MC	Captain's command	51MC	Aircraft maintenance and handling control
22MC	Electronic control	52MC	Unassigned
23MC	Electrical control	53MC	Ship administration
24MC	Flag command	54MC	Repair officer's control
26MC	Machinery control	55MC	Sonar service
27MC	Sonar and radar control	56MC	Unassigned
29MC	Sonar control and information	57MC	Unassigned
30MC	Special weapons	58MC	Hangar deck damage control
31MC	Escape truck	59MC	SAMID alert
32MC	Weapons control		

The following is an example of how to operate the intercom. You are on the signal bridge at the 24MC transmitter (fig. 4-13), and you want to call conn. First, you push the selector button marked CONN on the designation plate. We will assume the line is clear for your message, which means that a steady red light appears over the signal bridge selector button at the conn transmitter. When the operator at conn pushes the signal bridge button, the signal lights at both stations begin to flash. Now you can operate the PRESS-TO-TALK switch and start your message. Any other station attempting to cut in gets the flashing busy signal.

Student Notes:

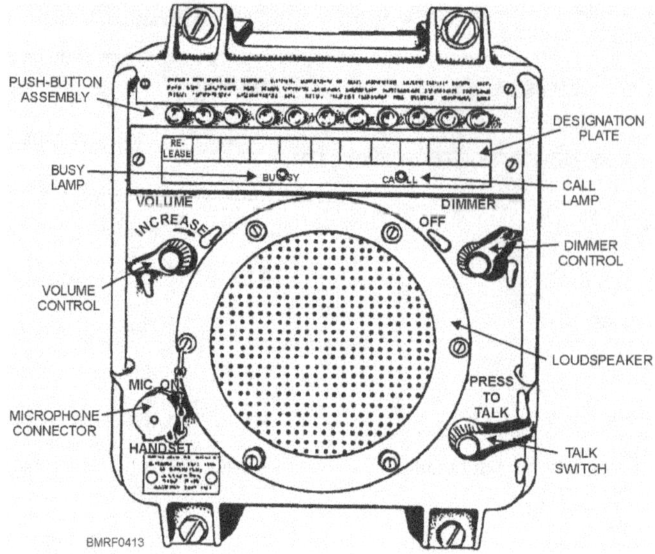

Figure 4-13.—A 24MC transmitter.

The chief disadvantage of the intercom is that it raises the noise level in any space in which it is located. For this reason, it seldom is used when sound-powered telephones are manned. Intercom circuits, which may be located on the bridge, are identified briefly as follows:

- 20MC, combat information announcing system, connects the same stations as the 1JS phones.
- 21MC, captain's command announcing system, is an approximate parallel to the JA phones.
- 22MC, radio room announcing system, is a substitute for the JX phones.
- 24MC, flag officer's command announcing system, is the intercom equivalent of the JF phones.

DAMAGE CONTROL WIREFREE COMMUNICATIONS (DC WIFCOM)

Learning Objective: When you finish this chapter, you will be able to—

- Recognize the purpose of DC WIFCOM.

DC WIFCOM is an improved means of damage control central (DCC) using modern hand-held radios specifically designed for shipboard needs. The system is initially installed in some ships and repair lockers with radios and antennas on a horizontal plane. An improvement in the system will eventually include additional radios and vertical antennas for other stations to include the bridge and electronics casualty control team.

Where installed, DC WIFCOM is the primary means of DCC within the repair locker area. Then hand-held portable transceivers, repair locker base stations, and a radiating antenna system provide instantaneous communications between repair lockers and repair locker personnel at the scene and investigators making damage reports. Each repair locker has an installed base station and four portable hand-held transceivers. Four to 12 channels are available for use. The first four channels have the following assignments:

Channel 1—Repair 5 area

Channel 2—Repair 2 area

Channel 3—Repair 3 area

Channel 4—Designated for ship-to-ship communications. Channel 4 may also be used for communications among ship control stations such as DCC, secondary DCC, secondary conn and the bridge major configurations.

Student Notes:

In the DC scenarios, WIFCOM hand-held transceivers are issued to the investigators and scene leader. They are the primary means of communication in the repair station area of responsibility. The 21J (or other designated) sound-powered telephone circuits are the primary means of communication between repair lockers and DCC. Personnel using WIFCOM must be aware of specific zones of reduced transmission capability or dead zones. Secondary communications, such as messenger or via second WIFCOM operator, must be used to communicate through dead zones. If emission control is necessary, special consideration must be given WIFCOM. In watertight areas during material condition ZEBRA, WIFCOM transmissions may be interrupted. These transmissions can be made only with command approval. In case of WIFCOM failure, repair locker personnel should establish effective communications as quickly as possible using other methods.

REVIEW 6 QUESTIONS

Q1. Your phone system is unsecured. When receiving a call, you should answer the phone by saying—

Q2. What system is tied into the IMC circuit?

Q3. What circuit is the damage control circuit?

Q4. What person(s) authorize(s) calls passed over the IMC?

a.

b.

c.

Q5. What is the difference between an IMC circuit and a 21MC circuit?

Q6. The first 4 channels of WIFCOM are assigned to—

a.

b.

c.

d.

FLAGS AND PENNANTS

Learning Objective: When you finish this chapter, you will be able to—

- Recognize the function and use of flags and pennants.

Flags and pennants serve various functions throughout the world. They have identified nations, governments, rank, and ownership and have conveyed messages for centuries. This section introduces flags and pennants that identify persons and ships and transmit information and orders. On special occasions, flags are used as a decoration, such as "dress ship."

The Navy uses the international alphabet flags; numeral pennants and a code/answer pennant; a set of numeral flags, special flags, and pennants; and four substitutes, or repeaters.

Each alphabet flag has the phonetic name of the letter it represents. A numeral flag takes the name of the numeral it represents; numeral pennants are used only in call signals. Special flags and pennants are used in tactical maneuvers to direct changes in speed, position, formation, and course; to indicate and identify units; and for specialized purposes. Flags and pennants are spoken and written as shown in figures 4-14 and 4-15.

Student Notes:

FLAG AND NAME	SPOKEN	WRITTEN	FLAG AND NAME	SPOKEN	WRITTEN	FLAG AND NAME	SPOKEN	WRITTEN
A	ALFA	A	M	MIKE	M	Y	YANKEE	Y
B	BRAVO	B	N	NOVEMBER	N	Z	ZULU	Z
C	CHARLIE	C	O	OSCAR	O	1	ONE	1
D	DELTA	D	P	PAPA	P	2	TWO	2
E	ECHO	E	Q	QUEBEC	Q	3	THREE	3
F	FOXTROT	F	R	ROMEO	R	4	FOUR	4
G	GOLF	G	S	SIERRA	S	5	FIVE	5
H	HOTEL	H	T	TANGO	T	6	SIX	6
I	INDIA	I	U	UNIFORM	U	7	SEVEN	7
J	JULIETT	J	V	VICTOR	V	8	EIGHT	8
K	KILO	K	W	WHISKEY	W	9	NINE	9
L	LIMA	L	K	XRAY	X	0	ZERO	0

BMRF0414

Figure 4-14.—Alphabet and numeral flags.

Student Notes:

PENNANT AND NAME	SPOKEN	WRITTEN
1	PENNANT ONE	p1
2	PENNANT TWO	p2
3	PENNANT THREE	P3
4	PENNANT FOUR	p4
5	PENNANT FIVE	p5
6	PENNANT SIX	p6
7	PENNANT SEVEN	p7
8	PENNANT EIGHT	p8
9	PENNANT NINE	p9
	PENNANT ZERO	p9
TACK LINE	TACK	—

PENNANT OR FLAG	SPOKEN	WRITTEN	PENNANT OR FLAG	SPOKEN	WRITTEN
CODE OR ANSWER	CODE OR ANSWER	CODE OR ANSWER	NEGATIVE	NEGAT	NEGAT
SCREEN	SCREEN	SCREEN	PREPARATIVE	PREP	PREP
CORPEN	CORPEN	CORPEN	PORT	PORT	PORT
DESIGNATION	DESIGNATION	DESIG	SPEED	SPEED	SPEED
DIVISION	DIVISION	DIV	SQUADRON	SQUADRON	SQUAD
EMERGENCY	EMERGENCY	EMERG	STARBOARD	STARBOARD	STBD
FLOTILLA	FLOT	FLOT	STATION	STATION	STATION
FORMATION	FORMATION	FORM	SUBDIVISION	SUBDIVISION	SUBDIV
INTER-ROGATIVE	INTER-ROGATIVE	INT	TURN	TURN	TURN

SUBSTITUTES

1st. SUBSTITUTE	FIRST SUB	1st.	3rd. SUBSTITUTE	THIRD SUB	3rd.
2nd. SUBSTITUTE	SECOND SUB	2nd.	4th. SUBSTITUTE	FORTH SUB	4th.

BMRF0415

Figure 4-15.—Numeral pennants, special flags, and pennants.

Student Notes:

EMERGENCY AND ADMINISTRATIVE SIGNALS

The flags and pennants (figs. 4-14 and 4-15) represent only a few of the thousands of signals that can be transmitted by flag hoist. Since they may be frequently seen displayed aboard Navy ships or stations, it would be to your advantage to learn to identify them and understand their meaning. Your own personal safety may someday depend on recognizing a particular signal flag.

Table 4-3 contains only those international signals most commonly used and having the same meaning as Navy signals.

Table 4-3.—Commonly Used International Signals

INTERNATIONAL SIGNALS	NAVY MEANINGS
EMERGENCY/WARNING FLAGS	
CODE ALFA	(International) I have a diver(s) down; keep well clear at slow speed.
BRAVO	I am taking in, discharging, or carrying dangerous materials.
KILO	Personnel working aloft.
MIKE 1	This ship has medical guard duty.
MIKE 2	This ship has dental guard duty.
OSCAR	Man overboard.
FLAG FIVE	Breakdown; the vessel is having engine or steering difficulty.
ADMINISTRATIVE FLAGS	
HOTEL	(International) This ship has a harbor pilot on board.
INDIA	Preparing to come alongside in-port or at anchor.
JULIETT	I have a semaphore message to transmit.
PAPA	General recall; all personnel return to the ship.
QUEBEC	Boat recall; all boats return to the ship.
ROMEO	In port; flown by ship having READY DUTY. At sea, flow by the ship PREPARING TO REPLENISH.
SIERRA	Holding flag hoist drill.
FIRST SUBSTITUTE	Indicates the absence of the flag officer or unit commander show personal flag or pennant is flying on the ship.
SECOND SUBSTITUTE	Indicates the absence of the chief of staff..
THIRD SUBSTITUTE	Indicates the absence of the captain. If the captain is absent over 72 hours, it indicates the absence of the executive officer.

Student Notes:

THE NATIONAL ENSIGN

Our national ensign (fig. 4-16) must always be treated with the greatest respect. It should never touch the ground or the deck. It should always be folded, stowed, and displayed properly. Our flag represents freedom to the world today and forever.

BMRF0416

Figure 4-16.—The United States national ensign.

When not under way, commissioned ships display the ensign from the flagstaff at the stern and the union jack from the jack staff at the bow from 0800 to sunset. While under way, the ensign is normally flown from the gaff. In ships having more than one mast, the gaff is usually positioned on the aftermast. In ships equipped with two macks (combination masts and stacks), the location of the flag depends on which mast is configured to accept halyards or a gaff.

When a U.S. naval ship enters a foreign port during darkness, at first light it briefly displays its colors on the gaff to make known its nationality. Other ships of war that are present customarily display their colors in return.

Our national ensign, along with the union jack, is referred to as colors. At commands ashore and on U.S. naval ships not under way, the ceremonial hoisting and lowering of the national flag at 0800 and sunset is known as morning and evening colors.

When the national ensign is hoisted and lowered or half-masted for any occasion, the motions of the senior officer present are followed. This is done by flying the PREPARATIVE pennant (called *PREP*) 5 minutes before morning and evening colors. Ceremonies for colors begin when PREP is hauled to the dip (the halfway point). The PREP pennant is shown in figure 4-17.

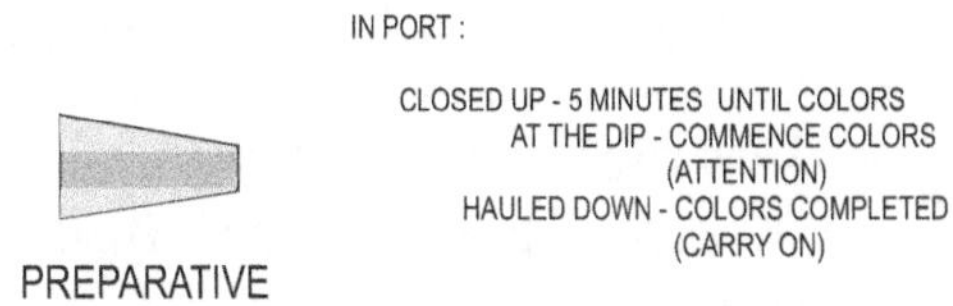

BMRF0417

Figure 4-17.—Preparative pennant.

If a band or recorded music is available for the colors ceremony, "Attention" is sounded, followed by the national anthem. At morning colors, the ensign is hoisted when the music begins. It is smartly hoisted to the top of the flagstaff. Remember, a furled (folded) ensign is never hoisted to the top of the flagstaff or gaff. At evening colors, lowering of the ensign also begins at the start of the music and is so regulated as to be completely lowered at the last note of the music. "Carry On" is sounded at the completion of the music. The national flag is always hoisted smartly and lowered ceremoniously.

If a band or music is not available for colors, "To the Colors" is played on a bugle at morning colors, and "Retreat" is played at evening colors. For ships having no band, music, or bugler, "Attention" and "Carry On" are signals for rendering and terminating the hand salute.

Sometimes the music for colors from another U.S. ship can be overheard aboard your ship. When this happens and no band, music, or bugler is aboard your ship, the command "Carry On" should not be given until the music being overheard is completed.

If foreign warships are present, the national anthem of each country represented is played after morning colors. If your ship is visiting a foreign port, the national anthem of that country is played immediately following morning colors, followed by the national anthems of any other foreign nations represented.

Student Notes:

There are times during the year that the ensign is flown at half-mast, or half-staff ashore. This is the internationally recognized symbol of mourning. Normally, the flag is half-masted on receiving information of the death of one of the officials or officers listed in *U.S. Navy Regulations*. Notification may be through the news media or by official message. The United States honors its war dead on Memorial Day by flying the flag at half-mast from 0800 until the last gun of a 21-minute gun salute that begins at noon (or until 1220 if no gun salute is rendered).

If the ensign is flown from the flagstaff and is half-masted, the union jack is also half-masted. In half-masting the national ensign, it will, if not already hoisted, first be hoisted to the peak and then lowered to the half-mast position. Before lowering from the half-mast position, the ensign is hoisted to the peak, then lowered ceremoniously. Distinctive marks, such as commission or command pennants, are not half-masted except when the ship's commanding officer or the unit commander dies.

U.S. Navy Regulations stipulates that when any ship under United States registry or the registry of a nation formally recognized by the United States salutes a U.S. Navy ship by dipping its flag (hauled halfway down and then raised), the courtesy is to be returned dip for dip. A U.S. Navy ship never dips to a foreign ship (flag) first. U.S. naval ships (USNS) of the Military Sealift Command do not dip the national ensign to Navy ships since they are public ships of the United States.

Formal recognition of a foreign country does not mean that diplomatic relations must exist. The fact that diplomatic relations have been severed does not mean that the United States no longer recognizes the existence of the state or the government concerned. However, the United States does not return the dip to countries such as Albania, North Korea, Vietnam, and South Yemen. If in doubt, ask the duty Signalman.

UNION JACK

The union jack is the rectangular blue part of the United States flag containing the stars. It is shown in figure 4-18. It symbolizes the union of the states of the United States. Each star represents a state.

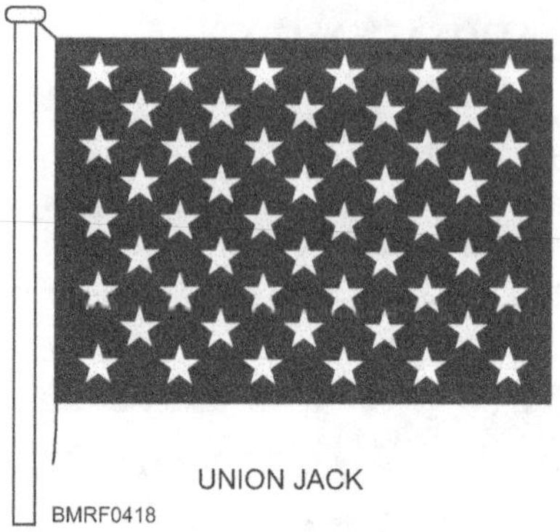

Figure 4-18.—Union jack.

When a naval ship is in port or at anchor, the union jack is flown from the jackstaff from 0800 to sunset. In addition to flying from the jackstaff, the union jack is hoisted at the yardarm to indicate that a general court-martial or a court of inquiry is in session.

The union jack is flown in boats as follows:

1. When a diplomatic official of the United States, at or above the rank of charge d'affaires, is embarked in a boat of the U.S. Navy and is within the waters of the country which that person represents

2. When a governor general, or a governor commissioned as such by the President, is embarked in a boat in an official capacity and the boat is within the governor's area of jurisdiction (for example, the Governor of the Virgin Islands)

When displayed from the jackstaff, the union jack is half-masted if the national ensign is half-masted. It is raised and lowered in the same manner as the national ensign. The union jack is not dipped when the national ensign is dipped.

The union jack is issued in several sizes; but, when flown at the jack staff, it must be the same size as the union of the ensign flown at the flagstaff. To make sure it is not flown upside down, always have the single point of the stars pointing toward the sky.

Student Notes:

U.S. NAVY FLAG

On 24 April 1959, the President, on the recommendation of the Secretary of the Navy, established an official flag for the United States Navy. That was done to fulfill a need for an official flag to represent the Navy in displays and on a variety of occasions, such as ceremonies and parades. Figure 4-19 shows the Navy flag.

BMRF0419

Figure 4-19.—U.S. Navy flag.

The U.S. Navy flag represents the Navy as follows:

- At official ceremonies
- In parades
- In displays during official Navy occasions
- At public gatherings when the Navy is an official participant
- On other occasions as may be authorized by the Secretary of the Navy

When used for the purposes listed above, the Navy flag accompanies, and takes the place of honor after, the national flag. However, when other branches of the armed forces are participating, the flags take precedence in the order of seniority of the services represented.

PERSONAL FLAGS AND PENNANTS

Every Navy ship in commission flies the commission pennant except when it is replaced by a personal flag, command pennant, or Red Cross flag. The commission pennant, shown in figure 4-20, is flown at the after truck of a naval vessel and at the highest and most conspicuous point of hoist on a fixed mastless ship (submarines in particular). It is also flown from the bow of a boat when a commanding officer, not entitled to a personal flag, is embarked on an official visit.

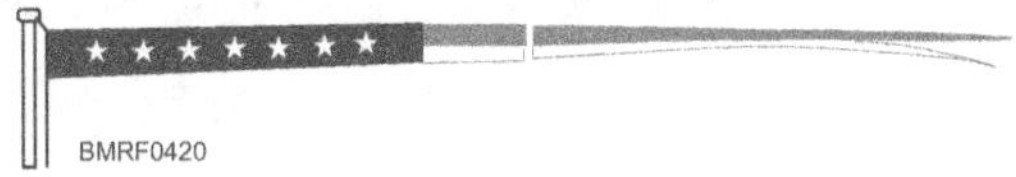

BMRF0420

Figure 4-20.—Commission pennant.

The commission pennant is not a personal flag, but sometimes it is regarded as the personal symbol of the commanding officer. Along with the ensign and union jack, it is half-masted upon the death of the commanding officer of a ship.

The Red Cross (Geneva Convention) flag, shown in figure 4-21, is the distinctive mark flown from the after truck of a commissioned hospital ship of the Navy. In general, the Red Cross flag is regarded as an international guarantee of amnesty from attack. None of the military services, however, fly it on the same halyard as the national ensign. Boats engaged in sanitary service and landing party hospital boats display the Red Cross flag in the bow.

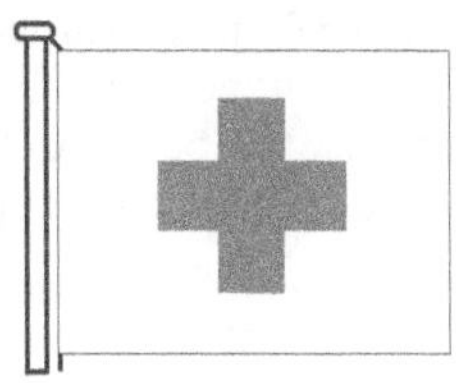

BMRF0421

Figure 4-21.—Red Cross flag.

Some nations in the Middle East regard the cross as a symbol contrary to their religious beliefs. Therefore, they use a design such as a red crescent on a white field or a red lion and sun on a white field to indicate a mission of mercy or amnesty from attack.

No flag or pennant may be flown above or, if on the same level, to the right of our national flag. One exception is the display of flags at the United Nations headquarters, where special rules apply. The only other

Student Notes:

exception is during church services aboard ship conducted by Navy chaplains or visiting church dignitaries. Then the church pennant (fig. 4-22) or the Jewish worship pennant (fig. 4-23) is flown above the ensign. Many ships are fitted with two halyards to the same point of hoist at both the staff and gaff to permit display of the church pennant and ensign simultaneously.

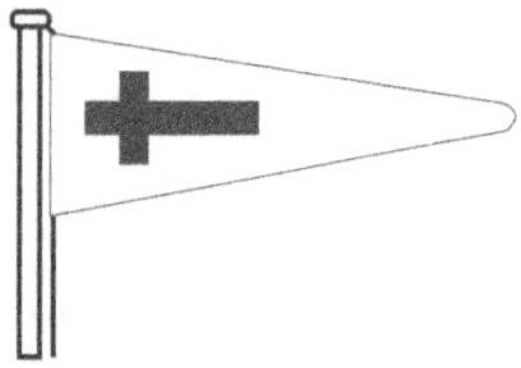

BMRF0422

Figure 4-22.—Church pennant.

Aboard ships under way, the church pennant is displayed by hoisting it to the peak or truck and then dipping the ensign just clear of it. If services are being conducted at the time of morning colors aboard ships not under way, the ensign is hoisted to the top of the flagstaff at the prescribed time. The church pennant is then hoisted and the ensign dipped just clear of the pennant. If the ensign is half-masted, the church pennant is hoisted just above the ensign. When the church pennant is lowered, the ensign is closed up (hoisted to the truck, peak, or top of the flagstaff) before the pennant is lowered. Although the church pennant may not be flown above the national flag ashore, it may be displayed separately.

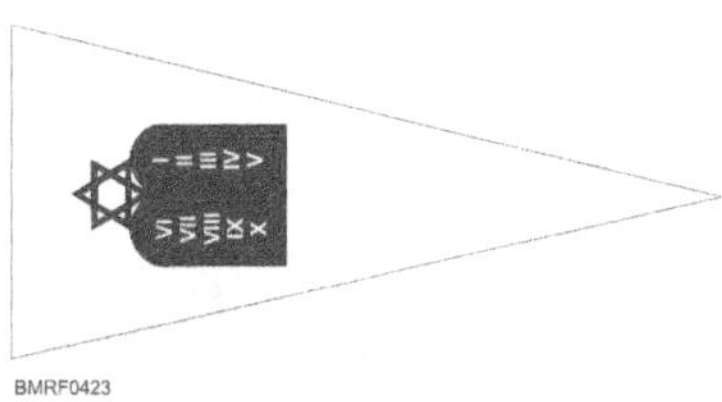

Figure 4-23.—Jewish worship pennant.

The Jewish worship pennant, shown in figure 4-23, is displayed during Jewish religious services afloat and ashore. This pennant was authorized by the Secretary of the Navy in 1975. The same rules governing the display of the church pennant apply to the display of the Jewish worship pennant.

The flag of the Chief of Naval Operations (fig. 4-24) is a blue and white rectangle, divided diagonally from lower hoist to upper fly. In its center is the official seal of the Chief of Naval Operations—an eagle clutching an anchor and encircled by 50 gold links of chain. The CNO's flag is displayed in the same manner as required for displaying flags of any flag officer.

Figure 4-24.—Chief of Naval Operations flag.

Aboard ships not under way, the absence (for a period of 72 hours or less) of various officers is indicated by the display of SUBSTITUTE pennants. These are general signal pennants. The pennants are assigned as shown in figure 4-25.

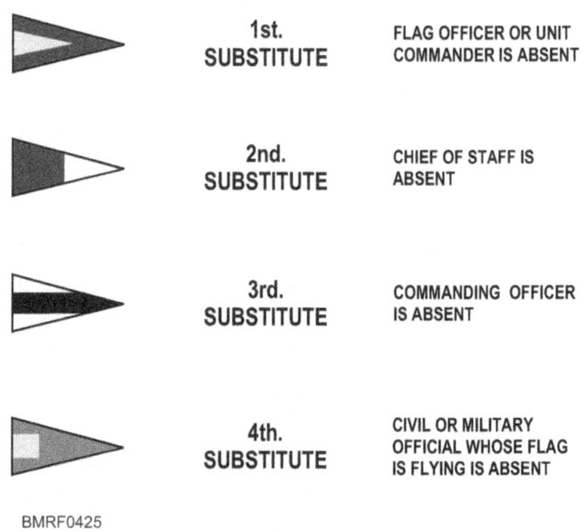

Figure 4-25.—Substitute pennants.

Student Notes:

On many small ships, it is the responsibility of the quarterdeck watch to hoist and haul down the absentee pennants. They are flown only between sunrise and sunset.

Whenever the ship is taking aboard, transferring, or handling dangerous commodities, such as ammunition and fuel, the BRAVO flag is hoisted and the smoking lamp is put out. BRAVO is hauled down when the dangerous condition no longer exists. The BRAVO flag (fig. 4-26) is a general signal flag.

Figure 4-26.—Bravo flag.

While standing watch, you will have many duties. One of them is to make sure special flags or pennants are displayed as required to indicate changing events aboard ship. Usually on a large ship, this is the responsibility of the duty signalman. On small ships, such as submarines, it is the duty of the topside watch (POOW). These flags or pennants are important because they tell other units what is happening within their area at any given time. A list of special flags and pennants is normally posted within the quarterdeck area for the ready reference of watch standers.

There are many more flags and pennants that have special meanings. You will have to know the meaning of some of these. They are called general signals, and those not previously discussed are shown in figure 4-27.

When flag officers of the Navy (admiral, vice admiral, rear admiral [upper half], rear admiral [lower half]) assume command of a fleet or a unit of a fleet, their personal flag (fig. 4-28) is hoisted and kept flying until they turn over their command to their successor. If the officer is absent from command for a period exceeding 72 hours, the flag is hauled down until the officer returns.

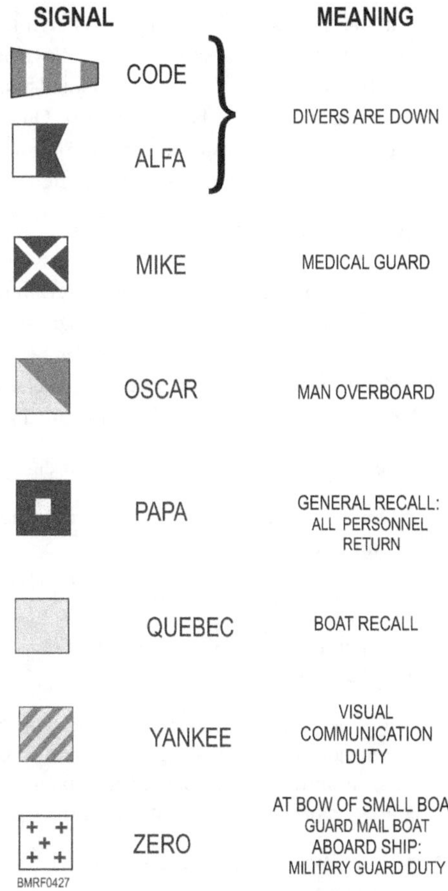

Figure 4-27.—General signals.

A flag officer's flag is never displayed simultaneously from more than one ship. It is flown at the main-truck of the ship the officer is aboard. Normally, no personal flag or pennant is shown at the same masthead with the national ensign. When a double display is required, the personal flag or pennant should

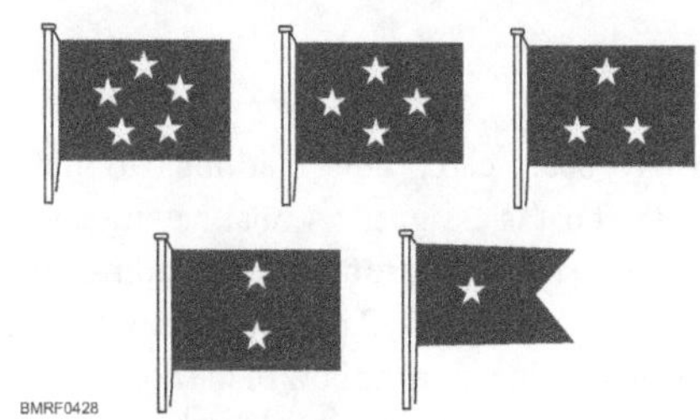

Figure 4-28.—Personal flags.

Student Notes:

be flown at the foretruck and the national ensign flown at the main-truck. When a single masted flagship is dressed or full-dressed, however, the personal flag or pennant is hoisted at the starboard yardarm. During a gun salute, the ensign is displayed at the main-truck. Any personal flag is lowered clear of the ensign.

FLAG DISPLAYS IN BOATS

The ensign is flown from the stern of naval boats. The ensign should never be so large that it hangs in the water when the boat is afloat. When the ensign becomes soiled, it should be changed for a clean ensign. Our flag should be flown from boats during the following times:

- When under way during daylight in a foreign port
- When ships are required to be dressed or full-dressed
- When going alongside a foreign vessel
- When an officer or official is embarked on an official occasion
- When a flag or general officer, a unit commander, a commanding officer, or a chief of staff, in uniform, is embarked in a boat of the command or in one assigned for personal use
- At other times when prescribed by the senior officer present

When an officer in command (or chief of staff) entitled to a personal flag or pennant is embarked in a boat on an official occasion, the appropriate flag or pennant is flown at the bow. (If not entitled to a personal flag or pennant, a commission pennant is displayed.) On other than official occasions, a miniature personal flag or pennant is displayed near the coxswain's station.

Bow Markings

Many boats carry bow markings to indicate to whom the boat is assigned. A boat having an arrow in the bow is assigned for use by a commanding officer or a chief of staff who is not a flag officer. A miniature of the command pennant is on the bow of the boat assigned to a unit commander. A boat assigned for the personal use of a flag or general officer has on each bow the number of stars corresponding to the officer's rank.

Flagstaff Insignias

Boats assigned to officers for personal use or boats in which a civil official is embarked on official business are marked with special devices on the flagstaff. The flagstaff for the ensign and for the personal flags or pennants is fitted at the peak with these special devices, shown in figure 4-29, as follows:

- Spread eagle: For any civilian official or flag officer whose official salute is 19 guns or more
- Halbert: For a flag or general officer whose official salute is less than 19 guns or for a civil official whose salute is 11 guns or more but less than 19
- Ball: For an officer of the grade, or relative grade, of captain in the Navy, and for certain diplomatic officials
- Star: For an officer of the grade, or relative grade, of commander
- Flat truck: For an officer below the grade, or relative grade, of commander, and for civil officials entitled to honors of a lesser nature than those previously described

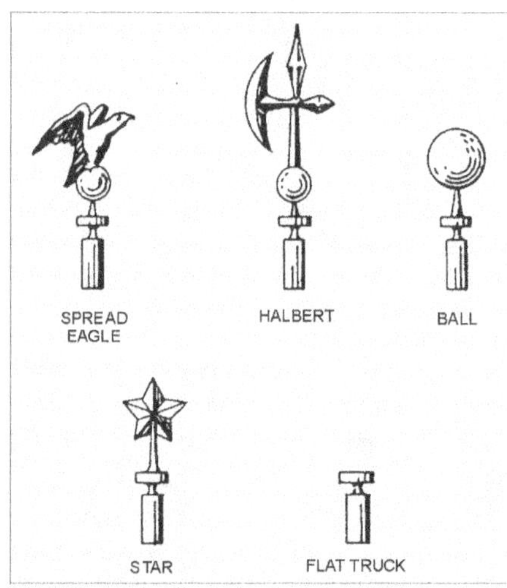

Figure 4-29.—Flagstaff insignias.

Student Notes:

Boat landings for officers usually are separate from those for enlisted personnel; but there may be times, especially overseas, when they are in the same location. Aboard ship, the bridge watch usually tells the quarterdeck that an officer's or enlisted's liberty boat is approaching the ship.

REVIEW 7 QUESTIONS

Q1. List some of the flags and pennants used by the Navy.

Q2. In the space provided, list the flag flown for the conditions described on the right.

CONDITIONS	FLAG
a. There is a man overboard.	
b. There are divers in the water.	
c. A general court-martial is in session.	
d. Worship service(s) in progress.	
e. The captain is absent.	

Q3. In port, commissioned ships display the national ensign and the union jack from what locations?

Q4. In large ships, what person is usually responsible for making sure that special flags and pennants are displayed?

Q5. What is the flagstaff insignia for a captain?

Q6. A boat with a halbert insignia on the flagstaff is approaching your ship. What is the rank of the person on the ship?

SIDE HONORS

Learning Objective: When you finish this chapter, you will be able to—

- Identify the purpose of and use of side honors.

Side honors, rendered to officers and officials boarding and departing the ship, are part of the honors stipulated for an official visit. The honors consist of parading the proper number of side boys and piping the side by the honors boatswain's mate. Officers appropriate to the occasion also attend the side. Side boys are not paraded on Sunday or on other days between sunset and 0800 or during meal hours of the crew, general drills and evolutions, and periods of regular overhaul, except in honor of civil officials and foreign officers; then they may be paraded at any time during daylight hours. Side boys are paraded only for scheduled (official) visits.

Student Notes:

The term *official* means a formal visit of courtesy requiring special honors and ceremonies. An informal visit of courtesy requiring no special ceremonies is a *call.*

HONORS FOR OFFICIAL VISITS

The honors specified for an official visit are rendered on arrival as follows:

- When the rail is manned, personnel are spaced uniformly at the rail on each weather deck, facing outboard. The command "Attention" is sounded as the visitor's boat or vehicle approaches the ship.

- If a gun salute is prescribed on arrival, it is fired as the visitor approaches and is still clear of the side. The proper flag or pennant is broken on the first gun and hauled down on the last gun except when it is to be flown for the duration of the visit. Other ships firing a concurrent salute also haul down, on the last gun, the flag or pennant displayed in honor of the visitor.

If the ship visited is moored to the pier in such a position that it is impractical to render the gun salute before arrival on board, the salute is rendered (provided local regulations don't forbid gun salutes) after the official arrives on board and the commanding officer is sure that the dignitary and party are moved to a position in the ship that is well clear of the saluting battery.

- The boat or vehicle is piped as it comes alongside.
- The visitor is piped over the side, and all persons on the quarterdeck salute and the guard presents arms until the termination of the pipe, flourishes, music, or gun salute, depending on which is rendered last.
- If the gun salute is not prescribed on arrival and a flag or pennant is to be displayed during the visit, it is broken at the start of the pipe.
- The piping of the side, the ruffles and flourishes, and the music are executed in the order named. In the absence of a band, "To the Colors" is sounded on the bugle, instead of the national anthem, when required.
- The visitor, if entitled to 11 guns or more, is invited to inspect the guard upon completion of the gun salute or such other honors as may be accorded.

On departure, the honors prescribed for an official visit are as follows:

1. The rail is manned, if required.
2. The command "Attention" is sounded as the visitor arrives on the quarterdeck.
3. When the visitor is ready to leave the ship, the guard presents arms, all persons on the quarterdeck salute, and ruffles and flourishes, followed by music, is sounded. The visitor then is piped over the side. The salute and present arms terminate with the call. If no gun salute is fired, the flag or pennant displayed in honor of the visitor is hauled down.
4. The boat or vehicle is piped away from the side.
5. If a gun salute is directed upon departure, it is fired when the visitor is clear of the side. If a flag or pennant is displayed in honor of the visitor, it is hauled down with the last gun of the salute.

When possible, the same honors and ceremonies are rendered for an official visit to a naval station.

SIDE BOYS

When required for attending the side, the required number of side boys will be on deck in the uniform of the day. Side boys are mustered, inspected, and instructed in their duties by the OOD and BMOW. They are stationed on either side of the route across the quarterdeck taken by arriving and departing high-ranking officers or civilian officials who are making official calls to the ship. When the side is piped by the BMOW on the boatswain's pipe, from two to eight side boys, depending on the rank of the honored official, will form a passageway to or from the gangway. They salute on the first note of the pipe and drop the salute together on the last note.80

Side boys must be particularly smart in appearance and groomed with polished shoes and immaculate uniforms. Enlisted women detailed to this duty are also called side boys.

Student Notes:

REVIEW 8 QUESTIONS

Q1. When are side boys paraded?

Q2. When a gun salute is prescribed, when is it fired?

SUMMARY

In this chapter, you have learned about communications equipment, telephones and telephone talker responsibilities, and how this equipment and responsibilities relate to you. You also learned about the importance of security and why the following correct procedures are important. This chapter also introduced you to the use of flags, pennants, and honors accorded various military and civilian personnel.

Student Notes:

REVIEW 1 ANSWER

A1. To give the serial number (23DBCX14) of the pump over the phone, you would say—**too, tree, delta, bravo, charlie, xray, wun, fo-wer.**

REVIEW 2 ANSWERS

A1. **True**, the mouthpiece and earpiece of a sound-powered phone are interchangeable.

A2. If you pick up the mouthpiece or the headpiece by itself, **delicate wires could break**.

A3. If you leave the headset plugged in, the **earpieces pick up background noises and transmit them over the circuit**.

A4. To talk or listen through the phone, **depress the button located between the transmitter and receiver**.

REVIEW 3 ANSWERS

A1. The three categories of sound powered phone circuits are—

a. **Primary**

b. **Auxiliary**

c. **Supplementary**

A2. An XJZ circuit is an **auxiliary** circuit.

A3. Some of the practices that make a good phone talker include—

a. **Speak clearly and directly into the phone**

b. **Don't have food or gum in your mouth**

c. **Don't paraphrase messages; repeat them word for word**

d. **Speak slowly**

e. **In an emergency, speak calmly and precisely**

f. **Don't use local accents**

A4. Some disciplines that must be followed when talking over sound-powered phone circuits include—

a. **Transmit official message only**

b. **Keep the button in the OFF position except when transmitting**

c. **Use standard terms and phrases**

d. **Don't use slang or profanity**

A5. To clear a sound-powered phone circuit to transmit an important message, the sender should say "**silence on the line**."

A6. The proper response upon receipt of a message is as follows: **"Repeat message, identify yourself, and then acknowledge the message."**

REVIEW 4 ANSWER

A1. The four elements you should include when taking a message are—

a. **Name of caller**

b. **The message**

c. **Time and date of message**

d. **Your name**

REVIEW 5 ANSWERS

A1. The terminal devices used with the IVCS are—

a. **Network**

b. **Dial**

A2. Within the IVCS, the **ICSC acts like a switchboard and connects the caller with the person who called.**

Student Notes:

REVIEW 6 ANSWERS

A1. Your phone system is unsecured. When receiving a call, you should answer the phone by saying **this line is unsecured.**

A2. The **alarm system** is tied into the 1C circuit.

A3. The **4MC** circuit is the damage control circuit.

A4. Calls passed over the 1MC are authorized by the—

a. **OD,**

b. **XO, or the**

c. **CO**

A5. The **1MC is a one-way system** and the **21MC is an intercom with two-way communication**.

A6. The first 4 channels of WIFCOM are assigned to—

a. **Channel 1—Repair 5**

b. **Channel 2—Repair 2**

c. **Channel 3—Repair 3**

d. **Channel 4—Ship-to-ship communications**

REVIEW 7 ANSWERS

A1. Flags and pennants used by the Navy include **the international alphabet flags; numeral pennants and a code/answer pennant; a set of number flags, special flags, and pennants; and four substitutes or repeaters.**

A2. The flag flown for the conditions is as follows:

CONDITIONS	FLAG
a. There is a man overboard.	**Oscar**
b. There are divers in the water.	**Code Alfa**
c. A general court-martial is in session.	**The Union Jack**
d. Worship service(s) in progress.	**Church pennant/ Jewish worship pennant**
e. The captain is absent	**The third pennant**

A3. When in port, commissioned ships display the national ensign and the union jack. The **national ensign is flown from the flagstaff at the stern**, and the **union jack is flown from the jackstaff at the bow**.

A4. On large ships, the **signalman** is usually responsible for making sure that special flags and pennants are displayed.

A5. A **ball** is the flagstaff insignia for a captain.

A6. A boat with a halbert insignia on the flagstaff is approaching your ship. **There is a flag or general officer on board, whose official salute is less than 19 guns**.

REVIEW 8 ANSWERS

A1. Side boys are paraded for **scheduled official visits**.

A2. A gun salute is fired when **the visitor approaches and is still clear of the side**.

Student Notes:

CHAPTER 5

NAVAL HISTORY

Why should I bother to learn or read history? Isn't it dead and gone? Read on....The quotes from George Santayana and Alfred Thayer Mahan tell the story!

Those who cannot remember the past are condemned to repeat it.
—George Santayana, American Philosopher, 1863 - 1952

The study of history lies at the foundation of all sound military conclusions and practice.
— Alfred Thayer Mahan, Philosopher of Naval Strategy, 1840-1914

These two quotes tell you the reasons **why** you should know what happened in the past. By studying history, you can avoid the mistakes made in the past. By studying both the failures and successes of the past, you can plan for future success.

There is another reason to study history—history is an adventure story. History is full of daring deeds, good luck and bad, heroes, cowards, and spies. The history of a country or an organization is like the biography of a person. A biography is the story of a person's life. Naval history is the story of the life of the Navy. Since this chapter is the biography of the life of the United States Navy, the logical place to start is with the birth of the Navy.

THE BIRTHDAY OF THE UNITED STATES NAVY

Learning Objectives: When you finish this chapter, you will be able to—

- Identify the important events of naval history.
- Recognize the importance of naval actions and traditions

In school, you learned about the birthday of the United States. You were told about the events that happened on July 4, 1776. The United States Navy had its birth on October 13, 1775. How could this be? How could the Navy be older than the United States?

Just as there wasn't a United States of America on July 4, 1776, there wasn't a United States Navy on October 13, 1775. But, what led to the formation of the United States Navy happened on October 13, 1775.

Remember when the Second Continental Congress met on May 10, 1775, the colonists were already fighting the British. Before long, it was clear that if the Colonies were to survive, a Navy was necessary. Therefore, on October 13, 1775, the Second Continental Congress authorized the purchase of two vessels; the United States Navy was born.

THE CONTINENTAL NAVY

Learning Objective: When you finish this chapter, you will be able to—

- Identify the ships of the Continental Navy to include the importance of their actions.

Navies are created from the spirit of independence and under the threat of war. They become mature by defending their country. This is the way it was with the first American Navy.

The American Colonies depended on the sea for their livelihood. All along the coast, harbors and shipbuilding docks offered work to many and provided income to thousands more. When the conflict between the Americans and the British began, these were the first ports the British attacked. These were also the ports from which the Continental Congress and the States sought to send out ships of a tiny and hastily organized naval force to harass the mightiest sea power in the world and its merchant fleet. This tiny naval force sought to capture enemy supply and munitions vessels.

What was life like in that first Navy? Where did its ships and men come from? How was it organized? And, importantly, what role did it play in building the proud tradition of the United States Navy today?

Like its beginnings, the Navy of the American Revolution was fragmented into many parts, each acting independently of the others. For instance, several naval engagements between the Americans and the British actually occurred before the Continental Congress authorized a Navy. Though the American Navy officially began in October 1775, some time passed before the new Navy had any effect on the mighty British Navy.

SHIPS OF THE CONTINENTAL NAVY

What constituted a *warship* in the late 1700s? During the revolutionary war and into the 19th century, naval vessels were grouped into three major classes—

1. Ships-of-the-line. These were the battleships of the sailing days. These ships were the largest of all sailing warships and carried 64 to over 100 guns of various sizes. However, our Navy's ships-of-the-line didn't come into existence until years later, long after the Revolutionary War was over.
2. Frigates. These were the cruisers of the 18th century. These cruisers were next in size, usually smaller and faster than average ship-of-the-line. They generally carried 28 to 44 guns.
3. Sloops-of-war. These were the small sailing warships. They carried 10 to 20 guns.

Another group of naval vessels were the privateers. Privateers were commissioned by the Continental Congress and by individual states to capture enemy merchant ships as prizes of war.

Typical of the independent "fleet" of privateers was the schooner. The schooner was a small, fast, flexible, flush-deck ship that carried smooth-bore cannon. With small ships like these schooners, the colonists broke the British stranglehold on main New England harbors by slipping past the Royal Navy's men-of-war and hiding in inlets. Unable to meet the British head-on, the American ships outmaneuvered them and jabbed here and there instead of standing full force and slugging it out.

Navy ships in the Continental Navy included the *Providence*, a 12-gun sloop; the *Lexington*, a 16-gun brig (converted from a merchantman); and the *Bonhomme Richard,* a loan from the French, an old East Indiaman. Later in this chapter, you will find out how other ships bearing some of these names made history in their own right.

THE FIRST UNITED STATES SUBMARINE

A young American experimented with a subsurface craft he hoped would help drive the British out of New York harbor and away from American shores for good. David Bushnell was a Yale medical student who had been working on a small submarine for some 4 years and finally completed it in 1775.

This first warfare submarine, named the *Turtle*, was described by Bushnell as having "some resemblance to two upper tortoise shells of equal size, joined together..." It was 7.5 feet deep, and under ideal conditions had a maximum speed of 3 knots. A single operator could stay down for 30 minutes.

The *Turtle* was armed with an oak casing filled with 150 pounds of explosives. This charge could be attached to the bottom of an enemy ship where it was intended to remain until detonated by a simple clockwork mechanism.

After completing the submarine, Bushnell took it for several dives to prove its seaworthiness. Finally, in September 1776, he was ready to try it against the British in New York harbor. Sergeant Ezra Lee, a volunteer from the Connecticut militia, maneuvered the *Turtle* through the use of hand-driven screw propellers. His mission was to attach a time-fuse charge of gunpowder to the hull of HMS *Eagle*. However, the mission was aborted when the auger failed to penetrate the copper sheathing of the *Eagle*.

Bushnell made a few more attempts to use the *Turtle* against the British in the Delaware River. He attached mines to the *Turtle* and floated the mines against ships. These attempts failed. The submarine was finally sunk by the British in New York harbor—the first recorded instance of an antisubmarine attack.

CONTINENTAL NAVY ACTIONS

The new Navy ordered to be established by the Continental Congress came into being in the last months of 1775. To build a fleet, Congress authorized the construction of 13 new frigates (ranging from 24 to 32 guns) and the conversion of 6 merchant ships (ranging from 10 to 24 guns). These merchant ships included the USS *Hornet* and the USS *Alfred.* The USS *Alfred* had the distinction of being the U.S. Navy's first flagship and is said to be the first U.S. naval vessel on which the "Flag of Freedom" was hoisted (by John Paul Jones). All were solidly constructed ships with a number of guns. Even so, they were at a serious disadvantage because they were pitted against the established and superior British force—then the finest Navy in the world.

Student Notes:

NOTE

As you read along, check the maps at the back of the chapter.

The first commander in chief, Esek Hopkins, put the first squadron of the Continental Navy to sea in February 1776. Under the guns of the USS *Providence* and the USS *Wasp* and with the squadron headed by the USS *Alfred*, over 200 Sailors and Marines landed on New Providence Island in the Bahamas. John Paul Jones served as first lieutenant aboard the USS *Alfred*.

Hopkins' raid on New Providence Island was the first amphibious operation carried out by the American Navy and Marines. The squadron captured a number of cannons and supplies from the fort.

Because the British blockaded the American coast, it was difficult for the newly outfitted ships to reach the sea. The USS *Montgomery* and the USS *Congress*, ships of 28 and 24 guns, were built at Poughkeepsie, NY on the Hudson River. When the British occupied the port of New York, these ships were bottled up. To prevent their capture by the enemy, the U.S. government had to destroy them. Two more ships built in Philadelphia suffered a similar fate. Some of the others were also blockaded in their home ports, and one ship, the USS *Trumbull*, was bottled up for 3 years because it couldn't clear the sandbar in the Connecticut River.

The new frigates of the Continental Navy had their moments. The USS *Hancock* and the USS *Boston*, both built in Massachusetts, set out together in mid-1777. They captured two British brigs and were then involved in separate actions with the British warships *Somerset* and *Fox*. After escaping from the *Somerset* on May 30, 1777, they met the *Fox* a week later and successfully captured it. Later, the two Continental ships were pursued by the powerful HMS *Rainbow*. Following a 39-hour pursuit, the *Rainbow* bore down on the USS *Hancock* and captured it. The USS *Boston* escaped and continued to serve in various actions over a period of some 3 years. Its last action was in the defense of the Charleston, South Carolina, harbor where it was captured by the British in May 1780.

After its capture by the British, the *Hancock* went on to serve in the Revolution, but on the enemy's side. By a twist of fate, it was the *Hancock* (renamed the *Iris*) that captured a sister frigate, the USS *Trumbull*, one of the original 13 frigates built for the Continental Navy. (The British crew was said to have called the American built ship one of the finest frigates in which it had sailed.)

Among the names associated with this new made-in-America fleet of frigates are John Barry, who courageously commanded many ships; John Manley, who captured the *Nancy* while in Washington's Navy; and Abraham Whipple.

The skipper of the USS *Providence*, Whipple, was a member of a three-ship force that found itself on the edge of a huge, heavily guarded, enemy convoy off Newfoundland during a fog. Sending armed boarding parties to the merchant ships, the Americans managed to take 11 ships as prizes without being detected by the ships protecting the convoy. Cargoes and captured ships worth a million dollars were dispatched back to the States.

John Paul Jones

Among the most daring commanders bringing the war to British waters was John Paul Jones (fig. 5-1). As skipper of the USS *Ranger*, he left France on April 10, 1778, for raids against the British. After capturing a number of ships, he actually landed on British soil, raiding Whitehaven, England.

Figure 5-1.—John Paul Jones, father of our highest naval traditions, represents the seaman, leader, officer, and gentleman at their best.

Student Notes:

The tiny new Navy played a significant role in the first official recognition by a foreign nation of the American "Stars and Stripes" flag. On February 14, 1778, John Paul Jones sailed into Quiberon Bay, France, in the USS *Ranger* and saluted the French fleet anchored there. A nine-gun salute was given in return. A gun salute given to a revolutionary government was a signal of that country's recognition. France became one of the first foreign powers to recognize the struggling government of the American Colonies. (In 1776, the Dutch had given recognition to an American flag [not the Stars and Stripes] at St. Eustatius, an island in the West Indies belonging to Holland.)

In 1779, John Paul Jones took command of an old, decaying French merchant ship that he renamed the USS *Bonhomme Richard*, honoring Benjamin Franklin. It carried 42 relatively light guns (some in doubtful condition). Jones headed for the coast of Ireland, capturing some ships and destroying others. On September 23, 1779, Jones met the British warship *Serapis* (with 50 guns), and a furious battle ensued near the headland of Flamborough Head. As Jones wrote later:

> *Every method was practiced on both sides to gain an advantage, and rake each other; and I must confess that the enemy's ship, being more manageable than the Bonhomme Richard, gained thereby several times an advantageous situation, in spite of my best endeavors to prevent it.*

The two ships, lashed together with grappling hooks so neither could escape, pounded away at one another. The USS *Bonhomme Richard* began taking the worst of the beating. The ship began to fill with water and fire broke out in several places. According to one story, a gunner in a state of panic was about to strike the colors when Jones hurled his pistol at him, striking him down. The battle continued and the fighting was furious. The outcome was uncertain until the end. The highlight of the battle came when, after being asked if he had struck colors, Jones replied, *Struck, sir? I have not yet begun to fight!* These words inspire Sailors to this day.

What turned the tide of victory for Jones? It was his forces aloft. Armed with muskets and climbing along the interlaced rigging of the two ships, Jones's men kept the deck of the *Serapis* clear by shooting and dropping chains and other material down on the enemy. A member of Jones' crew climbed to the *Serapis*' maintop and managed to drop a hand grenade on to the gundeck, which ignited the gunpowder and scattered cartridges. In that man-to-man sea battle, the British were finally forced to surrender. The battle of the USS *Bonhomme Richard* versus the *Serapis* went down as one of the great naval battles in history.

By the time the war was over, the official Continental Navy operated some 56 vessels at one time or another. However, it only managed to reach a peak of 27 ships, averaging 20 guns, that operated at the same time. This tiny Continental Navy, hurriedly assembled when the Colonies declared their independence, served not only to inflict damage on the proud ships of the Royal Navy but also to lift American morale with each of its victories. John Paul Jones, Gustavus Conyngham, and Lambert Wickes were among those who brought the battle to the British on their own waters. The news of daring raids and victorious battles at sea was acclaimed in the 13 youthful Colonies of the United States.

Privateers

American privateers harassed British shipping over lengthy sea-lanes. At first, ships of all types were converted for harassment purposes. Later, ships were specially built to do this job. These ships were fast and reasonably well armed. Men from all walks of life signed up to serve on these ships. Private financing to arm and fit the vessels was needed, but that was rarely a problem because a share in a privateer could mean a fortune almost overnight.

The British Navy began a system of convoys to protect its merchant shipping, but it was far from foolproof. The moment a merchantman dropped behind, it was in immediate danger because a warship couldn't leave the convoy to protect just one ship. Then, too, convoys could protect only so many ships.

It's estimated that Congress issued more than 1,600 commissions for privateers during the Revolutionary War. The privateers operated not only along the American coastlines, but also far out into the Atlantic and even into the English Channel and the Irish Sea.

According to one reasonable estimate, the British were said to have lost some 2,000 merchant ships, manned by crews totaling 16,000, to the American privateers. The merchant ships captured as prizes were manned by prize crews from the privateers and sailed to a friendly port where the ships and cargo were sold.

REVIEW 1 QUESTIONS

Q1. What was the reason for the formation of the United States Navy?

Q2. During the late 18th century, battleships were classified as—

Q3. During 1775, a craft was completed to fight the British Navy. What type of craft was this, what was it named, and who was its inventor?

Q4. What is the significance of the 1776 raid on the island of New Providence in the Bahamas?

Q5. The first official recognition of the American Stars and Stripes flag by a foreign nation was given by (a) what nation in (b) what location?

a.

b.

THE U.S. NAVY FROM 1783 TO THE CIVIL WAR

Learning Objectives: When you finish this chapter, you will be able to—

- Recognize the roles and responsibilities of the Navy from 1783 to the Civil War to include the War of 1812.

At the end of the Revolutionary War, a new federal government was established. In 1783, the Navy was down to five ships. The Navy was disbanded, and the last frigate, the USS *Alliance*, was sold in 1785.

Soon, Congress saw the need for a Navy. America's small merchant fleet was being molested on the high seas. In 1794, a Navy-conscious Congress authorized the construction of six frigates. They were to be of a new design—long and strong. These ships had a combination of firepower and class. One of these was the USS *Constitution* (fig. 5-2), which was completed in 1798. This ship was equipped with 44 guns, could sail at 13 1/2 knots, was 175 feet long (at its gundeck), and had a tonnage rating of 1,576 tons. Its mainmast towered 105 1/2 feet above its decks.

Figure 5-2.—The new and radical USS *Constitution*, built for speed and firepower, helped to rid the Mediterranean of the Barbary pirates.

NOTE

The USS *Constitution* is still in commission and can be seen at the Boston Navy Yard.

The USS *Constitution* fulfilled the thoughts and dreams of President John Adams, who did so much to form the U.S. Navy. John Adams established the Navy Department in 1798.

THE EARLY YEARS

Between America's first two wars with Great Britain (the Revolutionary War and the War of 1812), the early U.S. Navy was involved in two other conflicts—the Quasi War and the Barbary States War.

Student Notes:

Quasi War

The "Quasi War" with France, 1798 - 1801, was entirely a naval war. It followed worsening diplomatic relations with France, including a refusal by the French Secretary of Foreign Affairs to receive U.S. representatives unless a bribe was paid and a loan granted. The famous expression "Millions for defense, but not one cent for tribute" originated at this time. The Quasi War was the baptism of fire for the United States Navy under the new Constitution.

Barbary States War

The U.S. Navy was sent to the Mediterranean to deal with the Barbary States, who were forcing other nations to pay ransom for safe passage through the Mediterranean Sea. During the campaign, Lieutenant Stephen Decatur and 84 seamen slipped into the harbor at Tripoli on February 16, 1804, and burned the captured frigate USS *Philadelphia* (fig. 5-3). Not a single American Sailor was lost. Britain's Admiral Lord Nelson described the raid as "one of the most bold and daring acts of the age."

Figure 5-3.—Lieutenant Stephen Decatur and 84 seamen slipped into Tripoli harbor and burned the captured American frigate USS *Philadelphia*.

THE WAR OF 1812

The War of 1812 was brought on, in part, because the British were impressing (forcing Americans to serve in the British Navy) American seamen. England impressed American seamen to make its presence felt and demonstrate its power on the American continent.

On August 19, 1812, Captain Isaac Hull aboard the USS *Constitution* defeated the British frigate *Guerriere* (fig. 5-4), and the USS *Constitution* earned its nickname "Old Ironsides." The victory convinced Congress and President Madison that a stronger Navy was needed to win the war and protect the country.

Figure 5-4.—During the War of 1812, Captain Isaac Hull, aboard the USS *Constitution*, defeated the British frigate *Guerriere*.

Almost a year after Hull's important victory, another famous event in our naval history occurred. On September 10, 1813, Captain Oliver Hazard Perry defeated a British squadron on Lake Erie and wrote his dispatch, "We have met the enemy and they are ours." Perry's win cut British supply lines on the Great Lakes, gained control of Lake Erie, and strengthened the American claim to the Northwest Territory.

The Barbary States War and the War of 1812 saw bigger ships coming into the Navy. Typical was our first ship-of-the-line, the USS *Independence*, followed by the 74-gun USS *North Carolina*.

THE YEARS FROM 1813 TO 1815

Following the War of 1812, our Navy underwent technological changes. Before the Civil War, new scientific advances foreshadowed the incredible technological revolution that continues into today's world.

One change was the use of steam. The Navy entered a new era, an era of the "steam-driven warship." Harnessing the power of steam was the most important development in the surface Navy during the first half of the 19th century. Steam began to replace wind as a means of propulsion. It promised to eliminate some of the hazards and delays caused by ships being blown off course or left dead in the water.

Student Notes:

The principles of steam power were known for centuries. But, it was Robert Fulton who successfully used steam to power a commercial steamboat. After making a number of important modifications to James Watt's basic steam engine, Fulton sailed his riverboat *Clermont* up the Hudson River in 1807. Fulton helped build USS *Demologos*, the Navy's first warship to use steam. It was originally intended to defend the port of New York during the War of 1812. The USS *Demologos* was rechristened the USS *Fulton* in Robert Fulton's honor.

1815 TO THE CIVIL WAR

From 1815 to 1840, the Navy continued to expand its sailing fleet. In fact, more than 74 ships-of-the-line were built. In 1837 the Navy launched the 3,104-ton USS *Pennsylvania*, the largest of America's ships-of-the-line.

In 1841, the Navy launched the USS *Missouri* and the USS *Mississippi*. These were our first ocean-going, steam-driven capital ships. At the same time the US Navy was building bigger ships, it was developing steam powered ships and iron clad ships.

At the same time it was harnessing steam power for ship propulsion, the Navy was making advances in ship construction. The Navy began making its ships with iron instead of wooden hulls. In 1843, the Navy launched its first iron-hulled warship—the paddle sloop USS *Michigan*. This side-wheeler was 163 feet long and displaced 685 tons. It was powered by a 170-horsepower, two-cylinder, steam engine. Without using its sails, the USS *Michigan* was capable of making 8 knots.

Through the efforts of farseeing men like Commander Matthew Calbraith Perry, USN, the Navy was becoming more steam conscious. Perry is referred to as the "Father of the Steam Navy." He was enthusiastic about the possibilities of steam, and was in charge of construction and in command of the Navy's second steam frigate the USS *Fulton*. The harnessing of steam power was considered the most important naval development since the cannon.

The newly built steamships posed problems if engaged in battle. Their paddle wheels and steam engines could be easily damaged by enemy fire. This problem was fixed by changing the design of the ships so that the paddle-wheel housing was enclosed behind 5-foot-thick walls and set in an inboard channelway.

Steamship development overcame problems one by one. For example—

- Stronger engines were developed;
- Screw propellers replaced the paddle wheel; and
- Coal as a fuel was recognized as more efficient than wood.

These changes didn't happen overnight; they required long periods of trial and error. But in the 1840s, new ideas were being explored by their proponents. On September 5, 1843, the Navy's first successful steamship, the USS *Princeton*, was launched. Its new type of propeller eliminated the vulnerable paddle wheels and permitted the ship's engines to be placed below decks in protected spaces.

Other actions between 1815 and the Civil War included the following:

- The Navy took the first steps in Antarctic exploration. Notably, Lieutenant Charles Wilkes visited the subpolar region in January 1840 and proved conclusively that the icy land was, in fact, a continent.

- Following Texas' admission to the U.S. as the 28th state, Mexican troops crossed the Rio Grande. War broke out. The Mexican-American War was primarily a land war. However, the Navy did get involved. It blockaded port cities in the Gulf and provided protective action by the "Mosquito Fleet" during the first large-scale amphibious operation in U.S. military history—the landing of some 10,000 U.S. troops at Vera Cruz. (The Navy itself was not equipped to carry out such an operation at that time.) Marines were also involved in this war—they marched with Scott to Mexico City, coining the phrase "...from the halls of Montezuma..." in the famed Marines' song.

- The Navy was involved in diplomatic relations. Commodore Matthew C. Perry signed a treaty with Japan on March 31, 1854. This was the treaty that opened Japan's ports to American trade and provisioning of ships. England and Russia soon followed with their own treaties, all modeled after Perry's.

Student Notes:

REVIEW 2 QUESTIONS

Q1. After the Revolutionary War, what was the next significant role of the U.S. Navy?

Q2. List the two conflicts that the American Navy was involved in between the Revolutionary War and the War of 1812.

a.

b.

Q3. Describe the event during the Barbary States War that Lord Nelson thought of as *one of the most bold and daring acts of the age.*

Q4. List two events that the U.S. Navy was involved in during the War of 1812

a.

b.

Q5. The Mexican-American War was primarily a land war. However, the Navy provided what service during this war?

THE U.S. NAVY FROM THE CIVIL WAR TO THE 20TH CENTURY

Learning Objective: When you finish this chapter, you will be able to—

- Recognize the roles and responsibilities of the Navy from the Civil War to the 20th century to include the Civil War and the Spanish-American War.

The last half of the 19th century was a time of change for the United States. Marked by two wars and the first assassination of a United States President, it was a time of rapid change for the country and its Navy.

THE CIVIL WAR

This bloody struggle between the States was the stage for many events in U.S. naval history. Both Union and Confederate navies engaged in shipbuilding programs. These programs brought the ironclad era into being. Launched by the Union in 1862, USS *New Ironsides*, a powerful ironclad, had the armor that allowed it to survive 50 hits in one battle.

Ironclads

The Civil War saw the development of two famed ironclads—the USS *Merrimack* (renamed the CSS Virginia by the Confederacy) and the Union's USS Monitor (which sported a turret). The USSMonitor was ungainly, called a *cheese box on a raft*; however, it and its Confederate counterpart began the ironclad era. The battle of the ships was indecisive; both sides claimed victory.

Also appearing on the scene were riverboats, rams, and gunboats. Probably more changes and advances were made in ship designs during the 4 years of the Civil War (1861 - 1865) than during any period since our Navy had its start in 1775.

Submarines

The Confederate Navy took the next steps forward in the development of the submarine. The USS *Hunley* was built with funds provided by Captain H. L. Hunley, a man blessed with imagination but lacking in practicality. The ends of this 25-foot craft were loaded with ballast tanks that could be filled for descent but had to be hand pumped for ascent. Power was supplied by a propeller fitted to a camshaft that ran the length of the ship and was turned by as many as eight men.

The CSS *Hunley* was a jinx to the Confederate Navy. On its first voyage, it nosed into the mud and refused to surface, killing its seven crew members. It was hauled up and moored at James Island, where a passing steamer swamped it and six more crewmen were lost. It was hauled up once more and manned with another crew, but was swept over by another steamer and another three men were killed.

Student Notes:

A young Confederate lieutenant, George Dixon, was convinced that the boat could be useful to the South. The CSS *Hunley* was moored off Charleston's Sullivan Island, just a few hundred yards from the USS *Housatonic*. In the first true submarine attack in naval history, Dixon cast off toward the large warship. The CSS *Hunely* attacked the USS *Housatonic* in calm waters in the dark of night. The submarine was sighted by lookouts on the USS *Housatonic*; however, it didn't have the time or the opportunity to strike back or set sail.

The CSS *Hunley* hit the USS *Housatonic* driving its shaft deep into the ship's hull. The heavy charge of gunpowder the submergible was carrying went off prematurely, and the CSS *Hunley* never had a chance to escape. It and all of its crew went down. The USS *Housatonic* had the same fate. It was hit on the starboard side and went down in just 4 minutes. Another northern vessel moved to its rescue, and only a few of its seamen were lost. Even though he lost his life, Lieutenant Dixon had demonstrated that submarines could be useful weapons of war.

Other Innovations

Some people associated with the Navy during the 19th century were interested in the air above the ocean. The USS *George Washington Parke Custis* of the Civil War days might be labeled as the Navy's first "aircraft carrier." Actually, it was a balloon boat used to launch observation balloons over enemy installations. It was 122 feet long, and its total cost was $150.

Other Civil War Actions

Capture of Vicksburg. On the Mississippi River, the capture of Vicksburg, Mississippi, by the combined naval forces of Rear Admiral David G. Farragut, Acting Rear Admiral David D. Porter, and the commander of the Army in the West, General Ulysses S. Grant gave the North control of the entire river. The capture of Vicksburg cut off important Confederate supplies of food and clothing coming from Louisiana, Texas, and Arkansas.

Battle of Mobile Bay. On August 5, 1864, David Farragut, the Navy's first admiral, gave his famous order "Damn the torpedoes! Full speed ahead!" (Torpedo was the name used at the time for mines.) Farragut's order won the Battle of Mobile Bay (fig. 5-5). This victory closed the South's most important port (since New Orleans had already fallen) and tightened the Union blockade.

Figure 5-5.—Farragut (in rigging) at Mobile Bay.

The Civil War produced many men whose names are still famous in the Navy:

- Andrew Foote, whose gunboats helped General Grant capture the Mississippi River fortresses
- John Dahlgren, the father of modern naval ordnance (fig. 5-6)

Figure 5-6.—Rear Admiral Dahlgren, standing next to one of the guns he designed, was a leading pioneer in modern naval ordnance and gunnery.

Student Notes:

- David D. Porter (son of the captain of the *Essex*), who commanded the mortar flotilla in the capture of New Orleans

POST CIVIL WAR NAVY

Alfred T. Mahan (fig. 5-7) was one of the first instructors at the Naval War College, and he influenced naval strategy. In 1890, the first of his many books and articles on sea power was published. One of his books (*The Influence of Sea Power Upon History, 1660 - 1783*) stressed that without control of the seas, a nation at war could not expect victory. He defined sea power; showed the importance of understanding naval needs; and advocated a large, powerful Navy capable of assembling an overwhelming force to defeat the enemy's Navy. His books on sea power became the "bible" of many navies, and for many years, they influenced the thinking of naval strategists. Part of our Navy's readiness for the war with Spain was a result of the influence of his works.

Figure 5-7.—A philosopher of naval strategy, Mahan researched military history and proved that the nation controlling the oceans is the nation that maintains its supremacy in war or peace.

Surface Ships

Steam power was the major development in ship propulsion during the first half of the 19th century. Iron construction of ships was the outstanding development of the second half. The two developments went hand in hand—all the navies of the world recognized the advantages of steam power, and iron warships needed large steam engines to power them. The engines, in turn, called for bigger ships to accommodate them.

Shipbuilders used iron first as framing and then as a material for the entire ship. Iron was first used as framing to reinforce ships so that they could be used to ram their opponents as well as fire on them. It was several years before an economical way to process iron strong enough for the entire construction could be found. (Wooden ships had the advantage of being cheaper to build than iron ships.)

After the Civil War, the Navy began a drawdown period. A year and a half after the war, the total number of Navy ships was 236, with only 56 in active service.

World conditions made our Country aware that the Navy was small. Therefore, in 1882 and 1883, Congress authorized the construction of the "protected cruisers" USS *Atlanta*, USS *Boston*, and USS *Chicago* and the dispatch boat USS *Dolphin*, which had both masts for sails and stacks for smoke. They were steel hulled and signaled the end of the ironclads introduced only 40 years earlier. These new cruisers were in the 13- to 14-knot class. They sported new guns, new types of turrets, and armor.

Once more, the Navy began to rebuild its strength. Continued changes were made as the new steel Navy took on new shapes. Still clinging to the past, the USS *Newark*, a 4,098-ton protected cruiser, was the last of the Navy's warships to be fitted with sails. It was launched in 1890 and commissioned the following year. Because of its many improvements, the USS *Newark* has been labeled as *the first modern cruiser in the U.S. Fleet*.

With the development of the self-propelled torpedo, long-range torpedo boats made their debut. In 1890, one of the first torpedo boats joined the fleet—the 22.5-knot USS *Cushing*. The Navy acquired 16 fast torpedo boats and three 185-ton boats capable of speeds of 27 knots.

The development of torpedo boats caused the shape of ships to change. An example was the USS *Truxtun*, which led to the design of our present-day destroyers. These ships were designed to combat torpedo boats. Later improvements resulted in destroyers themselves carrying torpedoes.

Subsurface Ships

Since surface ships were driven by steam, why not submarines? Steam requires air, fire, and heat, and those were in limited supply aboard a submarine. During the 19th century, the internal combustion engine was developed. Use of this engine on ships had drawbacks.

Student Notes:

However, many of its problems were overcome by two inventors—John Holland and Simon Lake. Holland and Lake had opposite theories about the submarine.

- Holland thought submersion should be made by power-diving, using the force of the propeller and the angle of the bow planes.

- Lake said boats should descend on an even keel with slight negative buoyancy.

Lake was more interested in underwater exploration than naval warfare. He thought a submarine could be equipped with wheels and driven along the ocean's floor, although he did not pursue that idea. Holland was more practical; his design included a workable torpedo tube, which Lake's did not.

Holland received a $150,000 contract from the Navy for a subsurface vessel. His first attempt failed, but the Navy was impressed enough to award him another contract. By 1898, he had built USS *Holland*, a cigar-shaped craft, 52 feet long and 10 feet in diameter. The USS *Holland* was equipped with a gasoline engine for surface power and generators that charged batteries for underwater power. It was armed with a torpedo tube that fired an 18-inch torpedo and a bow gun recessed into the hull. A New York newspaper commented that "...the offensive powers of the *Holland* are, considering the size and method of attack, far greater than any other engine of war."

The submarine's problem of running blind when submerged was corrected after Simon Lake experimented with a set of prisms and lenses. Before that, the USS *Holland* had to surface to permit the crew to look out the conning tower; causing it to lose its greatest advantage—surprise. Lake and a professor from Johns Hopkins University worked out a design for the periscope. The periscope, with various improvements, remained the submarine's basic visual aid until 1958.

THE SPANISH-AMERICAN WAR

At the end of the 19th century, the United States and Spain became involved in diplomatic disputes about Cuban independence, trade, and U.S. citizens living there. On the evening of February 15, 1898, a terrific explosion suddenly tore through the battleship USS *Maine* at anchor in Cuba's Havana harbor. The explosion killed 250 American Sailors. The explosion was a major reason for the start of the Spanish-American War...*Remember the Maine* became our battle cry.

One event stood out in this short war—Commodore George Dewey's seizure of Manila Bay in the Philippines. On May 1, 1898, he steamed into Manila Bay and ordered, "You may fire when you are ready, Gridley." Dewey's resounding victory destroyed Spain's naval power in the East and was instrumental in quickly ending the war.

Shortly after the Battle of Manila Bay (fig. 5-8), U.S. naval forces at Cuba cornered the Spanish Atlantic Squadron at Santiago Bay. On the morning of July 3, 1898, the Spanish squadron tried to break out of the bay and was completely destroyed. Cuba and Puerto Rico fell shortly afterwards, effectively ending the war.

Figure 5-8.—Battle of Manila Bay.

Student Notes:

REVIEW 3 QUESTIONS

Q1. List the naval developments during the last part of the 19th century.

a.

b.

c.

d.

e.

f.

Q2. After the development of the ironclad, what was the Confederate Navy's next achievement?

Q3. During the Civil War, the U.S. Navy's first admiral gave the famous order, "Damn the torpedoes! Full speed ahead." List the admiral's name and battle where he gave the order.

Q4. Describe how Alfred T. Mahan influenced naval strategy.

Q5. What was the major cause of the Spanish-American War?

Q6. Who was instrumental in quickly ending the Spanish-American War?

THE NAVY FROM 1900 THROUGH WORLD WAR I

Learning Objective: When you finish this chapter, you will be able to—

- Recognize the roles and responsibilities of the Navy during World War I.

The 20th century began with a world at uneasy peace. Between the end of the 19th century and WWI, the U.S. Navy developed some new weapons. For example, in April 1900 the Navy accepted its first operational submarine, USS *Holland.*

SUBMARINES

The Navy continued to experiment with the development of submarines throughout the next decade. One of the main problems continued to be the gasoline engine—it heated up and gave off fumes that overcame many of the crew.

The gasoline engine was replaced by the diesel engine. The first diesel engines were installed in the USS *Skipjack* (SS 24) and the USS *Sturgeon* (SS 25). These new engines required no complicated ignition or sparking systems, produced fewer fumes, and were cheaper to operate. The diesel engine and electric battery remained as the main propulsion systems for submarines until nuclear power emerged in the 1950s.

DESTROYERS

Destroyers had been used primarily to deliver torpedo attacks. With the development of the submarine, they became submarine hunters. Construction of our first destroyer, which displaced 420 tons, began in 1899. Destroyers proved so successful that building these ships began on a large scale. From 1892 to 1914, the start of World War I, over 50 destroyers were built; and 273 were ordered during the war.

CRUISERS AND BATTLESHIPS

The battleship resulted from the major changes in ship design that took place during the 19th century. Battleships carried heavy guns and corresponding armor protection. The United States had begun building its *battlewagons* in the late 1880s; each succeeding class had more firepower than the one before.

Student Notes:

By 1895, the heavy elements of the U.S. Fleet consisted of 15 steel cruisers, the heavy cruiser USS *New York*, and three battleships. The first two battleships were the USS *Texas*, commissioned on August 15, 1895, and the USS *Maine*, commissioned on September 17, 1895. Both were listed as "second-class" battleships. The third ship, the USS *Indiana* (BB 1), was commissioned in 1895. It was our first "first-class" battleship.

In 1906, the United States began a large battleship-building program. Five battleships were of the same class as the USS *New Mexico* and USS *Colorado*; however, they weren't completed until after World War I. Based on lessons learned from wartime experiences, many improvements were incorporated into their design. For example, battleships of the same class as the USS *Colorado* were the first ones equipped with 16-inch guns.

NAVAL AVIATION

As the 19th century drew to a close, the Wright brothers were working on their flying craft. The Wright brothers' first flights at Kitty Hawk, North Carolina, began the vision of the future. Most people thought of flying as a stunt or a sport, while others talked about crossing the ocean by airplane. One European wrote in part,

> *...flights over the ocean will be made possible by a new type of ship...[its] deck will be clear of all obstacles, flat and wide as possible ... [it will] have the aspect of a landing field...its speed shall equal that of a cruiser...housing of planes will be arranged below deck and planes will have folding wings...and to one side there will be the service personnel workshop.*

Others saw the potential of aircraft serving as an extension of the might and range of a naval force at sea. They were convinced that airplanes wouldn't be used just for circus sideshows and crop-dusting. They believed aircraft would transport troops across oceans and be equipped to strike offensively.

The Navy was again looking upward. As the Assistant Secretary of the Navy, Theodore Roosevelt recommended that the Secretary of the Navy appoint two officers "...of scientific attainments and practical ability..." to examine Professor Samuel P. Langley's flying machine and report on its potential for military use.

One such man was Navy Captain Washington Irving Chambers, the U.S. Navy's first officer in charge of aviation. Captain Chambers' initial involvement was to answer letters from air-minded citizens and observe and report on aviation developments of particular concern to the Navy. What started as a collateral duty soon was a full-time job, and Chambers became a strong supporter of those who wanted to see the sea service add an air arm.

In April 1911, the Office of Aviation in Washington, D.C., consisted of only Captain Chambers. In May, he wrote requisitions for two machines made of wood, canvas, bamboo, rubber, and metal—two airplanes, the A-1 and the A-2. Earlier in the year, a civilian, Eugene Ely, had successfully taken off from and later landed a biplane on a platform rigged aboard USS *Pennsylvania* (ACR 4), demonstrating the practical use of naval aircraft.

Shortly thereafter, the Navy accepted delivery of its first airplane, the A-1. The A-1 was first flown by Lieutenant T. G. Ellyson, the Navy's first aviator. The A-1 was followed by the A-2; naval aviation had gotten off the ground.

By October 1911, the Navy was ready to try durability flights. Lieutenants Ellyson and J. H. Towers attempted a flight from Annapolis to Fort Monroe, Virginia. After flying 112 miles in 122 minutes, the pair was forced down somewhat short of their goal by mechanical problems. Although a failure in part, the flight paved the way for successful durability tests in the following months.

Based on tremendous headway made in a few short years, in 1914, Secretary of the Navy Josephus Daniels prophesied "...the science of aerial navigation has reached that point where aircraft must form a large part of our naval force for offensive and defensive operations." It had become evident that the airplane was no longer merely a plaything of the rich or eccentric—it had become a vital part of our nation's weaponry.

OTHER DEVELOPMENTS FROM 1900 THROUGH WWI

Meanwhile, the Navy was switching from coal to oil as fuel for its ships. USS *Nevada* (BB 36) was the first of the battleships to use oil. The day of the coal passer was on the way out.

Student Notes:

Navy involvement in exploration continued during the first decade of the century. On April 6, 1909, Commander Robert E. Peary, accompanied by Matthew Henson, reached the North Pole.

In pre-World War I days, the Navy also carried out its role as a diplomatic arm of the government. On December 16, 1907, the Great White Fleet left Hampton Roads, Virginia, for a round-the-world cruise to show the flag. The exercise demonstrated the strength of the U.S. Navy.

Although the United States entered World War I late, the Navy had plenty of time to make history. On May 4, 1917, six American destroyers commanded by Commander Joseph K. Taussig steamed into Queenstown, Ireland. They became the first U.S. Navy ships to operate in European waters during World War I. The event, billed as the "return of the *Mayflower*," was a great morale booster and aid for the Allied forces. The incident is probably best remembered by Commander Taussig's simple remark upon reporting to the British admiral in charge: "I shall be ready when refueled, sir."

Destroyers became a primary symbol of British-American cooperation during WWI. Destroyers were the main defense against German U-boats, which were practicing unrestricted warfare and terrorizing the seas. U-boat attacks were one reason for our entry into the war.

The British and Americans exchanged signals, codes, and inventions in combining their destroyer forces to seek out and attack the German submarines. Destroyers served as escorts for troopships and supply convoys for the Allies, helping to ensure their safety. On November 17, 1917, the destroyers USS *Nicholson* and USS *Fanning* were the first U.S. ships to sink an enemy submarine.

When the United States entered World War I, naval aviation assets were limited. The nation had only 54 aircraft, 1 air station, and 287 personnel assigned to aviation. The nation had no armed forces or operations abroad.

In spite of its size, the air arm proved its value as a supporting unit to surface antisubmarine (ASW) forces. Navy pilots served with Allied units in France and England. The airplane created a new breed of hero, the ace. Nineteen year-old Lieutenant David Ingalls, later Assistant Secretary of the Navy (Air), flew a Sopwith Camel to become the Navy's first ace.

In World War I, the women's role in the Navy came into its own. In 1811, a Navy surgeon recommended employing women in hospitals to care for the Navy's sick and wounded. The idea was not acted upon at that time.

NOTE

> In the Civil War, women nurses, although not part of the Navy, served aboard the hospital ship USS *Red Rover* in the medical department. In the war of 1898, the first trained nurses in the Navy, though not an official unit, were stationed at the Norfolk Naval Hospital to care for the injured. A decade later (in 1908), the Nurse Corps was officially born.

As the nation readied itself for World War I, it needed Yeomen and personnel in related jobs to handle the growing demand from headquarters and naval shore stations. Josephus Daniels, Secretary of the Navy, asked his legal advisors, "Is there any law that says a Yeoman must be a man?" The answer was no, but until that time only men had been enlisted. "Then enroll women in the Naval Reserve as Yeomen," the Secretary said. In such jobs, he added, they would offer the best "assistance that the country can provide."

Immediately after the United States entered World War I, women were enlisted on a large scale "in order to release enlisted men for active service at sea." By the time the armistice was signed, 11,275 women were enlisted in service as Yeomen (F). They handled most of the clerical work at the Navy Department, in addition to many highly important special duties. Yeomen (F) were stationed in Guam, the Panama Canal Zone, and Hawaii, in addition to the United States and France. About 300 "Marinettes," as the female enlisted personnel of the Marine Corps were designated, were on duty during the war. Most of them were stationed at Marine Corps Headquarters at the Navy Department, although a number were assigned with Marine Corps recruiting units.

All Yeomen (F) were released from active duty by July 31, 1919. Secretary Daniels sent the following message to the Yeomen (F): "It is with deep gratitude for the splendid service rendered by the Yeomen (F) during our national emergency that I convey to them the sincere appreciation of the Navy Department for their patriotic cooperation."

Student Notes:

REVIEW 4 QUESTIONS

All questions in this review concern WWI.

Q1. Before WW I, the Navy built surface ships and submarines. What other development occurred then that is still a big part of today's naval arsenal?

Q2. What was one reason why the U.S. Navy was deployed?

Q3. Describe the role of Navy destroyers.

Q4. Describe the role of the air forces.

Q5. What was the role of Navy women?

THE NAVY FROM 1920 TO 1950

Learning Objective: When you finish this chapter, you will be able to—

- Recognize the roles and responsibilities of the Navy from 1920 through 1950 to include World War II and the post-war years.

The world was changing rapidly from the end of WWI to 1950. During the 1920s, the world economy boomed, then fell. In the 1930s, there was the "Great Depression." In 1939, World War II began. In this section, you will learn about some of the developments made by the U.S. Navy.

1920 TO 1940

Between 1920 and 1940, the U.S. Navy was developing its aviation arm to include aircraft carriers and airships and airplanes. Also, it was building up its destroyer strength.

Aviation

Great strides in aviation had been made during World War I, and the end of the war did not slow the pace of progress. On May 8, 1919, three Navy Curtiss (NC) flying boats taxied into the bay of Far Rockaway, New York, and took off for Europe. Plagued by mechanical difficulties, two NCs failed to make it. The NC-4, piloted by Lieutenant Commander Albert C. Read, became the first airplane to fly the Atlantic. LCDR Read's message from Lisbon, Portugal, to the President read, "We are safely across the pond. The job is finished." The NC-4 is now located at the National Museum of Naval Aviation, Pensacola, Florida.

With transoceanic aircraft a reality, the Navy continued to research the use of rigid airships in its air arm. In 1923, *Shenandoah* was launched. During a severe squall in 1925, the *Shenandoah* broke in half and killed 14 men. At that time, some authorities questioned the safety of the airship since it was fueled with highly flammable hydrogen. In spite of some opposition, the Navy continued to test rigid airships throughout the next decade. In 1931, USS *Akron* was launched. The *Akron* crashed in 1933 during a thunderstorm, killing the entire crew.

In November 1929 a Ford trimotor aircraft, named the *Floyd Bennett*, carried Commander Richard E. Byrd and his crew on the first flight over the South Pole. Commander Byrd thereby became the first man to fly over both poles.

In 1933, *Macon* was commissioned. Two years later the *Macon* also crashed into the sea. The Navy then abandoned research and construction of rigid airships.

Aircraft Carriers

In 1934, the USS *Ranger*, the first carrier designed from the keel up, joined the fleet. Also in the 1930s and prewar 1940s, the large aircraft carriers USS *Enterprise*, USS *Wasp*, USS *Hornet*, and USS *Yorktown* were commissioned.

Student Notes:

Those carriers played an important role in the prewar years. They were used in exercises to test the possibility of launching air attacks from their decks. During fleet maneuvers, naval aviators received excellent training in mock attacks on Pearl Harbor. Flying predawn missions from carriers, flyers theoretically destroyed the U.S. Fleet and its aircraft there. Fleet commanders were impressed by the flexibility of the air arm, but no one else seemed to pay much attention to the exercises.

Destroyers

Between the two world wars, the United States built the Navy's destroyer fleet to 184 ships. Destroyers also became prime factors in America's policy to turn over older destroyers (fig. 5-9) to Britain after the British entered the war against Germany. When the Japanese attacked Pearl Harbor, a destroyer, USS *Ward* (DD 139), was among the first American forces to fire against the enemy, sinking a Japanese midget submarine. Destroyers went on to distinguish themselves in fighting enemy submarines both in the Atlantic and Pacific theaters.

Figure 5-9.—Destroyer built shortly after World War I.

WORLD WAR II

On the morning of December 7, 1941, the "Rising Sun" came out of the west when the Japanese pounced on Pearl Harbor. On that morning, over 15 U.S. Navy ships were sunk or damaged, including all 8 battleships of the Pacific Fleet (fig. 5-10). Some 3,400 Navy and Marine Corps personnel were killed or wounded. The United States declared war on Japan the next day.

Figure 5-10.—The day of infamy.

Pacific Arena

The Japanese attack on Pearl Harbor was the first attack in history conducted solely from aircraft carriers. The attack proved beyond a doubt that aircraft had become an essential part of naval armament. Fortunately, no United States carriers were lost during the attack on Pearl Harbor. The USS *Yorktown,* USS *Wasp*, and USS *Ranger* were in the Atlantic, and the USS *Saratoga* was in San Diego. The USS *Lexington* was about 425 miles south of Midway, and the USS *Enterprise* was 200 miles west of the Pearl Harbor.

The Japanese Imperial Navy captured island after island in the South Pacific as it advanced toward Australia. The U.S. Navy's air arm finally stopped that advance in early May 1942, which set the scene for the turning point of the war in the Pacific.

At the **Battle of Coral Sea**, the two fleets never saw each other—the battle was fought entirely with aircraft launched from carriers. The USS *Lexington* and USS *Yorktown*, jointly under the command of Admiral F. J. Fletcher, launched 93 attack planes against the Japanese carriers *Shoho, Shokaku*, and *Zuikaku*. Within 5 minutes, the *Shoho* was hit with 10 heavy bombs and 15 torpedoes. The USS *Lexington's* radio crackled with the voice of Lieutenant Commander Dixon of the air group, "Scratch one flattop. Dixon to carrier, scratch one flattop!" The other two enemy carriers were so badly damaged that their services to the Japanese fleet were lost for weeks. The United States suffered the loss of an oiler, an escort, and the USS *Lexington*. Even though American losses were heavy in tonnage and men, Australia had been saved from invasion.

Student Notes:

The turning point of the war in the Pacific came the next month at the **Battle of Midway**. The Japanese had concentrated on the central Pacific with the intention of occupying Midway Island. The four-carrier Japanese task force was met by a U.S. carrier force. The U.S. force included the carriers USS *Yorktown,* USS *Hornet*, and USS *Enterprise*, plus Navy, Marine, and Army air units from Midway.

Dive bombers proved to be the downfall of the Japanese carrier force. When the battle ended, the Japanese had lost four carriers, one heavy cruiser, and 258 aircraft. The United States had lost 132 aircraft, the destroyer USS *Hammann* (DD 412), and the aircraft carrier USS *Yorktown* (CV 5). In April 1943, another USS *Yorktown* was commissioned; and it continued in the proud tradition established by its predecessor.

In November 1942. the Navy fought the **Battle of Guadalcanal**. After 3 days of bitter fighting, the Japanese naval forces retreated, and U.S. Marines were able to secure the island. With the fall of Guadalcanal, the southern Solomons came under Allied control and Australia was in less danger of attack.

On June 19, 1944, U.S. Task Force 58 (fig. 5-11) caught the combined Japanese fleet steaming out of Tawi in the southern Philippines. The **Battle of the Philippine Sea** ended with the Japanese carrier forces short of ships, planes, gas, and pilots. Unable to replace these, the Imperial Navy was never able to recover from losses, although many desperate battles were to follow.

Figure 5-11.—Part of Task Force 58 at anchor in the Marshall Islands, April 1944.

The final blow to the Japanese Navy came October 23, 1944. In a last-chance effort to salvage the Philippines, the Japanese sent a naval force to the **Leyte Gulf** to attack the U.S. Fleet. Their plan backfired and the operation was a complete failure: It was the deciding catastrophe for their Navy. The loss of the Philippines severed their empire, and the homeland was cut off from its main source of supply from the south. With the losses at **Okinawa** and **Iwo Jima**, the war in the Pacific was approaching its final days.

European Arena

On the Atlantic side of World War II, the U.S. Navy had been fighting off U-boats in the long-running Battle of the Atlantic. The Navy protected convoys bound for Europe. Small escort carriers dubbed "jeeps" were operating with convoys; and their aircraft were successfully attacking German submarines as they surfaced to recharge their batteries. Limited range of land-based airplanes was no longer a significant factor; and distance offered no sanctuary for the U-boat. Eventually, the German submarine menace was contained, and England and Europe got vital supplies and troops.

The Navy's most notable Atlantic action may have been its part in the June 6, 1944, invasion of Normandy—the largest amphibious operation in history. The greatest armada ever assembled carried out minesweeping, shore-bombardment, amphibious operations, and transported supplies and troops. Those operations let the Allies complete D-Day landings successfully and eventually push on to Germany.

Widespread fighting on the oceans brought about the building of a fleet unlike any in history. This was a swift striking force. It had the advantages of speed, mobility, and surprise, yet it possessed the firepower and protective armor to stand and slug it out with enemy forces. Such a fleet was made up of ships with names synonymous with heroism, such as the USS *Tarawa,* USS *Missouri* (fig. 5-12), USS *Tucson,* USS *Higbee*, and USS *O'Bannon*.

Other Events during WWII

During the 5-year period ending in late 1944, 9 million tons of vessels had been added to the U.S. Navy. One novel development was the large assortment of landing ships that began appearing in the early stages of the war.

Student Notes:

Figure 5-12.—V-J Day aboard USS *Missouri.* Fleet Admiral Nimitz signs the Japanese surrender document on 2 September 1945.

Possibly the most versatile of the many new types of ships built during World War II were the destroyer escorts, now called *frigates*. Other types built during that time included attack cargo ships, transports, barracks ships, net tenders, repair ships, radar pickets, minelayers, and mine sweepers. Those ships, as well as many other types of ships too numerous to mention, changed the shape of the U.S. Navy almost overnight.

When the Japanese attacked Pearl Harbor, 111 American submarines were in commission, 60 in the Atlantic Fleet and 51 in the Pacific. After the invasion of North Africa, U.S. efforts were concentrated in the Pacific, leaving submarine operations in the Atlantic to U.S. Allies. The Pacific became the hunting grounds for American submarine forces.

The number of American submarines during the war peaked at 247. During the war, the United States lost 52 of these boats along with 3,505 submariners. The number of vessels sunk by U.S. submarines played a major part in the American victory in World War II. American submarines sank 1,750 Japanese merchant ships and more than 200 combatants. Those vessels represented 55 percent of the total Japanese tonnage sunk in the war. For an island nation such as Japan, those figures represented a fatal impact.

Radar and sonar came into full use during World War II. The English used them initially to combat German U-boats, but they were also incorporated into the submarine as an attack aid. Sonar has become the most important of the submarine's senses. Hydrophones listen for sounds from other ships and the echoes of sound waves transmitted from the submarine itself.

Women in the Navy

Twenty-one years after the Yeomanette era, women were needed to fill an acute shortage of personnel caused by rapid expansion of the Navy for World War II. On July 30, 1942, Congress authorized establishment of the Women's Reserve, with an estimated goal of 10,000 enlisted women and 1,000 officers. This new organization had certain congressional limitations. Women could not serve at sea or outside the continental United States and could not exercise military command over men. They could not go beyond lieutenant commander on the promotion ladder. On August 4, 1942, Mildred Helen McAfee was sworn in as Lieutenant Commander, U.S. Naval Reserve, to become Commander of the Women's Reserve.

A boot camp for women volunteers was established at Hunter College in New York City. It was promptly dubbed USS Hunter. Since basic training lasted from 6 to 8 weeks, every other week some 1,680 women seamen had to be housed, fed, and uniformed. (The housing was provided in 17 apartment buildings near the college taken over by the Navy.)

At about the same time, three other schools were commissioned in the Middle West to train enlisted women as Yeomen, Storekeepers, and Radiomen. In July 1943, the Navy Japanese Language School in Boulder, Colorado, opened to women.

Navy women came to work the same hours as Navy men, standing both day and night watches. They stayed in uniform at all times except in the barracks or when engaged in active sports. They were called on to meet the same standards of neatness and good behavior as those required of men in uniform. In short, women were fitted into the Navy as an integral part of the service. They slipped into the same spot in the chain of command as the men they replaced and performed the same duties. This system gave Navy women the same status, responsibilities, and restrictions as men.

The first Reserve classification for women officers was W-V(S), meaning Woman-Volunteer (Specialist).

Student Notes:

Professor Elizabeth Reynard (later LT Reynard) came up with the term *Women Appointed for Voluntary Emergency Service* (*WAVES*). That term was later changed to *Women Accepted for Voluntary Emergency Service*. The initials WR and the term *Women's Reserve* were official, and some women preferred these terms to the equally official, but less formal, term *WAVES*.

As the Women's Reserve observed its second anniversary on July 30, 1944, it could look back upon a brief but glowing record of expansion and achievement. During its 2 years of existence, its members had freed enough officers and men to crew a fleet of 10 battleships, 10 aircraft carriers, 28 cruisers, and 50 destroyers.

During World War II, WAVES were directly eligible for 34 different ratings. They performed nearly every conceivable type of duty at 500 naval shore establishments.

THE POSTWAR YEARS

Unlike the placid years following World War I, the postwar period from 1945 to 1950 was a busy one. The United States emerged from the war with an awareness that it couldn't afford any major cutbacks in military strength. The United States had become a nation committed to trading with and protecting other countries. The only way that responsibility could be discharged was by the maintenance of a strong and ready Navy.

Navy women. Since the WAVES had proved their worth during the war, the Navy was reluctant to give up its programs for women. After the war, a number of Navy women were retained in service. However, by the fourth anniversary of the program, only 9,800 remained on active duty.

The Women's Armed Services Integration Act, Public Law 625, was passed by the Senate and the House and signed by the President. It became law June 12, 1948, marking another step forward. That was perhaps the most significant milestone to date in the history of women in the Navy. That act gave women full partnership on the Navy team and abolished the Women's Reserve. For the first time, women became a part of the Regular Navy.

At the same time the Regular Navy opened to women, the Reserves established a program for women volunteers. The new laws authorized the transfer of all members to appropriate components of the permanent Naval Reserve.

Antarctic exploration. Following World War II, the U.S. Navy turned its attention once again to the exploration of Antarctica. In 1946, Operation Highjump got underway. Seaplanes flying from the open sea and the airstrip at Little America photographed the interior and coastline of the "white continent."

Naval aviation. Naval researchers continued to develop new, specialized ships and new planes capable of providing swift aid to Allies in a world of uneasy peace. All naval aircraft, featuring the most advanced radar and sonar systems, were redistributed into patrol, attack, and fighter squadrons.

Jet aircraft were perfected during the postwar years. In June 1948, a squadron of FH-1 *Phantoms* qualified for carrier operations aboard USS *Saipan* (CVL-48). Carrier flight decks were redesigned to launch and recover jets.

Submarines and nuclear power. During this time, the Navy was speeding development of the most revolutionary advancement in the history of submarines—nuclear power. Early in World War II, as part of the Navy's initial research on the atom, proposals were made to develop atomic power for use afloat. However, most of that work was diverted to development of the atomic bomb.

Nuclear power was the long-awaited propulsion source for the submarine. It turned the submersible surface ship into a true submarine, capable of almost indefinite operation. It was no longer bound to the earth's atmosphere.

In September 1947, Captain H. G. Rickover informally requested the first study of the application of a high-pressure, water-cooled reactor for a submarine. Personnel of the Daniels Pile Division at Oak Ridge, Tennessee, undertook that study.

In January 1948, the Department of Defense requested that the Atomic Energy Commission undertake the design, development, and construction of a nuclear reactor that would propel a naval submarine. In December 1948, the Commission contracted the Westinghouse Electric Corporation to develop design, construct, operate, and test a prototype nuclear propulsion plant. The outcome of those efforts was USS *Nautilus*.

Student Notes:

REVIEW 5 QUESTIONS

Q1. What was the significance of the Battle of Coral Sea?

Q2. List the other major naval battles in the Pacific during World War II and describe their significance.

a.

b.

c.

d.

e.

Q3. Describe the role of the U.S. Navy in the Atlantic Ocean during World War II.

Q4. The shape of the Navy changed during World War II because of new ships introduced during this period. List some of the types of ships that were introduced during this period.

Q5. What is the significance of the date 30 Jul 43?

Q6. What is the significance of the Women's Armed Services Integration Act?

THE NAVY FROM 1950 TO 1990s

Learning Objective: When you finish this chapter, you will be able to—

- Recognize the roles and responsibilities of the Navy from 1950 to 1990 to include the Korean Conflict, Vietnam, and the Persian Gulf.

As the second half of the 20th century arrived, the United States had been at peace for 5 years, and the Navy was involved in many scientific pursuits. However, scientific and exploratory pursuits were interrupted by the outbreak of the Korean Conflict.

THE KOREAN CONFLICT

Supported by the United Nations, the United States agreed to give the Republic of Korea air and naval assistance. Three days after that decision, June 29, 1950, the cruiser USS *Juneau* and the destroyer USS *Dehaven* fired the first shots of the war.

When North Korea attacked south of the 38th parallel, the Navy was called on for close air support to knock out bridges and block enemy supply routes. Navy

Student Notes:

jets flew from carriers for the first time in a war situation. Unlike World War II, the enemy didn't have the capability to strike our carriers, so pilots launched their Corsairs and Banshees on the first sustained ground-support missions in history.

The helicopter also came of age during the Korean Conflict. First studied and developed in 1942 when the Navy received four Sikorskys, the choppers were spotters for artillery. In Korea, they flew emergency supply runs and took part in direct combat duties. Later, the helicopter was used as a cargo transport between ships during underway replenishment, search and rescue missions, and ASW exercises. Korea was the testing ground for the helicopter and many other innovations our forces currently use.

On September 15, 1950, under massive shore bombardment by U.S. Navy ships, the amphibious landings at Inchon began. The successful operation cut enemy communications, split enemy forces, and dissolved enemy resistance in that area. The shelling of supply roads far inland by the battleship USS *Missouri* demonstrated a new tactical concept. That concept was the Navy's ability to intervene successfully in a ground operation far ashore.

The Korean Conflict (fig. 5-13 and fig. 5-14) lasted until July 1953. Other events were happening in the Navy while the war was being waged. For example, a program was established giving outstanding enlisted women the opportunity to receive commissions in the Regular Navy.

Photograph courtesy of Mr. Francis Jeffery.

Figure 5-13.—Korean War Memorial.

Photograph courtesy of Mr. Francis Jeffery.

Figure 5-14.—Korean War Memorial—soldier.

KOREA TO VIETNAM

The 1950s was a time of change. By the end of the decade, most operational aircraft in the attack and fighter arsenals of the sea service were jets. More and more angled-deck carriers were authorized, and new deck-edge elevators allowed simultaneous takeoffs and landings.

The USS *Nautilus*, the first nuclear submarine, was first put to sea on January 17, 1955. Under Commander Eugene P. Wilkinson, the USS *Nautilus* transmitted the historic signal, "Underway on nuclear power." On its shakedown cruise in May 1955, the USS *Nautilus* steamed submerged from New London, Connecticut, to San Juan, Puerto Rico. It traveled over 1,300 miles in 84 hours—a distance 10 times greater than the record for continuously submerged travel by any previous submarine.

After more than 2 years of operation and evaluation, the USS *Nautilus* was refueled in April 1957. On its first nuclear core, it steamed a total of 62,562 miles; it made more than half of that cruise while totally submerged. A conventionally powered submarine the size of the USS *Nautilus* would have required over 2 million gallons of fuel oil to duplicate that feat. A train of tank cars over a mile and a half long would have been necessary to transport that amount of fuel.

Student Notes:

On August 12, 1958, the USS *Nautilus* completed a history-making transpolar voyage from Pearl Harbor, Hawaii, to Portland, England. After diving under the ice near Point Barrow, Alaska, on August 1, 1958, it became the first submarine to reach the geographic North Pole.

Nuclear submarines produced after the USS *Nautilus* continued to pioneer new areas of submarine operations. The USS *Seawolf*, the Navy's second nuclear-powered submarine, operated as an active unit of the Atlantic Fleet. On October 6, 1958, it completed a record-breaking 60-day run, traveling a distance of 13,761 miles submerged.

While the USS *Nautilus* was still undergoing operational testing, the Navy began development of a ballistic missile of intermediate range. Brought from conception to initial operation in 5 years' time, the Polaris fleet ballistic missile (FBM) weapons system was mated with nuclear propulsion. That development produced a virtually invulnerable missile-firing submarine. Today, the missile-firing submarine constitutes one of the highest priority elements of the United States' deterrent capability; that is, a deterrent to nuclear conflict.

Each Polaris submarine could launch 16 two-stage ballistic missiles powered by solid-fuel rocket motors, containing a self-contained inertial guidance system. The Polaris provided a combined explosive power greater than the total of all the bombs dropped by all aircraft during World War II. Nuclear propulsion enabled these Polaris submarines to remain on patrol for extended periods, hidden beneath the surface of the sea, ready to launch their missiles.

On station, a Polaris submarine maintained complete radio silence, receiving radio messages while submerged, but not transmitting to prevent giving away its location. Each ship had two complete crews, the Blue and the Gold, of about 130 people each. The Polaris operated on a system that reflected a major change in the Navy's traditional ship-manning methods. The crews alternated on approximately 3-month-long deployments, providing maximum on-station time for the submarine. Its endurance was limited only by the limitations of its personnel.

Submarines were followed by the world's first nuclear-powered surface warships. They were the guided-missile cruiser USS *Bainbridge*, launched April 15, 1961; the guided-missile cruiser USS *Long Beach*, commissioned September 9, 1961; and the carrier USS *Enterprise*, commissioned November 25, 1961. On October 3, 1964, those three ships ended Operation Sea Orbit, a 64-day long, around-the-world, unreplenished cruise.

It was during this time that space exploration (fig. 5-15) began. The *Vanguard*, a 3 1/2-pound payload, was developed by the Naval Research Laboratory. On March 17, 1958, it was placed into orbit to test a system designed to launch earth satellites during the international geophysical year (IGY). Now the oldest man-made satellite in orbit, it is expected to remain aloft for 2,000 years.

Figure 5-15.—Seven original NASA astronauts.

Naval officers also participated in space exploration. On May 5, 1961, Commander Alan B. Shepard, Jr., made America's first suborbital flight. The 15-minute shot in *Freedom* 7 went 116.5 miles into space.

VIETNAM

Although the United States was at peace following the Korean Conflict, events were building that would plunge the country into another conflict. Since 1959, the French had been involved in fighting in a country most Americans had never heard of—Vietnam.

Student Notes:

Photograph courtesy of Mr. F4rancis Jeffery.
Figure 5-16.—The Wall .

Photgraph courtesy of Mr. Francis Jeffery.
Figure 5-18.—Vietnam memorial—soldier.

Americans were introduced to Vietnam in 1965. In that year, the United States entered the Vietnam Police Action. This police action, which caused conflict at home as well as on the battlefield, lasted until January1973. Figures 5-16 through 5-19 commemorate American actions in Vietnam.

The Navy's operations in support of South Vietnam's struggle against communist military aggression consisted mainly of gunfire support and carrier aircraft operations. These operations included coastal interdiction patrols against North Vietnamese ships moving troops and supplies to the south. They also included riverine operations by a swarm of various types of patrol craft in the maze of waterways in South Vietnam's delta area. (By early 1972 all boats and the responsibility for delta operations had been turned over to the South Vietnamese Navy.) Naval construction battalions (Seabees) built several military bases and constructed water and sanitary facilities for local communities. Often, as in World War II, they engaged in fighting as they worked. Navy medical personnel served in the field with Marine Corps and Seabee units, as they did in World War II and in the Korean Conflict. They often performed their duties under fire and often sacrificed themselves to protect their charges from further harm. As in previous wars, U.S. Navy service and amphibious forces transported over 90 percent of the personnel and supplies used in support of that conflict.

Photograph courtesy of Mr. Francis Jeffery.
Figure 5-17.—Vietman memorial.

Photograph courtesy of Mr. Francis Jeffery.
Figure 5-19.—Women in war—memorial.

Student Notes:

During the Vietnam era, five new attack carriers joined the fleet, including the world's first nuclear-powered carrier, USS *Enterprise* (CVN 65).

Vietnam was a different kind of war, a war in which the Navy's role was ever changing. The Navy used both new and old aircraft—OV-10 *Broncos*, propeller-driven *Skyraiders*, attack planes like A-4 *Skyhawks* and A-7 *Corsairs*, and fighter planes like F-8 *Crusaders*. It used various support aircraft for ASW, early warning, and advance communications links.

OTHER DEVELOPMENTS

Even during the Vietnam Police Action, the Navy was involved in exploration and development. Former Navy pilot Neil Armstrong became the first man to set foot on the moon on July 20, 1969. On November 14, 1969, the all-Navy *Apollo 12* crew lifted off from the Kennedy Space Center on the second lunar expedition.

The crew consisted of Commanders Charles Conrad and Richard Gordon and Lieutenant Commander Alan Bean. Another all-Navy crew (Captain Charles Conrad, Jr., and Commanders Joseph P. Kerwin and Paul J. Wietz) splashed down on the first Skylab mission on June 22, 1973. The crew set numerous records and accomplished virtually all of its objectives.

Space. The Navy stands tall in the first 10 years of manned space exploration. Records show that five of the six men to walk on the surface of the moon during that time had formerly been trained as naval aviators.

Research. In the 1960s, Navy scientific undersea research resulted in the USS *Alvin*. The USS *Alvin* was the Navy's first deep diving vehicle. It was successfully tested at 6,000-foot depths on July 20, 1965. The next month, 10 aquanauts, including astronaut Commander M. Scott Carpenter, entered the Sealab II capsule, 205 feet below the surface of the sea off the coast of La Jolla, California. Carpenter remained underwater for 30 days in a successful experiment of submerged living and working conditions. On January 25, 1969, the first nuclear-powered, deep-submergence research and ocean-engineering vehicle, NR-1, was launched. That five-man vessel can operate for weeks at a time at great depths.

Weapons. In early 1965 came the announcement of the proposal to develop a new missile for the fleet ballistic missile system—the Poseidon. The growth potential of the ballistic missile submarine launching system has enabled the Poseidon to fit into the same 16-missile tubes that carried the Polaris. Like the Polaris A-3, it is able to reach any spot on earth from its nuclear-powered hiding place. Its increased accuracy, greater payload, and improved ability to penetrate enemy defenses make the Poseidon more effective than the Polaris.

On July 19, 1974, construction of the new Trident undersea nuclear weapons system commenced. The Trident system consists of three principal elements: a nuclear-powered fleet ballistic missile submarine (SSBN), a strategic weapons system (the missile), and an integrated logistics support system. The first Trident submarine was the USS *Ohio* (SSBN-726), a nuclear powered fleet ballistic missile submarine. The USS *Ohio* was delivered to the Navy in 1981. Since then, the Navy has accepted delivery of 10 more Trident submarines.

THE PERSIAN GULF

As with other wars, conflicts, or areas of military aggression, U.S. naval forces operate in the hostile area of the Persian Gulf. U.S. naval forces have been present in this vital oil-rich region for many years.

The events leading to an increased number of U.S. naval units in the Persian Gulf (fig. 5-20) began in the mid 1980s. Iran and Iraq were at war. Iraq had begun attacking Iranian oil facilities and tankers; in response, Iran began attacks against ships flying flags of countries sympathetic to Iraq. U.S. Navy ships quickly began escort and protection operations for U.S.-flagged tankers.

BMRF0520

Figure 5-20.—Persian Gulf award.

Student Notes:

As the war between Iran and Iraq widened, so did the dangers to U.S. Navy ships operating in the Gulf. Iran started laying mines in the Gulf and began using small suicide boats to raid U.S. tankers and naval units. Iraq also possessed weapons that could cause tremendous damage and casualties. These weapons proved costly to the United States. In May 1987, an Iraqi aircraft mistakenly fired two missiles that struck USS *Stark* (FFG-31), killing 37 sailors and wounding many more. In April 1988, Iran's use of mines caused considerable damage to USS *Samuel B. Roberts* (FFG-58). Until that time, the U.S. Navy's presence was largely defensive. When forced to take offensive action, the United States acted quickly. U.S. Navy ships bombarded an Iranian oil platform being used as a command post and sank a mine-laying vessel carrying out operations.

DESERT SHIELD/DESERT STORM

On 2 August 1990, the president of Iraq Saddam Hussein, ordered the world's fourth largest army from Iraq to invade the country Kuwait. The United States deployed a major joint force which served as the foundation for a powerful 33-nation military coalition to stem Iraq's brutal aggression. Operation Desert Shield/Desert Storm was born. The United States Navy provided the sea control and maritime superiority that paved the way for the introduction of U.S. and allied air and ground forces. The United States offered strong leadership for the multinational naval force.

Desert Shield/Desert Storm brought together the largest force of Navy warships assembled in a single theater since World War II, adding a powerful punch to Navy forces already on scene the night of Iraq's invasion of Kuwait. Long-established maritime superiority facilitated the largest, fastest strategic sealift in history, with more than 240 ships carrying more than 18.3 billion pounds of equipment and supplies to sustain the forces of Desert Shield/Desert Storm.

Under the Navy's Total Force concept more than 21,000 naval reservists were called to active duty in support of Desert Shield/Desert Storm. Serving in specialties from medicine to mine warfare, reservists worked alongside their active duty counterparts in the Persian Gulf. Others filled critical vacancies on the home front.

Saddam Hussein's rejection of diplomatic efforts to solve the crisis led to the final decision to restore Kuwait's sovereignty by military force. The ensuing air war and the effects of the economic embargo decimated Iraq's military infrastructure, severed communication and supply lines, smashed weapons arsenals, and destroyed morale. Some of the first shots fired were from Navy ships in the Persian Gulf and Red Sea, as they launched salvos of Tomahawk cruise missiles against pre-programmed targets in Iraq.

After an impressive 38-day air campaign, the ground offensive began with allied forces sweeping through Iraqi defenses in blitzkrieg fashion. The allied push into Kuwait and southern Iraq was made easier by the amphibious forces on station in the Persian Gulf. The threat they posed forced tens of thousands of Iraqi troops to maintain positions along the Kuwaiti coastline to defend against attack from the sea. The Iraqi army was crushed after a mere 100 hours. Iraqi troops—tired, hungry and war-weary from 6 months of economic blockade and more than a month of relentless allied bombing—surrendered by the thousands. Less than 7 months after the Iraqi invasion, Kuwait was once again free.

It is likely that Navy ships will continue to represent and protect U.S. interests in the region for the foreseeable future.

REVIEW 6 QUESTIONS

Q1. List some of the Navy's roles during the Korean Conflict.

a.

b.

c.

d.

Student Notes:

Q2. List some of the Navy's missions during the Vietnam Police Action.

a.

b.

c.

Q3. What are other actions the Navy was involved with during the same timeframe as the Vietnam Police Action?

a.

b.

Q4. What service did the Navy provide during the Iraq – Iran War?

Q5. List the Navy's contributions during Operation Dessert Storm.

a.

b.

c.

SUMMARY

The United States Navy began more than 200 years ago with two ships, but today we are the finest naval force in history. The history of the Navy is a big story and an exciting one. We've only rippled the surface here, but maybe we've stimulated your curiosity enough that you will want to take a closer look at your Navy's past. If so, visit your ship or station library. You will find many fine books on naval history there.

From Flamborough Head to the Persian Gulf, the U.S. Navy has always been "on station" in time of trouble. The U.S. Navy's mission of preparedness to conduct prompt and sustained combat operations at sea means the U.S. Navy will be present at the first sign of conflict.

U. S. Navy ships continued to change with even greater momentum, ushering in another new era—that of nuclear propulsion, jet power, rockets, and guided missiles. New types of ships have emerged—ships such as guided-missile cruisers, tactical command ships, and helicopter flattops. The era of the 50s, 60s, 70s, 80s, and on into the 90s has seen the emergence of the nuclear Navy.

The heart of today's nuclear fleet is a highly complicated unit known as the nuclear reactor, which offers the following advantages:

- Almost unlimited steaming endurance at high speed. Nuclear ships have increased flexibility; an ability to obtain ammunition, aviation fuel, and other supplies from remote places in a minimum amount of time; and an attack ability in a much greater area.
- Reduced vulnerability. Nuclear ships need not remain exposed as long as nonnuclear vessels during replenishment. They can maneuver to avoid attack.
- Reduced dependence on logistic support. Nuclear ships require fewer mobile support forces.
- Greater attack effectiveness. Nuclear ships can remain in battle areas for a greater length of time and have a greater ability to exploit weather conditions to their advantage.
- Elimination of huge funnels. That provides more room for such items as a big, powerful radar.
- Power available upon command. Nuclear reactors eliminate the need to order "more boilers on the line" a half hour before full power is desired. Heat is produced in the nuclear reactor; in turn, steam and power is produced with little delay. Reduction from full power to one-third or stop is equally responsive.
- Reduced maintenance. The absence of corrosive stack gases cuts down on the wear and tear of the ships and a lot of at-sea and in-port repairs.

The Navy has been advancing in other areas of the surface fleet as well. An example is the new amphibious assault ships (LHAs). The LHAs are the largest and fastest amphibious ships in the Navy inventory and offer the greatest operational versatility in the history of amphibious warfare.

The size of the LHAs alone is impressive. The first of the LHAs, the USS *Tarawa*, is 820 feet long and 106 feet wide. The high point of its mast is 221 feet above the keel, and it has a full displacement of 39,300 tons. It can carry a large landing force with all its equipment and supplies, landing them either by helo or amphibious craft or both.

The primary advantage of these general-purpose assault ships is tactical integrity—getting a balanced force to the same point at the same time.

Spruance-class ships are the Navy's prime ASW destroyers. They are fitted with our most powerful sonar, helicopters, our best ASW weapons, and the Harpoon surface-to-surface missile system.

The most recent additions to the surface fleet are the Ticonderoga-class cruisers and the Arleigh Burke-class destroyers. Both are powered by gas turbines and are capable of high-speed transits. They are also outfitted with the Navy's new Aegis weapons system. That system has the capability to track and engage multiple targets, using a complex system of radars, missiles, guns, torpedoes, and self-defense systems. These capabilities make these cruisers and destroyers the most survivable units of today's surface fleet.

Our ability to quickly deploy large carrier battle groups and surface action groups quickly will assure our allies of our ability to exercise sea control. That ability, coupled with the U.S. submarine forces' strategic deterrence objective, will allow the United States and its allies the ability to deter further hostile action worldwide.

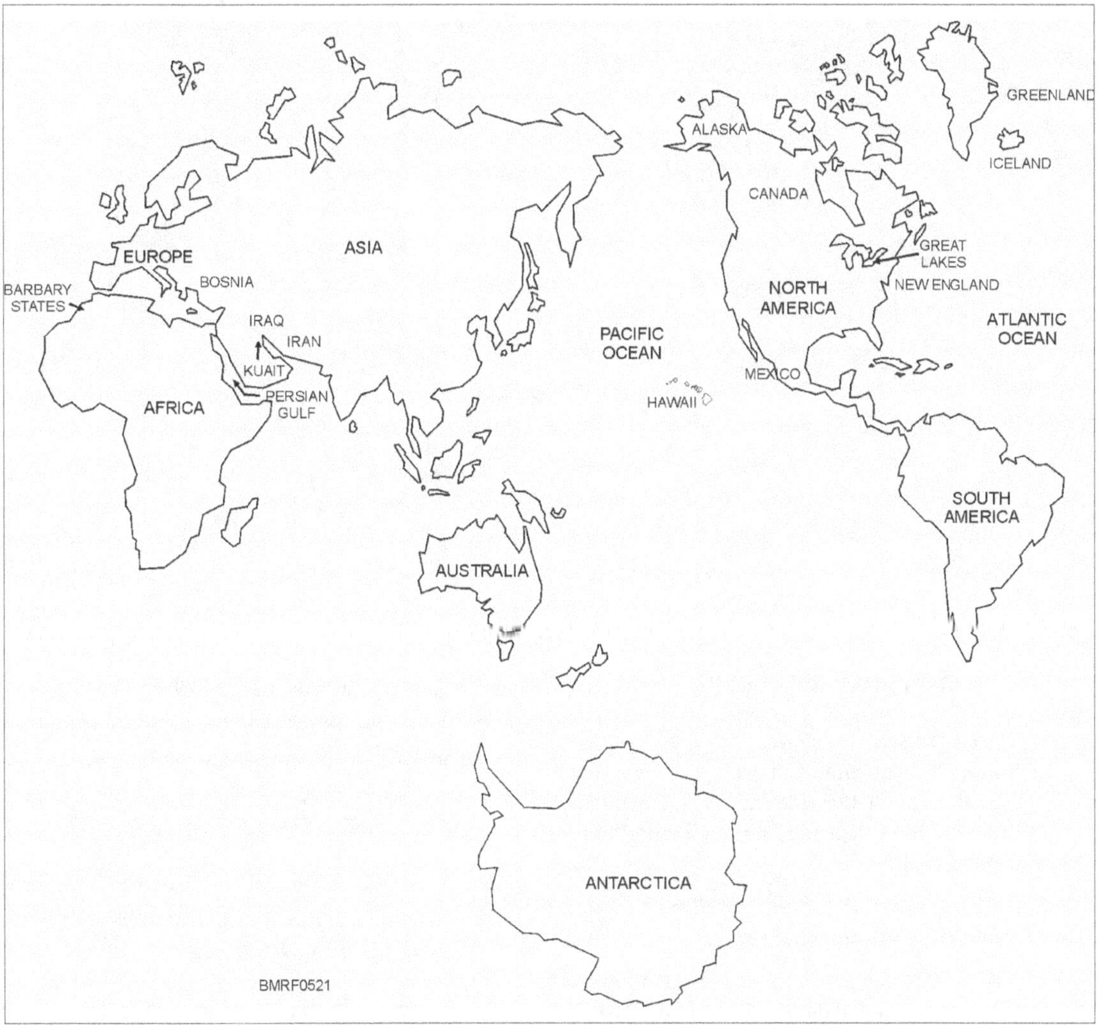

Student Notes:

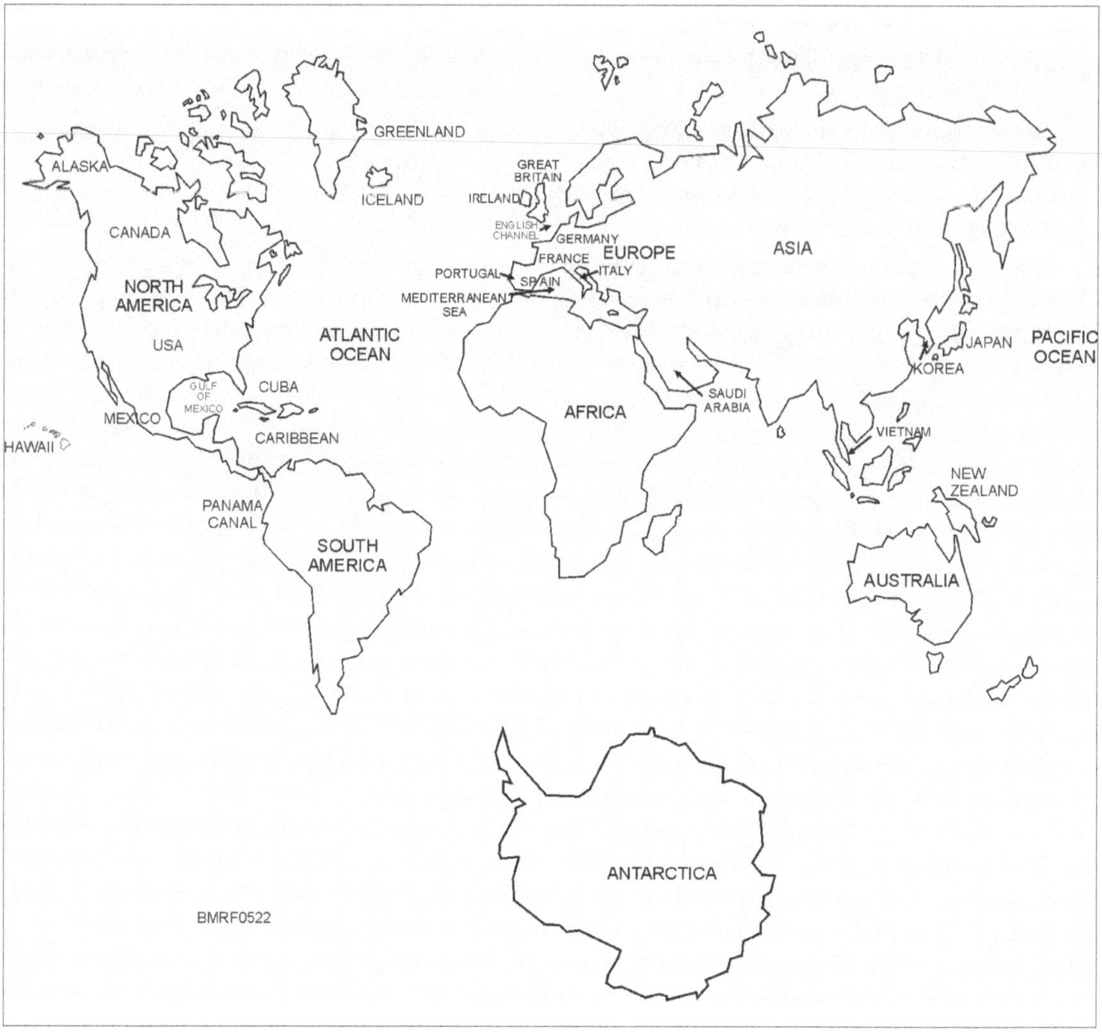

REVIEW 1 ANSWERS

A1. The United States Navy was formed because **the Second Continental Congress realized that the survival of the colonies as independent from England depended on the formation of naval forces.**

A2. During the late 18th century, battleships were classified as **ships-of-the-line**.

A3. The craft developed in 1775 was **a warfare submarine, named the *Turtle,* and invented by David Bushnell**.

A4. This raid was the **first amphibious operation carried out by the American Navy and Marines**.

A5. The first official recognition of the American Stars and Stripes flag by a foreign nation was given by (a) **France** (b) to the **USS *Ranger***.

REVIEW 2 ANSWERS

A1. After the Revolutionary War, **the U.S. Navy defended America's small merchant ship fleet from the Barbary pirates**.

A2. Between the Revolutionary War and the War of 1812, the U.S. Navy was involved with the—

a. **Quasi War with France** and

b. **Barbary States War**.

A3. Lord Nelson said that the **operation executed by LT Stephen Decatur and 84 seamen by slipping into the harbor of Tripoli and burning the captured frigate *Philadelphia*** was "one of the most bold and daring acts of the age."

A4. Two actions of the U.S. Navy during the War of 1812 were—

a. **The sea battle between the American frigate USS *Constitution* and the British frigate *Guerriere*** and

b. **The victory of Captain Oliver Hazard Perry over the British squadron on Lake Erie.**

A5. During the Mexican-American War, the Navy **blockaded the port cities on the Gulf** and the **"Mosquito Fleet" provided protective action during the first large-scale amphibious operation in U.S. military history.**

REVIEW 3 ANSWERS

A1. During the last part of the 19th century, naval developments included—

a. **Introduction of ironclad ships**

b. **Introduction of riverboats, rams, and gunboats**

c. **Development of submarines**

d. **Construction of steeled hull protected cruisers, signaling the end of the ironclads**

e. **Development of self-propelled torpedo and long-range torpedo boats**

f. **Development of the internal combustion engine for ships**

A2. After developing the ironclad, the Confederate Navy developed the **submarine.**

A3. During the Civil War **Battle of Mobile bay, Admiral Farragut** gave order, "Damn the torpedoes! Full speed ahead."

A4. Alfred T. Mahan influenced naval strategy through his books that stressed the idea that **without control of the seas, a nation couldn't expect victory. He was one of the first instructors at the Naval War College and shared his knowledge on sea power and the importance of understanding naval needs.**

A5. The Spanish-American war began when the ***Maine* was blown up and 250 Sailors were killed.**

A6. **Commodore George Dewey** was instrumental in quickly ending the Spanish-American War.

REVIEW 4 ANSWERS

A1. The development of **airplanes** occurred at this time.

A2. The U.S. Navy was deployed to **stop German U-boats from practicing unrestricted warfare and terrorizing the seas.**

A3. During this war, **destroyers were used as the main defense against German U-boats. They also served as an escort for troop ships and supply convoys for the allies.**

A4. During this war, the air forces **supported surface antisubmarine forces.**

A5. During this war, **women enlisted in the Navy as Yeoman (F), releasing enlisted men for active service at sea.**

REVIEW 5 ANSWERS

A1. The Battle of Coral Sea was fought by **aircraft, all of which were launched from carriers. This battle saved Australia from being invaded by the Japanese.**

A2. The major naval battles in the Pacific during World War II and their significance is as follows:

a. **Battle of Guadalcanal—The Solomon Islands came under allied control and the danger of Australia coming under Japanese attack was lessened**

Student Notes:

b. **Battle of the Philippine Sea—Heavy losses of ships, aircraft, and pilots paralyzed the Japanese Fleet**

c. **Battle of Leyte Gulf—Deciding blow to the Japanese Navy. Losing control of the Philippines meant that the Japanese homeland was cut off from its main source of supplies from the south.**

d. **Battle of Midway—The turning point of the war in the Pacific.**

e. **The Battles of Okinawa and Iwo Jima—Defeat of the Japanese in these battles signaled an approach to the end of the war.**

A3. During World War II, the U.S. Navy **protected convoys bound for Europe from German U-boat attack.**

A4. Some of the types of ships that changed the shape of the Navy changed during World War II include **landing ships, frigates, attack cargo ships, transport ships, barracks ships, net tenders, repair ships, radar pickets minelayers, and mine sweepers.**

A5. On 30 Jul 1943, **Congress authorized the establishment of the Women's Reserve to fill acute shortages of personnel during World War II.**

A6. The Women's Armed Services Integration Act **abolished the Women's Reserve and gave women full partnership in the Navy.**

REVIEW 6 ANSWERS

A1. Some of the Navy's roles during the Korean Conflict included—

a. **Providing close air support to knock out bridges and block enemy routes with the use of jets from carriers**

b. **Navy helicopters spotted enemy artillery**

c. **Navy ships supported the amphibious landing at Inchon through massive shore bombardment before ground forces landed**

d. **The Navy successfully used its battleships to intervene in ground operations far ashore.**

A2. Some of the Navy's missions during the Vietnam Police Action included—

a. **Surface ship-based gunfire support**

b. **Carrier-based aircraft operations**

c. **Coastal interdiction patrols against the enemy**

A3. Other actions the Navy was involved with during the same timeframe as the Vietnam Police Action include—

a. **The manned space exploration program**

b. **Manned undersea exploration, using deep submergence vehicles and underwater laboratories**

A4. During the Iraq – Iran War, **the Navy escorted and protected oil tankers in transit to and from the Persian Gulf against Iranian attacks.**

A5. The Navy's contributions during Operation Dessert Storm included—

a. **Providing sea control**

b. **Naval gunfire support for sea to ground forces**

c. **Surface and subsurface missile attacks on selected targets in Iraq**

CHAPTER 6

NAVAL ORGANIZATION

Organization is the element of administration which entails the orderly arrangement of materials and personnel by functions in order to attain the objective of the unit. Organization establishes the working relationship among unit personnel; establishes the flow of work; promotes teamwork; and identifies the responsibility, authority and accountability of individuals within the unit.

—*Standard Organization and Regulations of the U.S. Navy,*
OPNAVINST 3120.32C

The primary mission of the Navy is to support U.S. national interests. To do that, the Navy must be prepared to conduct prompt and sustained combat operations at sea. Each Navy unit must be prepared to engage in battle and support other units and forces in battle. Meeting the objectives of this mission requires organization. This chapter introduces you to naval organization, including the Department of Defense, the Department of the Navy, a typical unit organization, and the chain of command.

DEPARTMENT OF DEFENSE

Learning Objective: When you finish this chapter, you will be able to—

- Recognize the organization of the Department of Defense (DoD) to include the Department of the Navy (DoN) and the operating forces.

Originally, two executive departments managed the armed forces—the Department of War and the Department of the Navy. In 1947, the United States created the Department of Defense (DoD) as part of its security program by combining these two departments. The DoD consists of various agencies and three military departments—the Army, Navy, and Air Force. The DoD includes the Joint Chiefs of Staff, which consists of a chairman, the military heads of each department, and the Commandant of the Marine Corps. The DoD maintains and employs the armed forces to carry out the following missions:

1. To support and defend the Constitution of the United States against all enemies, foreign and domestic
2. To ensure, by timely and effective military action, the security of the United States, its possessions, and areas vital to its interests
3. To uphold and advance the national policies and interests of the United States
4. To safeguard the internal security of the United States

Figure 6-1 shows how the armed forces fit into the organization of the DoD to carry out these missions. The DoD is headed by the Secretary of Defense (SECDEF). The military departments that come under the DoD are the Department of the Army, the Department of the Air Force, and the Department of the Navy. Each department has a secretary as the head of department. By law, these secretaries (Army, Navy, Air Force, and SECDEF) are civilian appointees of the President. The Secretary of the Navy (SECNAV) heads the department of which you are a part.

DEPARTMENT OF THE NAVY

The mission of the Navy is to maintain, train, and equip combat-ready naval forces capable of winning wars, deterring aggression, and maintaining freedom of the seas. The Department of the Navy has two main objectives:

1. The first objective is to organize, train, equip, prepare, and maintain the readiness of Navy and Marine Corps forces to perform military missions. These forces carry out military missions as directed by the President through the Secretary of Defense, to the SECNAV, to the Navy Department.

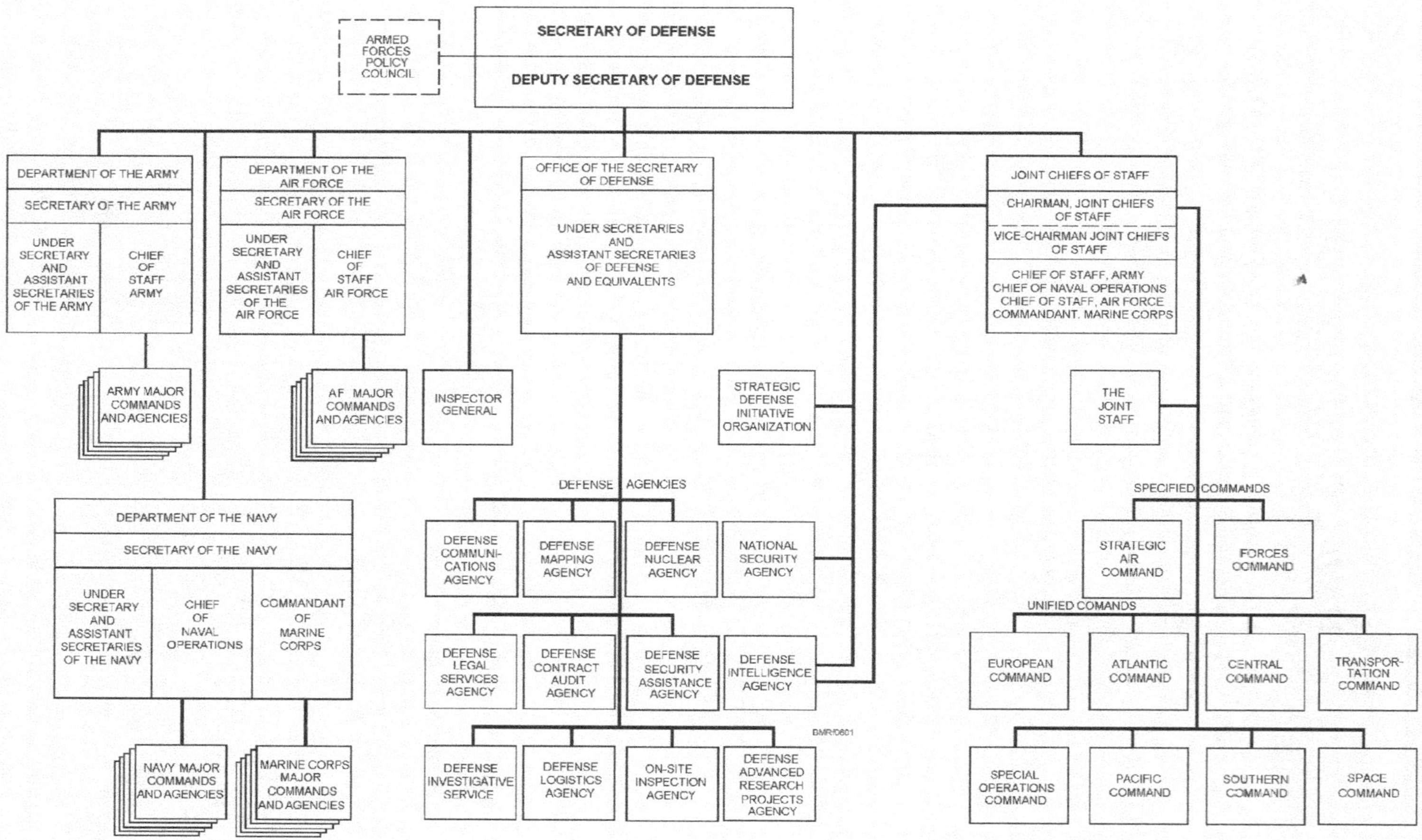

Figure 6-1.—Organizational chart of the Department of Defense.

2. The second objective is to support the Navy and Marine Corps forces as well as the forces of other military departments. The Department of the Navy supports these forces as directed by the Secretary of Defense.

Figure 6-2 shows the basic organization of the DoN. The SECNAV is responsible for, and has the authority under Title 10 of the United States Code, to conduct all the affairs of the Department of the Navy. The SECNAV has the following responsibilities:

- Conducts recruiting, organizing, supplying, equipping, training, and mobilizing, and demobilizing
- Oversees the construction, outfitting, and repair of naval ships, equipment and facilities
- Formulates and implements policies and programs that are consistent with the national security policies and objectives established by the President and the Secretary of Defense

The DoN consists of two uniformed Services—the United States Navy and the United States Marine Corps. You can find out more about these organizations by using the Internet. The Internet address is www.navy.mil. There, you can connect to a Navy organization's homepage.

Figure 6-3 shows you an overview of the organization of the DoN. The U.S. Navy was founded on 13 October 1775, and the Department of the Navy was established on 30 April 1798. The Department of the Navy has three principal components—

1. The Navy Department, consisting of executive offices mostly in Washington, D.C.
2. The operating forces, including the Marine Corps
3. The Shore Establishment

In this chapter, you will learn about the operating forces and shore commands. The reserve components are part of DoN. In times of war, the U.S. Coast Guard is also a part of the DoN; during peacetime, the U.S. Coast Guard is a part of the Department of Transportation.

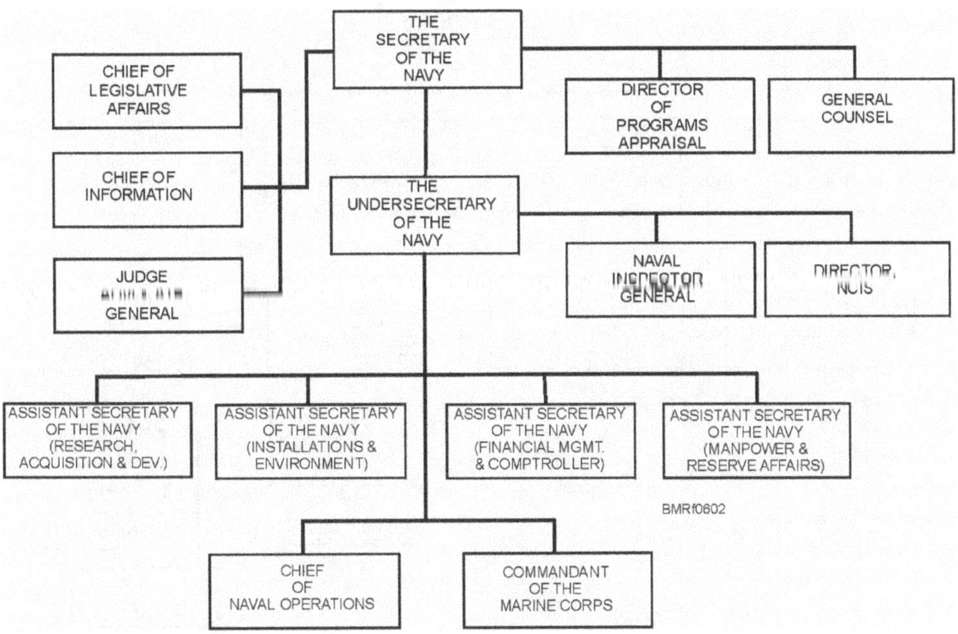

Figure 6-2.—Organizational chart of the Department of the Navy (DoN).

Student Notes:

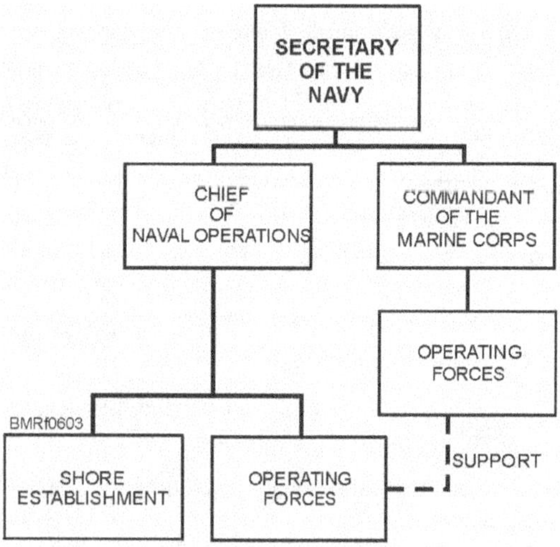

The dashed line marked "Support" indicates the cooperative support of the Navy-Marine Corps team. Each of the operating forces supports the other.

Figure 6-3.—Overview of the DoN.

THE OPERATING FORCES

The operating forces (fig. 6-4) consist primarily of combat and service forces, including several fleets and the Fleet Marine Forces. They include the Coast Guard (when operating as a part of the Navy) and other forces and activities under the command of the Chief of Naval Operations (CNO). You can get more information on the operating forces by going to the web page mentioned above. You'll also learn more about the operating forces when you complete the *Military Requirements for Petty Officer Third Class*, NAVEDTRA 12024.

THE SHORE ESTABLISHMENT

The shore establishment (fig. 6-5) provides support to the operating forces (known as *the fleet*). The support is in the form of facilities for the repair of machinery and electronics; communications centers; training areas and simulators; ship and aircraft repair; intelligence and meteorological support; storage areas for repair parts,

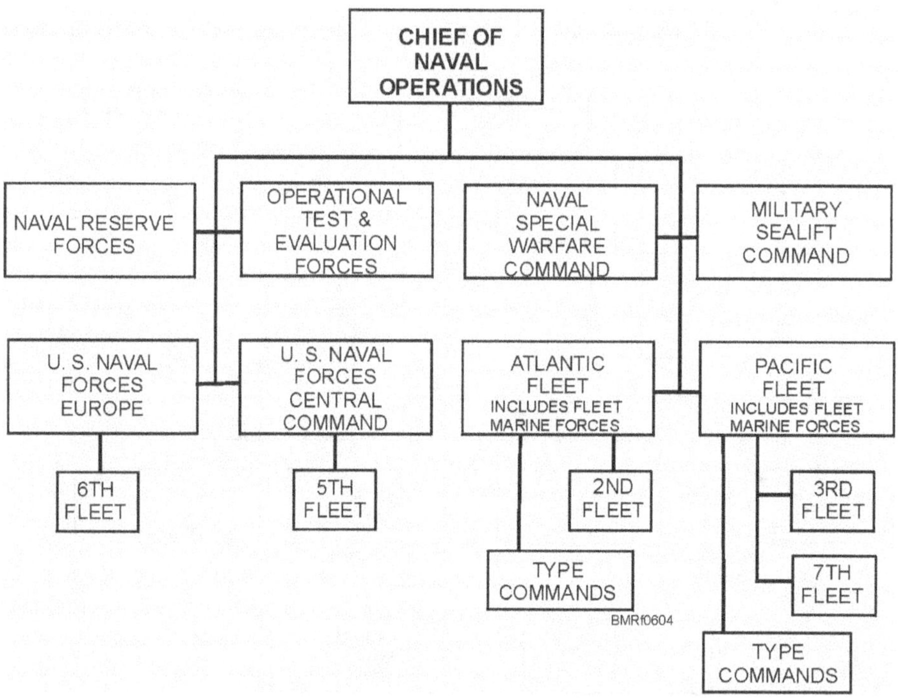

Figure 6-4.—Operating forces.

Student Notes:

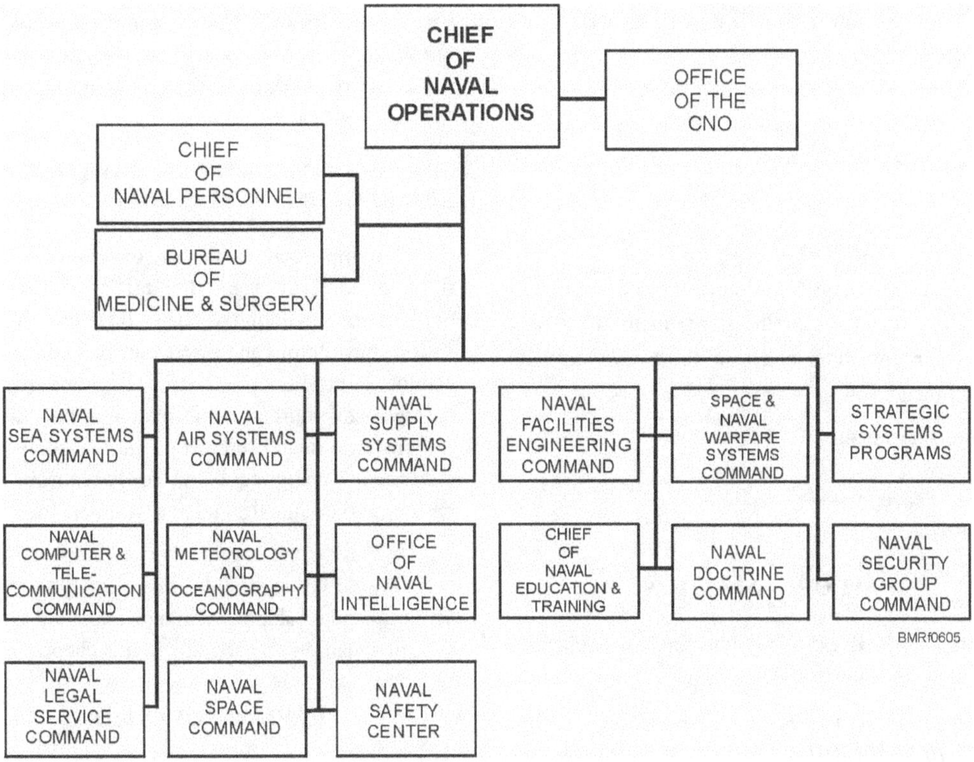

Figure 6-5.—Shore establishment.

fuel, and munitions; medical and dental facilities; and air bases. You can learn more about the commands shown here by going to the Internet web page mentioned above. Also, you'll learn more about the Shore Establishment when you complete the *Military Requirements for Petty Officer Third Class*, NAVEDTRA 12024.

REVIEW 1 QUESTIONS

Q1. List the three military departments of the Department of Defense (DoD).

a.

b.

c.

Q2. Describe the four missions of the DoD.

a.

b.

c.

d.

Q3. What are the two main objectives of the Navy?

a.

b.

Student Notes:

Q4. List the three basic components of the Department of the Navy.

a.

b.

c.

Q5. The U.S. Coast Guard operates under different departments. List the department the Coast Guard operates under in the two conditions shown below.

a. Wartime—

b. Peacetime—

UNIT ORGANIZATION

Learning Objectives: When you finish this chapter, you will be able to—

- Recognize the purpose and scope of the *Shipboard Organization and Regulations Manual*.
- Recall the application of the *Standard Organization and Regulations of the U.S. Navy* to unit organization to include ships'/commands' organization and regulations manual.

The purpose of a unit's organization is to help accomplish the mission of that unit. Each unit has its own mission in support of the overall mission of the Navy; however, each unit has the same basic organization. The *Standard Organizational and Regulations of the U.S. Navy*, OPNAVINST 3120.32, is used as a guide for unit organization.

STANDARD ORGANIZATION AND REGULATIONS OF THE U.S. NAVY

Standard Organization and Regulations of the U.S. Navy describes the many aspects of the standard unit organization. Each unit in the Navy has a ship's/command's organization and regulations manual based on the *Standard Organization and Regulations of the U.S. Navy*. Aboard ship, this manual is usually referred to as the *Shipboard Organization and Regulations Manual*.

SHIP'S/COMMAND'S ORGANIZATION AND REGULATIONS MANUAL

The ship's/command's organization and regulations manual governs the unit's administrative organization (including watches). It governs the coordination of evolutions and emergency procedures and the conduct of personnel in the unit. Its purpose is to provide a ready source of information about the duties, responsibilities, and authority of unit personnel. Ships/commands usually require all newly reporting personnel to read the manual and sign a statement to that effect.

Discussing the organization of every unit in the Navy would be impossible. Therefore, you will learn about a standard shipboard organization and a standard aircraft squadron organization in this chapter.

REVIEW 2 QUESTIONS

Q1. Aboard ship, you can find the ship's organization and regulations in what publication?

Q2. List some aspects of a ship's organization that are covered by the *Standard Organization and Regulations Manual*.

a.

b.

c.

SHIPBOARD ORGANIZATION

Learning Objectives: When you finish this chapter, you will be able to—

- Recall shipboard organization to include battle organization and administrative organization.

Student Notes:

- Identify the duties and responsibilities of the commanding officer, executive officer, department head, and division officer.

The officers and enlisted personnel make up a ship's wartime organization. They keep the ship in a state of readiness to fight a war. During peacetime operations, the ship's organization can be expanded if a wartime operation becomes necessary. The two elements of the ship's organization are the **battle organization** and the **administrative organization**.

THE BATTLE ORGANIZATION

The battle organization contains a list of the numbers and specialties of the personnel a unit will need to fulfill the wartime missions. The unit's battle organization depends on its armament, equipment, and personnel. As a part of the battle organization, you should know your assignments as posted on the Watch, Quarter, and Station Bill.

THE ADMINISTRATIVE ORGANIZATION

The administrative organization makes sure that the ship can fight or carry out its mission. Training, maintenance, and routine operations are covered by the administrative organization. The commanding officer (CO) is the head of the organization. He/she is assisted by the executive officer (XO) and other officers.

Look at figure 6-6. Each ship is organized into at least five departments—navigation, engineering, operations, supply, and a fifth department. For most ships, the fifth department is the weapons/deck department. Some ships have a separate deck department in addition to a weapons department, and some have a deck department instead of a weapons/deck department. Specially designated ships have a combat systems department instead of a weapons or weapons/deck department. Additional departments may be assigned according to ship type. Some of these departments are air, medical, dental, and repair.

Each department is under a department head. Departments are usually divided into divisions under a division officer. Each division is subdivided into sections, usually under senior petty officers. The following paragraphs describe the responsibilities of the five standard departments.

Navigation Department

The navigation department is responsible for the safe navigation and piloting of the ship. It is responsible for the operation, care, and maintenance of navigation equipment, charts, publications, and records.

Operations Department

The operations department has several divisions to carry out tasks such as collecting and evaluating

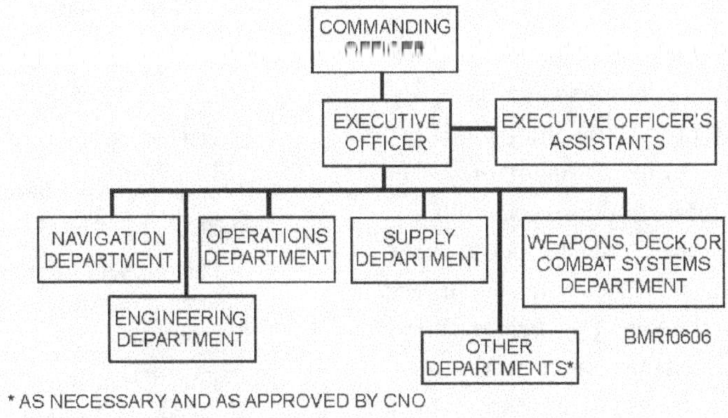

Figure 6-6.—Typical ship organizational chart.

Student Notes:

combat and operational information and conducting electronic warfare. Other tasks involve gathering and analyzing intelligence information, repairing electronic equipment, controlling aircraft, and forecasting weather. The operations department is usually in charge of all the radar, sonar, and communications equipment on the ship. The combat information center (CIC) is part of the operations department.

Supply Department

The supply department has many responsibilities. Some of these are—

- Operating the general mess, including preparing and serving food.
- Operating the ship's store, which provides personal articles for the ship's crew.
- Managing the clothing and small stores issue room, where crew members may buy uniform items.
- Maintaining the pay records of the crew (done by the disbursing office).
- Ordering and receiving general stores, supplies, spare parts, and equipment for the ship. In fact, just about everything that comes aboard the ship, other than people, is ordered by the supply department.

Engineering Department

The engineering department, under the engineer officer, is responsible for the operation, care, and maintenance of all propulsion and auxiliary machinery. It is responsible for the control of damage resulting from fire, explosion, collision, and so forth. The engineering department provides power, light, ventilation, heat, refrigeration, compressed air, and freshwater throughout the ship. The engineer officer may have several assistants, such as the main propulsion assistant, the damage control assistant, and the electrical officer.

Weapons/Deck/Combat Systems Department

The fifth department of ship's administrative organization varies. Some of the departments are listed below.

1. Surface combatants (ships) using ordnance (gun batteries, torpedoes, missiles, and so forth) have a **weapons department**, headed by a weapons officer.
2. Surface combatants (ships) with complex combat systems and some classes of submarines have **combat systems departments**, headed by combat systems officers.
3. Ships with offensive capabilities unrelated to ordnance have a **deck department**, headed by the first lieutenant.
4. Aircraft carriers and some other ships have a **weapons** or **combat systems department** in addition to a deck department.

In ships that have a weapons department or combat systems department, the **weapons** or **combat systems officer** is responsible for—

- The operation, care, and maintenance of the ship's armament and the weapons fire-control equipment.
- The care, handling, stowage, accountability, and issue of ammunition and pyrotechnics.
- The maintenance of magazines and the external security of the ship.
- If the ship doesn't have an air department the weapons department is responsible for

  —The launch and recovery of assigned aircraft

- If the ship doesn't have a deck department, the weapons department is responsible for

  —The preservation and cleanliness of the external areas of the ship not assigned to other departments.

  —The operation of the paint, sail, and boatswains' lockers and inspection and maintenance of survival equipment.

Student Notes:

—All decking seamanship operations and the care and use of deck equipment.

- If the ship has a deck department but no weapons or combat systems department

  —The first lieutenant (head of the deck department) is responsible for deck functions.

- If the ship has a combat systems department but no deck department

  —The operations department is responsible for deck functions.

COMMANDING OFFICER

The commanding officer (CO) has many and varied duties. The CO has so many duties that one entire chapter in *Navy Regulations*, consisting of nearly 70 articles, applies to commanding officers. In general, the CO is responsible for the safety, well-being, and efficiency of the command.

The commanding officer's responsibilities include the safe navigation of the ship and the condition and appearance of the material and personnel. The CO must also ensure the proper stationing of trained lookouts and the preparation of the ship for battle. The CO may delegate authority in these matters, but such delegation does not relieve the CO of responsibility. The officer of the deck (OOD), for example, has authority to run the ship; but if a collision occurs, the CO is still responsible.

The commanding officer must exert every effort to maintain the command in a state of maximum readiness for war. The commanding officer issues the necessary directions to the executive officer (XO). With the assistance of the various department heads, the XO then prepares and conducts exercises and drills needed to prepare the ship for battle.

During combat, the commanding officer directs the members of the crew in fighting to the best of their ability until action is complete. The CO's battle station is where the CO can best direct the fighting. If the ship should sink, both custom and regulations require the commanding officer to assure the completion of abandon ship procedures. All personnel should be off the ship before the commanding officer leaves.

The CO's power is authoritative and complete. With ultimate responsibility for the ship and everything pertaining to it, the commanding officer must have authority equal to the responsibility. To ensure efficiency, responsibility, and discipline, the commanding officer must have the power to enforce prompt obedience to orders. According to the *Uniform Code of Military Justice (UCMJ)*, the commanding officer has the power to impose limited punishment. This power is a part of a CO's command responsibility and may not be delegated.

Since the ship has only one CO but many crew members, a senior enlisted member gives advice on enlisted policies and informs the CO about the health, welfare, and general well-being of the crew. The senior enlisted member acts as a liaison between the officer and enlisted community. The senior enlisted member assigned to assist the CO is a master chief, a senior chief, or a chief petty officer (depending on the senior rate within the command). This person receives assignment as the command master chief (CM/C), command senior chief (CS/C), or command chief (CCh). This senior enlisted member transmits ideas and recommendations directly to the commanding officer.

If the CO is absent, disabled, relieved from duty, or detached without relief, another officer must assume the CO's responsibilities. This person is the next senior line officer that is eligible for command at sea, and who is attached to and aboard the ship. In most cases, this person is the executive officer.

EXECUTIVE OFFICER

The executive officer (XO) is the aide or "executive" to the commanding officer. The XO is usually the next ranking line officer aboard ship. As such, the XO is the direct representative of the commanding officer in maintaining the general efficiency of the ship. Some of the XO's responsibilities include the following:

- The command's assigned personnel. With the help of department heads, the XO arranges and coordinates all ship's work, drills, exercises, and policing and inspecting the ship.

Student Notes:

- Investigate matters affecting the discipline and conduct of the crew and makes recommendations concerning these matters to the commanding officer.
- Approve or disapprove liberty lists and leave requests.
- Inspect the ship and receive readiness reports from the various department heads when the ship is cleared for action; then report to the CO when the ship is ready for action.

If the captain is disabled during battle, the XO normally becomes the acting commanding officer. For this reason, the location of the XO's battle station (determined by the captain) is some distance from the captain's. This prevents disablement of both officers at the same time.

After a battle, the executive officer makes a detailed report to the commanding officer.

If the XO cannot fulfill the duties of the commanding officer, normally, the next senior line officer assigned to the ship assumes the duties of the commanding officer.

Depending on the size of the ship, the executive officer may have one or more assistants. Some of these assistants and their responsibilities are as follows:

Personnel officer. The personnel officer assigns personnel to the various departments, berthing arrangements, and to the task of maintaining enlisted service records.

Training officer. The training officer secures school quotas, schedules orientation courses for newly reporting personnel, and helps prepare long- and short-range training schedules.

Educational services officer (ESO). The educational services officer (ESO) receipts for, maintains, and distributes educational courses and training aids.

Drug and alcohol program advisor (DAPA). The drug and alcohol program advisor (DAPA) advises the CO and XO on all matters concerning the Navy's Drug and Alcohol Abuse Program. The DAPA provides onboard education, prevention, screening, command counseling, aftercare, probationary supervision, motivational training, and referral services.

Command master chief (CM/C). While serving as one of the XO's assistants, the CM/C has direct access to the CO and is the voice of all enlisted personnel.

In addition to these assistants, the executive officer may also have a legal officer, combat cargo officer, safety officer, and others as required. The master-at-arms force also works directly under the executive officer.

DEPARTMENT HEAD

As the representative of the commanding officer, the department head is responsible for and reports to the CO about all matters that affect the department. That includes administrative matters, the operational readiness of the department, and the general condition of equipment.

DIVISION OFFICER

The division is the basic unit of the shipboard organization. The CO assigns division officers to command the divisions of the ship's organization. Division officers are responsible to and, in general, act as assistants to department heads.

The number of divisions in a department varies among ships, with each division having only a few assigned personnel to as many as 200 personnel. The division officer is a major link in your chain of command, particularly in a small ship. At the working level, the division officer carries out command policies and personally sees that division tasks are completed in a timely manner. Some of the division officer's duties include—

- Making frequent inspections of division personnel, spaces, equipment, and supplies
- Maintaining copies of all division orders and bills and displaying them in a conspicuous place
- Training division personnel and preparing them for battle

Student Notes:

REVIEW 3 QUESTIONS

Q1. List the two elements of a ship's organization.

a.

b.

Q2. List the information contained in elements of a ship's organization.

a.

b.

Q3. Briefly describe the responsibility of the five departments listed below.

a. Navigation—

b. Operations—

c. Engineering—

d. Supply—

e. Weapons/Deck/Combat Systems—

Q4. What regulation has an entire chapter covering commanding officer duties?

Q5. Aboard ship, what officer is responsible for the safe navigation of the ship, the condition of the ship, and the appearance of material and personnel?

Q6. A CO must have authority equal to responsibility, including the power to impose limited punishment. Can the CO delegate this particular power?

Q7. What enlisted person transmits ideas and recommendations directly to the CO?

Q8. Describe four duties of the XO.

a.

b.

c.

d.

Q9. List the XO's assistants.

a.

b.

c.

d.

e.

Q10. What are the responsibilities of the department head?

Student Notes:

Q11. What is the basic unit of a shipboard organization?

Q12. List three division officer duties.

a.

b.

c.

AIRCRAFT SQUADRON ORGANIZATION

Learning Objective: When you finish this chapter, you will be able to—

- Recall aircraft squadron organization to include squadron departments and branch officer.

The organization of an aircraft squadron differs in some ways from that of a ship. *Standard Organization and Regulations of the U.S. Navy* specifies the basic organization required of an aircraft squadron. Figure 6-7 shows a standard organizational chart of an aircraft squadron.

Look at figure 6-7 again. As you can see, an aircraft squadron has different departments than a ship. This is one of the differences between the organization of an aircraft squadron and a ship. The departments have different names and responsibilities. However, the responsibilities of the commanding officer, executive officer, department heads, and division officers are the same in the organization of an aircraft squadron and a ship.

AIRCRAFT SQUADRON DEPARTMENTS

All aircraft squadrons have an administrative department and a safety department. Most squadrons also have an operations department and a maintenance department. Some squadrons have one or more departments in addition to the four already mentioned. Based on its mission, the squadron may have a training, a photographic, or an intelligence department.

Administrative Department

The administrative (ADMIN) department is responsible for all the administrative duties within the squadron. This department takes care of official correspondence, personnel records, and directives. Most of the jobs done by the XO's assistants in a shipboard organization are done by ADMIN in an aircraft squadron. The first lieutenant and command career counselor work as members of the ADMIN

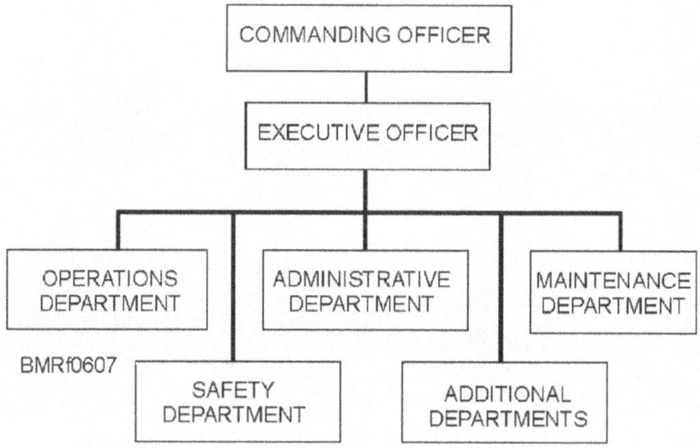

Figure 6-7.—Type aircraft squadron organization chart.

Student Notes:

department. Other parts of the administrative department include the following:

- Personnel office
- Educational services office
- Public affairs office
- Legal office

Safety Department

The safety department is responsible for all squadron safety program matters. This department is usually divided into the following:

- Ground safety.
- Aviation safety.
- Naval Air Training and Operating Procedures Standardization (NATOPS) divisions. (The NATOPS division makes sure standardized procedures are followed in the operation of the squadron's aircraft.)

Operations Department

The operations department is responsible for the operational readiness and tactical efficiency of the squadron. The operations department usually consists of the logs and records, schedules, training, communications, and navigation divisions.

Maintenance Department

The maintenance department is responsible for the overall maintenance of the squadron's aircraft. The maintenance department is usually divided into the following divisions:

- Maintenance/material control
- Quality assurance
- Maintenance administration
- Aircraft, avionics/armament, and airframes divisions

BRANCH OFFICER

A division on a ship is divided into watches or sections or both. In an aircraft squadron, the divisions are divided into branches. Each branch is headed by a branch officer. In aircraft squadrons, the branch officer is the officer with whom you will have the most direct contact.

The branch officer is directly responsible to the division officer. The branch officer has the same responsibilities for the branch that the division officer has for the division.

- Making frequent inspections of branch personnel, spaces, equipment, and supplies.
- Making sure branch tasks are completed in a timely manner.

REVIEW 4 QUESTIONS

Q1. List the four departments usually found in an aircraft squadron.

a.

b.

c.

d.

Q2. The safety department is responsible for the squadron safety program. What are the main divisions in this department?

a.

b.

c.

Q3. What is the responsibility of the operations department?

Student Notes:

Q4. List the four divisions of the maintenance department.

a.

b.

c.

d.

Q5. The branch officer is responsible to what officer?

CHAIN OF COMMAND

Learning Objectives: When you finish this chapter, you will be able to—

- Identify the purpose of the chain of command to include responsibilities, direction, communications, and solving work-related problems.
- Recognize a typical chain of command.

The chain of command is the relationship of juniors and seniors within an organization. The organizational charts you have seen in this chapter represent chains of command.

An effective chain of command is essential for the Navy to carry out its assigned mission. Good leadership is also essential for the Navy's mission. Good leadership supports an effective chain of command and vice versa; neither works well without the other.

The chain of command serves several purposes in the accomplishment of the Navy's mission. It defines responsibilities and identifies accountability. Properly used, it provides direction and smooth communications and ensures efficiency.

RESPONSIBILITY

Responsibility requires that an individual be accountable for the performance of assigned tasks within an organization. By defining responsibilities, the chain of command lets its personnel know what their responsibilities are and what they are expected to do.

The Navy expects its personnel to set good examples for their shipmates by doing their jobs quickly, correctly, and neatly. The Navy expects its members to instill a sense of pride in others to improve the efficiency of the command.

ACCOUNTABILITY

Every person in a chain of command is accountable to someone for professional performance and personal actions. Accountability is the ability of personnel to report, explain, or justify every action taken. They do this through two types of accountability—job accountability and military accountability.

1. *Job accountability* means you must answer to seniors in the chain of command for the way in which you carry out an assigned task.
2. *Military accountability* means you must answer to senior military personnel for your personal behavior and military appearance.

DIRECTION

The chain of command provides direction in the assignment of duties. All members of the chain of command know their specific duties. Seniors assign these duties, and juniors should carry them out to the best of their ability.

COMMUNICATION

The chain of command provides for smooth, rapid, and effective communication. Each person in the chain of command needs to clearly understand his/her status within that chain. Seniors should pass information down the chain of command about matters that may affect juniors. Juniors should pass information up the chain of command about problems that exist. In this way, communication flows in both directions.

Student Notes:

WORK-RELATED PROBLEMS

Work-related problems are situations that affect a person's job performance. A work-related problem might be a situation in which a person feels mistreated by a senior. It could also be a situation in which a person needs leave or liberty because of an illness in the family.

The chain of command is responsible to each Navy member for solving work-related problems. When a person's immediate senior is unable to resolve a problem, the next senior in the chain of command tries to solve the problem. If the senior at that level of command is unable to resolve the problem, it then goes to the next level in the chain of command. The problem continues to be referred to each level in the chain of command until it is resolved. **You must always use the chain of command when seeking solutions to work-related problems.**

TYPICAL CHAIN OF COMMAND

Figure 6-8 shows a typical shipboard, straight-line chain of command from the nonrated level to the commanding officer. An aircraft squadron's chain of command includes a branch officer.

For watch-standing assignments, the chain of command includes a section leader (not shown). The section leader may be from your division but often is a petty officer from another division. The section leader makes watch assignments for all personnel assigned to the section. Inform the section leader of situations, such as leave or special liberty, that affect your availability for watch assignments.

In most cases, the chain of command shown in figure 6-8 is complete. However, the chain of command does not stop with the commanding officer. **Remember, all people in the military are responsible to their seniors**!

The chain of command extends from nonrated personnel all the way to the President of the United States. Figure 6-8 shows the shipboard chain of command from the nonrated person to the commanding officer. Figure 6-9 shows a typical chain of command from the President to the commanding officer of a ship. To learn your chain of command, ask someone in the administrative (Admin) office to show you the command's organizational chart.

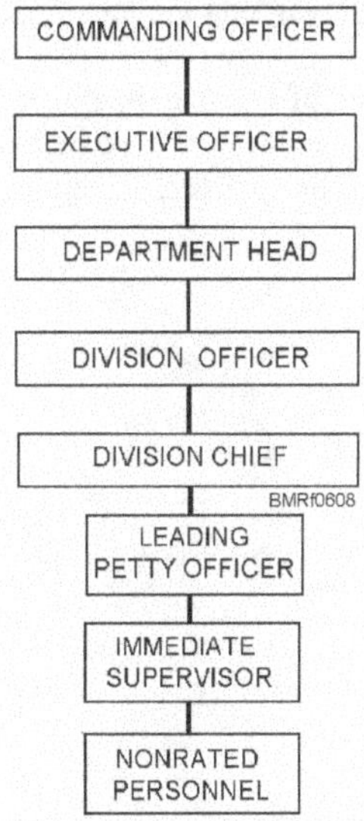

Figure 6-8.—Typical shipboard chain of command.

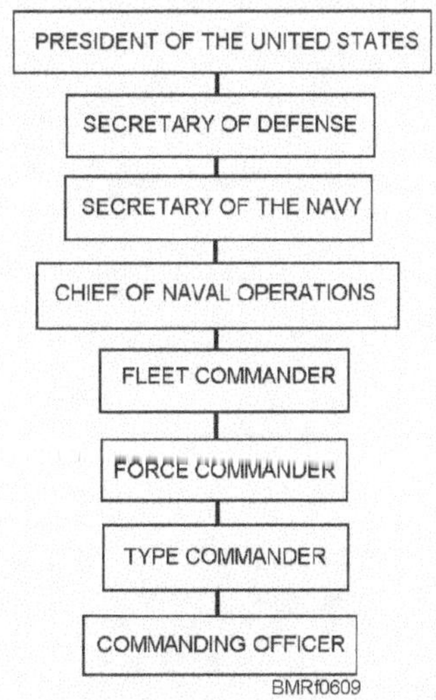

Figure 6-9.—Typical chain of command from the CO of a ship to the President.

Student Notes:

REVIEW 5 QUESTIONS

Q1. The chain of command defines the relationship of juniors and seniors in an organization. List the five areas that affect the chain of command.

a.

b.

c.

d.

e.

Q2. In your organization, you can find out about the chain of command by asking someone in what office?

SUMMARY

Where do I go? What do I do? Who is in charge? With the Navy's organizational structure, all personnel, from the CNO to the newest crew member, know what their job is, where they work, and who their supervisor is. Think about being at sea; its midnight and the general alarm sounds. Are the gun mounts manned by personnel trained to handle them or just by the people that showed up first? What about the repair lockers, the engine rooms, or the bridge?

Without an organization that ensures properly trained personnel manning each billet, our ships could not be in a continual high state of readiness.

What about a problem in the work space? Who do we tell about it? Do we tell the department head or the captain?

Our chain of command works in both directions, up and down. The upper level keeps us informed of the types of operations being conducted and what types of hazards we face. The people in the lower levels must keep the upper levels informed of all difficulties experienced in the performance of assigned duties. Every level in the chain of command is an integral part of a team. Members at each level must do their part to make sure their command functions effectively.

Flight deck operations are a good example of the effects of proper organization. Ships could never carry out these operations without superior organization. Every person knowing where to report, what job to do, and who to tell when things go wrong—that's organization.

Student Notes:

REVIEW 1 ANSWERS

A1. The three military departments of the Department of Defense (DoD) are the—

a. **Navy,**

b. **Army,** and

c. **Air Force.**

A2. The four missions of the DoD are—

a. **To support and defend the Constitution of the United States against all enemies, foreign and domestic**

b. **To ensure the security of the United States, its possessions, and areas vital to its interests**

c. **To uphold and advance the national policies and interests of the United States**

d. **To safeguard the internal security of the United States**

A3. The two main objectives of the Navy are to—

a. **Organize, train, equip, prepare, and maintain the readiness of the Navy and Marine corps forces to perform military missions**

b. **To support the Navy and Marine Corps forces as well as the forces of other military departments**

A4. The three basic components of the Department of the Navy are—

a. **Navy Department**

b. **Operating forces**

c. **Shore Establishment**

A5. The U.S. Coast Guard operates under the following two departments:

a. Wartime—**Department of the Navy**

b. Peacetime—**Department of Transportation**

REVIEW 2 ANSWERS

A1. Aboard ship, the ship's organization and regulations can be found in a publication known as the ***Shipboard Organization and Regulations Manual***, which is based on the ***Standard Organization and Regulations of the U.S. Navy***.

A2. Some aspects of a ship's organization covered by the *Shipboard Organization and Regulations Manual* include—

a. **Unit's admin organization, including watches**

b. **Coordination of evolutions and emergency procedures**

c. **Conduct of personnel**

REVIEW 3 ANSWERS

A1. The two elements of a ship's organization are the—

a. **Battle organization**

b. **Administration organization**

A2. The information contained in the—

a. Battle organization includes the **numbers and specialties the unit needs to fulfill its wartime missions**

b. Administration organization **makes sure the ship can fight or carry out its mission; training, maintenance, and routine operations are covered**

A3. The responsibilities of the departments are—

a. Navigation—**Safe navigation and piloting of the ship**

b. Operations—**In charge of all radar, sonar, and communications equipment on the ship. Operations collects and evaluates combat and operational information and conducts electronic warfare**

c. Engineering—**Operation, care, and maintenance of all propulsion and auxiliary machinery**

d. Supply—**Operates the general mess and ship's store; manages clothing and small stores issue room; maintains pay records; and orders and receives general stores, supplies, spare parts, and ship's equipment**

e. Weapons/Deck/Combat Systems—**Responsibilities depend on type of ship.**

1) Weapons department or combat systems department—**Operation, care, and maintenance of ships armament and weapons fire-control equipment**

2) Deck department—**Responsible for deck functions**

A4. The *Navy Regs* has an entire chapter covering commanding officer duties.

A5. Aboard ship, the **commanding officer** is responsible for the safe navigation of the ship, the condition of the ship, and the appearance of material and personnel.

A6. A CO has authority equal to responsibility, including the power to impose limited punishment. **This power can't be delegated.**

A7. The **senior enlisted person** transmits ideas and recommendations directly to the CO.

A8. Duties of the XO include—

a. **Arranging and coordinating ship/s work, drills, exercises, and policing and inspecting the ship**

b. **Investigating matters affecting crew discipline and conduct. Making recommendations to CO on these matters**

c. **Approving/disapproving liberty lists and leave requests**

d. **Inspecting the ship and receiving readiness reports from department heads; reporting to the CO when the ship is ready for action**

A9. Assistants to the XO include—

a. **Personnel officer**

b. **Training officer**

c. **Educational services officer (ESO)**

d. **Drug and alcohol program advisor (DAPA)**

e. **Command master chief (CM/C)**

A10. The department head is responsible for and reports to the CO on **matters that affect his/her department including administrative matters, operational readiness, and general condition of the equipment.**

A11. The **division** is the basic unit of a shipboard organization.

A12. Division officer duties include—

a. **Inspecting division personnel, spaces, equipment, and supplies**

b. **Maintaining copies of division orders and bills and displaying them in conspicuous places**

c. **Training division personnel and preparing them for battle**

REVIEW 4 ANSWERS

A1. The four departments usually found in aircraft squadrons include—

a. **Administration**

b. **Safety**

c. **Operations**

d. **Maintenance**

A2. The safety department is usually divided into

a. **Ground safety**

b. **Aviation safety**

c. **Naval Air Training and Operating Procedures Standardization (NATOPS)**

A3. The operations department is responsible for the **operational readiness and tactical efficiency of the squadron**.

A4. The four divisions of an aircraft maintenance department are the—

a. **Maintenance/material control**

b. **Quality assurance**

c. **Maintenance administration**

d. **Aircraft, avionics/armament, and airframes divisions**

A5. The branch officer is responsible to the **division officer**.

REVIEW 5 ANSWERS

A1. The five areas that affect the chain of command are—

a. **Responsibility**

b. **Accountability**

c. **Direction**

d. **Communication**

e. **Work-related problems**

A2. In your organization, you can find out about the chain of command by **asking someone in the administration office.**

CHAPTER 7

BASIC SEAMANSHIP

To ensure safety at sea, the best that science can devise and that naval organization can provide must be regarded only as an aid, and never as a substitute for good seamanship, self-reliance, and sense of ultimate responsibility which are the first requisites in a seaman...

—C.W. Nimitz
Letter to U.S. Pacific Fleet
13 February 1945

At this stage in your Navy career, you're learning thousands of things entirely new to you. You're probably finding it hard to assign relative importance to them. The importance of these skills and knowledge will become more obvious the longer you're in the Navy. This is true because the most important things will be emphasized in your day-to-day living. If you're not assigned to the deck force, you may think that seamanship is not important. Well, you're wrong!

Seamanship ties every member of the Navy together. The uniform worn by Navy members, from seaman to admiral, implies that the wearer has a certain degree of proficiency in the art of seamanship. The fact that you may later become an Electronics Technician doesn't change the fact that you're first a seaman and then a technician. Be as proud of your ability as a seaman as you are of your ability to perform your other duties.

Even though you don't work on deck everyday, there will be times, particularly on small ships, when you will be required to assist the deck force. You may have to carry stores, assist in replenishment-at-sea operations, assist in mooring to or untying a ship from the pier, and so forth. When working as part of the deck force, you'll be expected to have a general idea of what's going on, how and why a task is being accomplished, and be able to carry out orders intelligently. Therefore, take every opportunity to observe and learn as much as you can about seamanship. This chapter provides only basic seamanship information.

In this chapter, seamanship is divided into the following basic sections—deck, boat, and marlinespike seamanship.

- Deck seamanship concerns the general work that goes on about the ship's deck and the equipment used. Anchoring, mooring, rigging and handling heavy weights and cargo, underway replenishment, towing, and a host of other skills are considered deck seamanship.
- Boat seamanship, as the name implies, concerns the handling of boats.
- Marlinespike seamanship concerns the use and care of line and consists of forming knots, making splices, and fashioning useful and decorative articles from small stuff and twine.

DECK SEAMANSHIP

Learning Objective: When you finish this chapter, you will be able to—

- Identify deck equipment and recognize their purpose.

Deck equipment consists of all equipment used in the application of deck seamanship, which is work normally performed by the deck force. You need to know shipboard equipment and terminology because you may be called on to assist the deck force in various seamanship evolutions. To help you, some of the more familiar items of deck equipment are discussed in this section.

GROUND TACKLE

Ground tackle is the equipment used in anchoring and mooring with anchors. It includes anchors, anchor cables and chains, and associated equipment, such as chain stoppers, bending shackles, outboard swivel shots, and detachable links. Figure 7-1 shows a typical ground tackle arrangement on a forecastle.

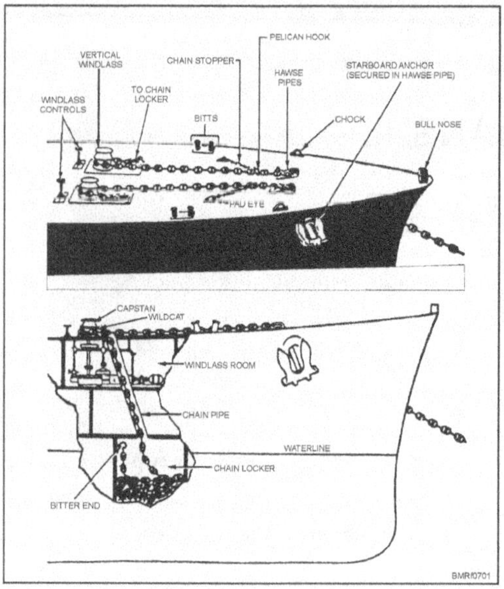

Figure 7-1.—Typical ground tackle and chain stowage.

ANCHORS

Anchors can be defined by their stowage locations aboard ship or by their type of construction. Bower anchors are carried on the bow and are secured (housed) in the hawsepipes. Stern anchors are carried on the stern. On landing ships and craft, stern anchors are secured to the stern and are used to help pull away from beaches.

The most common types of anchors used aboard ship are the stockless and the lightweight (or stock-in-crown) anchors. The two anchors shown in figure 7-2 are of Navy design. The stockless types are used chiefly as bow anchors (bowers) on most Navy ships. Originally, the lightweight types were used only on small boats and as stern anchors of landing ships and craft. However, recently they are carried as bowers for several types of vessels.

ANCHOR CHAIN

Modern Navy anchor chain consists of studded links of high strength steel. (Studs are crosspieces of metal forged or welded in the center of the links to prevent the chain from kinking.) Chains are made up of 15-fathom (90-foot) sections called *standard shots*. The number of shots per chain depends on the size of the ship. Shots are secured together by detachable links that can be readily disassembled whenever it is desirable to break the chain.

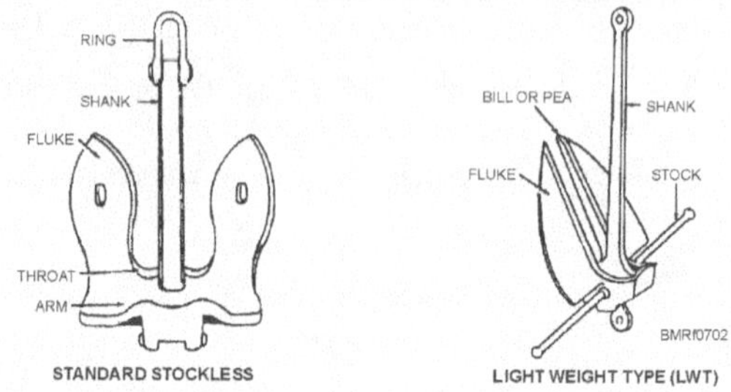

Figure 7-2.—Anchors.

STOWING CHAIN

As the chain comes aboard, it passes along the deck on metal flash plates, over the wildcat, and down into the chain locker. Each chain goes into a bin called a chain locker, as shown in figure 7-1. Its *bitter end* is secured to a ring bolt on the bulkhead of the chain locker.

ANCHOR WINDLASSES

The Navy uses two types of anchor windlasses for lifting the ship's anchor—the vertical shaft type and the horizontal shaft type (fig. 7-3). The vertical shaft type is used on most combatant ships. The horizontal shaft type is used on amphibious and auxiliary ships. Both types are equipped with wildcats, which engage the links of the anchor chain. The wildcat may be disengaged when it is desired to use the capstan (vertical type) or the gypsy heads (horizontal type) for handling lines or wire.

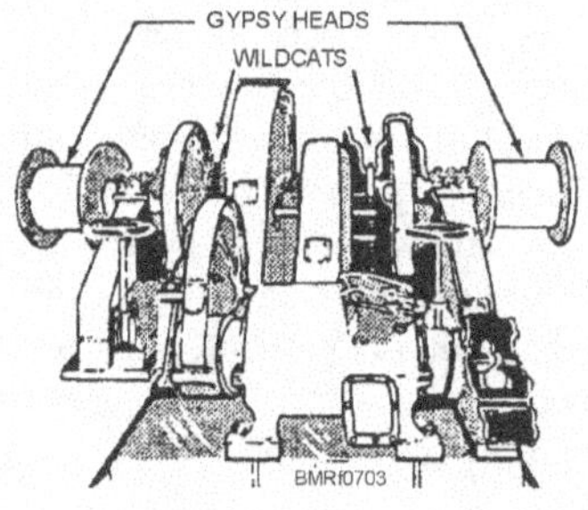

Figure 7-3.—Horizontal shaft anchor windlass.

Student Notes:

ACCOMMODATION LADDER

Frequently, the accommodation ladder is mistakenly called the gangway. However, gangway actually means *the opening in a bulwark or life rail that gives access to a brow or an accommodation ladder.* An accommodation ladder (fig. 7-4) consists essentially of an upper and a lower platform connected by a ladder. The lower end is supported, raised, and lowered by a block and tackle (called *falls*) and is usually suspended from a davit.

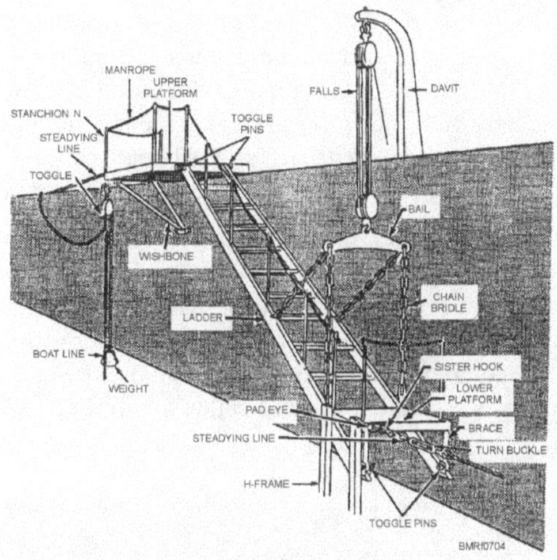

Figure 7-4.—A rigged accommodation ladder.

Brow is the Navy term for gangplank. Brows are ramps used between ships and between a ship and pier. They may be simply two or three wooden planks fastened together, or they may be elaborate affairs with handrails and wheels at one or both ends to prevent a ship's motion from unduly affecting the positioning of the brow.

MOORING LINES

A ship is moored when it's made fast to a buoy, when it's between two buoys, when it's between two anchors, or when it's secured by lines alongside a pier or another ship.

The lines used in mooring a ship alongside a pier are shown in figure 7-5. Well in advance of mooring, the lines should be faked down, fore and aft, each near the chock through which it passes in preparation for passing the line. You will learn about the procedure for faking a line and a description of deck fittings later in this chapter.

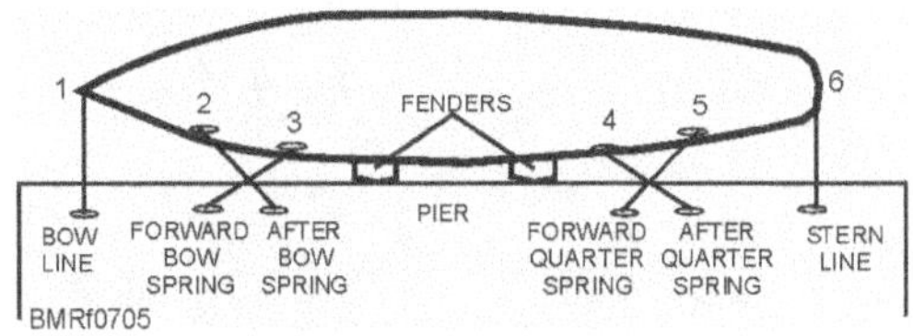

Figure 7-5.—Ship's mooring lines.

Rat guards are hinged conical metal shields secured around mooring lines. They are used to prevent rats from coming aboard ship.

The bowline and forward spring lines prevent the ship from drifting astern. The stern line and after spring lines prevent the ship from drifting forward. Look at figure 7-5. Here, lines 1, 3, and 5 are called *forward lines*; lines 2, 4, and 6 are called *after lines*. When secured, these lines tend to breast the ship in. The forward and after spring lines are used to prevent the ship from drifting forward or aft.

NOTE

The various types of line and wire rope are discussed in the "Marlinespike Seamanship" section of this chapter.

Teamwork is essential in carrying out the mooring operation. Lines must not be kinked or fouled. Keep control of the lines and avoid dipping them into the water. **Remember, observe all safety precautions!**

If the ship is to remain moored for a long period, lines are doubled up and bound together with marline hitches, and rat guards are placed on each line. Look at figure 7-6. To provide protection to the side of the ship while it is alongside a pier, *camels* (large wooden logs or rectangular structures) (views B and C) are often placed between the pier and the ship. *Fenders* (large cylindrical objects of rubber or fibrous material) (views A and D) are swung over the side of the ship to give bumper support against damage whenever a ship lies alongside another ship or a pier.

Student Notes:

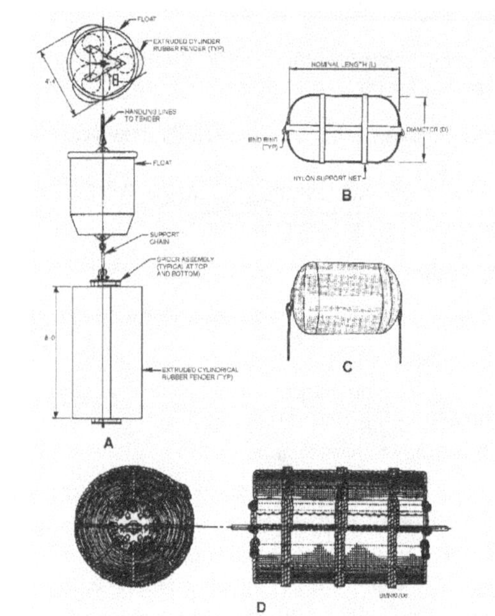

Figure 7-6.—Protection for the side of a ship.

DECK FITTINGS

Deck fittings are used aboard ships and boats mainly for the securing of mooring lines. All fittings shown in figure 7-7 are found aboard ship except the bollard, which is a pier fitting. The pad eye shown in the figure is not used for mooring but for towing other vessels. Different variations of the pad eye are used for securing heavy objects and equipment.

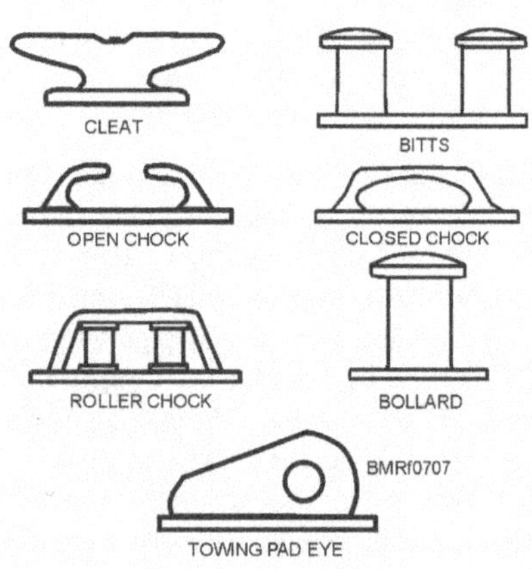

Figure 7-7.—Deck fittings.

DAVITS

Boats carried aboard ships usually are handled by powerful cranes and booms. These cranes and booms hook onto slings attached to hoisting points built into the strong parts of the boat's structure. Boats stowed at davits are lowered and hoisted by the davit machinery. Basically, a set of davits is nothing more than a special crane that is designed specifically for handling boats in a safe and timely manner.

BOAT BOOMS

Ships that are at anchor or moored to a buoy rig out their boat booms for the purpose of mooring their boats well clear of the side. This method of securing is known as *hauling out to the boom*. Forward booms are called *lower booms*; after booms are called *quarter booms*.

The boat boom shown in figure 7-8 is a spar that is secured to a gooseneck by a pin on the side of the ship. This arrangement allows free motion fore and aft. The outboard end of the boom hangs from a wire vang and tackle combination called the *topping lift*. Fore-and-aft motion is controlled by lines called *forward* and *after guys*.

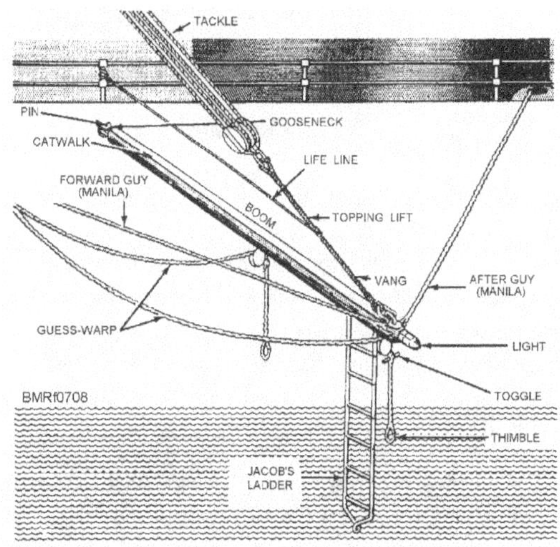

Figure 7-8.—Parts of a boat boom.

A strong line called a *guess-warp* runs from well forward on the ship out through a block or blocks on the boom and ends in a metal thimble through which boats can reeve (pass) their bowlines. A small piece of wood

Student Notes:

called a *toggle* is seized between strands of the guess-warp above the thimble to keep it from running up out of reach when a boat lets go. One or more *Jacob's ladders* (a rope ladder) from the boom permit boat crews to come aboard.

REVIEW 1 QUESTIONS

Q1. List six types of ground tackle used aboard ships.

a.

b.

c.

d.

e.

f.

Q2. Where are bower anchors located?

Q3. List the standard parts of the mooring line used to secure a normal sized ship at a pier.

a.

b.

c.

d.

e.

f.

Q4. Aboard ship, deck fittings are used for—

Q5. While anchored, what deck equipment is used to moor the ship's boat?

BOAT SEAMANSHIP

Learning Objectives: When you finish this chapter, you will be able to—

- Identify various types of boats, service craft, and combatant craft to include boat terms and nomenclature.
- Identify safety practices for boat passengers.

Boat seamanship is much more than a knowledge of the kinds of boats in operation in the Navy. Boat crews are responsible for the safe operation and upkeep of their craft and must receive training in a number of areas. Some of the techniques to be mastered require much practice and experience before a boat crew can become accomplished in their assigned duties. If you are assigned to duties as a member of a boat crew, you should study the *Seaman* and *Boatswain's Mate 3 & 2* training manuals and complete the required personnel qualification standards (PQS).

Boats used by the Navy are of three general groups—support craft, combatant craft, and boats in general. Each group may be determined by its assigned mission and by its type, design, and construction. Chapter 8 has detailed information about these craft.

BOATS

The term *boat* refers to a noncommissioned waterborne vessel that is not designated as a service craft. A boat is capable of limited independent operation. Officer/personnel boats, motor whaleboats, and utility boats fit into this group. Boats carried aboard ship that can be hoisted from and lowered into the water are known as *ship's boats*.

SERVICE CRAFT

The term *service craft* (figs. 7-9 and 7-10) is applied to waterborne craft that are designed for special use. Harbor tugs, ferryboats, various nonself-propelled barges, and floating dry docks are designated service craft.

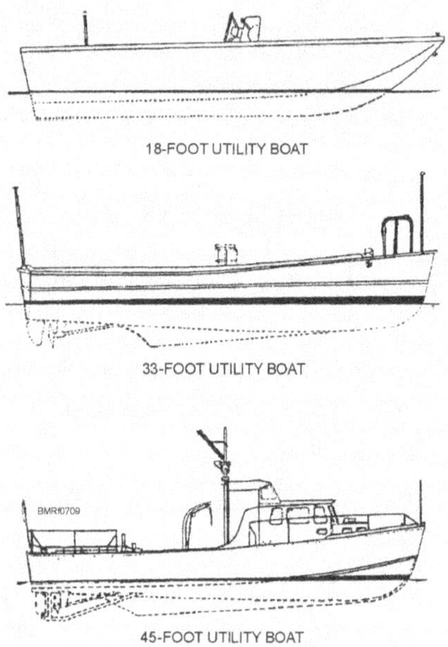

Figure 7-9.—Boats and service craft of the U.S. Navy.

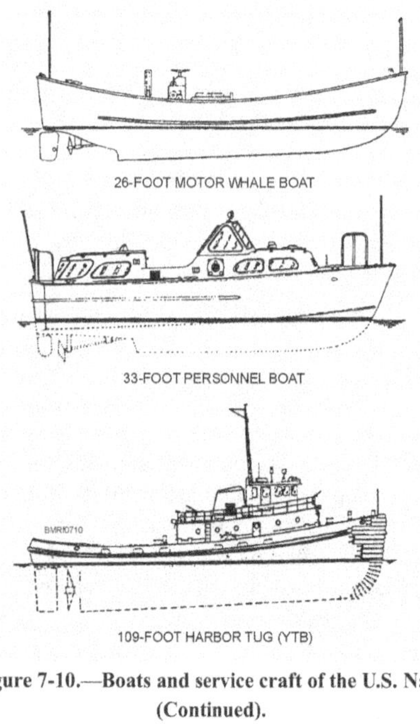

Figure 7-10.—Boats and service craft of the U.S. Navy (Continued).

COMBATANT CRAFT

Combatant craft are craft or boats specifically designed for combat roles. Figures 7-11 and 7-12 show a variety of patrol, riverine, amphibious warfare, and special combatant craft.

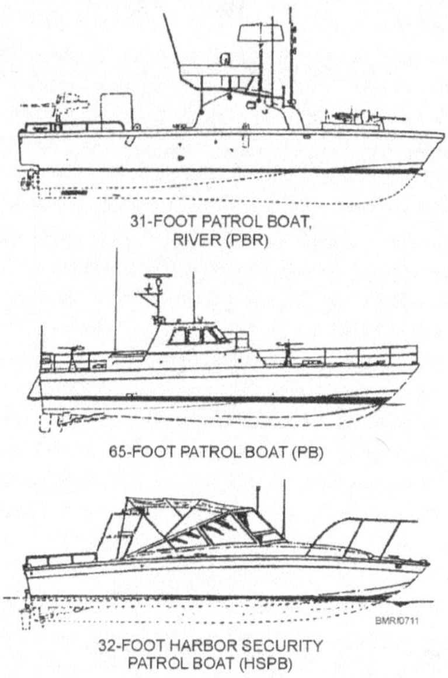

Figure 7-11.—Combatant craft of the U.S. Navy.

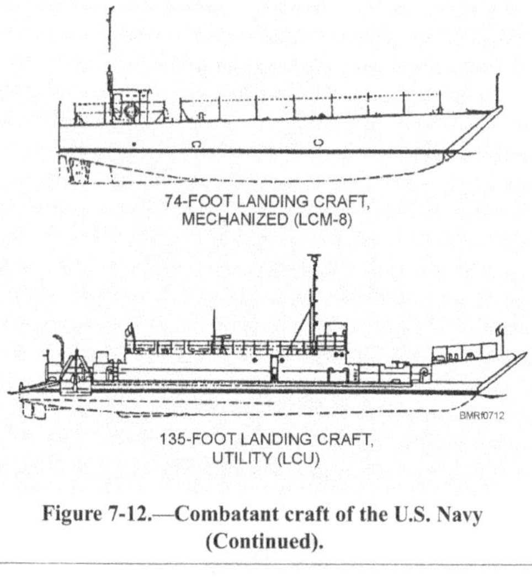

Figure 7-12.—Combatant craft of the U.S. Navy (Continued).

Student Notes:

BOAT SAFETY

Because the majority of Navy personnel are concerned with small boats only in the role of passengers, this section is written from the standpoint of passengers, rather than crew members. Every Sailor should be familiar with the following boat safety precautions:

- Obey all orders from the coxswain and boat officer.
- Embark in a quiet, orderly manner and move as far forward as possible. Once embarked, stay in place. Keep all parts of your body in the boat; do not perch on the gunwales.
- Don't engage in horseplay.
- Never needlessly distract the attention of crew members from their duties.
- Don't sit on life jackets—to do so mats the filler and reduces buoyancy.
- When told to do so, don your life jacket immediately.
- Don't smoke in a boat.
- During heavy weather, boat loads must be reduced.
- If told not to embark or requested to disembark, do so without argument
- If a boat swamps or capsizes, don't panic. Fear is transmitted easily from person to person, and a terrified individual drowns easily. Never strike out alone.
- Never strike out alone. Stay with the boat or huddle with other passengers because a large group can be found much more easily than individual swimmers.

BOAT TERMS AND NOMENCLATURE

Boat crew members often develop the habit of calling objects and the activities around them by their proper names. In times of emergency, your understanding and correct response to such terms could save valuable time.

Abaft. Any part of the boat aft of amidships.

After end (aft). The after end (aft) of a boat is the stern.

Amidships. Amidships is a point about halfway between the bow and stern and the sides of the boat.

Athwartships. When something is said to be *athwartships*, it's across the boat from side to side.

Forward end (fore). The forward end (fore) of the boat is the *bow.*

Inboard. Inboard usually describes the area inside the boat or an object nearer the centerline of the boat.

Outboard. Outboard describes the area furthermost from the boat's centerline or beyond the side of a boat.

Starboard. When facing forward of the boat, your right-hand side is the starboard.

Port. When you are facing forward of the boat, your left-hand side is the port.

Figure 7-13 shows a 26-foot personnel boat with features that are similar to most Navy boats.

By studying the nomenclature shown in figure 7-13, you will become familiar with much of the deck and hull equipment used on Navy boats. The glossary in appendix I of this training material will help you identify some of the terms.

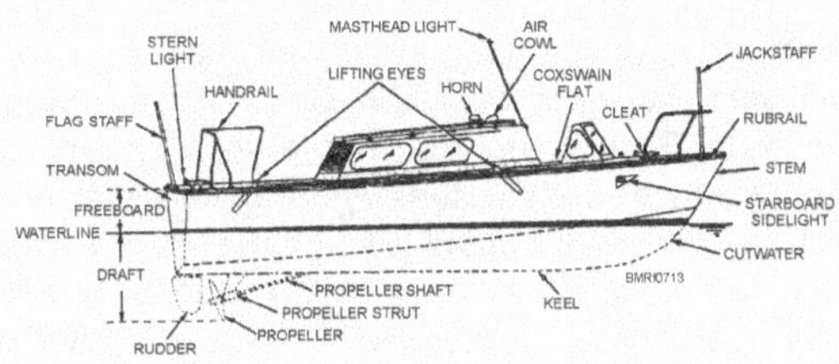

Figure 7-13.—Boat nomenclature.

Student Notes:

REVIEW 2 QUESTIONS

Q1. List two types of combatant craft .

a.

b.

Q2. As a boat passenger, you should obey the orders of what person(s)?

Q3. If a boat capsizes while you're a passenger, you shouldn't panic for what reason?

Q4. What does the term *athwartships* mean?

MARLINESPIKE SEAMANSHIP

Learning Objectives: When you finish this chapter, you will be able to—

- Identify the purpose of various types of line and rope.
- Recognize the procedures used to tie knots, bends and hitches, and to make splices.
- Identify the procedures for securing at sea.

Marlinespike seamanship is the art of handling and working all kinds of fiber and wire rope. *Rope* is a general term and can include both fiber and wire rope. In the Navy, Sailors generally refer to fiber rope as *line*, and wire rope is referred to as *rope*, *wire rope*, or *wire*. A better definition of a line is as follows: *A line is a length of rope, either fiber or wire, that is in use or has been cut for a specific purpose, such as a lifeline, heaving line, or lead line.* A few such lines have the word *rope* in their names, such as wheel rope, foot rope, and bell rope.

In sailing ships, the fiber ropes that gave athwartship support for the masts were so numerous that they actually shrouded the tops of the masts from the view of an observer on deck, hence, the name *shroud*. Stays, the fore and aft supports, were not so numerous, but there were several on each ship. Running rigging, tackles used to hoist and trim (adjust) the sails and handle cargo and other heavy weights, spanned the areas between sails, yards and decks, and yards and bulwarks. Lines secured the guns to the ship's sides and prevented them from rolling or recoiling across the gun decks. Gun tackles were used to haul the guns back into battery (firing position) after the guns were fired. Even the anchor cable was made of line. Obviously, line played a vital role in those ships.

In today's Navy, line isn't used as much as on sailing ships; however, it's still an important and expensive item. Therefore, every Sailor needs to learn the proper use and care of all kinds of line and wire rope. Today's Navy uses line made of fiber (natural and artificial); wire rope made of steel, phosphor bronze, and other metal; and a combination of wire and fiber (spring-lay).

Lines made from a variety of natural fibers have seen service in the Navy, but most have been replaced by lines made of synthetic fibers. The two most commonly used lines made of natural fibers are marline (tarred hemp fibers) and manila (abaca plant fibers). Manila line was formerly authorized for use only where great strength was required, such as mooring lines, towing lines, personnel transfers at sea and boatfalls. Fiber ropes made of tarred hemp are used in seizing, worming, serving ropes, and lashing. For most applications, nylon line (synthetic fiber) has replaced manila. Nylon line is about 2 1/2 times as strong as manila of the same size, has a greater strength and elasticity, and has a higher resistance to weather.

Wire rope usually is substituted for line where the line is subjected to a great deal of wear, weathering, or heat, and where greater strength is required. Spring lay is used for mooring lines, particularly at the bow and stern.

FIBER LINE

Any rope that is not wire is fiber rope. Except in a few instances where it has special uses, **fiber rope is never called anything but line aboard ship**.

Student Notes:

Lines are classified by both their construction and their material. Nearly all line used in the Navy is three-strand line.

Line is made by twisting fibers into threads (or yarns), threads into strands, and strands into rope. Taking the process further, ropes twisted together form a cable—an item seldom seen nowadays. Most of our lines are three-strand and right-laid; that is, as you look along a line, the twist is to the right. During construction of natural fiber line, a lubricant is added that also serves as a preservative.

Large line is measured by circumference. Line 1 3/4 inches and under in circumference, called *small stuff*, is identified by the number of threads in the line. A line with twenty-four thread is 1 1/2 inches in circumference. Inasmuch as the numbers of threads per strand are equal, thread numbers in a three-strand line are divisible by 3—24, 21, 18, and so on, down to the smallest—6 thread (3/4 inch). Line from 1 3/4 inches to about 4 inches is manufactured in 1/4-inch graduations. The length of all line and wire rope is usually measured in feet.

The chart shown below lists tips on the care of natural fiber line. You should be thoroughly familiar with them and observe them at all times.

NEVER	ALWAYS
Stow wet or damp line in an unventilated compartment or cover it so that it cannot dry. Mildew will form and weaken the fibers.	Dry line before stowing it.
Subject line to intense heat nor unnecessarily allow it to lie in the hot sun. The lubricant (natural oils) will dry out, thus shortening the useful life of the line.	Protect line from weather when possible.
Subject a line to loads exceeding its safe working load. To do so may not break the line, but individual fibers will break, reducing the strength.	Use chafing gear (canvas, short lengths of old firehose, and so on) where line (or wire) runs over sharp edges or rough surfaces.
Allow line to bear on sharp edges or run over rough surfaces. The line will be cut or worn, reducing the strength and useful life.	Slack off taut lines when it rains. Wet line shrinks, and if the line is taut, the resulting strain may be enough to break some of the fibers.
Scrub line. The lubricant will be washed away, and caustics in strong soap may harm the fibers.	Coil right-laid line to the right (clockwise).
Put a strain on a line with a kink in it.	Inspect a line before using it. Overworked or overstrained line will have a bristly surface. Mildew can be seen, and it has peculiar, unpleasant odor. Untwist the line so that the inner parts of the strands can be seen. If they have a dull, grayish look, the line is unsafe.
Try to lubricate line. The lubricant you add may do more harm than good.	Give line the care it deserves—someday your safety may depend on it.

Student Notes:

NYLON LINE

Most of the tips for the care of natural fiber line should be observed with nylon line. Nylon, however, is not subject to mildew. It should be scrubbed if it becomes slippery because of oil or grease.

A stretch of one third of its length is normal for nylon line under safe working loads. Nylon stretches about 50 percent before it will break. Because of its elasticity, nylon line breaks with a decided snapback; therefore, stand well clear when it is under a heavy strain.

CAUTION

Snapback is extremely dangerous and has caused severe injuries and death. The utmost caution must be observed when working with or around all synthetic lines.

WIRE ROPE

The basic unit of wire rope construction is the individual wire, which is made of steel or other metal and comes in various sizes. These wires are laid together to form strands. The number of wires in a strand varies according to the purpose of the rope. A number of strands are laid together to form the wire rope itself.

Wire rope is designated by the number of strands per rope and the number of wires per strand. For example, a 6 by 19 rope will have 6 strands with 19 wires per strand. It may have the same outside diameter as a 6 by 37 wire rope, which will have 6 strands with 37 wires of much smaller size per strand. The more wires per strand, the more flexible the rope. Rope with fewer and larger wires per strand is more resistant to external abrasion.

The strands of the wire rope are laid up around a central core, which may be only a single wire, a single strand of wire, or hemp. A hemp core contributes flexibility, cushions the strands as the wire rope contracts under strain, and holds a portion of lubricant for continuous lubrication. A wire core is stronger than hemp and can be used where conditions, such as high temperatures, would damage a hemp core.

WHIPPINGS

Whippings are bindings on the ends of rope that keep the rope from unlaying. On line, whippings are made with cord, such as sail twine or with marline. The ends of all line must be whipped because of the frequent need for passing the ends through rings and pad eyes and for reeving them through blocks. Unlaid and frayed ends of line are unsightly and unseamanlike and waste many feet of line. Knots or backsplices in the end of a line are not allowed, nor are friction tape or wire whippings. Knots and backsplices will jam in a block; friction tape will not hold for long; and wire may tear a line-handler's hands.

The most secure whipping is made with the aid of a sail needle and palm. However, an excellent whipping can be made without a needle if the procedure shown in figure 7-14 is followed. First, lay one end of the whipping along the line, bind it down with a couple of turns, and snug up the edges. Then lay the other end on in an opposite direction with the body portion of the whipping, continuing with several more turns from the bight of the whipping. The whipping length should be about equal to the diameter of the line being whipped. Snug up the edges and cut off the twine close to the line. This type of whipping is a temporary one. If the line is to be used frequently, a permanent whipping should be used.

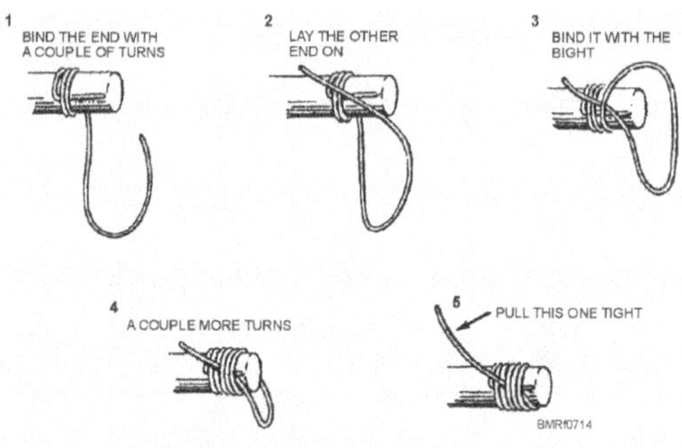

Figure 7-14.—Plain whipping a line.

Student Notes:

KNOTS, BENDS, AND HITCHES

Except among seamen, the word *knot* is ordinarily used as an all-inclusive term, covering the more specific use of knots plus bends and hitches. Even seamen find it hard to clearly define the terms *knot*, *bend*, and *hitch* because their functions overlap like the bowline knot and many other instances. In general, however, the terms may be defined as follows:

Knots. Knots are used to form eyes or to secure a cord or line around an object, such as a package. In other words, the line is bent to itself.

Hitches. Hitches are used to bend a line to or around an object, such as a ring, spar, or stanchion.

Bends. Bends are used to secure two lines together.

All Navy personnel should know the square knot, bowline, single- and double-becket bends, round turn and two half hitches, and clove hitch. Navy personnel should know when these knots, bends, and hitches are used. Before reading further, look at figure 7-15, which shows a few terms that make it easier for you to understand the following procedures.

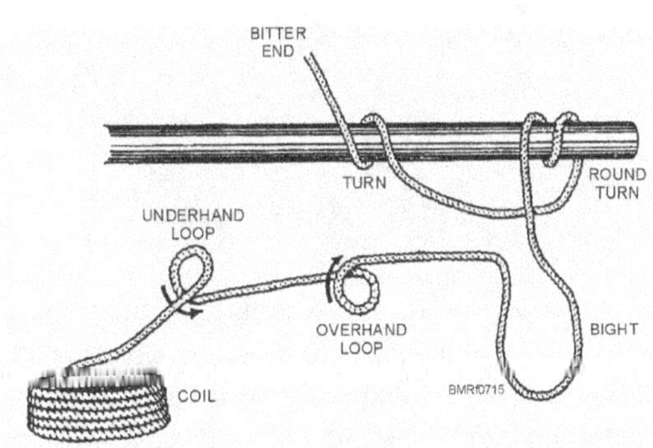

Figure 7-15.—Elements of knots, bends, and hitches.

Square Knot

The square knot, also known as the *reef knot* from its use in reefing sails, is quickly and easily made and has a great many uses. It will not slip, but it can jam under heavy strain. It can be loosened, however, by pulling on first one and then the other end. Figure 7-16 shows steps in making a square knot.

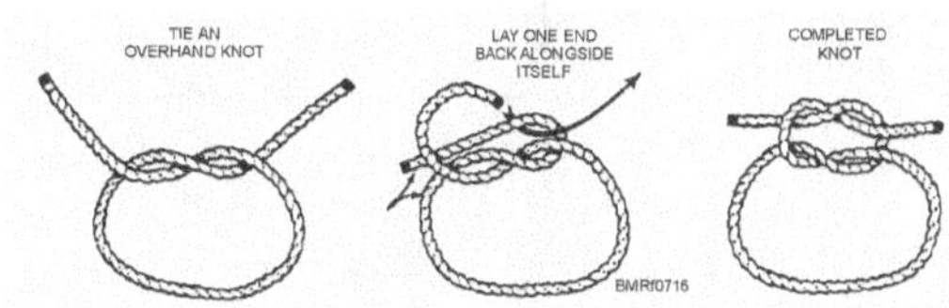

Figure 7-16.—Square knot.

Bowline

The bowline, with its many variations, has a lot of uses. Its chief use is to form an eye; but it can also be used to secure a line to a pad eye, to secure another ring around a stanchion or other object, or to bend two lines together.

To tie a bowline, you should—

1. Grasp the bitter end of the line in the right hand and the standing part in the left hand (opposite, if left-handed). Assuming you are using small stuff, the length of line between your hands should be about 2 feet.
2. Throw an overhand loop counterclockwise near your left hand (clockwise near your right hand, if you are left-handed).
3. Grasp the loop formed and hold it. Pass the bitter end up through the bottom of the loop, as shown in figure 7-17, view A.
4. Pull the bitter end up through the loop, and pass it around behind the standing part of the line (fig. 7-17, view B).
5. Pass the bitter end down through the loop beside the line that was pulled up through the loop (fig. 7-17, view C).
6. To tighten the knot, grasp the standing part in one hand and the two lines passed through the loop with the other hand, and pull.

Student Notes:

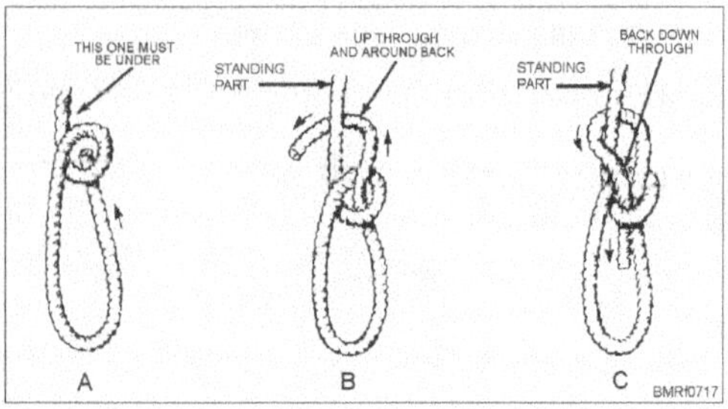

Figure 7-17.—Tying a bowline.

Becket Bend

The chief value of the becket bend is its use in binding together two lines of different sizes. If there is a great difference in sizes or the strain on the line is to be great, always use a double becket bend.

To fashion a single becket bend, you should—

1. Make a bight in one line and run the bitter end of the other line up through it, as shown in figure 7-18, view A.
2. Pass the end around behind both parts of the bight and back under itself (fig. 7-18, view B).

Figure 7-18, view C, shows how you make a double becket bend by simply taking another turn around the bight. (These bends are also known as *sheet bends*.)

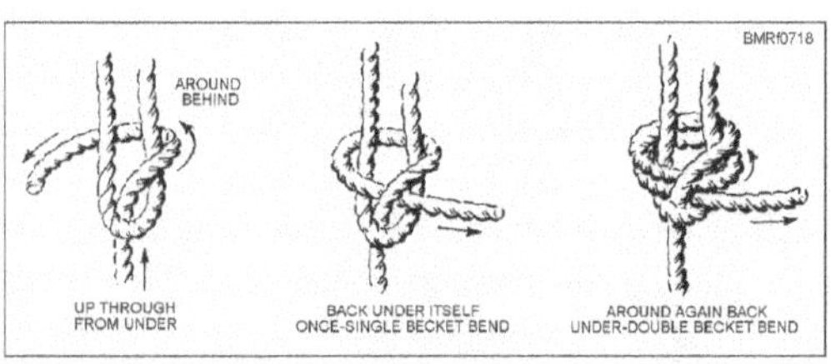

Figure 7-18.—Single- and double-becket bends.

Clove Hitch

The clove hitch can be quickly and easily tied in several ways, and it will hold as long as there is a strain on it. Once the strain is taken off, however, the hitch must be checked and tightened to prevent the bitter end from pulling out when the strain is reapplied. To make this checking and tightening unnecessary, you lash a clove hitch with a half hitch around the standing part.

To tie this hitch (fig. 7-19), you should—

1. Take a turn with the bitter end.
2. Pass the end across the standing part.
3. Take another turn. (Notice that both turns go around in the same direction.)
4. Pass the end under itself, and the hitch is complete.

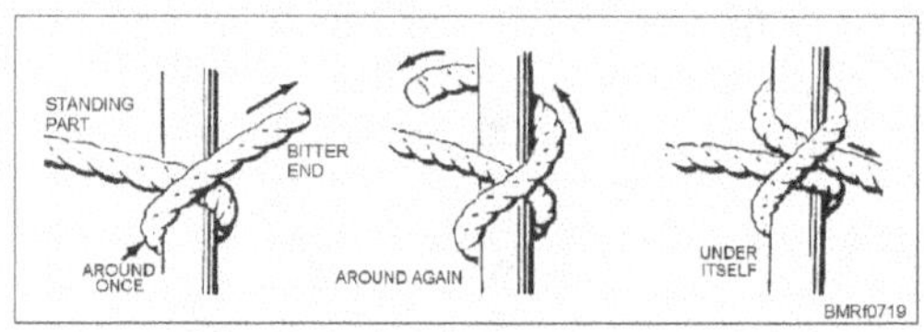

Figure 7-19.—Tying a clove hitch.

Another way to make this hitch is to form two underhand loops. Lay the second loop on top of the first. This method is the usual way to form the hitch when it can be slipped over the end of the object to which the line is to be secured.

Round Turn and Two Half Hitches

The chief advantage of the round turn and two half hitches over other hitches is that it won't slip along the object to which it is secured. It's made by taking a round turn and making two half hitches (fig. 7-20). (The two half hitches actually consist of a clove hitch taken around the line itself.)

Student Notes:

Figure 7-20.—Round turn and two half hitches.

Making Up a Line

Once line is removed from the manufacturer's coil, or spool, it may be made up (for ready use) by coiling down, faking down, or flemishing. Figure 7-21 shows the methods of coiling, faking, and flemishing lines.

"Coiling down a line" means laying it up in circles, roughly one on top of the other. Faking down a line is laying it up in the same manner as for coiling down, except that it is laid out in long, flat bights, one alongside the other, instead of in round coils. The main advantage of working with line that is faked down is that it runs off more easily. To flemish down a line, start with the bitter end, and lay on deck successive circles of line in the manner of a clock spring with the bitter end in the center. Right-laid line is laid down clockwise; left-laid line is laid down counterclockwise.

SPLICES

Splices are used to permanently join two lines or to form an eye or loop in the end of a line. When time permits, splices should be used instead of knots because splices are much stronger.

Eye Splice

To make an eye splice, unlay (untwist) the strands in the end of your line about 8 to 10 turns of lay. Whip the end of each strand to prevent the strands from unlaying while you splice.

NOTE

When splicing synthetic line, such as nylon, it is sometimes easier to use tape on the strand ends. Large line, such as mooring lines, should be seized or bound together at the point where unlaying stops.

To form the eye, bend the line back until the eye is the desired size. This is the point where your splicing begins.

Follow the steps shown in figure 7-22 by tucking each whipped strand under one strand of the line. Pull the slack out of each tuck and check the size of the eye. (If a thimble is to be used, insert it at this point.) Follow the "over one strand, under the next" procedure until you complete at least three tucks for natural fiber line or four tucks for synthetic line. (**NOTE:** The splice can be smoothed by rolling it on deck under your foot.)

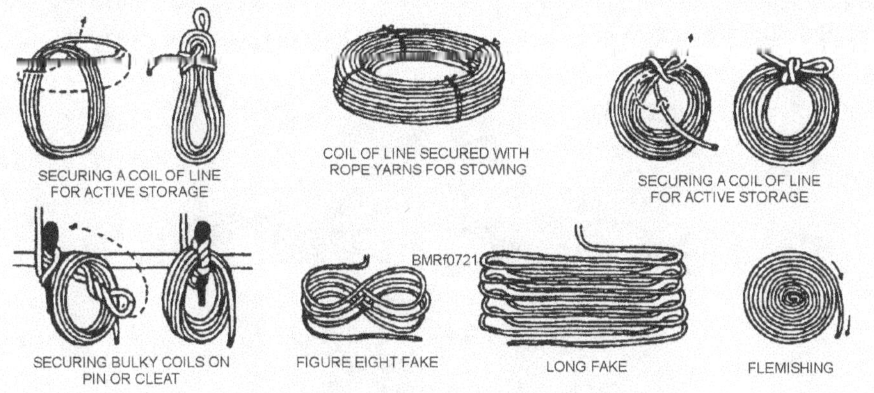

Figure 7-21.—Coils, fakes, and flemishes.

Student Notes:

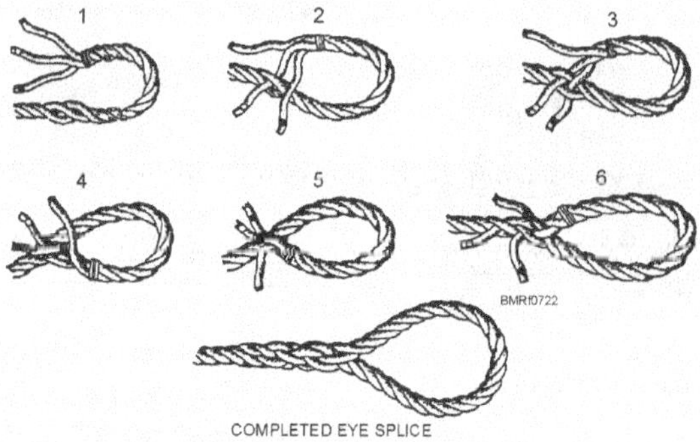

Figure 7-22.—Steps for making an eye splice.

Upon completion of the splice, the excess length of each strand must be cut off. When natural fiber line is used for the splice, the strands can simply be cut off near the line. With synthetic line, a short length of each strand should be left intact. The ends of the threads of each strand are then melted together over an open flame to prevent the strands from frazzling.

When you melt the ends of the strands, don't allow any of the melted synthetic line to drip on you, your clothing, another person, or anything that might present a fire hazard. Also, observe all safety precautions pertaining to the use of open flames aboard your ship or station.

Short Splice

A short splice is used where two lines are to be permanently joined, provided a slight enlargement of the diameter is not important. When properly made, the short splice is much stronger than any knot.

After unlaying and whipping the strands as described for the eye splice, seize each line where the unlaying stops. Now butt the two lines together so that they are interlaced, and follow the steps shown in figure 7-23.

With large lines, you must put on a temporary seizing where they join to keep them from suddenly coming apart. It's better to do that with small lines, too, until you get the hang of holding them together while you tuck.

Once your seizing is on, tuck over and under the same way you finish off an eye splice. Three tucks (natural fiber) or four tucks (synthetic fiber) on each side of the seizing are ample. Remove the seizing, cut off the ends of the strands, and melt them (if appropriate) as previously described.

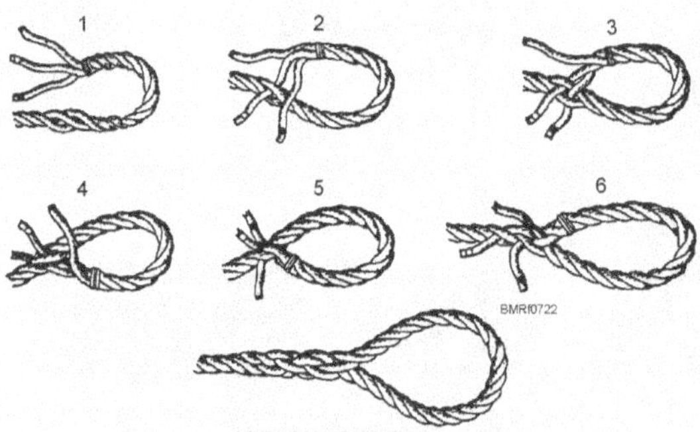

Figure 7-22.—Steps for making an eye splice.

Student Notes:

SECURING FOR SEA

You are required to learn these knots, bends, and hitches so that you can use them when and where necessary. Rest assured that a person who goes to sea will find frequent use for them in securing equipment to prevent damage from rough waters. How the equipment is secured depends on the gear and the places of stowage. By observing the few tips that follow and by using a little common sense, you should be able to do a good job of securing your equipment for sea.

- Use line strong enough to hold the gear.
- Make certain the line is in good condition. Make fast the belay objects from at least two points that, preferably, are near the object.
- Lash tightly all objects against something solid (such as the bulkhead).
- Make the lashings taut so that the object will not "work" with the pitch and roll of the ship. Frequently check all lashings, and tighten as necessary.
- Use chafing gears on sharp corners and rough surfaces.
- Never make fast your lashings to electric cables or small slightly secured pipes, lagged pipes, door and hatch dogs or hinges, electric motors, lifeline stanchions, or anything not solidly secured.
- Never block access to vents, fireplugs, switches, valves, doors, or hatches.

Never underestimate the force of the sea! Secure everything properly the first time and be safe.

REVIEW 3 QUESTIONS

Q1. List the three advantages of using nylon line over natural fiber line.

a.

b.

c.

Q2. When a natural fiber line will bear on sharp edges or run over rough surfaces, what action should you take to protect the line from damage?

Q3. Under what condition should you use a wire rope?

Q4. You are right-handed. When tying a bowline, the overhand loop should be in what direction?

Q5. You are going to join two separate lines together using a becket bend. What is the first step you should take?

SUMMARY

Becoming an accomplished seaman takes time, hard work, and patience. At some time in your career, you could be a member of a detail where handling lines will be required, or you may work with deck equipment and lines on a daily basis. Knowing how to use deck equipment and handle lines safely is essential.

Lines are used in the Navy for many reasons—from mooring aircraft carriers to securing bookshelves at sea. Without the wide variety of lines available to us, our way of doing our jobs would be extremely difficult.

Each piece of deck equipment or fitting has been designed for a specific purpose. A set of bits on a destroyer is used primarily for mooring, where a set of bits on a tug is used almost exclusively for towing. Becoming familiar with deck equipment and its use, and knowing how to makeup and use lines is not only a sign of good seamanship but could, in case of emergencies, make the difference between saving or losing the ship or your shipmates.

Each person in the Navy is first and foremost a seaman and then a technician in his or her specific rate. Become proficient in seamanship. It may help you in your daily duties and most certainly assist you in times of difficulty.

Student Notes:

REVIEW 1 ANSWERS

A1. Ground tackle used on board ship includes—

a. **Anchors**

b. **Anchor cable and chains**

c. **Chain stoppers**

d. **Bending shackles**

e. **Outboard swivel shots**

f. **Detachable links**

A2. Bower anchors are **carried on the bow of a ship and secured in the hawspipe**.

A3. The standard mooring lines used to secure a normal size ship at a pier include the—

a. **Bowline**

b. **Forward bow spring**

c. **After bow spring**

d. **Forward quarter spring**

e. **After quarter spring**

f. **Stern line**

A4. Aboard ship, deck fittings are used **in securing mooring lines, in towing operations, and in securing heavy objects and equipment**.

A5. **Boat booms** are used to moor the ship's boat while anchored.

REVIEW 2 ANSWERS

A1. Combatant craft include—

a. **Mechanized landing craft (LCM)**

b. **Utility landing craft (LCU)**

A2. As a passenger, you should **obey all orders from the boat officer and the coxswain**.

A3. If a boat you're a passenger in capsizes, you shouldn't panic because **fear is easily transmitted from one person to another**.

A4. The term *athwartships* refers **to the position of something that is across the boat from side to side at a right angle.**

REVIEW 3 ANSWERS

A1. Three advantages of using nylon line over natural fiber line include—

a. **Nylon line is 2 1/2 times stronger than natural fiber**.

b. **Nylon has greater strength and elasticity**.

c. **Nylon has greater resistance to weather**.

A2. To protect a natural fiber line from sharp edges and rough surfaces, you should **use a chafing gear between the contact point of the line and the damaging surface**.

A3. You should use a wire rope **when a great deal of wear and tear and weathering and heat is met, and greater strength is needed**.

A4. The overhand loop should be turned in a **counterclockwise direction**.

A5. The first step you should take when joining two separate lines together using a beck bend is to **make a bight on one line and run the bitter end of the other line up through the bight**.

Student Notes:

CHAPTER 8

SHIP/AIRCRAFT CHARACTERISTICS

This ship is built to fight. You'd better know how.

—Admiral Arleigh Burke

The air fleet of an enemy will never get within striking distance of our coast as long as our aircraft carriers are able to carry the preponderance of air power to the sea.

—Rear Admiral W. A. Moffett

The U.S. Navy has thousands of vessels and aircraft in its inventory. They range from small harbor patrol boats to huge super carriers and from helicopters to giant transport planes. You won't be expected to know the characteristics of each one, but you should be able to recognize the type of ship or aircraft you see. You should also be able to identify its mission and armament and have an idea about its size. In this chapter, you'll learn about the major classes and the major types of ships and aircraft the Navy operates and what their characteristics and missions are. You will also learn some of the more common terms used to identify structural features and the terminology used to express direction and locations aboard ship.

Before you learn about the types and classes of ships, you need some background information about ships in general. To take advantage of scientific advances, the fleet is making changes. Cruise missiles, close-in defense systems, and multirole radar units are replacing conventional electronic and weapons systems. The Navy's new submarines and aircraft carriers are nuclear-powered. Therefore, steaming endurance is limited only by the replenishment of necessary supplies and food.

Many ships have been modernized to perform a wide variety of missions and to accomplish old missions more efficiently. During overhaul, older ships are outfitted with new radar, fire control, and communications systems. The hulls are strengthened and power plants reworked to extend the lives of these ships. However, it's not economically sound to convert all ships to nuclear power.

SHIP TERMS

Learning Objectives: When you finish this chapter, you will be able to—

- Identify terms used aboard ship.
- Recall the names used for superstructures and components of ship's hulls to include decks and doors and hatches.
- Identify structural terms.

In civilian life you used terms such as *upstairs, downstairs, windows, floors, ceilings, walls*, and *hallways*. In the Navy, you must learn to use Navy language. To use civilian terminology aboard ships marks you as a *landlubber*—a scornful term used to describe those who know nothing of the sea.

GENERAL TERMS

Lengthwise direction on a ship is *fore* and *aft*; **crosswise** is *athwartships*. The front of the ship is the *bow*; the rearmost is the *stern*. To move forward toward the bow is to go *forward*; to move toward the stern is to go *aft*. Anything that is more toward the bow than another object is *forward* of it, and anything that is more toward the stern is *abaft* (behind) the other object.

A ship is divided in half lengthwise by a *centerline*. When you face forward along the centerline, everything to your right is to *starboard*; everything to your left is to *port*. Fixtures and equipment take the name of the side on which they are located, such as the *starboard gangway* and the *port anchor*.

When you go toward the centerline, you go *inboard*. An object nearer the centerline is inboard of another object and that object is *outboard* of the first. The section around the midpoint area is called *amidships* (also called the *waist*). The extreme width of a ship, usually in the midship area, is its *beam*.

You never go downstairs in a ship; you always go *below.* To go up is to go *topside.* However, if you climb the mast, stacks, rigging, or any other area above the highest solid structure, you go *aloft.* The *bridge* is topside and usually forward. It contains control and visual communication stations. Human beings live *in* a ship or *on board* a ship. Inanimate objects, stores, and equipment are *aboard* a ship. Similarly, you *board* a ship or *go on board.* Stores, ammunition, and so on are *taken aboard* and *struck below.*

An object hanging against the side, bow, or stern is *over* the side, bow, or stern. An object in the water but not touching the ship is *outboard* of or *off* the ship (off the starboard side, off the port bow, and so on). An object in front of a ship is *ahead* of it. An object to the rear is *astern*, never in back. Cooking is done in the *galley*, not in the kitchen.

The fore-and-aft inclination of a ship is the ship's *trim*—down by the head or down by the stern. To trim a submarine is to adjust water in the *variable ballast tanks*, or *trim tanks*. A ship is said to *list* if it has a permanent or semipermanent inclination to one side or the other. This is a less than optimum condition.

STRUCTURAL TERMS

In this section, you will learn some of the terms related to ship construction. These terms won't tell you "how to" build a ship; however, by learning the terms, you will understand the major structural characteristics of the hull, decks, and superstructure of a ship.

Hull

Figure 8-1 shows the hull structure of a cruiser. You should refer to this figure as you read this section. The *hull* is the supporting body of a ship. Think of the *hull* as an envelope. Inside the hull are *strengthening members* that prevent the envelope from collapsing. The hull also contains partitions that form machinery, berthing, messing, and other spaces.

The *keel* is the backbone of the ship. The keel of most steel ships does not extend below the ship's bottom; hence, it is known as a *flat keel.* Its usual shape is that of an I-beam. All other parts used in constructing the hull are attached, either directly or indirectly, to the keel.

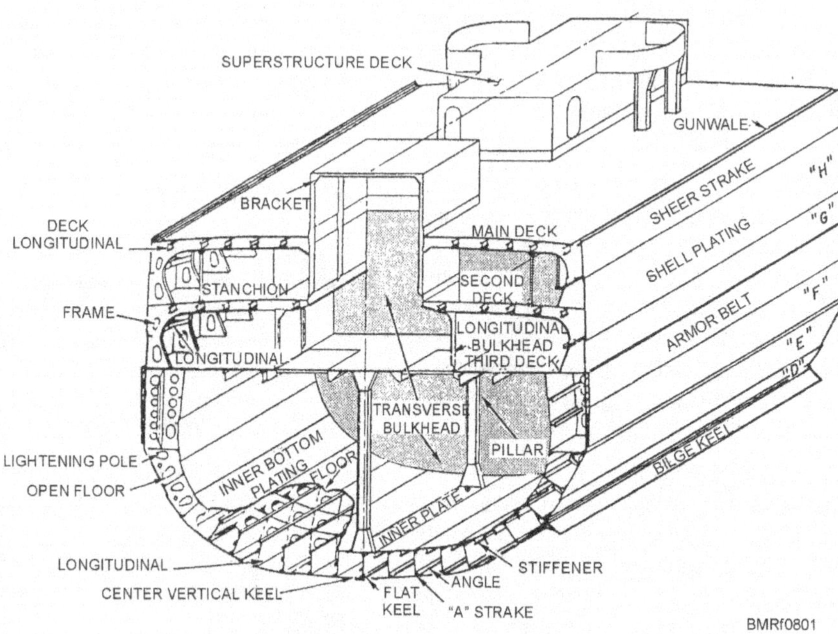

Figure 8-1.—Hull structure of a cruiser.

Student Notes:

The athwartships structure consists of *transverse frames* and *decks*. The decks run outboard from the keel to the turn of the bilge (where the bottom turns upward). Here, they are attached to the transverse frames, which then extend upward to the main deck.

Frames running parallel with the keel are called *longitudinals*. From the turn of the bilge up the sides they are also called *stringers*. The network of floors and longitudinals resembles a honeycomb (known as *cellular construction*), which greatly strengthens the bottom of the ship. When plating covers the honeycomb, double bottoms are formed. The space between the inner and outer bottoms (known as *tanks*) is used for liquid stowage. Planks laid upon the tank tops are called *ceilings*. The forward end of the keel, which is extended upward, is called the *stem*. The after end of the keel has a similar extension called the *sternpost*. The part of the stem above water is the *prow*; the forward edge of the stem is the *cutwater*.

The interior of a ship is divided into compartments by vertical walls, called *bulkheads*. Bulkheads run both transversely and longitudinally. Most bulkheads are merely partitions; but spaced at appropriate intervals, they are transverse watertight bulkheads. These bulkheads extend from the keel to the main deck and from side to side to provide extra stiffening and to partition the hull into independent watertight sections. Large ships have a series of longitudinal side bulkheads and tanks that provide protection against torpedoes. Usually, the outer tanks are filled with oil or water, and the inner tanks (called *voids*) are empty. The innermost bulkhead is called the *holding bulkhead*. If a torpedo were to hit the ship, the outer tanks, although ruptured, would absorb enough energy from the explosion that the holding bulkhead would remain intact, thus preventing flooding of vital spaces.

The plates that form the ship's hull are called *strakes*. Strakes are fastened to the framework in longitudinal rows. The keel forms the center strake. Strakes are lettered, beginning with the A strake on either side of the keel and extending up to the main deck. Some of the strakes also have names. The A strake is called the *starboard strake*; the strake along the turn of the bilge is the *bilge strake*; the uppermost strake is the *sheer strake*. A protecting keel running along the bottom near the turn of the bilge is called a *bilge keel*. Its purpose is to reduce rolling of the ship. (A ship *rolls* from side to side; it *pitches* when it goes up and down fore and aft; it *yaws* when the bow swings to port and starboard because of wave action.)

The upper edges of the sides, where the sheer strakes join the main deck, are called the *gunwales* (rhymes with funnels). The foremost part of the ship, where the gunwales join the stem, is known as the *eyes* of the ship. The port and starboard quarters are located where the gunwales curve inward to the sternpost.

The water level along the hull of a ship afloat is the *waterline*. The vertical distance from the bottom of the keel to the waterline is the ship's *draft*. *Freeboard* is the distance from the waterline to the main deck. Figures 8-2 and 8-3 show various parts of the hull and deck.

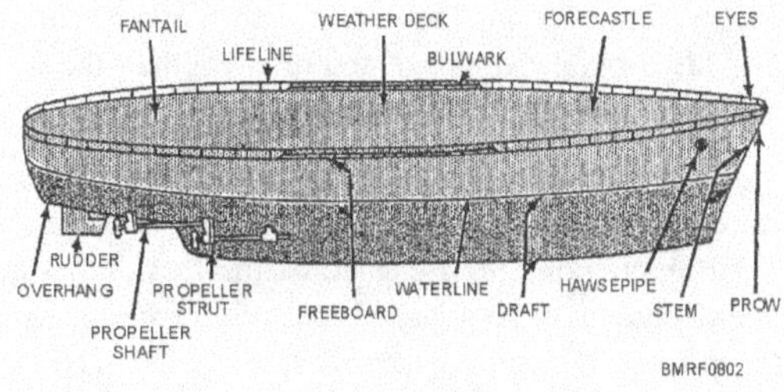

Figure 8-2.—The hull.

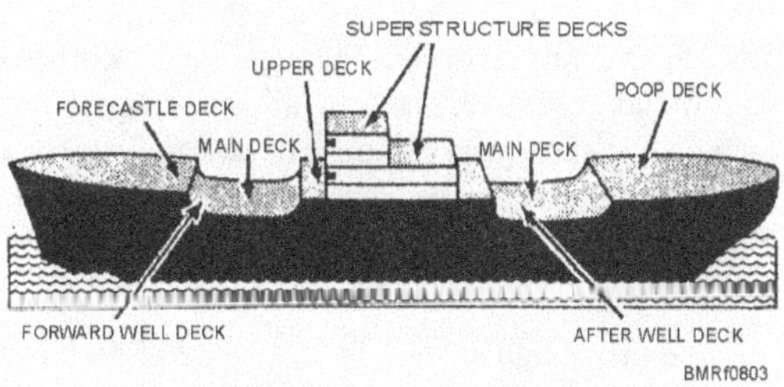

Figure 8-3.—The weather deck.

Decks

The floors of a ship are called *decks*. They divide the ship into layers and provide additional hull strength and protection for internal spaces. The undersurface of each deck forms the *overhead* (never the ceiling) of the compartment below. *Compartments* are the rooms of a ship. Some compartments are referred to as rooms, such as the wardroom, stateroom, and engine room.

Student Notes:

Generally speaking, you do not use the word *room*. For instance, you never refer to the space where you sleep as the bedroom or where you eat as the dining room. These spaces are called the *berthing compartment* or *space* and the *messdeck*.

A steel deck is made of steel plating (strakes) running fore and aft. The outboard strake in the deck plating is composed of stringer plates that are welded or riveted to the side plates of the ship adding additional strength to the ship's sides. Decks are supported by athwartships deck beams and by fore-and-aft deck girders. Further deck support is provided throughout the ship by vertical steel pillars called *stanchions*. Stanchions are mounted one above the other or one above a strength bulkhead. (The short posts used as lifeline supports also are called *stanchions*.) Look at figure 8-2. Decks are usually slightly bowed from the gunwale to the centerline to provide for water drainage and to strengthen the deck. The arch so formed is called *camber*.

A deck or part of a deck exposed to the weather is called a *weather deck* (fig. 8-3). *Bulwarks* are a sort of solid fence along the gunwale of the main (weather) deck. The bulwarks are fitted with *freeing ports* (scuppers) to permit water to run off during heavy weather.

A deck that extends from side to side and stem to stern is a *complete deck*. On an aircraft carrier, the uppermost complete deck is the *flight deck* from which aircraft take off and land. In all ships but aircraft carriers, the uppermost complete deck is the *main deck*. On an aircraft carrier, the *hangar deck* is the main deck. The hangar deck is the deck on which aircraft are stowed and serviced when not on the flight deck.

The first complete deck below the main deck is the *second deck*; the next, the *third deck*; the next, the *fourth deck*; and so on. *Half decks* or *'tween decks* take the number of the deck above and have the fraction 1/2 added to them.

A *strength deck* is just what the name implies. It is a complete deck (usually the main deck) designed to carry not only deck loads on it but also to withstand the hull stresses. A damage control deck (on most ships the second or third deck) is the lowest deck having access through the main transverse bulkheads, from forward to aft. This deck usually contains damage control main repair equipment in addition to the facilities for the control of flooding, sprinkling, and pumping if the ship is damaged.

The following are definitions that relate to decks in modern ships (the location of each deck is also given):

Companionways (ladders). Companionways, or ladders, lead from one deck level to another. They may or may not be covered by hatches.

Flats. Flats are plating or gratings installed only to provide working or walking surfaces above bilges.

Forecastle (pronounced folk' sel) *deck*. The forecastle deck is the deck above the main deck at the bow. Ships that don't have raised forecastles are called *flush-deckers*. In them, the part of the deck from the stem to just abaft the anchor windlass is the forecastle.

Gallery deck. The gallery deck is the first half deck or partial deck below the flight deck.

Half deck. The half deck is any partial deck between complete decks.

Levels. A level is a general term used to designate deck heights above the main deck. The first level above the main deck is the 01 (pronounced oh-one) level, the second the 02 level, and so on. Different decks at a particular level, however, carry different names. For example, both a poop deck and a boat deck (usually) are on the 01 level.

Platforms. Platforms are partial decks below the lowest complete deck. They are usually broken to admit machinery and are called *platform decks* or just *platforms*. They are numbered downward, as first platform, second platform, and so on.

Poop deck. The poop deck is a partial deck above the main deck located all the way aft. A flush-decker does not have a poop deck, so the stern area of the main deck on a flush-decker is called the main deck aft, or the *fantail*.

Quarterdeck. The quarterdeck is not an actual deck, but an area designated by the CO for the conduct of official functions. It is the station of the officer of the deck in port, and its location depends on how the ship is moored or which side of the ship is tied up to the pier.

Student Notes:

Superstructure deck. The superstructure deck is a partial deck above the main, upper, or forecastle deck that might not extend to the sides of the ship; or if it does, it does not have side plating carried up to it.

Upper deck. The upper deck is a partial deck extending from side to side above the main deck amidships. It is part of the superstructure, which is the part of a ship's structure above the deck. The superstructure does not include masts, yards, stacks, and related parts. The side plating extends upward to the upper deck.

Well deck. The well deck is the forward part of the main deck between upper deck and forecastle and aft between the upper deck and the poop deck.

Doors and Hatches

Access through bulkheads is provided by *doors* and through decks by *hatches*. *Watertight* (*WT*) *doors*, as the term implies, form a watertight seal when properly closed. All doors leading to weather decks are of the watertight variety, as are those in structural (watertight) bulkheads. (See fig. 8-4.) The doors are held closed by fittings called *dogs*, which bear up tight on wedges. A rubber gasket around the edge of the door presses against a knife-edge around the doorframe forming a watertight seal when all dogs are properly seated (*dogged down*). Some doors have individually operated dogs, as shown in figure 8-5. Other doors are quick acting types, for which a handwheel or lever operates all the dogs at once, as shown in figure 8-6. Some WT doors have openings, called *passing scuttles,* through which ammunition is passed. These scuttles (small tubelike openings) are flashproof as well as watertight.

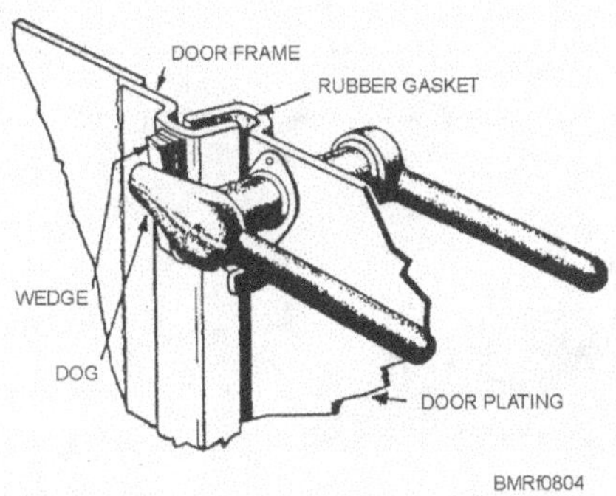

Figure 8-4.—Dogs and gasket for watertight door.

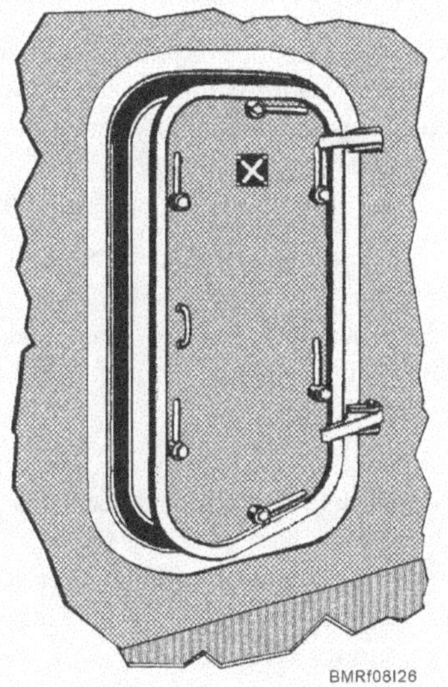

Figure 8-5.—Watertight door with individually operated dogs.

Nonwatertight (NWT) doors are used in NWT bulkheads and are of various types. Some slide, some fold, and others are similar to the regular house door (but made of metal). Some NWT doors have dogs, but fewer than those used on WT doors.

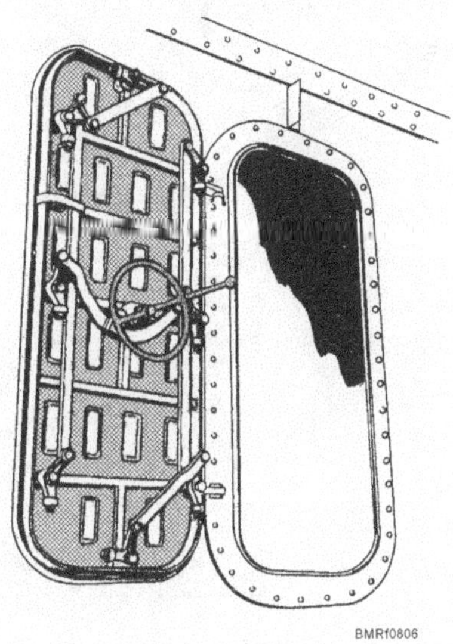

Figure 8-6.—Quick-acting watertight door.

Student Notes:

Hatches are horizontal openings for access through decks. A hatch is set with its top surface either flush with the deck or on a *coaming* (frame) raised above the deck. Hatches don't operate with quick-acting devices. They must be secured with individually operated dogs or drop bolts.

Figure 8-7 shows a typical hatch with an *escape scuttle*, which is a round opening with a quick-acting closure. An escape scuttle may also be found in the deck (or overhead) of a compartment that otherwise has only one means of access.

Manholes of the hinged type are miniature hatches provided in decks for occasional access to water, fuel tanks, and voids. Bolted manholes are sections of steel plate that are gasketed and bolted over deck access openings. Manholes are also found in bulkheads but are not as common as deck manholes.

A cargo hatch and hold are shown in figure 8-8. The hatch is a large opening in the deck that permits loading and unloading of equipment and materials. It is covered by hatch boards or a mechanical/hydraulic hatch cover. A cargo hatch is protected from the weather by a canvas tarpaulin (tarp for short). The tarp is pulled over the hatch boards and down the sides of the coaming around the hatch and then battened down. To batten down is to secure the tarp by wedging battens (slats of wood or steel) that hold it against the side of the coaming.

Superstructure

The solid part of a ship above the main deck is called the *superstructure* (fig. 8-9). The masts, stacks, and related gear above the superstructure are referred to as the ship's *top hamper* (fig. 8-10). Masts are of three general designs—pole, tripod, and cage. On a single-masted ship, the mast is called simply the *mast*. A two-masted ship has a *foremast* and *mainmast*. A three-masted ship has a *foremast, mainmast,* and *mizzenmast*, in that order from forward. *Stacks* (never chimneys or funnels) are the large pipes that carry off smoke and gases from the boilers. The wider lower section of a stack is an *uptake*.

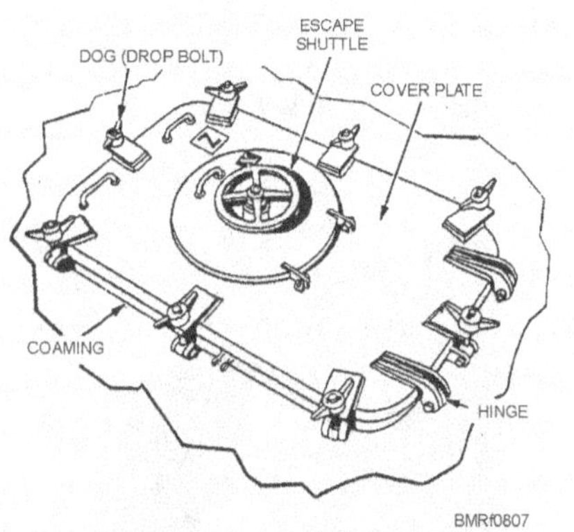

Figure 8-7.—Bolted hatch with escape scuttle.

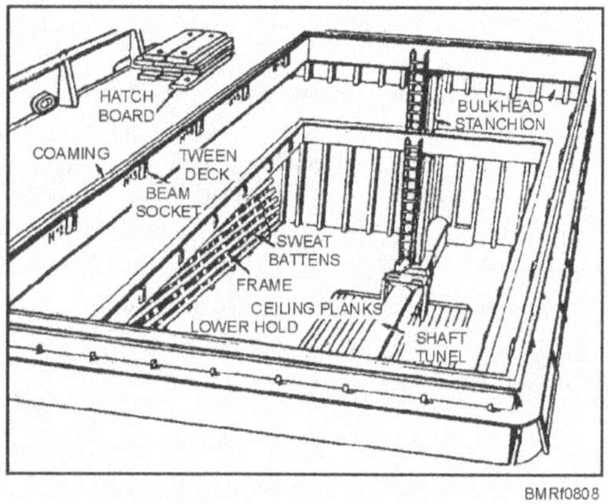

Figure 8-8.—Cargo hatch and hold.

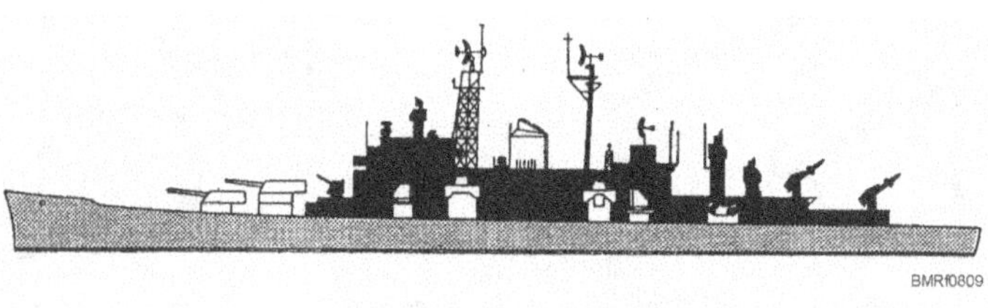

Figure 8-9.—Superstructure.

Student Notes:

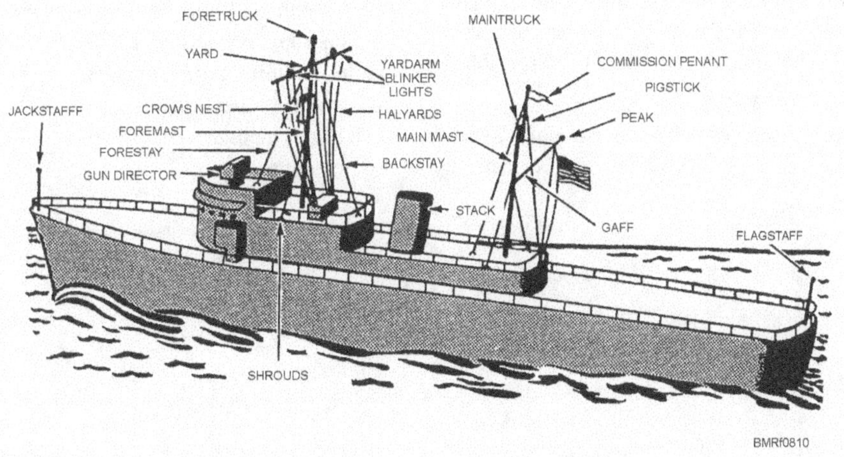

Figure 8-10.—Top hamper.

Masts are used to support radio and radar antennas, signal halyards (lines used for hoisting signal flags signal lights, and booms). Stays and shrouds, together with other wires used for similar purposes on stacks, masts, and so on, are known as the ship's *standing rigging*. Lines or wires used for hoisting, lowering, or controlling booms, boats, and so on, are known as *running rigging*.

Look at figure 8-10. The top of a mast is called the *truck*. A small sheave (a pulley, pronounced *shiv*) at the truck is used to run halyard lines for hoisting. The top of the foremast is the *foretruck*, and the top of the main- mast is the *main truck*. Commissioned ships of the U.S. Navy fly a commission pennant secured to a pigstick and hoisted to the truck. Ships that have radar antennas at the top of their masts fly the commission pennant from a sheave fixed in the highest convenient location.

Most foremasts have a light spar, called a *yard*, and mounted horizontally athwartships on their upper part. The port and starboard halves of a yard are the port and starboard *yardarms*. The yardarms carry a number of sheaves for signal halyards. Also, yardarms usually carry a set (two) of blinker lights, used (by means of a telegraph key) for signaling. The *gaff* is a light spar suspended at an angle abaft the upper part of the mainmast. The upper end of the gaff is the *peak*. The national ensign is flown at the peak while a ship is under way. When a ship is anchored or moored, the national ensign flies from the *flagstaff* at the stern, and the union jack flies from the *jackstaff* at the bow.

The *bridge*, from which the ship is controlled while under way, is located in the superstructure. The bridge contains the primary equipment used by the bridge watch personnel to control (*conn)* the movement of the ship: *helm* (steering control), *lee helm* (speed control), and *radar repeaters*. Ships also have a secondary conning station from which control can be maintained if the bridge is put out of commission. Some larger classes of ships have, in addition to the navigation bridge (conn), a *flag bridge* for the use of the squadron commander or admiral and staff.

The *signal bridge* (where Signalmen operate the signal lights, flags, and pennants) is normally located atop the bridge. On aircraft carriers, the signal bridge is abaft and usually one deck above the navigation bridge. Outboard, open ends of a bridge are called *bridge wings*. Located near the bridge is the *chart house*, where charts (maps) are stowed and worked on by the Quartermaster. Also nearby (on some ships) is the *combat information center* (CIC) manned by operations and combat systems department personnel.

Student Notes:

Main control is the station where the engineer officer controls the engineering functions of a ship. Main control is normally located below the main deck inboiler or mhinery spaces.

Each type of ship uses its superstructure spaces differently; hence, only generalitics can be made to describe them. Some of the spaces that may be found in the superstructures, in addition to the bridges, include administration and personnel offices, officers staterooms (berthing spaces), CPO quarters, a helicopter hangar, and radar and other electronic equipment rooms.

REVIEW 1 QUESTIONS

Q1. Label the following ship's parts.

a. Bow	d. Centerline
b. Beam	e. Port
c. Stern	f. Starboard

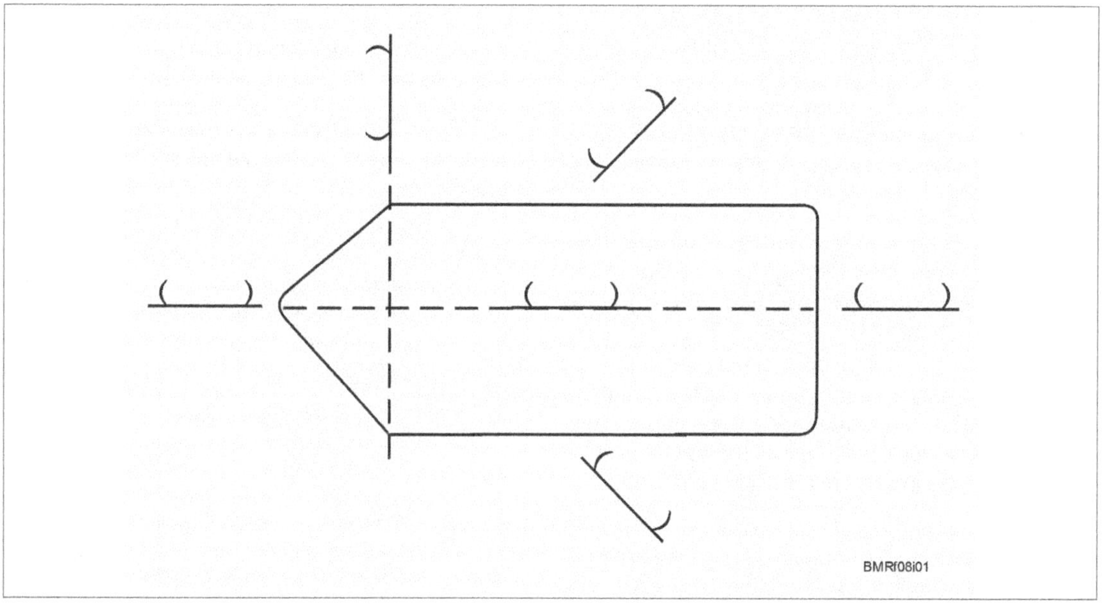

Student Notes:

Q2. Lable the following areas of a ship.

a. Bulkheads
b. Gunwale
d. Longitudinals
e. Stanchion

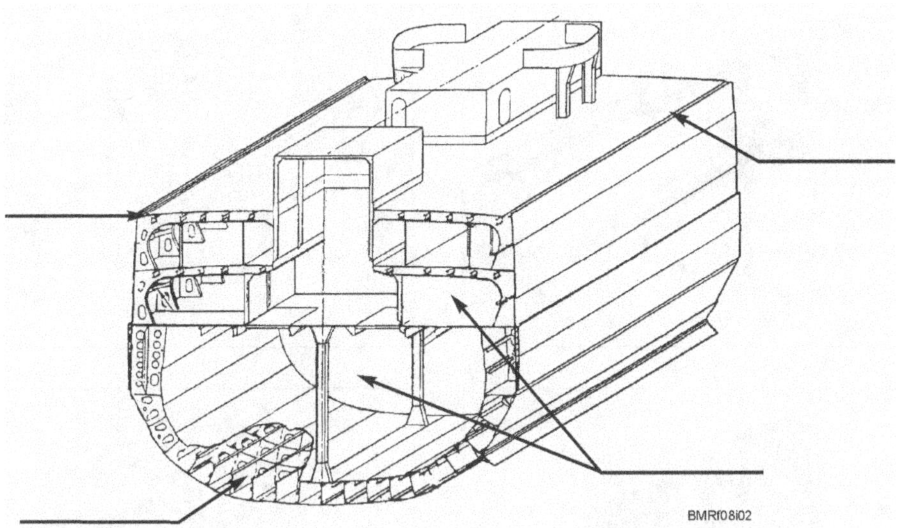

BMRf08i02

Q3. Label the following areas of a ship.

a. Forecastle
b. Poop deck
c. Main deck
d. Well decks
e. Superstructure decks
f. Upper deck

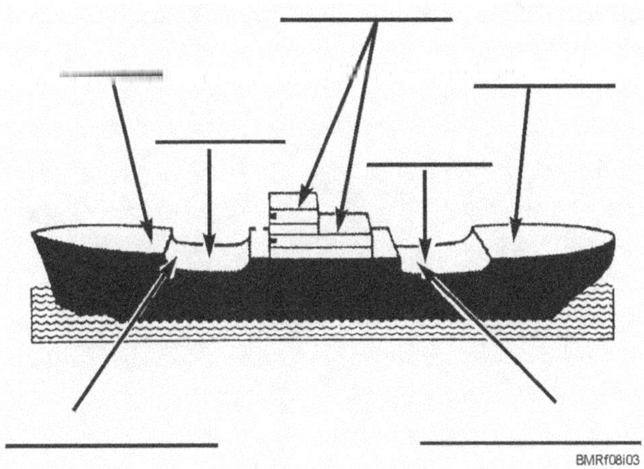

BMRf08i03

Student Notes:

Q4. Label the following doors/hatches.

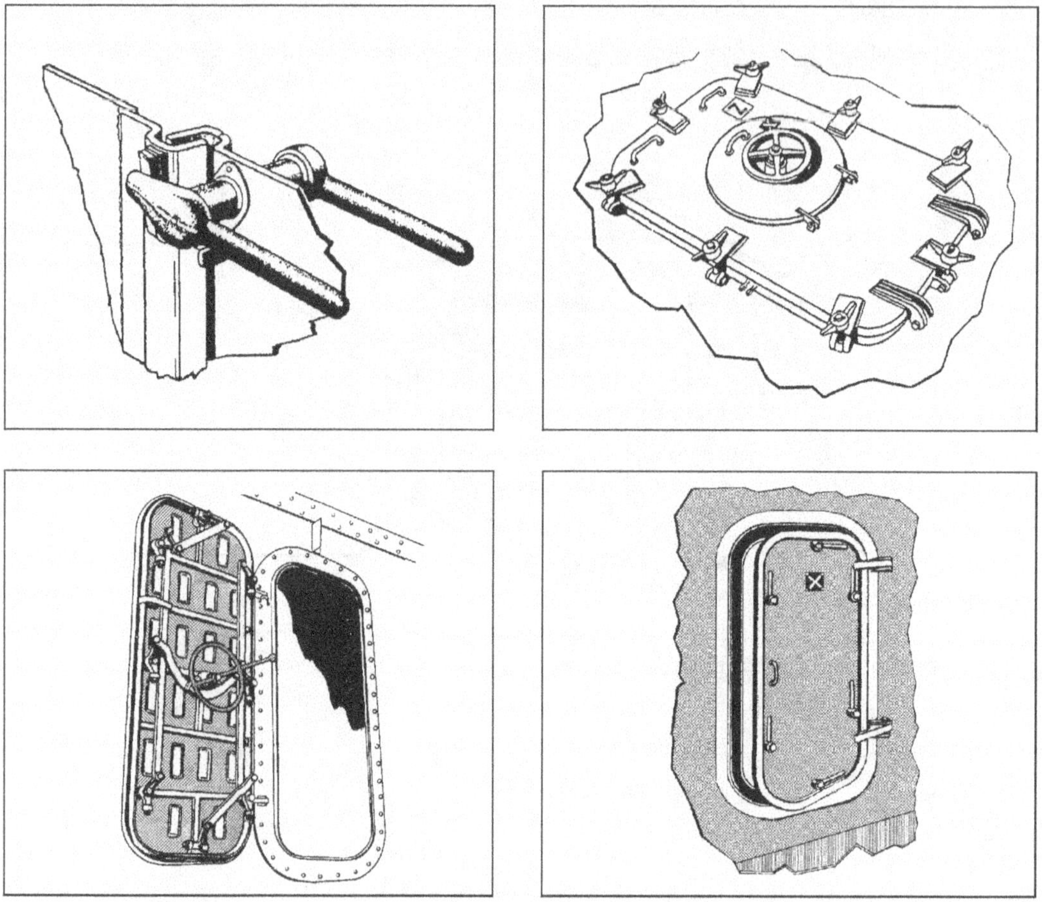

BMRf0860

Student Notes:

COMPARTMENT DESIGNATION/DECK NUMBERING SYSTEM

Learning Objectives: When you finish this chapter, you will be able to—

- Recall compartment designations.
- Recall deck lettering and numbering systems.

Every space in a ship (except minor spaces, such as pea coat lockers, linen lockers, and cleaning gear lockers) is assigned an identifying letter and number symbol. This symbol is marked on a label plate secured to the door, hatch, or bulkhead of the compartment. Compartments on the port side end in an even number and those on the starboard side end in an odd number (fig. 8-11). A zero precedes the deck number for all levels above the main deck. Figure 8-12 shows the system of numbering decks.

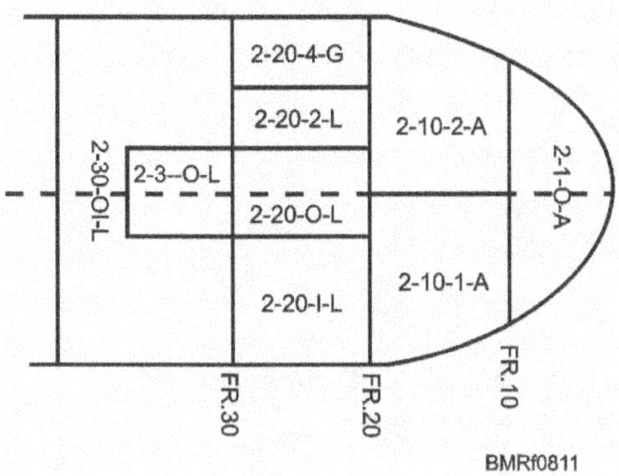

Figure 8-11.—Compartment designations.

Ship's compartment designations consist of a deck number, a frame number, the relationship of the compartment to the centerline, and a letter showing the use of the space. Where a compartment extends through two or more decks, the number of the lower deck is used. The frame number indicates the foremost bulkhead of the compartment. If the forward boundary is between frames, the frame number farthest forward within the compartment is used.

Compartments located on the centerline carry the number 0. Compartments to starboard are given odd numbers, and compartments to port are given even numbers. Where two or more compartments have the same deck and frame number, they have consecutively higher odd or even numbers, as applicable, numbering from the centerline outboard. For example, the first compartment to starboard is 1, the second is 3, and so on. To port of the centerline, they are numbered 2, 4, and so on. When the centerline passes through more than one compartment with the same frame number, the compartment having the forward bulkhead through which the centerline passes carries the number 0. Compartments above the main deck are numbered 01, 02, 03, as applicable, shown in figure 8-12.

The last part of the compartment number is the letter that identifies the primary use of the compartment. On dry and liquid cargo ships, a double letter is used for cargo holds to differentiate them from spaces containing the same commodity for use by the ship (for example, fuel oil). Compartment usage in the present system is shown in table 8-1.

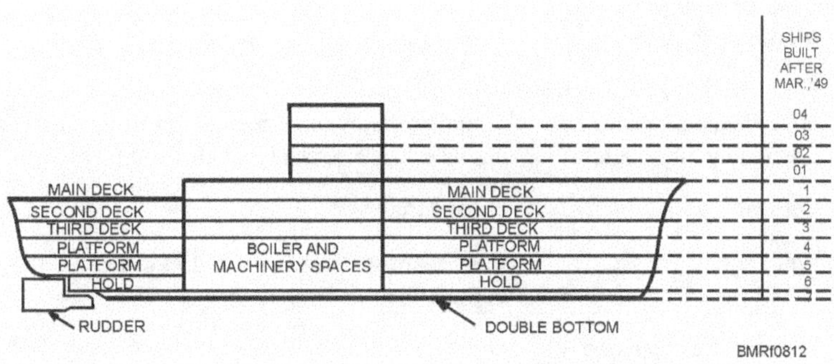

Figure 8-12.—Deck numbering system.

Student Notes:

Table 8-1.—Compartment Letters for Ships

Letter	Type of Compartment	Examples
A	Stowage spaces	Store and issue rooms; refrigerated compartments
AA	Cargo holds	Cargo holds and cargo refrigerated compartments
C	Control centers for ship and fire-control operations (normally manned)	CIC; plotting rooms; communications centers; pilothouse; electronic equipment operating spaces; IC rooms
E	Engineering control centers (normally manned)	Main machinery spaces; evaporator rooms; steering gear rooms; pump rooms; auxiliary machinery spaces; emergency generator rooms
F	Oil stowage compartments (for ship use)	Fuel- , diesel- , and lubricating-oil compartments
FF	Oil stowage compartments (cargo)	Compartments carrying various types of oil as cargo
G	Gasoline stowage compartments (ship use)	Gasoline tanks, cofferdams, trunks, and pump rooms
GG	Gasoline stowage compartments (cargo)	Spaces for carrying gasoline as cargo
J	JP-5 fuel (ship use)	Jet fuel stowage spaces
JJ	JP-5 fuel (cargo)	Spaces for carrying JP-5 fuel as cargo
K	Chemicals and dangerous materials (other than oil and gasoline)	Chemicals, semisafe materials, and dangerous materials carried as cargo or for ship's use
L	Living spaces	Berthing and messing spaces; staterooms; washrooms; heads; brig; sick bay; and passageways
M	Ammunition spaces	Magazines; handling rooms; turrets; gun mounts; shell rooms; ready service rooms
Q	Miscellaneous spaces not covered by other letters	Laundry; galley; pantries; wiring trunks; unmanned engineering; electrical and electronic spaces; shops; offices
T	Vertical access trunks	Escape trunks
V	Voids	Cofferdam spaces (other than gasoline); void wing compartments
W	Water stowage spaces	Drainage tanks; freshwater tanks; reserve feedwater tanks

Student Notes:

The following is an example of compartment designation for a ship:

Number 2-175-7-A

Second deck...................................2

Frame Number................................175

Fourth compartment to starboard
from the centerline........................7

Compartment usage (stowage)..............A

Access closures are numbered in the same manner as compartments, except that the letter designating usage is omitted.

REVIEW 2 QUESTIONS

Q1. Identify the following compartment designation number:

01-56-2-Q

Q2. Label the following areas on the ship.

a. 04
b. Main deck
c. Third deck
d. Platform
e. Machinery space

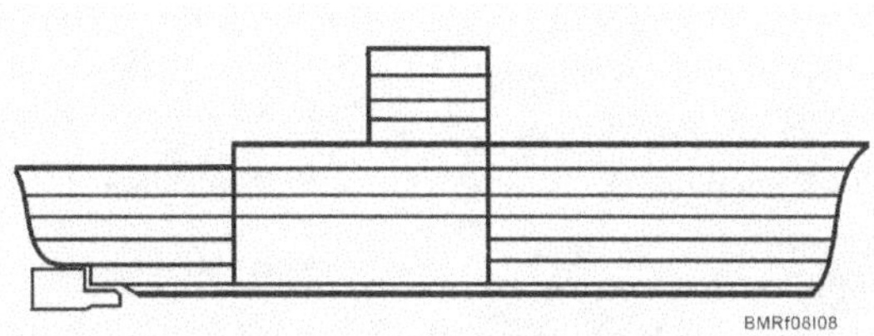

SHIP IDENTIFICATION

Learning Objectives: When you finish this chapter, you will be able to—

- Identify major types of ships to include their size, armament, armor, speed, class, and category.
- Identify types of warships to include aircraft carriers, surface combatants, submarines, and other types of combatants.
- Identify auxiliary types of ships to include replenishment-at-sea ships, material support ships, and fleet support ships.
- Identify the purpose and use of combatant craft.
- Identify the purpose and use of support craft.

Name and designation identify each Navy ship. In the name USS *Kitty Hawk* (CV-63), for example, USS means United States ship; CV is the designation—it indicates this type of ship is a multipurpose aircraft carrier. The ship's identifying or hull number is a general indication of the number of ships of the same type that have been built. (There are gaps in the sequence of numbers of most types because of the cancellation of shipbuilding orders, particularly at the end of World War II.) A ship's hull number never changes unless its designation also changes and not always then.

NOTE

Official designations for various types of ships are contained in appendix III, titled Ship's Classification.

TERMS USED IN SHIP IDENTIFICATION

The terms you will learn in this chapter will help you identify ships. Some of the terms you will learn are ship's size, armament, speed, class, and categories.

Student Notes:

Ship size. The size of a ship usually is given in terms of its displacement in long tons. Displacement means the weight of the volume of water that the ship displaces when afloat; in other words, the weight of a ship by itself. The Navy uses standard displacement, which is the weight of a ship when ready for sea. All weights given in this chapter are standard displacements, except where otherwise noted. Cargo ships usually are measured in light displacement (no cargo aboard) because of the wide difference in the weights of cargo carried.

Ship armament. Armament describes the offensive weapons a ship carries—guns, rockets, guided missiles, and planes.

Ship armor. Armor means protective armor—special steel installed along the sides of the ship, on a deck, and on some gun mounts and turrets.

Ship speed. The speed of a ship is stated in knots. A knot is 1 nautical mile per hour (mph) or about 1 1/8 statute miles per hour. When a ship goes 20 nautical miles an hour, its speed is said to be 20 knots (but never 20 knots per hour). A land (or statute) mile is 5,280 feet. A nautical mile is about 6,080 feet, or roughly 2,000 yards. A ship traveling at 20 knots is, therefore, traveling at the rate of about 23 mph.

Ship class. Ships are said to be of a particular class. Do not confuse this characteristic with type, which is shown by a ship's designation. The *Forrestal*, for example, was the first of several aircraft carriers of the same general advanced type and configuration to be completed. The next three carriers completed after the *Forrestal* are of the Forrestal class; however, later CVs or CVNs (nuclear-powered carriers) of other types are different classes (such as the Kitty Hawk class, Nimitz class, and so forth).

Ship categories. Ships of the U.S. Navy are divided into four categories that include combatant ships, auxiliary ships, combatant craft, and support craft.

REVIEW 3 QUESTIONS

Q1. How is the size of a ship usually given?

Q2. What is meant by a ship's armor?

Q3. What term is used to indicate the speed of a ship?

SHIPS CATEGORIES

Ships of the U.S. Navy are divided into four categories:

- Combatant ships
- Auxiliary ships
- Combatant craft
- Support craft.

Combatant Ships

Depending on size and type, combatant ships may have missions other than simply "slugging it out" with an enemy ship. Combatant ships are of two types—warships and other combatants.

WARSHIPS.—Most warships are built primarily to attack an enemy with gunfire, missiles, or other weapons. There are exceptions, however, that you will see as we go along. The following types of ships are included in the warship category:

- Aircraft carriers
- Battleships
- Cruisers
- Destroyers
- Frigates
- Submarines

Student Notes:

Aircraft Carriers.—There are three types of aircraft carriers—

1. Multipurpose aircraft carriers (CVs)
2. Multipurpose aircraft carriers (nuclear propulsion) (CVNs)
3. Training carriers

The job of the CV or CVN is to carry, launch, retrieve, and handle combat aircraft quickly and effectively. The aircraft carrier can approach the enemy at high speed, launch planes for the attack, and recover them. The attack carrier is an excellent long-range offensive weapon and is the center of the modern naval task force or task group. Figure 8-13 shows the USS Nimitz, and figure 8-14 shows aircraft flying over the USS Enterprise.

The displacement and aircraft capacity of the older CVs is less than the newer nuclear-powered CVNs. The older Forrestal class CVs displace about 79,000 tons and embark about 75 aircraft. The larger Nimitz class displaces about 96,000 tons and embarks about 85 aircraft. There is also a big difference in ships company and air wing complement (personnel assigned). The Forrestal class has about 5,400 personnel assigned, while the Nimitz class has about 5,700. Most carriers have the following equipment/capabilities:

- Angled flight decks
- Steam catapults
- Ability to launch and recover planes simultaneously
- Large hangar deck for plane stowage
- Deck-edge elevators to move aircraft rapidly between the hangars and flight decks
- Extensive repair shops and storerooms
- Fast-fueling equipment

Figure 8-13.—USS *Nimitz* (CVN 68).

Student Notes:

Photograph courtesy of LT Brian Douglas

Figure 8-14.—Various aircraft from Carrier Wing Three fly over the USS *Enterprise* (CVN 65).

The emphasis is on speed (all carriers can do over 30 knots), endurance, and sea-keeping ability (ability to stay at sea for long periods under all conditions), plane-carrying capacity, and maintenance capability.

Battleships.—The battleships have been decommissioned. However, they could be reactivated. Battleships participated in few surface engagements in World War II, but with their large number of antiaircraft guns, they proved to be excellent support ships in carrier task forces. Another major role was that of providing gunfire support of amphibious landings in both the Pacific and European theaters. Only their large-caliber guns could knock out heavily reinforced gun emplacements. They also provided gunfire support in the Korean conflict.

Several battleships (BBs) were modernized to include additional armament such as Tomahawk and Harpoon missile systems or the Phalanx close-in weapons system (CIWS). Battleships were given state names. However, since there is little likelihood of our building any more battleships, state names are being given to cruisers like the USS *South Carolina* (CGN 37) and to submarines (SSBNs) like the USS *Ohio* (SSBN 726) and USS *Michigan* (SSBN 727).

Cruisers.—Cruisers are medium-sized, general-utility ships. They have a large cruising range and are capable of high speeds (over 30 knots). They serve as protective screens against surface and air attacks and also provide gunfire support for land operations. The two basic types of cruisers are the guided-missile cruiser (CG) and guided-missile cruiser (nuclear propulsion) (CGN). Cruisers displace about 10,000 tons. The CGs include cruisers with missiles, but some of these also have guns that are 5"/54 caliber. CGNs are the same as the CGs except that their main engines are nuclear-powered. Figures 8-15 and 8-16 show two cruisers.

Student Notes:

Photograph courtesy of PH3 Jason D. Malcom

Figure 8-15.—USS *Philippine Sea* (CG 58) comes alongside USS *Enterprise during an under way replenishment.*

Photograph courtesy of PH3 Christopher Mobley

Figure 8-16.—USS *Port Royal* (CG-73).

The *Ticonderoga* (CG 47) class cruisers are built on the *Spruance* (DD 963) hull. Modern U.S. Navy guided-missile cruisers perform primarily a battle force role. These ships (fig. 8-16) are multimission surface combatants capable of supporting carrier battle groups, amphibious forces, operating independently, and as flagships of surface action groups. Because of their extensive combat capability, these ships have been designated as battle force capable.

Destroyers.—Destroyers (DDs) and guided-missiles destroyers (DDGs) are multipurpose ships that are useful in almost any kind of naval operation. They are fast ships with a variety of armaments, but little or no armor. For protection, they depend on their speed and mobility. Their displacement varies from 2,425 tons to 7,800 tons.

The principal mission of destroyers is to operate offensively and defensively against submarines and surface ships and to take defensive action against air attacks. They also provide gunfire support for amphibious assaults and perform patrol, search, and rescue missions.

The destroyers armament consists of 5-inch guns and a variety of antisubmarine weapons, such as torpedoes, ASROCs, and surface-to-air missiles.

Traditionally, destroyers have been named after Secretaries of the Navy and officers and enlisted personnel of the Navy and Marine Corps.

Destroyers make up the Navy's largest group of similar types of ships. Only a few are mentioned so you will have some idea of the several types and classes.

Spruance class destroyers. The Spruance (fig. 8-17) class destroyers displace 7,800 tons fully loaded. Each of these ships has two 5"/54-caliber guns, one Seasparrow missile launcher, one ASROC launcher, and two Mk 32 triple-torpedo tubes. They also have full helicopter facilities to accommodate the SH-2H or SH-60B helicopter, and the larger Sea King SH-3 helicopter. The Spruance class destroyers are the first large U.S. warships to use gas-turbine propulsion. This propulsion system was selected because of its smaller space requirements, rapid replacement capability, and cold start capability. (The engines can go from "cold iron" to full power in 12 minutes.)

Kidd class guided-missile destroyers. The Kidd class guided-missile destroyers are designed around the Spruance hull and engineering plant. Armament includes two Mk 26 Tartar/ASROC launchers; two Quad Harpoon canisters; two Mk 45, 5"/54-caliber gun mounts; and two Vulcan/Phalanx CIWSs. There are facilities for two SH-2 LAMPS or one LAMPS III. Displacement of these ships is 8,500 tons and propulsion is gas turbine.

Arleigh Burke class destroyers. The DDGs of the Arleigh Burke class (fig. 8-18 and fig. 8-19) are the most

Photograph courtesy of PH1 James Slaughenhaupt

Figure 8-17.—Spruance class destroyer, USS *Moosebrugger* (DD 980).

Student Notes:

Figure 8-18.—Guided-missile destroyer USS *The Sullivans* (DDG 68).

Figure 8-19.—USS *Hopper* (DDG 70).

powerful and survivable class of destroyers ever put to sea. They possess the following capabilities:

- AEGIS weapons system with the AN/SPY-1D multi-function radar, capable of detecting and tracking over one hundred targets simultaneously, while conducting multiple engagements
- The vertical launching system, capable of storing and rapidly firing 90 missiles
- The SQQ-89 antisubmarine warfare system with its SQR-19 towed array sonar and the SQS-53C digital hull-mounted sonar
- The Harpoon antiship cruise missile system
- The Tomahawk antiship and land attack cruise missile system, capable of hitting targets hundreds of miles away
- Improved versions of the 5-inch gun and the Phalanx close-in weapons system.

The Arleigh Burke class represents a return to all-steel construction and incorporates electromagnetic pulse hardening, enhanced firefighting features, and a collective protection system to provide protection against nuclear, chemical, or biological contamination. This vital equipment is distributed through the ship, giving the ship improved blast and fragmentation protection, which lets them to survive a hit and continue to fight.

Frigates.—The classification "frigate" designates ships used for open-ocean escort and patrol. Frigates resemble destroyers in appearance, but they are slower, have only a single screw, and carry less armament. Frigates are slowly being replaced by DDGs. The Oliver Hazard Perry class is the only class of guided-missile frigates still commissioned. The USS *Ingraham* (FFG 61) (fig. 8-20) carries the following armament:

- A single 76-mm, .62-caliber
- Dual-purpose gun
- A 20-mm Vulcan/Phalanx rapid-fire gun
- A single launcher for Harpoon missiles
- Two SH-60 LAMPS III helicopters
- Two Mk 46 triple-torpedo tubes

Submarines.—The Navy deploys two classes of submarines attack submarines (SSNs) and ballistic missile submarines (SSBNs). The mission of nuclear attack submarines (SSNs) is to locate and destroy enemy ships and submarines. They also act as scouts, deliver supplies and personnel to locations in enemy territory, and perform rescue missions.

Fleet ballistic missile submarines (SSBNs) deliver ballistic missile attacks against assigned targets from either a submerged or surfaced condition. Most of the SSBNs are being converted to carry Trident missiles, which have greater range and multiple warheads.

A new class of submarine, the Ohio class (fig. 8-21), has been developed for the Trident missile. The Ohio class is the largest undersea craft developed by the Navy. It displaces 16,600 to 18,700 tons. The size of the Trident submarine is dictated by the larger size missile required for ranges of 4,000 to 6,000 miles and by the larger reactor plant required to drive the ship. The submarine has 24 tubes for the Trident missile and 4 torpedo tubes located in the bow.

A nuclear-powered attack submarine, like that of the Sturgeon class, displaces 3,800 to 4,700 tons, can do more than 20 knots, and has four torpedo tubes. The newer Seawolf class fast-attack submarine displaces about 9,137 tons, has four torpedo tubes, and can attain speeds of over 35 knots (fig. 8-22). The Seawolf performs a variety of crucial assignments, from underneath the Arctic icepack to all regions anywhere in the world. Its missions include surveillance, intelligence collection, special warfare, covert cruise-missile strike, mine warfare, and anti-submarine and anti-surface ship warfare. The Seawolf's stealth characteristics make it the world's quietest submarine.

Early submarines were named after marine life. The first SSBNs, however, were given names of persons well known in American history, like USS *George Washington*, USS *Patrick Henry*, and USS *Lafayette*. The new fast-attack submarines (SSNs) are named after American cities, like the USS *Los Angeles*, USS *Albuquerque*, and USS *Memphis*. The Tridents (SSBNs) are being named after American states, like the USS *Ohio* and USS *Michigan*.

Student Notes:

Figure 8-20.—USS *Ingraham* (FFG 61).

Photograph courtesy of PH3 Erik Kenny

Figure 8-21.—USS *Maryland* (SSBN 773).

Figure 8-22.—USS *Seawolf* (SSN 21).

OTHER COMBATANTS.—Other ships classified as combatants are amphibious warfare ships and mine warfare ships.

Amphibious warfare ships.—An amphibious assault operation is the fastest means of landing large numbers of personnel, equipment, and supplies on enemy-held territory. The lessons learned during World War II, Korea, and Vietnam have resulted in the U.S. Navy having the largest and most capable amphibious force in the world. With the introduction of new classes of ships and new types of landing craft and helicopters, the U.S. Navy can conduct an amphibious operation almost anywhere in the world.

Amphibious assault ships. Tarawa-class amphibious assault ships (LHAs) are able to embark, deploy, and land a marine battalion landing team by helicopter, landing craft, amphibious vehicle, or by a combination of these methods. The Tarawa-class ships have 9 *Sea Stallions* and 12 *Sea Knight* helicopters plus 6 *Harrier* attack planes. It also carries 2 RAM launchers, two 5"/54 caliber Mk-45 lightweight guns, two *Phalanx* 20mm CTWS mounts and six 25mm Mk 38 machine guns. The USS *Belleau Wood* (LHA 3) (fig 8-23) and the USS *Peleliu* (LHA 45) (fig. 8-24) are examples of amphibious assault ships.

The Wasp-class LHDs are designed to embark, transport, and land 2,000 troops and their equipment using transport helicopters in conjunction with a beach assault. The Wasp-class ships are the largest amphibious ships in the world (fig. 8-25). Their vertical envelopment is more effective than older methods of amphibious landings. One feature of this class of ships is the ability to commit the landing force in an assault without being limited to favorable beaches. These ships allow establishment of beachheads in enemy territory more quickly than older methods. When not in used for amphibious assaults, LHDs have the capability to assist in antisubmarine warfare.

Amphibious transport dock. Amphibious transport docks (LPDs) are versatile ships. They perform the mission of amphibious transports, amphibious cargo ships, and older LSDs. The Navy's newest class of ships are scheduled to replace the Navy's amphibious fleet. The LPD (fig. 8-26) is a highly reliable, warfare-capable ship, as well as the most survivable amphibious ship ever put to sea. The LPD incorporates the latest quality of life standards for the embarked Marines and Sailors—they accommodate women as part of the crew and embarked troops.

Student Notes:

Photograph courtesy of PHC Justin A. Kahkosoko

Figure 8-23.—USS *Belleau Wood* (LHA-3) refuels USS *Vincennes* (CG 49) during an under way replenishment.

Photograph courtesy of PHC Larry Nixon

Figure 8-24.—Landing craft, utility (LCU-1663) back loads equipment and personnel to USS *Peleliu* (LHA 45).

Photograph courtesy of JO2(SW/AW) Tyler A. Swartz

Figure 8-25.—USS *Bonhomme Richard* (LHD 6).

Photograph courtesy of PH2 Felix Garza

Figure 8-26.—USS *Duluth* (LPD 6).

Dock landing ships. Dock landing ships (LSDs) (fig. 8-27) were designed to transport and launch a variety of amphibious craft and vehicles with embarked crews and troops. All landing craft operate from a well deck that is over 300 feet long and 50 feet wide. The types of amphibious craft vary from the newer LCAC (landing craft air cushion) (fig. 8-28) to the conventional LCU (landing craft utility) or LCM (landing craft mechanized). The number of amphibious craft embarked will vary, depending on the type of craft and class of ship.

Photograph courtesy of PH2 Todd R. Cromer.

Figure 8-27.—USS *Comstock* (LSD 45).

The newer class of LSD is capable of transporting and operating four LCACs while the older classes may embark only three. A newer variant of the LSD will be designed to handle only two LCACs but will have a

Student Notes:

Photograph courtesy of SGT Don L. Mayes

Figure 8-28.—Landing craft air cushion (LCAC).

larger cargo capacity. These ships also have a helicopter platform over the well deck that allows them to conduct limited helicopter operations.

Tank landing ships. Tank landing ships (LSTs) (fig. 8-29) were developed during World War II. The Navy required a ship capable of transporting troops, tanks, ammunition, and all sorts of supplies. The LSTs of today's fleet are fitted with bow doors and a bow ramp that give access to the tank deck. Another ramp and turntable in the tank deck enable vehicles to turn around and reach the main deck under their own power. They also have a stern gate that permits off-loading of amphibious vehicles directly into the water. In addition to transporting and landing equipment in amphibious assaults, these ships can transport and launch a pontoon causeway section in support of amphibious operations. With booms and winches mounted on the main deck forward, this class of ship is capable of numerous missions. They carry one 20mm *Phalanx* and two 25mm Mk3 machine guns.

Amphibious command ships. Amphibious command ships (LCCs) (fig. 8-30) provide amphibious command and control for major amphibious operations.

Figure 8-29.—USS *LA Moure County* (LST 1194).

Photograph courtesy of PH2 John Sokolowski

Figure 8-30.—USS *Mount Whitney* (LCC-20).

With the latest command and control facilities available, these ships have become fleet flagships. They are capable of supporting a naval amphibious task force, a landing force, and an air force simultaneously.

Mine Warfare Ships.—Mine countermeasures ships (MCM) are ships designed to clear mines from vital waterways. In the early 1980s, the U.S. Navy began development of a new mine countermeasures (MCM) force, which included two new classes of ships and minesweeping helicopters. The Iran-Iraq war and Operation Desert Shield/Desert Storm showed the importance of a state-of-the-art mine countermeasures force when the *Avenger* (MCM 1) and *Guardian* (MCM 5) ships conducted MCM operations.

Avenger class ships are designed as mine hunter-killers capable of finding, classifying, and destroying moored and bottom mines. These ships use sonar and video systems, cable cutters, and a mine-detonating device that can be released and detonated by remote control. They are also capable of conventional sweeping measures. The ships are of fiberglass sheathed, wooden hull construction. They are the first large mine countermeasures ships built in the United States in nearly 27 years. (See fig. 8-31.)

Figure 8-31.—USS *Chief*.

Student Notes:

Osprey (MHC 51) class ships are also designed as mine hunter-killers. The MHC 51 has a 15-day endurance and depends on a support ship or shore-based facilities for resupply. Ships under this class are named after birds.

REVIEW 4 QUESTIONS

Q1. List the four categories of ships.

a.

b.

c.

d.

Q2. List the six classes of warships.

a.

b.

c.

d.

e.

f.

Q3. What are battleships names after?

Q4. Name the two basic classes of cruisers.

a.

b.

Q5. For protection, the destroyer depends on their

_____________ and _____________.

Q6. What class of ship was developed for the purpose of open ocean escort and patrol?

Q7. Name the two classes of submarines.

a.

b.

Q8. What class of ship is used to land large numbers of personnel, equipment, and supplies on enemy held territory?

Auxiliary Ships

Today's fleet is highly mobile and can respond to an area of conflict quickly. However, its ships cannot remain on station indefinitely. There must be a means of resupply and repair. The auxiliary ships of today's fleet are the lifeline to the combatant force. These ships keep the fleet operating by furnishing vital supplies and repair facilities. They can deliver such items as fuel, food, ammunition, and repair parts.

The types of ships in the auxiliary force range from fast combat support ships (AOEs) to rescue and salvage ships (ARSs). The type of service an auxiliary provides determines its classification. The initial letter in each designation is the letter *A*. The second and subsequent letter indicates the service it performs. An *AE* indicates an ammunition (explosives) supply ship, while an *AO* is an oiler. These types of ships do not always receive the level of publicity a carrier or cruiser might receive, but they fight and work just as hard in times of emergency. Certain classes of auxiliaries have the capability to function in many roles. An AOE is capable of supplying not only fuel and ammunition but can supply dry stores and refrigerated stores.

Student Notes:

REPLENISHMENT-AT-SEA SHIPS.—*Replenishment* at *sea* is the term applied to the transfer of fuel, munitions, supplies, and personnel from one vessel to another while ships are under way. During World War II, replenishment at sea (fig. 8-32) was developed to a fine art of seamanship, which is taken as a matter of course today.

Replenishment at sea is accomplished with both the replenishment ship and the ship(s) being replenished steaming side by side on parallel courses at a

Figure 8-32.—Replenishment at sea enables the fleet to remain at sea and make successive strikes without returning to base for fuel, ammunition, and supplies.

Student Notes:

predetermined speed. In most cases, the replenishment ship maintains its course and speed while the other ship(s) maneuver(s) into position alongside. A separation of about 100 feet is maintained between ships, with the replenishing ship frequently serving ships both to port and starboard. Messenger lines are passed to the receiving ships, which send back telephone and distance measuring lines and then haul over cargo-handling gear or fuel hoses by means of the messengers.

Ships designed for that purpose do most of the replenishment, but major combatant ships are capable of refueling smaller ships. Even the smallest ships can, and do, transfer light freight, mail, and personnel by means of highlines.

In addition to the standard replenishment capabilities, all recently constructed, as well as many of the older auxiliary, ships have helicopter platforms for the transfer of munitions, personnel, cargo, and stores by vertical replenishment. Vertical replenishment permits a receiving ship to remain on station in combat formation, eliminating the necessity of temporarily immobilizing itself by going alongside another ship for replenishment.

Ammunition Ships.—Ammunition ships (AEs) (fig. 8-33) operate with replenishment groups to deliver ammunition and missiles to the fleet at sea. Their design incorporates a mechanical handling system for more rapid loading and off-loading of ammunition. The mechanical handling system includes such equipment as dual-cantilevered elevators in the holds; forklift trucks; and low-lift, power-operated transporters on the main deck for handling palletized ammunition from the elevators to the transfer stations. Universal portable metal dunnage provides maximum stowage with ready access to all types of ammunition. A tension highline system is built into the design along with new, improved electro-hydraulic cargo winches for replenishment at sea. These improvements provide for much more rapid and reliable transfers and conservation of deck space. These ships are capable of handling all types of missiles (fig. 8-34).

Oilers and Tankers.—Oilers (AOs), carrying Navy fuel oil, jet fuel, and other petroleum products, operate with replenishment groups and deliver their cargo to ships at sea. Oilers, as well as ammunition ships (fig. 8-35), can service ships on both sides simultaneously.

The AO (Jumbo) is a conversion of the AO that includes the installation of a new midsection in the hull. This midsection increases the payload and provides for an improved balance of cargo fuel products to meet the more recent demands placed upon the AO by the increase in fleet requirements for jet aircraft fuel.

BMRf0833

Photograph courtesy of PH3 Jason Branson

Figure 8-33.—USS *John C. Stennis* (CVN 74) off-loads ammunition onto ammunition ship USS *Mount Hood* (AE 29).

Student Notes:

Figure 8-34.—An AE highlines a missile to a guided-missile ship during replenishment at sea

Fast Combat Support Ships.—The fast combat support ship (AOE) is the largest and most powerful auxiliary ship in the Navy. Unlike other replenishment ships, the AOE is designed to operate as an integral force rather than as a unit of an underway replenishment group.

The AOE (fig. 8-36) is a multiple-product ship (missiles, fuel, ammunition, and general cargo) that has a cargo-fuel capacity greater than that of our largest fleet oilers plus a hold capacity equal to the largest ammunition ship. In addition, the ship carries a large load of both general supplies, materials, and refrigerated cargo.

Other than speed and capacity, this ship has two major areas of improvement over other replenishment vessels—material handling and replenishment at sea. Materials, other than missiles and special weapons, are moved vertically by elevators or conveyors. Horizontal movement of general cargo and ammunition is mechanized through the use of pallet transporters and forklift trucks. Cargo helicopters are available to

Student Notes:

Figure 8-35.—AOE conducting an evolution.

Figure 8-36.—A multiple-product AOE conducting under way replenishment.

replenish outlying units of the force with dry cargo and ammunition.

The missile and special weapons-handling system is separate from the cargo-handling system. This arrangement permits a continuous flow of missiles from the cargo holds to the missile-transfer system, port or starboard.

The fuel hoses on the AOE are designed to permit an average ship separation of 200 feet during replenishment instead of the normal 100 feet. The greater distance reduces the possibility of collision and makes increased replenishment speeds feasible. There are nine replenishment stations to port and six to starboard.

FLEET SUPPORT SHIPS.—While certain types of naval auxiliary ships are designed and equipped specifically for towing, for salvage, or for rescue operations, most of these types may, in an emergency and to a limited extent, perform all these operations. Among ships as versatile and as adaptable as the auxiliaries, there is bound to be an occasional overlapping of functions to meet an unexpected situation.

Rescue and Salvage Ships.—The mission of the rescue and salvage ship (ARS) has four parts—debeaching stranded vessels, heavy lift capability from ocean depths, towing other vessels, and manned diving operations. For rescue missions, these ships are equipped with fire monitors forward and amidships, which can deliver either fire-fighting foam or seawater. The salvage holds of these ships are outfitted with portable equipment to provide assistance to other vessels in dewatering, patching, and supplying of electrical power and other essential services required to return a disabled ship to an operating condition.

The Navy employs ARSs (fig. 8-37) to salvage U.S. government-owned ships and, when it is in the best interests of the United States, privately owned vessels. The rugged construction of these steel-hulled ships, combined with speed and endurance, make rescue and salvage ships well suited for rescue/salvage operations of Navy and commercial shipping throughout the world. The versatility of this class of ship adds to the capabilities of the U.S. Navy with regard to assisting those in need on the high seas.

Photograph courtesy of PH1 Todd P. Cichonowicz

Figure 8-37.—USS *Grapple* (ARS-53).

Oceangoing Tugs.—There is one major type of oceangoing tug—the ATF (fleet ocean tug) (fig. 8-38). It has a large cruising range and limited salvage capabilities. ATFs are equipped with firefighting equipment, including fire monitors. (A fire monitor is similar in appearance to a gun and permits water to be discharged through a horizontal arc of 360°.) They also are fitted with automatic towing machines and booms. In addition to hauling and towing, fleet tugs may be called on to patrol certain areas, lay smoke screens, and pull landing craft off beaches. They are often used in search and rescue (SAR) operations. Military Sealift Command (MSC) personnel now operate most fleet tugs.

Photograph courtesy of Robert J. Sitar

Figure 8-38.—USNS *Powhaten* (T-ATF-166).

Student Notes:

Combatant Craft

Combatant craft include patrol craft, amphibious warfare craft, and mine warfare craft.

Patrol craft. Surface patrol craft are intended for use relatively near the coast or in sheltered waters or rivers. These craft may be transported aboard larger units.

Amphibious warfare craft. All amphibious craft that have the organic capacity for amphibious assault, principally in coastal waters. They may be transported aboard larger units.

Mine warfare craft. All craft with the primary function of mine warfare that operate in coastal waters. They may be transported aboard larger units.

Support Craft

Among the hardest working ships of the Navy are the support craft. Not primarily fighting ships, they are for the most part unarmed. These are ships that serve a variety of purposes in continental and overseas harbors, sea frontiers, convoys, amphibious forces, and task forces. Many are small, but of incalculable use to the Navy.

With a few exceptions, support craft designations start with the letter *Y*. A few of the class names identify the many duties they perform:

- Auxiliary floating dry dock—large (AFDB) and small (AFDL)
- Floating crane (YD)
- Diving tender (YDT)
- Ferryboat or launch (YFB)
- Fuel oil barge (YO)
- Gasoline barge (YOG)
- Oil storage barge (YOS)
- Floating workshop (YR)
- Tug (YTL, YTM, or YTB)
- Water barge (YW)

REVIEW 5 QUESTIONS

Q1. What is the term used to describe the transfer of fuel and supplies between ships while underway?

Q2. Ships usually maintain a distance of _______ feet while taking on supplies at sea.

Q3. What type of replenishment allows a receiving ship to stay on station in combat formation?

Q4. What class of ship is the largest and most powerful auxiliary ship?

Q5. The mission of the rescue, salvage, and towing ships is to—

Q6. Support craft designators usually start with what letter?

NAVAL AIRCRAFT

Learning Objective: When you finish this chapter, you will be able to—

- Recognize fixed-wing and rotary-wing aircraft, to include aircraft nomenclature and characteristics.

The history of naval aviation goes back to 1911 when the Navy acquired its first aircraft, a pusher-type biplane with no cockpit. The only covered surfaces were the wings and tail, and flight speed was less than

Student Notes:

50 mph. By contrast, today's high-performance planes have speeds in excess of 2,000 mph.

AIRCRAFT NOMENCLATURE

In this section, you will learn the basic parts of aircraft and how the Navy identifies aircraft.

Fixed-Wing Aircraft Nomenclature

A fixed-wing aircraft (fig. 8-39) may be divided into three basic parts—fuselage, wings, and empennage (tail).

FUSELAGE.—The fuselage is the main body of the aircraft, containing the cockpit and, if there is one, the cabin. On virtually all naval fighter and attack aircraft operational today, the engines and some of the fuel tanks are mounted within the fuselage.

WINGS.—Wings are the primary lifting devices of an aircraft, although some lift is derived from the fuselage and tail. Located on the trailing (rear) edge of the wings are flaps that may be used to give extra lift on takeoff or to slow the aircraft in flight or landings; ailerons that control the roll or bank of the aircraft; and trim tabs used to aerodynamically unload the control surfaces to relieve some of the pilot's work. On the leading (front) edge of the wing may be found auxiliary lifting devices, resembling flaps, which are used to increase camber (curvature) of the wing for added lift on takeoff. Most Navy jet aircraft carry their bomb loads on pylons (called *stations*) under the wings and, in some cases, under the fuselage. Some jets have missile stations on the sides of the fuselage. Fuel cells are located in the wings; additional external tanks can be fitted for extra range. Larger jets may have their engines slung beneath the wings in pods. Some low-wing aircraft have their main landing gear retract into the wings, while the nose wheel retracts into the fuselage. On most high-wing aircraft all gear retracts into the fuselage.

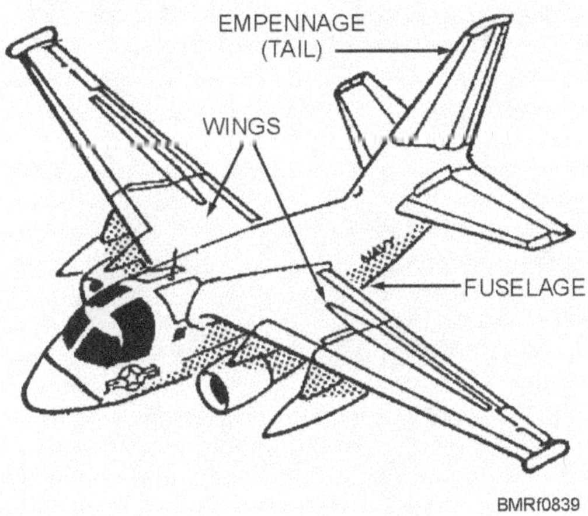

Figure 8-39. Fixed-wing aircraft.

EMPENNAGE.—The empennage consists of the stabilizing fins mounted on the tail section of the fuselage. These include the vertical stabilizer on which is generally mounted the rudder that is used to control yaw, or direction of the nose about the vertical axis; and the horizontal stabilizer, on the trailing edge of which are the elevators that determine the pitch (climb or dive). Some supersonic aircraft may have a full delta wing. In that case, there is no horizontal stabilizer and the elevators and ailerons are combined into control surfaces called *elevons*.

In aircraft with internally mounted jet engines, exhausts normally are in the tail. High-performance jets have afterburners that give additional thrust at the cost of greatly increased fuel consumption.

Rudder, ailerons, and elevators are collectively grouped as control surfaces. The "stick" or a similar device in the cockpit controls these surfaces, while foot pedals control the rudder. On high-performance aircraft, aerodynamic pressures on these surfaces become too great for a pilot to overcome manually; hence, all high-speed models today have power-assisted controls.

Rotary-Wing Aircraft Nomenclature

The aerodynamics of rotary-wing aircraft (fig. 8-40) are considerably more complex than those of fixed-wing aircraft. A helicopter essentially consists of a fuselage, main rotor or rotors, and often a tail rotor.

FUSELAGE.—As in fixed-wing aircraft, the fuselage contains the cockpit and cabin.

MAIN ROTOR.—The main rotor is the approximate equivalent of the wing of a fixed-wing aircraft. Each rotor blade is an airfoil, like a wing, and

Student Notes:

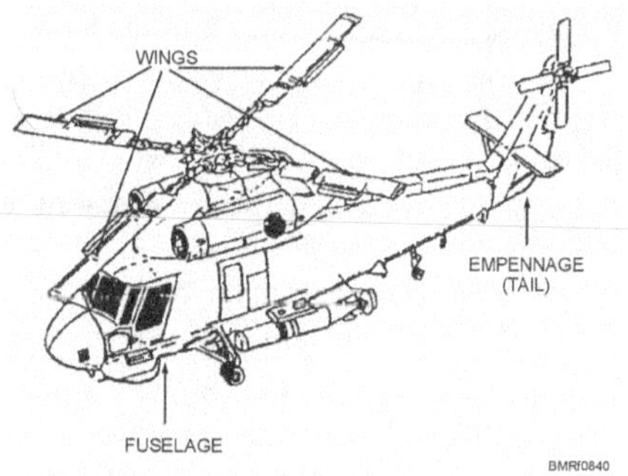

Figure 8-40.—Rotary-wing aircraft.

the lift is generated by the rotation of the assembly, which creates a flow of air over the blades.

A helicopter is lifted into the air by the aerodynamic forces on the rotor and not pushed up by the downwash. Some helicopters have twin rotors in tandem at either end of the fuselage; but most have a single, main rotor with a tail rotor mounted at right angles. A few have tandem intermeshing rotors.

TAIL ROTOR.—The tail rotor is used for directional control and stability. It is mounted at right angles to the main rotor to counteract the torque of that system. By varying the pitch of the tail rotor blades, the pilot controls yaw.

Helicopter engines are connected to the rotor shaft(s) by a transmission, which may be disengaged. That permits the engine(s) to be operated on the ground without engaging the rotor system and also permits a mode of flight known as *autorotation*. If the engines should stop while in flight, they can be disengaged; the freewheeling action of the rotor will allow a slower descent.

AIRCRAFT MODEL DESIGNATIONS

All aircraft have tri-service designations; that is, a given aircraft has the same alphanumeric identification symbol, regardless of which service uses the aircraft. Look at table 8-2. Here, you can find the four basic parts of an aircraft model designation.

Table 8-2.—Aircraft Model Designations

Mission/type modification symbol		Basic mission/type symbol		Aircraft series number	Model series letter
A	Attack	A	Attack	These numbers are assigned sequentially within each basic mission category. The number is separated from the basic mission symbol by a dash.	This letter, added to the series number, indicates an improvement or alteration of the basic mode. These are assigned in sequence; for example: F-4A, F-4B, F-4C, and so forth.
C	Cargo/transport	B	Bomber		
D	Drone control	C	Cargo/transport		
E	Special electronics	E	Special electronics		
H	Search and rescue	F	Fighter		
K	Tanker	H	Helicopter		
L	Cold weather operations	K	Tanker		
M	Missile capability	O	Observation		
O	Observation	P	Patrol		
Q	Drone	S	Antisubmarine		
R	Reconnaissance	T	Trainer		
S	Antisubmarine	U	Utility		
T	Trainer	V	Vertical takeoff and landing (VTOL)/short takeoff and landing (STOL)		
U	Utility	X	Research		
V	Staff transport				
W	Weather reconnaissance				

1. Mission/type modification symbol
2. Basic mission/type symbol
3. Aircraft series number
4. Model series letter

Now, let's try out this system of aircraft designation. For example:

EA-6B *Prowler*

1. Find the letter "E" in the first column of the table. This aircraft has special electronics.
2. Find the letter "A" in the second column of the table. The basic mission symbol tells you that this is an attack aircraft.
3. The third column of the table explains the number after the dash. This is the sixth aircraft of this series.
4. The fourth column explains the last letter of the aircraft designation. This is improvement/alteration B.

Let's try another one:

CH-46 *Sea Knight*

1. First letter is "C." This is a cargo aircraft.
2. Second letter is "H." This is a helicopter.
3. 46. This is the forty-sixth of the series.
4. No letter. There have been no improvements/alterations.

CURRENT FIXED-WING NAVY AIRCRAFT

This section briefly describes some of the fixed-wing aircraft (fig. 8-41) currently operational within the Navy.

Attack Class

Attack planes are used for low-level bombing, ground support, or nuclear strikes. They do not need the speed of fighters, but should be capable of heavy payloads, have good stability, and be able to carry enough fuel to remain on station long enough to render extended support to troops, if needed. Attack aircraft normally operate under conditions of good visibility, but some have the equipment needed for all-weather and night attacks.

EA-6B *PROWLER*.—The *Prowler* (fig. 8-42) is an all-weather tactical electronic warfare aircraft, based on the A-6 airframe. The *Prowler* provides jamming coverage to prevent missile engagement of U.S. or allied aircraft during strike operations. The *Prowler* also carries the high-speed antiradiation missile (HARM).

AV-8B *HARRIER*.—The *Harrier* (fig. 8-43) is the western world's only operational fixed-wing vertical short takeoff or landing (V/STOL) strike aircraft. It is an integrated V/STOL weapons system incorporating the inertial navigation and attack system (INAS) with an electronic display. The aircraft is used by the Marine Corps and is operated from the decks of aircraft carriers and amphibious support ships.

Fighter Class

Fighters are high-performance aircraft generally employed to gain air superiority. They may be deployed defensively as interceptors, offensively as escorts for bombers or on ground support missions, or independently to counter enemy aircraft. Some are capable of carrying sufficient payloads for bombing missions.

F-14 *TOMCAT*.—The F-14 *Tomcat* (fig. 8-44) is an aircraft-carrier-based, jet-powered fighter aircraft. The aircraft is mainly missile oriented, carrying the new air-to-air missile, Phoenix, and capable of carrying the older Sidewinder and Sparrow. The *Tomcat* can be configured for bombing and rocketry.

F/A-18 *HORNET*.—The *Hornet* (fig. 8-45) is a sonic, single-seat, twin-engine jet. The fighter and attack versions are identical, except for selected interchangeable external equipment. Conversion from the fighter to attack mode (and vice versa) takes less than 1 hour. The aircraft is designed for aerodynamic agility, high reliability, high survivability, and reduced manpower maintenance requirements.

Student Notes:

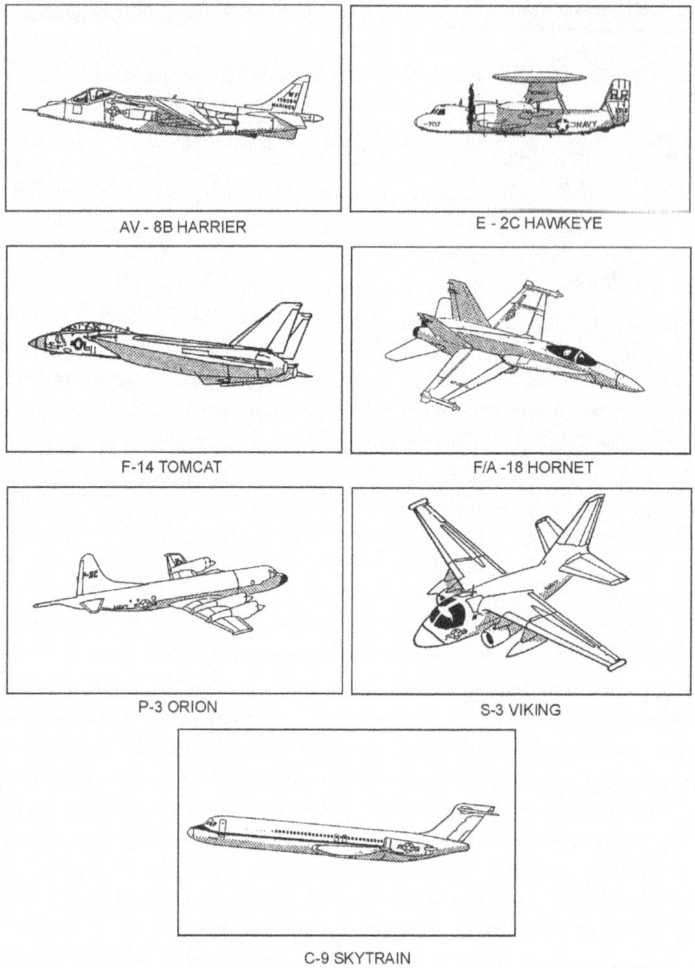

Figure 8-41.—Representative of fixed-wing aircraft.

Photograph courtesy of Senior Airman Greg L. Davie

Figure 8-42.—EA-6B *Prowler.*

Photograph courtesy of PH3 Timothy C. Ward

Figure 8-43.—AV-8B H*arrier* lands and launches for deck qualifications on USS *Constitution (CV 64)*.

Photograph courtesy of LCDR Mike Harrison

Figure 8-44.—F-14 *Tomcat*.

Photograph courtesy of Randy Hepp

Figure 8-45.—F/A-18 *Hornet*.

Patrol Class

Patrol craft are land-based, long-range, multiengine aircraft used primarily for antisubmarine warfare (ASW) patrol. Patrol squadrons operate from the continental United States and overseas bases. The P-3 *Orion* is the Navy's primary ASW patrol aircraft

The P-3 *Orion* (fig 8-46) is equipped with magnetic anomaly detection (MAD) gear, sonobuoys, radar, and other submarine detection systems. It is armed with torpedoes, bombs, missiles, and depth charges for kills. It has the primary mission of detecting, locating, and destroying enemy submarines. The P-3 *Orion* can respond quickly to hunt down submarine contacts long before surface units can arrive. Other duties include convoy escort, photographic missions, and aerial mining.

Photograph courtesy of PH2 Damon J. Mortiz

Figure 8-46.—P-3 *Orion*.

Antisubmarine Class

Antisubmarine aircraft operate from CVs in conjunction with hunter-killer group helicopters and surface craft. The S-3 *Viking* is an example of such an aircraft.

The *Viking* (fig. 8-47) is a high-wing, jet-powered, twin-engine, carrier-based ASW aircraft. It carries surface and subsurface search equipment with integrated target-acquisition and sensor-coordinating systems that collect, interpret, and store ASW sensor data. It has direct attack capability with a variety of armaments.

Photograph courtesy of LCDR Mike Harrison

Figure 8-47.—S-3 *Viking* refueling an F-14 *Tomcat*.

Warning Aircraft

Carrier-based airborne early warning (AEW) aircraft maintain station at some distance from a task

Student Notes:

force to provide early warning of approaching enemy aircraft and direct interceptors into attack position.

E-2C *HAWKEYE*.—The E-2C *Hawkeye* (fig. 8-48) has long-range antennas that are enclosed in a saucer-shaped, rotating disk atop the fuselage. The *Hawkeye* is manned by a crew of five.

Photograph courtesy of PH2 Damon J. Mortiz

Figure 8-48.—E-2C *Hawkeye* waits for direction prior to flight operations on the USS *Enterprise* (CVN 65).

ES-3 *SHADOW*.— The ES-3 *Shadow* (fig. 8-49) is a jet aircraft used to collect and disseminate tactical aircraft resembles the S-3 *Viking*, with the addition of numerous antennas and antenna housings. The ES-3 *Shadow* is a carrier-based, subsonic, all-weather, long-range, electronic reconnaissance aircraft. It

Photograph courtesy of LCDR Mike Harrison

Figure 8-49.—ES-3 *Shadow* landing aboard USS *George Washington* (CVN 73).

operates primarily with carrier battle groups providing indications and warning support to the battle group and joint theater commanders. It carries an electronic sensors and communications gear.

C-2A *GREYHOUND*.— The C-2A *Greyhound* (fig. 8-50) is a twin-engine cargo aircraft, designed to land on aircraft carriers. The C-2A *Greyhound* provides logistics support to aircraft carriers. It's powered by two PT-6 turboprop engines and can deliver a payload of up to 10,000 pounds. The cabin can carry cargo, passengers, or both. It's also equipped to accept litter patients in medical evacuation missions. Cargo such as jet engines can be transported from shore to ship in a matter of hours. A cage system or transport stand provides cargo restraint for loads during carrier launch or landing. The large aft cargo ramp and door and a powered winch allow straight-in rear cargo loading and downloading for fast turnaround. The C-2A's open-ramp flight capability allows airdrop of supplies and personnel from a carrier-launched aircraft. This, plus its folding wings and an on-board auxiliary power unit for engine starting and ground power self-sufficiency in remote areas, provide an operational versatility found in no other cargo aircraft.

Photograph courtesy of Larry Smith

Figure 8-50.—C-2A *Greyhound* landing aboard USS *John F. Kennedy* (CV 67).

C-2 *SKYTRAIN*.—The C-9 *Skytrain* (fig. 8-51) fleet is located throughout the continental United States, Europe, and Asia. The Navy and Marine Corps C-9 aircraft provide cargo and passenger transportation as well as forward deployment logistics support. The Air Force C-9s are used for medical evacuation, passenger transportation, and special missions. The C-9 *Skytrain*

Student Notes:

is the military version of the McDonnell Douglas DC-9 used for many years by commercial airlines.

Photograph courtesy of Senior Airman Delia Castillo

Figure 8-51.—C-2 *Skytrain*.

C-12 *HURON*.—The C-12 *Huron* is a twin-engine logistics aircraft that carries passengers and cargo between military installations. The C-12F provides logistics support between Navy air stations. It's powered by two PT-6A-42 turboprop engines and can deliver a total payload of up to 4,215 pounds. The cabin can carry cargo, passengers, or both. It is also equipped to accept litter patients in medical evacuation missions.

C-130 H*ERCULES*.—The C-130 *Hercules* (fig. 8-52) is a four-engine turboprop aircraft. It's the workhorse of the military services, capable of landing and taking off from short, rough dirt runways. It's a people and cargo hauler that's used in a wide variety of other roles, such as gunships, weather watchers, tankers, firefighters and aerial ambulances. There are more than 40 versions of the *Hercules*, and it is widely used by more than 50 nations.

Photograph courtesy of SGT David W. Richards

Figure 8-52.—C-130 *Hercules*.

T-45A *GOSHAWK*.—The T-45A *Goshawk* (fig. 8-53) is a tandem-seat, carrier capable, jet trainer. The T-45A aircraft is used for intermediate and advanced portions of the Navy/Marine Corps pilot training program for jet carrier aviation and tactical strike missions. There are two versions of T-45 aircraft currently in operational use at this time.

- The T-45A has an analog design cockpit.
- The T-45C is built around a new digital "glass cockpit" design.

Photograph courtesy of Bob Lawson, USN (Ret)

Figure 8-53.—T-45A *Goshawk*.

T-34C TURBOMENTOR.—The T-34C *Turbomentor* is an unpressurized two-seat, tandem cockpit low-wing turboprop trainer. The T-34C is used to provide primary flight training for student pilots attached to the Chief of Naval Air Training. As a secondary mission, approximately 10 percent of the aircraft provide pilot proficiency and other aircraft support services.

Student Notes:

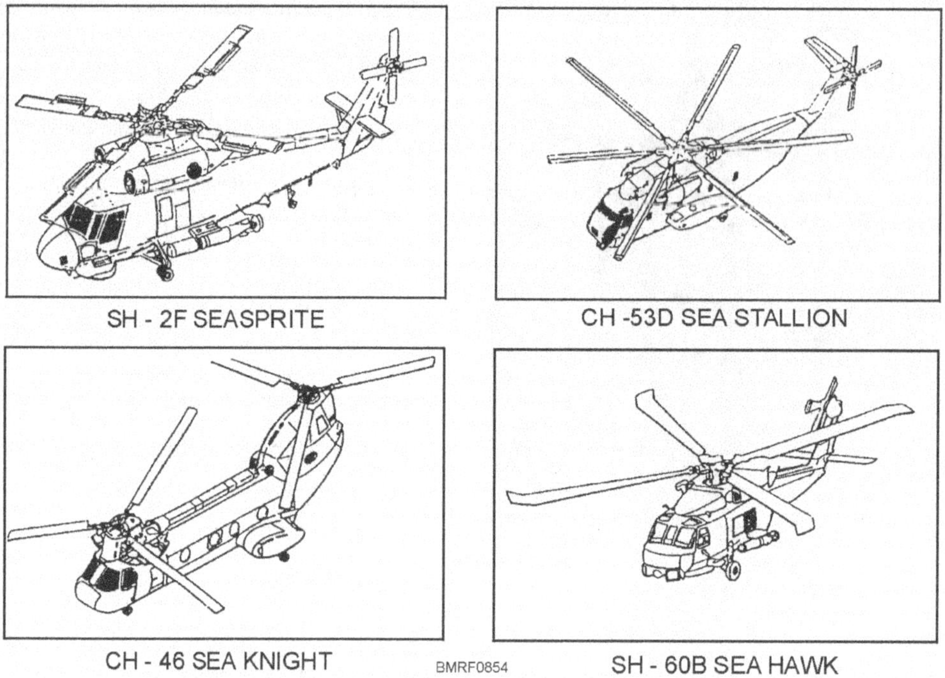

Figure 8-54.—Representative types of rotary-wing aircraft.

CURRENT ROTARY-WING NAVY AIRCRAFT

Since World War II, the helicopter has become an indispensable part of naval warfare. Its applications seem limitless—ASW; pilot rescue; transfer of supplies, mail, and personnel within dispersed forces; amphibious warfare; evacuation of wounded; counterinsurgency; minesweeping; and others. Figure 8-54 shows representative types of rotary-wing aircraft.

CH-46 *Sea Knight*

The *Sea Knight* (fig. 8-55) is a twin-turbine transport helicopter that provides the fleet with a day/night underway replenishment capability. It is used primarily for supply missions at sea and for casualty evacuation. Its carrying capacity is 25 troops, 15 litters and attendants, or 4,000 pounds of cargo. Rotor blades fold for shipboard use. The CH-46 is a small version of the Army's *Chinook.*

Photograph courtesy of PHAN Mahaya Jordan

Figure 8-55.—CH-45 *Sea Knight* transports ordnance from flight deck of the USS *Independence* (CV 62).

Student Notes:

SH-2 *Seasprite*

The *Seasprite* (fig. 8-56), an ex-utility helicopter, is now serving in the LAMPS (light airborne multipurpose system) program with the destroyer Navy.

Photograph courtesy of PH2(NAC) Jeff Viano

Figure 8-56.–SH-2 *Seasprite*

CH-53D *Sea Stallion*

The *Sea Stallion* (fig 8-57) tows and operates various mine countermeasure devices designed to detect and neutralize submerged naval mines. CH-53D squadrons are capable of rapid worldwide deployment.

Photograph courtesy of SSGT D.W. Mobley

Figure 8-57.—CH-53D *Sea Stallion* airlifts grain for Somalia.

SH-60B *Seahawk*

The *Seahawk* SH-60B (fig. 8-58) is placed aboard frigates and destroyers. The *Seahawk* is the airborne platform segment of the LAMPS Mk III weapons system. It can carry personnel as well as weapons to detect, localize, and destroy submarines at long range. It is designed to be in constant voice and data link contact with the ship's CIC. In addition to its primary mission of seeking and engaging submarines many miles from the ship, the *Seahawk* helicopter is able to provide targeting information for over-the-horizon, surface-to-surface missiles. The secondary mission of the *Seahawk* helicopter is search and rescue, medical evacuation, vertical replenishment, and communications relay.

Photograph courtesy of PH3 Anthony Haley

Figure 8-58.—SH-60B *Seahawk* aboard USS *Carney* (DDG 64) during VERTREP.

MH-53E *Sea Dragon*

The MH-53E (fig. 8-59) is used primarily for airborne mine countermeasures, with a secondary mission of shipboard delivery. The MH-53E *Sea Dragon* is heavier and has a greater fuel capacity than its ancestor, the CH-53E *Super Stallion.* MH-53s can operate from carriers and other warships. The *Sea Dragon* is capable of carrying up to 55 troops or a 16-ton payload 50 nautical miles or a 10-ton payload 500 nautical miles. The MH-53E is capable of towing a variety of mine-sweeping countermeasures systems, including the Mk 105 minesweeping sled, the ASQ-14

Student Notes:

Figure 8-59.—MH-53E *Sea Dragon.*

side-scan sonar, and the Mk 103 mechanical mine-sweeping system.

V-22A *Osprey*

The V-22 *Osprey* is a joint-service, multimission aircraft with vertical take-off and landing (VTOL) capability. It performs VTOL missions as effectively as a conventional helicopter while also having the long-range cruise abilities of a twin turboprop aircraft. The Marine Corps is the lead service in the development of the *Osprey*. The Marine Corps version, the MV-22A, will be an assault transport for troops, equipment and supplies, and will be capable of operating from ships or from expeditionary airfields ashore. The Navy's HV-22A will provide combat search and rescue, delivery and retrieval of special warfare teams along with fleet logistic support transport. The Air Force CV-22A will conduct long-range special operations missions.

The *Osprey* is a tiltrotor aircraft with a 38-foot rotor system and engine/transmission nacelle mounted on each wing tip. It can operate as a helicopter when taking off and landing vertically. Once airborne, the nacelles rotate 90 degrees for horizontal flight, converting the V-22 to a high-speed, fuel-efficient turboprop airplane. The wing rotates for compact storage aboard ship. The first flight occurred in March 1989. The V-22 is the world's first production tiltrotor aircraft. Planned purchases include 360 for the Marine Corps, 48 for the Navy, and 50 for the Air Force.

TH-57 *Sea Ranger*

The TH-57 *Sea Ranger* is a derivative of the commercial Bell Jet Ranger 206. Although primarily used for training, these aircraft are also used for photo, chase, and utility missions. The Jet Ranger was initially designed to compete in a U.S. Army light observation helicopter competition. Bell lost that competition; but, the 206 was commercially successful. The TH-57 Sea Ranger provides advanced (IFR) training to several hundred aviation students a year at Naval Air Station Whiting Field in Milton, Florida.

Student Notes:

REVIEW 6 QUESTIONS

Q1. When did the Navy acquire its first aircraft?

Q2. Label the three basic parts of a fixed-wing aircraft.

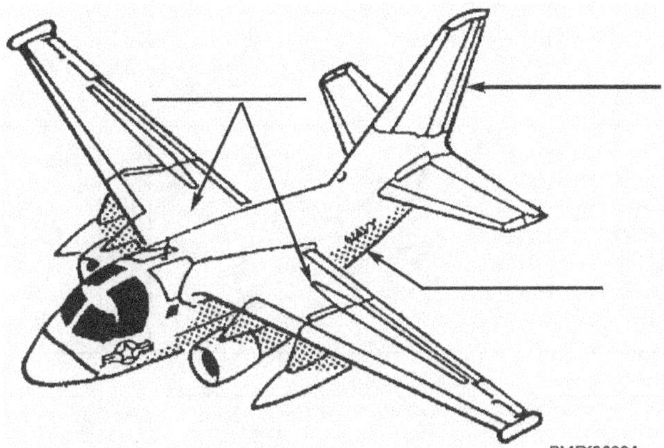

BMRf0839A

Q3. Label the three basic parts of a rotary-wing aircraft.

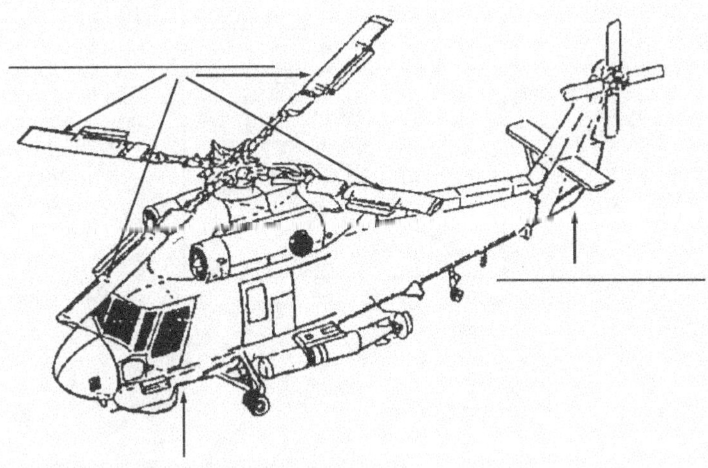

BMRf0840A

Q4. All aircraft have what type of designation?

Q5. Identify the following aircraft.

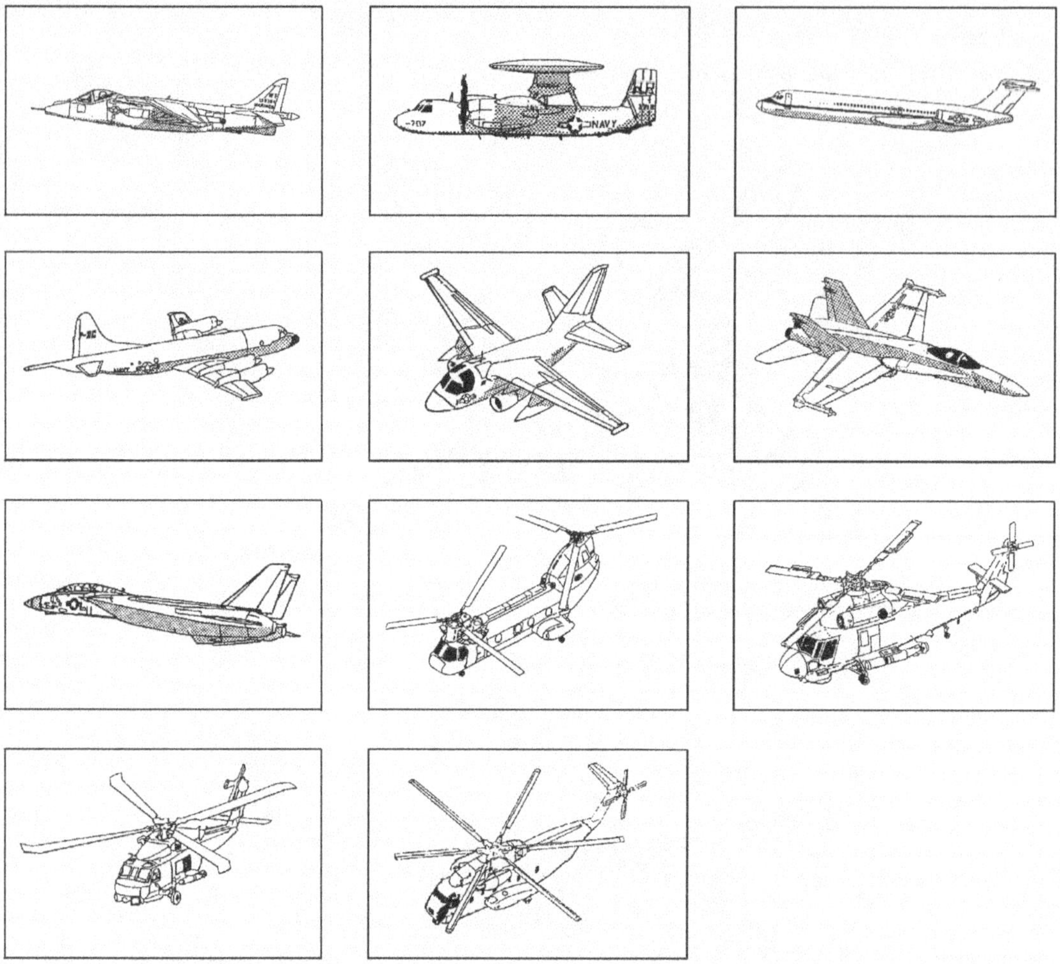

BMRf0841

Student Notes:

SUMMARY

In today's world, the United States requires military power adequate to strengthen national security objectives. The United States Navy is an integral component of this nation's military forces. Freedom of the seas is not a gift; it must be won through naval presence or engagements. Naval forces provide our nation with the ability to provide a significant presence in crisis areas, or, if required, a rapid offensive capability.

The U.S. Navy has the ability to control enemy naval forces in three areas—air, surface, and subsurface. It can also conduct amphibious and mine warfare operations.

One of the most important aspects of naval warfare is the ability to provide supply and support operations. With the Navy's wide range of underway replenishment and supply ships, we can keep U.S. Navy battle groups under way in crisis areas for long periods of time. The most recent example of this ability is the Persian Gulf War. Today's Navy consists of a new generation of cruisers, destroyers, fighter and strike aircraft, high-speed amphibious assault ships, mine countermeasures ships, replenishment ships, submarines, and weapons systems. With these craft, vessels, and weapons systems, our nation employs the most modern and capable naval force in existence.

REVIEW 1 ANSWERS

A1. Ship's parts are labeled as shown.

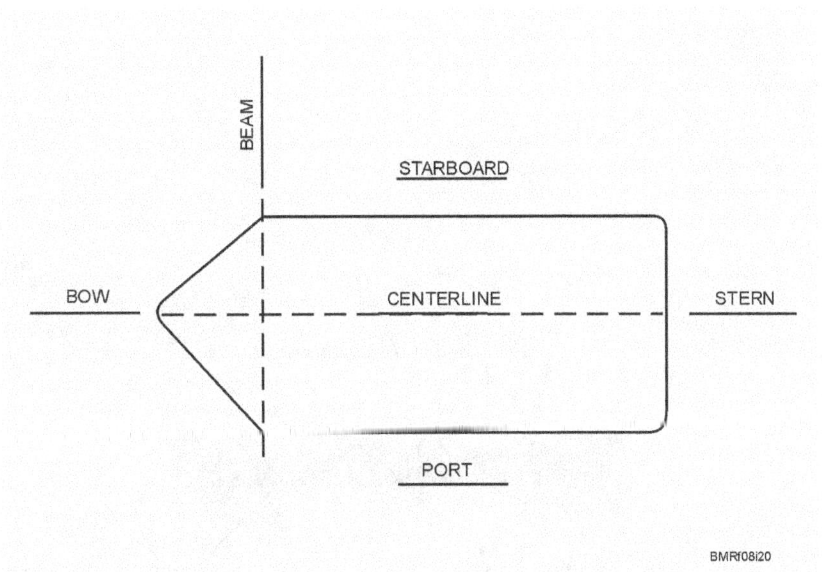

Student Notes:

A2. Some of the areas of a ship are labeled as shown.

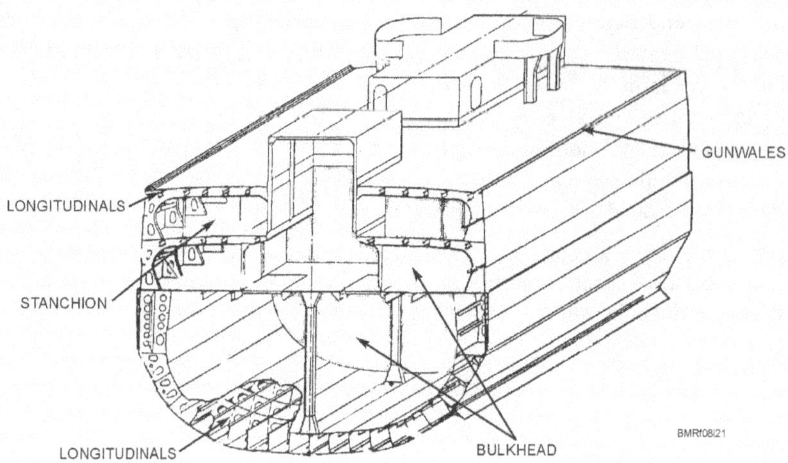

A3. Some of the decks of a ship are labeled as shown.

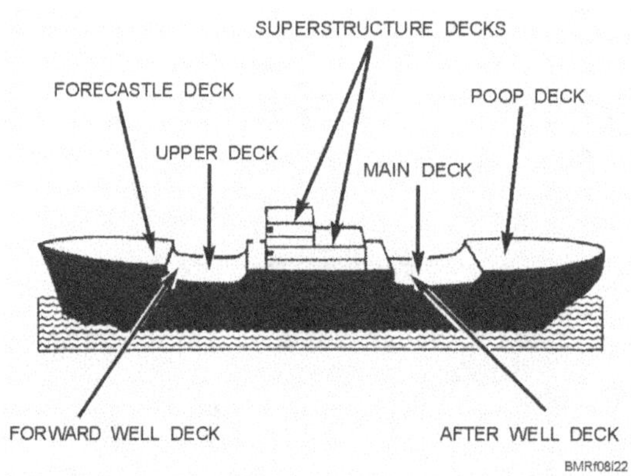

Student Notes:

A4. Doors and hatches.

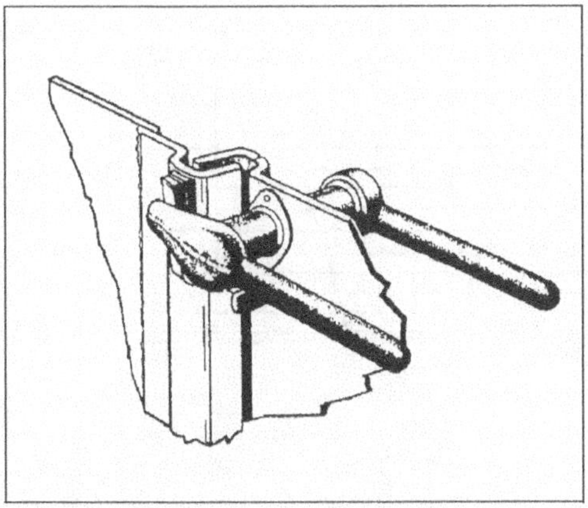

DOGS AND GASKET FOR WATERTIGHT DOOR

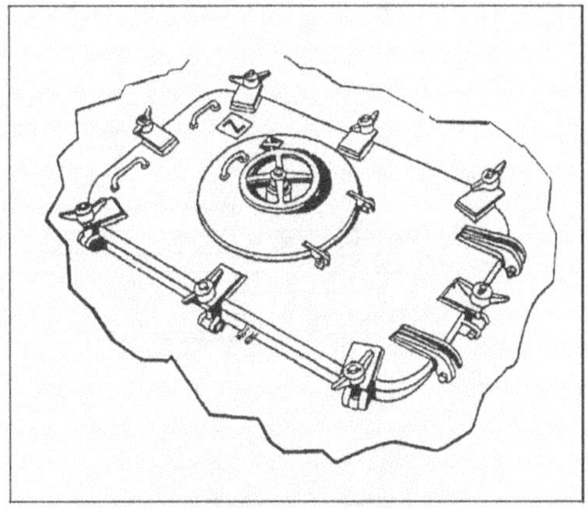

BOLTED HATCH WITH ESCAPE SCUTTLE

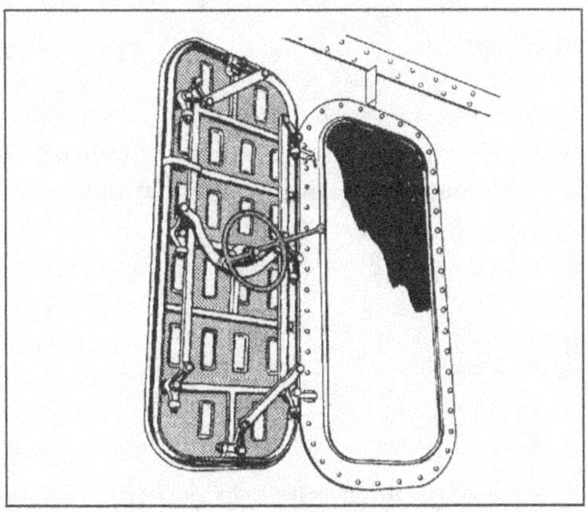

QUICK-ACTING WATERTIGHT DOOR

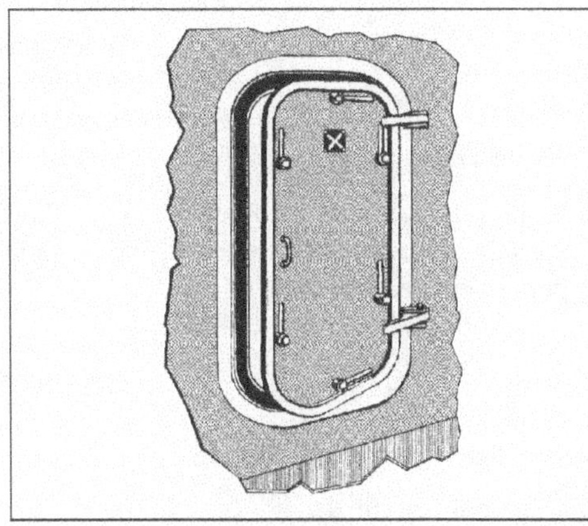

WATERTIGHT DOOR WITH INDIVIDUALLY OPERATED DOGS

BMRf0860A

REVIEW 2 ANSWERS

A1. Compartment designation number 01-56-2-Q is identified as follows:

01— **Main deck**

56— **Frame number**

2— **First compartment on the portside**

Q— **Miscellaneous or office space**

Student Notes:

A2. The following spaces of a ship are shown.

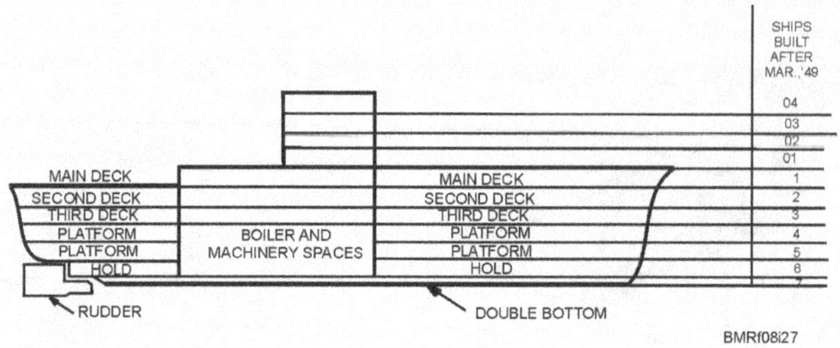

REVIEW 3 ANSWERS

A1. The size of a ship is usually given as **displacement in long tons**.

A2. A ship's armor is **the protective armor along the sides of the ship, on the deck, and on some gun mounts and turrets**.

A3. The term used to indicate the speed of a ship is the **knot, which is 1 nautical mile per hour or about 1 1/8 statute miles per hour.**

REVIEW 4 ANSWERS

A1. The four categories of ships are—

a. **Auxiliary ships**

b. **Combatant craft**

c. **Combatant ships**

d. **Support craft**

A2. The categories of warships include—

a. **Aircraft carriers**

b. **Battleships**

c. **Cruisers**

d. **Destroyers**

e. **Frigates**

f. **Submarines**

A3. The battleships are named after **states**.

A4. The two basic classes of cruisers are—

a. **Guided-missile cruisers (CG)**

b. **Guided-missile cruisers nuclear (CGN)**

A5. For protection, the destroyer depends on its **speed and mobility**.

A6. The class of ship developed for the purpose of open ocean escort and patrol was the **frigates**.

A7. The two classes of submarines are the—

a. **Attack submarine**, and the

b. **Ballistic missile submarine**

A8. The class of ship used to land large numbers of personnel, equipment, and supplies on enemy held territory is the **amphibious war ship**.

Student Notes:

REVIEW 5 ANSWERS

A1. The term used to describe the transfer of fuel and supplies between ships while under way is **replenishment at sea**.

A2. Usually, ships maintain a distance of **100 feet** while taking on supplies at sea.

A3. A receiving ship can stay on station in combat formation while undergoing **vertical replenishment**.

A4. The largest and most powerful auxiliary ship is the **fast combat support ship (AOE)**.

A5. Rescue, salvage, and towing ships provide **rapid firefighting, dewatering, battle damage repair, and rescue towing assistance.**

A6. Support craft designators usually start with the letter ***Y***.

REVIEW 6 ANSWERS

A1. The Navy acquired its first aircraft in **1911**.

A2. The three basic parts of a fixed-wing aircraft are shown below.

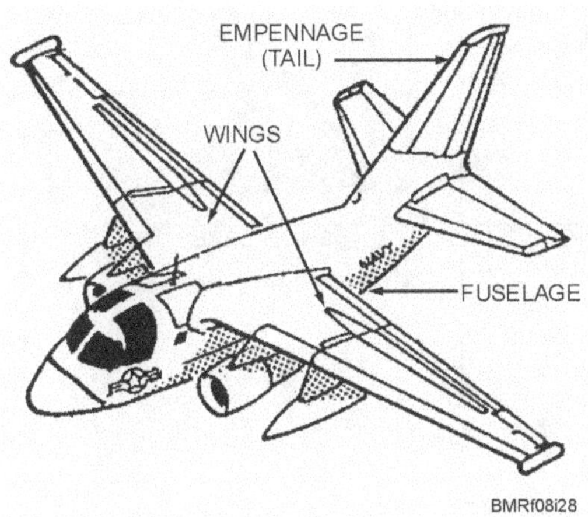

A3. The three basic parts of a rotary-wing aircraft are shown below

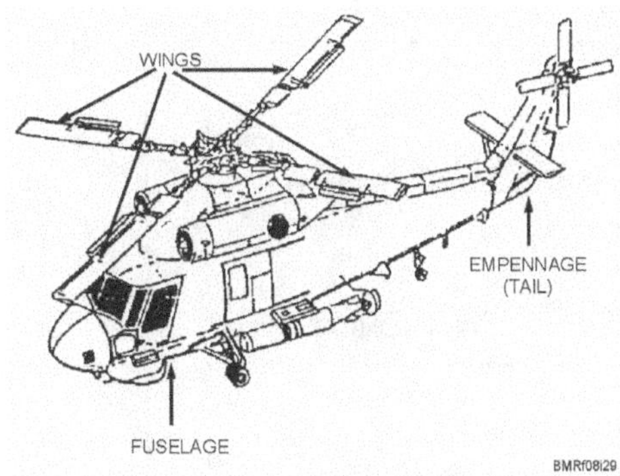

A4. All aircraft have **tri-service designations.**

Student Notes:

A5. Aircraft identification.

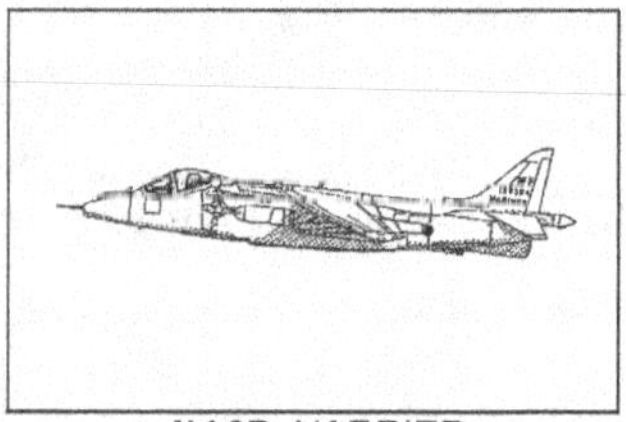
AV-8B HARRIER

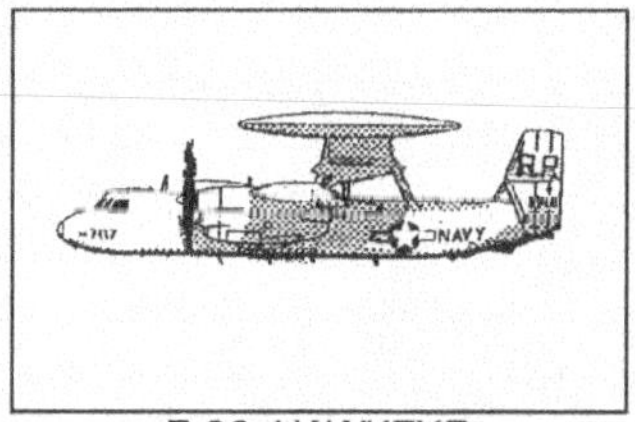
E-2C HAWKEYE

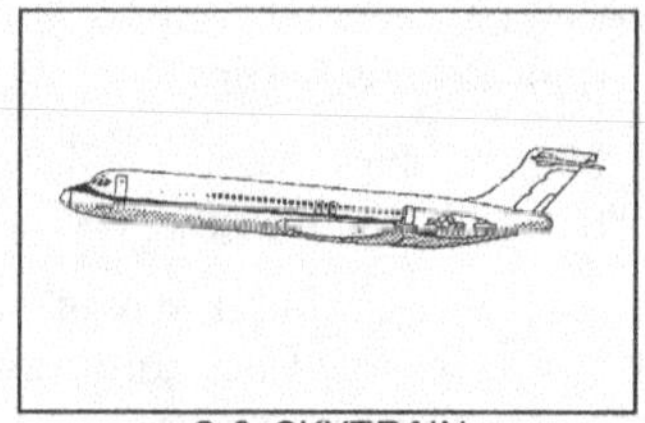
C-9 SKYTRAIN

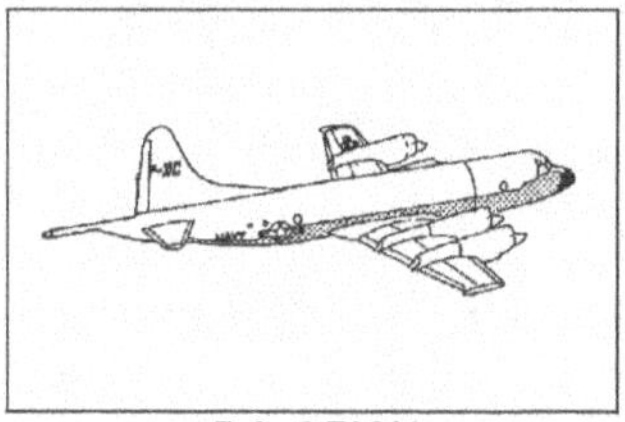
P-3 ORION

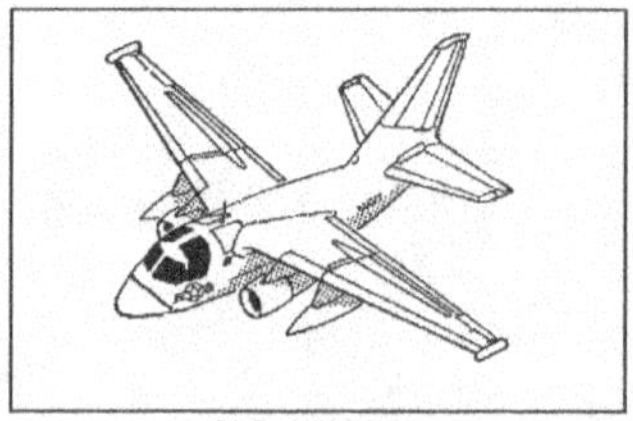
S-3 VIKING

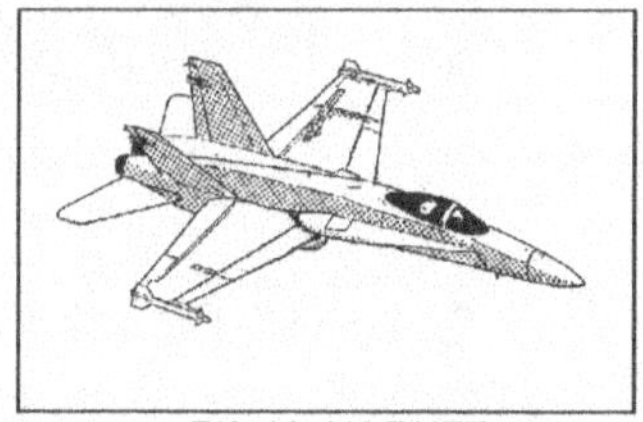
F/A-18 HORNET

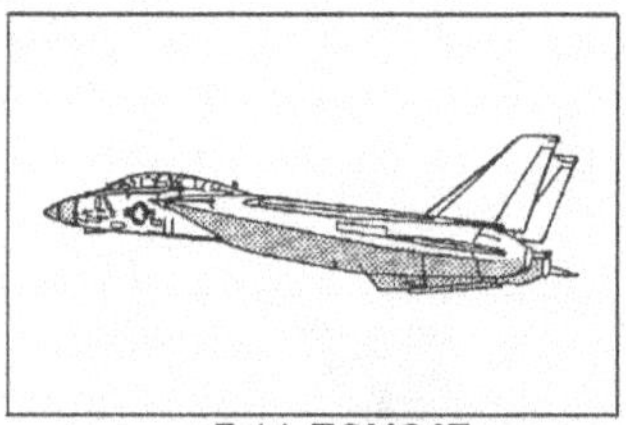
F-14 TOMCAT

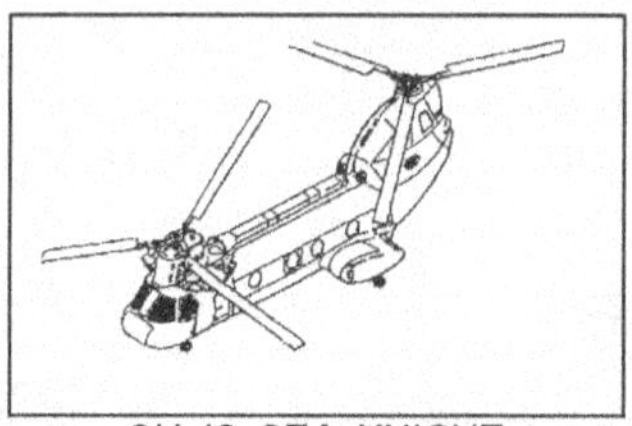
CH-46 SEA KNIGHT

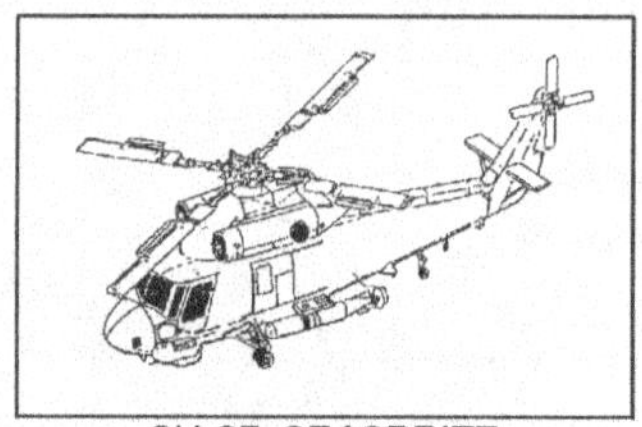
SH-2F SEASPRITE

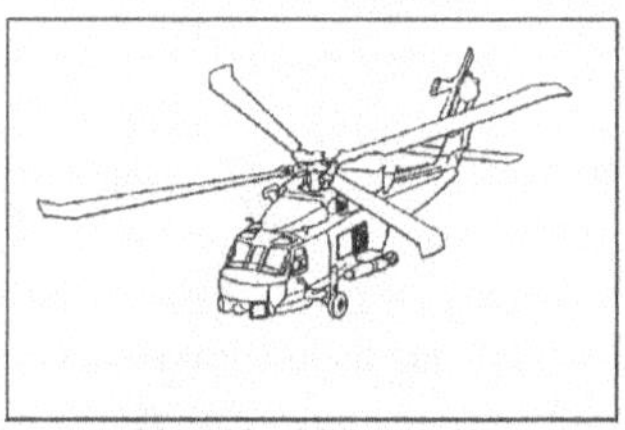
SH-60B SEA HAWK

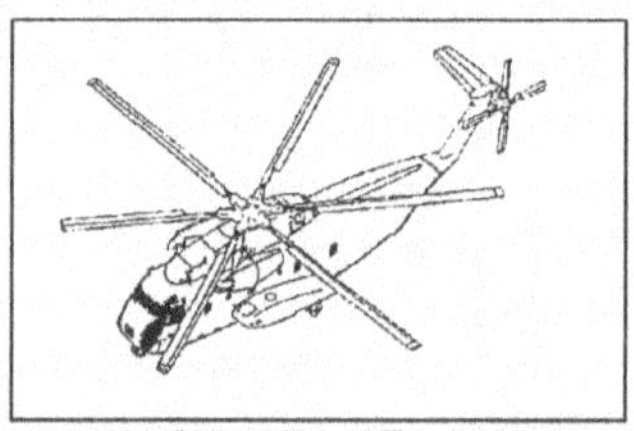
CH-53D SEA STALLION

BMRf0841A

Student Notes:

CHAPTER 9

CUSTOMS AND COURTESIES

It rests with us to make the traditions and to set the pace for those who are to follow and so upon our shoulders rests a great responsibility.

—Esther Voorhess Hasson,
First Superintendent, Navy Nurse Corps, 1908

The military services have a long history. Many traditions have been established as a result of this long history. If you are familiar with some of these traditions, you will understand the military better. These traditions can be broken down into various customs and courtesies.

A custom is a way of acting—a way that has continued consistently over such a long period that it has become like law. A courtesy is a form of polite behavior and excellence of manners. You will find that Navy life creates many situations, not found in civilian life, that require special behavior on your part. Customs and courtesies help make life orderly and are a way of showing respect.

Customs are regular, expected actions. They have been repeated again and again and passed from one generation to the next. Courteous actions show your concern and respect for others and for certain objects or symbols, such as the American flag.

The use of customs, courtesies, and ceremonies helps keep discipline and order in a military organization. This chapter will give you some of the more common day-to-day customs and courtesies and ways to deal with them.

MILITARY CUSTOMS

Learning Objective: When you finish this chapter, you will be able to—

- Recognize the purpose of military customs.

From time to time, situations arise that are not covered by written rules. Conduct in such cases is governed by customs of the service. Customs are closely linked with tradition, and much esprit de corps of the naval service depends on their continued maintenance. (Custom has the force of law; usage is merely a fact. There can be no custom unless accompanied by usage.)

A custom is a usual way of acting in given circumstances. It is a practice so long established that it has the force of law. An act or condition acquires the status of a custom under the following circumstances:

- When it is continued consistently over a long period
- When it is well defined and uniformly followed
- When it is generally accepted so as to seem almost compulsory
- When it is not in opposition to the terms and provisions of a statute, lawful regulation, or order

MILITARY COURTESIES

Learning Objectives: When you finish this chapter, you will be able to—

- Identify how to, when to, and to whom to render the hand and rifle salute.
- Identify the military courtesies when ship and boat passing honors are rendered.

Courtesy is an act or verbal expression of consideration or respect for others. When a person acts with courtesy toward another, the courtesy is likely to be returned. We are courteous to our seniors because we are aware of their greater responsibilities and authority. We are courteous to our juniors because we are aware of their important contributions to the Navy's mission.

In the military service, and particularly in the Navy where personnel live and work in close quarters, courtesy is practiced both on and off duty. Military courtesy is important to everyone in the Navy. If you know and practice military courtesy, you will make favorable impressions and display a self-assurance that will carry you through many difficult situations. Acts of respect and courtesy are required of all members of the naval service; the junior member takes the initiative, and the senior member returns the courtesy.

SALUTING

One required act of military courtesy is the salute. Regulations governing its use are founded on military custom deeply rooted in tradition. The salute is a symbol of respect and a sign of comradeship among service personnel. The salute is simple and dignified; but, there is great significance in that gesture. It is a time-honored demonstration of courtesy among all military personnel that expresses mutual respect and pride in the service. Never resent or try to avoid saluting persons entitled to receive the salute. (The privilege of saluting is generally denied prisoners because their status is considered unworthy of the comradeship of military personnel.)

The most common form of salute is the hand salute. However, there are other types, such as gun and rifle salutes, which are discussed later in this chapter.

The Hand Salute

The hand salute began in the days of chivalry when it was customary for knights dressed in armor to raise their visors to friends for the purpose of identification. Because of the relative position of rank, the junior was required to make the first gesture. Another school of thought traces the salute back to a custom at the time of the Borgias. Assassinations by dagger were not uncommon at that time, and it became the custom for men to approach each other with raised hand, palm to the front, to show that there was no weapon concealed.

In the U.S. Navy, it's reasonable to believe that the hand salute came from the British navy. There is general agreement that the salute as now rendered is really the first part of the movement of uncovering. From the earliest days of military units, the junior uncovered when meeting or addressing a senior. Gradually, the act of taking off one's cap was simplified into merely touching the cap or, if uncovered, the head (forelock), and finally into the present form of salute.

The way you render the hand salute depends on whether you are in civilian clothes or in uniform. Personnel in civilian clothes render the salute in two ways:

1. Hat in front of the left shoulder (men only)
2. Right hand over the heart (men without hats; women with or without hats)

NOTE

These forms of saluting are used only to salute the flag or national anthem, never to salute officers.

In this chapter, the hand salute usually refers to a salute rendered by personnel in uniform. Except when walking, you should be at attention when saluting. In any case, turn your head and eyes toward the person you're saluting (unless it is inappropriate to do so, such as when a division in ranks salutes an inspecting officer on command). Navy personnel salute the anthem, the flag, and officers as follows:

- Raise the right hand smartly until the tip of the forefingers touches the lower part of the headgear or forehead above and slightly to the right of the eye (fig. 9-1).
- Extend and join the thumb and fingers.
- Turn the palm slightly inward until the person saluting can just see its surface from the corner of the right eye.
- The upper arm is parallel to the ground; the elbow is slightly in front of the body.
- Incline the forearm at a 45° angle; hand and wrist are in a straight line.
- Complete the salute (after it is returned) by dropping the arm to its normal position in one sharp, clean motion.

Student Notes:

Figure 9-1.—Hand salute.

NOTE

The salute should not be ended as though the person is waving to someone or trying to get something off the fingers. Navy custom permits left-hand saluting when a salute cannot be rendered with the right hand. Army and Air Force customs permit only right-hand salutes.

Under naval customs, the hand salute is accompanied by a word of greeting. The junior stands at attention, looks the senior straight in the eye, and says (depending upon the time of day) the following:

- From first rising until noon "Good morning, ..."
- From noon until sunset "Good afternoon, ..."
- From sunset until turning in "Good evening, ..."

It is preferable to call the senior by grade and name; that is, "Commander Jones," rather than by the impersonal "sir" or "ma'am."

The following are some of the major points you should remember when rendering a salute:

1. If possible, always use your right hand. Use your left hand only if your right hand is injured. Use your left hand to carry objects and to leave your right hand free to salute.

2. Accompany your salute with a cheerful, respectful greeting; for example, "Good morning, sir"; "Good afternoon, Commander [Jones]"; "Good evening, Chaplain [Smith]".

3. Always salute from the position of attention. If you are walking, you need not stop; but hold yourself erect and square. If on the double, slow to a walk when saluting.

4. Look directly into the officer's eyes as you salute.

5. If you are carrying something in both hands and cannot render the hand salute, look at the officer as though you were saluting and render a verbal greeting as previously described.

6. Remove a pipe, cigar, or cigarette from your mouth or hand before you salute.

7. Salute officers even if they are uncovered or their hands are occupied. Your salute will be acknowledged by a verbal greeting, such as "Good morning," "Good afternoon," or something similar.

8. Army and Air Force policy, unlike the Navy's, is to salute when uncovered. Suppose you are in an office with several Army personnel, and all of you are uncovered. An officer enters and the soldiers rise and salute. You should do likewise; to do otherwise would make you seem ill-mannered or disrespectful.

9. If you are walking with or standing by a commissioned officer and the occasion for a salute arises, do not salute until the officer salutes. Assume that you are walking with a lieutenant. A commander approaches. Do not salute the commander until the lieutenant salutes; but as soon as the lieutenant starts to salute, you should quickly do the same.

10. When approaching an officer, start your salute far enough away from the officer to allow time for your salute to be seen and returned. This space can vary; but a distance of about six paces is considered good for this purpose. Hold your salute until it is returned or until you are six paces past the officer.

11. Salute all officers who are close enough to be recognized as officers. It is unnecessary to identify an officer by name; however, ensure that he/she is wearing the uniform of an officer.

Student Notes:

12. Salute properly and smartly. Avoid saluting in a casual or perfunctory manner. A sharp salute is a mark of a sharp Sailor.

WHOM TO SALUTE.—Enlisted personnel salute all officers, and officers salute their seniors. Salutes are returned by persons saluted except when they are uncovered—the person saluted should acknowledge the salute with an appropriate greeting or a nod of the head.

Salutes are rendered to all of the following officers:

- Navy.
- Army.
- Air Force.
- Marine Corps.
- Coast Guard.
- National Oceanic and Atmospheric Administration.
- Public Health Service.
- Foreign military services.
- Officers of the Navy, Army, Air Force, Marine Corps, and Coast Guard Reserves.
- Officers of the National Guard when they are on active duty. When not on active duty, they rate a salute only when they are in uniform.

Civilians who are entitled (rate), by reason of their position, gun salutes, or other honors are also entitled (by custom) to the hand salute. The President, as Commander in Chief of the armed forces, is always saluted. Other civilians may be saluted by persons in uniform when appropriate, but the uniform hat or cap must not be raised as a form of salutation.

WHEN TO SALUTE.—On occasion, you might be uncertain whether the person approaching you in uniform is an officer, thus rating a salute. The safest course of action is to salute immediately and not wait for the person approaching you to disclose his/her rank. In other words, when in doubt, salute. Figures 9-2 and 9-3 show some examples of when to salute officers.

Aboard Ship

When boarding a ship that is flying the national ensign, all persons in the naval service must do the following:

1. Stop on reaching the upper platform on the accommodation ladder or the shipboard end of the brow,
2. Face the ensign,
3. Salute, and
4. Then salute the officer of the deck (OOD).

On leaving the ship, personnel render the salutes in reverse order—first to the OOD and then to the national ensign. These salutes also are rendered aboard foreign men-of-war.

You are required to salute all flag officers (officers above the grade of captain), the commanding officer, and visiting officers (senior to the commanding officer) on every occasion of meeting, passing near, or being addressed. On your first daily meeting, you salute all senior officers attached to your ship or station. Many ships consider salutes rendered at quarters sufficient for this first salute of the day. When the progress of a senior officer may be blocked, officers and enlisted personnel clear a path by calling out "Gangway" and stand at attention facing the senior officer until he/she passes.

In Boats

When a boat is not under way, the person in charge salutes officers that come alongside or pass nearby. If there is no one in charge, all those in the boat render the salute. Boat coxswains salute all officers entering or leaving their boats. (Although it is customary to stand when saluting, if the safety of the boat is endangered by standing, remain seated.) When boat awnings are spread, enlisted personnel sit at attention while saluting; they should not rise. Officers seated in boats rise when rendering salutes to seniors who are entering or leaving.

When boats pass each other with embarked officers or officials in view, hand salutes are rendered by the senior officer and coxswain in each boat. Coxswains rise to salute unless it is dangerous or impracticable to do so.

Student Notes:

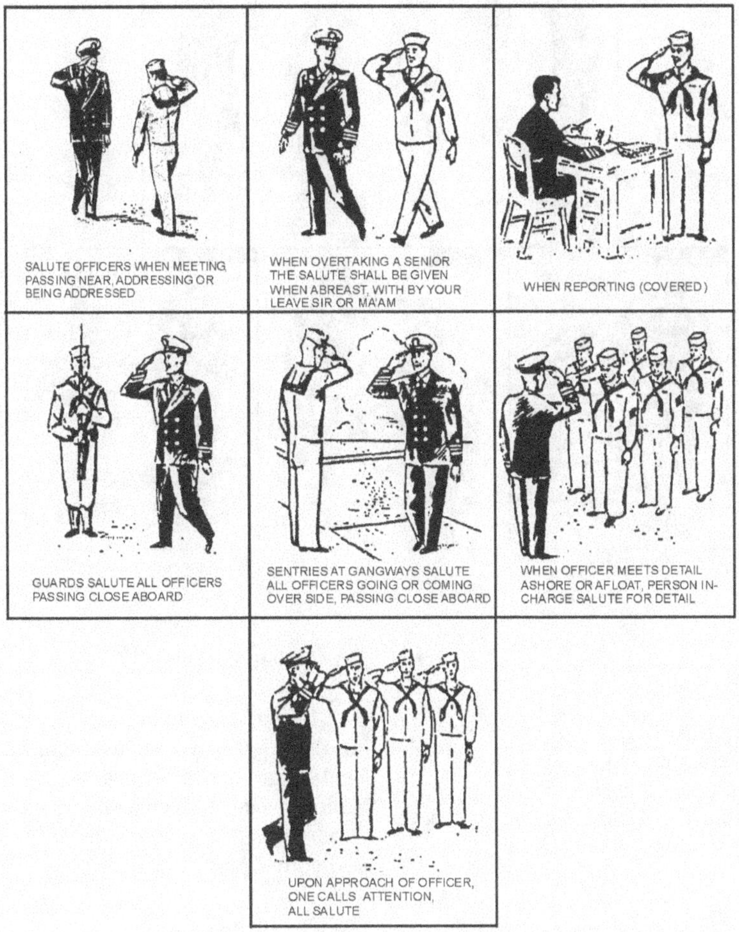

Figure 9-2.—When to salute officers.

In a Group

If enlisted personnel and officers are standing together and a senior officer approaches, the first to see the senior should call out "Attention," and all face the officer and salute.

Overtaking

Never overtake and pass an officer without permission. If it becomes necessary for you to pass, you should do so to the left, salute when abreast of the officer, and ask, "By your leave, sir/ma'am?" The officer should reply, "Very well," and return the salute.

Student Notes:

Figure 9-3.—When to salute officers (Continued).

Reporting

When reporting on deck or out-of-doors ashore, you should remain covered and salute accordingly. When reporting in an office, you should uncover upon approaching the senior; therefore, you should not salute.

Sentries

Sentries at gangways salute all officers going or coming over the side and when passing or being passed by officers close aboard in boats.

In Vehicles

You salute all officers riding in vehicles, while those in the vehicle both render and return salutes, as required. The vehicle's driver salutes if the vehicle is stopped; to do so while the vehicle is in motion endangers the safety of the occupants and may be omitted.

In Civilian Clothes

If you are in uniform and recognize an officer in civilian clothes, you should initiate the proper greeting and salute. In time of war, however, an officer not in uniform may be deliberately avoiding disclosure of his/her identity, so you should be cautious in following the normal peacetime rule.

At Crowded Gatherings

At crowded gatherings or in congested areas, you normally salute only when addressing or being addressed by officers.

Rifle Salutes

When armed with a rifle, you should use one of the three rifle salutes described in this section instead of the hand salute. (**NOTE**: The salute at sling arms shown in fig. 9-4 is simply a hand salute and is not considered a rifle salute.) The occasions for rendering each type of rifle salute are as follows:

Student Notes:

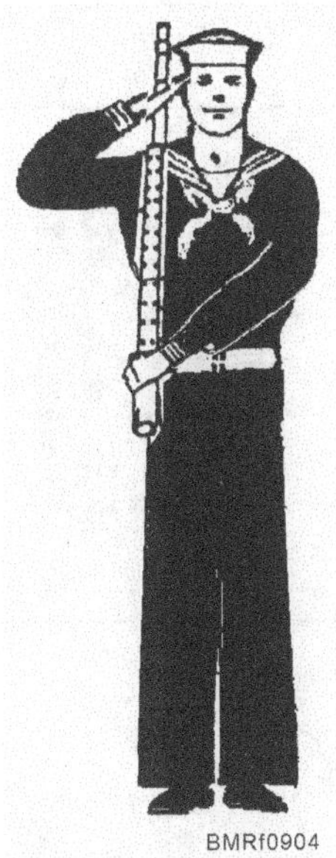

Figure 9-4.—Hand salute at sling arms.

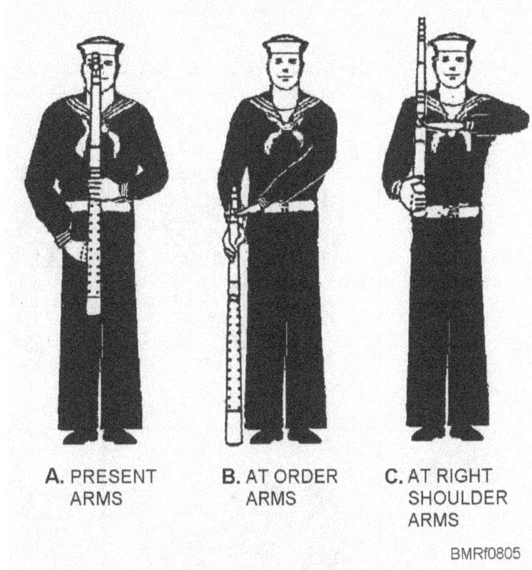

Figure 9-5.—Rifle salutes.

1. Present arms (fig. 9-5, view A)

   - When standing in a sentry box or on a post and addressed or approached by any person entitled to a salute.
   - When halted while on patrol (such as an area security patrol) to reply to or to address an officer.
   - When in ranks and so commanded; for example, at colors.

2. Rifle salute at order arms (fig. 9-5, view B)

   - When standing sentry or guard duty by a door inside a building. Present arms may also be required by competent authority; but where there is considerable traffic, the salute at order arms is usually prescribed.
   - When reporting individually to an officer indoors. For example, you would approach an officer's desk at trail arms, come to order arms, and render the rifle salute at order arms.
   - When reporting a roll call (if already at order arms).

3. Rifle salute at right shoulder arms (fig. 9-5, view C)

   - When on patrol and passing, without halting, a person entitled to a salute.
   - When leading a detail past an officer.
   - When reporting a roll call (if already at shoulder arms). When reporting at shoulder arms to an officer outdoors.
   - When going individually to and from drill or place of duty and you pass any person entitled to a salute.

WHEN NOT TO SALUTE

There are some situations in which it is improper for you to salute (fig. 9-6 and fig. 9-7). These are as follows:

- When uncovered, except where failure to salute might cause embarrassment or misunderstanding.

Student Notes:

Figure 9-6.—When not to salute officers.

Figure 9-7.—When not to salute officers (Continued).

Student Notes:

- In formation, except on command.
- On a work detail (the person in charge of the detail salutes).
- When engaged in athletics or assembled for recreation or entertainment.
- When carrying articles with both hands, or otherwise occupied making saluting impracticable.
- In public places where saluting is obviously inappropriate (theaters, restaurants, elevators, etc.).
- In public transportation.
- In action or under simulated combat conditions.
- When a member of a guard is engaged in performance of a duty that prevents saluting.
- At mess (when addressed, stop eating and show respectful attention).
- When guarding prisoners.

Honors

Honors are salutes rendered to individuals of merit, such as recipients of the Medal of Honor, to high-ranking individuals, to ships, and to nations. The type of honors rendered depends upon who or what is being saluted. Passing honors are rendered by a ship to other ships and to boats having officials embarked. Side honors are rendered to officials or officers as they board and depart a Navy ship. Gun salutes are rendered to high-ranking individuals, to nations, and to celebrate national holidays. Honors are not rendered to nations or officials of nations not recognized by the United States. Officials and officers who request that the honors be dispensed with do not receive them.

PASSING HONORS.—Passing honors are honors (other than gun salutes) rendered on occasions when ships, officials or officers pass in boats or gigs, or are passed (flag officers or above) close aboard. "Close aboard" means passing within 600 yards for ships and 400 yards for boats. Passing honors between ships, consisting of sounding "Attention" and rendering the hand salute by all persons in view on deck and not in ranks, are exchanged between ships of the Navy and between ships of the Navy and the Coast Guard passing close aboard.

Signals for the actions required to be performed by personnel are as follows:

- One blast—Attention (to starboard)
- Two blasts—Attention (to port)
- One blast—Hand salute
- Two blasts—End salute
- Three blasts—Carry on

NOTE

Signals are given by police whistle on small ships and by bugle on large ships.

On the signal of "Attention," all hands in view on deck (starboard or port as indicated by number of blasts) come to attention and face outboard. At the sound of one blast, all hands in view and not in ranks salute. (When personnel are in ranks, only the division officer and the division petty officer salute; all other persons stand at attention.) At two blasts, persons saluting bring their hands back to their sides but remain at attention until three blasts are sounded.

For boats passing honors, flag officers, unit commanders, or commanding officers in uniform embarked in boats are saluted by all persons on the quarterdeck.

Passing honors for the President of the United States and for rulers of foreign nations include manning the rail. Manning the rail consists of the ship's company lining up at regular intervals along all weather deck rails. Normal saluting procedures are followed.

Having the crew at quarters when the ship is entering or leaving port is a less formal ceremony than manning the rail. The crew is paraded at quarters on ceremonial occasions, such as—

Student Notes:

- When the ship is entering or leaving U.S. ports at times other than operational visits,
- When the ship is visiting foreign ports, or
- When the ship is departing for or returning from extended deployments, and other special occasions as determined by a superior.

When the ship is entering or leaving U.S. ports on operational visits or home port on local operations, the normal procedure is to parade only an honor guard.

SIDE HONORS.—Side honors, rendered to officers and officials boarding and departing the ship, are a part of the honors stipulated on the occasion of an official visit. The honors consist of parading the proper number of side boys and piping the side.

Acting as a side boy may be one of your shipboard duties. When you are assigned to side boy duty, you must remain in dress uniform and in the vicinity of the quarterdeck at all times, ready to fall in when required. Your uniform must be clean and neat, and you must be especially neat and military in appearance. Enlisted women may be detailed to this duty, but they are still called *side boys*.

Side boys are paraded between 0800 and sunset daily except on Sunday. Normally, side boys are not called away during meal hours, general drills, all hands evolutions, or periods of regular overhaul except in honor of civil officials or foreign officers; then, they may be called away at any time during daylight. The number of side boys paraded varies from two to eight (always an even number), depending on the rank of the individual being saluted.

When called away, side boys form two ranks facing each other to form a passageway at the gangway. When the Boatswain's Mate (BM) begins to pipe the call "Over the Side," the side boys salute in unison, hold the salute until the last note of the call, and then drop their hands smartly to their sides.

Gun Salutes

Gun salutes are used to honor individuals, nations, and certain national holidays. Practically all shore stations have saluting batteries, but not all ships are so equipped. Whether aboard ship or ashore, you must be able to act properly whenever you hear a gun salute being rendered.

The salutes always consist of an odd number of guns, ranging from 5 for a vice consul to 21 for the President of the United States and for rulers of foreign nations recognized by the United States. Military officers below the rank of commodore are not entitled to gun salutes. Normally, only one gun is fired at a time at intervals of about 5 seconds. During the salutes, persons on the quarterdeck, in the ceremonial party, or if ashore, render the hand salute. All other personnel in the vicinity (in the open) should stand at attention and, if in uniform, render the hand salute.

Gun salutes also mark special occasions in our country's history. On President's Day, Memorial Day, and Independence Day, a standard 21-gun salute is fired at 1-minute intervals, commencing at 1200. Thus, on these holidays, the salute ends at 1220.

REVIEW 1 QUESTIONS

Q1. Define the following terms.

a. Courtesy—

b. Custom—

Q2. Name a required act of military courtesy.

Q3. What is the most common form of saluting?

Q4. When in uniform, Navy personnel salute which of the following persons/things?

a. Flag

b. Anthem

c. Officers

d. Each of the above

Q5. True or false. Under naval customs, the hand salute is accompanied by a word of greeting.

Student Notes:

Q6. As an enlisted person, who should you salute?

Q7. What procedure should you follow when boarding a ship that is flying the national ensign?

a.

b.

c.

Q8. As a sentry at a gangway, when should you render a salute?

a.

b.

Q9. List the three rifle salutes.

a.

b.

c.

Q10. Define the term *honors*.

Q11. Name the passing honors for the President of the United States?

Q12. A ruler of a country recognized by the United States rates which of the following gun salutes?

a. 5

b. 17

c. 21

MILITARY CEREMONIES

Learning Objective: When you finish this chapter, you will be able to—

- Identify the procedures for conducting colors, performing military courtesies, handling the Ensign and Union Jack, and boarding.

Ceremonies are formal acts performed on public occasions. There are too many types of ceremonies and too many occasions when they are performed to include them all here. Instead, you will learn about some of the common situations involving a formal ceremony and the behavior required of you during the event.

COLORS

At commands ashore and aboard ships of the Navy not under way, the ceremonial hoisting and lowering of the national flag at 0800 and sunset are known as *morning* and *evening colors*. Every Navy shore command, and every ship not under way, performs the ceremony of colors twice a day.

You will render honors as follows:

- If you are in ranks, you will be called to attention or order arms.
- If you are in uniform but not in ranks, face the colors and give the hand salute.
- If you are driving a vehicle, stop and sit at attention but do not salute.
- If you are a passenger in a boat, remain at attention, seated or standing. The boat officer or coxswain salutes for the boat.
- If you are in civilian clothes or athletic uniform, face the colors at attention and salute by placing your right hand over your heart.

Aboard Navy ships or naval shore activities, when the national ensign is hoisted and lowered or half-masted for any occasion, the motions of the senior officer present are followed. Five minutes before morning and evening colors, the PREPARATIVE

Student Notes:

pennant (called *PREP*) is hoisted. Ceremonies for colors begin when PREP is hauled to the dip (the halfway point).

Navy ships not under way also hoist and lower the union jack on the jackstaff, at the ship's bow, and at morning and evening colors. The union jack is also flown from a yardarm to denote that a general courts-martial or court of inquiry is in session. The union jack is the rectangular blue part of the United States flag containing the stars (fig. 9-8).

If a band is available for color ceremonies, "Attention" is sounded, followed by the band playing the national anthem. At morning colors, hoisting the ensign begins when the music starts. It is hoisted smartly to the top of the flagstaff. At evening colors, lowering of the ensign also starts at the beginning of the music. Hoisting and lowering of the ensign are completed at the last note of the music. The national flag is always hoisted smartly and lowered ceremoniously. "Carry on" is sounded at the completion of the music.

If a band is not available for colors, "To the Colors" is played on the bugle at morning colors, and "Retreat" is played at evening colors. For ships without a band or a bugler, "Attention" and "Carry on" are signals for beginning and terminating the hand salute.

Sometimes the music for colors from another U.S. ship can be overheard aboard your ship. When this happens, and no band or bugler is aboard your ship, the command to "Carry on" should not be given until the music being overheard is completed.

After morning colors, if foreign warships are present, the national anthem of each country

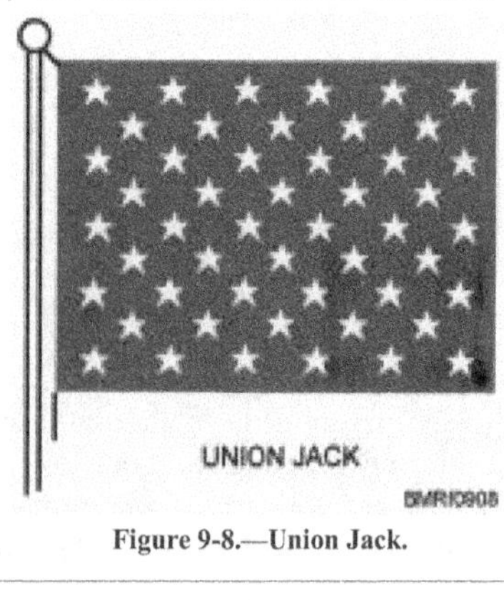

Figure 9-8.—Union Jack.

represented is also played. If your ship is visiting a foreign country, the national anthem of that country is played immediately following morning colors, followed by the national anthems of any other foreign nations represented. You should show the same respect for national anthems of foreign countries as you do for our own.

On Sundays, authorized holidays, and other days proclaimed by the President, the largest national ensign in the ship's or station's allowance is flown. This ensign is referred to as *holiday colors*. When the holiday colors are flown on a Navy ship not under way, the union jack flown is the same size as the blue field in the holiday colors.

Ships that are under way do not hold morning or evening colors because the ensign usually is flown day and night. Just as the ship gets under way, the ensign is shifted from its in-port position on the stern to its at-sea position at the mainmast. This is called *shifting the colors*.

HALF-MASTING THE ENSIGN

National flags flown at half-mast (or half-staff ashore) are an internationally recognized symbol of mourning. The United States honors its war dead on Memorial Day by half-masting the flag from 0800 until the last gun of a 21-minute-gun salute that begins at noon (until 1220 if no gun salute is rendered).

Normally, the flag is half-masted on receiving information of the death of one of the officials or officers listed in *U.S. Navy Regulations*. Notification may be received through news media reports or by an official message.

In half-masting an ensign already flying at the peak or truck aboard ships under way, lower it ceremoniously to half-mast. If the ensign is not flying, hoist it smartly to the peak or truck before lowering it to half-mast. In lowering a half-masted ensign, raise it first to the peak or truck, then lower it ceremoniously.

When the national anthem, "To the Colors," or "Retreat" is played at morning or evening colors aboard ships not under way, all hands should hold the salute during the raising or lowering of the flag. In half-masting during morning colors, "Carry on" should

Student Notes:

not be sounded until the flag is lowered to half-mast. At evening colors, "Attention" is sounded and the salute rendered before raising the flag to the top of the flagstaff from its half-mast position.

If the ensign is flown from the flagstaff and is half-masted, the union jack is half-masted also. Distinctive marks, such as commission or command pennants, are not half-masted except when the ship's commanding officer or the unit commander dies.

A special ceremony calling for half-masting the ensign is required of ships passing Washington's tomb between sunrise and sunset. A full guard and band are paraded (if aboard), the ship's bell is tolled, and the ensign is half-masted as the ship comes in sight of Mount Vernon, Virginia. When the ship is opposite the tomb, the guard and all persons on deck face the tomb and salute. When the bugler begins to sound taps, the ensign is raised to the peak; tolling of the bell ceases on the last note of taps. The band then plays the national anthem, followed by the command to "Carry on."

You may have the duty of raising or lowering the ensign at some time in your career. You should remember that the ensign is raised smartly but lowered ceremoniously. After the ensign is lowered, it is folded properly and placed in safekeeping until morning colors. Figure 9-9 shows the correct way to fold the ensign. The union jack is folded and handled in the same manner as the national ensign.

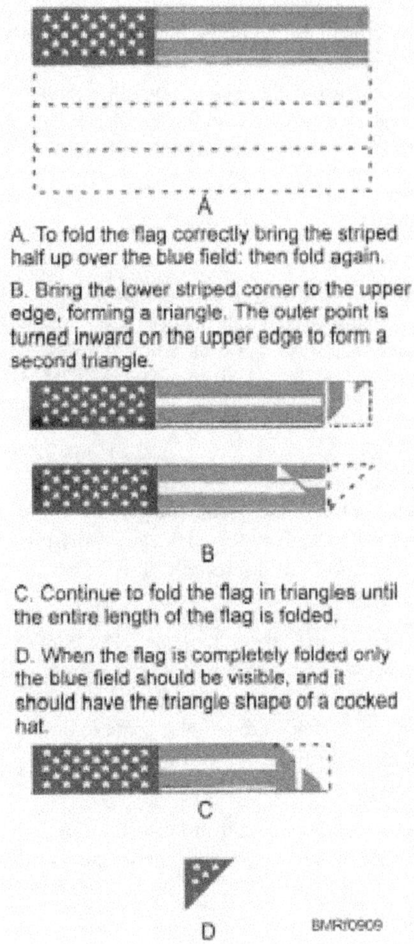

Figure 9-9.—Folding the ensign.

NATIONAL ANTHEM AND FLAG HONORS

Honors to the ensign or national anthem are rendered on occasions other than at colors. In this section, you will learn about procedures for rendering honors when the anthem is played indoors and outdoors, with or without the flag present. Foreign anthems and ensigns are shown the same mark of respect as our own anthem and ensign. All salutes are held from the first note of the anthem to the last.

All the following rules for saluting the national anthem apply only when you hear it played as part of a public ceremony at which you are present. If you hear a broadcast or recording of the anthem on a radio, tape or CD player, television, or as you pass a store, you don't need to stop or salute.

Indoors

If the flag is **not** displayed when the anthem is played inside a building, you stand at attention facing the source of the music. If you are in uniform and covered, you render the hand salute; if not covered, you stand at attention. If you are in civilian clothes, render the hand-over-the-heart salute.

Student Notes:

If the flag is displayed when the anthem is played, you face the flag and stand at attention. If in uniform and covered, render the hand salute; if in civilian clothes or if in uniform and uncovered, you place your right hand over your heart. Persons in formation stand at attention, and those in charge of the formation salute.

Outdoors

With some exceptions, saluting procedures when the anthem is played outdoors (with or without the flag displayed) are the same as when indoors. Marching formations are halted at attention, and the person in charge faces and salutes the flag or music, as appropriate.

Personnel in boats, whether in uniform or in civilian clothes, do not salute during the playing of the anthem. Only the boat officer (or coxswain if there is no boat officer) stands and salutes; all other personnel remain seated at attention.

You are not likely to hear "The Star Spangled Banner" played in a parade, but most marching units do carry the national ensign. The rules for saluting the flag passing in a parade are simple: come to attention, face the flag, and salute. (If you are in a vehicle, remain seated at attention.)

The musical selection "Hail to the Chief" is performed to honor the President of the United States. When "Hail to the Chief" is played, stand at attention and salute.

BOARDING AND LEAVING A NAVAL VESSEL

You cannot just walk on and off a ship as you would enter and leave your home. You must follow certain procedures.

When you are in uniform and boarding **any** ship and the national ensign is flying, you halt at the gangway, face aft, and salute the ensign. You then turn to the OOD and salute. If you are returning to your own ship, you say, "I request permission to come aboard, sir/ma'am." The OOD returns both salutes and says, "Come aboard" or a similar expression.

When you salute the OOD in boarding a ship other than your own, you say, "I request permission to come aboard, sir/ma'am." You should then add the purpose of your visit: "to visit a friend" or "to go to small stores."

When you leave a ship, the order of saluting is reversed. You salute the OOD first and say, "I request permission to leave the ship, sir/ma'am." After receiving permission, you then face and salute the ensign (if it is flying) and depart. If you are not in the liberty uniform, state your reason for wanting to leave the ship: "I request permission to go on the pier to check the mooring lines, sir/ma'am."

When boarding a ship in civilian attire and the national ensign is flying, you will halt at the gangway, at attention, and face aft. You then turn to the OOD at attention. If you are returning to your own ship, you say, "I request permission to come aboard, sir/ma'am." The OOD salutes and says, "Very well" or a similar expression.

When you board a ship other than your own, you say, "I request permission to come aboard, sir/ma'am." You should then add the purpose of your visit. The OOD will then say, "Permission granted" or "Permission not granted."

When you are leaving a ship in civilian attire, the procedure is reversed. You stand at attention in front of the OOD first and say, "I request permission to leave the ship, sir/ma'am." After receiving permission, you then stand at attention facing the ensign (if it is flying) and depart.

Sometimes it is necessary for destroyers, submarines, and other ships to tie up in nests alongside a repair ship, tender, or pier. In this case, you may have to cross several ships to go ashore or return to your own ship. When you have to cross one or more ships to reach the pier, to reach another ship or to return to your own ship, you should use the following procedure: Upon boarding a ship that you must cross, salute the colors (if flying), then turn toward and salute the OOD, and request permission to cross. After receiving permission, proceed to cross without delay. When you depart that ship, it is not necessary to salute the colors or OOD again. Repeat this crossing procedure until you reach your destination.

Student Notes:

On many ships, particularly those of destroyer size and smaller, there may be a first-class or chief petty officer instead of an officer on the quarterdeck. Although you do not salute enlisted personnel, you must salute an enlisted person who is the OOD because you are saluting the position and authority represented—not the individual. If you are part of a working party that will be using the quarterdeck when loading supplies, you normally salute only when first leaving the ship.

REVIEW 2 QUESTIONS

Q1. Hoisting the national flag at 0800 and lowering at sunset are known as

Q2. Define the term *shifting the colors.*

Q3. What is the significance of national flags flown at half-mast (half-staff ashore)?

Q4. You are indoors for an event, and the anthem is being played but the flag is not displayed. What should you do—

a. When in uniform and covered—

b. When in uniform and uncovered—

c. When in civilian clothes—

MILITARY ETIQUETTE

Learning Objectives: When you finish this chapter, you will be able to—

- Recognize military etiquette when addressing and introducing enlisted personnel.
- Recognize military etiquette when passing, meeting, addressing, replying to, walking or riding with an officer of the U.S. or foreign armed forces to include aboard ship or in a boat.

The rules of behavior to be observed by Navy personnel at certain times, in specified places, and on certain occasions are described in this section. "Behavior," in this case, means social conduct rather than strict military behavior, though the two sometimes are related.

For passing through doorways, let the senior go first; if possible, hold the door for him or her. On meeting an officer in a passageway, step aside so the officer may pass. If other enlisted persons and/or junior officers are present, call out "Gangway" so everyone can make way for the senior officer.

Juniors should show respect to seniors at all times by recognizing their presence and by being courteous and respectful in speech and manner. Juniors take the leftmost seat in a vehicle and walk on the left side of seniors whom they are accompanying.

ABOARD SHIP

There are rules of etiquette to follow during divine services, on the quarterdeck, or in officer's country.

When **divine services** are held on board ship, the following word is passed: "Divine services are being held in (such and such a space). The smoking lamp is out. Knock off all games and unnecessary work. Maintain quiet about the decks during divine services."

If you enter the area where divine services are being held, you must uncover even though you are on watch and wearing a duty belt. (Remain covered during Jewish ceremonies.)

Another area in which special rules apply is the **quarterdeck**. The quarterdeck is not a specific deck; it is an area designated by the commanding officer to serve as the focal point for official and ceremonial functions. The quarterdeck, consequently, is treated as a "sacred" part of the ship; and you should obey the following rules:

1. Don't be loud or sloppy in its vicinity.
2. Never appear on the quarterdeck unless you are in complete uniform.

Student Notes:

3. Never smoke or have coffee cups and soda cans or bottles on the quarterdeck.

4. Never cross or walk on the quarterdeck except when necessary.

5. Don't lounge on or in the vicinity of the quarterdeck.

6. When on the quarterdeck, salute whenever the quarterdeck watch salutes (as during a gun salute).

Shore stations, as well as ships, have areas designated as the quarterdeck. The same rules apply in all cases.

A **messing compartment** is where enlisted personnel eat; the **wardroom** is where officers eat. If you enter any of these areas while a meal is in progress, you must uncover.

Officers' country is the part of the ship where officers have their staterooms and wardrooms; **CPO country** is where the chief petty officers have their living spaces and mess. You must avoid entering these areas except on official business. Never use their passageways as thoroughfares or shortcuts. If you enter the wardroom or any compartment or office of an officer or a CPO, you must remove your hat, unless you are on watch and wearing the duty belt. Always knock before entering an officer's or a chief petty officer's room.

IN A BOAT

The basic rule in Navy etiquette, as in civilian etiquette, is to make way for a senior. Thus the rule for entering boats, airplanes, and vehicles is seniors in last and out first. (Enlisted personnel board a boat first, leaving room, of course, for officers.) The reason is that the captain should not have to wait in a boat for a less senior person to amble down the accommodation ladder. When the destination is reached, the senior is allowed to disembark first as a mark of respect from juniors.

In general, seniors take the seats farthest aft. If officers are present, enlisted personnel should not sit in the stern seats unless invited to do so. Enlisted personnel maintain silence as long as officers are in the boat. (For reasons of safety, personnel should never become noisy or boisterous in a boat regardless of the hour, condition of the sea, or who is present.)

The boat coxswain salutes all officers entering or leaving the boat. Enlisted personnel seated well forward do not rise when officers enter or leave the stern seats. Personnel in the after section, however, rise and salute when an officer enters or leaves. (Although it is customary to stand when saluting, this formality is dispensed with if the safety of the boat crew would be endangered.) When boat awnings are spread, enlisted personnel remain seated at attention while saluting; they do not rise under these circumstances.

A boat assumes rank according to the rank of the highest grade officer embarked in the boat. The coxswain and senior officer in each boat salute, with the person in the junior boat saluting first. Other crew members stand at attention; passengers sit at attention. The rules of etiquette for personnel aboard airplanes and other vehicles are the same as for boats.

Boats passing a ship during colors must lie to, or proceed at the slowest safe speed. The boat officer, or in his or her absence, the coxswain, stands (if safe to do so) and salutes. Other persons in the boat remain seated or standing, but do not salute.

ADDRESSING AND INTRODUCING NAVAL PERSONNEL

Custom, tradition, and social change determine how members of the naval service are introduced. Although tradition and military customs generally hold true, there are some differences in methods of addressing and introducing military personnel, depending on whether you are in civilian or military circles.

The proper forms of addressing and introducing naval personnel are summarized in table 9-1. Except as provided in the paragraphs that follow, all officers in the naval service are addressed or introduced with the titles of their grades preceding their surnames.

Officers of the Medical or Dental Corps, and officers of the Medical Service Corps having a doctoral degree, may be addressed as "doctor." Likewise, an officer of the Chaplain Corps may be addressed as "chaplain." However, if a doctor or chaplain prefers to

Student Notes:

be addressed by the title of his or her grade, such preference should be honored. When you are addressing an officer whose grade includes a modifier (for example, lieutenant commander), the modifier (lieutenant) may be dropped.

As a general rule, use the officer's title and name. It is better to say, "Yes, Ensign Smith"; "No, Doctor Brown"; or "Yes, Lieutenant Jones"; than to say, "Yes, sir" or "No, ma'am." However, in prolonged conversation where repetition would seem forced or awkward, the shorter "sir" or "ma'am" is used more often.

Aboard ship, the regularly assigned commanding officer is addressed as "captain" regardless of grade. The regularly assigned executive officer (if of the grade of commander) may be addressed as "commander" without adding the name. In some ships it is customary to address the executive officer as "commander" even though the grade is that of lieutenant commander.

Naval officers are introduced to civilians by title, and the method of introduction should give a clue as to how the person should be addressed from then on. You might say, "This is Lieutenant Jones. Mr. Jones is a shipmate of mine." This serves a double purpose; it gives the civilian to whom you are introducing the officer knowledge of the naval person's grade, and it also gives the correct method of address, "Mr. Jones."

Military and civilian practices differ in the introducing and addressing of enlisted personnel. Under military conditions, petty officers are addressed and introduced by their respective title followed by their last name. Petty officers in paygrades E-7, E-8, and E-9 are introduced and addressed as "Chief ______ ," prefixed by "Senior" or "Master," if appropriate. Petty officers in paygrades E-4 through E-6 are addressed and introduced as "Petty Officer _______ ." Persons in paygrades E-3 and below are addressed by their last names only in informal situations. However, in the formal situation or introductions, their last names are preceded by "Seaman," "Fireman," "Airman," and so forth, as appropriate.

Civilians sometimes feel uncomfortable in social gatherings when addressing enlisted personnel as described in the preceding paragraph. It is customary, therefore, for those outside the service to extend to enlisted personnel the same courtesies they would extend to them in civilian life and to prefix their names with Mr., Mrs., Miss, or Ms., as the case may be. In introducing them, one should give their titles and names, then the mode of address: "This is Petty Officer Smith. Mr. Smith will be visiting us for a while." Thereafter, he will be addressed as "Mr. Smith."

In civilian life you are supposed to introduce men to women and youth to age; that is, a young man to a young woman or a young woman to an older woman. If the person is a male member of the clergy, however, you introduce women of any age to him; or, if a man is aged or distinguished, you introduce the woman to him.

The same general rules are followed in military life, except that in most instances naval rank establishes the order of introduction. Thus, you introduce the junior to the senior, whether male or female. An exception is that you always introduce others, regardless of the rank or sex, to a chaplain. If one of the persons is a civilian, you follow the rules of youth to age and male to female.

The only proper response to an oral order is "Aye, aye, sir/ma'am." This reply means more than yes. It indicates "I understand and will obey." Such responses to an order as "O.K., sir" or "All right, sir" are taboo. "Very well" is proper when spoken by a senior in acknowledgment of a report made by a junior, but a junior never says "Very well" to a senior.

"Sir" or "Ma'am" should be used as a prefix to an official report, statement, or question addressed to a senior. It should also be used when addressing an official on duty representing a senior. For example, the OOD, regardless of grade, represents the commanding officer and should be addressed as "Sir" or "Ma'am."

If you are a junior addressing a senior, you should introduce yourself unless you are certain the senior knows you by sight.

REVIEW 3 QUESTIONS

Q1. What is the quarterdeck?

Q2. Aboard ship, how is the CO addressed?

Student Notes:

Table 9-1.—Introducing and Addressing Naval Personnel

PERSON ADDRESSED OR INTRODUCED	TO MILITARY		TO CIVILIAN	
	INTRODUCE AS:	ADDRESS AS:	INTRODUCE AS:	ADDRESS AS:
COMMANDER or above	COMMANDER (or appropriate rank) SMITH	COMMANDER (or appropriate rank) SMITH	COMMANDER (or appropriate rank) SMITH	COMMANDER (or appropriate rank) SMITH
LIEUTENANT COMMANDER or below	LIEUTENANT COMMANDER (or appropriate rank) SMITH	COMMANDER SMITH	LIEUTENANT COMMANDER SMITH	Mr. (Mrs., Miss, Ms.) SMITH
MEDICAL And/or DENTAL CORPS OFFICER	DR. SMITH	DR. SMITH	LIEUTENANT SMITH OF THE NAVY MEDICAL CORPS	DR. SMITH
CHAPLAIN CORPS OFFICER	CHAPLAIN SMITH	CHAPLAIN SMITH	CHAPLAIN SMITH	CHAPLAIN
NAVY NURSE CORPS OFFICER	COMMANDER (or appropriate rank) SMITH	COMMANDER SMITH	COMMANDER SMITH OF THE NAVY NURSE CORPS	COMMANDER (Mr., Mrs., Miss., Ms.) SMITH
CHIEF WARRANT OFFICER	CHIEF WARRANT OFFICER SMITH	CHIEF WARRANT OFFICER SMITH	CHIEF WARRANT OFFICER SMITH	MR. (Mrs., Miss., Ms.) SMITH
MIDSHIPMAN	MIDSHIPMAN SMITH	MIDSHIPMAN SMITH	MIDSHIPMAN SMITH	MR. (Mrs., Miss., Ms.) SMITH
CHIEF PETTY OFFICER	CHIEF SMITH	CHIEF or SMITH	CHIEF YEOMAN SMITH	MR. (Mrs., Miss., Ms.) SMITH
AVIATION CADET	AVIATION CADET SMITH	MR. SMITH	AVIATION CADET SMITH	MR. (Mrs., Miss., Ms.) SMITH
PETTY OFFICER	PETTY OFFICER SMITH	PETTY OFFICER SMITH	PETTY OFFICER SMITH	MR. (Mrs., Miss., Ms.) SMITH
SEAMAN	SEAMAN SMITH	SMITH	SEAMAN SMITH	MR. (Mrs., Miss., Ms.) SMITH

SUMMARY

Customs and courtesies play an integral part of the seafarer's life. Through them, we show respect for each other and for certain symbols or objects, such as the American flag. When we address the commanding officer as *captain,* even though he or she holds the rank of commander, we are using a time-honored custom of respect for the person in command.

The courtesy of the salute is not only the required recognition of a senior but also the expression of mutual respect and pride in service. The courtesy of rendering honors to the Arizona Memorial and Washington's tomb is a sign of respect. The custom of officer personnel boarding ships' boats after enlisted personnel is another sign of respect. The customs and courtesies of removing your hat indoors or in the presence of a lady, of rendering honors to the national ensign, and of playing the national anthem at morning and evening colors are also signs of respect. Our customs and courtesies will continue to be a part of our daily routine as long as we maintain pride and respect in our Nation, our service, and ourselves.

REVIEW 1 ANSWERS

A1. Define the following terms.

a. Courtesy—**an act or verbal expression of consideration or respect for others**

b. Custom—**a usual way of acting in a situation that has been practiced so long that it has the force of law**

A2. A **salute** is a required act of military courtesy.

A3. The **hand salute** is the most common form of salute.

A4. When in uniform, Navy personnel salute the **flag, anthem, and officers.**

A5. **True**, under naval customs, the hand salute is accompanied by a word of greeting.

A6. As an enlisted person, you should salute **all officers.**

A7. When boarding a ship in which the national ensign is flying, you should—

a. **Stop on reaching the upper platform of the accommodation ladder or end of brow**

b. **Face the ensign and salute**

c. **Salute the OOD**

A8. As a sentry at a gangway, you should render a salute—

a. **To all officers going or coming over the side**

b. **When passing or being passed by officers close aboard in boats**

A9. The three rifle salutes are—

a. **Present arms**

b. **Rifle salute at order arms**

c. **Rifle salute at right shoulder arms**

A10. Honors are **salutes rendered to ships, high-ranking individuals, and nations.**

A11. **Manning the rail** is a passing honor rendered to the President of the United States.

A12. A ruler of a country recognized by the United States rates a **21-gun** salute.

REVIEW 2 ANSWERS

A1. Hoisting and lowering the national flag at 0800 and sunset are known as **morning colors** and **evening colors.**

A2. Shifting the colors—**as a ship gets underway, the ensign is shifted from its in-port position on the stern to its at-sea position on the mainmast.**

A3. National flags flown at half-mast are **internationally recognized symbols of mourning.**

Student Notes:

A4. When indoors at an event and the anthem is being played but the flag is not displayed, you should—

a. In uniform and covered—**render a hand salute**

b. In uniform and uncovered—**stand at attention**

c. In civilian clothes—**place your hand over your heart**

REVIEW 3 ANSWERS

A1. The quarterdeck is **an area designated by the CO that serves as the focal point for official and ceremonial functions.**

A2. Aboard ship, the CO is addressed as **captain, regardless of rank**.

CHAPTER 10

UNIFORMS AND FORMATIONS

You shall wear your uniforms properly as described in these regulations. Naval personnel must present a proud and professional appearance that will reflect positively on the individual, the Navy, and the United States. The uniforms of the United States Navy and the indications of rank and specialty displayed thereon, are but outward symbols of naval organization and military rank or rating. As such, the Navy uniform is a visibly important element in the morale, pride, discipline and effectiveness of the organization.

—*U.S. Navy Uniform Regulations*, NAVPERS 15665

Today's Navy has narrowed the gap between men's and women's career paths. Women now perform many of the same tasks and have the same specialties as their male counterparts. These changes caused Navy uniform policy to change, bringing the uniforms of both men and women more closely in line with each other. Navy uniforms are distinctive visual evidence of the authority and responsibility vested in their wearer by the United States.

Because Navy ways are new to you, many questions probably have crossed your mind, such as, "What is that officer's rank"? "What does that petty officer's insignia mean"? "What does that pin stand for"? This chapter explains officer grades; precedence; authority; the enlisted rating structure; and how to wear, mark, and exchange uniforms.

The United States Navy has had a basic uniform policy for many years. The purpose of the uniform policy is to ensure that naval personnel have attractive, distinctive, and practical uniforms. *U.S. Navy Uniform Regulations,* NAVPERS 15665, provides the basic naval uniform policy. You can download the *U.S. Navy Uniform Regulations* from the BUPERS homepage at www.bupers.navy.mil. When you get to this page, select the "services" button and look under "uniform matters."

WEARING THE UNIFORM

Learning Objective: When you finish this chapter, you will be able to—

- Recognize uniform components.
- Identify the proper procedures for wearing uniforms.

Your dress and conduct should always reflect credit upon yourself, the Navy, and the United States. It should be a matter of personal pride to present the best possible appearance in **your** manner of dress and grooming.

You are given a complete issue of regulation clothing when you enter the Navy. The maintenance of your uniform and the replacement of articles of clothing are **your** responsibility. You cannot use the excuse, "I have nothing fit to wear," or "I'm out of clean uniforms."

ENLISTED MEN

Look at figures 10-1 and 10-2. Here, most enlisted (E-6 and below) men's uniforms are shown. Uniforms must be tailored according to the specifications in the following paragraphs:

The dress blue and dress white jumpers must hang straight and be the same length. The sleeves of the dress white jumper are cut square at the cuff openings.

The **dress white trousers** have a zipper front and are the same length as the blue trousers. A white belt with a regulation buckle is worn with the white uniform. The belt should be of the same fabric as the uniform.

The **neckerchief** is 36 inches square and made of black silk, acetate, or other suitable material. It is folded diagonally to form a triangle, then rolled (fig.10-3) and placed around the neck under the collar. It is tied (fig. 10-4) in a square knot with both ends of equal length and falling naturally. The upper edge of the knot must be even with the lowermost point of the collar opening. Neckerchiefs are worn with both the dress blue and the dress white uniforms. (**NOTE**: For enlisted women neckerchiefs are worn with dress white uniform only.)

Dress blue and dress white jumpers. The dress

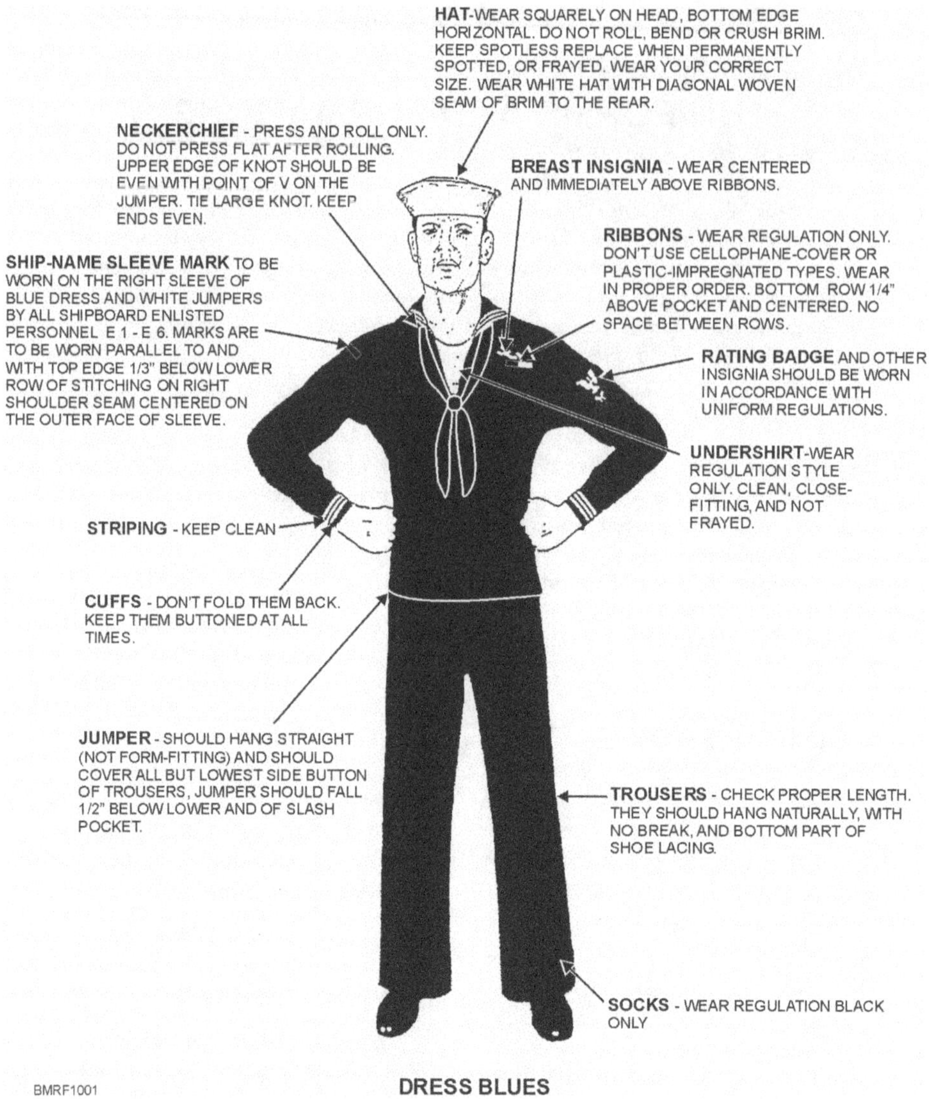

Figure 10-1.—Proper wearing of uniforms.

blue jumper should hang straight and be long enough to cover all but the lowest button of the 13-button broadfall front trousers. Sleeves have two button cuffs, the edges of which reaches to the knuckles when the cuffs are unbuttoned. The collar has three stripes of white tape (piping). The dress blue trousers are plain and cuffless and have a 13-button broadfall front. The old story that the 13 buttons represent the 13 original colonies is false.

Government-issue dress blue jumpers and trousers are made of wool serge. The dress white jumpers and trousers are made of a polyester material designated certified Navy twill.

Dungarees and winter working blues. Dungarees and winter working blues are considered to be the working uniforms. Dungarees consist of a blue

Student Notes:

Figure 10-1. Proper wearing of uniforms (continued).

chambray shirt and blue denim trousers. Command or navy ball caps are only authorized for use with dungarees. The winter working blue uniform consists of blue winter shirt and blue trousers. Enlisted men E-6 and below have the option of wearing the garrison cap instead of the white hat with the winter working blue uniform (command ball cap may be authorized as with dungarees and worn within unit premises or working areas). Both uniforms are to be worn in working spaces where other uniforms would be unsafe, inconvenient, or would become excessively soiled.

Shoes. Black dress shoes and black socks are worn with all uniforms. Shoes may be of leather or a synthetic material with a plain toe.

Peacoat. The peacoat is a blue winter-weight coat that may be prescribed with an appropriate winter uniform. It is an easy-fitting, double-breasted coat with a convertible collar. The peacoat should reach the hips. The sleeves should reach to about three fourths of the distance from the wrist to the knuckles of the hand when the arms hang naturally at the sides. The peacoat must

Student Notes:

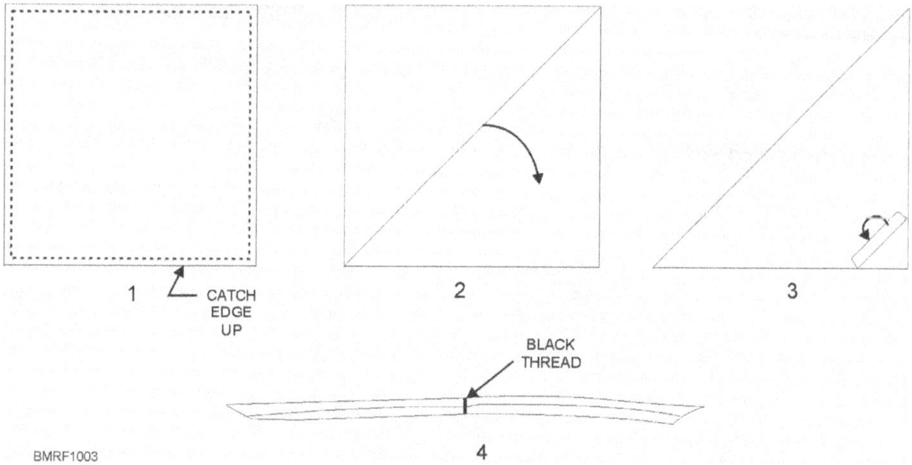

Figure 10-3.—Rolling the neckerchief.

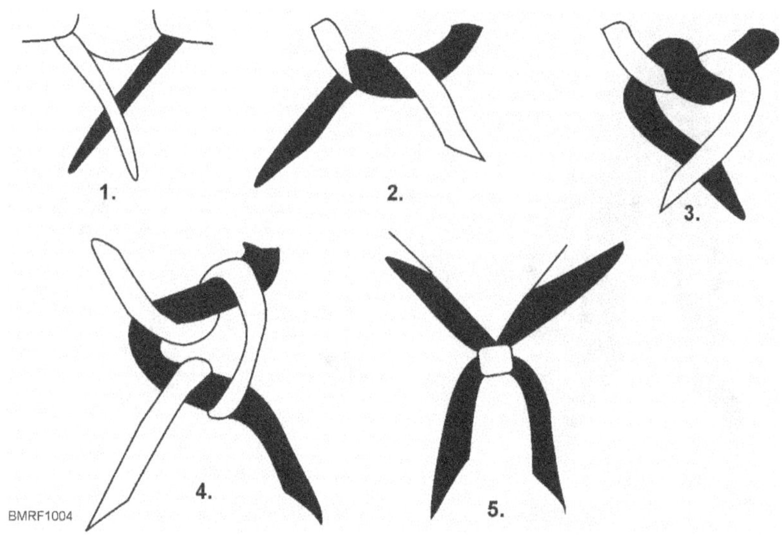

Figure 10-4.—Tying the neckerchief.

be worn buttoned with the three lower buttons on the right side. In foul weather, the coat may be buttoned to the neck.

All-weather coat. A black, single, breasted, beltless, water-repellant coat. It's made from 65% polyester and 35% cotton poplin fabric. The coat has epaulets, tabs on the cuffs, and a zip-out lining. Women's coats button to the left and men's coats button to the right. When worn, button all the buttons except the collar button (in bad weather, the collar button may be buttoned). When wearing the all-weather coat, a clear plastic, combination cap rain cover may be worn.

Blue working jacket. The blue working jacket is made of navy blue material. It is fully lined with a zipper front. Two military organization patches may be worn on the jacket on an optional basis, subject to the following restrictions:

Student Notes:

- The patch of the command to which you are assigned must be centered on the left breast.
- A second unit patch of your choice, acceptable to the command, will be centered on the right breast.

The blue working jacket may be worn with the dungarees uniform.

Hat. The white hat should be worn squarely on the head, as shown in figure 10-1.

Medals. Large medals may be prescribed for wear with the dress blue jumper for special occasions, such as change of command or formal inspections. Medals and neckerchief may be prescribed for wear with the dress white jumper. When large medals are worn with either uniform, the uniform is called *full dress*.

ENLISTED WOMEN

Enlisted (E-6 and below) women's uniforms and their proper wear are shown in figures 10-5 and 10-6 (not all uniforms are shown). Uniforms must be tailored according to the specifications in the following paragraphs:

Service dress blues. The enlisted women's service dress blues (fig. 10-5) is worn with ribbons. The necktie or necktab is worn with the outer edges parallel to the outer edges of the collar. An equal amount of the necktab should show on each side of the collar. For prescribed (required) and optional items, refer to NAVPERS 15665.

Full dress blue. The full dress blue uniform is the same as the service dress blue uniform. Large medals and ribbons are worn (fig. 10-5) with this uniform.

Dress white jumper. The enlisted women's dress white jumper (fig. 10-6). The jumper should fit comfortably with no binding. The sleeves will have inverted creases at the inside and outside edge, hang straight, and be long enough to cover the wristbone. The bottom of the jumper should be loose fitting at the hips. Normally, a white belted skirt is worn with the dress white jumper. However, white jumper slacks can be worn with the dress white jumper. (**NOTE**: When white jumper slacks are used with the white jumper, the hem of the jumper should fall to within 1 inch above the bottom of the pocket opening of the slacks.) The slacks have side seam pockets and inverted creases. The slacks

Figure 10-5.—Enlisted women's uniforms.

Student Notes:

Figure 10-6.—Enlisted women's uniforms (continued).

should cover the shoes at the heel by 1 inch.

Working uniforms. Working uniforms consist of belted blue slacks or skirt and blue winter shirt. Dungarees (fig. 10-6) are also authorized. Black socks, a garrison cap, black service shoes, and a black handbag may be worn with each. Command ball caps, berets, and the blue cardigan are other items that may be prescribed.

Shoes. Black dress shoes are pumps made of smooth leather, calf, or synthetic leather. They will be of plain design with closed heels and toes. The heels will be no higher than 2 5/8 inches nor less than 5/8 inch when measured from the forward edge of the heel. Wedge heels are not authorized. The black service shoes will be laced leather or synthetic oxfords with one line of black stitching around the top of the toe.

Hat. The combination hat may be worn with all dress uniforms. It is oval in shape, with a stiffened crown. The brim is rolled at the sides and straight in front and back. The hat is worn with a detachable white hat cover. Enlisted women E-6 and below may wear the garrison cap as an option with blues only when wearing the black v-neck sweater instead of the service dress blue coat. As with the enlisted male E-6 and below, command or navy ball caps are only authorized with dungaree uniforms.

UNIFORM OF THE DAY

The uniform of the day is that uniform prescribed by proper authority to be worn on occasions such as work, liberty, and inspections. The prescribed uniform of the day is published in the Plan of the Day or the Plan of the Week. Wear your uniform with pride in self, the Navy, and the United States.

REVIEW 1 QUESTIONS

Q1. What is the length of the dress blue jumper for men?

Q2. What uniform should enlisted women E-1 through E-6 wear with the black silk neckerchief?

Student Notes:

Q3. When wearing a peacoat over a dress blue jumper uniform, the uniform collar should be—

Q4. When added to a service dress white uniform, what uniform component makes the service dress white uniform a full dress uniform?

PROPER CARE OF UNIFORMS

Learning Objectives: When you finish this chapter, you will be able to—

- Identify the methods for caring for and maintaining uniforms to include marking enlisted clothing and transferring clothing.
- Identify the contents and layout of the seabag.
- Identify the uniform components that can be worn with civilian clothes and recognize the rules for wearing civilian clothes.

All Navy personnel must maintain their full requirement of authorized uniforms and are forbidden to possess or wear unauthorized uniforms. Division officers are required to inspect the uniforms of all nonrated personnel at regular intervals. The purpose of the inspection is to make sure that each person has the prescribed outfit. Insignia, decorations, medals, badges, and ribbons are worn as prescribed. All uniforms must be kept scrupulously clean; gold bullion lace, devices, and insignia must be kept free of tarnish and corrosion. Shoes should be kept well shined and in good repair.

You should **not** wear a uniform if any of the following conditions exist:

- Frayed, torn, ill fitting, badly wrinkled, badly stained, or dirty uniforms. (A little leeway is permitted in the dungaree uniform, but if it is damaged beyond the possibility of a professional-looking repair, the item of clothing should be discarded.)
- Discolored or frayed piping.
- Missing buttons.
- Cracked, badly stained, or nonregulation shoes.
- Badly faded, discolored, or frayed ribbons or rate/rating badges; tarnished or corroded metal devices.

You may not wear any uniform, article, insignia, or decoration that is not yours or to which you are not entitled.

Further, you should **not** wear a uniform in the following manner:

- Unbuttoned coats
- Rolled up sleeves, hat not squared, cuffs unbuttoned, or shirttails hanging out
- Incomplete or nonregulation uniforms

MARKING ENLISTED CLOTHING

You should mark your clothing legibly with your name and social security number. Use black marking fluid for marking white clothing and chambray shirts. Use white marking fluid for marking blue clothes and dungaree trousers. Where labels are provided, mark them with indelible ink.

If available, men's and women's clothing should be marked with a 1/2-inch stencil or stamp. If a 1/2-inch stamp is not available, a stencil no larger than 1 inch may be used.

Required items of men's and women's clothing (tables 10-1 and 10-2) are marked according to current Navy uniform regulations and as indicated in the chart shown below. Where the word *right* or *left* appears in marking instructions, it means the owner's right or left when wearing the garment. On towels and similar articles, it means the owner's right or left when standing behind the article.

Student Notes:

Table 10-1.—Marking Men's Clothing

ITEM	LOCATION
Shirts	1. Blue chambray—Vertically, beginning 1 inch from the bottom on the inner side of the right front facing on which the buttons are sewn; embroider last name on right front, 1 inch above pockets. 2. Winter blue —Vertically, beginning 1 inch from the botton on the lower side of the right front facing. 3. White summer—Same as winter blue.
Jumper (blue or white)	1. Turn inside out, front down, collar away from you, stencil three initials, ¾ inch below collar seam to left of center, and last four digits of SSN ¼ inch below horizontal seam using white ink; fill in manufacturer's tag, using ball- point pen.
Trousers	1. Dress blue—On designated nameplate, turn trousers inside out, fly down, waistband away from you, stencil three initials and last four digits of SSN on rear pocket, 1/4 inch below horizontal seam using white ink; fill in manufacturer's tag, using ball-point pen. 2. Dungaree—On waistband inside front at the right of center line, last name will be embroidered in white on the outside, 1 inch above right hip pocket, centered. 3. White long—Turn inside out, fly down, waistband away from you, stencil three initials and last four digits of SSN on left rear pocket in between the horizontal seams; fill in manufacturer's tag, using ball-point pen.
Caps	1. Command ball—Initials only on sweatband. 2. Knit (watch)—Initials only, on a label on the inside. 3. White hat—On the inside of the hem at the right of the center line on the back.
Jackets	1. Blue windbreaker—On the inside of the hem at the right center line on the back. 2. Blue working—On the inside of the hem at the right on the center line on the back; last name only on the left 1 inch above the pocket; centered. Will be white.
Peacoat	1. On label located on the inside breast pocket.
All-weather coat	1. Inside lining, 3 inches below collar seam.
Sweater	1. On label on the inside below the back of the collar.
Shoes	1. Initials only inside, near top.
Socks	1. Initials only on the foot.
Drawers	1. On the outside of the right half of the waistband, or immediately underneath the waistband on drawers with elastic waistbands.
Undershirts	1. On outside of the front, 1 inch from the bottom of the shirt, right of the center.
Belts	1. Inside near tab.
Gloves	1. Initials only on inside, near the top.
Neckerchiefs	1. Diagonally across the center before folding, initials only.

Table 10-2.—Marking Women's Clothing

ITEM	LOCATION
Coat (blue)	1. On designated name label.
Jumper (white)	1. Same as for men.
Shirts	1. Blue chambray—Vertically, beginning 1 inch from the bottom on the inner side of the right front fold on which the buttons are sewn. The chambray shirt is embroidered in black, last name only on right front, 1 inch above the pocket, centered. Maternity chambray shirts that do not have pockets are embroidered in the same relative position as the blue chambray shirt, with pockets. 2. Blue winter—Vertically, beginning 1 inch from the bottom on the inner side of the right front fold on which the buttons are sewed. 3. White—Same as blue winter.
Skirts	1. Blue, belted—Center front, inside waistband. 2. Blue, unbelted—Center front, inside on waistband. 3. Blue, formal—Center front, inside on waistband. 4. White, belted—White-certified Navy twill skirts will be marked on the name tag sewn on the liner directly underneath the right pocket.
Slacks (blue, white, or dungarees)	1. Blue and white—Center back, inside on waistband. 2. Dungarees—Center back, embroidered inside on waistband
Caps	1. Command ball—Initials only on sweatband. 2. Knit (watch)—Initials only on label on the inside. 3. Beret—On designated nameplate. 4. Combination white—On designated nameplate. 5. Garrison blue—On designated nameplate.
Jackets	1. Blue windbreaker—On the inside of the hem at the right on the center line on the back. 2. Blue working—On the inside of the hem at the right of the center line of the back and the last name only on the left front, 1 inch above the pocket; centered. Will be in white
Overcoat	1. On designated nameplate; and inside left front panel.
Peacoat	1. On the label located on the inside breast pocket.
All-weather coat	1. Inside the lining, 3 inches below collar seam.
Sweater	1. On the manufacturer's tag.
Scarf (blue or white)	1. Center back, inside.
Shoes (black dress or black service)	1. Initials only inside, near top.
Socks (black)	1. Initials only on the foot.
Undershirt	1. On the outside of the front, 1 inch from the bottom of the skirt and at the right on the center.
Belts (black or white)	1. Inside, last name and the first and middle initials only.

Table 10-2.—Marking Women's Clothing (continued)

ITEM	LOCATION
Handbag	On the manufacturer's tag inside large pocket.
Necktie	Inside, center, back.
Neckerchief	Same as for men.

NOTE

Embroidered name/nametags are required on the chambray shirt and dungaree trousers.

Your clothing is marked for good reasons. When you send your clothing to the laundry, there must be a method to identify it. If your clothing is lost or misplaced, the only way it can be recovered is if it has been properly marked.

TRANSFER OF CLOTHING

No transfer or exchange of an enlisted person's uniform clothing will be made without the commanding officer's authorization. When such transfers or exchanges are authorized or when clothing belonging to deserters is sold, obliterate (make the name unreadable) the former owner's name with a red "D.C." stamp. The purchaser's name will be placed above, below, or next to it.

SEABAGS

The CO requires clothing of all nonrated personnel to be inspected at regular intervals to make sure that each person has the required seabag items (table 10-3 and table 10-4). Also, before a nonrated person is transferred to another ship or station, another seabag inspection is made. Petty officers clothing may be inspected on an individual basis, as appropriate. All personnel are required to have at least the following items and quantities in their seabag:

Table 10-3.—Seabag Requirements for Enlisted E-1 through E-6—Men

ENLISTED MEN			
ITEM	QUANTITY	ITEM	QUANTITY
All-weather coat (blue)	1	Shirt, winter blue	2
Bag, duffel	1	Shirt, blue chambray, long sleeve	1
Belt, web, black, with silver clip	2	Shirt, blue chambray, fire retardant, long sleeve	4
Belt, web, white, with silver clip	3	Shirt, white short sleeve	2
Buckle, silver	4	Shoes, dress black	1 pr
Cap, ball	2	Shoes, safety chukka	1 pr
Cap, knit	2	Socks, cotton/nylon, black	6 pr
Gloves, leather, black	1 pr	Sweater, wool, blue	1
Group rate mark, black	3	Towel, bath	4
Group rate mark, white	4	Trousers, broadfall, blue	1 pr
Hat, white	3	Trousers, poly/wool, dress blue	2 pr
Jacket, blue working	1	Trousers, dungaree/denim	1 pr

Student Notes:

Table 10-3.—Seabag Requirements for Enlisted E-1 through E-6—Men (continued)

ENLISTED MEN			
ITEM	QUANTITY	ITEM	QUANTITY
Jumper, blue working	1	Trousers, dungaree, fire retardant	4 pr
Jumper, blue dress	1	Trousers, white	2 pr
Jumper, white dress	2	Trousers, white jumper (polyester)	2 pr
Neckerchief	1	Undershirts	8
Necktie, black	1	Undershorts, white	8

Table 10-4.—Seabag Requirements for Enlisted E-1 through E-6—Women

ENLISTED WOMEN			
ITEM	QUANTITY	ITEM	QUANTITY
All-weather coat (blue)	1	Shirt, cotton/poly, blue chambray, long sleeve	1
Bag, duffel	1	Shirt, blue chambray, fire retardant, long sleeve	4
Belt, web, black, with silver	2	Shirt cotton/poly white short sleeve	3
Belt, web, white, with silver	3	Shirt, winter blue	2
Buckle, silver	2	Shoes, black dress pumps	1 pr
Cap, ball	2	Shoes, black safety	1 pr
Cap, combination with 2 crowns	1	Shoes, service black	1 pr
Cap, garrison, blue	2	Skirt, blue belted	1
Cap, knot, blue	1	Skirt, blue unbelted	1
Coat, service dress blue	1	Skirt, CNT, poly white belted	1
Gloves, leather, black	1pr	Skirt, poly/cotton, white belted	2
Group rate mark, black	3	Slacks, blue belted	2 pr
Group rate mark, white	5	Slacks, blue unbelted	1 pr
Handbag, black	1	Slacks, cotton/poly, dungaree/denim	1 pr
Hosiery, nylon	As needed	Slacks, dungaree, fire retardant	4 pr
Insignia, service, hat, cap	1	Slacks, dress, CNT, white	2 pr
Jacket, utility (blue working)	1	Slacks, poly/cotton, white	2 pr
Jumper, white dress	1	Socks, black	5 pr
Lingerie	As needed	Sweater, blue	1
Neckerchief, black	1	Towel, bath	4
Necktab, black	1	Undershirts	8

CIVILIAN CLOTHING

You may be permitted to have civilian clothing in your possession aboard ship or at a naval activity ashore. You may wear such clothing while leaving or returning to your ship or station, while awaiting transportation after permission to leave the ship has been given, while on authorized leave of absence, liberty, or in any off-duty status ashore.

When wearing civilian clothing, you must ensure that your dress and personal appearance are appropriate for the occasion and won't bring discredit upon the naval service. Current styles and fashions are authorized. Tank-top shirts, white undershirts worn as outer garments, cutoff shorts, and shower sandals are considered appropriate civilian attire for occasions such as picnics, athletic events, and other daytime activities of an extremely casual nature. The above items will not otherwise be worn within the confines of a military installation.

In cases of individuals who do not wear civilian clothing as outlined or who fail to maintain proper and adequate uniforms, individual commands may suspend the privilege of wearing civilian clothing to and from the command.

Military personnel may wear the following military uniform articles with civilian clothing:

- All-weather coat/raincoat (without insignia)
- Belts with civilian buckles
- Knit watch cap
- Command/Navy ball cap (without insignia)
- Gloves
- Handbag
- Blue windbreaker jacket (without insignia)
- Khaki windbreaker jacket (without insignia)
- Shoes
- Socks/hosiery
- Cardigan and blue pullover sweaters (recruit issue)
- Black V-neck pullover sweater (without nametag)
- Underwear

REVIEW 2 QUESTIONS

Q1. DELETE

Q2. Men and women stencil what uniform the same way?

Q3. What person can authorize the transfer of an enlisted person's clothing to another enlisted person?

Q4. You are required to have fire retardant dungaree shirts and trousers/slacks in your seabag. How many pair(s) should you have?

Q5. List five uniform items that you can wear with civilian clothes.

a.

b.

c.

d.

e.

Student Notes:

OFFICER AND ENLISTED INSIGNIA

Learning Objective: When you finish this chapter, you will be able to—

- Identify rating insignia to include service stripes, paygrade insignia of U.S. armed forces enlisted personnel, grade insignia and corps devices of naval officers, and special insignia.

In the enlisted branch of the Navy, a field of work or an occupation is called a *rating*. Levels within the rating are called *rates*. In the case of a Boatswain's Mate second class (BM2), for example, Boatswain's Mate is the rating and second class is the rate. In other words, rating is the job or occupation, while rate is the paygrade of the person.

As a newcomer without previous military experience, you probably entered the service as a recruit in paygrade E-1. This is the basic paygrade in the armed forces grading structure. From the recruit rate, you began to train in one of the six broad occupational groups. Your occupational group is identified by a group rate mark, which is worn on the left sleeve of jumpers and white summer shirts. Group rate marks are shown in figure 10-7.

Personnel in paygrades E-1, E-2, and E-3 who have qualified in a particular rating will wear the specialty mark of that rating. This is called a *striker mark*. The striker mark is worn immediately above the group rate mark. If you were a qualified striker and in paygrade E-1, you would wear the striker mark, even though you wouldn't wear the group rate mark.

After advancing to Seaman, Fireman, Airman, Constructionman, Hospitalman, or Dentalman, you'll want to qualify for the lowest petty officer rate—petty officer third class. The rating you are trying for will depend on your personal qualifications and desires. At this time, you will enter the occupational field that you will normally follow for the rest of your Navy career. Subject to standard instructions, changes from one field to another are allowed quite freely in the lower paygrades before a person has been intensively trained in one particular field. This lets you have enough time to find the choice of work you want in the Navy. However, once you have advanced to a senior petty officer level, changes to another field are seldom permitted.

As mentioned before, every enlisted person in the Navy has a rate. You must be able to identify a person's rate. To enable you to do this, every enlisted person in the Navy (with the exception of E-1) is required to wear an insignia indicating rate on the left arm of the service uniform. This is usually called a rating badge. Figures 10-8 through 10-10B identifies all enlisted rating insignia.

SERVICE STRIPES

Service stripes (called *hash marks*) indicate length of service. One stripe is worn on the left sleeve of jumpers for each **full** 4 years of active or Reserve service in any of the armed forces, or any combination thereof, such as 2 years in the Army and 2 years in the Navy. Scarlet hash marks and rating badges are worn on blue uniforms; blue hash marks on white uniforms.

Gold rating badges and service stripes are worn when good conduct in the naval service totals 12 years. This 12 years may be active or drilling reserve time in the Navy, Navy Reserve, Marine Corps, or Marine Corps Reserve. For more information about authority to wear gold rating badges and service stripes, refer to the *U.S. Navy Uniform Regulations*, NAVPERS 15665.

RATE INSIGNIA OF THE U.S. NAVY AND OTHER U.S. ARMED FORCES ENLISTED PERSONNEL

Look at figure 10-11, which shows the rate insignia of all the U.S. armed forces enlisted personnel.

The most senior enlisted person in the U.S. Navy is the master chief petty officer of the Navy (MCPON). (**NOTE**: The senior enlisted insignia for the other armed services are shown directly under the MCPON in fig. 10-11.) The rating insignia of the MCPON is similar to that of all other master chief petty officers, except that it has three gold stars in line above the eagle and a gold star in the space between the eagle and the upper chevron that replaces the specialty mark.

Major commands have a Command Master Chief Petty Officer (CM/C). The CM/C insignia differs from the MCPON. There are two silver stars above the eagle, and there is a silver star that replaces the specialty mark.

Student Notes:

GROUP	PAYGRADE	RATE	ABBR	UPPER SLEEVE
DECK/ADMINSTRATION MEDICAL/DENTAL	E-1	SEAMAN RECRUIT HOSPITALMAN RECRUIT DENTALMAN RECRUIT	SR HR DR	
	E-2	SEAMAN APPRENTICE HOSPITALMAN APPRENTICE DENTALMAN APPRENTICE	SA HA DA	
	E-3	SEAMAN HOSPITALMAN DENTALMAN	SN HN DN	
ENGINEERING/HULL	E-1	FIREMAN RECRUIT	FR	
	E-2	FIREMAN APPRENTICE	FA	
	E-3	FIREMAN	FN	
AVIATION	E-1	AIRMAN RECRUIT	AR	
	E-2	AIRMAN APPRENTICE	AA	
	E-3	AIRMAN	AN	
CONSTRUCTION (SEABEES)	E-1	CONSTRUCTION RECRUIT	CR	
	E-2	CONSTRUCTION APPRENTICE	CA	
	E-3	CONSTRUCTION	CN	

BMRF1007C

Figure 10-7.—Group rate marks for paygrades E-1 through E-3.

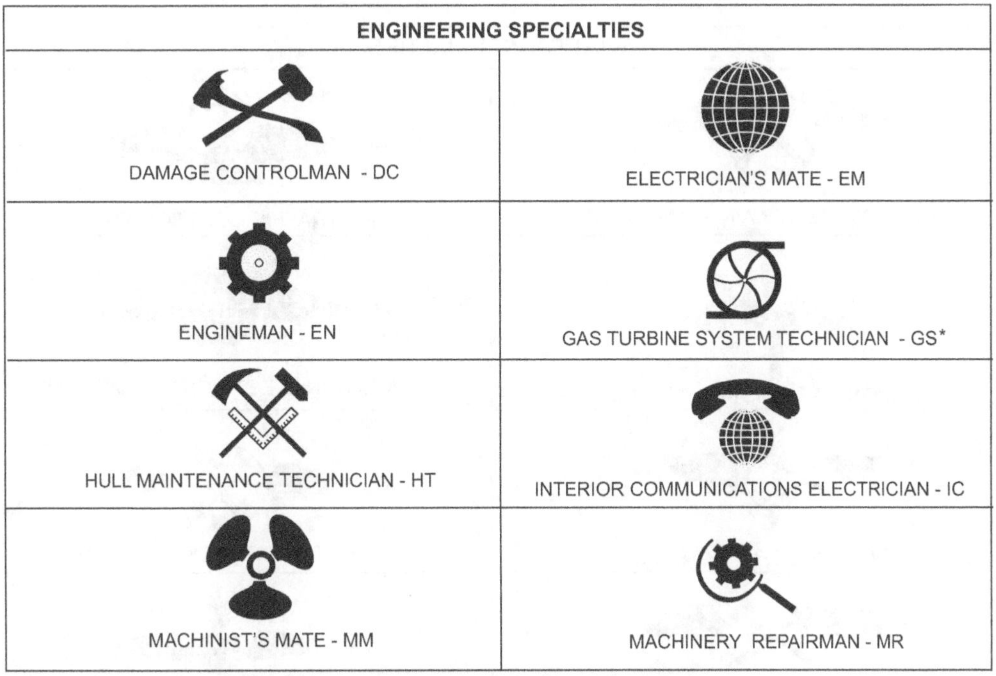

*GAS TURBINE TECHNICIAN IS USED AT PAYGRADE E-9 ONLY. LEADING TO GSCM IS THE GAS TURBINE SYSTEM TECHNICIAN - ELECTRICAL (GSE) AND GAS TURBINE ELECTRICIAN - MECHANICAL (GSM)

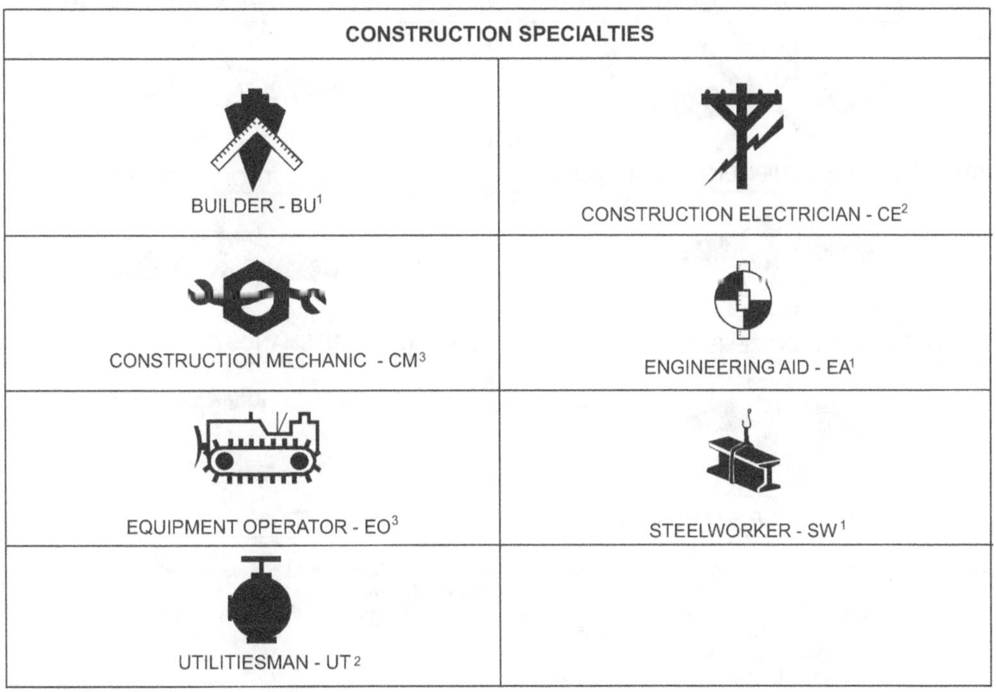

[1]BU, EA, AND SW BECOME CUCM AT PAYGRADE E-9
[2]CE AND UT BECOME UCCM AT PAYGRADE E-9
[3]CM AND EO BECOME EQCM AT PAYGRADE E-9

BMRF1008

Figure 10-8.—Rating insignias for enlisted personnel.

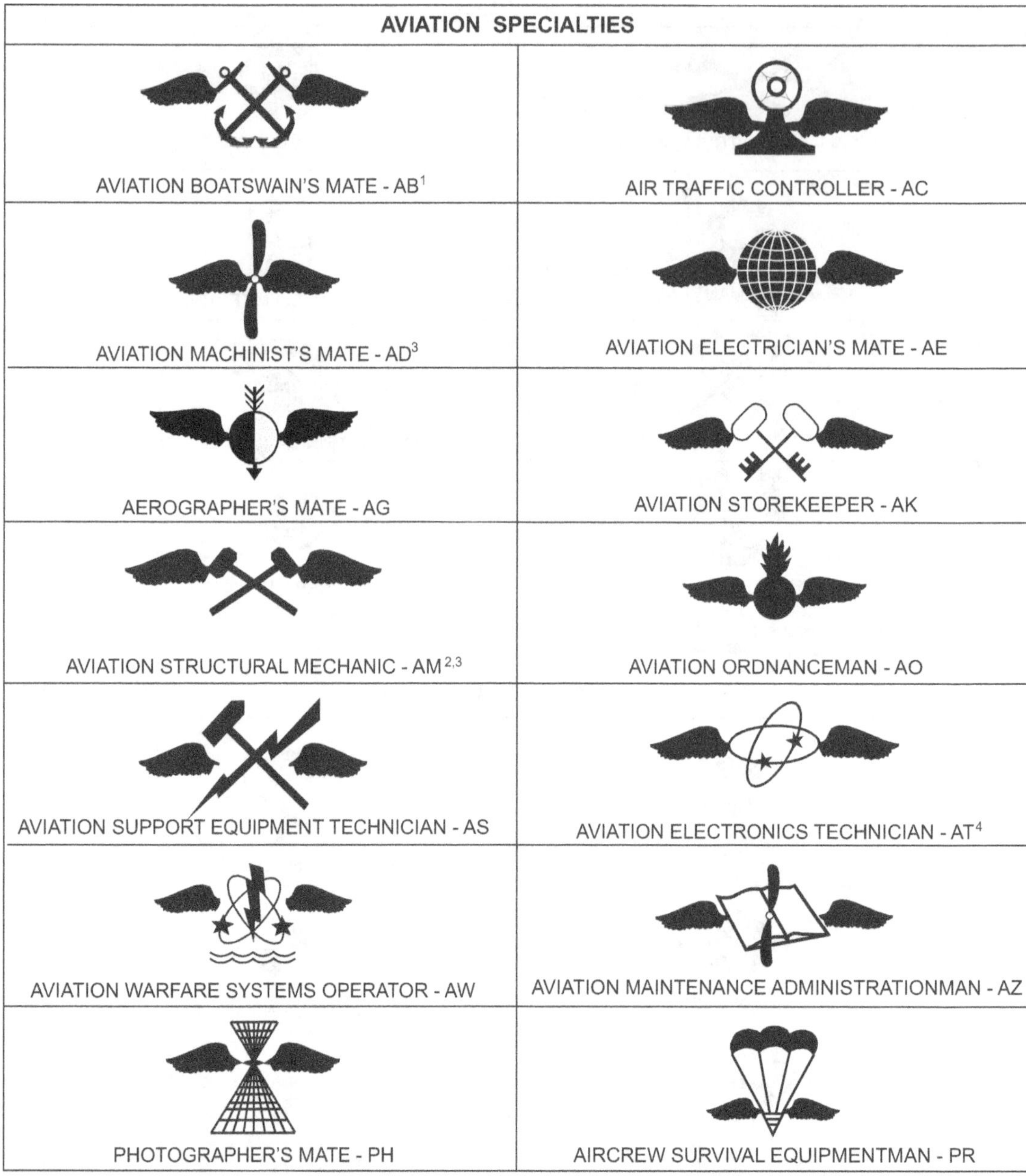

[1] AVIATION BOATSWAIN'S MATE IS USED AT PAYGRADE E-9 ONLY. LEADING TO ABCM IS LAUNCHING AND RECOVERY EQUIPMENT (ABE), FUELS (ABF), AND AIRCRAFT HANDLING (ABH).
[2] STRUCTURAL MECHANIC IS USED AT PAYGRADE E-8 ONLY. LEADING TO AMCS IS SAFETY EQUIPMENT (AME), HYDRAULICS (AMH), AND STRUCTURES (AMS).
[3] ADCS AND AMCS LEADING TO AFCM.
[4] ATCS AND AECS LEADING TO AVCM.

BMRF1009

Figure 10-9.—Rating insignias for enlisted personnel (continued).

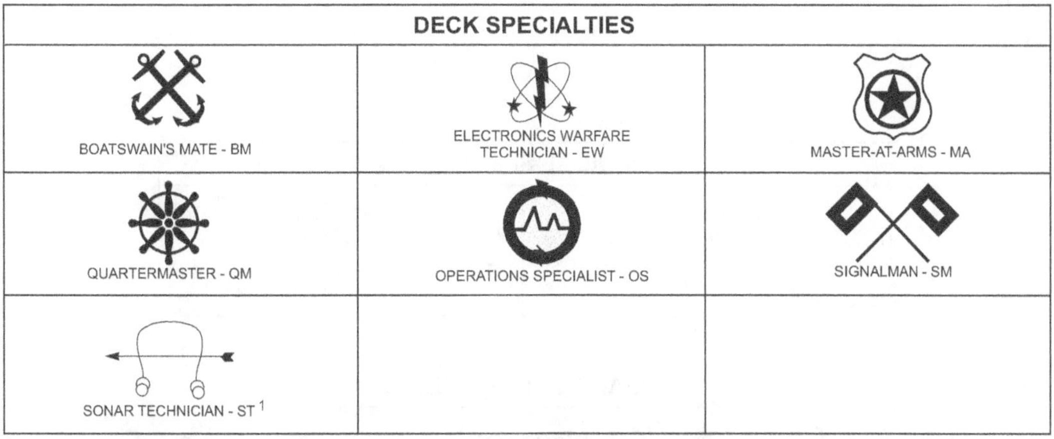

[1] SONAR TECHNICIAN HAS TWO BRANCHES: SONAR TECHNICIAN-SURFACE (STG) AND SONAR TECHNICIAN-SUBMARINE (STS)

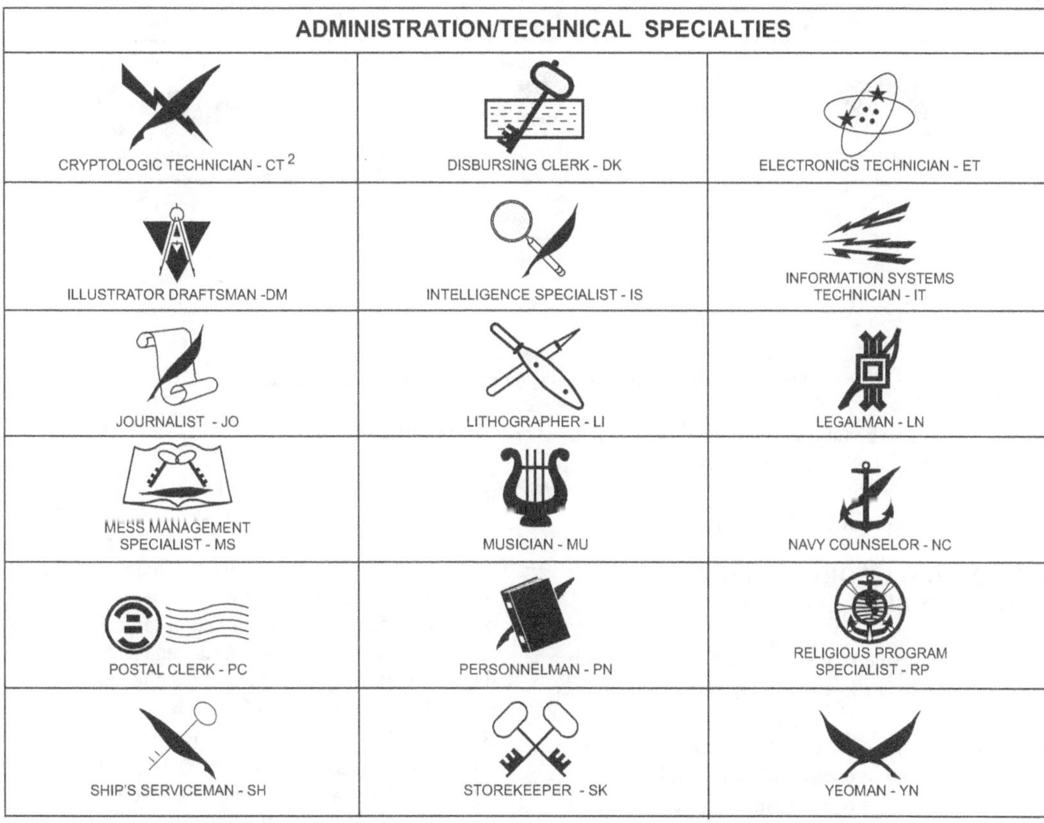

[2] CRYPTOLOGIC TECHNICIAN HAS OF SIX BRANCHES: CRYPTOLOGIC TECHNICIAN-ADMINISTRATION (CTA); CRYPTOLOGIC TECHNICIAN-INTERPRETIVE (CTI); CRYPTOLOGIC TECHNICIAN-MAINTENANCE (CTM); CRYPTOLOGIC TECHNICIAN-COMMUNICATIONS (CTO); CRYPTOLOGIC TECHNICIAN-COLLECTION (CTR); AND CRYPTOLOGIC TECHNICIAN-TECHNICAL (CTT)

BMRF1010A

Figure 10-10A.—Rating insignias for enlisted personnel (continued).

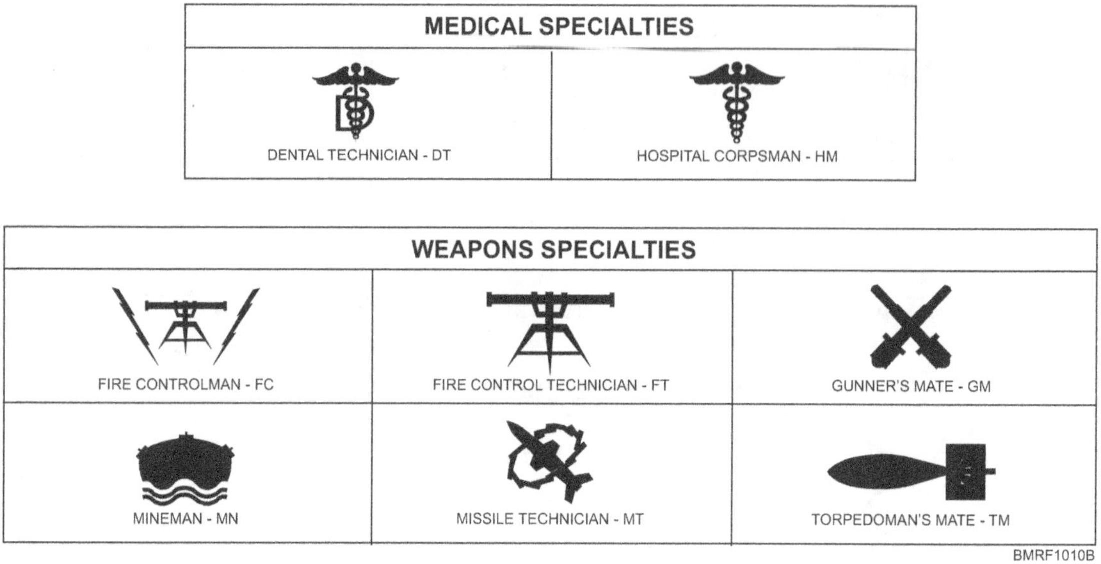

Figure 10-10B.—Rating insignias for enlisted personnel (continued).

You should also be familiar with the rating structure of the other branches of the U.S. military. Refer to figure 10-11 for a comparison of the Navy, Marine Corps, Army, and Air Force enlisted structure.

RANK INSIGNIA/CORPS DEVICES—U.S. NAVAL OFFICER

The paygrade of an enlisted person is referred to as a rate. For example, an E-4 is a petty officer third class. The paygrade of a commissioned officer is called a *rank*. A lieutenant is an 0-3. Commissioned officers hold their positions of command by the authority given them by the President of the United States.

Naval Officers

An officer of a given grade is junior to all officers of higher grades and senior to all officers of lower grades. Any commissioned officer is senior to any warrant officer or enlisted person. When officers are of the same grade, then the officer first commissioned in that grade outranks officers commissioned at later dates. In other words, officers of the same grade rank according to their date of commission—the earlier the date, the more senior the officer. If the officers have the same date of rank, then they take precedence according to a numerical listing kept by the Navy Department. Figures 10-12, 10-13, and 10-14 show the rank insignia of U.S. naval officers as well as the insignia of other branches of the U.S. armed forces.

Student Notes:

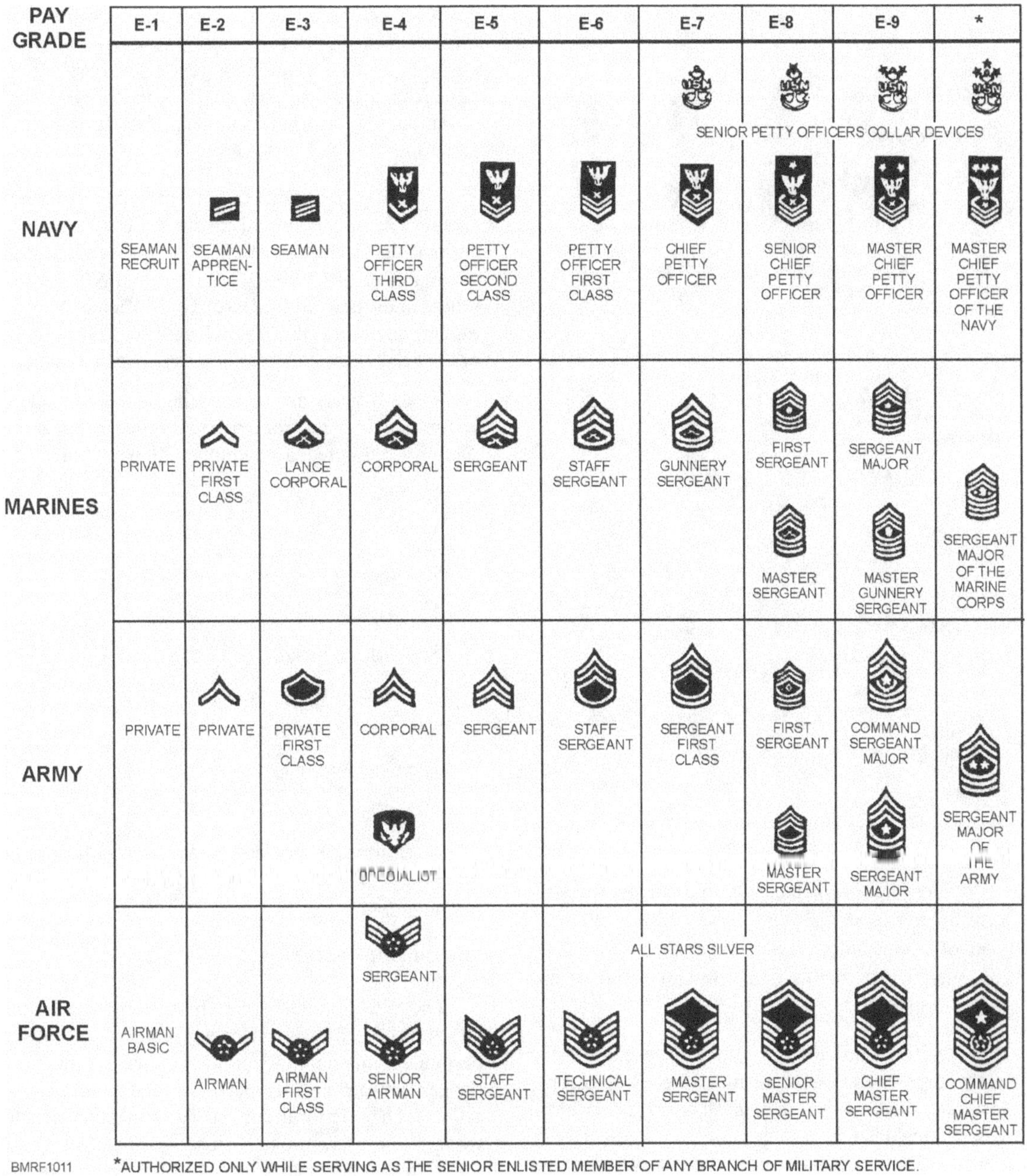

Figure 10-11.—Rate insignia of U.S. armed forces enlisted personnel.

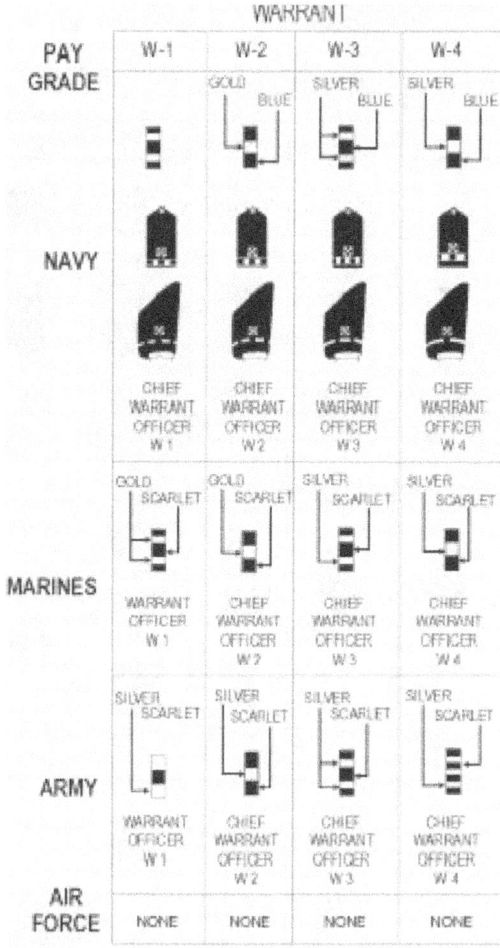

BMRF1012

Figure 10-12.—Rank insignia of warrant officers of the U. S. armed services.

Officers are also divided into line officers and staff corps officers. A star is worn on the sleeve or shoulder board of the line officer, depending on the uniform. The line category is subdivided into unrestricted and restricted line.

- Only unrestricted line officers are eligible for command at sea and the command of aircraft squadrons, fleets, and shore bases such as naval bases and naval air stations. Included in this category are limited duty officers (LDOs), who have been specifically authorized to assume such command and certain naval aviators.

- Restricted line officers are those designated for engineering or other special duty, such as communications, naval intelligence, photography, and other technical fields. They aren't eligible for command at sea but may assume command of designated shore facilities.

- Staff corps officers are specialists in certain areas, such as supply and medicine. They wear staff corps insignia (fig. 10-15). A doctor can become the commanding officer (CO) of a hospital or a medical school or the chief of the Bureau of Medicine. A supply officer can become the CO of a supply depot or a school or the head of the Navy Supply Systems Command.

When officers are of the same grade and date of commissioning, but from different officer's categories, seniority takes precedence in the following order:

1. Line
2. Medical
3. Supply
4. Chaplain
5. Civil Engineer
6. Judge Advocate General
7. Dental
8. Medical Service
9. Nurse

In addition to regular commissioned officers, the Navy has another group of officers called *commissioned warrant officers*.

Warrant Officers

The Navy needs specialists to supervise the operation of equipment and weapons and needs enlisted personnel to maintain them. Chief warrant officers fill the gap between enlisted personnel and commissioned officers. They are former enlisted personnel selected for warrant status because of their professional ability and for their demonstrated qualities of leadership, loyalty,

Student Notes:

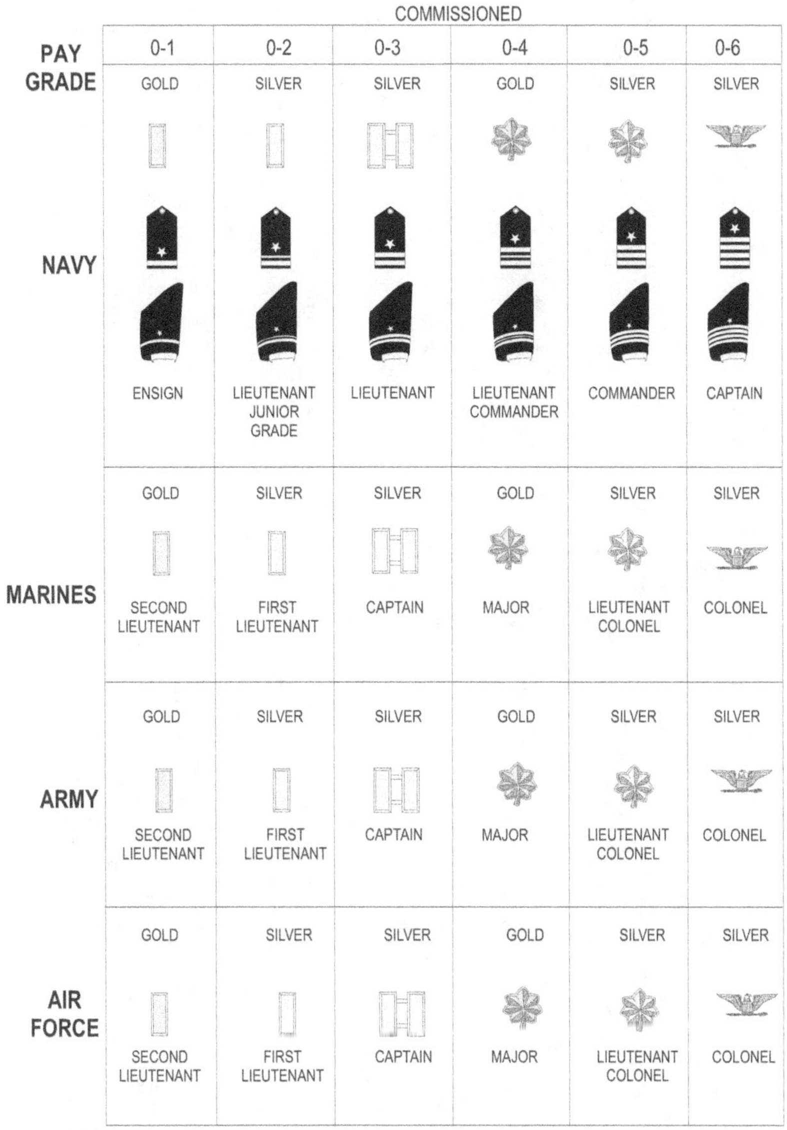

BMRF1013

Figure 10-13.—Rank insignia of commissioned officers of the U. S. armed services.

and devotion to duty. All warrant officers are commissioned as W-2s.

Chief warrant officers wear collar devices or sleeve insignia symbolic of their specialty in the same manner as staff corps officers, as shown in figures 10-16 and 10-17.

SPECIAL INSIGNIA

Special insignia are worn on the breast to indicate special qualifications or designations (figs. 10-18 and 10-19). (**NOTE**: Not all insignias are shown.) Some examples of special insignia are as follows:

Student Notes:

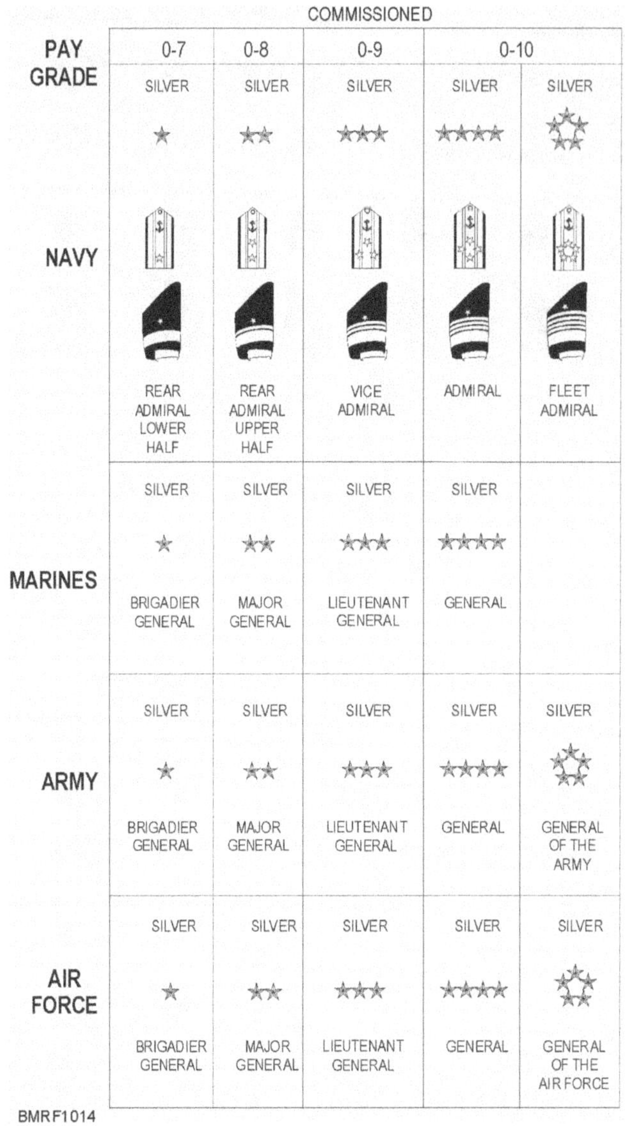

Figure 10-14.—Rank insignia of commissioned officers of the U. S. armed services (Continued).

The **Command at Sea** insignia is worn by persons below flag rank who have or have had command of commissioned ships or aviation squadrons at sea. Officers currently in command wear the insignia on the right breast. Those not presently in command, but who have held command, wear it on the left breast below any ribbons, medals, or other insignia.

The **Command Ashore/Project Manager** insignia is worn by officers below flag rank who have or have had command ashore or served as a project manager. It is worn in the same manner as the Command at Sea insignia.

The **Small Craft OIC/POIC** insignia is worn by

Student Notes:

Figure 10-15.—Line and staff corps officers' insignia.

enlisted and officer personnel currently serving or who have previously served as an officer in charge of a small craft. Enlisted personnel wear silver and officers wear gold when they are authorized to wear these special insignias. This insignia is worn in the same manner as the Command at Sea insignia.

The **Surface Warfare** insignia is worn by officers and enlisted personnel who have qualified in all phases of surface warfare.

The **Submarine Warfare** insignia is worn by personnel who have qualified to serve in submarines. In addition to the basic insignia, other submarine insignia include those for submarine medical, engineer, and supply officers and for all who participated successfully in combat patrols.

The **Aviation Warfare** insignia is worn by personnel qualified to serve in flight. In addition to the Aviation insignia, other insignia are worn by flight officers, flight surgeons, flight nurses, aircrewmen, astronauts, and air warfare specialists.

The **Special Warfare** insignia is worn by personnel qualified in underwater and beach reconnaissance, demolition, and special warfare tactics. They are usually associated with underwater demolition or

Student Notes:

Figure 10-16.—Warrant officers' specialty insignia.

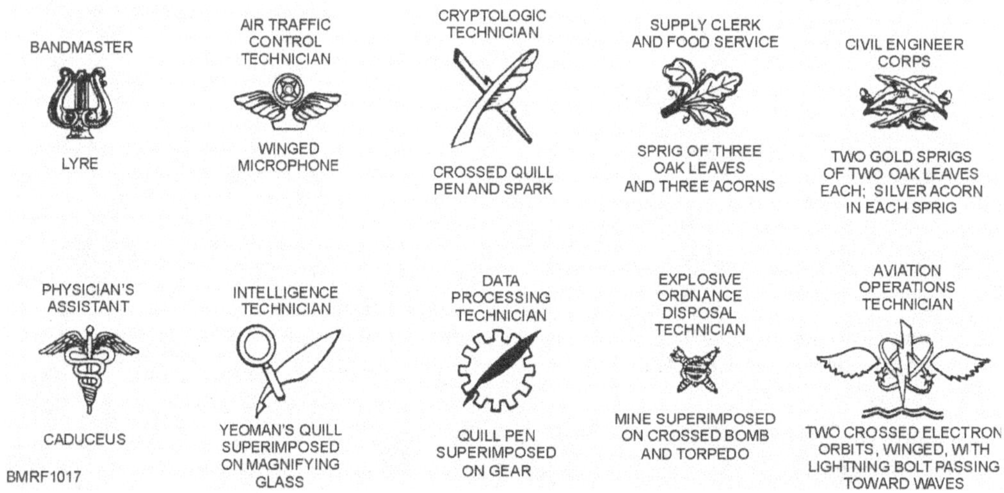

Figure 10-17.—Warrant officers' specialty insignia (continued).

Figure 10-18.—Breast insignia.

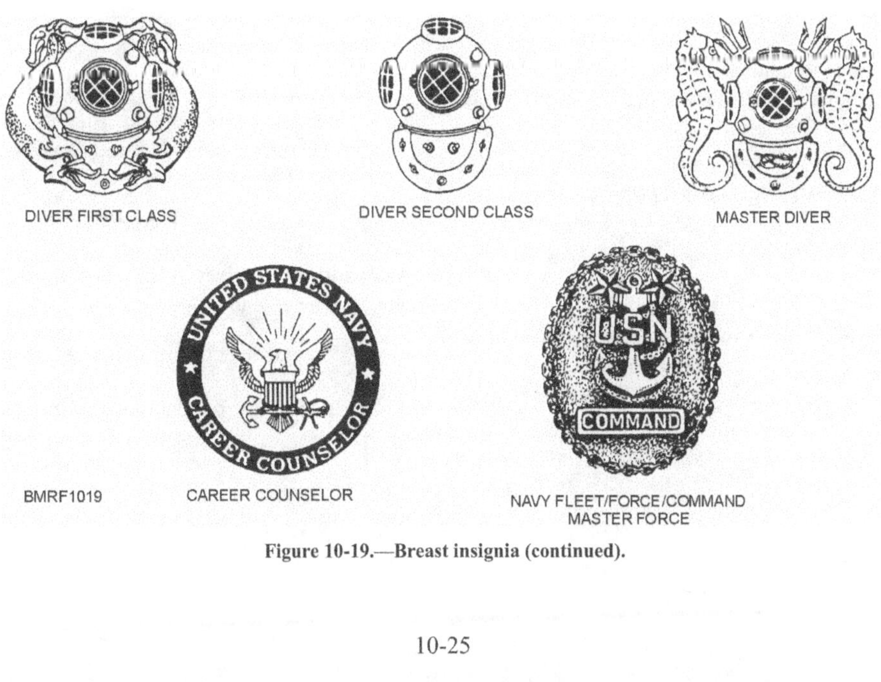

Figure 10-19.—Breast insignia (continued).

sea-air-land (SEAL) team detachments.

The **Explosive Ordnance Disposal Warfare** insignia is worn by personnel who are qualified in the identification and safe disposal of many different types of ordnance produced by the United States, our allies, and our enemies.

The **Underwater** insignia is worn by officer and enlisted personnel qualified in various classes of diving. The Underwater insignia is silver with the exception of the Diving Officer and the Diver Medical insignia, which are gold.

The **Seabee Combat Warfare Specialist** insignia is worn by personnel who are qualified in all phases of Naval Mobile Construction Battalion operations including mobilization, combat operations, disaster recovery, and combat readiness.

Most insignia worn by officers and enlisted personnel are identical, with one exception—those worn by officers are gold, while those worn by enlisted personnel are silver. Examples of some of these are **Submarine**, **Small Craft**, and **Explosive Ordnance Disposal**. Two exceptions to this are the **Enlisted Aircrew** insignia and the **Naval Parachutist's** insignia, which are gold.

Another type of special insignia worn by naval personnel is an identification badge, which is displayed by those engaged in Presidential service or assigned to certain staffs, such as the organization of the Joint Chiefs of Staff (JCS) or the Office of the Secretary of Defense.

EMBLEM

Different commands within the Navy have special emblems. Figure 10-20 shows one example of an emblem.

REVIEW 3 QUESTIONS

Q1. What is a striker mark?

Q2. To be eligible to wear a service stripe, you must have served 4 full years of active or Reserve service in what branch of the U.S. Armed Forces?

BMRF1020

Figure 10-20.—Chaplain corps emblem.

Q3. A naval officer is wearing one silver star insignia on the uniform collar. What rank does this insignia show?

Q4. When worn as a sleeve insignia, what device designates the officer as a line officer?

Q5. What kind of special insignia is worn on the breast by personnel qualified in underwater and beach reconnaissance and demolition?

AWARDS

Learning Objectives: When you finish this chapter, you will be able to—

- Recognize authorized military awards.
- Identify the regulations governing their wear.

Student Notes:

An award is any decoration, medal, badge, ribbon, or letter of commendation given to an individual or unit in recognition of outstanding acts or service performed. There are seven broad categories of awards:

1. Military decorations
2. Unit awards
3. Nonmilitary decorations
4. Campaign and service awards
5. Foreign decorations and non-U.S. service awards
6. Marksmanship awards
7. Awards of military societies and other organizations

Military decorations. A military decoration is an award given to an individual for a specific personal act of gallantry or meritorious service. Some examples of military decorations are the Medal of Honor, Navy Cross, Distinguished Service Medal, and Purple Heart.

The first military decoration awarded to individuals by this country was the Purple Heart. The Purple Heart was founded by George Washington in 1782. It was awarded for unusual gallantry and/or extraordinary fidelity and essential service. Only three people were awarded the original Purple Heart. The awards were made for action during the Revolutionary War. One hundred and fifty years later, in 1932, the President Franklin Deleno Roosevelt revived the Purple Heart decoration.

The basis for this award was changed from the original idea. As reestablished, the Purple Heart is now awarded for wounds received as a result of enemy action.

Unit awards. An award presented to an operating unit worn only by members of the unit who participated in the cited action. Unit awards include the Presidential Unit Citation, the Navy Unit Commendation, and the Navy "E."

Nonmilitary decorations. Nonmilitary decorations are awarded for various actions by an individual. A few examples are the Presidential Medal of Freedom, the Gold and Silver Life Saving Medals, and the National Sciences Medal.

Campaign and service awards. Campaign and service awards are issued to personnel who have participated in designated wars, campaigns, expeditions, or who have fulfilled creditable, specific, service requirements. Examples of campaign and service awards are the Prisoner of War Medal, Good Conduct Medal, Antarctica Service Medal, Armed Forces Expeditionary Medal.

Foreign decorations and non-U.S. service awards. The foreign decorations and non-U.S. service awards that may be worn are listed in the *U.S. Navy Uniform Regulations,* NAVPERS 156650.

Marksmanship awards. Some examples of the marksmanship awards include the Navy Pistol Marksmanship Medal/Ribbon, the Distinguished Marksman Badge, and the Navy Rifleman Excellence In Competition Badge.

Awards of military societies and other organizations. Awards of military societies and other organizations are awards from the following organizations: Regular Army and Navy Union, Medical Scientific Societies, Naval Reserve Association, and the Moreell Medal.

Normally, awards are medals suspended from a pin by a distinctive ribbon; but there are exceptions, such as the Presidential Unit Citation, Navy Unit Commendation, Meritorious Unit Commendation, and Combat Action Ribbon. These awards are not medals but ribbons that denote the citation.

Medals are attached to the uniform just above the left breast pocket. Up to three medals are worn side by side. When there are more than three medals, they are attached to a bar in an overlapping fashion with a maximum of five medals to a row (in their order of precedence).

Except on special occasions (such as a personnel inspection), replicas of the medal suspension ribbons are worn. These ribbons are part of the service dress uniform. They are worn centered 1/4 inch above the left breast pocket, with no space between ribbons or between rows. Each row may contain no more than three

Student Notes:

ribbons. A row of fewer than three ribbons is centered above a full row. Persons possessing four or more ribbons must wear a minimum of three but may wear all if desired. The ribbons, which may be either sewn on the uniform or attached to bars, are worn in their order of precedence—from top to bottom and inboard to outboard within rows. Transparent covering or the use of preservatives isn't permitted. No alteration may be made that would change the appearance of the ribbons. When medals are prescribed as part of the uniform, ribbons that don't have corresponding large medals are worn on the right breast.

REVIEW 4 QUESTIONS

Q1. If a person has four or more ribbons, what is the minimum number of medals that can be worn on the uniform?

Q2. List four examples of military decorations.

a.

b.

c.

d.

Q3. List two types of nonmilitary decorations.

a.

b.

Q4. When wearing medals or ribbons on a uniform, what is their order of precedence?

IDENTIFICATION CARDS

Learning Objective: When you finish this chapter, you will be able to—

- Identify the regulations governing military identification (ID) cards and identification tags.

The armed forces identification card is used to identify you as a member of the U.S. armed forces. It isn't a pass. It remains the property of the United States. Anyone altering, damaging, lending, counterfeiting, or using the card in an unauthorized manner is subject to disciplinary action.

You must carry the card at all times. You can't give your ID card as security for the return of property or equipment provided by civilian or naval recreational activities.

The active-duty ID card **must be** surrendered by the holder for the following reasons:

- When it is replaced. It must be replaced for the following reasons:
  - — To show a change in rank or rate.
  - — To show a change in the card's expiration date
  - — To replace a lost, stolen, or destroyed card.
  - — To correct an error.
  - — To replace a mutilated card.
  - — To change data that makes the card questionable as a means of identification.
  - — To effect a name change.
- When the holder is released from extended active duty.
- When it is required by proper military authority for identification or investigation purposes, or while in disciplinary confinement.

Since the armed forces ID card (active) meets all the requirements of article 17 of the Geneva Convention pertaining to the treatment of prisoners of war, it serves as identification for that purpose. If you're captured as a prisoner of war, you **may** show your ID card to the capturing authorities but you **may not** surrender it to

Student Notes:

them.

Identification tags (dog tags) are designed for the identification and casualty reporting of members who become casualties and for grave registration of members who die in a combat zone. As soon as possible after reporting for active duty, each Navy member is issued two complete identification tags.

Identification tags are made of metal, approximately 2 inches long by 1 1/8 inches wide, and attached to a 25-inch necklace. These tags are a prescribed part of your uniform and must be kept in your possession. When prescribed by directives, they are worn suspended from the neck under the clothing. When not required to be worn, they should be regarded as part of your equipment and will be regularly inspected as such.

Identification tags must be worn while you are on active duty in the Navy under the following conditions:

- In time of war
- In time of national emergency
- When engaged in flight operations
- When traveling in aircraft
- When reporting to an armed forces medical facility for treatment
- When prescribed by the Chief of Naval Operations (CNO)
- When prescribed by competent authority

Each tag is embossed with the following information:

First line—Last name, first name, and middle initial of the wearer; that is, DOE, John R. When the space provided for the first line is insufficient for the name as prescribed above, the first line will contain only the last name.

Second Line—The first name and middle initial, if needed.

Third line—Military personnel identification number (SSN), the letters USN, and the blood type and Rh factor.

Fourth line—The religious preference of the wearer.

REVIEW 5 QUESTIONS

Q1. Name five actions that you should **not** take with regard to your ID card.

a.

b.

c.

d.

e.

Q2. Describe the difference between the purpose for the military ID card and dog tags in wartime.

Q3. List the information contained on the dog tags.

a.

b.

c.

d.

GROOMING STANDARDS

Learning Objective. When you finish this chapter, you will be able to—

- Recognize grooming standards for men and women.

Grooming standards are based on several elements—including neatness, cleanliness, safety, military image, and appearance in uniform. The standards aren't intended to be overly restrictive or designed to isolate Navy men and women from society. The limits are reasonable; they make sure that personal appearance contributes to a favorable military image,

Student Notes:

yet they allow a degree of individuality. The seeming difference between the policy on grooming for male and female members is simply recognition that there is a difference between the sexes—mustaches and sideburns for men, longer hair and cosmetics for women.

Remember! Grooming promotes pride in who and what you are, pride of being in the Navy, and pride in being in the United States.

GROOMING STANDARDS FOR MEN

Hair will be neat and clean and present a groomed appearance. Hair above the ears and around the neck will be tapered from the lower hairline upward at least 3/4 inch and outward to greater than 3/4 inch to blend with the hairstyle. Hair on the back of the neck may not touch the collar. Hair will be no longer than 4 inches and groomed so that it does not touch the ears or collar, extend below the eyebrows when headgear is removed, or interfere with proper wearing of the headgear. The primary consideration remains a neatly groomed appearance for the hairstyle and the type of hair that the individual has.

Sideburns are permitted, but they are to be of even width (not flared), end with a clean-shaven horizontal line, and cannot extend below the middle of the ear.

In most instances, mustaches also are permitted, but must be kept neatly trimmed so that they don't appear ragged. No eccentricities, such as long drooping mustaches, are permitted. Beards are not allowed.

Articles such as pencils, pens, watch chains/fobs, pins, jewelry, handkerchiefs, combs, cigars, cigarettes, or pipes must not be worn or carried exposed on the uniform. Wristwatches, bracelets, and rings (one ring per hand) are permitted. While in uniform, men may wear one necklace or choker, but it must not be visible. Men must not wear earrings while in uniform or while in civilian clothes, while on a military installation, or when attending a military function.

Navy personnel assigned to Marine Corps units, **must** abide by the grooming standards established for Marines; otherwise, when wearing Navy uniforms (including fatigues), they must abide by Navy regulations.

GROOMING STANDARDS FOR WOMEN

Hairstyles will be neat and present a groomed appearance. They will not be outrageously multicolored or faddish, to include shaved portions of scalp or designs cut or braided into the hair. Haircuts and styles will present a balanced appearance. Lopsided and extremely asymmetrical (not balanced) styles aren't authorized.

Braids

Braided hairstyles must be conservative and conform to the guidelines stated in NAVPERS 15665. Some of the guidelines include the following information:

- Ponytails, pigtails, widely spaced individual hanging locks, and braids that protrude from the head **aren't** authorized.
- Multiple braids are authorized.
- When a hairstyle of multiple braids is worn, braids will be
  - — Uniform in dimension
  - — Small in diameter (approximately 1/4 inch)
  - — Tightly interwoven to present a neat, professional, well-groomed appearance.
  - — Foreign material (such as beads or decorative items) will not be braided into the hair.
  - — Short hair may be braided in symmetrical fore and aft rows (corn rowing) that minimizes scalp exposure.
  - — Corn row ends will not protrude from the head and must be secured with inconspicuous rubber bands that match the hair color.

Hairstyles

Appropriateness of the hairstyle is judged by its appearance when headgear is worn. All headgear must fit snugly and comfortably around the largest part of the head without distortion or excessive gaps. Hair shouldn't show from under the front of the brim of the

Student Notes:

combination hat, garrison, or command ball cap. Hairstyles that don't allow headgear to be worn in this way, or that interfere with the proper wear of protective masks or equipment, are prohibited.

Hair Length

When in uniform, the hair may touch, but not fall below a horizontal line level with the lower edge of the back of the collar. With jumper uniforms, hair may extend a maximum of 1 1/2 inches below the top of the jumper collar. Long hair, including braids, that falls below the lower edge of the collar must be neatly and inconspicuously fastened, pinned, or secured to the head. No portion of the bulk of the hair as measured from the scalp will exceed approximately 2 inches.

Barrettes/Combs/Clips

A maximum of two small barrettes/combs/clips, similar to hair color, may be used in the hair. Additional bobby pins or rubber bands matching the hair color may be used to hold hair in place, if necessary. Fabric elastics and colored bands/pins aren't authorized. Hair ornaments will not present a safety or foreign object damage (FOD). Hair nets will not be worn unless authorized for a specific type of duty.

Fingernails

Fingernails must not exceed 1/4 inch measured from the tip of the finger. Nail polish must be a soft shade, complementary to the skin tone.

Cosmetics

Cosmetics should be conservative in color and applied sparingly. No eccentricities or faddishness of dress, jewelry, or grooming is permitted. No pencils, pens, pins, handkerchiefs, or jewelry may be worn or exposed on the uniform. Earrings may be worn with the uniform and must be the 6mm-ball (approximately 1/4 to 1/8 inch) type with a plain brushed matte finish or a shiny finish; either the screw-on or post type. E-6 and below must wear silver earrings; CPOs and officers must wear gold. Small single pearl earrings are authorized for dinner or formal dress uniforms. While in uniform, women may wear one necklace or choker, but it must not be visible.

Maternity Uniforms

The wearing of the maternity uniform is mandatory for all pregnant women in the Navy when a uniform is prescribed, and regular uniforms no longer fit.

REVIEW 6 QUESTIONS

Q1. While in uniform, how many (a) wristwatches and (b) bracelets can be worn?

a.

b.

Q2. To what Navy publication should you refer for further explanation of grooming standards?

Q3. When in uniform, what type of earrings is authorized for E-6 and below and CPOs and above?

a. E-6 and below:

b. CPO and above:

DRILL AND FORMATIONS

Learning Objectives: When you finish this chapter, you will be able to—

- Recognize the purpose of formations.
- Identify facing movements within a formation.

Understanding and correctly following the basic drill positions and facing movements are a necessary part of your military life. Drills teach discipline and instill habits of precision and automatic response to orders.

What is the purpose for formations? Is it to see if you know your right from our left? Not really!

Student Notes:

One of the purposes for formations is to move a large number of personnel from one place to another in an orderly manner. Another purpose is to make sure people receive correct up-to-date information. People listen more closely to and better understand what is being said when they are alert and paying attention. Just imagine a group of people standing around, their hands in their pockets, daydreaming or talking while someone is trying to relay important information. How many people will actually hear and understand what is being said? Probably not many! Formations also help teach a group of individuals to act as members of a team instead of "doing their own thing."

POSITIONS WITHIN A FORMATION

Many military functions, such as morning quarters and personnel inspections, require that you assemble in formation. The terms used to identify these formations may vary at different commands. For example, the term *squad* or *platoon* at one command may be a *detail, division*, or *class* at another. Here, the term *squad* is used to represent a basic formation. Remember that the members of any formation must respond in unison (together) to the commands given. By studying the following terms and referring to the diagram in figure 10-21, you can easily learn the basic positions within a formation:

Distance. A space of 40 inches between the chest of one person and the back of the person ahead within ranks.

Element. An individual, squad, section, platoon, company, or some other unit that is part of a larger unit.

File or column. A formation of elements or persons placed one behind the other.

Flank. The extreme right or left of a unit, either in line or in column. The element on the extreme right or left of the rank. A direction at a right angle to the direction an element or a formation is facing.

Formation. An arrangement of elements in line, in column, or in any other prescribed manner.

Guide. The individual on whom a formation or element regulates its alignment. The guide is usually positioned to the right.

Interval. The space between individuals from shoulder to shoulder, normally one arm's length.

Pace. The length of a full step (30 inches for men and 24 inches for women).

Rank or line. A formation of elements or persons abreast of each other or side by side.

Step. The distance from heel to heel between the feet of a marching person. The half step and back step are 15 inches. The right and left steps are 12 inches.

POSITION AND FACING COMMANDS

The two types of commands are the preparatory command, such as RIGHT, which indicates the type of movement to be made, and the command of execution, such as FACE, which causes the desired movement to be

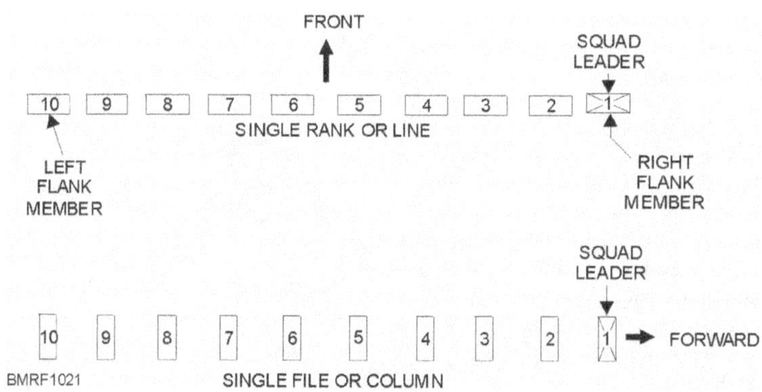

Figure 10-21.—Positions within a formation.

Student Notes:

made. **In this chapter, preparatory commands are bolded and the first letter is capitalized. In commands of execution, each letter is capitalized. When both types of commands are combined, such as FALL IN, AT EASE, and REST, they are capitalized just like commands of execution**.

The command AS YOU WERE cancels a command or an order started but not completed. On this command, you resume your former position.

POSITIONS

Assume the following positions only when you are at a halt. One person or an entire formation may execute them.

ATTENTION. The position of attention is the basic military position. It indicates you are alert and ready for instruction. In this position, stand with your heels together, feet forming an angle of 45°, head and body erect, hips and shoulders level, and chest lifted. As shown in figure 10-22, allow your arms to hang naturally—thumbs along skirt or trouser seams and fingers joined and in their natural curl. Keep your legs straight, but not stiff at the knees. Direct your head and eyes to the front. Keep your mouth closed, and pull your chin in slightly. When called to attention, bring the heel of your left foot to the heel of your right foot.

PARADE REST. The command Parade REST is given only when the formation is at attention; the movement is executed in one count (fig. 10-23).

AT EASE. On the command AT EASE, you can relax and shift about, but keep your right foot in place. Do not talk. This command may also be given when you are not in ranks, as in a classroom. You must not talk, but you may remain relaxed.

REST. On the command REST, follow the same movement restrictions as you would when at ease, but you may talk.

FALL OUT. (This command is not a dismissal order.) Upon the command FALL OUT, leave your position in ranks but remain nearby. On the command FALL IN, resume your place in ranks, and come to attention.

BMRF1022

Figure 10-22.—Attention.

BMRF1023

Figure 10-23.—Parade rest.

Student Notes:

To bring a formation to attention again when it is in any one of the four positions of rest, the person issuing commands gives a preparatory command (such as **Company**) before the command ATTENTION. If at rest or at ease, come to the position of parade rest on the preparatory command.

FACINGS

Facings are movements that can be made either to the right or left, with the exception of about face. While facing, your arms should remain at the position of attention. The following commands describe only the movement to the right. To perform a movement to the left, simply substitute "left" for "right" and "right" for "left."

RIGHT FACE. Right face (fig. 10-24) is a two-count movement started on the commands Right FACE. On the command FACE, (1) raise your left heel and right toe slightly and turn 90° to the right. Keep your left leg straight but not stiff; (2) bring your left heel smartly alongside the right heel and stand at attention.

EYES RIGHT. When given the commands **Eyes** RIGHT, smartly turn your head 45° to the right on the command RIGHT. The commands to turn your head back to the position of attention are **Ready** FRONT. On the command FRONT, snap your head to the front.

ABOUT FACE. About face is a two-count movement performed on the commands **About** FACE. On the command **About**, shift your weight to your left leg without noticeable movement. On the command FACE, (1) place your right toe about 6 inches behind and slightly to the left of your left heel (fig. 10-25; (2) on the ball of the right foot and the heel of the left foot, turn smartly to the right until you are facing the rear. Your feet will be in the position of attention when the turn is completed if you place your right toe properly behind your left heel.

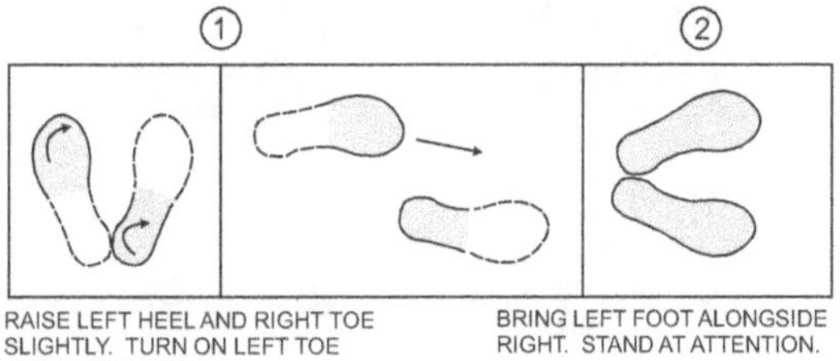

Figure 10-24.—Right Face.

FALLING INTO FORMATION

Up to this point, you have learned about movements that can be made by one person or by a group. In a sharp military formation, each member must correctly respond to commands as a team. Always listen carefully to the person in charge since formation movements are usually made up of both preparatory and execution commands. In the following movements, you must pay special attention to the duties of the left and right flank members since their response to a command is slightly different from the other members in formation.

FALL IN. On the command FALL IN, the squad forms in line on the left of the right flank member (squad leader). Each member of the squad, except the left flank member, raises the left arm shoulder high in line with the body. Fingers are straight and touching each other palm down. Each member (except the right flank member) turns their head and looks to the right. To obtain a normal interval (fig. 10-26), move in line so that your right shoulder touches the fingertips of the person to the right. As soon as you are in line with the person to your right and the person on your left has obtained normal interval, return smartly and quickly to the position of attention.

Close interval (fig. 10-27) is the horizontal distance between the shoulder and elbow when the left hand is

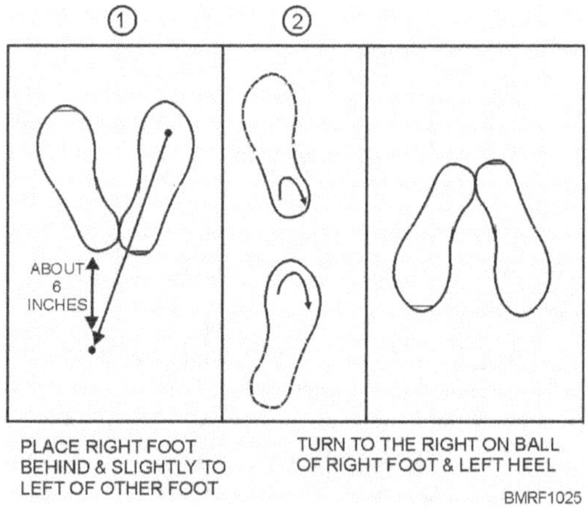

Figure 10-25.—About Face.

Student Notes:

Figure 10-26.—Normal interval.

placed on the left hip. The command **At Close Interval** requires the same movements as for normal interval. The only exception is that each member places the left hand on the beltline above the left hip with the elbow in line with the body. The heel of the hand rests on the hip with fingers straight, touching each other, and pointing down. The left flank member makes the adjustment without moving the arms.

TO ALIGN THE SQUAD. On the commands **Dress Right**, DRESS (normal interval) or **At Close Interval Dress Right**, DRESS (close interval), members of the squad align themselves with each other.

On the command DRESS, all members, except the right flank member, smartly turn their heads, look, and align themselves to the right. At the same time, all members, except the left flank member, smartly raise their left arm shoulder high (normal interval) or place their left hand on their hip (close interval). The right flank member stands fast and looks to the front. Using the right flank member as a guide and taking short steps, the other members align themselves and obtain the proper interval. Whether commanded to dress to the right or to the left, use only the left arm to obtain the interval, and hold that position until the next command is given.

Figure 10-27.—Close interval.

Student Notes:

When the alignment is correct, the commands **Ready**, FRONT are given. On FRONT, heads snap to the front and arms drop to the side.

TO COVER OFF. This command is given when the formation is in column or in two or more ranks. On the command COVER, the forward member or forward rank stands fast. You will then move, left or right, to position yourself directly behind the person in front of you while maintaining a 40-inch distance.

FROM NORMAL TO CLOSE INTERVAL. The commands **Close**, MARCH tell members to move from normal interval to close interval while in line. On MARCH, all members, except the right flank member, pivot to the right on the ball of the right foot and step off on the left foot (one count). They march forward until they obtain an approximate close interval, halt, and face to the left. They then form at close interval, as already described. All members lower their arms when the member on their left has obtained the proper interval.

FROM CLOSE TO NORMAL INTERVAL. The commands **Extend**, MARCH tell members to change from close interval to normal interval while in line. On MARCH, all members, except the right flank member, pivot to the left on the ball of the right foot and step off on the left foot (one count). They march forward until they obtain an approximate normal interval, halt, and face to the right. Then they form at normal interval, as previously described . Each member drops the left arm when the member to the left has obtained the proper interval.

DOUBLE-ARM INTERVAL. The commands **Take Interval To The Left** MARCH, tell members at either close or normal interval to form a double-arm interval. At the command MARCH, members move as when extending ranks, except that each member raises both arms and touches the fingertips of the members on either side to obtain the double-arm interval. (The right flank member raises only the left arm, and the left flank member raises only the right arm.) Each member smartly lowers the right arm after obtaining proper interval to the right and lowers the left arm when the member on the left lowers the right arm.

FROM DOUBLE-ARM TO NORMAL INTERVAL. The commands **Assemble To The Right** MARCH, instruct members to obtain normal interval from double-arm interval. Execute this movement as you would in closing, but form at normal interval.

TO COUNT OFF. The commands **Count** OFF instruct members to count off while in a rank or line. On the command OFF, all members, except the right flank member, smartly turn their heads and look to the right. The right flank member shouts ONE, the next member in rank or line shouts TWO, and so on, in quick cadence on down the line through the left flank member. As each member shouts the appropriate number, he or she turns the head smartly to the front.

Members in a file or column count off when given the commands **From Front To Rear Count** OFF. Each member, starting with the squad leader, turns the head to the right and shouts the appropriate number while turning the head back to the front.

The commands **Open Ranks** MARCH are given when more distance between ranks is required; for example, for a personnel inspection. On the command MARCH, the front rank takes two paces forward, the second rank takes one pace (30 inches) forward, and the third rank stands fast. Each succeeding rank takes two, four, or six (15-inch) steps backward. Each rank automatically dresses right as soon as it halts. When the alignment is completed, the commands **Ready** FRONT are given.

TO CLOSE RANKS. The commands **Close Ranks** MARCH tell members to close ranks. On the command MARCH, the front rank stands fast, the second rank takes one pace forward, the third rank takes two paces forward, the fourth rank takes three paces forward, and so on. You will halt and cover without command.

HAND SALUTE. Begin a salute on the commands **Hand** SALUTE, and complete the salute on the command TWO. On the command SALUTE, raise your right hand smartly. At the command TWO, return to attention by moving your hand smartly in the most direct manner back to its normal position at your side. (If you are in formation, the preparatory command Ready will be given before the command of execution, TWO.)

UNCOVER. Many religious ceremonies, and usually inspections, require you to remove your hat

Student Notes:

when given the commands **Uncover** TWO. On the command **Uncover**, raise your right hand as in the hand salute, but grasp the brim of your hat with your fingers instead of touching your forehead. Hold this position until the command TWO is given (you may lift your hat slightly so as not to muss your hair); then return your hand and your hat to your side in the most direct manner, but do not remove it with an exaggerated or sweeping motion. On the command **Cover**, grasp your hat with both hands and place it squarely on your head. Drop your left hand holding the hat brim. On the command TWO, drop your right hand to your side.

DISMISSED. The single command DISMISSED is used to secure an individual or an entire formation.

REVIEW 7 QUESTIONS

Q1. List three reasons for military formations.

a.

b.

c.

Q2. What is the meaning of the term *rank* or *line* with reference to military formations?

Q3. List the two parts of a facing command.

a

b.

Q4. When in a formation, what are the four commands for rest?

a.

b.

c.

d.

Q5. What command tells you to smartly turn your head 45° to the right?

SUMMARY

All branches of the military establishment take pride in the appearance and sharpness of their personnel during ceremonies. The Navy is no exception. During your naval career, you will probably receive some type of award in recognition of an accomplishment. The recognition will probably take place at a command function, such as an inspection or awards ceremony. Ship's company or command personnel standing at attention in formation during such ceremonies makes the award more meaningful and the ceremony more impressive.

REVIEW 1 ANSWERS

A1. The men's dress blue jumper should **hang straight and cover all but the lowest button of the 13-button front of the trouser**.

A2. Enlisted women E-1 through E-6 **wear the dress white jumper with the black silk neckerchief**.

A3. When the peacoat is worn over the dress blue jumper, **the uniform's collar is worn inside the outergarment (peacoat)**.

A4. When **large medals are worn with a service dress white uniform, the uniform is considered a full dress uniform**.

Student Notes:

REVIEW 2 ANSWERS

A1. DELETE.

A2. Both men and women Sailors **stencil the dress white jumper in the same way**.

A3. The **commanding officer** can authorize the transfer of enlisted person's uniform clothing to another individual.

A4. You are required to have **four dungaree shirts and four pairs of dungaree trousers/slacks in your seabag**.

A5. You may wear any of the following with civilian clothes:

a. **All-weather raincoat**

b. **Underwear**

c. **Shoes**

d. **Belt with civilian buckle**

e. **Socks/hosiery**

f. **Sweaters**

g. **Neckties**

REVIEW 3 ANSWERS

A1. A striker mark is **a specialty mark of a rating** worn by qualified E-1 through E-3 personnel.

A2. To be eligible to wear a service stripe, you must have **served 4 full years of active or reserve service in any branch of the U.S. Armed Forces**.

A3. A **rear admiral (lower half)** wears one silver star metal grade insignia on the uniform collar.

A4. A line officer wears **a five-pointed gold star on the sleeve insignia**.

A5. Personnel qualified in underwater and beach reconnaissance and demolition wear the **Special Warfare insignia**.

REVIEW 4 ANSWERS

A1. If a person has four or more ribbons, that person **must wear three on the uniform**.

A2. Military decorations include the following:

a. **Medal of Honor**

b. **Navy Cross**

c. **Distinguished Service Medal**

d. **Purple Heart**

A3. Nonmilitary decorations include the following:

a. **Presidential Medal of Freedom**

b. **Gold and Silver Life Saving Medals**

c. **National Sciences Medal**

A4. The person wears the medals from **top to bottom inboard to outboard, within rows**.

REVIEW 5 ANSWERS

A1. A person shouldn't take any of the following actions with regard to their ID card:

a. **Alter it**

b. **Intentionally damage it**

c. **Lend it to someone**

d. **Counterfeit it**

e. **Use it in an unauthorized manner**

A2. The military ID card **identifies you to the capturing authorities** if held as a prisoner of war. The ID tags (dog tags) are designed for **identification purposes in casualty reporting and for grave registration** of members who die in combat.

A3. The following information is embossed on the dog tags:

a. **Full name**

b. **Social security number (SSN) and the letters USN**

c. **Blood type and Rh factor**

d. **Religious preference of the wearer**

REVIEW 6 ANSWERS

A1. While in uniform, you may wear—

a. **One wristwatch**

b. **One bracelet**

A2. To find an explanation of grooming standards, you should refer to the ***U.S. Navy Uniform Regulations Manual***.

A3. When in uniform, female Sailors may wear **6mm ball-type earrings (either post or screw on) with a brushed matte finish.** The following information applies:

a. E-6 and below **wear silver earrings.**

b. CPOs and officers **wear gold earrings**.

REVIEW 7 ANSWERS

A1. Military formations have the following purposes:

a. **To move a large number of personnel from one place to another in an orderly fashion.**

b. **To make personnel alert and to pay attention to information or instructions being put out.**

c. **To teach a group of individuals to act as a team.**

A2. With reference to military formations **rank or line is a formation of elements or persons abreast of each other or side by side.**

A3. The two parts of a position and facing command are—

a. **Preparatory command**

c. **Execution Command**

A4. The four positions of rest for a formation are—

a. **Parade rest**

b. **At ease**

c. **Rest**

d. **Fall out**

A5. The command given to turn your head 45° to the right is **Eyes** RIGHT.

CHAPTER 11

SMALL ARMS

A good Navy is not a provocation to war. It is the surest guarantee of peace.

—Theodore Roosevelt

A number of duty assignments in the Navy may require you to be armed with a rifle or pistol. Examples of shipboard duty assignments include the forecastle, fantail, and pier sentry watches, quarterdeck, and magazine security watches. Examples of ashore duty assignments include base security forces and duties of Seabee personnel. Although none of these assignments may be your normal watch-standing duties, you may be required to support these or other security forces at any time. For this reason, you must be familiar with the proper use and care of small arms.

Strictly defined, the term *small arm* means any firearm of .60-caliber, 15-mm, or smaller bore. However, the term is generally considered to mean a weapon intended to be fired from the hand or shoulder, such as a rifle or a pistol.

At most naval commands, the small arms carried by security watches are maintained by armory personnel. Armory personnel should give you instructions on the proper use and handling of small arms.

SMALL ARMS SAFETY PRECAUTIONS

Learning Objective: When you finish this chapter, you will be able to—

- Recognize the purpose for and identify the safety precautions to follow when using small arms.

Before you learn to use any firearm, you must learn to handle it safely. **Remember, firearms are dangerous**. Their purpose is to kill or to cause injury.

NOTE

When at the firing range, follow all safety precautions.

Every firearm used by Navy personnel has some type of built-in safety device, and some have more than one. The safety device guards against accidental discharge of a firearm. In almost every case of accidental shooting, negligence or carelessness is the prime cause. A weapon is only as safe as the person using it. Learn to respect each firearm as a deadly weapon.

You should observe the following general precautions when handling any type of firearm:

1. Treat every weapon with respect. Consider it loaded.
2. Never point a weapon at anything or anyone you do not intend to shoot.
3. Always make sure that the bore is clear and that all oil and grease have been removed from the barrel and chamber before firing.
4. Use only the proper size of ammunition.
5. Unload firearms before transporting them to and from a shooting area.
6. Always carry the firearm so as to control the direction of the muzzle. Keep the muzzle pointed in a safe direction until ready to fire.
7. Keep the safety on until you are ready to shoot.
8. Never shoot until you have positively identified the target.
9. Unload unattended weapons. At home, store firearms (with trigger locks installed) and ammunition out of the reach of children.
10. Do not climb trees or fences with a loaded firearm.
11. Do not pull a firearm toward you by the muzzle.
12. Avoid shooting a rifle over a hard, flat surface or body of water because of possible erratic and lengthy bullet ricochets.
13. Like oil and water, firearms and alcohol do not mix. Do not drink alcoholic beverages or partake of any narcotic or drug before or during shooting activities.

14. Know your weapon—its shooting characteristics, its safeties, and its loading and unloading procedures.

15. Never indulge in horseplay when carrying a firearm.

In addition to observing these safety precautions when handling and using firearms, you should take steps to protect your hearing and sight, particularly when you are exposed to repeated small arms fire such as that on a rifle or pistol range. Blast noise from small arms fire may cause a temporary or permanent hearing loss. The extent of injury depends on a number of factors, such as intensity of the noise, length of exposure, and your own sensitivity to noise hazards.

Two general types of personal ear protective devices are used to reduce noise and thus protect the ear. These are the insert type (earplug) and the circumaural type (covers the entire outer ear). The circumaural type is sometimes referred to as *Mickey Mouse ears*.

If you work in a noise hazard area, you may be fitted with a pair of earplugs. It's important to have the plugs fitted by a qualified member of the medical department because each person's auditory canals are a different size and structure.

The Mickey Mouse ears (fig. 11-1) are made with rigid plastic ear cups lined with foam, plastic, or rubber to provide a comfortable seal around the outer ear. The cups are connected over the head with an adjustable spring type of headband for a snug fit.

Figure 11-1.—Circumaural (Mickey Mouse) type of ear protector.

Protecting your sight is as important as protecting your hearing. The Navy has several types of safety glasses and goggles that provide adequate protection from the danger of small arms. They range from the standard safety goggles used in everyday work to prescription safety glasses.

When you are on the range, use these protective devices so that you will always be able to see and hear the full spectrum of sounds and sights you see and hear today.

REVIEW 1 QUESTIONS

Q1. When handling a firearm, you should NOT take which of the following actions?

a. Always treat a weapon a if it were loaded

b. Never point a weapon at anything or anyone you don't intend to shoot

c. Carry the weapon by the muzzle as long as the safety is on

d. Never engage in horseplay when carrying a firearm

Q2. List the protective equipment you should wear when firing a weapon.

a.

b.

THE M14 RIFLE

Learning Objective: When you finish this chapter, you will be able to—

- Identify the M14 rifle.

The 7.62-mm M14 (fig. 11-2) rifle is a lightweight, air-cooled, gas-operated, magazine-fed shoulder weapon. It was widely used during Vietnam. Since then, it has since been replaced by the M16. It is designed for semiautomatic or automatic fire at the rate of 750 rounds per minute. The rifle is chambered for the 7.62-mm NATO cartridge and is

Student Notes:

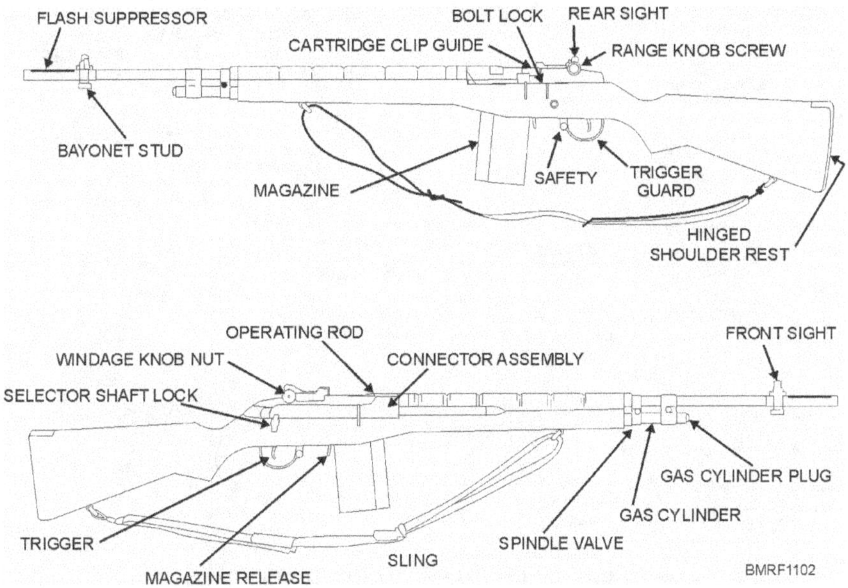

Figure 11-2.—The 7.62-mm M14 rifle.

designed to accommodate a 20-round magazine, the M76 grenade launcher, and the M6 bayonet.

The overall length of the rifle (with a flash suppressor) is 41.31 inches. The weapon has a muzzle velocity of 2,800 feet per second and a maximum range of 4,075 yards. Empty, the rifle weighs about 9 pounds. Fully loaded and ready to fire, the rifle weighs about 11 pounds.

THE M16A1/A2 RIFLE

Learning Objectives: When you finish this chapter, you will be able to—

- Identify the differences between the M16A1 and M16A2 rifles.
- Recognize procedures to load and unload the magazine.
- Identify procedures for cleaning and maintaining the M16A1 and M16A2 rifles.

The M16A1 and the M16A2 rifles (fig. 11-3 and fig. 11-4) are magazine-fed, gas-operated shoulder weapons. They are chambered for a 5.56-mm (about .22-caliber) cartridge. The magazine has a capacity of 20 or 30 rounds and may be loaded with any amount, up to capacity. The caliber may seem small, but the bullet has a muzzle velocity of more than 3,000 feet per second and a muzzle energy of more than 13,000 foot-pounds. A forward assist assembly, located on the right rear of the upper receiver, permits the operator to ensure the bolt is locked in the forward position. They have a maximum effective range of 460 meters.

DIFFERENCE BETWEEN THE M16A1 AND THE M16A2

The steps you take when field-stripping and cleaning the M16A1 and M16A2 weapons are basically the same. However, there are some major differences between the two weapons. These differences include the barrel, sights, selector assembly, ammunition, and hand guards of the two rifles.

WARNING

Ensure you are using the proper ammunition for the M16A1 and the M16A2. Improper use may cause serious damage.

Student Notes:

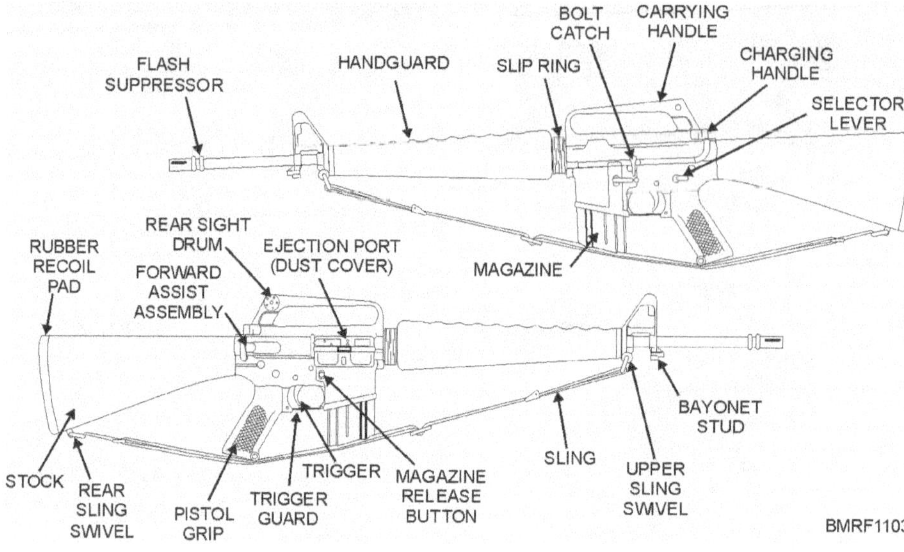

Figure 11-3.—M16A1 service rifle, 5.56 mm, left and right side views.

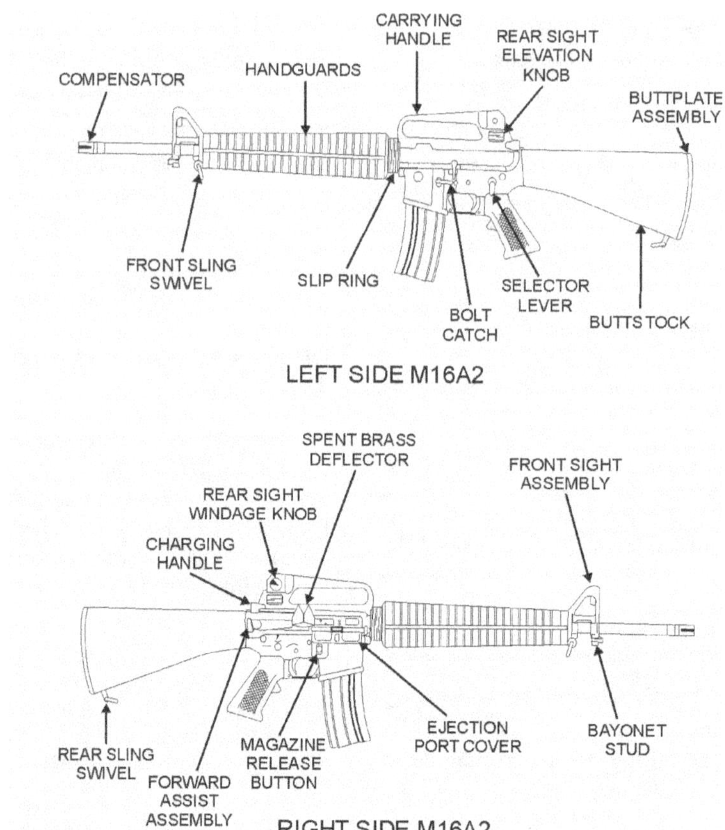

Figure 11-4.—M16A2 service rifle, 5.56 mm, left and right side views.

COMPONENT	M16A1	M16A2
Barrel	Lighter barrel because of ammunition fired	Heavier barrel because of ammunition fired
Sights	Rear sight uses a windage drum for windage (fig. 11-5) Front sight is same as the M16A2 (fig.11-6)	Rear sight is adjusted by means of a windage knob for windage and an elevation knob for elevation (fig.11-7)
Selector assembly	Has semiautomatic or fully automatic features	Has a semiautomatic or burst (three rounds) feature
Ammunition	Fires a 5.56-mm round and a 62-grain NATO round	Fires a 5.56-mm round and a 62-grain NATO round
Hand guards		Round and ridged, making them stronger and easier to grip than the M16A1 hand guards. They are also interchangeable.

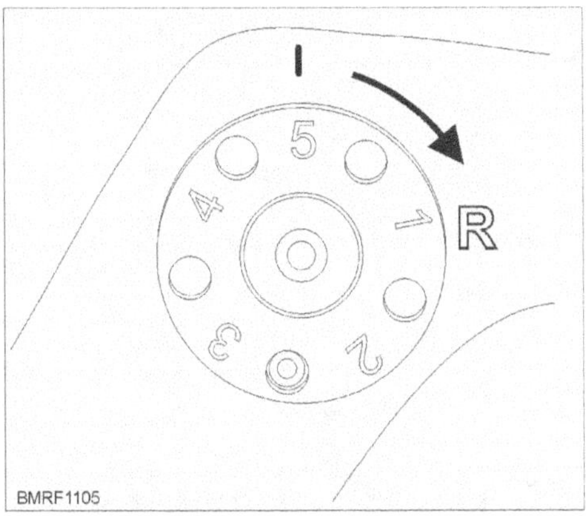

Figure 11-5.—M16A1 windage drum.

CLEARING THE RIFLE

The following maintenance procedures (clearing, field-stripping, assembling, etc.) for the M16 service rifles are written for the right-handed Sailor. The left-handed Sailor can reverse the hand directions for these procedures for the M16

NOTE

Unless specifically stated otherwise, the following discussion of the M16 rifle applies equally to both the M16A1 and M16A2.

The first precaution you should take in handling any weapon is to make it safe by clearing it. To clear the M16 rifle, place the butt against the right thigh and proceed as follows:

1. Point the selector lever toward SAFE (fig. 11-8, view A). If the weapon is not cocked, the selector lever cannot be pointed toward SAFE. If that is the case, do

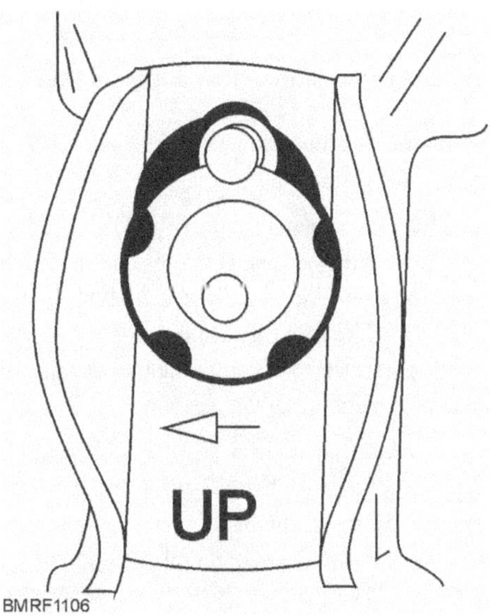

Figure 11-6.—Front sight for the M16A1 and the M16A2 rifle.

Student Notes:

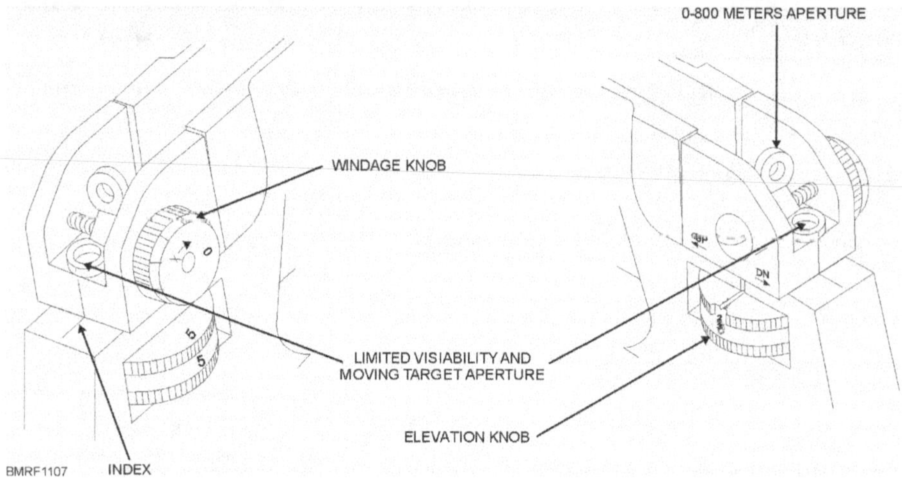

Figure 11-7.—M16A2 rear sight.

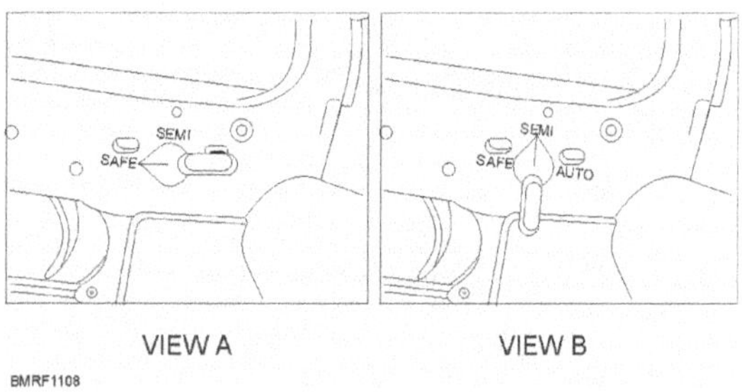

Figure 11-8.—Selector lever pointing to SAFE.

not cock the weapon at this time; instead, go on to the next step in clearing.

2. Remove the magazine (fig. 11-9). Grasp the magazine with the left hand (fingers curled around the front of the magazine, thumb placed on the magazine catch button). Use your right hand and apply pressure on the magazine catch button with the thumb and pull the magazine straight out of the weapon.

3. Lock the bolt open (figs. 11-10 and 11-11). Grasp the charging handle with the thumb and forefinger of the right hand, depress the charging handle, latch it with the right thumb, and pull to the rear (fig. 11-10). When the bolt is fully rearward, press the bottom of the bolt catch with the thumb or forefinger of the left hand (fig. 11-11). Allow the bolt to move slowly

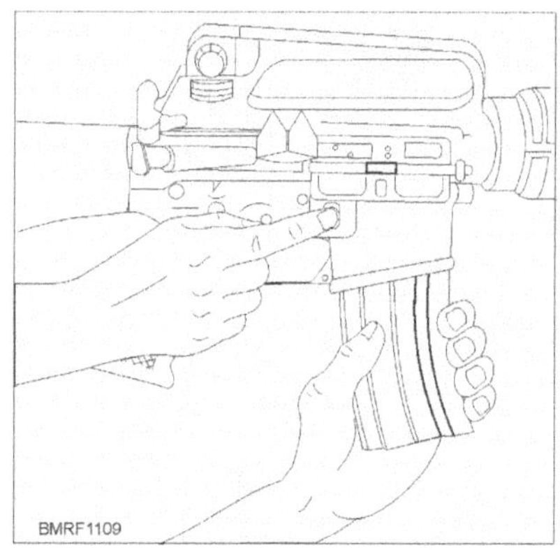

Figure 11-9.—Removing the magazine.

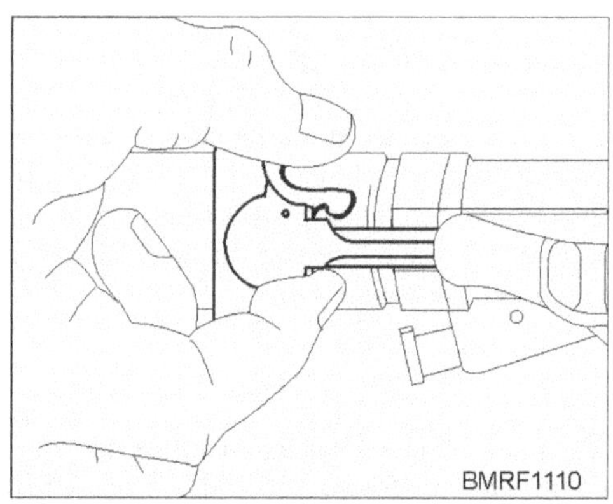

Figure 11-10.—Pulling the charging handle rearward.

Student Notes:

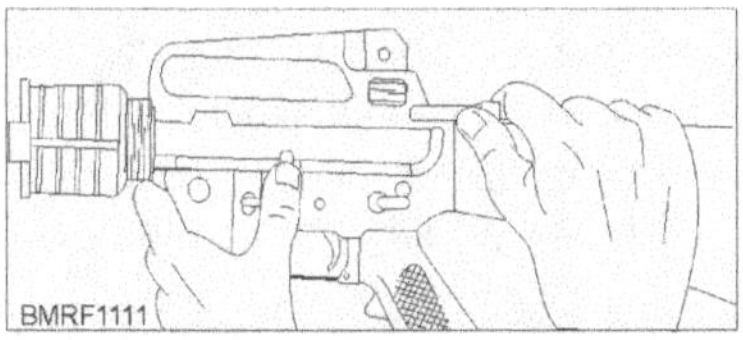

Figure 11-11.—Locking the bolt open.

forward until it engages the bolt catch, and return the charging handle to its forward position.

4. Inspect the receiver and chamber of the weapon by looking through the ejection port to ensure these spaces contain no ammunition.

5. Check the selector lever to ensure it points toward SAFE; then allow the bolt to go forward by depressing the upper portion of the bolt catch.

CAUTION

The selector must be on SAFE to prevent damage to the automatic sear during assembly and disassembly.

FIELD-STRIPPING THE RIFLE

Before you can field-strip a rifle, you must be qualified to do so. **You will use instructions when you actually field-strip a rifle**. Follow the procedures in the instruction that deals with the weapon that you are field stripping.

LOADING THE MAGAZINE

A magazine can come in either a 20- or 30-round capacity and may be loaded with any amount up to that capacity. The magazine follower has a raised portion generally resembling the outline of a cartridge. Cartridges are loaded into the magazine so the tips of the bullets point in the same direction as the raised portion of the follower (fig. 11-12).

CAUTION

Do not load or attempt to load more rounds than what the magazine was designed for. Overloading deforms the lips of the magazine and can cause malfunctions.

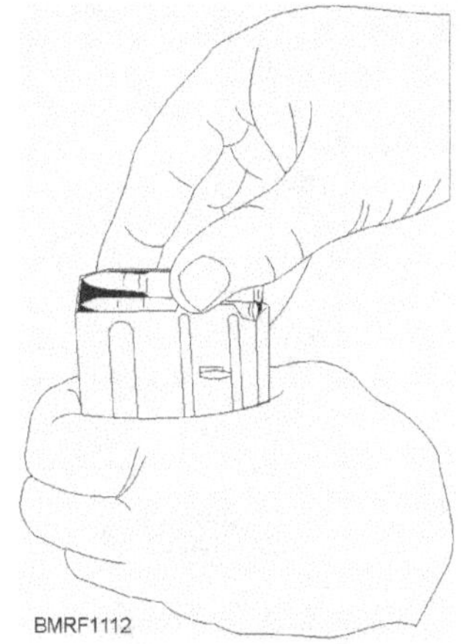

Figure 11-12.—Loading cartridges into the magazine, 20 -round capacity.

UNLOADING THE MAGAZINE

To prevent damage to the lips of the magazine, you should remove the ammunition in the following manner:

1. Hold the magazine in your left hand with the open end away from your body and with the nose of the cartridge down (fig. 11-13, view A).

2. Depress the center of the second round in the magazine using the nose of the cartridge, allowing the first round to drop out of the magazine (fig. 11-13, view B). Repeat this procedure until you remove all the rounds from the magazine except the last one.

3. Use the nose of the cartridge to depress the follower to remove the last round, allowing the last round to drop out of the magazine (fig. 11-13, view C).

LOADING THE RIFLE

With the hammer cocked, place the selector lever on SAFE. (Refer to figure 11-8, view A.) Notice that you can't place the selector lever on SAFE unless the rifle is cocked. You may insert the magazine with the bolt and bolt carrier open or closed; however, you should learn to load with the bolt open. Loading with the bolt open reduces the possibility of first round stoppage and saves the time needed to pull the charging handle to the rear.

Hold the stock of the rifle under your right arm with your right hand. Grasp the pistol grip; then point the

Student Notes:

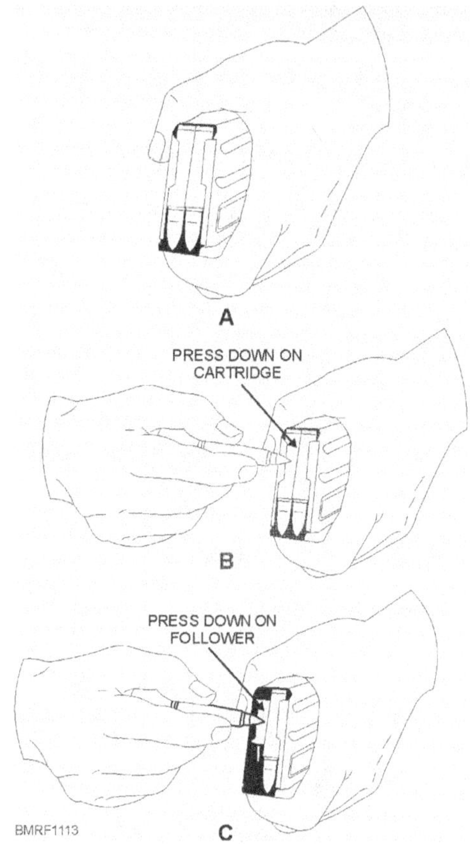

Figure 11-13.—Unloading the magazine with the nose of a cartridge.

muzzle in a safe direction. With your left hand, insert the loaded magazine into the magazine housing. Push upward until the magazine catch engages and holds the magazine. Rap the base of the magazine sharply with the heel of your hand to ensure positive retention. If the action is open, release the bolt by depressing the upper portion of the bolt catch with the thumb of your left hand, allowing the action to close, chambering the round. If the action is closed when the magazine is inserted, pull the charging handle fully to the rear with your right hand and release it. (Refer to figure 11-10.)

WARNING

Don't *ride* the charging handle forward with the right hand. If the charging handle is eased forward from the open position, the bolt may fail to lock. If the bolt fails to go fully forward, use the bolt closure forward assist assembly (fig. 11-4) with the heel of your right hand. The rifle is now loaded and is ready to fire when you place the selector lever in the automatic or semiautomatic position. If it is not ready to fire, make sure the selector lever is on SAFE.

After the last round has been fired, the bolt catch holds the bolt carrier to the rear. To change the magazine for reloading, press the magazine catch button and remove the empty magazine from the weapon.

FIRING THE RIFLE

The rifle fires semiautomatic or burst (automatic for the M16A1) when you move the selector lever to the desired position. (Refer to figure 11-8.) With the selector lever in the semiautomatic position, the rifle fires one round each time you pull the trigger. With the selector lever in the burst position, the M16A2 fires in short bursts of three rounds. (**NOTE**: The M16A1 rifle fires fully automatically and will continue to fire until the magazine is empty or you release the trigger.) When the rifle is fired in either SEMI or BURST or AUTOMATIC, the bolt locks in the open position when the last round from the magazine has been fired.

UNLOADING AND CLEARING THE RIFLE

As you read this section, refer back to figures 11-8, 11-9, 11-10, and 11-11. To unload the rifle and make it safe, place the selector lever on the SAFE position (fig. 11-8); and remove the magazine by pressing the magazine release button (fig. 11-9). Pull the charging handle to the rear (fig. 11-10), ejecting any round from the chamber. Inspect the chamber and receiver to ensure that it is clear. Releasing the charging handle will allow the bolt to close. To keep the bolt open, depress the lower portion of the bolt catch before returning the charging handle forward (fig. 11-11). The rifle is clear only when the following conditions exist:

- No case or round is in the chamber.
- The magazine is out.

Student Notes:

- The bolt carrier is to the rear.
- The selector lever is on the SAFE position.

CARING AND CLEANING OF THE RIFLE AND AMMUNITION

A clean, properly lubricated and maintained rifle that is loaded with clean ammunition will fire when needed. To keep the rifle in good condition, you need to take care of it and clean it. Under bad weather conditions, some key parts may need care and cleaning several times a day. The cleaning material (fig. 11-14) used for the care of the rifle is carried in the rifle stock. Special attention must be given to the barrel bore and chamber, bolt carrier group, upper receiver group, lower receiver group, and the ammunition magazines.

Figure 11-14.—M16Al and M16A2 rifle cleaning material.

PART	ACTION
Barrel bore and chamber	1. Dip a bore brush in the bore cleaner; then brush from the chamber to the muzzle, using straight-through strokes. **Don't reverse the brush while it is in the bore: it may jam.** A jammed brush is hard to remove, and removing the brush might damage the bore. 2. Dip the brush in bore cleaner; then clean the chamber with the bore brush. 3. Replace the bore brush with a slotted cleaning patch tip, and push the dry patches through the bore and chamber until they come out clean. 4. After you clean the bore, lightly lubricate the bore and chamber to prevent corrosion and pitting. Use the recommended lubricant on a patch. 5. Lightly lubricate the lugs in the barrel extension.

Student Notes:

PART	ACTION
Bolt carrier group	1. Dip the bore brush in the bore cleaner, and clean the inside of the carrier key. 2. Dry with a pipe cleaner 3. Clean the locking lugs, bolt, extractor ejector, and bolt rings with the bore brush. 4. Remove any accumulation of dirt, carbon, or oil from the firing pin and the external and internal surfaces of the bolt and bolt carrier. 5. Wipe all parts dry; then lubricate them with the recommended lubricant.
Upper receiver group	1. Coat the bore brush or a swab with bore cleaner; then remove the powder fouling collected on the group. 2. Clean the protruding gas tube inside and outside. 3. After cleaning these components, wipe them dry. 4. Apply a light coat of the recommended lubricant.
Lower receiver group	1. Coat the bore brush or a swab with bore cleaner; then remove dirt, carbon, and sand from the lower receiver group. 2. Dry and apply a light coat of the recommended lubricant.
Ammunition magazines	1. After removing all cartridges from the magazine, depress the spring steel lock band on the bottom of the magazine, using the nose of a cartridge. 2. Slide the base until it is free of the tabs, and remove it from the magazine body. 3. Remove the magazine spring and follower, but do not remove the follower from the spring. 4. Clean the exterior and interior of the magazine with a dry rag or swab. 5. Apply a light coat of the recommended lubricant to the magazine spring only; otherwise, keep the magazine dry. • Assemble the magazine in reverse order and test it to ensure that the follower is free to move without binding. • If the magazine and the ammunition in it gets wet, be sure to wipe them dry as soon as possible. • When left wet, both the magazine and the ammunition can become corroded and are dangerous to use. • Remember not to use oil or grease on any cartridge. If you do this, injurious abrasives can collect in the weapon or produce excessive and hazardous chamber pressures when the weapon is fired. Whenever practical, ammunition should be stored under cover. This applies particularly to tracer ammunition.

Student Notes:

REVIEW 2 QUESTIONS

Q1. What type of ammo is the M-14 rifle chambered to fire?

Q2. When fully loaded, the M-14 weighs __________.

Q3. Identify the following rifle components as either M16A1 or M16A2.

Component	Rifle
a. Has a lighter barrel because of the ammunition fired	
b. Rear sights are adjusted by means of a windage knob and elevation knob	
c. Has a semiautomatic or burst feature	
d. Fires a 5.56-mm round and a 62-grain round	

Q4. What is the first step in cleaning the M16 rifle?

Q5. How many rounds does an M16 rifle carry?

Q6. When loading the M16 rifle, why shouldn't you ride the charging handle forward?

Q7. When the selector is set for burst, how many rounds will the M16A2 fire?

Q8. List the conditions that must exist to consider the M16 rifle clear.

a.

b.

c.

d.

Q9. Where are the cleaning materials for the M16 stored?

REVOLVERS AND SERVICE PISTOLS

Learning Objectives: When you finish this chapter, you will be able to—

- Recognize the operating characteristics of the .38-caliber revolver.
- Recognize the procedures to follow when loading and unloading the magazine, and unloading and cleaning the .45-caliber pistol.
- Identify the procedures for cleaning and maintaining the .45-caliber pistol.
- Identify the safety devices of the 9mm caliber pistol.
- Recognize the procedures to follow when loading and unloading the 9mm pistol.

NOTE

The .38-caliber revolver and the .45-caliber service pistol have been replaced by the 9mm pistol. However, small units may still carry these revolvers and pistols in their allowance.

Student Notes:

THE .38-CALIBER REVOLVER

The .38-caliber revolver is a cylinder-loading, single- or a double-action, manually operated hand weapon (fig. 11-15). Several barrel lengths and weights are available.

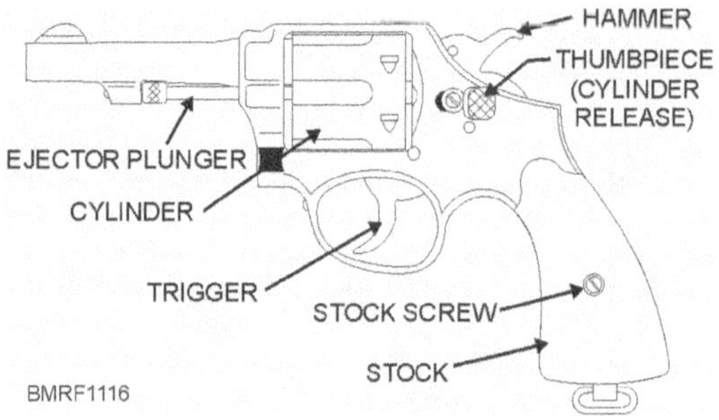

Figure 11-15.—The .38-caliber revolver.

THE .45-CALIBER SERVICE PISTOL

The .45-caliber service pistol (fig. 11-16) is an individual weapon intended for use in close combat. The .45-caliber pistol is a semiautomatic, recoil-operated, magazine-fed hand weapon. The pistol fires one round each time the trigger is squeezed. The pistol can be carried in either a hip or shoulder holster.

The magazine holds seven cartridges. The forward movement of the slide strips the upper cartridge from the magazine into the chamber. After the last cartridge from the magazine has been fired, the slide remains in the rear.

Only your ability to change magazines, aim, and squeeze the trigger rapidly limits the rate of fire of the .45-caliber service pistol.

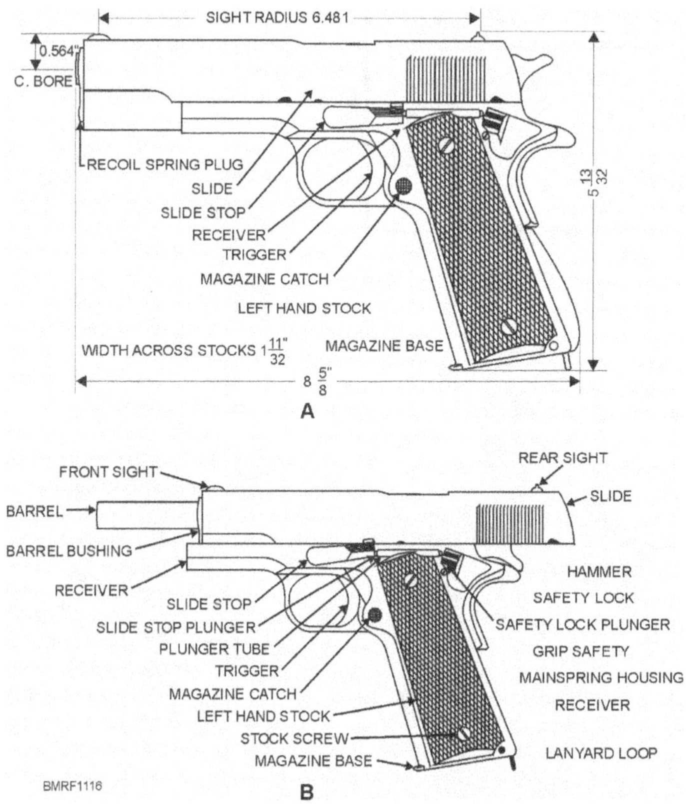

Figure 11-16.—.45-caliber semiautomatic service pistol. (A) Assembled; (B) Sectional view in recoil position.

Student Notes:

The pistol is 8 5/8 inches in length and weighs 3 pounds fully loaded, with a maximum range of 1,500 yards, and a maximum effective range of 50 yards.

THE 9MM SERVICE PISTOL

In 1985, the armed forces selected a 9mm pistol to replace the .45-caliber pistol. The pistol selected is a single- or double-action semiautomatic hand weapon. As soon as the pistol is fired, either in single or double action, the slide automatically comes back and cocks the hammer. To fire the pistol again, all you have to do is pull the trigger. The 9mm pistol has a large magazine capacity—it can hold 15 rounds in the magazine. Slots in the magazine help the user know the number of rounds that remain.

Description and Technical Data

As you know, the 9mm service pistol is a semi-automatic, magazine-fed, recoil operated, double-action pistol. The pistol fires one round each time the trigger is pulled. The energy needed to operate the pistol comes from the recoil, which is created by the rearward force of expanding gases of a fired round. The double-action feature lets you fire a weapon when the hammer is in the forward position, the safety is in the fire position, and the trigger is pulled. The magazine holds 15 cartridges. When the last cartridge from the magazine is fired, the slide remains locked to the rear. Look at the following chart for the technical data of the 9MM service pistol.

COMPONENT	TECHNICAL DATA
Caliber	9 x 9mm (9mm NATO)
Diameter of bore	.356 in.
Length	217 mm (8.54 in.)
Width	38 mm (1.50 in.)
Height	140 mm (5.51 in.)
Weight with 15-round magazine	1,134 g (2.50 lb.)
Weight with empty magazine	960 g (2.12 lb.)
Barrel length	125 mm (4.92 in.)
Muzzle velocity	375 m/sec (1,230 ft/sec)
Maximum effective range	50 mm (45.7 yd)
Cyclic rate of fire	1,100 rpm
Trigger pull	4 to 6.5 lb single action and7.5 to 16.6 lb double action

Safety Devices

The 9mm service pistol is equipped with three types of safety features—the ambidextrous slide safety, the firing pin block, and the half cock notch.

AMBIDEXTROUS SAFETY.—This safety allows for safe operation of the pistol by both right- and left-handed users. It lowers the hammer safely without causing an accidental discharge. When the hammer is cocked, it may be safely lowered by moving the safety to the safe (down) position. When the safety is in the fire (up) position, a red dot will be visible indicating that the pistol should be handled with caution (**red you're dead**).

FIRING PIN BLOCK.—This safety prevents any motion of the firing pin and is only overcome by pulling on the trigger. Both the firing pin and the firing pin block have a notch cut into them. The firing pin block remains in the firing pin notch and prevents any motion of the firing pin. When you pull the trigger, the firing pin block is pushed upward and aligns its notch with the firing pin so that the firing pin can move forward to strike the primer of a round.

HALF COCK NOTCH.—This safety prevents accidental discharge. The half cock notch is located on the hammer. If a cocked hammer should fall forward because of a mechanical failure, the half cock notch would catch on the sear before the hammer strikes the firing pin and prevent an accidental discharge of the pistol.

Loading

The 9mm service pistol incorporates single- and double-action modes of fire. With the safety in the **FIRE** position, in the double-action mode, squeezing

Student Notes:

the trigger will automatically cock and fire the pistol. Always keep your finger away from the trigger unless you intend to fire. The safety should be in the down position (the red dot not visible), which indicates that the pistol is in a safe condition before loading. With the pistol pointing in a safe direction and the slide in its forward position, follow the steps listed below.

STEP	ACTION
1	Insert a loaded magazine into the magazine well of the pistol until you hear a click. This ensures a proper catch engagement.
2	Grasp the serrated portion of the slide with the nonshooting hand (fig. 11-17).
3	Pull the slide all the way to the rear (fig. 11-18).
4	Release the slide. This will strip a cartridge from the magazine and chamber a round (fig. 11-19).

WARNING

Although rare, it is possible that the safety may become disengaged during the loading procedure. Make sure the safety is on after loading.

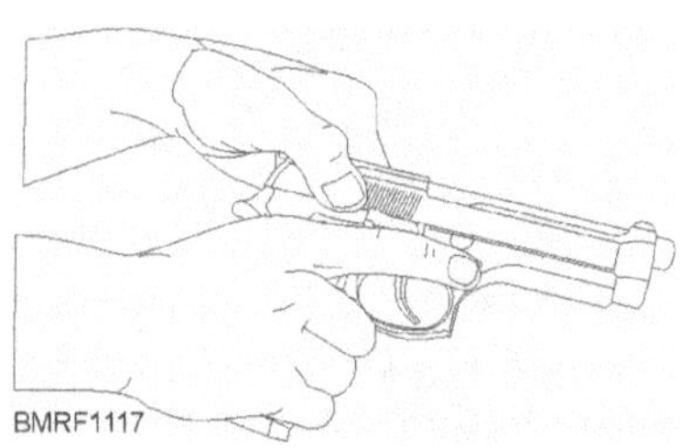

Figure 11-17.—Grasping the serrations of the slide.

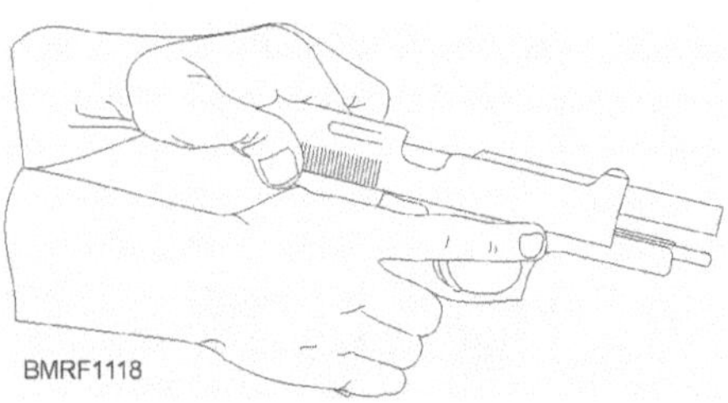

Figure 11-18.—Pulling the slide to the rear.

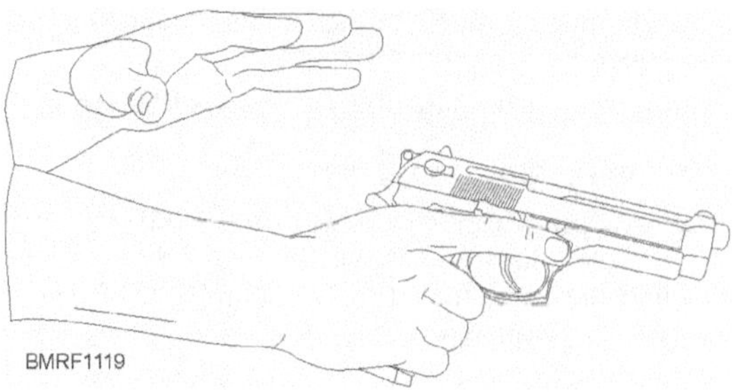

Figure 11-19.—Releasing the slide.

Unloading

To unload the 9MM service pistol, follow the steps listed below.

STEP	ACTION
1	Place the safety in the **safe** position.
2	Depress the magazine release button to remove the magazine from the pistol.
3	With the pistol pointing in a safe direction, grasp the slide serrations and fully retract the slide to remove the chambered cartridge.
4	Use the right thumb and push upward on the slide stop to lock the slide to the rear, and visually inspect the chamber to ensure that it is empty.

Unloading the Magazine

To unload the magazine, hold the magazine using one hand, with the front end forward. With your thumb, press firmly on the cartridge rim and push forward. Repeat this procedure until the magazine is empty.

REVIEW 3 QUESTIONS

Q1. How many rounds will a .45-caliber pistol magazine hold?

Student Notes:

Q2. The .45 caliber pistol was designed to be used as

____________________________.

Q3. List the three safety features of the 9mm service pistol.

a.

b.

c.

Q4. You are loading your pistol. The safety should be in what position?

SHOTGUNS

Learning Objective: When you finish this chapter, you will be able to—

- Identify the safety practices to follow when using shotguns.

Shotguns (fig. 11-20) are being used by the Navy in security guard work. Shotguns are the weapons of choice for short-range work, which includes the requirement to protect vital nuclear propulsion systems and nuclear weapons. The advantage of shotguns over pistols/revolvers is that sight alignment is not so critical. Each trigger pull of a shotgun expels anywhere from nine to hundreds of projectiles (shot). These projectiles cover a wide area. The characteristics of the M870 shotgun are shown in the following chart:

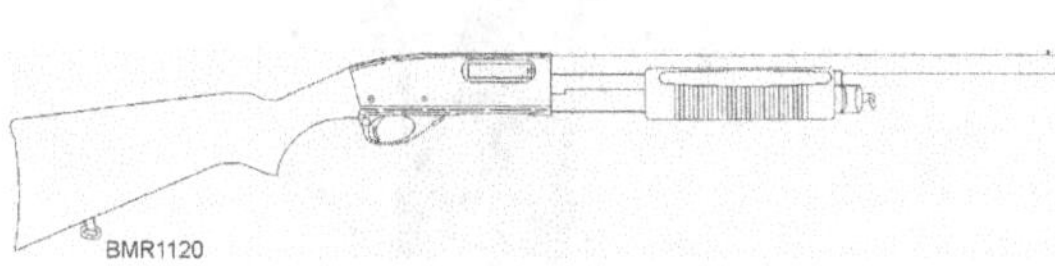

Figure 11-20.—M870 shotgun.

NOTE

Make certain you load the M870 with the proper ammunition. For example, many people have been injured by a shotgun loaded with a smaller gauge shell. This smaller shell goes part way down the barrel and cannot be fired. The user thinks a misfire has occurred and chambers the proper size shell. Firing the weapon causes the gun to explode because the smaller shell is an obstruction. Serious injury or death could occur.

COMPONENT	CHARACTERISTIC
M870 shotgun	A manually operated, magazine-fed (tubular), pump-action shoulder weapon.
Length–Overall	39 inches
Barrel	20 inches
Ammunition	Four rounds of 12-gauge , 2 ¾ inch in the magazine. Normal ammunition issue is 12 gauge, 00 buck
Crossbolt Safety	

REVIEW 4 QUESTION

Q1. What advantage does a shotgun have over other firearms?

MARKSMANSHIP

Learning Objective: When you finish this chapter, you will be able to—

- Recognize the firing techniques for rifles and pistols.
- Identify the shooting positions.

Marksmanship training gives you proper information and instruction on how to be a safe and effective shooter. Good shooting, whether on the firing range or in combat, depends on your knowledge and use of basic marksmanship principles. These principles are interrelated and must be practiced each time you fire a shot.

Student Notes:

FIRING TECHNIQUES—RIFLE

The most important factors for you to remember about firing a rifle are sight alignment and achieving a correct aiming point. Together they make up the sight picture.

Sight Alignment

Sight alignment involves looking through the rear sight aperture, focusing the eye on the front sight post (or blade), and centering the front sight post exactly in the rear sight aperture, both vertically and horizontally. The top of the front sight is level with the horizontal center line of the rear sight, and the body of the front sight is centered between the rear sight aperture (fig. 11-21).

REAR SIGHT.—When you are in different firing positions (standing, kneeling, or sitting), your aiming eye is at a slightly different distance from the rear sight. This distance is referred to as *eye relief.* Eye relief makes the opening (peep) of the rear sight appear larger or smaller, depending on the firing position. **Regardless of the apparent size of the rear sight opening, you must align the front sight in the center of the opening.**

It is important to keep your eye the same distance from the peep sight in any particular firing position. To ensure this distance is always the same, you must hold the rifle in the same exact location for each shot. This location is commonly called the *spot weld or anchor.* There are several tricks shooters use to help maintain this distance. For example, you can place a small piece of tape on the stock of the rifle where it touches the cheek. In this way, you can feel whether your cheek has the proper *eye* relief.

FRONT SIGHT.—The front sight always appears to be the same size. However, depending on the distance your eye is from the rear sight, more or less of the front sight may be visible in the sight picture. The front sight, not the target, is the point of focus for the eye. As such, the front sight will be sharp and distinct in outline. For this reason, keep the front sight square, level, and blackened.

AIMING POINT.—The aiming point is the point on the target where the sights of the weapon are brought to bear. The correct aiming point is at 6 o'clock; that is, the bottom of the bull's-eye of a type "A" target (fig. 11-22) or the silhouette of a type "D" target (fig. 11-23). Any location on the target face is always given relative to a similar position on a clock face regardless of the target shape. Therefore, a vertical line in the exact center of the target would be described as running from 12 o'clock (top) to 6 o'clock (bottom).

SIGHT PICTURE.—To obtain the correct sight picture, you need to align the rear sight, the front sight, and the bull's-eye (figs. 11-22 and 11-23). Each of these three elements affects the sight picture. As you can see from figure 11-24, any error in sight alignment will increase as the range increases. An error in the aiming point remains constant as the range increases.

At close ranges, the bull's-eye or silhouette appears larger in relation to the front sight than it will at longer

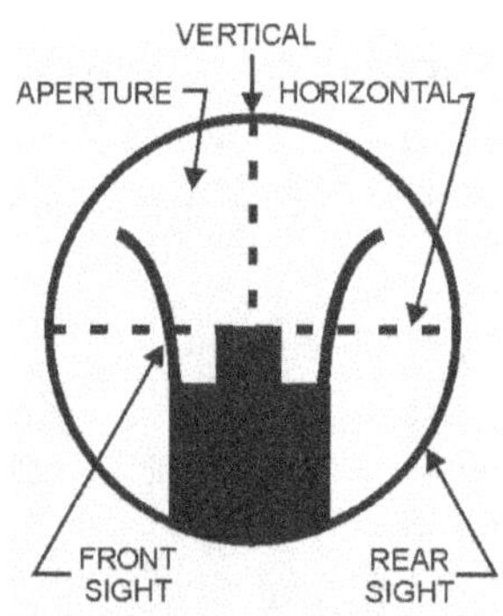

Figure 11-21.—Proper sight alignment.

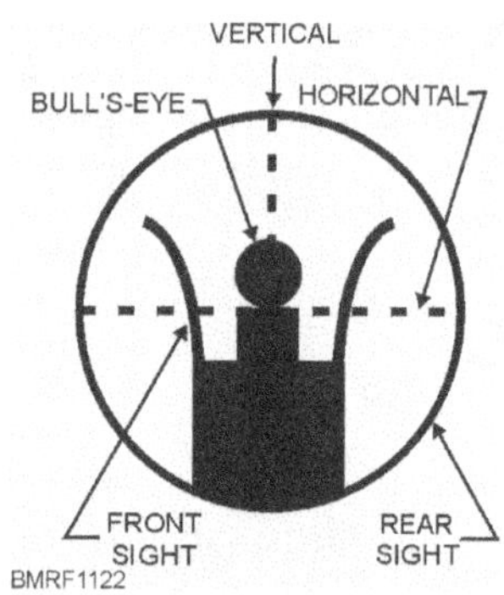

Figure 11-22.—6 o'clock sight picture held on "A" target at a range of 200 yards.

Student Notes:

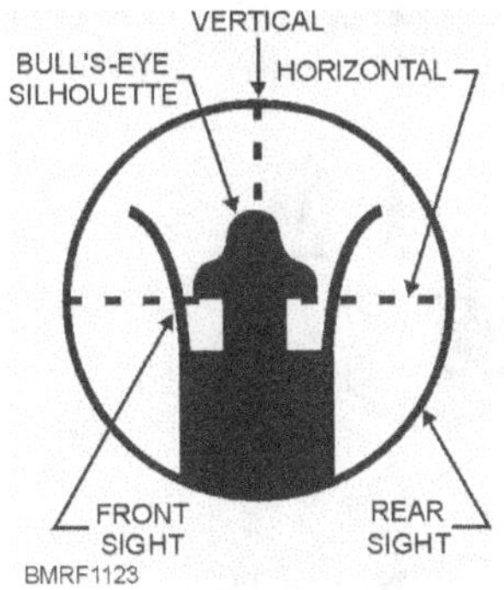

Figure 11-23.—6 o'clock sight picture held on a "D" target at a range of 200 yards.

ranges. This means that the sight picture will vary not only from one firing position to another but also from one firing line to another (fig. 11-25).

TRAINING.—You will receive training in aiming along with the position and trigger squeeze before actually firing on the rifle range. You do this by aiming at a series of small bull's-eyes at least 20 feet away on a *dry firing range*; this training is known as *snapping in.*

BLACKENING SIGHTS.—You should blacken the sights during sighting and aiming exercises to help eliminate light reflection or glare. Blacken all sights, both front and rear, on the base of the receiver and the top of the barrel. Usually, sights are blackened by using a smudge pot, carbide lamp, oily patch, candle, cigarette lighter, or ordinary match. Be sure to remove all oil from the sight before blackening it.

Shooting Positions

For the best results in rifle shooting, you need to shoot in the correct shooting position. The better the position, the easier it is to hold the rifle and squeeze the trigger while the sights are properly aligned on the target. However, shooting position **won't** compensate for lack of practice. You may have difficulty in assuming a correct position until sufficient practice has limbered up your muscles. Once your muscles are limber, you will find the positions both comfortable and steady.

A standard qualification course requires you to learn and use three standard positions while shooting—standing, kneeling, and sitting. Experience has proved that these positions produce excellent results with men and women of all physical types.

Once you master the correct positions, you must combine sighting and aiming with your practice. Learn to get into the correct position and align the sights without moving the rifle. If the target isn't properly aligned with the sights, you must move your body

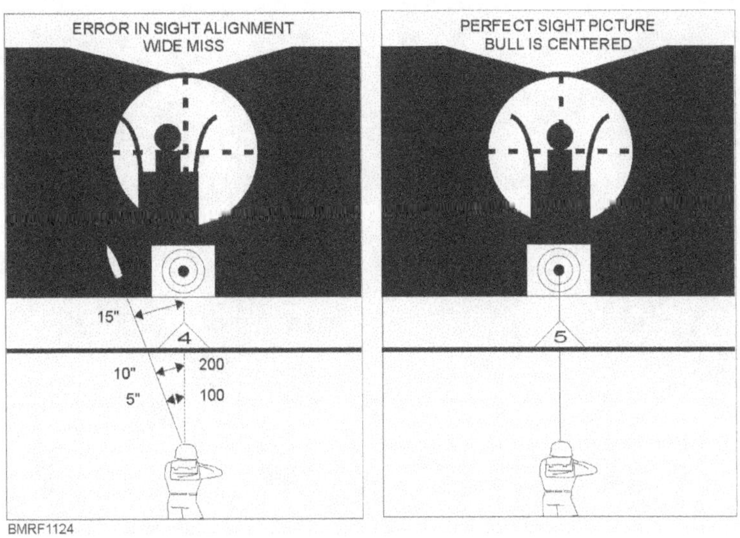

Figure 11-24.—Error in sight alignment increases as range increases.

Student Notes:

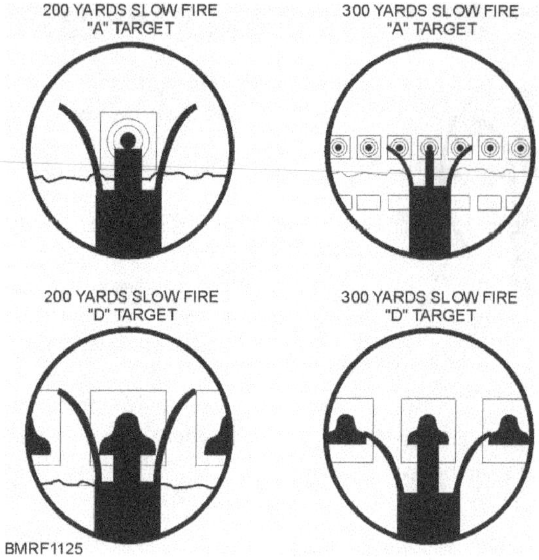

Figure 11-25.—Variation in sight picture for each range of fire.

instead of the rifle until you obtain the proper sight picture.

STANDING (OFF-HAND) POSITION.—The standing position (fig. 11-26) is used to engage surprise targets that appear at close ranges. Normally, you use this position when engaging targets less than 100 yards in range and when you are constantly firing and moving.

Figure 11-26.—Standing position.

Figure 11-27.—Kneeling position.

Student Notes:

KNEELING POSITION.—The kneeling position (fig. 11-27) is a natural position that can be assumed quickly. It is suitable for use on level ground or on ground that slopes upward.

SITTING POSITION.—There are three variations of the sitting position:

1. Open leg
2. Cross leg
3. Cross ankle

The position used depends entirely on the shooter. The open-leg position (fig. 11-28) is especially suited

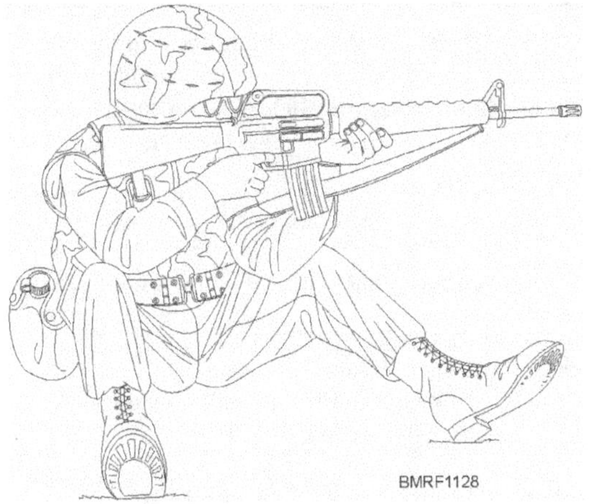

Figure 11-28.—Open-leg position.

Figure 11-29.—Cross-leg position.

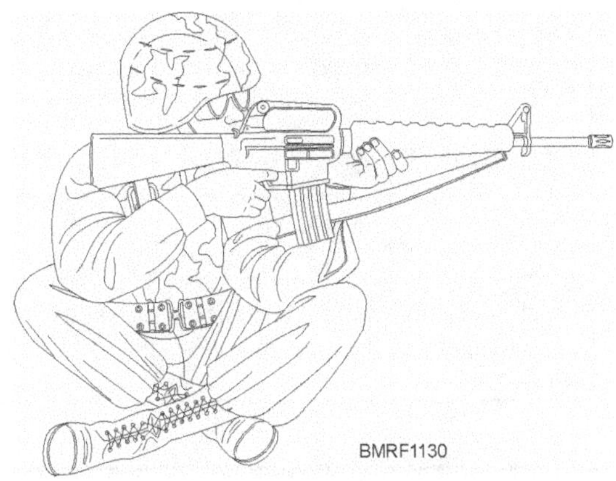

Figure 11-30.—Cross-ankle position.

for use on ground that slopes downward. The other two alternate sitting positions are the cross-leg position (fig. 11-29) and the cross-ankle position (fig. 11-30).

Trigger Control

The most important single factor in marksmanship is trigger control. Everything about your position and aim may be perfect; but if you do not squeeze the trigger properly, your shot will not go where you aimed it.

The key to trigger control is that the trigger must be squeezed smoothly, gradually, and evenly straight to the rear. Any sideward pressure, however slight, applied to the trigger during its rearward movement will likely result in a wide shot. Similarly, upward or downward pressure on the trigger will result in high or low shots. Trigger control can be done as shown in the chart on the following page:

When you fire from the standing position, coordinating the trigger squeeze and proper aim is critical. You must start and continue the squeeze **only** when the front sight is momentarily at rest or is slowly moving in the smallest area of the bull's-eye. Inexperienced shooters usually tend to *snap shoot* in this position; that is, they attempt to complete the trigger action instantly as the front sight moves across the aiming point. This invariably results in jerking the rifle and producing a wild shot.

Student Notes:

Trigger hand	Grasp the stock or pistol grip firmly, but without strain, so the trigger finger has the proper support to overcome trigger weight. An unnatural, straining grasp causes excessive muscle tension in the hand, which results in a tremor that is transmitted to the weapon.
Trigger (index) finger	Make contact with the trigger where the contact produces a movement straight to the rear (usually between the first joint and the tip). (**NOTE**: The trigger finger must not touch the receiver or rifle.)
Line up the sights and apply pressure on the trigger	Gradually increase the pressure until the hammer releases and the shot fires. (**NOTE**: If, during this process, the sights drift off the target, interrupt the trigger squeeze but maintain the pressure. When the sight picture is correct, continue the squeeze until you fire the shot.)

B	**BREATH**	Proper breathing is essential. It helps you relax, steadies your aim, and clears your vision. First, take a normal breath; then release part of it (enough to be comfortable); and hold the remainder. **Do not hold your breath for more than 10 seconds before shooting**. This may tense your muscles and blur your vision. If you do not shoot during this breathing period, take another normal breath and repeat the procedure.
R	**RELAX**	You must relax. The more relaxed you are, the better your shot will be.
A	**AIM**	Concentrate on the proper sight alignment of the correct sight picture. Focus your eye on the front sight post (blade).
S	**SLACK**	Some rifles have a certain amount of slack in the trigger. Take up this slack before starting your squeeze to the rear to fire. The M16 trigger slack is insignificant, and this step is generally omitted when firing that weapon. Knowing your weapon is important.
S	**SQUEEZE**	Squeeze the trigger as previously described. If you squeeze it properly, you will not know when the round will fire. This will prevent flinching, caused by anticipation of the shock, or recoil, from the exploding cartridge.

Squeezing the trigger correctly is not as easy as it may appear; the technique must be fully mastered. To help you remember the correct technique, the acronym BRASS was developed (see chart above).

FIRING TECHNIQUES—PISTOL

Good pistol shooting, like rifle shooting, depends on your ability to master and apply certain basic marksmanship skills. You must practice these skills—aiming, position, and trigger squeeze—often. Apply these fundamentals of marksmanship! If your life ever depends on how well and accurately you shoot the pistol, you might walk away alive.

Aiming the Pistol

Aiming the pistol consists of combining proper sight alignment with the correct aiming point to obtain a correct sight picture.

SIGHT ALIGNMENT.—Sight alignment is best defined as placing the front and rear sights into correct alignment with the eye. The top of the front sight is level with the top of the rear sight, and the body of the front sight is centered between the rear sight aperture (fig. 11-31). Correct sight alignment is essential for accuracy because of the short sight radius (about 6 1/2 inches). For example, if a 1/10-inch error is made in aligning the

Student Notes:

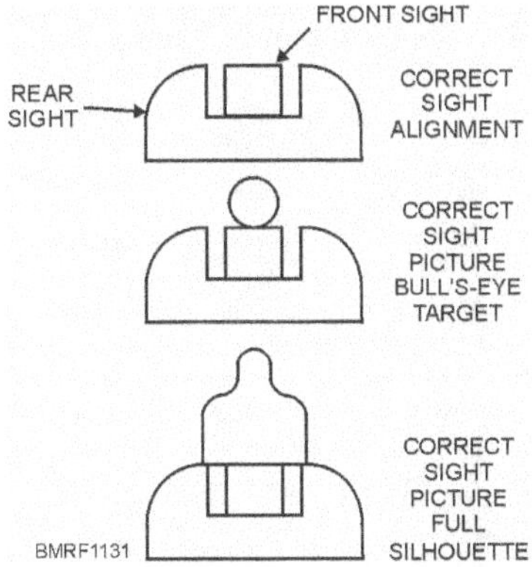

Figure 11-31.—Correct sight alignment and sight picture.

front sight in the rear sight, the bullet will miss the point of aim by almost 15 inches at 25 yards of range.

AIMING POINT.—The correct aiming point, when you fire at a bull's-eye target at 25 yards, is a 6 o'clock sight picture. At 15 yards, bring the aiming point well up into the black. When you fire at an "E" type of silhouette target, the aiming point is in the center of the target.

CORRECT SIGHT PICTURE.—A sight picture is the pattern of the pistol sights in relation to the target as you aim the pistol. A correct sight picture combines correct sight alignment and correct aiming point (fig. 11-31). When you are aiming, your eye cannot focus simultaneously on three objects (rear sight, front sight, and bull's-eye) at different ranges. Therefore, the last focus of the eye should always be on the front sight. You will see the front and rear sights sharp and clear, but the bull's-eye will appear to be a bit hazy.

NOTE

If sight alignment is correct, the bullet will strike the bull's-eye even if the sight picture is partially off center but still touching the bull's-eye.

Since it is physically impossible to hold the weapon perfectly still, you must learn to apply trigger squeeze and to maintain correct sight alignment while the weapon is moving around the bull's-eye. This movement is referred to as the *wobble area*. You must accept this wobble area, or movement, trying to keep it to a minimum.

Position

To position yourself properly for firing the pistol, you need to know how to grip the pistol correctly and how to position your body in relation to the target. Only the standing position will be covered in this section because it is the one used in qualification. However, the pistol can also be fired accurately from the kneeling, standing, and sitting positions. The pistol may be gripped with either a one-hand grip or a two-hand grip.

STANDING POSITION ONE-HAND GRIP.—To assume the standing position using the one-hand grip (fig. 11-32), face the target squarely and then execute an exaggerated half-left face (about 50°). Spread your feet about shoulder width apart until you're standing comfortably. Your legs should be straight, but not stiff, and your hips should be level. Extend the index finger of your shooting hand and point it at the target, forming a V with the thumb and forefinger. Adjust your stance until your finger points naturally, without muscle tension, at the center of the target. Pick the pistol up with your other hand and place it in the V of your shooting hand.

Figure 11-32.—Standing position (pistol).

Student Notes:

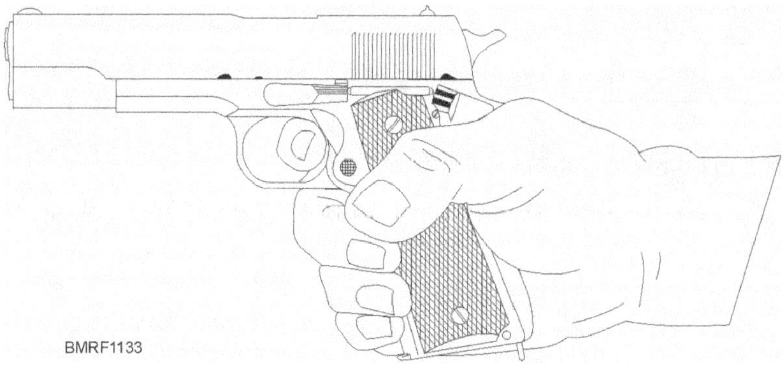

Figure 11-33.—Gripping the pistol (one hand).

As you read this paragraph, refer to figure 11-33. Grip the receiver firmly with the hand and fingers. Wrap the three lower fingers around the receiver (grip), and place the trigger finger on the trigger between the tip and second joint so that the trigger can be squeezed to the rear. Hold your thumb up and along the side of the pistol with enough pressure to steady the pistol and to equalize any pressure from the other side by the palm and forefinger. Once you have a firm grip, maintaining the same degree of firmness throughout firing is important. A change in your grip will change the location of the shot group on the target. A tight grip tends to cause the bullet to strike low and a loose grip to strike high.

With a proper grip on the pistol, the muscles of your arm should be firm, but not rigid. Your arm should be straight with your wrist and elbow locked. This will prevent excessive up-and-down movement of the weapon. When the weapon is fired, the recoil will be absorbed through the arm to the shoulder. If you are in the correct position, the pistol will return to approximately the same sight picture after each shot.

STANDING POSITION TWO-HAND GRIP.—In this position (fig. 11-34), you face the target squarely with your feet placed comfortably about shoulder width apart. Keep your legs straight without stiffness and your hips level and slightly forward. Relax the muscles of your diaphragm, and make no effort to hold in your abdomen.

Grasp the pistol in the same manner as if you were firing one-handed. Place the nonfiring hand under the firing hand, wrapping the nonfiring fingers around the back of the firing hand. Place the thumb over the middle finger of the firing hand. Lock the weapon firmly in both hands (fig. 11-35). Bring the weapon onto the center of the target by shifting your feet.

Figure 11-34.—Standing position (pistol), two-handed grip.

MISCELLANEOUS.—In both the one-hand and two-hand positions, position your head so that you are

Student Notes:

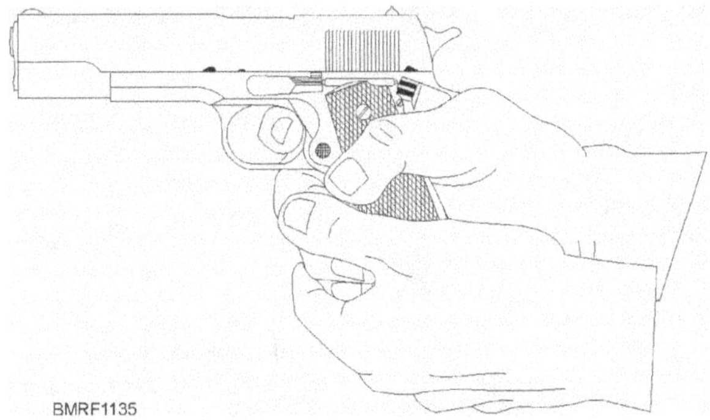

Figure 11-35.—Two-hand grip.

looking straight out through your shooting eye. Keep your shooting arm fully extended.

In the one-handed position, the shoulder of your shooting hand should be slightly raised. Turn your head in order to see the target through the sights. The ease with which your head can be turned is another determining factor in how far you must turn to the right or left. There should be no strain on the neck muscles with your head held upright. The whole position, with the exception of your shooting arm, is one that can be maintained with the least muscular effort. Your body is balanced, rather than held in position. The muscles of your shooting arm and shoulder should be tightened somewhat to sustain the weight of the pistol and to maintain a correct grip. Excessive tightening of the muscles of your shooting arm and hand should be avoided. The tension in the muscles of your shooting arm and hand should be maintained after the hammer falls. This will assist in getting off your second shot quicker.

Because of the differences in the body structure of individuals, the standing position will vary slightly. However, regardless of your size, your position should be relaxed and comfortable. The pistol should point at the center of the target or you will be tense while firing. If you are tense, there will be excessive muzzle movement.

Trigger Squeeze

Poor shooting is most often caused by disturbing your aim as the bullet is leaving the barrel. This is usually the result of jerking the trigger or flinching. The trigger does not have to be jerked violently to spoil your aim; even a slight off-center pressure of your finger while squeezing the trigger is enough to move the strike of the bullet several inches.

- Flinching is a subconscious reflex caused by anticipating the recoil from firing.
- Jerking results from attempting to fire the pistol at the precise time that you align the sights with the target.
- Heeling causes the bullet to strike the target high and to the right.

You can correct all these shooting errors by understanding and using the correct trigger squeeze. Both flinching and jerking will cause the bullet to strike the lower left section of the target. An attempt to correct flinching and jerking by tightening the large muscle in the heel of the hand may cause heeling. An improper trigger squeeze will cause more misses on the target than any other single step of preparatory marksmanship training.

You obtain a correct trigger squeeze by applying a uniformly increasing pressure on the trigger straight to the rear without disturbing the sight alignment until the pistol fires. The trigger slack, or free play, is taken up first, and the correct squeeze continues steadily until the hammer falls. If the trigger is squeezed properly, you will not know when the hammer will fall. This is the best way to prevent jerking, flinching, and heeling.

To help you squeeze the pistol trigger properly, use the acronym BRASS as you did with the rifle. You must also learn to call your shots. If you cannot call your shots

Student Notes:

correctly, you are not concentrating properly on sight alignment and trigger squeeze.

REVIEW 5 QUESTIONS

Q1. When you have properly aligned your sights, what is the relationship between the front sight and the rear sight?

Q2. The distance between your eye and the rear sight is referred to as the ____________.

Q3. Why is it sometimes necessary to blacken sights during sighting and aiming exercises?

Q4. List the three standard positions for shooting.

a.

b.

c

Q5. List the three sitting positions for shooting.

a.

b.

c.

Q6. What is the single most important factor in good marksmanship?

Q7. What does the acronym BRASS mean?

Q8. When sighting with a pistol, how should the top of the front sight align with the rear sight?

SUMMARY

In this chapter, you have learned about various small arms used by the Navy. You may never have the occasion to use small arms; then again you may use them every day in your routine aboard a ship. It is imperative that when using small arms, you ALWAYS abide by all prescribed safety precautions. The use of safety equipment is also essential. Respect small arms, but don't be afraid of them. They can be your best source of personal defense in times of trouble if handled properly. All the small arms used by the U.S. Navy are manufactured to have a high degree of reliability if they are cared for and maintained properly. Proper maintenance could mean the difference between your small arm working properly or jamming at critical moments.

Take advantage of every opportunity you may have of going to a rifle or pistol range. The more familiar you become with small arms, the better you will be able to handle them safely. The more you shoot, the better you will become.

REVIEW 1 ANSWERS

A1. When handling a firearm, you should never **carry the weapon by the muzzle**.

A2. The protective gear you should wear when firing a weapon includes—

a. **hearing protection and**

b. **safety goggles**.

Student Notes:

REVIEW 2 ANSWERS

A1. The M-14 rifle is chambered to fire the **7.62-mm NATO cartridge**.

A2. Fully loaded, the M-14 weighs **11 pounds**.

A3. Identify the following rifle components as either M16A1 or M16A2:

Component	Rifle
a. Has a lighter barrel because of the ammunition fired	**M16A1**
b. Rear sights are adjusted by means of a windage knob and elevation knob	**M16A2**
c. Has a semiautomatic or burst feature	**M16A2**
d. Fires a 5.56-mm round and a 62-grain round	**M16A1** **M16A2**

A4. The first step in cleaning the M16 rifle is to **point the selector lever toward SAFE**.

A5. The M16 rifle carries **20 or 30 rounds** of ammunition.

A6. When loading the M16 rifle, you shouldn't ride the charging handle forward because the **bolt may fail to lock.**

A7. When the selector is set for burst, the M16A2 fires **three rounds.**

A8. The conditions that must exist to consider the M16 rifle clear include—

a. **No case or rounds is/are in the chamber**

b. **The magazine is out**

c. **The bolt carrier is to the rear**

d. **The selector lever is on the SAFE position**

A9. The cleaning materials for the M16 are stored **in the rifle stock.**

REVIEW 3 ANSWERS

A1. A .45-caliber pistol magazine will hold **seven cartridges.**

A2. The .45-caliber pistol was designed to be used as **an individual weapon for close use**.

A3. The three safety features of the 9mm service pistol are—

a. **Ambidextrous safety**

b. **Firing pin block**

c. **Half cock notch**

A4. When loading your pistol, you should make sure the safety is in the **down position**.

REVIEW 4 ANSWER

A1. The shotgun has the advantage of being able to **expel anywhere from nine to hundreds of projectiles covering a wide area.**

REVIEW 5 ANSWERS

A1. When you have properly aligned your sights, **the top of the front sight is level with the horizontal center line of the rear sight, and the body of the front sight is centered between the rear sight aperture.**

A2. The distance between your eye and the rear sight is referred to as ***eye relief***.

A3. At times, it's necessary to blacken sights during sighting and aiming exercises **to reduce reflection and glare.**

A4. The three standard positions for shooting are—

a. **Standing**

b. **Kneeling**

c. **Sitting**

A5. The three sitting positions for shooting are—

a. **Open leg**

b. **Cross leg**

c. **Cross ankle**

A6. The single most important factor in good marksmanship is **trigger control**.

A7. The acronym BRASS means **breath, relax, aim, slack, squeeze**.

A8. When sighting with a pistol, **the front sight is level with the top of the rear sight**.

CHAPTER 12

DAMAGE CONTROL

...the events of 14-15 April 1988 have proven that solid damage control, good training, and sound leadership based on experience can save a ship that is on fire and sinking, to fight another day.

—Paul X. Rinn, CAPT
CO, USS *Samuel B. Roberts* (FFG-58)

Damage control is the responsibility of ALL HANDS, from the commanding officer to the newest recruit. Without proper damage control training, your shipmates may be injured or killed, or perhaps your ship may sink. During World War II, many ships were saved because of the positive damage control practices.

For instance, the German battleship *Bismarck* was kept afloat and remained in a fighting status for an extended period in spite of the British pounding it with heavy gunfire. You can see that survivability through positive damage control is not only important for the United States, but on an international level.

Effective damage control (DC) requires the correct use of equipment and techniques to prevent or minimize the damage effects caused by battle, fire, collision, grounding, explosion, and so forth. DC also includes defensive measures used to reduce the effects of weapons of mass destruction, such as chemical, biological, and radiological (CBR) warfare. Remember, our Navy gives you the best damage control training in the world. As a ship's crew member, it's your responsibility to properly safeguard yourself and your ship. You learn how to do this through training.

You may feel the information in this chapter doesn't apply to you because of your chosen occupational field. Remember that as a Sailor, no matter what your rating, you may serve aboard ship at any point in your career. Therefore, as a crew member, you must know your damage control responsibility. For more in-depth information about the administration and organization of damage control, you should refer to the following manuals:

- *Surface Ship Survivability,* Naval Warfare Publication (NWP) 3-20.31
- *Naval Ship's Technical* Manual (*NSTM*), chapter 079, vol. 2, "Practical Damage Control"
- *Naval Ship's Technical* Manual (*NSTM*), chapter 555, vol. 1, "Surface Ship Firefighter"

Our Navy has a complete organization in place relating to damage control. This organization consists of the following two main parts:

1. The administrative organization, and
2. The battle organization.

ADMINISTRATIVE ORGANIZATION

Learning Objectives: When you finish this chapter, you will be able to—

- Identify the chain of command within the damage control administrative organization.
- Recognize the functional purpose of each part of the organization and the interrelationships between the parts.

The damage control administrative organizational chain of command is shown in figure 12-1. As you can see, the responsibility for damage control begins with the commanding officer (CO) of a naval ship and runs to the most junior Sailor in the DC division and damage control petty officer community.

Look at figure 12-1. As you can see at the level of the ship's damage control assistant (DCA), other personnel report to the DCA. Each person has a different administrative responsibility and tasking. Each person reports back up the chain via the DCA.

PURPOSE OF ADMINISTRATIVE ORGANIZATION

The administrative organization requires the efforts of all hands to establish and maintain material readiness conditions. Material readiness is

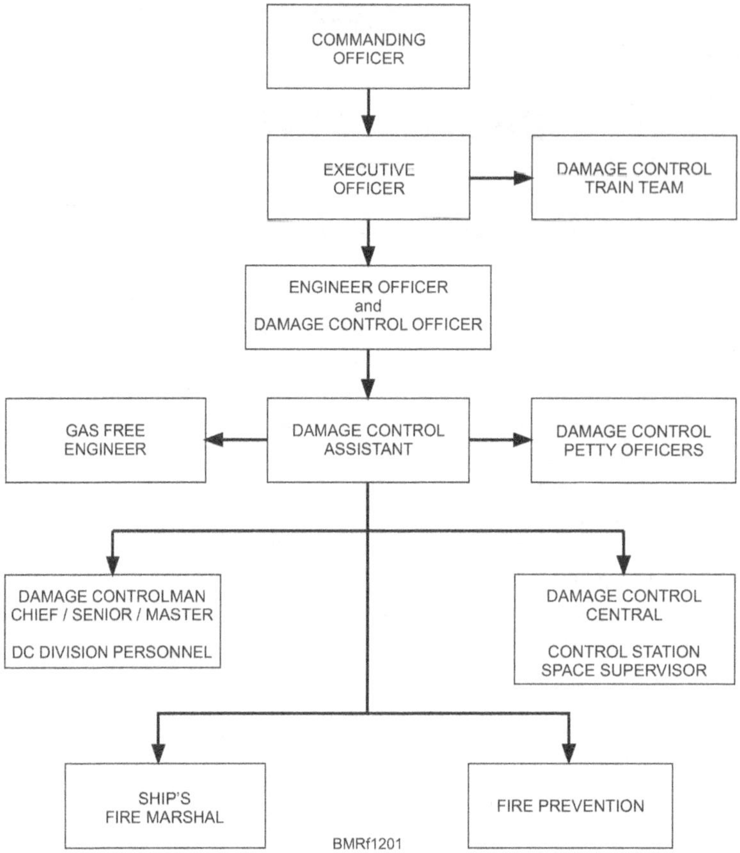

Figure 12-1.—Administration organization chain of command.

accomplished when DC fittings and equipment aboard ship are available and in proper working order to combat any emergency. Aboard ship, there is a damage control petty officer (DCPO) for each department or division. The DCPO makes sure that all emergency equipment and fittings are maintained and work properly. The DCPO also maintains the compartment checkoff list.

Each ship has a specified number of damage control lockers with a repair party assigned. Repair parties and repair locker personnel, along with the DCPOs, receive periodic training so they can perform their assigned duties. The engineer officer is responsible for maintaining properly trained DCPOs, repair parties, and repair locker personnel. Under the engineer officer, the DCA coordinates the efforts of the repair parties to control damage. Also, the DCA oversees all DC training. When properly maintained, the administrative organization reduces and confines any initial damage.

The survivability of a ship depends on the level of preparedness of its personnel. The condition of the equipment, shipboard systems, and the amount of training are factors that affect the ship's survivability. Ship's bills, records and schedules for maintenance, written doctrine, and procedures relating to damage control are all part of the administrative organization of damage control.

For the list of responsibilities for various ship's personnel, refer to OPNAVINST 3120.32, or ask your LPO or LCPO.

Student Notes:

COMMANDING OFFICER

United States Navy Regulations states the various broad responsibilities of the commanding officer (CO). These regulations require the commanding officer to maintain his/her command in a state of maximum effectiveness for war. They also require that he/she repair as much damage as possible immediately after a battle or action.

To carry out this charge, the CO must ensure that the ship's officers and crew are trained and continually exercised in all aspects of survivability. The CO should be fully aware of the adequacy and operability of all survivability systems and equipment. Shortages and deficiencies must not be tolerated, and they should be immediately rectified (fixed) or reported as casualties in the ready reporting system.

EXECUTIVE OFFICER

The executive officer (XO) keeps the command informed of the ship's survivability readiness. The XO carries out the requirements regarding the ship's survivability training, the readiness to manage casualties, and the readiness to control and recover from damage. As the team leader, the XO is the senior person assigned to the damage control training team (DCTT). The DCTT is responsible for all shipboard drills and exercises in the area of damage control.

DAMAGE CONTROL TRAINING TEAM

Each ship has a DCTT. The DCTT trains the ship's DC organization in ship equipment, ship systems, and procedures and techniques relating to the ship's survivability. The DCTT is organized to train all types of DC-related exercises.

The DCTT evaluates and critiques the DC team's abilities as well as their own. The team is made up of a variety of ratings, which provides the ship with a wide range of experience. All team members are qualified, as a minimum, to at least the level of personnel they are training and/or evaluating.

ENGINEER OFFICER

The engineer officer is designated as the ship's damage control officer. The ship's engineer officer fills the billet as the department head of the engineering department, and the DC division is one of the divisions in that department. The engineer officer has the following responsibilities:

- The operation, care, and maintenance of the main propulsion plant, auxiliary machinery, and piping systems
- The control of damage
- The operation and maintenance of electric power and distribution systems
- Repairs to the hull
- Repairs to material and equipment in other departments that require the attention of engineering department

Also the damage control officer (DCO), the engineer officer reviews all administrative items drafted by the DCA and DCA personnel. This is a primary administrative duty that occurs before items are routed to the CO for approval.

DAMAGE CONTROL ASSISTANT (DCA)

The DCA is the primary assistant to the damage control officer in the areas of damage control; firefighting; and chemical, biological, and radiological defense. For the purpose of administrative organization, the DCA is in charge of many programs relating to damage control on the ship and monitors the effectiveness of other programs.

The DCA is responsible for the overall administration and training of the ship's DC organization. Under the guidance of the DCA, individual departments and divisions are responsible for administration and training as related to damage control for their personnel. All ships afloat must meet the minimum prescribed formal training requirements to include training for the following personnel:

Student Notes:

- All DCTT members
- In-port emergency teams (IETs)
- Rescue and assistance (R&A) detail
- Repair party personnel
- Damage control petty officer (DCPO) organization
- All hands

Shipboard training by the DCTT is scheduled, executed, and documented to meet the type commander's (TYCOM's) requirements. The DCA maintains records of all DC personnel qualification standards (PQS) accomplishment for all hands.

All ship's instructions relating to casualty prevention and response are either drafted by or drafted for and reviewed by the DCA. This review is made before seeking the approval and signature of the ship's CO. The DCA is the resident expert for the engineer officer and the command on instructions relating to casualty prevention and response. The DCA administers the following billets and programs: DCPO, gas free engineer (GFE), fire marshal and fire prevention, and the operation of damage control central (DCC) and supporting watches.

DAMAGE CONTROL DIVISION

The damage control division is responsible for the damage control systems and equipment on board ship. The DCA is the senior member (division officer) of the DC division. As such, the DCA oversees all the paper work and the scheduling for maintenance of these systems and equipment.

Damage Control Petty Officer (DCPO)

A petty officer, who has completed the PQS, is designated as the damage control petty officer (DCPO) based on TYCOM instructions. Each work center has a DCPO. Division officers nominate DCPOs for endorsement by the chain of command. The XO gives the final approval for the nomination, replacement, and rotation of all DCPOs. DCPOs normally serve for a period of 6 months; they check with the fire marshal and DCA when first assigned to or relieved from this duty. The DCPO is responsible for—

- Acquainting themselves with all phases of the ship's damage control, firefighting, and defense procedures
- Assisting with the instruction of division personnel in damage control
- Firefighting, egress, and CBR procedures
- Ensuring the preparation and maintenance of damage control checkoff lists for all spaces under their cognizance
- Supervising the setting of specified damage control material conditions within division spaces and making all required reports
- Weighing portable CO_2 bottles, inspecting and testing damage control and firefighting equipment, and preparing all required reports for approval of the division officer based on current ship's instructions and planned maintenance system (PMS) requirements
- Ensuring all battle lanterns, dog wrenches, spanners, and other damage control equipment are in place and in a usable condition in all division spaces
- Ensuring all compartments, piping, cables, and damage control equipment are properly stenciled or identified by color codes based on *NSTM*, chapter 079, volume 2, and *General Specifications for Ships of the United States Navy*, NAVSEA S9QQO-AA-SPN-010/Gen-Spec, chapters 505, 507, and 602
- Ensuring the posting of safety precautions and operating instructions in required division spaces
- Assisting the division officer in inspection of division spaces for cleanliness and preservation and assisting in the preparation of required reports

Student Notes:

- Conducting daily inspections of division spaces for the elimination of fire hazards
- Performing any other actions in regard to damage control and maintenance of spaces that are directed by the division leading petty officer, division officer, fire marshall, DCA, and executive officer

Gas Free Engineer (GFE)

The gas free engineer (chief petty officer or above) decides when it's safe for personnel to enter closed or poorly ventilated spaces. The GFE makes sure that all necessary measures have been taken to eliminate the risk of fire, explosion, exposure to toxic substances, suffocation, or asphyxiation. The GFE has a working knowledge of all definitions, instruments, and equipment listed in the GFE program. The shipboard GFE meets the qualifications and is capable of performing the duties and responsibilities specified in *NSTM*, chapter 074, volume 3, "Gas Free Engineering." All Navy ships must be adequately staffed with personnel qualified to perform gas free engineering services for normal and emergency situations.

Fire Marshal

All ships designate a fire marshal to assist the engineer officer. The fire marshal helps the DCA train personnel to prevent and fight fires. The ship's fire marshal and duty fire marshals are qualified to ensure continuity of inspections for improperly stored or inoperative DC equipment. Fire marshals are free of duties that would not let them instantly respond to emergencies. The fire marshal conducts daily inspections throughout the ship, paying particular attention to the following areas:

- Housekeeping
- Firefighting equipment, both portable and fixed
- Safety precautions
- Flammable liquids
- Smoking
- Welding and burning
- Stowage of compressed gas
- Fume/spraytight light covers
- Grease filters/traps, Gaylord hoods
- Ventilation/air conditioning ducts/flammable stowage ventilation
- Material condition

The fire marshal has the following responsibilities:

- Submitting reports citing fire hazards and making recommendations for correction

NOTE

Immediate steps should be taken to eliminate hazards resulting from poor housekeeping, welding or burning, smoking, and equipment deficiencies. The fire marshal submits discrepancy reports to the DCA with copies to the XO and the appropriate department head(s) and conducts follow-up inspections to ensure compliance.

- Conducting training for fire teams, rescue and assistance teams, and departmental/divisional DCPOs stressing fire hazard consciousness
- Setting up a fire watch team before regular/interim overhauls or availabilities
- Training and assigning fire watches
- Being overall in charge at the fire scene until relieved by a general quarters (GQ) scene leader and keeping DCC and/or the OOD informed with an accurate status of the situation

Student Notes:

REVIEW 1 QUESTIONS

Q1. The damage control organization is divided into what two main parts?

a.

b.

Q2. What is the purpose of the administrative phase of DC?

Q3. Who is responsible to maintain properly trained DCPOs, repair parties, and repair locker personnel?

Q4. Who is the senior person assigned as team leader on DCTT?

Q5. What person is responsible for making sure that all battle lanterns, dog wrenches, spanners, and other damage control equipment are in place and in a usable condition within their division?

Q6. Before anyone can enter a poorly ventilated space or void, who must certify the space as safe?

BATTLE ORGANIZATION

Learning Objectives: When you finish this chapter, you will be able to—

- Recognize the battle organization of damage control parties.
- Recall the location and contents of shipboard damage control lockers.

The battle phase starts when the ship has received actual damage. The DCA coordinates the efforts of the repair parties from damage control central (DCC). These efforts may include fighting fires, flooding, controlling the ship's stability, and repairing battle damage. Repair party personnel also use CBR defense measures (CBR is covered in chapter 13).

PURPOSE OF THE BATTLE ORGANIZATION

Once the ship has been damaged, the ship's damage control battle organization is responsible for restoring the ship to as near normal operation as possible. The organization varies somewhat from one ship to another, depending on the size, type, and mission of the ship. However, the same basic principles apply to all battle organizations.

The DCA is responsible, under the engineer officer, for the ship's survivability systems. The DCA's responsibilities include control of damage; control of stability, list, and trim; fighting fires; restoration from damage; medical casualty response; and CBR countermeasures. The battle station for the DCA is DCC. The primary damage control battle organization units are repair parties or teams. Battle dressing stations (satellite medical stations) should be located near the repair parties.

DAMAGE CONTROL CENTRAL/CENTRAL CONTROL STATION

The primary purpose of damage control central (DCC) is to collect and compare reports from various repair stations to determine the ship's condition and the corrective action to be taken. DCC is the nerve center and directing force of the entire damage control organization. Personnel from various shipboard divisions man DCC. On newer class ships, the central point for reporting is the central control station (CCS). The CCS has the added capability of being able to control the closing of fire zone (FZ) doors and certain valves electronically. Also, CCS can remotely activate fire-fighting systems.

Student Notes:

Reports from the repair parties are carefully checked in DCC/CCS. This is done so that immediate action can be taken to isolate damaged systems and to make effective emergency repairs. As reports are received, graphic records of the damage are made on damage control diagrams and status boards under the direction of the DCA. For example, reports on flooding are used to update the status boards showing the liquid distribution (fuel and water) before the damage occurred. With this information, the stability and buoyancy of the ship can be estimated and necessary corrective measures taken.

In the unlikely event DCC/CCS is destroyed or is unable to retain control, a plan is in place to designate repair stations to take over the responsibilities of damage control central.

REPAIR PARTIES

The following chart shows a variety of repair lockers that are found on an aircraft carrier. All ships have a minimum of repair lockers 2, 3, and 5. Each locker has an officer or senior petty officer in charge. The makeup of each repair party in these lockers depends on the type of ship and the area they serve.

Repair Locker	Location or Function
Repair 1	Main deck repair
Repair 2	Forward repair
Repair 3	After repair
Repair 4	Amidships repair
Repair 5	Propulsion repair
Repair 6	Ordnance
Repair 7	Gallery deck and island structure
Repair 8	Electronic casualty control

Repair Party Personnel

For a repair party to control battle damage effectively, PQS-qualified personnel are assigned to specific functions and duties. Some personnel may have more than one assignment, depending on the number and qualifications of personnel available. Repair party personnel must know their own area of responsibility along with the areas of other repair parties. The following functions are common to all repair parties:

- Make repairs to electrical and communication circuits and rig casualty power.
- Give first aid and transport injured personnel to battle dressing stations without seriously reducing the party's damage control capabilities.
- Detect, identify, and measure radiation dose and dose rate intensities.
- Decontaminate the affected areas of nuclear, biological, and chemical attacks.
- Identify, control, and extinguish all types of fires.
- Control and remove flooding water.
- Evaluate and correctly report the extent of damage in the repair parties' area of responsibility, to include maintaining—
  - — Deck plans showing location of CBR contamination, location of battle dressing stations, casualty collection and decontamination stations, and safe routes to them.
  - — Graphic display boards showing damage and action taken to correct disrupted or damaged systems, using standard DC symbology and plotting techniques.
- Make emergency repairs to various piping systems.
- Be familiar with all damage control fittings in the assigned area, such as watertight doors, hatches, scuttles, ventilation systems, and various valves.
- Control and clean up hazardous material spills.

Student Notes:

In-Port Emergency Team (IET)

In-port emergency teams are made up of duty section personnel. IET members are emergency team member qualified and stand ready to respond to any type of casualty.

At times, it may be necessary to provide additional personnel and material support to the IET. These additional personnel come from a pool of excess duty personnel and assemble at a designated location. They can assist in setting fire or flooding boundaries and are available to back up the primary IET, if needed.

Rescue and Assistance (In Port/At Sea)

All ships are required to have a rescue and assistance detail in port and at sea. The primary missions of this detail are to rescue personnel at sea, assist another unit in distress, and assist persons or activities in distress ashore. All personnel assigned to the rescue and assistance detail must, as a minimum, be qualified as an emergency team member and in first aid.

Rapid Response Team

When in port or under way, each ship has a designated rapid response team. This team proceeds directly to the scene when a fire or flooding is called away. The team attempts to quickly extinguish or contain the fire or minimize or stop a flooding. At a minimum, this team is made up of four personnel qualified as fire team members and the ship's fire marshal. Because immediate response is required, protective clothing or OBAs aren't donned. If they are unable to gain control of the casualty within a preset time frame, a full IET or GQ team relieves them and takes over.

General Quarters (GQ)

General quarters is an all hands evolution. It is the highest state of readiness of the ship, and all repair lockers are manned and fully equipped to combat casualties. When responding to GQ, all hands adjust their clothing to battle dress and proceed to their GQ stations.

Battle dress—Bottom of pants tucked into boots or socks, long sleeves pulled down and buttoned, top button on shirt buttoned, and flash hood and gloves donned. All exposed skin covered.

GQ route—Move forward in passageways and up ladders on starboard side, move aft in passageways and down ladders on port side. Since all hands will be moving at the same time, you have to move with the flow of traffic.

DAMAGE CONTROL LOCKERS

The equipment and materials required for making battle damage repairs vary according to the nature of the damage. Since many different kinds of damage can occur aboard ship, you must know how to use a variety of equipment and materials.

Checks should be made to see that all damage control equipment tools and materials on the allowance list are actually on board and in working order. Comparing the ship's allowance list with an accurate and up-to-date inventory of onboard damage control equipment does that.

Damage control equipment should be stowed or installed in its designated location and be readily accessible. Emergencies can be handled much more effectively if equipment is available than if you have to waste time looking for it.

Damage control equipment must not be used for any purpose other than damage control. Because damage control equipment is located throughout the ship, some people are tempted to use it merely because it is handy. That must not be allowed. It is important to make all hands realize their lives may literally depend on the ready availability of damage control equipment if an emergency should arise.

REVIEW 2 QUESTIONS

Q1. When does the battle phase of damage control start?

Student Notes:

Q2. Where is the battle station for the DCA?

Q3. During GQ, DCC receives reports and casualty status from—

Q4. What is the minimum requirement to be a member of the in-port emergency team?

Q5. What are the primary missions of the rescue and assistance team?

Q6. What is the ship's highest state of readiness?

COMMUNICATIONS

Learning Objectives: When you finish this chapter, you will be able to—

- Identify the communications devices used during damage control evolutions.
- Recognize the purpose of communications devices used during damage control evolutions.

Communications are vital to the damage control organization. Without good communications, the entire organization could break down and fail in its primary mission. The size and complexity of the surface ships govern the design of DC communications systems found aboard ship. Major shipboard DC communications systems are as follows:

NOTE

All the systems listed below are explained in detail in chapter 4 of this TRAMAN.

- Sound-powered telephones
- IVCS
- Ship's service telephones
- Announcing systems
- Intercommunications systems
- DC WIFCOM

EMERGENCY COMMUNICATIONS

The X40J is an emergency damage control communication system. It's employed in the event of primary, auxiliary, and supplementary communications circuit failure. The X40J is composed of both portable ("salt and pepper" line) and permanently installed (risers) components.

ALARMS

The general announcing system (1MC) is integrated with a system of alarm signals. The signals override the microphone control stations and are intended to notify the ship's crew of imminent danger. These alarms, in order of priority, are as follows:

1. Collision
2. Chemical attack
3. General
4. Flight crash

Collision Alarm

The OOD/conning station sounds this alarm signal when there is a possibility that the ship will run into a pier, run aground, or another waterborne unit will strike the ship. All hands should move away from the area of impact and brace for shock. After a collision, all hands set material condition ZEBRA and are prepared to control fires and flooding.

Chemical Attack Alarm

The chemical attack alarm signal is sounded by the OOD/conning station, DCC, and automatically by the chemical agent point detection system (CAPDS) on

Student Notes:

ships so equipped. This alarm is sounded when there has been a CBR attack on or in the vicinity of the ship. All hands must exercise protective measures and procedures to reduce exposure and personnel injuries.

General Alarm

The general alarm signal is sounded by the OOD to notify the crew of a battle condition (GQ). Immediately after the alarm is sounded, some ships pass the word, "General quarters, general quarters, all hands man their battle stations." All hands report to preassigned stations following the correct GQ traffic routes and set material condition ZEBRA.

Flight Crash Alarm

The flight crash alarm is sounded by the OOD or PreFly to notify ship's company of a pending or actual flight deck emergency.

MESSENGERS

Messengers are used to deliver written messages between repair lockers and DCC as a matter of record. If all methods of communications have failed, messengers are used to relay orders and information. Messengers deliver messages between repair lockers and DCC, between DCC and the bridge, or between other locations. Messengers should be familiar with the ship to get from one place to the other without delay. Messengers must be able to deliver oral messages without error.

REVIEW 3 QUESTIONS

Q1. The X40J circuit is commonly referred to as what kind of line?

Q2. List the names of the alarms that can override the microphone control of the 1MC.

a.

b.

c.

d.

Q3. When all communications systems fail, how are messages and orders relayed between repair lockers and DCC?

SHIP'S INTEGRITY THROUGH SUBDIVISION

Learning Objective: When you finish this chapter, you will be able to—

- Identify material conditions of readiness and recognize their purpose.

Naval ships are subdivided into many small watertight rooms or as we say *watertight compartments*. This breaking up process of a very large area is called the "compartmentation process." It limits the spread of toxic or noxious gases, fire and flooding, and other battle damage when it occurs. This system when combined with material conditions of readiness (discussed below) increases a ship's ability to survive.

Each space or compartment on a ship is accessed through a door to a new space or passageway connecting other spaces on the same deck or to the weather deck area (outside). The space may have a hatch or scuttle that allows transit from one deck to another via a ladder. In some spaces you may have many doors, hatches, and scuttles. These accesses are either watertight, airtight, fumetight, oiltight, or nontight by design.

Piping and ventilation systems work on the same concept as compartment access fittings. They have valves at strategic points that will isolate sections of a system, limiting the spread of damage to smaller areas. Because of all the fittings and closures that are on a ship, you might be asking yourself the question, "How will I know which ones to close or leave open and when to do it?" Those answers are covered in the paragraphs to follow.

Student Notes:

MATERIAL CONDITIONS OF READINESS

There are three material conditions of readiness. Those conditions are named XRAY, YOKE, and ZEBRA. When set, each condition affords the ship with a level of protection, with XRAY being the least and ZEBRA the most. Some fittings or closures on a ship may not be classified although the majority of them are classified. They are identified by a classification plate affixed to the closure or fitting, or sometimes the classification is painted on the bulkhead next to the item. Ship's instructions (and Navy instructions) specify what type of classification an item should hold, how it will be marked, and when that condition should be set. The commanding officer of the ship is authorized by instruction to alter this to a small extent if he/she desires but basically this system is standard Navywide.

XRAY. Provides the least watertight integrity and the greatest ease of access throughout the ship. It is set when the threat to the ship is minimal. Condition XRAY is set during working hours when the ship is in port, when there is no danger of attack, and when there is no threat from bad weather. All fittings marked with a black X and circle X are closed when condition XRAY is set.

YOKE. Provides a greater degree of watertight integrity than condition XRAY but to a lesser degree than the maximum condition. YOKE is normally set at sea and in port during wartime. All fittings marked with Xs and Ys, Circle X, and Circle Y are closed when condition YOKE is set.

ZEBRA. Provides the greatest degree of subdivision and watertight integrity to the ship. It is the maximum state of readiness for the ship's survivability system. Condition ZEBRA is set when the following situations occur:

1. Immediately when GQ is sounded
2. When entering or leaving port in wartime
3. To localize damage and control fire and flooding when the crew is not at GQ
4. At any time the CO deems the maximum condition of survivability should be set

All fittings marked with X or Y, Circle X, Circle Y, Z, Circle Z, and DOG Zs are closed when condition ZEBRA is set.

The setting of material conditions is normally carried out by the division or department responsible for the compartment and is accomplished by using the compartment checkoff lists (CCOLs). In an emergency the repair party personnel responding to the casualty assisted by those Sailors in the area will set the required condition to restrict the spread of damage. CCOLs are a tool used by repair party personnel to find the damage control fittings and closures in each compartment. The fittings and closures are identified by name (type) and a number that represent the location of the fitting by deck, the frame it is located at, and the side of the space similar to the compartment identification process. The CCOL list is posted in each space near the entrance and should be kept up to date by the division who owns the space. An example of a CCOL is shown in (fig. 12-2).

The normal setting of a material condition should be logged in the Damage Control Closure Log by each division or repair party completing the setting. Any exceptions to the setting should also be noted in the log so DCC/OOD are aware of the reduction in the ship's ability to restrict the spread of potential damage. This log and the Inoperative Fitting Log are maintained in either DCC or on the quarterdeck by the OOD. If you have a need to break (open) a fitting that should be normally set in a specific condition, you must first request it open and then log it in before opening the fitting. If you don't do this, you are in violation of the setting and have jeopardized your shipmate's survival. When you have completed work in a space that you needed a fitting/closure open, close it first, then log it closed. Now those who maintain the log will have a clear picture of what is open and closed in a given area of the ship. In addition to the closure log, a log for inoperative fittings and closures exists. As its name implies, this log lists all DC-related fittings and closures on the ship that don't work properly. It becomes a tracking tool for future maintenance.

A modified condition YOKE or modified condition ZEBRA is sometimes set, rather than the normal setting of XRAY, YOKE, or ZEBRA. These are settings of convenience at the discretion of the CO.

Student Notes:

COMP'T NO. 2-108-1-L	NAME Crews Berthing (LSD Wing Wall)				
ITEM	FITTING	NUMBER	LOCATION AND PURPOSE	CLASSIFICATION	DIVISION RESPONSIBLE
	ACCESS				
1	WT DOOR	2-108-1	Access to: 2-96-1-L	Z	REPIII
2	WT DOOR	2-129-3	Access to: 2-120-1-L	X	REPIII
3	WT HATCH	2-108-1	Access to: 3-108-1-L	X	S
	MISCELLANEOUS CLOSURES				
4	ATC	2-108-1	In WITH 2-108-1 used to test: 3-108-1-L 3-103-3-A 3-115-1-A	X	E
5	ATC	2-108-1	In WTD 2-108-1 used to test: 2-95-1-L	X	E
	DRAINAGE				
6	DECK SOCKET (remote)	2-112-1	Bilge eductor overboard discharge valve 5-112-1	X	M
7	STC	2-118-1	Sound Ball 6-108-1-W	X	R
8	GAGGED SCUPPER	2-109-1	Plumbing drain from 1-110-1-L	Z	REPIII
	FIRE MAIN & SPRINKLING SYSTEM AND WASH DOWN				
9	FMCOV	2-109-1	Cut out to FP 1-109-1	W	REPIII
10	FMCOV	2-110-1	Cut out to Group IV magazine sprinkler	W	REPIII
	FUEL OIL				
11	STC	2-116-1	Sound F.O. & Ball. 6-108-3-F	X	B
	REMOTE OPERATION				
12	Remote start/stop switch	2-119-1	For exhaust blower 2-108-1	Z	REPIII
	MISCELLANEOUS UNCLASSIFIED				
13	Loud speaker		General announcing 1 MC		
14	C.P. Riser Terminal	2-114-1	Casualty Power Outlet		
15	15 lb CO_2	2-119-1	Portable fire extinguisher		
16	One OBA		In box at Fr. 110 stbd.		

Figure 12-2.—Compartment checkoff list.

Student Notes:

A modified YOKE is sometimes used at sea when cruising independently in good weather and calm seas and in port in peacetime. In the modified condition, YOKE fi+ttings above the waterline are left open to improve ventilation and habitability. All other XRAY and YOKE fittings shipwide will be closed unless logged on a case-by-case basis as discussed earlier.

An alternative to the setting of ZEBRA shipwide is the setting of modified ZEBRA. An example of the setting of modified ZEBRA would be setting Zebra on the DC deck and below only if the area the ship was transiting was a known or suspected hazardous navigational area. This would provide a higher survivability stance than condition YOKE. At the same time modified ZEBRA is less restrictive in the rest of the ship and will more readily allow the accomplishment of other operational requirements. Modified ZEBRA may result as an upgrade from condition YOKE or as a downgrade from condition ZEBRA.

SPECIAL CLASSIFICATIONS

Circle XRAY and YOKE. These fittings are access fittings to battle stations and compartments containing equipment that require periodic checks or fittings that must be opened for ammunition transfer, as well as some systems.

Circle XRAY and YOKE fittings are normally closed when condition XRAY or YOKE are set. They may be opened without special authority when going to or securing from general quarters, transferring ammunition, during periodic space or equipment checks, or during operation of damage control equipment. Circle XRAY and YOKE fittings are marked with a black X or Y in a black circle. They should be secured once passage is complete.

Circle ZEBRA. These fittings are closed when condition ZEBRA is set. They may be opened with the permission of the CO only. This would be during extended periods of general quarters for the preparation and distribution of battle messing, to provide access to sanitary facilities, to ventilate battle stations, to transit from squadron ready rooms to the flight deck, and to allow limited access throughout the ship as the CO prescribes.

Open Circle ZEBRA fittings must be guarded so they can be closed immediately because the ship is still in a battle station position when ZEBRA is set. Circle ZEBRA fittings are marked with a red Z surrounded by a red circle.

DOG ZEBRA. These fittings are secured when condition ZEBRA is set and whenever the ship is set for "darken ship," which is a night steaming condition, not a material condition. During darken ship these fittings and closures are closed to prevent light inside showing to the outside. DOG ZEBRA fittings are marked with a red Z surrounded by a black letter D.

NOTE

Darken ship is a night steaming condition, not a material condition.

WILLIAM. These fittings are vital sea suctions, ventilation fittings valves serving vital equipment, and valves that must be open to maintain mobility and fire protection. WILLIAM fittings are open during all material conditions. They are secured only to control damage, contamination, or to repair equipment served. WILLIAM fittings are marked with a black colored letter W.

Circle WILLIAM. These fittings, like WILLIAM fittings, are normally open but are secured for protection in CBR attack. Fittings that are marked with this classification are those that provide ventilation opening to the outside of the ship. Circle WILLIAM fittings are marked with a black W surrounded by a black circle.

REVIEW 4 QUESTIONS

Q1. What is the most important feature of a ship to ensure its survivability?

Student Notes:

Q2. List the three basic material conditions of readiness.

a.

b.

c.

Q3. What material condition provides the least amount of watertightness?

Q4. When is YOKE normally set?

Q5. What material condition is set when GQ is sounded?

Q6. What fittings are closed to set darken ship?

Q7. To find a list of all DC fittings within a compartment, you should refer to—

LIFE SUPPORT EQUIPMENT

Learning Objective: When you finish this chapter, you will be able to—

- Identify life support equipment used aboard ship.

All life support devices discussed in this chapter are designed to allow the wearers to breath (and thereby to escape), continue work, and assist in saving the ship and their fellow crew members. Remember that the crew must save the ship or no one will be saved. Therefore, the purpose of this section of the chapter is to provide you with information on the emergency escape breathing device (EEBD), supplemental emergency egress device (SEED), oxygen breathing apparatus (OBA), and self-contained breathing apparatus (SCBA).

EMERGENCY ESCAPE BREATHING DEVICE (EEBD)

Studies of fire casualties have proven that most casualties are the result of smoke and toxic fumes and not from the fire itself. For this reason, the EEBD (fig. 12-3) was developed for emergency escape. It provides the wearer with 15 minutes of breathable air. It is to be worn until you can get topside during evacuation from below deck spaces. The EEBD is designed to provide respiratory and eye protection in an atmosphere that will not support life. With the proper training you should be able to activate and don an EEBD in less than 30 seconds.

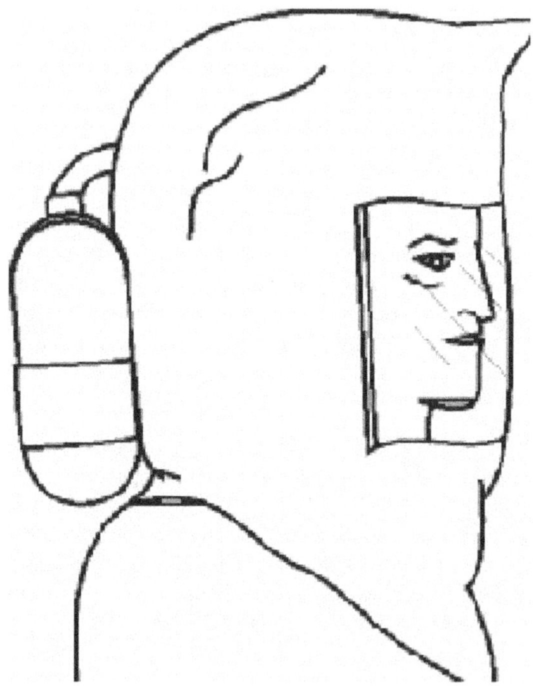

Figure 12-3. —emergency escape breathing device (EEBD)

Student Notes:

WARNING

EEBDs must not be used for firefighting purposes.

WARNING

Remember, when donning the EEBD, if you don't hear the hissing sound of oxygen being generated, discard the unit and get another.

DANGER

Don't smoke immediately after using the EEBD because your hair is saturated with oxygen and could catch on fire.

SUPPLEMENTAL EMERGENCY EGRESS DEVICE (SEED)

Personnel working in engineering spaces wear supplemental emergency egress devices (SEEDs) on their belts for easy access. When a main space fire is called away, the watch stander should use the SEED (fig. 12-4) to proceed to an EEBD. The watch stander should obtain an EEBD and don it when not in danger of immediate harm from heat or flames. Because the SEED lacks protection for the eyes and nose and has a short operational time, it is a supplemental device. However, it is immediately available for the engineering watch standers and is easily operated on the run. Factors to consider when using these devices include the following:

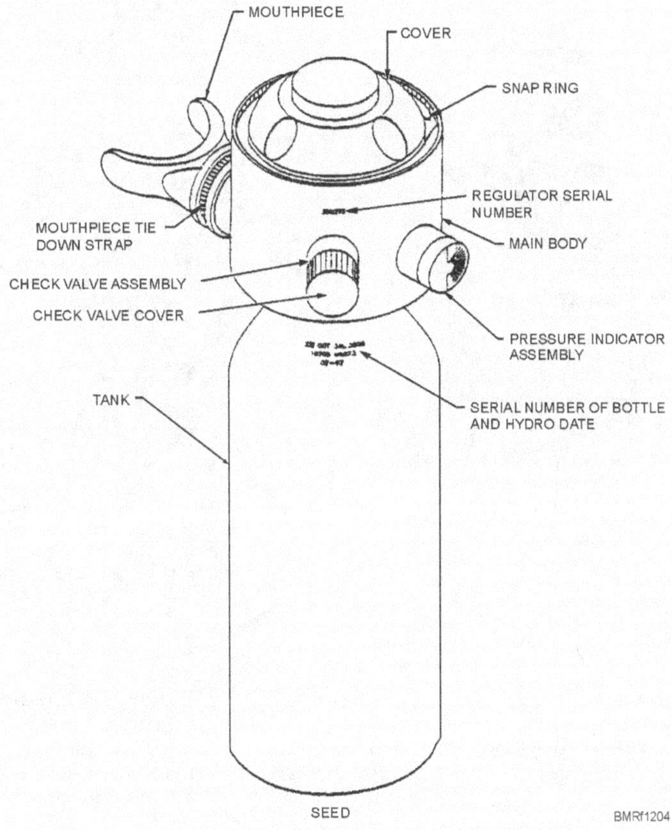

Figure 12-4.—Supplemental emergency egress device (SEED).

Student Notes:

- How quickly conditions are deteriorating
- Ease of egress, including travel time to a breathable atmosphere
- Operating times for each device
- Capabilities and limitations of each device

WARNING

Don't breath through the nose—breath only through the mouth when using the SEED.

WARNING

SEEDs must not be used for firefighting purposes.

OXYGEN BREATHING APPARATUS (OBA)

The oxygen breathing apparatus (OBA) (fig. 12-5) is a self-contained device that generates oxygen through a chemical process and lets the wearer breathe independently of the surrounding atmosphere. Currently, the OBA is the primary tool used by firefighting teams for respiratory protection. The effective time limit of the oxygen supply is in excess of

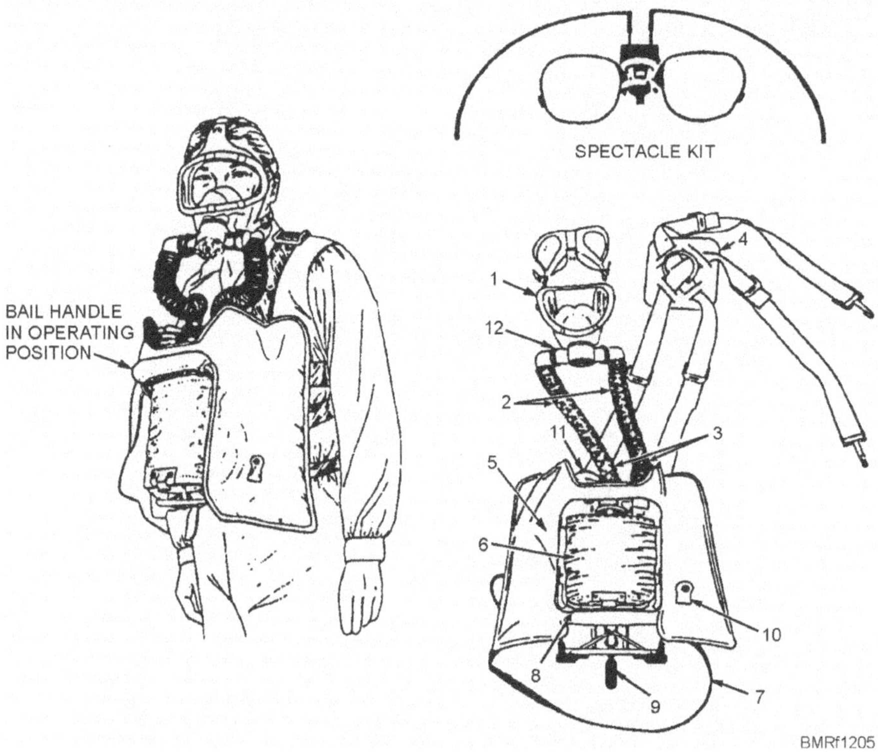

1. Facepiece
2. Breathing tubes
3. Breathing tube couplings
4. Body harness and pad
5. Breathing bag
6. Breastplate
7. Waist strap
8. Bail assembly handle (standby position)
9. Canister release strap
10. Pressure relief valve and pull tab timer
11. Timer
12. Valve housing

Figure 12-5.—Navy oxygen breathing apparatus (OBA).

Student Notes:

45 minutes. For your personnel protection you should set the timer on the OBA for 30 minutes, allowing 15 minutes to leave the area and return to fresh air.

When in operation, the air within the apparatus is continuously replenished with oxygen while the chemicals in the canister remove exhaled carbon dioxide (CO_2) and water vapor. As a result of this chemical action, the OBA wearer may survive and work in a toxic atmosphere, such as a smoke-filled compartment. With the facepiece and canister in place, using the OBA forms a closed self-sustaining system. For personnel having eyeglasses, a spectacle kit is available for the installation of prescription lenses in the facepiece.

WARNING

When removing the candle cover, do not remove the cotter pin while pulling the lanyard. Removal of the cotter pin fires the candle and starts generating oxygen. If that happens while the copper foil is intact, internal pressure in the canister will build up, causing the copper foil or canister seam to rupture. Do not pull the cotter pin until the canister is inserted in the OBA and the bail assembly is up and locked.

CAUTION

Do not pull the breathing bag tab during normal use because the oxygen in the bags will leak into the atmosphere, causing loss of your breathing oxygen.

WARNING

Never allow grease, oil, or water to enter the neck of the canister. Any of these liquids may cause a violent chemical reaction or explosion. The chemicals contained in the canister are injurious to skin and equipment.

SELF-CONTAINED BREATHING APPARATUS (SCBA)

The self-contained breathing apparatus (SCBA) (fig. 12-6) is replacing the OBA throughout the Navy. The unit's main components consist of a harness, high-pressure bottle, pressure regulator, full-face mask, and the high- and low-pressure hoses as shown below High-pressure air cylinders are filled with compressed grade D breathing air and then stored until needed. The cylinders themselves are rechargeable, replacing the need for additional canisters in the OBA system.

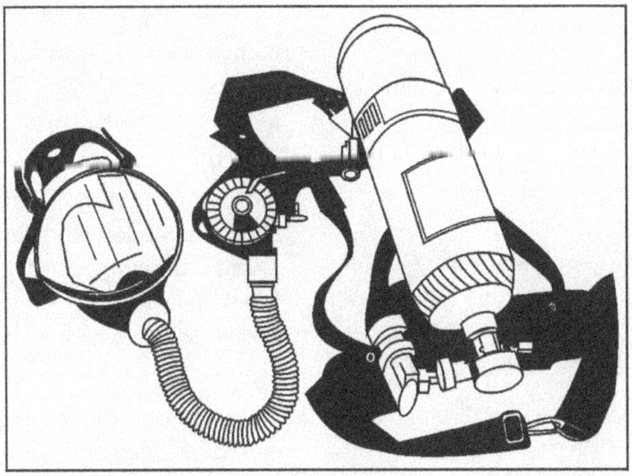

BMRF1206

Figure 12-6.—Self-contained breathing apparatus (SCBA).

Student Notes:

Once the cylinder valve has been turned on, it provides a continuous air supply to the system regulator via the high-pressure hose. The pressure is reduced by the regulator for use by the wearer, and the low-pressure hose carries the breathable air into the facepiece. As a safety feature, these are positive pressure style regulators, which means that the air supply to the mask is just above any demand requirements eliminating any possibility of toxic fumes from entering the wearer's mask.

The time each cylinder lasts will be determined by a number of factors. The main factors will be the physical condition and size of the wearer and the work to be done. Generally, each bottle will last approximately 30 minutes from the time it is activated.

REVIEW 5 QUESTIONS

Q1. How many minutes of air is the EEBD designed to provide?

Q2. Why should you stay away from open flames when you first remove an EEBD?

Q3. What should you listen for when donning an EEBD?

Q4. What device is worn by engineering watch standers?

Q5. What is the effective time of an OBA?

Q6. When do you pull the cotter pin on an OBA canister?

FIREFIGHTING

Learning Objectives: When you finish this chapter, you will be able to—

- Recognize the properties of a fire triangle and fire tetrahedron.
- Identify the conditions necessary for spontaneous combustion.
- Identify the types of heat transfer.

Fire is a constant threat aboard ship. All possible measures must be taken to prevent a fire, or if one is started, to extinguish it quickly. Fires may start from several causes—spontaneous combustion, carelessness, hits by enemy shells, or collision. If the fire is not controlled quickly, it may cause more damage than the initial casualty and could cause the loss of the ship.

FIRE AND FIREFIGHTING

You cannot win against a fire. You can fight the fire and you can hold down its damage; but some property will be destroyed and, all too often, people will be injured or killed. Time is always lost, productive work is stopped, and additional effort and materials are required to make repairs and to clean up the mess.

The objective of fire prevention, therefore, is to prevent fires from starting. Fire prevention is an all-hands, all-day, all-night, heads-up effort. A cigarette tossed in the wrong direction can cause as much damage as an enemy bomb. An oily wiping rag or a sparking tool can be as dangerous as an open flame in a gasoline depot.

Each ship is required to institute and maintain a fire prevention program. Your part in the fire prevention program is as follows:

- Ensure that all gear is stowed properly.

Student Notes:

- Practice good housekeeping procedures, such as the daily removal of trash from spaces.
- Practice safety precautions when working with flammable materials.
- Report all potential fire hazards.
- Keep firefighting equipment handy and in good working order.
- Ensure closures and fittings are working properly and report any discrepancies.

FIRE TRIANGLE

The entire chemistry and physics of fire and burning, or combustion, can be simplified into a relationship between three components—fuel, heat (temperature), and oxygen (air). To have a fire in any combustible substance, each one of these components must be present to help each other. Picture these components in the form of a triangle, as shown in figure 12-7.

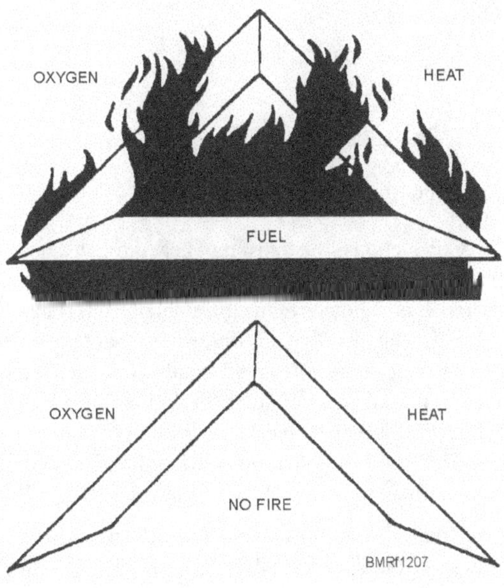

Figure 12-7.—Requirements for combustion.

Look at figure 12-7. Here, you can see that if the oxygen reacts with the fuel, it creates heat, which causes a draft or some other condition that takes in more oxygen and creates still more heat, and so on. Or the heat may cause more fuel to become available (such as causing gasoline to boil into vapor), which then takes more oxygen to burn and creates more heat, which then produces still more fuel, and so on. The burning reaction can go in many different directions.

The modern science of firefighting and fire extinguishment is based on the sides of the fire triangle and an uninhibited chain reaction of burning. Obviously, the firefighter can remove one or more of the components to cause the burning to stop. The type of firefighting agent the firefighter has at hand determines which component or components of the triangle will be removed.

Another way the firefighter can stop the fire (and the combustion) is to place a screen between any two components of the triangle. If the fighter uses an agent as a temporary screen that breaks the triangle, the fire goes out. Obviously, the fire can quickly start up again if this method is used because each of the three necessary components is still there waiting to start the fire again once the screen is gone.

FIRE TETRAHEDRON

The fire triangle describes the requirements for surface glowing or smoldering, but it doesn't completely describe flaming combustion requirements. A fourth requirement, an uninhibited chain reaction, is needed for flames to exist. This is shown by the fire tetrahedron (fig. 12-8). A tetrahedron is a solid figure with four triangular faces. It is useful for illustrating the flaming combustion process because it provides for the chemical chain reaction requirement and each face touches the other three sides. As described for the fire triangle, flaming combustion stops when one of the four sides of the fire tetrahedron is removed.

SPONTANEOUS COMBUSTION

Fire, also called *burning* or *combustion*, is a rapid chemical reaction that results in the release of energy in the form of light and heat. Most spontaneous combustion involves very rapid oxidation; that is, the

Student Notes:

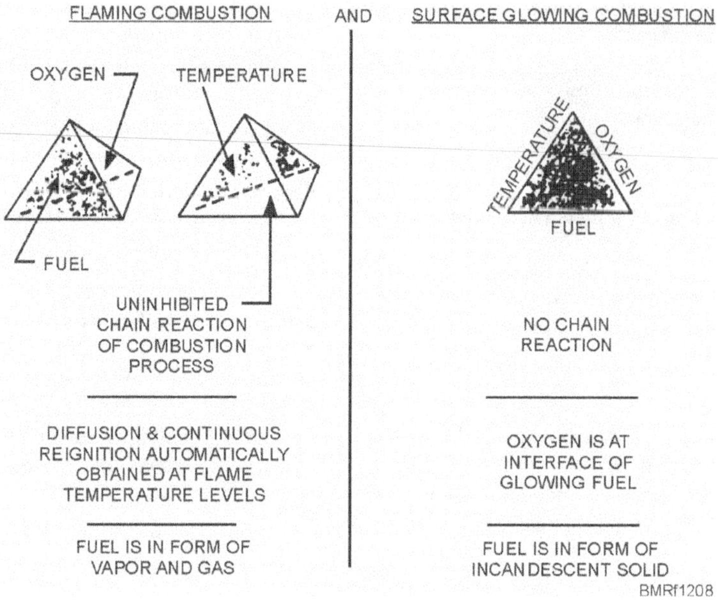

Figure 12-8.—Tetrahedron and fire triangle.

chemical reaction by which oxygen combines chemically with the burning material.

Such things as rags or paper soaked with oil or with paints or solvents are particularly subject to spontaneous combustion if they are stowed in confined spaces where the heat caused by oxidation cannot be dissipated rapidly enough.

A fire involving combustible fuel or other material must have an ignition source, and the material must be hot enough to burn. The lowest temperature at which a flammable material gives off vapors that will burn when a flame or spark is applied is called the *flash point*. The *fire point*, which is usually a few degrees higher than the flash point, is the temperature at which the fuel will continue to burn after it has been ignited. The *ignition* or *self-ignition point* is the lowest temperature to which a material must be heated to give off vapors that will burn without the aid of a spark or flame. In other words, the ignition point is the temperature at which spontaneous combustion occurs. The ignition point is usually at a much higher temperature than the fire point.

METHODS OF HEAT TRANSFER

Heat from a fire is transferred by one or more of the following methods:

1. Conduction
2. Convection
3. Radiation

Conduction

Conduction is the transfer of heat through a body or from one body to another by direct physical contact. For example, on a hot stove, heat is conducted through the pot to its contents. Wood is ordinarily a poor conductor of heat, but metals are good conductors. Since most ships are constructed of metal, heat transfer by conduction is a potential hazard. Fire can move from one fire zone to another, one deck to another, and one compartment to another by heat conduction.

Often, the skillful application of water, using fog patterns to rapidly coat and recoat surfaces with a film of water, will slow or halt the transmission of heat by conduction. Fog patterns coat surfaces more efficiently

Student Notes:

than solid streams, reducing run off and the effect on ship stability.

Convection

Convection is the transfer of heat through the motion of circulating gases or liquids. Heat is transferred by convection through the motion of smoke, hot air, and heated gases produced by a fire.

When heat is confined (as within a ship), convected heat moves in predictable patterns. The fire produces lighter than air gases that rise toward high parts of the ship. Heated air, which is lighter than cooler air, also rises. As these heated combustion products rise, cool air takes their place; the cool air is heated, in turn, and then rises to the highest point it can reach.

Hot smoke originating at a fire on a low deck will travel horizontally along passageways, and then upward by way of ladder and hatch openings, heating flammable materials in its path. To prevent fire spread, release the heat, smoke, and gases to the atmosphere. However, the structural design of a ship makes it difficult to rapidly cut openings through decks, bulkheads, or the ship's hull for ventilation. It is imperative that the fire be confined to the smallest possible area. Doors and hatchways should be kept closed when they are not in use. If a fire is discovered, attempts should be made to close off all openings to the fire area until firefighting personnel and equipment can be brought into position to fight the fire.

Radiation

Heat radiation is the transfer of heat from a source across an intervening space; no material substance is involved. The heat travels outward from the fire in the same manner as light; that is, in straight lines. When it contacts a body, it's absorbed, reflected, or transmitted. Absorbed heat increases the temperature of the absorbing body. For example, radiant heat that is absorbed by an overhead will increase the temperature of that overhead, perhaps enough to ignite its paint.

Heat radiates in all directions unless it's blocked. Radiant heat extends fire by heating combustible substances in its path, causing them to produce vapors, then igniting the vapor.

Within a ship, radiant heat raises the temperature of combustible materials near the fire and, depending on the ship's design, at quite some distance from the fire. Ship fires can spread as a result of radiating bulkheads and decks. Intense radiated heat can make an approach to the fire extremely difficult. For this reason, protective clothing should be worn by firefighters.

REVIEW 6 QUESTIONS

Q1. Fire prevention is the responsibility of—

Q2. List the three components that make up a fire.

a.

b.

c.

Q3. What process is involved in most cases of spontaneous combustion?

Q4. List the three methods of heat transfer.

a.

b.

c.

CLASSES OF FIRE

Learning Objective: When you finish this chapter, you will be able to—

- Recognize the four classifications of fire and identify the means used to extinguish them.

Fires are divided into four classifications, each indicating the type of material burning. By knowing the

Student Notes:

CLASSES OF FIRE	TYPES OF MATERIALS INVOLVED	METHOD(S) OF EXTINGUISHING
A	Wood and wood products, cloth, textiles and fibrous materials, and paper and paper products.	Water in straight or fog pattern. If the fire is deep-seated, aqueous film forming foam (AFFF) is more effective than seawater and can be used as a wetting agent.
B	Flammable liquids, such as gasoline, diesel fuel (F-76), jet fuels, hydraulic fluid, and lube oil. Also, involves flammable gases.	Extinguished with AFFF, Halon 1211, Halon 1301, or potassium bicarbonate (PKP). Never extinguish flammable gases unless there is a good chance that the flow of gas can be secured. **Securing the flow of gas is the single most important step in controlling a gas fire**.
C	Energized electrical fires.	Use nonconductive agents, such as CO_2, Halon 1211, or water spray. Most effective tactic is to de-energize and handle the fire as a class A fire. If fire isn't deep-seated, clean agents that don't a cleanup problem are preferred (CO_2 or Halon 1211).
D	Combustible metals, such as magnesium and titanium.	Water in quantity, using fog patterns. When water is applied to burning class D materials, explosions may occur. Apply water from a safe distance or from behind shelter.

Figure 12-9.—Classes of fire.

class of fire, the primary agent and best method for extinguishing the fire can be determined as shown in figure 12-9.

Many substances, such as liquids, gases, and solids, are used as firefighting agents. The selection and use of these agents varies with the class of fire, its location, and the extent of the fire involvement. Although seawater is the most valuable firefighting agent available based on its endless supply, considerations should always be taken to determine if water is the best agent to put out the type of fire being fought.

REVIEW 7 QUESTIONS

Q1. Match the class of fire with the types of materials involved.

MATERIALS	CLASSES OF FIRE
a. Hydraulic fluid	A
b. Energized circuit board	B
c. Paper	C
d. Gasoline	D
e. Magnesium	
f. Cloth	
g. Titanium	

Student Notes:

Q2. What is the most available heat-removing agent for shipboard use?

Q3. AFFF was developed to combat what class of fire?

Q4. CO_2 is the primary agent to disrupt the fire triangle of what class fire?

Q5. What is an effective agent against a class D fire?

DAMAGE CONTROL EQUIPMENT

Learning Objective: When you finish this chapter, you will be able to—

- Identify the portable equipment and fixed systems that make up the shipboard inventory of protection equipment.

As you've learned in this chapter, damage control is an all hands job. The time it takes to get a fire or flooding under control is vital. You should become familiar with the portable equipment and fixed systems that make up your ship inventory of protection equipment. Because of the large amount of explosives, fuels, and other flammable materials aboard ship, you must know where equipment is located and how to use it before an emergency exists.

Knowing where equipment is and how to use it applies not only to firefighting equipment but also the equipment used for flooding or CBR protection. The equipment and its storage location varies from ship to ship, so you must know your ship.

You may find yourself the first person on the scene, so knowing the basic rules of damage control and equipment usage may make the difference in the loss of many spaces or a shipmate's life. Perhaps you will be called to serve on a fire party or flooding detail. As a team member, keep in mind, the lack of equipment and procedures could result in a disastrous outcome.

Figures 12-10 through 12-15 show some examples of the equipment that's common to all ships. These figures show only a small amount of what you'll find in a ship's repair locker.

THE FIREFIGHTER ENSEMBLE

The firefighter's ensemble (fig. 12-10) is used to protect the firefighter from short duration flame (flash) exposure, heat, and falling debris.

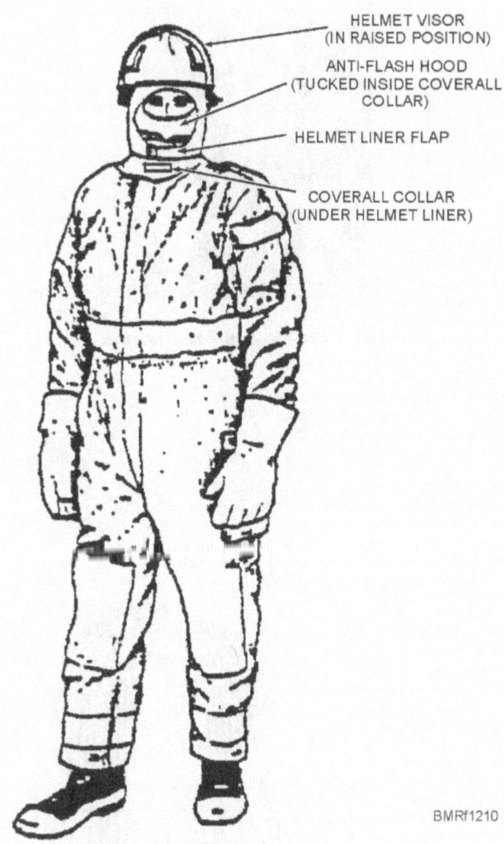

Figure 12-10.—Firefighter's ensemble.

Student Notes:

Figure 12-11.—Anti-flash clothing.

ANTI-FLASH CLOTHING

Anti-flash clothing (fig. 12-11) is used to protect personnel from high temperatures resulting from the use of explosive weapons and from burns caused by fire. The clothing consists of the following two items:

1. Anti-flash hood
2. Anti-flash gloves

FIRE HOSE STATION

A fire hose station (fig. 12-12) is commonly referred to as either a *fire station* or a *fireplug*. The fire hose station is the location of a fireplug and associated equipment. Branches of the firemain system supply water to the fire hose stations throughout the ship. Generally, fire hose stations aboard frigates and larger ships have 1 1/2-inch fireplugs and fire hose stations aboard ships larger than frigates have 2 1/2-inch fireplugs.

P-100 PUMP

The P-100 pump is a diesel-engine-driven portable pump unit. It's designed for firefighting (fig. 12-13) and limited dewatering (fig. 12-14) functions aboard ships.

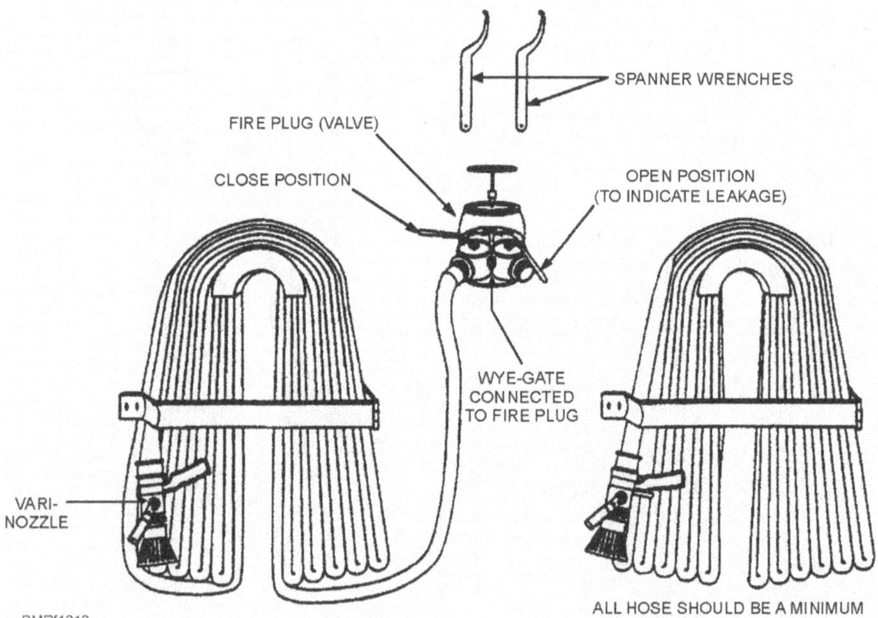

Figure 12-12.—Fire hose station.

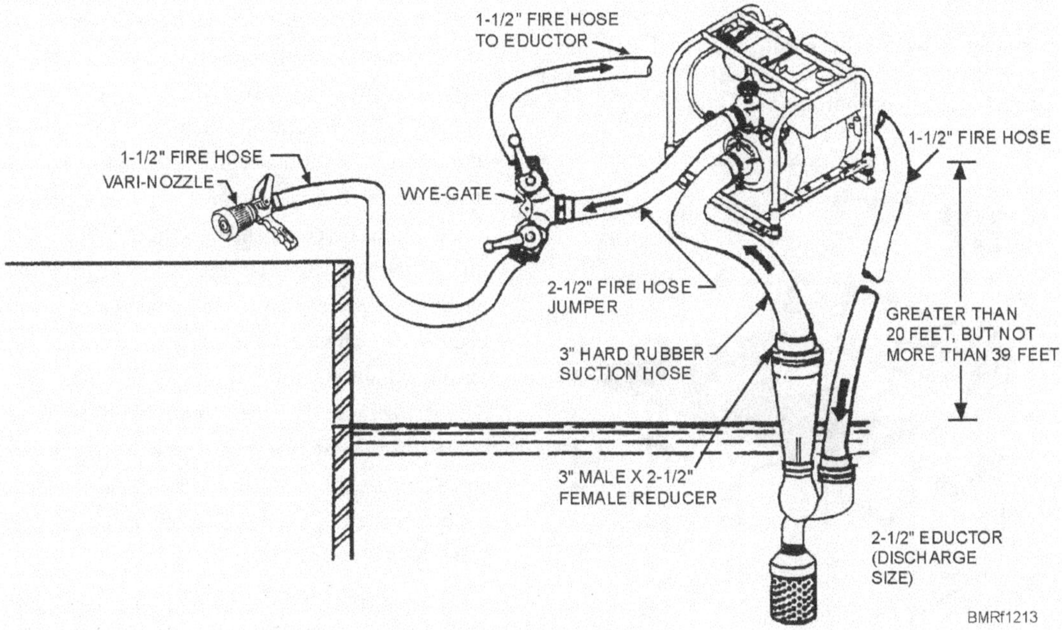

Figure 12-13.—Firefighting hookup.

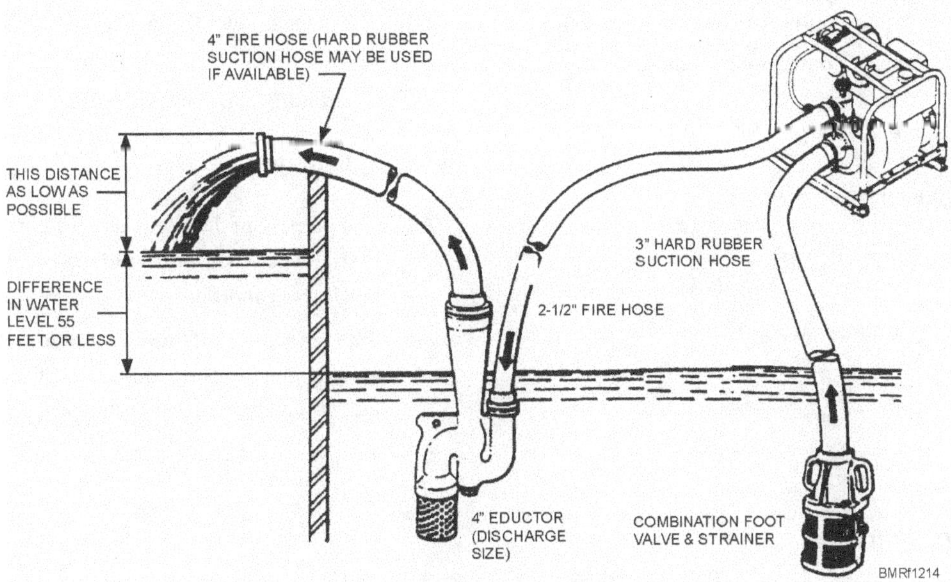

Figure 12-14.—Dewatering hookup.

DAMAGE CONTROL SHORING CHEST

Each repair station has a damage control shoring chest (fig. 12-15). The chest is divided into three compartments to separate the different tools and materials needed for shoring.

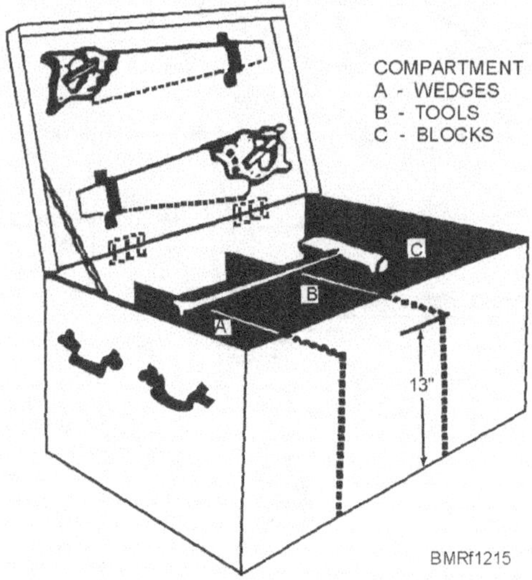

Figure 12-15.—Damage control shoring chest.

REVIEW 8 QUESTION

Q1. Describe the purpose of the following items.

a. Damage control shoring chest

b. P-100 pump

c. Firefighter ensemble

d. Anti-flash clothing

e. Fire hose station

SUMMARY

A ship lives or dies depending on its crew's ability to combat a wide range of casualties, and these casualties are not restricted to times of conflict. Casualties can occur while moored in your homeport, at anchorage overseas, or during a routine peacetime deployment. The requirement for every person on board, from the commanding officer to the fireroom messenger, to have a good, basic knowledge of damage control procedures and constant training and drills in combating shipboard casualties is essential to the ship's survival. This training and preparation demonstrated during the Persian Gulf crisis saved several U.S. Navy ships that sustained considerable damage. If these crews had not been trained in damage control, these ships and most likely a large portion of their crews may have been lost.

The proper use of the closure log is just as important as knowing how to don an OBA or use a fire hose. The business of damage control is serious. Learn it well; your life, the lives of your shipmates, and your ship depend on it.

REVIEW 1 ANSWERS

A1. The damage control organization is divided into the—

a. **Administrative organization**

b. **Battle organization**

A2. The purpose of the administrative phase of DC is **to establish and maintain material readiness conditions.**

A3. **The engineer officer** is responsible for maintaining properly trained DCPOs, repair parties, and repair locker personnel.

A4. The **XO** is the senior person assigned as team leader on DCTT.

Student Notes:

A5. The **DCPO** is responsible for making sure that all battle lanterns, dog wrenches, spanners, and other damage control equipment are in place and in a usable condition within their division.

A6. Before anyone enters a poorly ventilated space or void, the **gas free engineer** (**GFE**) certifies the space as safe.

REVIEW 2 ANSWERS

A1. The battle phase of damage control starts **when the ship receives actual damage**.

A2. The battle station for the DCA is **DCC/CSS**.

A3. DCC receives reports and casualty status from **repair parties**.

A4. The minimum requirement to be a member of the in-port emergency team is to be **repair party qualified**.

A5. The primary missions of the rescue and assistance team are **to assist persons from the water, other units in distress, and other persons or activities in distress ashore**.

A6. **General quarters** is the highest state of readiness.

REVIEW 3 ANSWERS

A1. The X40J circuitry is commonly referred to as the **salt and pepper line**.

A2. The alarms that can override the microphone control of the IMC are the

a. **Collision**

b. **Chemical attack**

c. **General**

d. **Flight crash**

A3. When all communications systems fail, **messengers** are used to deliver messages and orders between repair lockers and DCC.

REVIEW 4 ANSWERS

A1. Survivability of the ship can be ensured by **compartmentation**.

A2. The three basic material readiness conditions are—

a. **XRAY**

b. **YOKE**

c. **ZEBRA**

A3. Material condition **XRAY** provides the least amount of watertightness.

A4. Normally, **YOKE** is set at sea and in port during wartime.

A5. Material condition **ZEBRA** is set when GQ is sounded.

A6. **DOG ZEBRA** fittings are closed to set darken ship.

A7. To find a list of all DC fittings within a compartment, you should refer to the **compartment checkoff list**.

REVIEW 5 ANSWERS

A1. The EEBD is designed to provide **15 minutes** of air.

A2. You should stay away from open flames when you first remove an EEBD **because your hair is saturated with oxygen and could catch fire**.

A3. When donning an EEBD, you should hear a **hissing sound**.

A4. Engineering watch standers wear a **SEED**.

A5. The effective time of an OBA is **45 minutes**.

A6. You pull the cotter pin on an OBA **after the canister has been inserted in the OBA and locked in place.**

REVIEW 6 ANSWERS

A1. Fire prevention is the responsibility of **all hands**.

A2. The three components that make up a fire are—

a. **Heat**

b. **Fuel**

c. **Oxygen**.

A3. Most cases of spontaneous combustion involve the process of **rapid oxidation.**

A4. The three methods of heat transfer are—

a. **Conduction**

b. **Radiation**

c. **Convection**

REVIEW 7 ANSWERS

A1.

MATERIALS	CLASSES OF FIRE
a. Hydraulic fluid	**B**
b. Energized circuit board	**C**
c. Paper	**A**
d. Gasoline	**B**
e. Magnesium	**D**
f. Cloth	**A**
g. Titanium	**D**

A2. **Sea water** is the most available heat-removing agent for shipboard use.

A3. AFFF was developed to combat **class B fires**.

A4. **CO_2** is the primary agent to disrupt the fire triangle of class C fires.

A5. **Water fog** is an effective agent against a class D fire.

REVIEW 8 ANSWERS

A1. The purposes of the following items are as follows:

a. Damage control shoring chest. **A chest that is divided into compartments for storing materials used for shoring.**

b. P-100 pump. **The P-100 pump can be used to fight fires or to dewater spaces.**

c. Firefighter ensemble. **The firefighter ensemble protects personnel from short duration flames, heat, and falling debris.**

d. Anti-flash clothing. **Anti-flash clothing protects personnel from heat caused by high explosive weapons and from burns caused by fire. There are two items that make up anti-flash clothing—anti-flash gloves and the anti-flash hood.**

e. Fire hose station. **The fire hose station is where the fireplug and associated equipment is located.**

CHAPTER 13

CHEMICAL, BIOLOGICAL, AND RADIOLOGICAL DEFENSE

If we do not stem the proliferation of the world's deadliest weapons, no democracy can feel secure... One of our most urgent priorities must be attacking the proliferation of weapons of mass destruction, whether they are nuclear, chemical or biological.

—President Bill Clinton
1993

As a member of the Navy, you need a good working knowledge of chemical, biological, and radiological defense (CBR-D). CBR-D is defined as *defensive measures taken against the effects of a chemical, a biological, or a nuclear weapons attack.* Training in defensive measures lets the command maintain operational readiness and complete its mission.

Throughout history, countries and factions have developed and used chemical, biological, and radiological (CBR) warfare. In WWI, chlorine gas was used, which forced the development of the gas mask. In WWII, nuclear weapons were used for the first time. During WWII, many nations began developing chemical and biological agents as a warfare tool. More recently, a nerve agent was used in Japan's subway system, and blister agents were used on the Kurds during the Iraq and Iran war. The use of CBR clearly demonstrates the need for a positive defensive posture within our Navy.

United States national policy prohibits our being the first to use chemical agents against an attacking nation. The approval for our initial use of chemical weapons must come from the President of the United States.

The United States **will not use** biological agents, including toxins, regardless of source or manner of production, or other methods of biological warfare under any circumstances. The United States strictly limits its biological and toxin research program to defensive measures, such as production of vaccines, antidotes, treatment, and protective equipment.

An aggressive nation might decide using a chemical or biological (CB) weapon first is to its advantage, perhaps as a surprise attack. Therefore, all members of the U.S. Navy must be highly trained in CBR-D. CBR-D training allows Navy units to survive CBR attacks and continue to fight and defend their ship or unit under CBR-contaminated conditions.

The need for training in CBR-D is never ending. The Navy has developed and continues to develop different countermeasures against many possible CBR applications that an enemy might think of. Because possible enemies continually develop CBR applications, training programs using information about the greatest CBR threats are developed to train naval personnel. These programs include disaster preparedness drills and personnel qualification standards (PQS). Don't take this training lightly; it may save your life.

CHEMICAL, BIOLOGICAL, AND NUCLEAR WARFARE OPERATIONS

Learning Objectives: When you finish this chapter, you will be able to—

- Recognize chemical, biological, and radiological (CBR) attack methods.
- Recognize the need for CBR defense.
- Identify terms used with CBR.

The primary purpose of nuclear weapons is the mass destruction of targets and personnel. The primary purpose of biological and chemical attacks is the mass casualties of personnel, livestock, and/or crops. These weapons are called *weapons of mass destruction* because they destroy large areas or kill and disable large segments of the population.

In chemical warfare (CW) operations, chemical agents can kill or disable personnel by affecting their blood, nerves, eyes, skin, lungs, or stomach. Biological warfare (BW) operation agents include microorganisms, fungi, toxins, and microtoxins to cause diseases that will kill or produce other casualties. Nuclear weapons produce explosions of great force and heat and release nuclear radiation.

Many types of weapons and methods may be used to deliver chemical and biological agents and nuclear bombs. The use of two or more different types of weapons to deliver these agents may be used at the same time. Missiles having long-range attack capability provide a means of delivering chemical, biological, and nuclear weapons that can be launched in almost any manner by land, sea, and/or air units.

Chemical agents have been placed in projectiles and used effectively. A similar possibility exists with biological agents; however, for technical reasons it appears that the most probable method of delivery is by aerosol.

CHEMICAL WARFARE

Learning Objectives: When you finish this chapter, you will be able to—

- Identify terms used with chemical warfare (CW).
- Identify types of CW.
- Identify the effects of CW agents.
- Identify self-aid and first-aid methods for countering nerve, blister, and choking agents.

CW agents are used to produce death, injury, temporary incapacitation, or irritation effects. Broadly speaking, there are two types of antipersonnel agents—casualty and incapacitating. Some of the types of CW agents are described in table 13-1.

Table 13-1.—Characteristics of Selected CW Agents

Agent Name	Agent Type	Physical Properties	Physiological Effects	Relative Rate of Action
Sarin	Nerve	Colorless odorless, volatile liquid	Difficulty breathing, excessive contraction of the pupil of the eye (miosis), blurred vision, headache and nausea leading to respiratory distress, convulsions, and eventually death.	Rapid (within minutes)
VX	Nerve	Colorless odorless, low volatility, oily liquid	Difficulty breathing, miosis, blurred vision, headache and nausea leading to respiratory distress, convulsions, and eventually death.	Relatively rapid (within 30 minutes)
Mustard	Blister	Garlic odor, medium volatility, oily liquid	Blisters or irritates skin, eyes, and lungs.	Delayed onset (4-6 hours)
Hydrogen cyanide	Blood	Almond odor, highly volatile gas	Prevents the normal transfer of oxygen from the blood to body tissue resulting in respiratory paralysis.	Rapid (within minutes)

Student Notes:

CASUALTY CW AGENTS

Casualty CW agents can cause death or severely incapacitate personnel for long periods of time. Casualty agents can be either persistent or nonpersistent. They are classed as blood, choking, nerve, and blister agents, all of which can inflict serious injury or death.

Some casualty agents have a cumulative effect, which means that successive doses add to the effect of each preceding dose. You might receive a nonlethal dose of a nerve agent, for example, followed within a few hours by another nonlethal dose. However, the cumulative effects of the two exposures could kill you.

INCAPACITATING CW AGENTS

Incapacitating CW agents temporarily disable personnel but do not create permanent injury. They can produce physiological and/or psychological effects. These effects make individuals incapable of performing duties for hours or days even after exposure has ceased.

Some incapacitating agents have effects that typically last for significant periods of time but do not seriously endanger life; for example, riot control agents. Riot control agents produce only temporarily irritating or incapacitating effects when used in normal concentrations. Complete recovery is usually expected without medical treatment.

EFFECTS OF CW AGENTS

CW agents will make you a casualty when your body comes in contact with a bigger dose than it can withstand. The limits of tolerance of the human body vary from short periods of exposure and low concentrations of certain agents to extended periods of exposure and high concentrations of certain other agents. Furthermore, the limits of tolerance to specific agents vary with individuals. Your principal concern is recognizing the symptoms and relieving the effects of exposure before the limit of exposure is exceeded.

Nerve Agents

Poisoning by nerve agents affects bodily functions. The disruption of nerve impulses produces different effects on different body systems. It's important for you to recognize both mild and severe signs and symptoms of nerve agent poisoning. Mild symptoms will become severe if personnel are repeatedly or continually exposed to low concentrations of a nerve agent. High concentrations of nerve agent poisoning will cause rapid onset of severe symptoms, possibly without any mild symptoms at all. The symptoms of nerve poisoning are shown in the following chart:

MILD SYMPTOMS	SEVERE SYMPTOMS
• Unexplained runny nose	• Strange or confused behavior
• Unexplained sudden headache	• Wheezing, difficult, or labored respiration and cough
• Excessive sudden drooling	• Severely pinpointed pupils
• Difficulty seeing (reduced vision or miosis)	• Red eyes with tearing
• Tightness in chest, difficulty breathing	• Vomiting
• Localized sweating and muscular twitching in the area of contaminated skin	• Severe muscular twitching and general weakness
• Stomach cramps	• Involuntary urination and defecation
• Nausea	• Convulsions
	• Unconsciousness
	• Respiratory failure

Student Notes:

NOTE

Some symptoms of heat stress are similar to symptoms of nerve agent poisoning.

The rapid action of nerve agents calls for immediate administration of the antidotes atropine and pralidoxime chloride (2-PAM C1). Atropine acts to dry up secretions in the respiratory tract and to stimulate the central respiratory functions, and 2-PAM C1 simultaneously relieves muscle paralysis, especially in the respiratory tract. Both antidotes are self-injected into the lateral thigh muscle by the use of automatic injectors.

Blister Agents

Blister agents act on the eyes, mucous membranes, lungs, and skin. Blister agents include mustard vapors and mustard liquids. Mustards burn and blister the skin they contact, damage the respiratory tract when inhaled, and cause vomiting and diarrhea when absorbed. The degree of damage depends on the type and concentration of the agent, the weather, the amount of activity of the individual, and amount of exposure time. Blister agents are effective even in small quantities and produce both immediate and delayed injuries.

Mustard vapors burn any area of the skin; but, the burn is most severe in moist areas, such as the neck, genitals, groin, armpits, bends of knees, and elbows. Redness of the skin follows in 1/2 to 36 hours after exposure. This condition may be accompanied by intense itching, and blisters may then appear. Stiffness, throbbing pain, and swelling may also occur.

A few hours after breathing the mustard vapor, a victim experiences irritation of the throat, hoarseness, and coughing. After severe exposure, the lining of the respiratory system swells and interferes with breathing. Frequently, pneumonia develops.

If the whole body is exposed to mustard vapor, the body goes into a state of shock. This reaction is accompanied by nausea and vomiting.

Personnel who suspect contamination of their eyes or face must seek overhead shelter and flush the eyes with potable (drinkable) water from a canteen or shower. Mild exposure to skin can be treated by applying calamine lotion or topical steroid creams. All blisters should be opened, drained, and cleansed with tap or saline water. Any exposure to mustards require medical care by a corpsman or medical personnel.

Blood Agents

Blood agents inhibit the action of an enzyme responsible for transferring oxygen from the blood to the cells of the body. Thus the cells become starved for oxygen. Inhalation is the usual route of entry for blood agents.

The symptoms produced by blood agents depend on the concentration of the agent and the duration (length of time) of the exposure. Typically, either death occurs rapidly or recovery takes place within a few minutes after removal of the victim from the toxic atmosphere. High concentrations of blood agent cause labored breathing within a few seconds, violent convulsions, followed by cessation (stoppage) of breathing completely. Finally, the heart stops only a few minutes after initial exposure. The symptoms of exposure to blood agents are shown in the following chart:

INITIAL SYMPTOMS	ADVANCED SYMPTOMS
• Increased respiration • Headache • Giddiness • Dizziness • Increased pulse rate • Red, flushed skin	• Convulsions • Coma • Death

Student Notes:

If you're exposed to a blood agent, immediately don (put on) a protective mask. Speed is essential! Blood agents act so rapidly that within seconds, the effects of exposure can make it impossible for individuals to don their own mask. If this happens, the nearest person should help those who can't don their mask. Medical personnel should administer medications.

Choking Agents

In low concentration, choking agents produce an action on the respiratory system that results in the accumulation of fluid in the lungs. Accumulation of fluid in the lungs can cause death. High concentrations produce death for the same reason, but the upper respiratory tract may be involved as well. Exposure to choking agents may produce immediate dryness of the throat, coughing, choking, tightness across the chest, headache, nausea, and at times, irritated and watery eyes. However, symptoms are usually delayed, and it's possible that no immediate symptoms will appear when exposed to a fatal dose.

Even a mild exposure to a choking agent that is accompanied by immediate symptoms may cause fluid to accumulate in the lungs within 2 to 24 hours after exposure. Shallow and rapid breathing, a hacking and painful cough, frothy saliva, and an ashen gray color of the skin indicate the presence of fluid in the lungs.

After exposure to a high dose of a choking agent, it's important to begin medical treatment quickly to prevent accumulation of fluid in the lungs. It's important to keep the victim at rest and warm. Cough suppressant and pain relievers can be given as long as the doses don't interfere with respiratory functions.

Riot Control Agents (RCAs)

RCAs are classified as either tear agents or vomiting agents and are characterized by very low toxicity and brief action. They are used to produce temporary misery and harassment. Most personnel exposed to RCAs don't require medical attention and casualties are rare. Tear agents act rapidly on nerve ends in the cornea and mucous membranes of the eye. Vomiting agents cause local inflammation of the respiratory tract, sinuses, and eyes. The symptoms of exposure to RCAs are shown in the following chart.

First aid for personnel exposed to tear agents includes providing a supply of fresh air as soon as possible and changing exposed clothing. If symptoms

SYMPTOMS OF TEAR AGENTS	SYMPTOMS OF VOMITING AGENTS
• Violent burning sensation of the eyes • Reddening of the eyelids • Uncontrollable winking • Excessive tearing • Intolerance to light • Burning sensation of the throat, with developing pain and a sensation of choking • Sneezing • Nausea • Diarrhea • Headache • Burning sensation of the skin	• Irritation of the eyes, mucous membranes of the mouth and nose • Runny nose, sneezing, and coughing • Headache • Tightness and pain in the chest • Nausea and vomiting

Student Notes:

continue, the eyes, mouth, and skin should be flushed with large amounts of water. Although the effects of vomiting agents can be dramatic, personnel can usually perform duties despite their effects. Personnel should continue to wear a face mask even though coughing, sneezing, salivating, and nausea occur. (The mask can be lifted from the face briefly to allow for vomiting and to drain saliva from the face piece). Analgesics can be given to relieve headache and general discomfort.

REVIEW 1 QUESTIONS

Q1. What term is used to describe weapons that destroy large areas or kill and disable large segments of a population?

Q2. What is the most probable delivery method for chemical or biological weapons?

Q3. List the two types of antipersonnel agents.

a.

b.

Q4. The use of nerve agents produces symptoms that are similar to what other, more common condition?

Q5. What part of the body is most affected by blister agents?

Q6. If you are exposed to a blood agent, what action should you take first?

Q7. True or false. Cough suppressant and pain relievers can be given to a victim of a choking agent.

BIOLOGICAL WARFARE

Learning Objectives: When you finish this chapter, you will be able to—

- Recall the terms used with biological warfare (BW).
- Identify the types of BW.
- Identify the effects of BW.

Biological warfare (BW) is the intentional use of living organisms, toxins, and microtoxins to disable or to destroy people and domestic animals, damage crops, or deteriorate supplies. BW might be used on a large scale; therefore, biological immunizations of military forces and the development of detection equipment, such as the Interim Biological Agent Detection System (IBADS), are being used. Some of the types of BW agents and their symptoms are described in table 13-2.

Do not underestimate BW as a weapon. BW agents can be produced on a scale not considered possible in the past. Even small nations with modern, adequate research facilities can produce large quantities of BW toxins and microtoxins more cheaply than they can produce other types of weapons. These toxins and microtoxins are hundreds to thousands times stronger than today's chemical weapons. The disadvantage of BW agents is that many are rapidly degraded when exposed to certain environmental conditions, such as ultraviolet radiation, visible radiation, heat, dryness, or humidity.

Animals, insects, and rodents can be used as carriers to spread BW agents. Saboteurs can also infect large numbers of people by contaminating a water supply. Infecting water, milk, and food supplies with microorganisms can spread diseases, such as anthrax, typhoid fever, cholera, and influenza.

In the early stages of any biological disease, the general symptoms include fever, malaise, and inflammation.

Student Notes:

Table 13-2.—Characteristics for Selected BW Agents

Disease (common name)	Causative Agent	Physiological Effects	Time to effect
Anthrax	Bacillus anthracis	Mild fever and fatigue, worsening to severe respiratory disorders, high fever and excessively rapid pulse rate. Death can occur within 5-12 days of exposure if left untreated. Pulmonary anthrax is fatal more than 90% of the time.	1-5 days
Plague	Yersinia pestis	Fever, headache, and rapid heart rate, followed by pneumonia and hemorrhaging in the skin and mucous membranes. Untreated plague pneumonia fatalities approach 100%, but early treatment can reduce mortality to as low as 5%.	2-3 days
Tularemia	Francisella tularensis	Symptoms include fever, chills, headache and muscular pain. Untreated tularemia can result in 30-60% mortality; treated, the mortality rate is reduced to 1%.	3-5 days
Botulinum Toxin	Clostridium botulinum	Initial symptoms include extreme weakness, nausea, headaches, and intestinal pain leading to respiratory paralysis that may cause death.	2-26 hours

The degree of **fever** varies with the individual, depending on the person's resistance. However, fever does serve as a rough guide to the severity of infection. Often a violent chill precedes the fever. Whether the chill occurs or not, fever is usually one of the earliest symptoms.

Malaise is a feeling of bodily discomfort and weakness. There may be nausea, dizziness, loss of appetite, and general aches and pains.

Inflammation is caused by the reaction of body tissues combating and sealing off an infection. In almost every case there is pain, redness, and swelling. Some types of infection result in a characteristic rash, making it possible for a doctor to make an early diagnosis.

REVIEW 2 QUESTIONS

Q1. BW is the intentional use of

(a) ________________ to disable or

destroy (b) ________________.

Q2. What is the disadvantage an enemy has when using BW agents?

Q3. List the symptoms of biological disease in its early stages.

a.

b.

c.

NUCLEAR WARFARE

Learning Objectives: When you finish this chapter, you will be able

- Recall the terms used with nuclear warfare.
- Identify the types of nuclear warfare and the effects of nuclear weapons.
- Identify self-aid and first-aid methods for countering the effects of nuclear radiation.
- Recall the difference between radiological and radiation contamination.

Student Notes:

In one way, nuclear weapons are no different from ordinary high-explosive bombs—both are designed to cause destruction by blast and shock effects. Of course, nuclear weapons have a much greater destruction capability than conventional high-explosive weapons, with the added effects of nuclear radiation.

Nuclear explosions are classed according to the point of detonation with relationship to the surface of the earth—a high altitude blast, an air blast, a surface blast, and a subsurface blast.

HIGH ALTITUDE BLAST

A high altitude blast (fig. 13-1) is defined as a blast that takes place above 100,000 feet. The major aim of this blast is to destroy or interrupt satellites and communication systems through the effect of an electromagnetic pulse (EMP). Basically, the EMP is an intense electrical surge that affects electronic or electrical equipment in a burnout that's equivalent (equal) to that caused by a lightning strike.

AIR BLAST

An air blast (fig. 13-2) is one in which the fireball is below 100,000 feet and doesn't touch the earth's surface. The radiation effects from an air blast are minimal. The main reason for using an air blast is its destructive value produced in the expansion and compression phases of weapon detonation. The blast causes an over pressurization that crushes everything in its path. The front of the blast is called the *mock front.* An air blast would be most effective to use against a battle group at sea because it would structurally damage and/or sink many ships.

SURFACE BLAST

A surface blast (fig. 13-3) is one in which the fireball touches the earth's surface. Most of the damage caused by a surface blast is due to the shock (or blast) wave that accompanies the explosion. Large amounts of surface materials are vaporized and taken into the fireball. As the fireball rises, more debris is sucked up by the strong after winds. Much of this debris returns to earth as radioactive fallout.

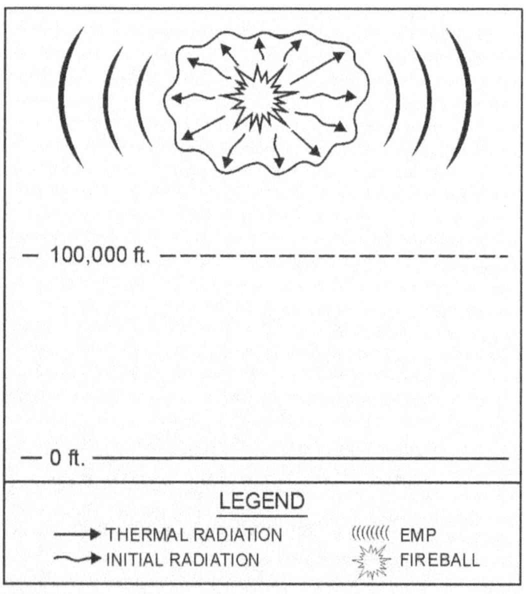

Figure 13-1.—A high altitude blast.

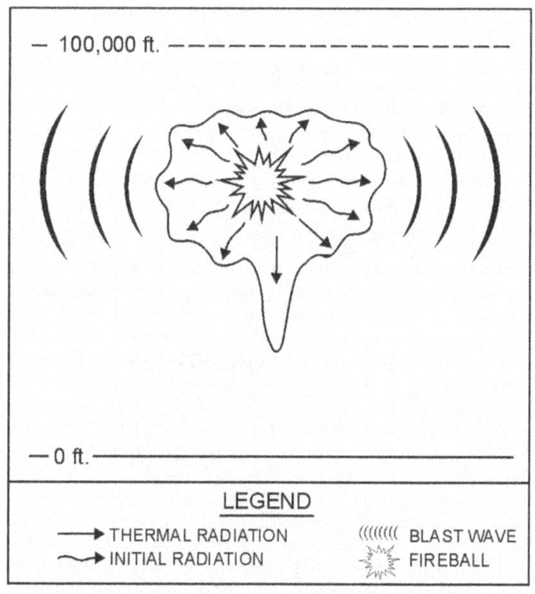

Figure 13-2.—An air blast.

Student Notes:

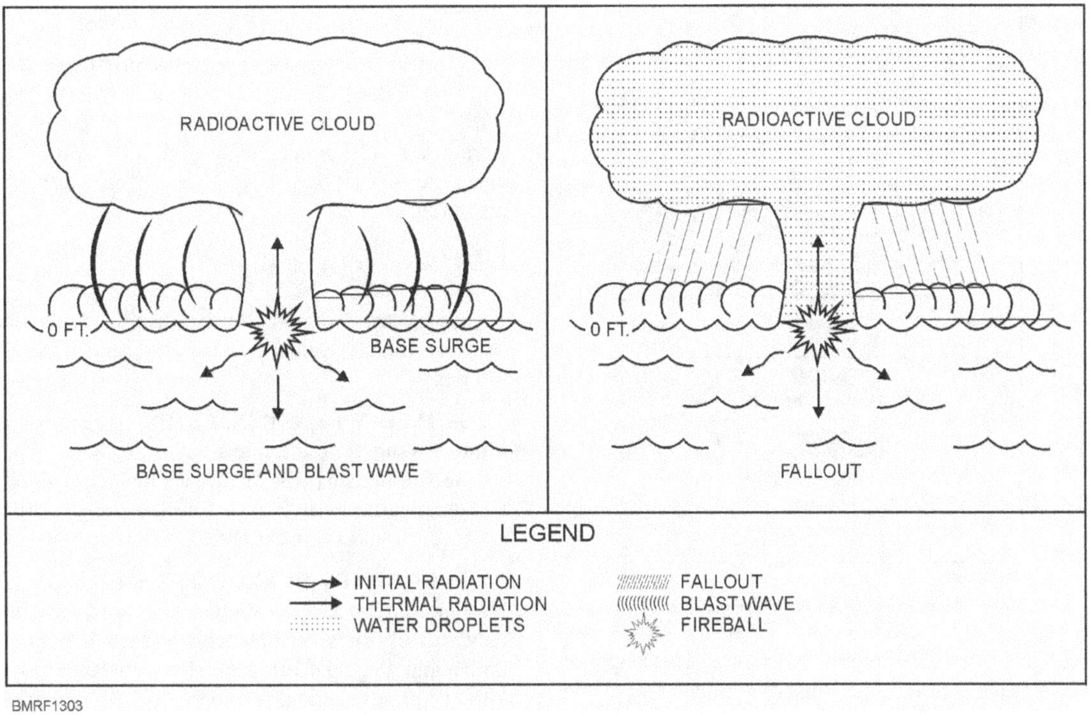

Figure 13-3.—A surface blast.

The effective range of blast damage is less than that from an air blast because much of the energy is transmitted in the form of a ground or water shock wave. Near ground zero, however, the severity of the shock wave is greater than that of the blast wave. The distance at which thermal radiation (heat) is hazardous is slightly less than that from an air blast.

Nuclear radiation is of two types—initial and residual.

Initial Radiation

Initial radiation occurs within the first minute after an explosion; residual radiation occurs thereafter. In most instances, initial radiation is of little consequence because the lethal range of its effects is less than that of the blast wave.

Residual Radiation

Residual radiation for a surface burst is dangerous because the large amount of surface material drawn into the cloud is heavy enough to fall while still highly radioactive. Additionally, the fallout area of a surface blast is much larger than the area affected by heat and shock.

SUBSURFACE (UNDERWATER) BURST

In an underwater burst (fig. 13-4) a fireball is formed. However, it's smaller than the fireball of an air burst and is not normal. The explosion creates a large bubble (cavity) that rises to the surface where it expels steam, gases, and debris into the air. Water rushing into the cavity is thrown upward in the form of a hollow column that may reach a height of several thousand feet. When the column collapses, a circular cloud of mist, called the *base surge*, is formed around the base of the

Student Notes:

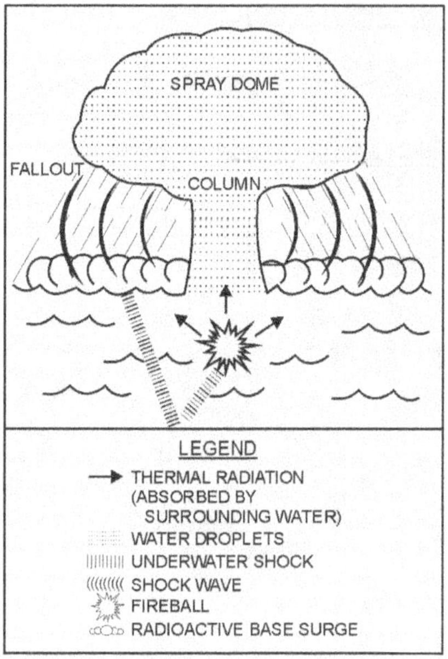

Figure 13-4.—A subsurface burst.

column. The base surge billows upward to a height of several hundred feet and expands rapidly outward to a distance of several thousand yards. Then it gradually rises from the surface and merges with the cloud formed by the escaping fireball.

EFFECTS OF NUCLEAR WEAPONS

Detonation of the nuclear bomb creates a blast wave that travels outward in all directions at an initial speed much greater than the speed of sound. When the wave strikes the earth's surface, another wave is formed by reflection. At some distance from ground zero (depending on the height of the blast), the primary and reflected waves combine to form a reinforced blast wave. Pressure at the wave front, called *overpressure*, is many times that of normal atmospheric pressure and is what causes most of the physical damage. Additionally, underwater bursts create large water waves, some of which reach heights of over 90 feet within a few hundred feet from the blast. The water waves travel outward at high speed for a distance of several miles, gradually diminishing in size. The overpressure decreases as the distance from the blast increases, but it can cause damage many miles from the blast.

Nuclear weapons produce explosions of great force and heat and release nuclear radiation. Their primary purpose is the mass destruction of property and personnel. Their effects are divided into three categories—blast waves or shock waves, incendiary, and radiation.

Blast Waves or Shock Waves

Injuries caused by blast waves can be divided into primary (direct) injuries and secondary (indirect) injuries.

PRIMARY BLAST INJURIES.—Primary blast injuries result from the direct action of the air shock wave on the human body. The greater the weapon's size, the greater the blast wave's effective range, with a subsequent increase in casualties.

SECONDARY BLAST INJURIES.—Secondary blast injuries are caused by strong blast winds reaching hundreds of miles per hour collapsing buildings and timber and flinging debris about. Personnel may also be hurled against stationary objects or thrown to the ground by high winds accompanying the explosion.

At sea, the shock wave accompanying an underwater burst produces various secondary injuries. Causalities resemble those caused by more conventional underwater weapons, such as mines and depth charges. Instead of being localized, the casualties extend over the entire ship. Also, injuries result from personnel being thrown against fixed objects or structures. Unsecured objects can act as missiles and cause many injuries.

Incendiary

There are two general ways fires can originate in a nuclear explosion.

1. First, kindling fuels can be ignited as a direct result of the absorption of thermal radiation.
2. Second, fires can be started from electrical short circuits, broken gas lines, or other interrupted heat sources as an indirect effect of the blast wave.

Student Notes:

Interaction of the blast wave, fire, and extent of blast damage are important factors in determining fire spread.

Flash burns are likely to occur on a large scale as a result of an air or surface blast of a nuclear weapon. Because thermal radiation travels in straight lines, it burns primarily on the side facing the explosion. But under hazy atmospheric conditions a large proportion of the thermal radiation may be scattered, resulting in burns received from all direction. Depending on the size of the weapons, second-degree burns may be received at distances of 25 miles or more.

The intense flash of light that accompanies a nuclear blast may produce flash blindness, even at a range of several miles. Flash blindness is normally temporary, though, the eyes can recover in about 15 minutes in the daytime and in about 45 minutes at night. A greater danger lies in receiving permanent damage to your eyes caused by burns from thermal radiation, which may occur 40 miles or more from a large-yield nuclear weapon.

Under some conditions, individual fires created by a nuclear explosion can come together into mass fires with great potential for destruction. The most significant types of mass fires are divided into two categories—firestorms and conflagrations.

FIRESTORMS.—In a firestorm, many fires merge to form a single column of hot gas that rises from the burning area. Strong, fire-induced, radial winds are associated with the column. Therefore, the fire front is essentially stationary and the outward spread of fire is prevented by the in-rushing wind. Virtually everything combustible within the firestorm area is destroyed.

CONFLAGRATIONS.—Conflagrations have moving fire fronts driven by the wind. Conflagrations can spread as long as there is fuel. Unlike firestorms, conflagrations can develop from a single ignition.

Radiation

Nuclear radiation hazards consist of alpha and beta particles, gamma rays, and neutrons.

ALFA PARTICLES.—Alpha particles have little skin-penetrating power and must be taken into the body through ingestion or cuts to be injurious.

BETA PARTICLES.—Beta particles can present a hazard to personnel if the emitters of these particles (carried in contaminated dust, dirt, or bomb residue) come into contact with the skin or get inside the body. Beta particles with enough intensity cause skin burns (radiation burns).

GAMMA RAYS.—Gamma rays are pure energy and not easily stopped. They can penetrate every region of the body. In fact, many gamma rays will pass right through a body without touching it. However, gamma rays that do strike atoms in the body cause the atoms to ionize. The ionization may result in any number of possible chemical reactions that damage the cells of the body.

NEUTRONS.—Of all the nuclear radiation hazards, neutrons have the greatest penetrating power. When the neutron is captured in the atoms of various elements in the body, atmosphere, water, or soil, the elements become radioactive and release high-energy gamma rays and beta particles.

Initial radiation contains both gamma and neutron radiation. Residual radiation, our greatest concern, contains both gamma and beta radiation.

EFFECTS ON SHIPS AND SHIPBOARD SYSTEMS

Ships close to a detonation point may sustain considerable material damage from air blast, underwater shock, water waves, and possibly thermal radiation. There will be a ship kill zone around ground zero. Outside ground zero, there will be a much larger damage-survival zone. Here, ships will receive severe, moderate or light topside damage as well as operational and equipment damage.

Damage from an Air Blast

Depending on the weapon yield, the blast wave from nuclear detonations can cause damage to ships miles from the blast. Damage will be inflicted primarily on the superstructure and the hull above the waterline. Some examples of damage from an air blast might include the warping or buckling of the flight deck; a distortion of airplane elevators, hull girders, deck

Student Notes:

machinery and radar antennas; and the cracking of seams above and below the waterline.

Damage from Underwater Shock

The pressure pulse created in water by an explosion on or below the surface is called an *underwater shock*. It travels much faster than an air blast and can inflict damage to ships at a distance of several miles. Possible effects include damage to the hull, heavy machinery, gun mounts, and electronics systems.

Damage from Water Waves

An underwater nuclear burst may result in waves over a hundred feet in height, but water waves are seldom the primary source of ship damage. The impact of water waves may cause distortion of the superstructure, carry away deck gear, or flood through damaged weather doors.

Damage to Ships Tactical Systems

Nuclear detonation can cause considerable damage to tactical systems, including electrical and electronic systems, sonar, radar and communications. Such damage can be a result of an electromagnetic pulse (EMP), transient radiation effects on electronics, blueout, or blackout.

ELECTROMAGNETIC PULSE (EMP).—Shipboard damage occurs when metal conductors, such as electrical cables, antennas, and sensors, absorb EMP. Computers and other equipment using solid-state components are most vulnerable to EMP. Vacuum-tube equipment is less susceptible. Personnel aren't directly injured by EMP, but they may suffer electrical shock if they are in contact with a large conductor of electrical energy.

Preventive measures to protect or *harden* equipment against damage by EMP include metal shielding, good grounding, use of surge arresters, and the proper arrangement of electrical wiring.

TRANSIENT RADIATION EFFECTS ON ELECTRONICS (TREE).—TREE occurs in electronics systems as a result of exposure to gamma or neutron radiation. The actual effects are determined by the characteristics of the circuits in the electronics package, the components in the circuits, and the construction techniques and materials used to make the components. In general, radios, radar, computers, cables and wiring, and inertial guidance systems are susceptible to TREE. The response of such systems to radiation depends on the nature of the radiation and on the specific components and operating status of the systems.

BLUEOUT.—Blueout is the prolonged disturbance of an underwater nuclear detonation and is caused by ocean basin shock reverberations that interfere with passive sonar systems. The noise resulting from the initial nuclear weapon detonation (the interaction of steam and water and the pulsations of the steam bubble) masks out all other sound for a short period of time making it impossible for sonar operators to listen for target data. The effects of blueout are temporary.

BLACKOUT.—Blackout, caused by an atmospheric nuclear explosion, is the interference of radio transmissions through ion fields formed in a detonation. In a tactical situation, straight-line communications (radar and radio transmissions) between ships on opposite sides of the fireball will be lost. Following a high altitude detonation, satellite communications may be affected or lost. Blackout alters or inhibits radar or radio waves and affects all frequency bands.

Procedures to counteract the effects of blackout include providing alternate paths for communications, shifting radio operating frequencies, changing transmission modes, and waiting for blackout effects to diminish.

REVIEW 3 QUESTIONS

Q1. List the four types of nuclear weapon explosion classification.

a.

b.

c.

d.

Student Notes:

Q2. Describe why residual radiation is more dangerous than initial radiation.

Q3. Describe how a secondary blast can cause injuries.

Q4. List the nuclear radiation hazards.

a.

b.

c.

d.

Q5. List the measures that should be taken to protect electronic equipment from the effects of EMP.

a.

b.

c.

d.

CONTAMINATION, DETECTION, AND IDENTIFICATION

Learning Objectives: When you finish this chapter, you will be able to—

- Identify the purpose of CBR monitoring and decontamination teams.
- Identify the markers used to indicate CBR contamination.
- Recall the purpose of the markers used to indicate CBR contamination.

For a ship or station to retain its offensive power and carry out its mission, immediate detection and identification of radiation and BW and CW agents are of great importance. However, the nature of radiation and BW and CW agents makes it difficult to detect and identify them. Here are some examples.

You know a nuclear attack is taking place because you can see it, hear it, and feel it. But, you can't see the nuclear radiation. Nuclear radiation is just as deadly over a period of time as the blast itself. A biological and chemical attack can be just as invisible. You might not know about them until it's too late. Because CBR attacks might be invisible, you need to recognize symptoms of radiation and BW and CW contamination.

SURVEY TEAMS

After a CW, BW, or nuclear attack, survey teams go through the ship to determine the extent and location of any contamination. Rapid detection and identification are vital so that effective defense measures may be taken immediately. A survey team, or monitoring party, consists of a minimum of three people—a monitor, a recorder, and a messenger.

The **monitor** is in charge of the party. The monitor carries high-range and low-range survey meters. The monitor is responsible for the safety of the team and for determining intensities and locations of contamination.

The **recorder** maintains a record of intensity readings (obtained by the monitor), time of the readings, location of the hazardous areas, and specific hazards. Also, the recorder may act as a marker, using line to rope off hazardous areas and chalk to mark on bulkheads and decks the intensities of contamination found during the survey.

The **messenger** reports to damage control central (DCC) the contaminated areas and the readings obtained by the monitor. In DCC, personnel plot the reports from the various teams to get a general outline of contaminated areas, to pinpoint hot spots (areas of higher-than-average intensities), and to establish stay times for specific areas (fig. 13-5).

Student Notes:

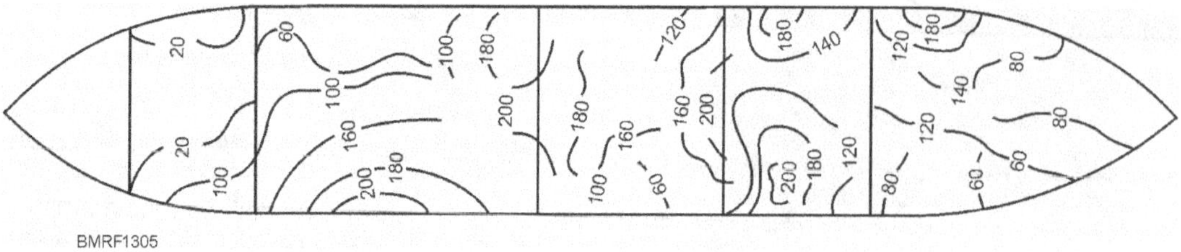

Figure 13-5.—General outline of contaminated areas on weather decks.

Two types of surveys are usually conducted—a rapid, or gross, survey and a detailed survey.

The **rapid survey** is a preliminary reconnaissance. Limited numbers of readings are taken in a minimum amount of time. The purpose of the rapid survey is to obtain a quick estimate of radiation levels at specified locations to determine the possibility of keeping stations manned.

A **detailed survey** is used to determine the effectiveness of decontamination measures. All accessible areas and equipment are surveyed in a slow, methodical manner. Special attention is paid to areas that tend to hold contamination (rust spots, caulking in wood decks, canvas, rope, and so on).

Each member of a monitoring team wears a protective mask and clothing and is equipped with both a pocket dosimeter and a high-range casualty dosimeter. No member with an open cut or wound should enter any contaminated area. Smoking, drinking, and eating are prohibited in contaminated areas.

CBR CONTAMINATION MARKERS

A standard system is used to mark areas contaminated by CW, BW, or nuclear agents. Look at figure 13-6, which shows CBR contamination markers. The markers are triangular in shape, with a base of approximately 11 1/2 inches and sides of about 8 inches. Each type of contamination is readily identified by the color of the marker. Additionally, they are labeled GAS, BIO, or ATOM, as appropriate. The front of the marker indicates the safe limits of the contaminated area. **Never go beyond the markers without permission**. The front of each marker also contains information about the contaminated area, such as the date and time of detection and the type of agent.

NUCLEAR RADIATION

When a ship is exposed to radiation or is radiologically contaminated (such as from a base surge or fallout), surveys are made to determine the degree of contamination.

During surveys, two types of measurement are made—intensity (dose rate) of the radiation field and the total amount (dose) received. This information is used to calculate (find) the safe entry time (time after exposure when an area may be entered safely) and stay time (length of time a person may remain in an area without exceeding permissible radiation exposure levels). Dose rate is expressed in roentgens (gamma ray measurement only). The total dose is expressed in rads (any type of radiation).

One measurement instrument is the radiac meter (radiac stands for *radioactivity detection, indication, and computation*). Usually, only qualified damage control (DC) personnel use the radiac meter; therefore, only the personnel dosimeter is covered here.

Measurements are made using two basic types of personnel dosimeters—self-reading and nonself-reading.

The self-reading pocket dosimeter (fig. 13-7) is about the size and shape of a fountain pen and comes in the following ranges:

0 to 5 roentgens

0 to 200 roentgens

0 to 600 roentgens

0 to 200 milliroentgens

Self-reading instruments measure exposure to radiation over a period of time, not dose rates at any given time. Hold the dosimeter up to a light source and

Student Notes:

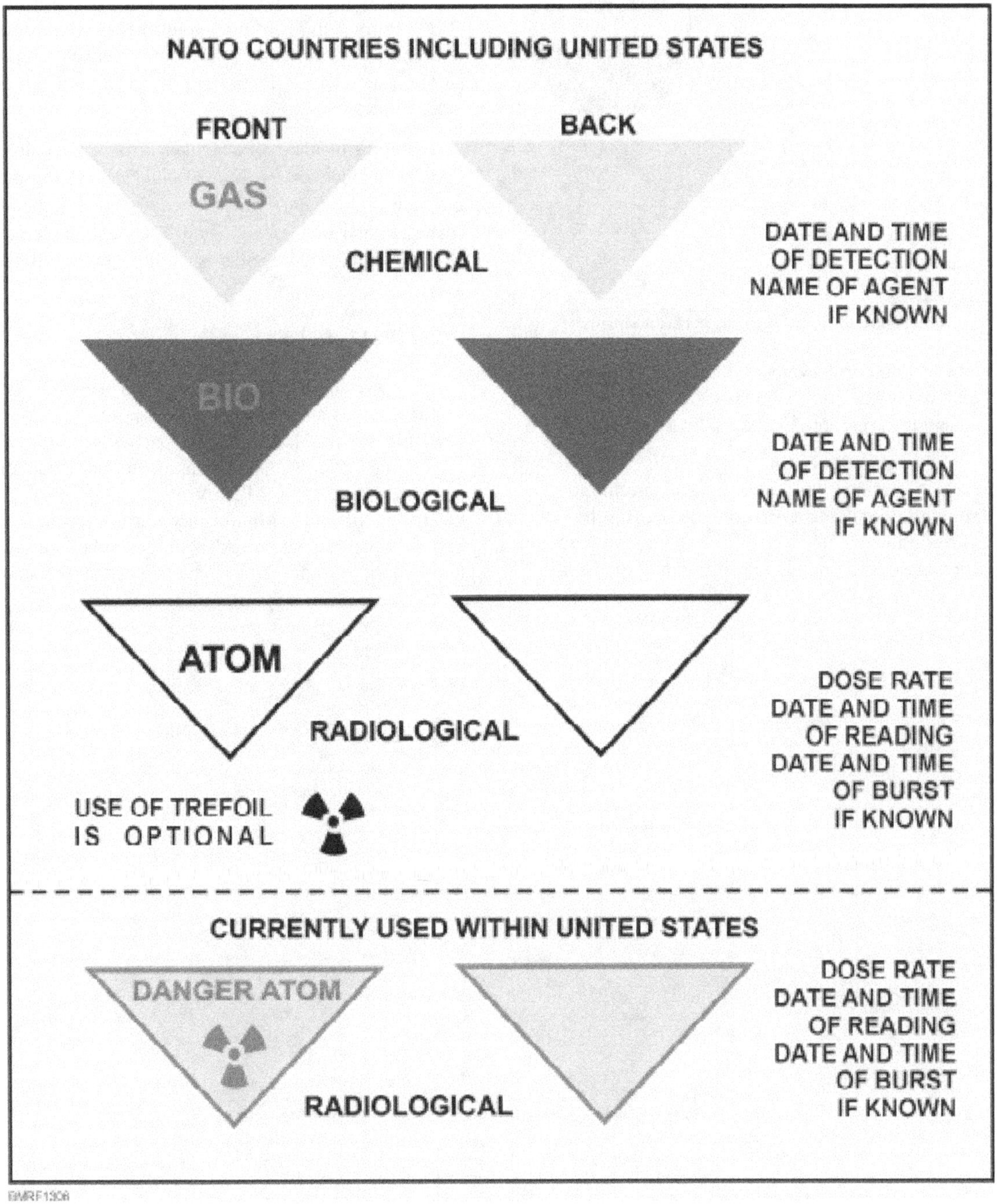

BMRF1306

Figure 13-6.—CBR contamination markers.

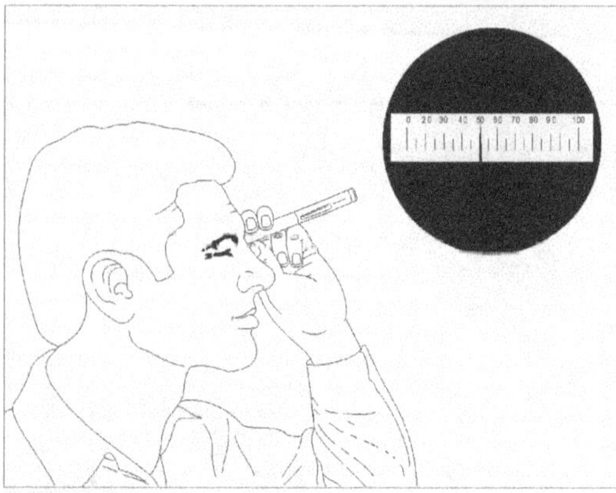

BMRF1307

Figure 13-7.—Self-reading pocket dosimeter.

look through the eyepiece; the total radiation dose received is read directly on the scale. After each use, the dosimeter is recharged and the indicator line set to zero.

The nonself-reading category is a high-range casualty dosimeter (fig. 13-8). To determine the total amount of gamma radiation the wearer has been exposed to, it's put in a special radiac computer-indicator. Its range is 0 to 600 roentgens.

BIOLOGICAL AGENTS

No simple or rapid method can be used to detect BW contaminants. The only known method consists of two phases—a sampling phase conducted by a CBR survey team and a laboratory stage conducted by medical personnel.

Samples of material are taken from a wide area. Samples include air, surfaces of bulkheads and decks, clothing, equipment, water, food, or anything else suspected of being contaminated. Then the samples are shipped to a medical laboratory for identification of the agent.

CHEMICAL AGENTS

Warning of a CW attack based on detection by the physical senses alone is not only dangerous but would probably be too late. This is particularly true if fast-acting nerve agents were used. Special detection equipment, such as the M256A1 vapor sample detector kit and the M8 and M9 liquid chemical agent papers, is used to detect CW agents. Also, draeger tubes are used to detect the presence of phosgene gas. Other pieces of

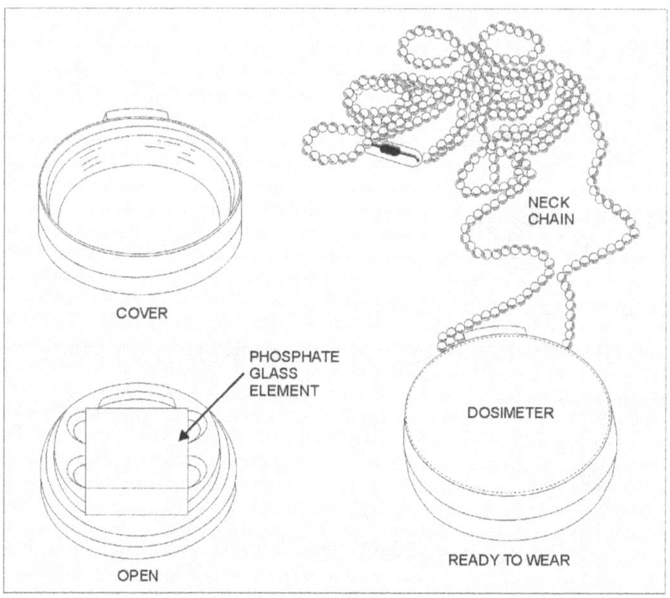

BMRF1308

Figure 13-8.—High-range casualty dosimeter, DT-60/PD.

Student Notes:

CW detection equipment used by Navy personnel include the portable AN/KAS-1A chemical warfare directional detector (CWDD) and the permanent chemical agent point detector system (CAPDS). No one piece of equipment can detect all CW agents, which is why the Navy uses several different methods of CW detection.

M256A1 detector kits are used to check areas suspected of being contaminated, to test an area after decontamination operations, and to indicate when masks might be removed. The kits are not designed to indicate when it is necessary to don (put on) gas masks.

REVIEW 4 QUESTIONS

Q1. A survey team consists of—

a.

b.

c.

Q2. What are the two types of surveys?

a.

b.

Q3. Biological markers are (a) what color with (b) what color inscription?

a.

b.

Q4. To calculate safe entry time and stay time in a radiologically contaminated area, you need what two measurements?

a.

b.

Q5. Describe the only known method for detecting BW contaminants.

Q6. To check areas suspected of being contaminated by CW agents, you should use what kit?

CBR DEFENSE PROTECTIVE MEASURES

Learning Objective: When you finish this chapter, you will be able to—

- Recognize the procedures to follow in case of a CBR attack.

For a ship or shore activity to be able to continue its mission after a CW, BW, or nuclear attack, personnel must be protected. Protective measures include both individual and group actions. Individual protection is an immediate concern. What you do in the first few moments of a CW, BW, or nuclear attack may keep you alive!

WHAT TO DO IN A CBR ATTACK

In a nuclear attack, defensive measures are much the same as the general damage control precautions taken against any explosion. These measures are to keep things squared away, maintain watertight integrity, make repairs as quickly as possible, protect yourself with your clothing and protective mask, be ready to fight fires ignited by the blast, and be ready to administer first aid to shipmates who are injured or burned. What you learned about damage control and firefighting in chapter 12 also applies to the damage and fires caused by nuclear weapons.

General Precautions to Follow in a CBR Attack

As soon as the initial effects of the explosion are over, you should then take the following precautions:

1. Put on your mask immediately or cover your nose and mouth with a handkerchief or cloth.

Student Notes:

2. Adjust your clothing to cover exposed skin (battle dress).
3. Slip on a protective cover, if you have one, or cover yourself with anything at hand.
4. Keep upwind of the explosion, if possible.
5. Administer first aid to yourself and to others.
6. If you are not a casualty, report to your duty station or to the designated area where you can take a shower and get clean clothes.
7. Keep your hands away from your face, particularly your mouth.
8. Don't eat, chew, drink, or smoke until the items are checked by a medical officer.
9. Don't stir up dust or step into puddles.
10. Don't brush against or touch decks, bulkheads, structures, or objects in the contaminated area.

Breathing radioactive particles is dangerous. Take shelter from dust clouds raised by wind, by aircraft, or by moving vehicles. Otherwise, use a protective mask or a handkerchief for protection.

Nuclear Attack

If there is warning of a nuclear attack, the word is passed to take cover. When the word is passed, go to your designated shelter as quickly as possible. At the sound of the alarm, get your protective mask ready. If you are ordered to a shelter, remain there until the all-clear signal is given.

In general, the farther you are below the main deck (deep shelters), the better the protection from nuclear radiation. To reduce the contamination from the base surge and from fallout, secure the appropriate Circle W fittings. All topside openings will be closed for as long as the ship is in the danger area.

BW or CW Attack

If you are in a BW or CW attack, avoid the spray, mist, or cloud if you can. Wear your mask, cover your body as much as possible, and seek shelter. Assume that all surfaces in the vicinity of the attack are contaminated; leave the area quickly, and follow the route to the closest decontamination area. Report any sickness promptly, and do not eat, drink or smoke. Since BW and CW agents can sometimes enter your body through the skin, cover any cuts or scratches. As with nuclear warfare protective measures, if you have no mask with you, cover your nose and mouth with your handkerchief or cloth (such as a rag or shirt).

PROTECTIVE EQUIPMENT

The protective equipment described here includes the MCU-2P mask, clothing, and antidotes for certain chemical agents.

- You should know how to use a MCU-2/P protective mask and how to apply antidotes.
- Two types of clothing are useful, to varying degrees, in CBR defense—wet-weather clothing and ordinary work clothing

MCU-2/P Protective Mask

The protective mask, or gas mask, is your most important piece of protective equipment against CBR agents. It protects your face, eyes, nose, throat, and lungs. The reason it is so important is because inhaling CBR agents is much more dangerous than getting them on the outside of your body. Without filtration of some kind, a large amount of contamination could be inhaled in a short time.

The mask serves two functions:

1. It filters the air, removing particles of dust that may be radioactive or otherwise contaminated.
2. It purifies the air of many poisonous gases.

The mask does not produce oxygen. Therefore, it doesn't provide protection against smoke or against toxic gases, such as carbon monoxide, carbon dioxide, and ammonia. Therefore, it may be used for emergency escape only as a last resort. When entering a compartment containing such gases, you must use an oxygen breathing apparatus or an air hose mask.

The operation of the mask is simple. On inhalation (breathing in), the air passes through a filter system that filters and absorbs the CBR agents. Exhaled (breathing out) air is expelled through a one-way valve.

Student Notes:

From the moment you hear the alarm or suspect a CBR attack, hold your breath until you can put on the mask. You should be able to don (put on) and adjust your mask within 10 seconds. If your eyes or face becomes contaminated before you can get the mask on, the contamination should be taken care of first, provided the necessary materials are readily available. The most important action is to don the mask immediately; then, proceed with decontamination.

The MCU-2/P protective mask is designed to provide full protection. It provides protection against tactical concentrations of chemical and biological agents, toxins, and radiological fallout particles. The MCU-2/P mask also accommodates the use of the tri-service/NATO canisters.

The MCU-2/P protective mask (fig. 13-9) is built with a silicone rubber facepiece. It has the following features:

- Two voice emitters
- A drinking tube
- A flexible lens that lets you use binoculars, gunsights, and other optical equipment; and the option to put the filter canister on either side

The mask can be worn over approved mask-compatible glasses. You can order compatible glasses through your medical department. The large lens size provides the user with a good all-around view.

CBR Protective Clothing

Basically, any clothing or coverall that covers the body can provide a degree of protection from CBR contaminants. However, the type of clothing and its proper wear determine the amount of protection.

CHEMICAL-PROTECTIVE OVERGARMENT.—The chemical-protective overgarment consists of two pieces—a smock and trousers (fig. 13-10). The smock has two layers of materials: inner (antigas) and outer (monacrylic/nylon). The smock is generously cut to allow complete freedom of movement. It has a large front flap pocket for gloves, and so forth, and a sleeve patch where you can place detector paper for easy visibility. You can make quick and easy adjustments with hook-and-pile fasteners at the wrist and waist. The trousers are made of the same two layers of material and have suspender-type fittings located at the waist and across the shoulders. Hook-and-pile fasteners are located at the base of each leg for adjustment. The chemical-protective overgarment is issued in a plastic envelope that is

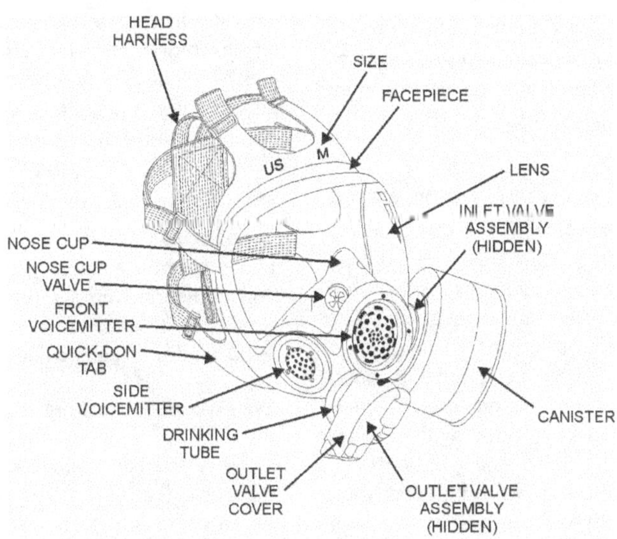

BMRF1309

Figure 13-9.—MCU-2/P protective mask.

Student Notes:

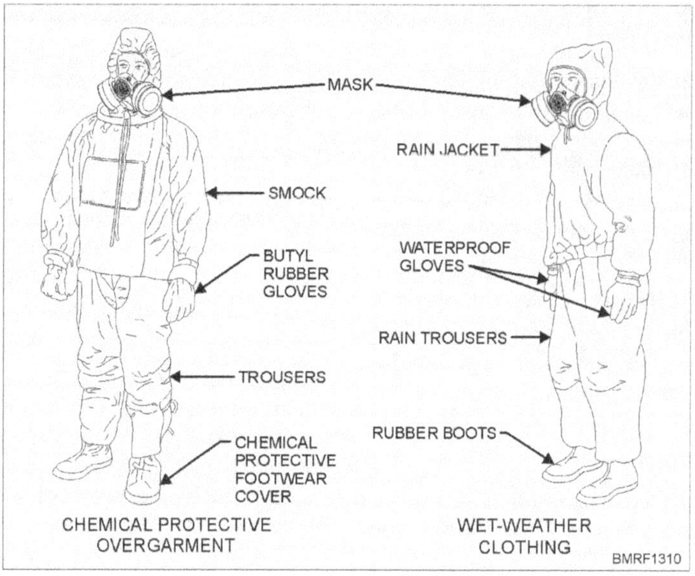

Figure 13-10.—Types of chemical-protective overgarments.

pressure packed, air evacuated, and heat sealed. It is then placed in a polyethylene bag and heat sealed. The overgarment has a shelf life of 5 years when unopened.

The protective overgarment protects against all CBR agents and is permeable to water vapor. Once removed from its protective envelope, it has a shelf life of 14 days in a nonchemical environment. If it is opened but uncontaminated, keep it for training purposes. Once exposed to chemical contamination, the overgarment provides 6 hours of continuous protection, after which it should be discarded.

CHEMICAL-PROTECTIVE FOOTWEAR COVERS.—The chemical-protective footwear covers (overboots) are worn over the standard work shoe and provide protection to the feet against exposure to all known concentrations of nerve and blister agents. The overboots are made of loose-fitting, impermeable, butyl sheet rubber and have a premolded, nonslip, butyl rubber sole. The overboot is approximately 16 inches high with a grommet lace closure, including five eyelets to allow lacing around the foot. The overboots are available in two sizes and can be worn on either foot. They are issued in a polyethylene bag with two pairs of laces and an instruction sheet. Upon contamination, the overboots provide 6 hours of protection from agent penetration.

CHEMICAL-PROTECTIVE GLOVE SET.—The chemical-protective glove set is worn to protect the hands against nerve and blister agents, liquids, and vapors. The set consists of an outer glove to provide chemical protection and an inner glove to assist in absorption of perspiration. The five-finger outer glove is made of impermeable, unsupported, black butyl rubber and is manufactured for both the right and left hand. The thin, white cotton inner glove can be worn on either hand. The glove set is issued in a clear polyethylene bag with an instruction sheet.

The black outer glove protects against chemical agent vapors, aerosols, and small droplets. Upon contamination, the set provides at least 6 hours of protection from agent penetration. These gloves, in good condition, can be decontaminated and reissued.

Wet-Weather Clothing

Wet-weather clothing (refer back to fig. 13-10) is often described as impermeable or rubberized clothing. Its value results from the fact that the previously described impregnated/protective clothing can be

Student Notes:

partially penetrated by all but the smallest droplets of liquid agents, especially in relatively high winds. Moreover, the impregnated/protective clothing is not equally efficient in neutralizing all liquid CW agents. On the other hand, wet-weather clothing is resistant to all liquid CW agents for a limited amount of time, provided that the closures at the neck, wrists, and protective mask are well adjusted or taped.

Wet-weather clothing provides a measure of protection against CBR contaminants when worn over ordinary clothing; but it provides the most complete protection when worn over impregnated or protective clothing. Gradual penetration of the synthetic rubber layer of the wet-weather clothing will eventually occur unless CW agent contaminants are promptly removed. The contaminants are removed by frequent and thorough flushing of the surface with a seawater washdown or an equivalent, such as jury-rigged topside seawater showers, or by swabbing with liquid hypochlorite.

In warm weather or during periods of increased physical activity, wet-weather clothing has a major disadvantage in that it can only be tolerated for relatively short periods of time. Tolerance is limited because no air can pass through the clothing to cool the wearer's body by the evaporation of perspiration.

Perspiration is normally accumulated inside an impermeable suit. Underclothing, gloves, socks, and shoes may become saturated. Sweating can be reduced and tolerance times lengthened by reducing the exercise rate, by using water-spray cooling, and by reducing exposure to direct sunlight.

Ordinary Work Clothing

Special protective clothing is not required for all personnel. Ordinarily, only the personnel of monitoring and decontamination teams who must work in or near hazardous areas wear it. All other personnel working near these areas should wear two layers of ordinary clothing, which provide partial protection against agents and radioactive particles.

REVIEW 5 QUESTIONS

Q1. True or false. Eating food after a CBR attack is okay as long as the food was in a sealed container before the attack.

Q2. Aboard ship, the safest place to be during a nuclear attack is ____________________.

Q3. What are the two functions of an MCU-2/P mask?

a.

b.

Q4. How long should it take you to don and adjust an MCU-2/P mask?

Q5. List the types of clothing that are useful for CBR defense.

a.

b.

MISSION ORIENTED PROTECTIVE POSTURE (MOPP)

Learning Objective: When you finish this chapter, you will be able to—

- Recall the procedures for protection at each level of mission oriented protective posture (MOPP).

Mission oriented protective posture (MOPP) is a means of establishing levels of readiness. MOPP is a flexible system of protection against chemical agents

Student Notes:

and is used in CW defense to help accomplish the mission.

The MOPP doesn't require that personnel wear protective clothing all the time. Duty requirements, body heat buildup, and basic human needs will prevent you from using full protective equipment for an infinite period of time. The MOPP does, however, give the CO the option of no protection to full protection, depending on the threat to the ship.

All operations are conducted under the MOPP system, even when there is no threat. There are four levels of MOPP—from Level-1, the least protection, to Level-4, the most protection.

MOPP Level-1

1. Individual protective equipment and medical supply items are issued to shipboard personnel and maintained at respective battle stations. Protective masks are fitted for immediate use.
2. Inventory stowed chemical/biological defense equipment and supplies.
3. Set readiness Condition III and material condition YOKE, if not already set.

MOPP Level-2

1. For both chemical and biological threats, protective mask is in a carrier and worn on the person.
2. Preposition decontamination supplies in decon stations and at repair lockers. Preposition stowed detection and monitoring equipment, supplies, and empty canteens as specified in the ship's CBR Defense Bill.
3. Set material condition ZEBRA (modified).

MOPP Level-3

1. Install new filter canisters on protective masks, maintain in a carrier and on the person. Provide wet-weather gear for donning over other protective clothing and equipment for weather deck activities. Don overgarment trousers and coat with hood down. Don chemical-protective overboots. Stow personnel decontamination kit in mask carrier. Stow chemical-protective glove set and medical supply items in pocket on overgarment coat. Initiate pyridostigmine pretreatment regimen.
2. Go to general quarters (GQ) (readiness Condition I may be relaxed and readiness Condition II set at CO's discretion); set material condition ZEBRA.
3. Fill prepositioned canteens with potable water.
4. Activate decontamination stations and contamination control areas (CCAs) and assure operability. Post detection and monitoring teams.
5. Post and monitor detection equipment and materials as designated by the ship's CBR Defense Bill.
6. Activate countermeasures washdown system intermittently.

MOPP Level-4

1. Don protective mask and secure hood over head and around mask. Don chemical-protective glove set.
2. Direct ship to GQ (if not previously in effect).
3. Initiate continuous monitoring and operation of detection equipment.
4. Set CIRCLE WILLIAM.
5. Activate countermeasures washdown system to operate continuously.

The setting of MOPP levels may be different at various locations around the ship. This depends on the mission, work rate, and heat buildup in these battle station areas (engine rooms, combat information center, flight deck, and so on).

REVIEW 6 QUESTIONS

Q1. What does the acronym MOPP stand for?

Q2. There are how many levels of MOPP?

Student Notes:

Q3. What MOPP level provides the most protection?

Q4. At what MOPP level is material condition ZEBRA (modified) set?

DECONTAMINATION

Learning Objectives: When you finish this chapter, you will be able to—

- Recall the procedures for area decontamination afloat.
- Identify the purpose of the decontamination station.

There are four levels of decontamination—emergency personnel decontamination, limited operational decontamination, operationally complete decontamination, and complete decontamination.

LEVELS OF DECONTAMINATION

Level 1—Emergency personnel decontamina- tion. Emergency personnel decontamination is decontamination necessary to save your life. It is your responsibility. The primary purpose of emergency personnel decontamination is to safeguard you in protective gear that includes the following items:

- Mask
- Protective overgarment
- Boots
- Gloves

However, if a chemical attack takes place before you don all of the protective gear, you **need to** destroy, neutralize, or remove the chemical agents from inside your protective gear and from exposed skin area. Personnel decontamination kits (M291) give you the ability to decontaminate skin surfaces. The cleansing/decontamination stations used for entering and leaving the ship's interior provide soap, detergent, and shower facilities.

Level 2—Limited operational decontamination. Limited operational decontamination is decontamination necessary to let you, while in protective clothing and/or masks, do your job with a minimum risk of contact, pickup, and transfer of chemical agent contamination. Initial contamination is most likely to be on the upper-outer surfaces of structures and equipment. Further contamination may be picked up and/or transferred to noncontaminated areas. The two types of contamination hazards are—

1. Pickup hazards. A chemical agent on a surface that is touched by an individual, contaminating himself/herself.
2. Transfer hazards. A chemical agent picked up, transferred, and then deposited on an otherwise uncontaminated area.

The objective of limited operational decontamination is to destroy, neutralize, or remove persistent chemical agents that are located on structures and/or equipment in places where they constitute a contact hazard.

Level 3—Operationally complete decontamination. Operationally complete decontamination (also known as *full decontamination*) is decontamination so that the contamination of personnel, structures, and equipment is reduced to a level that results in a significant operational benefit. Level 3 decontamination reduces contamination to the lowest level possible. However, it should only be conducted when there is a reasonable chance that work can be performed without masks or gloves for limited periods, and the ship's mission can be completed without undue hazards to personnel.

One hundred percent decontamination can't be accomplished on each and every item suspected of being contaminated. Level 3 decontamination isn't a fixed level of decontamination. It depends on the ship's operating schedule and the urgency of the assigned mission. Decontamination at sea or by ship's personnel will be of this type.

Level 4—Complete decontamination. Complete decontamination is a degree of decontamination where

Student Notes:

appropriate chemical tests fail to give a positive response for a residual agent. Decontamination at naval shipyards, advanced bases, or by shore-based personnel will normally be of the 100% chemically complete type. This level is **not** mission essential for shipboard units.

DECONTAMINATION OF THE SHIP

The purpose of decontamination is to remove or reduce CBR contamination so that the ship can carry out its mission without danger to the life or health of its crew. Each type of contamination requires different decontamination procedures. Radiological (nuclear) contamination may be removed by washing it over the side; CW agents may be neutralized; BW agents must be destroyed.

Nuclear Radiation Decontamination

Complete decontamination of a ship usually requires the service of a shipyard. However, radiation levels can be reduced by shipboard personnel to the point where radiation no longer presents a serious hazard to the crew. Most of the radioactive particles can be removed by washing down the ship. Two washdown methods are used—mechanical and manual.

MECHANICAL METHOD.—The mechanical method, called the *ship's water washdown system*, consists of a system of piping and nozzles that spray water over all weather surfaces. Water is supplied by the fire main.

NOTE

The washdown system actually is a preventive measure against fallout, rather than a decontamination method, because normally the system is activated before the ship enters the fallout area.

The water spray carries away the radioactive particles as they fall on the ship. At the same time, the flowing water fills in the cracks and crevices so that the particles that do get through the spray cannot settle into the cracks and crevices.

MANUAL METHOD.—If parts of the ship are contaminated before the washdown system is turned on, water from the sprinklers may not effectively reduce the radioactivity because the slowly flowing water doesn't have enough force to wash away the particles. The areas of heavy contamination must be hosed down with water under pressure. Hosing and scrubbing down the ship is the manual method.

Decontamination teams are formed to hose and scrub down the ship. A team usually consists of six people—the monitor, who is in charge; two hosemen; and three other team members. The hosemen wash down the hot spots with fire hoses, moving from the areas of less contamination toward areas of greater contamination, and working from top to bottom. Then the areas are scrubbed by the remaining team members with soap or detergent and water and are rinsed by hosing (fig. 13-11). The hosing-scrubbing-hosing continues until monitoring shows that contamination is removed or at least reduced to a safe level. Keep the contaminated water away from vent systems, doors, and hatches, because washing away the particles does not destroy them; they are simply being moved over the side.

BW and CW Decontamination

BW decontamination means eliminating the sources of infection. Using a chemical disinfectant is the most effective way to decontaminate BW agents. The type of disinfectant depends on the agent, the material to be decontaminated, and sometimes the area. Other methods include burning, dry heat, and moist heat. Burning usually is unsatisfactory because it naturally destroys surface material. An example of dry heat is a hot air oven set at 180°. Moist heat includes hot water or steam under pressure. Sunlight also is effective in reducing BW contamination. The ultraviolet rays of the sun kill most BW agents.

In **CW decontamination**, weather alone is the simplest means. Bright sunlight is a decontaminant, even in cold weather. However, lack of time, unfavorable weather, or contamination of critical areas may require a faster method. Enclosed spaces can be steamed. All spaces can be treated with liquid detergents. Water alone is often satisfactory as a flushing agent; hot water or steam is better than cold water.

Student Notes:

Figure 13-11.—A decontamination team at work.

PERSONNEL DECONTAMINATION

Each ship has a special area designated as a decontamination (decon) station. Personnel exposed to CBR agents are processed through these decon stations. Aboard ship, the decon stations are shower rooms, one forward and one aft. Large ships have more stations. Each decon station is divided into three parts—

1. A contaminated or an unclean area
2. A washing area
3. A clean area

Whenever practicable, the clean and unclean sections have separate access routes and entrances. Undressing is done in the unclean area, and containers are located there for the disposal of contaminated clothing. A box containing a mixture of sand and bleach may be located at the entrance to the undressing area; if so, scuff your feet in the box before entering the station.

The following are some general decontamination procedures you should remember:

1. Enter the undressing area after scuffing your feet in the box. Then, sit on a bench with both feet on the unclean side. Remove your shoes, swing your legs to the clean side of the bench, and remove your outer clothing only. In case of a BW or CW attack, keep your protective mask on. (Remove your protective mask only when told to do so.) Carefully remove your clothing to prevent the possible rise of a secondary aerosol. After placing your outer clothing in the containers, proceed to another section, remove your underwear and socks, and place them in the appropriate containers.

2. Proceed to the washing area. You should spend at least 5 minutes soaping, scrubbing, and rinsing. Give special attention to the hair, nails, skin creases, and ears, using a brush on the nails. You should rinse, soap, and scrub; then rinse again.

3. Proceed to the dressing area and dry off. (If nuclear contamination is involved, you will be monitored and required to repeat the shower until you are free of contamination.) Dress in clean clothing and proceed as directed.

Remember that showering doesn't destroy nuclear agents or many of the BW agents—it merely washes them away. Therefore, you should immediately report any illness (however minor) to medical personnel.

Student Notes:

M291 Decontamination Kit. The purpose of the M291 decontamination kit is to decontaminate skin and selected personnel equipment contaminated with chemical agents. The kit contains six sealed foil packets, enough for three complete skin applications.

Each packet contains a folded applicator pad with a handle on one side. The pad is filled with the black decontaminating powder, which is a reactive and absorbent resin that is not toxic but may be slightly irritating to the skin or eyes.

New and/or improved CBR defense and decontamination kits, clothing, and equipment are being introduced rapidly. Check with your supervisor to see if any new or improved articles are available.

Atropine and oxime are used to counteract the effects of and to relieve the symptoms of nerve agents only. At the appropriate level of readiness, each crew member will be issued three atropine autoinjectors and two oxime autoinjectors. To use the injectors, remove the safety caps and press the injectors against the thigh or buttocks. The pressure on the end of the injector causes the automatic injection of the contents. As soon as the symptoms of nerve agent poisoning are noticed, immediately inject one atropine autoinjector and one oxime autoinjector. Wait 10 to 15 minutes; if symptoms are still present, inject another atropine and oxime autoinjector.

CAUTION

Use atropine and oxime only against nerve agents.

THE COLLECTIVE PROTECTION SYSTEM

Learning Objective: When you finish this chapter, you will be able to—

- Identify the purpose and use of the collective protection system (CPS) decontamination station.

The collective protection system (CPS) protects specific areas of the ship from the effects of CBR contamination by filtering the air supply and maintaining an overpressure to prevent the penetration of contaminants. The system is divided into two protection zones:

1. The total protection (TP) zone, which provides a pressurized, toxic-free environment
2. The limited protection (LP) zone, which isn't pressurized and doesn't provide protection against gaseous chemical agents

The extent of CPS coverage varies. Some ships have only one or two TP zones and no LP zones. Other ships may have different numbers of protection zones. The level of protection is determined by the ship's mission, operational requirements, and the overall cost of installation. The following are the three levels of protection:

Level I—the shelter envelope. Level I provides protection for messing, berthing, sanitary, and battle dressing functions for 40% of the crew.

Level II—the minimum operational protection envelope. Level II provides at least the same protection as level I, but also includes protection for key operational functions.

Level III—the maximum operational protection envelope. Level III provides sufficient protection of the ship for mission requirements, but does not include launching aircraft or troops.

REVIEW 7 QUESTIONS

Q1. There are how many levels of decontamination?

Q2. Describe the primary purpose of level 1 decontamination.

Q3. What are the two types of contamination hazards?

a.

b.

Student Notes:

Q4. What is the most effective way to remove radioactive particles from the ship?

Q5. What is the most effective way to decontaminate areas exposed to BW agents?

Q6. Decon stations are divided into what three areas?

a.

b.

c.

SUMMARY

In this chapter, you have learned about CBR defense. During a major conflict, an enemy who uses weapons of mass destruction will find a way to get these weapons through our defenses. The U.S. Navy has spent many years and a considerable amount of money developing protective systems, equipment, and measuring devices that are available to us today. These systems give us the ability to defend ourselves and our units against CBR attacks and the ability to continue as a combat-capable force. These systems, devices, and equipment will work if used properly and at the right time. The continued training on procedures, techniques, systems, and equipment will ensure the maximum protection available.

REVIEW 1 ANSWERS

A1. **Weapons of mass destruction** are weapons that can be used to destroy large areas or kill and disable large segments of a population.

A2. The most probable delivery method for chemical or biological weapons is by **aerosol**.

A3. The two types of antipersonnel agents are—

a. **Casualty**

b. **Incapacitating**

A4. The use of nerve agents produces symptoms that are similar to **heat stress**, which is a more common condition.

A5. **Moist areas** of the body are most affected by blister agents.

A6. The first action you should take if exposed to a blood agent, is to **don (put on) a protective mask.**

A7. True, **cough suppressant and pain relievers can be given to a victim of a choking agent.**

REVIEW 2 ANSWERS

A1. BW is the intentional use of (a) **living organisms, toxins, and microtoxins** to disable or destroy (b) **people, domestic animals, crops, or supplies**.

A2. The disadvantage an enemy has when using BW agents is that **BW agents degrade rapidly when exposed to environmental conditions such as ultraviolet light, radiation, heat, dryness, or humidity**.

A3. The symptoms of biological disease in its early stages include—

a. **Fever**

b. **Malaise**

c. **Inflammation**

REVIEW 3 ANSWERS

A1. The four types of nuclear weapon explosion classification are—

a. **High altitude blast**

b. **Air blast**

c. **Surface blast**

Student Notes:

d. **Subsurface burst**

A2. Residual radiation is more dangerous than initial radiation because **residual radiation is caused by large amounts of surface material drawn up into the cloud, which falls back to earth as radioactive fallout and affects a large area**.

A3. A secondary blast can cause injuries by **its strong winds that collapse structures and trees**.

A4. Nuclear radiation hazards include—

a. **Alpha particles**

b. **Beta particles**

c. **Gamma rays**

d. **Neutrons**

A5. The measures that should be taken to protect electronic equipment from the effects of EMP are—

a. **Metal shielding**

b. **Good grounding**

c. **Surge arresters**

d. **Proper arrangement of electrical wiring**

REVIEW 4 ANSWERS

A1. A survey team consists of a—

a. **Monitor**, a

b. **Recorder**, and a

c. **Messenger**

A2. The two types of surveys include—

a. **Rapid** and

b. **Detailed**

A3. Biological markers are (a) **blue** and have (b) a **red inscription**.

A4. To calculate safe entry time and stay time in a radiologically contaminated area, you need to know the—

a. **Dose rate**

b. **Dose**

A5. The only known method for detecting BW contaminants is **to gather samples and ship them to a laboratory**.

A6. To check areas suspected of being contaminated by CW agents, you should use an **M256A1 kit**.

REVIEW 5 ANSWERS

A1. **False**, eating food after a CBR attack is **not** okay.

A2. Aboard ship, the safest place to be during a nuclear attack is **below the main deck.**

A3. The two functions of an MCU-2/P mask are to—

a. **Filter air**

b. **Purify**

A4. It should take you **10 seconds** to don and adjust an MCU-2/P mask.

A5. The types of clothing that are useful for CBR defense are—

a. **Wet-weather clothes**

b. **Ordinary work clothes**

REVIEW 6 ANSWERS

A1. The acronym MOPP stands for **Mission Oriented Protective Posture**.

A2. There are **four** MOPP levels.

A3. **MOPP level 4** provides the most protection.

A4. At **MOPP level 2** material condition ZEBRA (modified) is set.

REVIEW 7 ANSWERS

A1. There are **four levels** of decontamination.

A2. The primary purpose of level 1 decontamination is **to safeguard you in protective gear that includes mask, overgarment, boots, and gloves.**

A3. The two types of contamination hazards are—

a. **Pick up hazards**

b. **Transfer hazards**

A4. **Washdown** is the most effective way to remove radioactive particles from the ship.

A5. Chemical disinfectant **is the most effective way to decontaminate areas exposed to BW agents.**

A6. Decon stations are divided into an—

a. **Unclean area**, a

b. **Washing area**, and a

c. **Clean area**

CHAPTER 14

FIRST AID AND HEALTH

If you do something once, people call it an accident. If you do it twice, they call it coincidence. But do it a third time and you've just proven a natural law.

—Rear Admiral Grace Murray Hopper

In this chapter, you will learn some guidelines on giving first aid in an emergency. You won't be an expert or even qualify to administer first aid. You will learn why first aid is important and the results of properly administered first aid. You will also learn the measures you should take for the treatment of shock, bleeding, burns, and fractures; methods of resuscitation; and methods of moving injured persons.

Personal hygiene is also important, not only to you, the individual, but to the entire ship's company. In this chapter, you will receive pointers for maintaining cleanliness of the body, clothing, and bedding. You will also learn the effects of sexually transmitted diseases.

FIRST AID—ITS PURPOSE, LIMITATIONS, AND GENERAL RULES

Learning Objective: When you finish this chapter, you will be able to—

- Recognize the purpose, general rules, and limitations of first aid.

First aid is the emergency care you give to sick or injured persons until medical care is available. In addition to knowing what to do for a victim, it's just as important to know what not to do.

Your knowledge of first-aid measures and their proper application may mean the difference between life and death, between rapid recovery and long hospitalization, or between temporary disability and permanent injury.

PURPOSE AND LIMITATIONS

The objectives of first aid are to save life, prevent further injury, and limit infection. However, first aid isn't a substitute for proper medical treatment. Keep in mind the objectives of first aid. Everyone in the Navy must know when and how to apply first-aid measures and must be prepared to give assistance to persons injured in battle, collision, fire, and other mishaps.

In administering first aid, you have three primary tasks:

1. Maintain breathing
2. Stop bleeding/maintain circulation
3. Prevent or treat for shock

The first step, of course, is to determine the victim's injuries. When you treat a victim, first consideration usually must be given to the most serious injury. In general, the order of treatment is to restore breathing, stop bleeding, and treat for shock.

Work quickly, but don't rush around frantically. Don't waste time looking for ready-made materials. Do the best you can with whatever is at hand. Send for medical help as soon as possible.

GENERAL FIRST-AID RULES

Although each case involving injury or sickness presents its own special problems, some general rules apply to practically all situations. Before you go on to learn first-aid treatment for specific types of injuries, learn with the following basic rules:

1. Keep the victim lying down; head level with the body, until you have found out what kind of injury has occurred and how serious it is. However, if the victim shows one of the following difficulties, follow the rule given for that specific problem:

    a. Vomiting or bleeding about the mouth and semiconsciousness: If the victim is in danger of sucking in blood, vomited matter, or water, place the victim on his or her side or back with the head turned to one side and lower than the feet.

b. Shortness of breath: If the victim has a chest injury or breathing difficulties, place the victim in a sitting or semisitting position.

c. Shock: If the victim is in shock, place the victim on his or her back with the head slightly lower than the feet. (Shock is explained later in this chapter.)

2. Move the victim no more than is absolutely necessary. To determine the extent of the victim's injuries, carefully rip or cut the clothing along the seams. Removal of clothing in the normal way may make injuries worse, especially if fracture injuries are involved. Shoes may also be cut off to avoid causing pain or increasing an injury. When the clothing is removed, make sure the victim does not become chilled.

3. Keep the victim reassured and as comfortable as possible. If possible, don't let the victim see his or her injuries. The victim can endure pain and discomfort better if he or she is confident of your abilities.

4. Don't touch open wounds or burns with fingers or other objects except when sterile compresses or bandages aren't available and it's absolutely necessary to stop severe bleeding.

5. Don't try to give an unconscious person any solid or liquid substance by mouth. The person may vomit and get some of the material into the lungs when he or she breathes, causing choking and possibly death.

6. If a bone is broken or you suspect that one is broken, don't move the victim until you have immobilized the injured part. That may prove lifesaving in cases of severe bone fractures or spinal cord injuries, for the jagged bone may sever nerves and blood vessels, damage tissues, and increase shock. Of course, threat of fire, necessity to abandon ship, or other similar situations may require that you move the victim. But always keep in mind the principle that moving the victim could do further damage; always weigh the risk of moving the victim against other factors.

7. When transporting an injured person, always see that the litter is carried feet forward no matter what the injuries are. Carrying the litter this way lets the rear bearer observe the victim for any respiratory obstruction or stoppage of breathing.

8. Keep the injured person comfortably warm—warm enough to maintain normal body temperature.

Very serious and mutilating injuries may require heroic first-aid measures on your part. However, the greater the number of injuries, the more judgment and self-control you must exhibit to prevent yourself and well-intentioned bystanders from trying to do too much.

REVIEW 1 QUESTIONS

Q1. Describe the primary purpose of first aid.

Q2. List the primary tasks when administering first aid.

a.

b.

c.

Q3. Describe the general first-aid rule for the following conditions:

a. Shock

b. Broken bones

c. Transporting injured personnel

ARTIFICIAL VENTILATION

Learning Objective: When you finish this chapter, you will be able to—

- Recall the procedures used to administer artificial ventilation.

Student Notes:

A person who has stopped breathing may not be dead but is in immediate critical danger. Life depends on oxygen that is breathed into the lungs and then carried by the blood to every body cell. Since body cells can't store oxygen and the blood can hold only a limited amount (and only for a short time), death will result from a continued lack of oxygen.

The heart may continue to beat and the blood may still be circulated to the body cells for some time after breathing has stopped. For a short time, blood will contain a small supply of oxygen; therefore, the body cells won't die immediately. **For a very few minutes, there's a chance that the person's life may be saved**. A person who's stopped breathing but who is still alive is in a state of *respiratory failure*. The first-aid treatment for respiratory failure is *artificial ventilation*.

Artificial ventilation provides air exchange until natural breathing is reestablished. Artificial ventilation should be given only when natural breathing has stopped. **Never give artificial ventilation to any person who is still breathing**.

Don't assume breathing has stopped if a person is unconscious or if a person has been rescued from the water, from poisonous gas, or from contact with an electrical wire. **Remember, never give artificial ventilation to a person who is breathing naturally**. If the victim doesn't begin spontaneous breathing (breaths by himself/herself) after using the head or jaw tilt techniques (discussed later) to open the airway, give artificial ventilation immediately. If a blocked airway prevents ventilation, one of the "thrust" methods (discussed later) to clear the airway must be performed, followed by another attempt at artificial ventilation.

MOUTH TO MOUTH

To perform mouth-to-mouth ventilation, take the following steps:

1. Clear the victim's mouth of obstructions (false teeth and foreign matter).
2. Place the heel of one hand on the victim's forehead, and use the other hand placed under the chin to tilt back the head to open the airway.
3. Using the thumb and index finger, pinch the nostrils shut.
4. Take a deep breath, cover the victim's mouth with your own, and blow.
5. Then remove your mouth from the victim to allow him or her to exhale.

Observe the victim's chest for movement. If the victim hasn't started to breathe normally, start artificial ventilation with four quick ventilations in succession, letting the lungs inflate only partially. If the victim still doesn't respond, then you must fully inflate the victim's lungs at the rate of 12 to 15 ventilations per minute, or one breath every 5 seconds.

MOUTH TO NOSE

Mouth-to-nose ventilation is effective when the victim has extensive facial or dental injuries or is very young. Mouth-to-nose ventilation creates an effective air seal.

To administer this mouth-to-nose ventilation—

1. Place the heel of one hand on the victim's forehead and use the other hand to lift the jaw.
2. After sealing the victim's lips, take a deep breath, place your lips over the victim's nose, and blow.

Observe the chest for movement and place your ear next to the victim's nose to listen for or feel air exchange. Again, you must continue your efforts at the rate of 12 to 15 ventilations per minute, or one breath every 5 seconds, until the victim can breathe without assistance.

Sometimes during artificial ventilation air enters the stomach instead of the lungs. This condition is called *gastric distention*. It can be relieved by moderate pressure exerted with a flat hand between the navel and the rib cage. Before applying pressure, turn the victim's head to the side to prevent choking on the stomach contents that are often brought up during the process.

BACK PRESSURE/ARM LIFT

The back pressure/arm lift method is an alternate technique used when other methods are not possible. To

Student Notes:

perform the back pressure/arm lift method, do the following steps:

1. Place the victim on the stomach, face to one side, neck hyper extend, with hands under the head.
2. Quickly clear the mouth of any foreign matter.
3. Kneel at the victim's head and place your hands on the victim's back so that the heels of the hands lie just below a line between the armpits, with thumbs touching and fingers extending downward and outward.
4. Rock forward, keeping your arms straight, and exert pressure almost directly downward on the victim's back, forcing air out of the lungs.
5. Then rock backward, releasing the pressure and grasping the arms just above the elbows.
6. Continue to rock backward, pulling the arms upward and inward (toward the head) until resistance and tension in the victim's shoulders are noted. That expands the chest, causing active intake of air (inspiration).
7. Rock forward and release the victim's arms. That causes passive exiting of air (expiration).

Repeat the cycle of **press, release, lift**, and **release** 10 to 12 times a minute until the victim can breathe naturally.

CARDIAC ARREST AND CARDIOPULMONARY RESUSCITATION

Learning Objective: When you finish this chapter, you will be able to—

- Recall the procedures to administer cardiopulmonary resuscitation (CPR).

Cardiac arrest is the complete stoppage of heart function. If the victim is to live, action must be taken immediately to restore heart function. The immediate administration of cardiopulmonary resuscitation (CPR) by a rescuer using correct procedures greatly increases the chances of a victim's survival. CPR consists of external heart compression and artificial ventilation. The compression is performed on the outside of the chest, and the lungs are ventilated either by mouth-to-mouth or mouth-to-nose techniques. **To be effective, CPR must be started within 4 minutes of the onset of cardiac arrest**. The victim should be lying on a firm surface.

CAUTION

A rescuer who has not been properly trained should not attempt CPR. (To learn CPR, you should take an approved course from a qualified CPR instructor.) Improperly done, CPR can cause serious damage. Therefore, it is **never** practiced on a healthy individual for training purposes; a training aid is used instead.

ONE-RESCUER TECHNIQUE

In an unwitnessed cardiac arrest, don't assume that an arrest has occurred solely because the victim is lying on the floor and appears to be unconscious. Before beginning CPR, you should—

1. Try to arouse the victim (shake the victim's shoulders and shout to try to obtain a response).
2. Lie the unconscious victim on his/her back.
3. Kneel at the shoulders and establish an open airway, using the procedures outlined previously in artificial ventilations.
4. Check for breathing by looking, listening, and feeling.
   a. Look to see if the chest is rising and falling.
   b. Listen for air coming from the mouth.
   c. Check close to the victim's mouth and feel for air coming out.
5. If the victim isn't breathing, seal the nose, take a deep breath, and blow four quick breaths into the victim without allowing time for the lungs to deflate fully.
6. Quickly remove your mouth and allow the victim to exhale by himself/herself.
7. Check the carotid pulse as shown in figure 14-1. If no pulse is present, start CPR immediately.

Student Notes:

Figure 14-1.—Feeling for the carotid pulse.

To start **external cardiac compression**—

1. Place the victim on his/her back, establish an open airway, and kneel at right angles to the victim's body.
2. Then locate the victim's sternum (breastbone) by—
   a. Baring the chest and locating the sternum by drawing an imaginary line from one nipple to the other to identify the proper area of the sternum, which is darkened in figure 14-2.
   b. Locating the lower tip of the sternum with the index and middle fingers, placing the heels of your hands above your fingers in the darkened area.

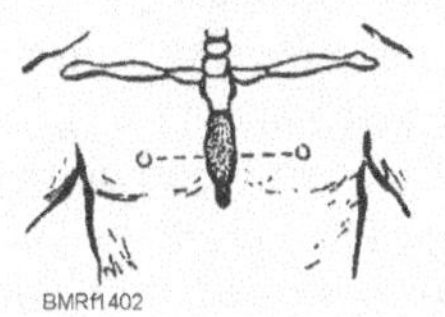

Figure 14-2.—Locating the sternum.

NOTE

There is a small piece of cartilage at the lower end of the sternum (fig. 14-2). A fracture of this area can damage the liver, causing hemorrhage (heavy bleeding) and death. When you place the heels of your hands on the victim's chest, stay above the tip of the sternum.

3. Place the heel of one hand directly on the sternum, and the heel of the other on top of the first. Figure 14-3, view A, shows this technique. Interlock your fingers, and **keep them off the victim's chest**!
4. Lean or rock forward with elbows locked, and apply vertical pressure to depress the sternum (adult) 1 ½ to 2 inches (fig. 14-3, view B).
5. Then release the pressure, keeping the hands in place.
6. Administer 60 to 80 compressions per minute.

You won't get as tired if you use the proper technique, and you will be more effective. Ineffective compression occurs when the elbows are not locked, the rescuer is not directly over the sternum, or the hands are improperly placed on the sternum.

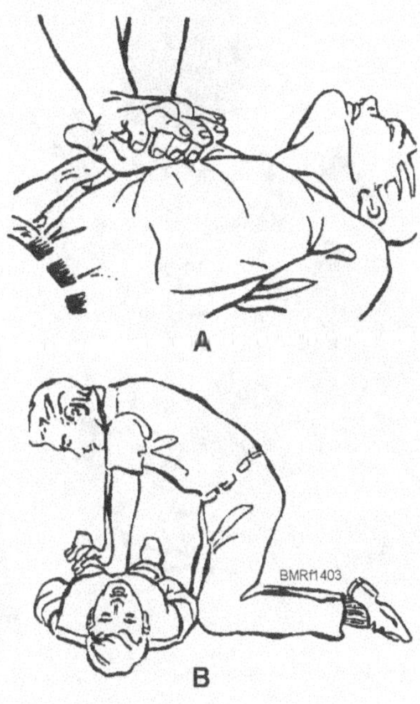

Figure 14-3.—Position for cardiac compression.

Student Notes:

When one rescuer performs CPR, as shown in figure 14-4, the ratio of compressions to ventilations is 15 compressions to 2 ventilations (or 15 to 2). This ratio must continue for four full cycles. Then check for pulse and breathing. If there are still no signs of recovery, continue CPR until the victim can breathe unassisted or you are relieved by medical personnel.

Before reviewing the next technique, let's go over the steps to take in an unwitnessed cardiac arrest involving one rescuer.

1. Determine whether the victim is conscious.
2. Check the vital signs.
3. Ventilate four times. (You may have to remove an airway obstruction at this time.)
4. Again check the vital signs; if none—
   a Begin compression-ventilation rate of 15 to 2 for four complete cycles;
   b. Check pulse, breathing, pupils; if no change,
   c. Continue compression—ventilation rate of 15 to 2 until victim is responsive or you are relieved by medical personnel.

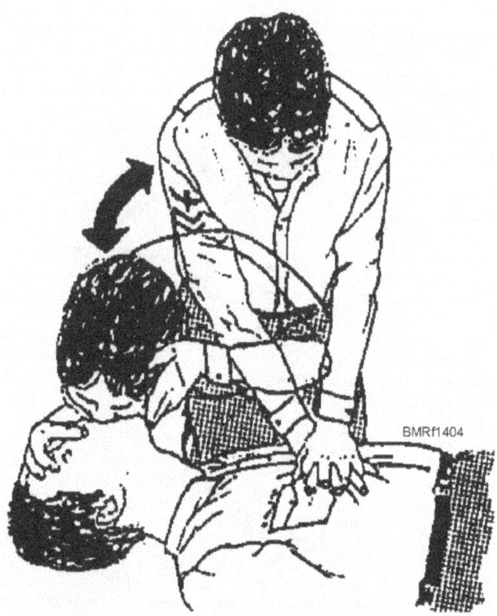

Figure 14-4.—One-rescuer CPR technique.

TWO-RESCUER TECHNIQUE

If two people trained in CPR are on the scene, one performs compressions while the other performs artificial ventilation. The compression-ventilation ratio for two-person CPR is 5 compressions to 1 ventilation (5 to 1). One rescuer is positioned at the chest area and the other beside the victim's head. The rescuers should be on opposite sides of the victim.

To avoid confusion, one rescuer is designated the leader. The leader makes the preliminary checks of the victim's vital signs and performs the initial four ventilations. The second rescuer will perform the compressions.

When CPR is started, the compressions should be given in a constant, methodical rhythm. The rescuer giving the compressions counts them out loud. As the fifth compression is released, the other rescuer ventilates the victim. Do not stop the compressions while ventilation is being given.

AIRWAY BLOCKAGE

Learning Objective: When you finish this chapter, you will be able to—

- Recall the procedures used to clear an airway passage.

Obstruction in the upper airway (throat) is often caused by attempting to chew food and talk at the same time. One of the most reliable indications of an airway obstruction is the inability of the victim to speak. Other indicators are the victim's grasping or pointing at his or her throat, exaggerated breathing efforts, and the skin turning a bluish color. Your first action upon encountering a victim with this problem is to clear the mouth of any food particles, foreign objects, or loose dentures. If that is not effective, you should use one of the following procedures:

Student Notes:

PROCEDURE		STEPS
Standing abdominal thrust	1.	Stand behind the victim and wrap your arms around the victim's waist (fig. 14-5).
	2.	Grasp your wrist and place the thumb side of your fist against the victim's abdomen, above the navel and just below the rib cage (fig. 14-6).
	3.	Give four quick upward thrusts to the victim. The obstruction should pop out like a champagne cork. If unsuccessful, repeat until the obstruction is dislodged.
Reclining abdominal thrust	1.	Position yourself for the thrust by either straddling the victim at the hips, straddling one leg, or kneeling at the victim's hips.
	2.	Place your hands one on top of the other in the area between the lower end of the sternum and the navel, and give four quick upward thrusts into the abdomen, as shown in figure 14-7.
Standing chest thrust	1.	Bring your arms under the arms of the victim and encircle the lower chest, as shown in figure 14-8.
	2.	Grasp your wrist, keeping the thumb side close to the victim's chest. (Keep your fist on the middle, not the lower part, of the sternum.)
	3.	Press the chest with a sharp, backward thrust.
Reclining chest thrust	1.	Kneel at either side of the victim, place hands in same position as used for CPR.
	2.	Deliver thrusts slowly and downward on the sternum (fig. 14-9).

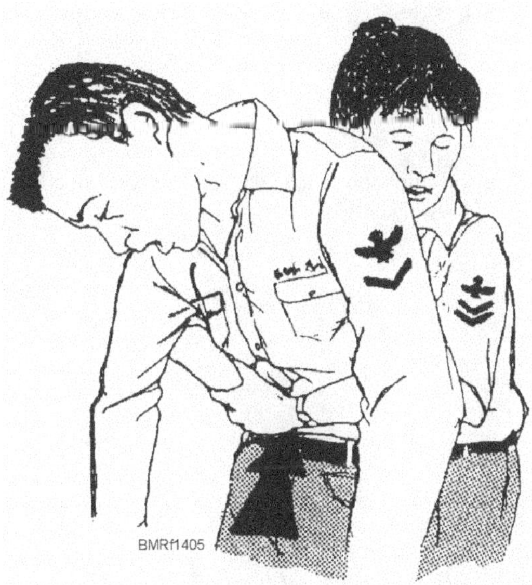

Figure 14-5.—Position for standing abdominal thrust.

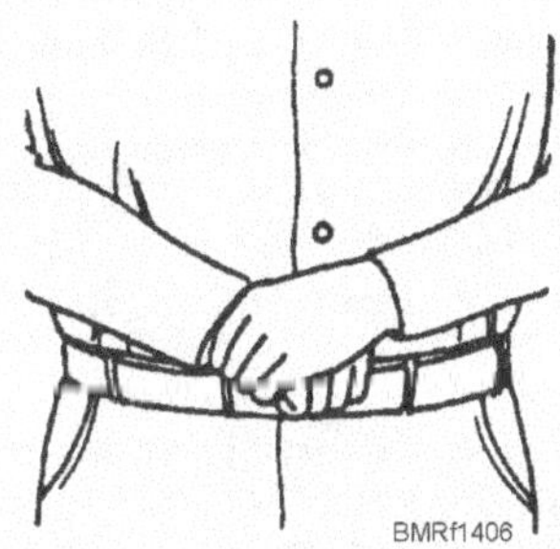

Figure 14-6.—Correct hand positioning.

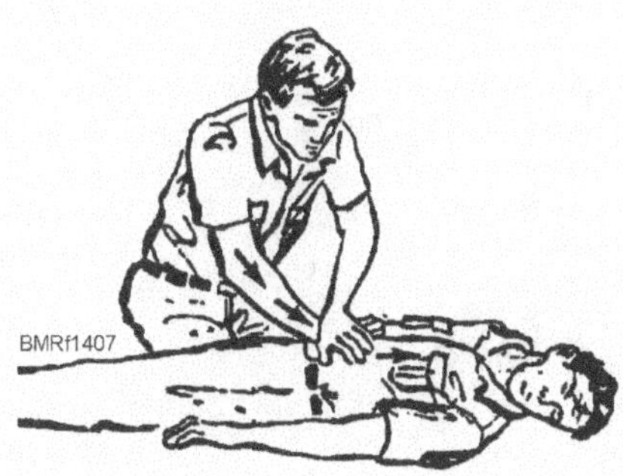

Figure 14-7.—Position for reclining abdominal thrust.

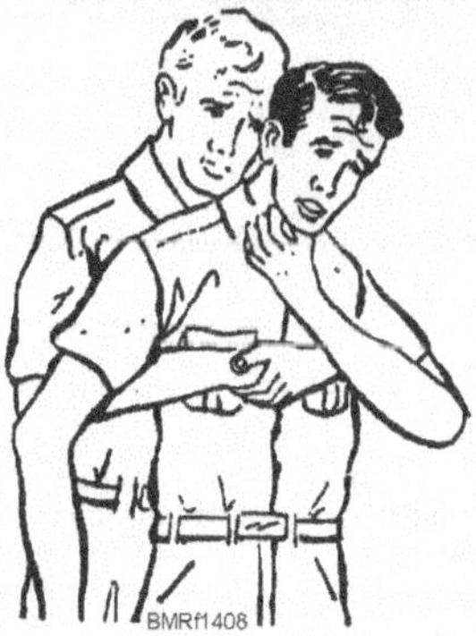

Figure 14-8.—Position for standing chest thrust.

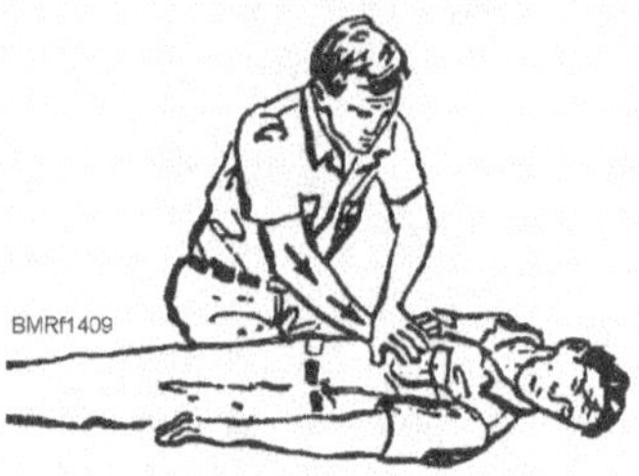

Figure 14-9.—Position for reclining chest thrust.

REVIEW 2 QUESTIONS

Q1. What is the first-aid treatment for respiratory failure?

Q2. When should artificial ventilation be administered?

Q3. List the three types of artificial ventilation.

a.

b.

c.

Q4. What is cardiac arrest?

Q5. To be effective, CPR must be started within how many minutes of the onset of cardiac arrest?

Q6. When you use the one-rescuer technique of CPR, what is the ratio of compressions to ventilations?

Q7. When you use the two-rescuer technique of CPR, what is the ratio of compressions to ventilations?

Q8. List the symptoms of airway blockage.

a.

b.

c.

Q9. List the four methods you can use to clear a person's airway.

a.

b.

c.

d.

Student Notes:

HEMORRHAGE AND METHODS OF CONTROLLING BLEEDING

Learning Objective: When you finish this chapter, you will be able to—

- Recall the procedures used to control external bleeding.

Blood is circulated throughout the body by three different kinds of blood vessels.

1. Arteries, which are large vessels that carry the blood away from the heart
2. Veins, which are large vessels that carry the blood back to the heart
3. Capillaries, which form a connecting network of smaller vessels between the arteries and the veins

Hemorrhage (escape of blood) occurs whenever there is a break in the wall of one or more blood vessels. In most small cuts, only capillaries are injured. Deeper wounds result in injury to veins or arteries. Bleeding severe enough to endanger life seldom occurs **except** when arteries or veins are cut.

The average adult body contains about 5 quarts (4.75 liters) of blood. One pint of blood can usually be lost without harmful effect—in fact, that's the amount usually given by blood donors. However, the loss of 2 pints (0.95 liter) will usually cause shock, and shock becomes greater as the amount of blood loss increases. (Shock will be discussed later in this chapter.) If half the blood in the body is lost, death almost always results.

Capillary blood is usually brick red in color. If capillaries are cut, the blood oozes out slowly. Blood from the veins is dark red. If a vein is cut, the blood escapes in a steady, even flow. If an artery near the surface is cut, the blood will gush out in spurts that are synchronized with the heartbeats; but if the cut artery is deeply buried, the bleeding will appear to be a steady stream. Arterial blood is usually bright red in color.

In actual practice, you might find it difficult to decide whether bleeding was from a vein or an artery; but the distinction is not usually important. A person can bleed to death quickly from a cut artery; prolonged bleeding from any large cut can, of course, have the same effect. The important thing to know is that **all** bleeding must be controlled as quickly as possible.

The only way to stop serious bleeding is by the application of pressure. In practically all cases, bleeding can be stopped if **pressure** is applied **directly to the wound**. If direct pressure doesn't stop the bleeding, pressure should be applied at the appropriate pressure point. In those very rare cases where bleeding is so severe that it cannot be controlled by either of these methods, pressure can be applied by a tight constricting band. The actual procedures you should use to stop bleeding are shown in chart on pages 14-10 and 14-11.

CAUTION

Never put on a constricting band unless the hemorrhage is so severe that it cannot be controlled in any other way. Once a constricting band has been applied, it should be released only by medical personnel.

BATTLE DRESSINGS

Learning Objective: When you finish this chapter, you will be able to—

- Recall the procedures used to apply battle dressings.

A battle dressing is a combination compress and bandage, in which a sterile gauze pad is fastened to a gauze, muslin, or adhesive bandage. Most Navy first-aid kits contain both large and small battle dressings. Battle dressings are also supplied at battle dressing stations located throughout the ship. Any part of a dressing that is to come into direct contact with a wound should be absolutely sterile. The dressing you find in Navy first-aid kits have been sterilized. Never touch a battle dressing with your fingers, clothing, or any other unsterile object.

When applying a battle dressing, make sure the dressing is the proper size so that it covers the wound completely. Some wounds, such as protruding abdominal wounds, require the dressing to be moistened in sterile water. Battle dressing should be applied so it doesn't allow the dressing to move or slip

Student Notes:

PROCEDURE	STEPS
Direct pressure	In most cases, bleeding can be stopped by the application of pressure directly on the wound. • Place a dressing (sterile or clean, if possible) over the wound and firmly fasten it in position with a bandage. • If bleeding doesn't stop, firmly secure another dressing over the first, or apply direct pressure with your hand to the dressing (fig. 14-10). • In cases of severe hemorrhage, don't worry too much about the danger of infection. The basic problem is to stop the flow of blood. If no material is available, simply place your hand firmly on the wound. Remember, direct pressure is the first method to use when you are trying to control hemorrhage.
Pressure points	Bleeding from a cut artery or vein may often be controlled by applying pressure to the appropriate pressure point. A pressure point is a place where the main artery to the injured part lies near the skin surface and over a bone. Pressure at such a point is applied with the fingers (digital pressure) or with the hand; no first-aid materials are required. The object of the pressure is to compress the artery against the bone, shutting off the flow of blood from the heart to the wound. There are 10 principal points (fig. 14-11) on each side of the body where hand or finger pressure can be used to stop hemorrhage. You should memorize these pressure points so that you will know immediately which point to use for hemorrhage from a particular part of the body. The correct pressure point you should use is the one that is— 1. Nearest the wound. 2. Between the wound and the main part of the body, or between the wound and the heart. Applying finger pressure is very tiring, and it can seldom be maintained for more than 15 minutes. Pressure points are recommended for use while direct pressure is being applied to a serious wound. While pressure is being applied at the appropriate pressure point, an assistant can bandage the wound (or wounds). If available, a battle dressing should be used. After opening the dressing, be careful not to contaminate it. Place the compress portion over the wound, then bind it tightly in place with the attached straps (fig. 14-12). If bleeding continues to be severe even after direct pressure and pressure points have been used, you may have to apply a constricting band.
Constricting band	A constricting band is a band used to cut off the supply of blood to an injured limb. **It can't be used to control bleeding from the head, neck, or body** because its use in these locations would result in greater injury or death. Only use a constricting band when hemorrhage can't be controlled by other means. A constricting band consists of a pad, a band, and a device for tightening the band so that the blood vessels will be compressed. There are several different kinds of ready-made constricting bands. A variety of materials can be used to improvise constricting bands. Any round, smooth pressure object may be used for the pad (such as a compress, a roller bandage, a stone, or a rifle shell), and any long, flat material may be used as the band. Remember, the band must be flat! Belts, stockings, flat strips of rubber, or neckerchiefs can be used; but rope, wire, string, or very narrow pieces of cloth shouldn't be used because they will cut into the flesh. A short stick may be used to twist the band, tightening the constricting band.

PROCEDURE	STEPS
Constricting band (Continued)	A constricting band must always be applied above the wound; that is, toward the body, and it must be applied as close to the wound as practicable. The best object to be used for the pad is either a pad, compress, or similar pressure object. The pad goes under the band. Place it directly over the artery, or it will actually decrease the pressure on the artery and allow greater flow of blood. If a constricting band placed over a pressure object doesn't stop the bleeding, the pressure object is probably in the wrong place. If that occurs, shift the object around until the constricting band, when tightened, controls the bleeding. If no suitable pressure object is available, use the constricting band without it. To apply an emergency constricting band (fig. 14-13) made from something like a neckerchief— 1. Wrap the material (which is a minimum of 2 inches wide) at least twice around the limb and tie an overhand knot. 2. Place a short stick on the overhand knot and tie a square knot over it. Then twist the stick rapidly to tighten the constricting band. The stick may be tied in place with another strip of material. To be effective, a constricting band must be tight enough to stop the blood flowing to the limb. If the pressure from the constricting band is less than the arterial pressure, arterial bleeding will continue. Also, insufficient constricting band pressure may actually increase the amount of bleeding from the veins. So be sure to draw the constricting band tight enough to stop the bleeding. However, don't make it any tighter than necessary. After you have brought the bleeding under control with the constricting band, apply a sterile compress or dressing to the wound, and fasten it in position with a bandage. Some points to remember about using a constricting band are as follows: • Don't use a constricting band unless you can't control the bleeding by any other means. • Don't use a constricting band for bleeding from the head, face, neck, or body. Use one only on the limbs. • Always apply a constricting band above the wound and as close to the wound as possible. • Be sure you draw the constricting band tight enough to stop the bleeding, but don't make it any tighter than necessary. • Don't loosen a constricting band after it has been applied. Don't cover a constricting band with a dressing. If it's necessary to cover the injured person in some way, make sure all other people concerned with the case know about the constricting band. Using a crayon, skin pencil, or blood, make a large *T* on the victim's forehead or on a medical tag attached to the wrist, and note the time the constricting band was applied.

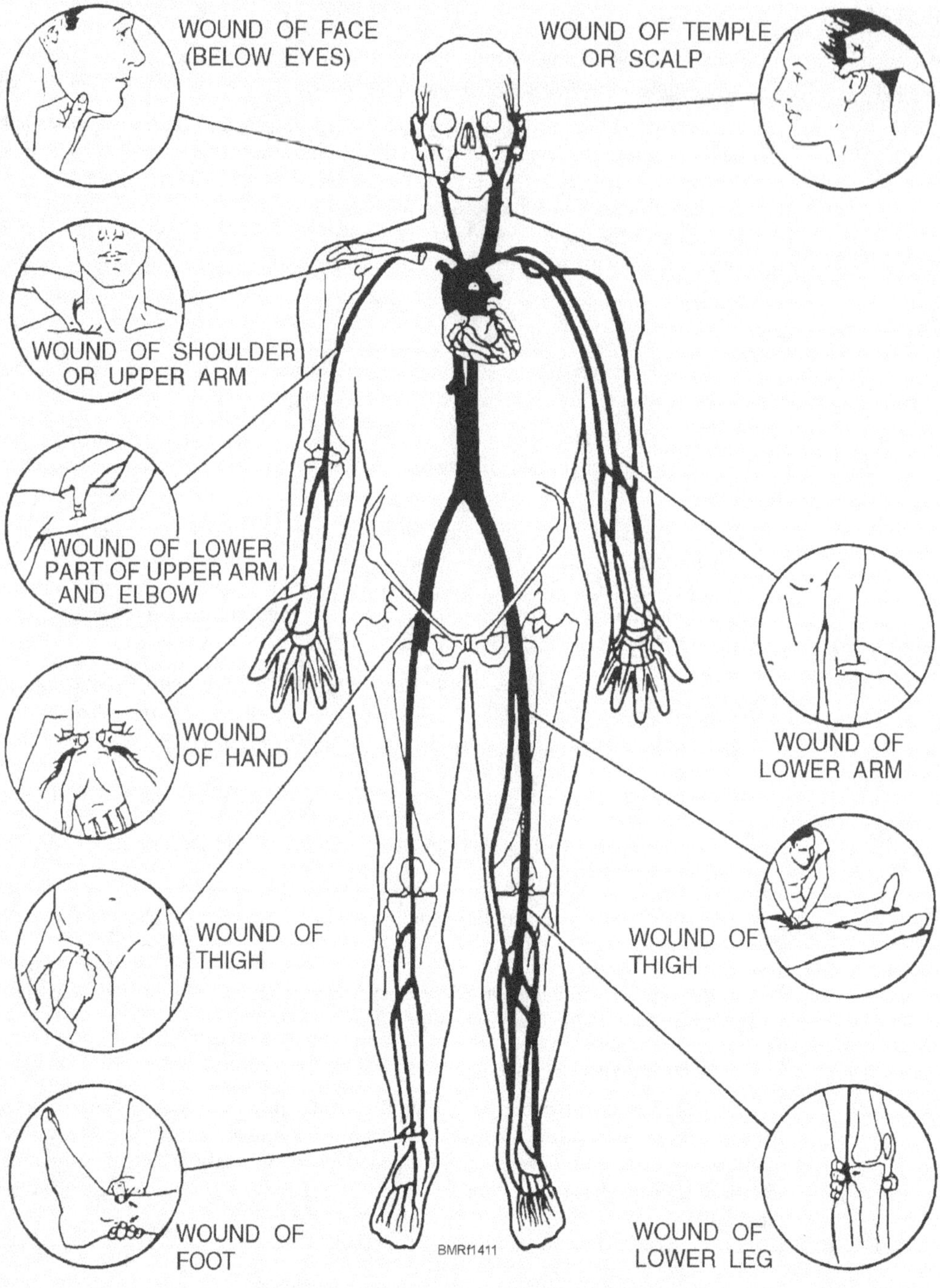

Figure 14-11.—Pressure points for control of bleeding.

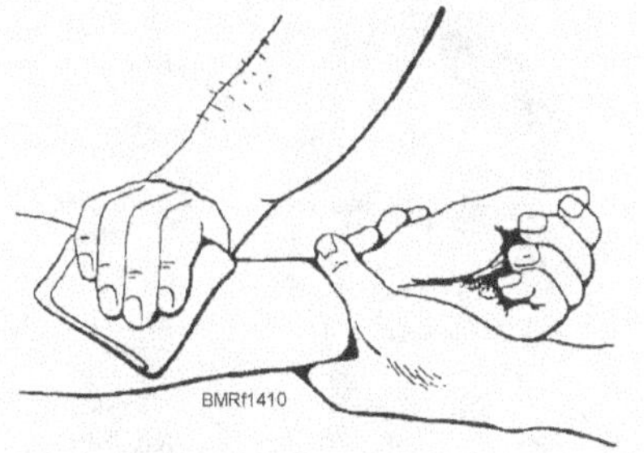

Figure 14-10.—Direct pressure.

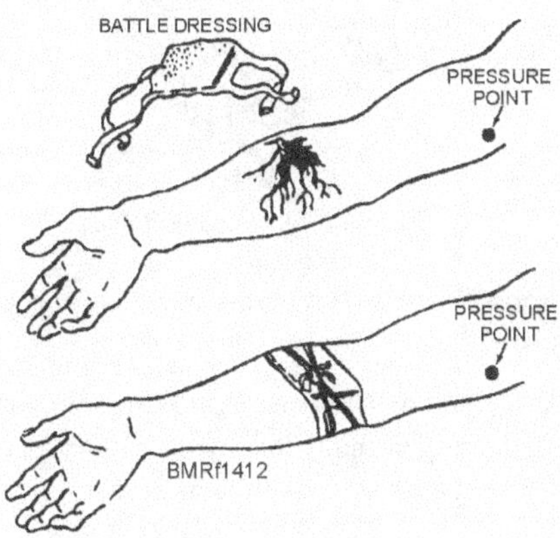

Figure 14-12.—Battle dressing.

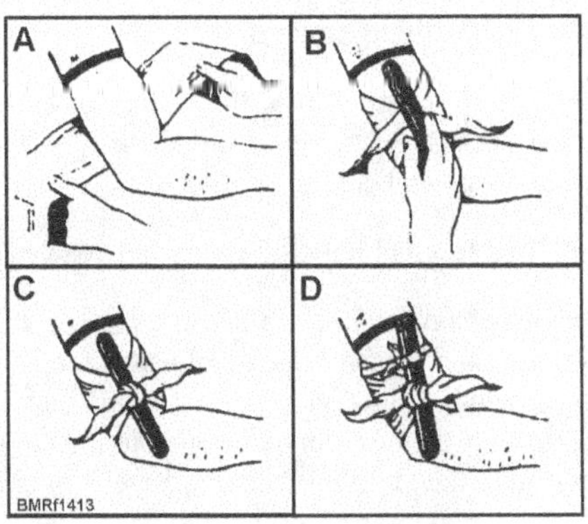

Figure 14-13.—Applying a constricting band.

off the wounded area. Once a battle dressing has been applied to a wound, it shouldn't be removed except by medical personnel. Each ship in the Navy holds periodic training on first aid. There are always new and updated techniques on how to administer first-aid procedures, including how to apply battle dressings. Pay particular attention to these training sessions and learn as must as you possibly can.

REVIEW 3 QUESTIONS

Q1. List the three types of blood vessels the body uses to circulate blood.

a.

b.

c.

Q2. Under what condition is hemorrhage (bleeding) severe enough to endanger life?

Q3. A loss of how many pints of blood will usually cause shock?

Q4. What color is blood carried by (a) capillaries, (b) veins, and (c) arteries?

a.

b.

c.

Q5. What is the only way to stop serious bleeding?

Student Notes:

Q6. What is a constricting band?

Q7. When a battle dressing is applied, what person should release or remove it?

Q8. What is a battle dressing?

Q9. How should you apply a battle dressing?

SHOCK

Learning Objective: When you finish this chapter, you will be able to—

- Recognize the symptoms, prevention, and treatment of shock.

If you've ever hit your finger with a hammer and felt—in addition to the pain—weak, dizzy, and nauseous, then you have experienced a mild form of shock. In this case, the symptoms appeared immediately after the injury, but they may not show up for several hours.

Shock is a condition in which blood circulation is seriously disturbed. Crushed or fractured bones, burns, prolonged bleeding, and asphyxia all cause shock. Shock may be slight or it may be severe enough to cause death. Because all traumatic injuries result in some form of shock, you should learn its symptoms and know how to treat the victim.

HOW TO RECOGNIZE SHOCK

A person who is going into shock may show quite a few signs or symptoms, some of which are indicated in figure 14-14, and are discussed in the following paragraphs. Remember, that signs of shock don't always appear at the time of the injury; and, in many very serious cases, symptoms may not appear until hours later.

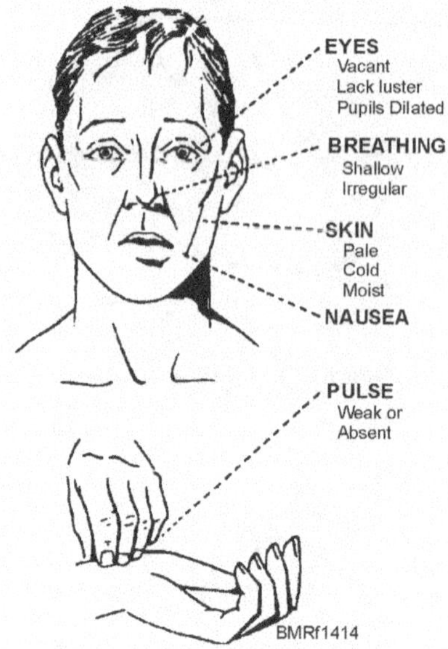

Figure 14-14.—Symptoms of shock.

The symptoms of a person suffering from shock are caused, directly or indirectly, by the disturbance of the circulation of the blood. Symptoms of shock include the following:

- The pulse is weak and rapid.

- Breathing is likely to be shallow, rapid, and irregular, because the poor circulation of the blood affects the breathing center in the brain.

- The temperature near the surface of the body is lowered because of the poor blood flow; so the face, arms, and legs feel cold to the touch.

- Sweating is likely to be very noticeable.

- A person in shock is usually very pale, but, in some cases, the skin may have a bluish or reddish color. In the case of victims with dark skin, you may have to rely primarily on the color of the mucous membranes on the inside of the mouth or under the eyelid or under the nail bed. A person in or going into shock has a bluish color to these membranes instead of a healthy pink.

Student Notes:

• The pupils of the eyes are usually dilated (enlarged).

• A conscious person in shock may complain of thirst and have a feeling of weakness, faintness, or dizziness. The victim may feel nauseous, restless, frightened, and/or anxious. As shock deepens, these signs gradually disappear and the victim becomes less and less responsive to what is going on. Even pain may not arouse the shock victim. Finally, the victim may become unconscious.

You will not likely see all the symptoms of shock in any one case. Some of them may appear only in late stages of shock when the disturbance of the blood flow has become so great that the person's life is in serious danger. Sometimes the signs of shock may be disguised by other signs of the injury. You must know what symptoms indicate the presence of shock, but don't ever wait for symptoms to develop before beginning the treatment for shock. **Remember, every seriously injured person is likely to develop serious shock!**

PREVENTION AND TREATMENT OF SHOCK

You should begin treatment for shock as soon as possible. Prompt treatment may prevent shock or, if it has already developed, prevent its reaching a critical point. Keep the victim lying down and warm. If conscious, the victim should be encouraged and assured that expert medical help will arrive soon.

Keep an injured person warm enough for comfort, but do not let the victim become overheated.

The best position to use to prevent or to treat shock is one that encourages the flow of blood to the brain. If possible, place the injured person on his or her back on a bed, a cot, or a stretcher. Raise the lower end of the support about 12 inches so that the feet are higher than the head (fig. 14-15). If you can't do that and it's possible, raise the feet and legs enough to help the blood flow to the brain. Sometimes it's possible to take advantage of a natural slope of ground and place the victim so that the head is lower than the feet.

Figure 14-15.—Position for the treatment of shock.

Of course in every case, you'll have to consider what type of injury is present before you can decide on the best position. Here are some examples:

- If a person has a chest wound, he/she may have so much trouble breathing that you will have to raise the head slightly.
- If the face is flushed, rather than pale, or if you have any reason to suspect a head injury, don't raise the feet. Instead, you should keep the head level with or slightly higher than the feet.
- If the person has broken bones, you will have to judge what position would be best both for the fractures and for shock. A fractured spine must be immobilized before the victim is moved at all, if further injuries are to be avoided.

If you have any doubts about the correct position to use, have the victim lie flat on his/her back. **The basic position for treating shock is one in which the head is lower than the feet.** Do the best you can under the particular circumstances to get the injured person into this position. In any case, never let a seriously injured person sit, stand, or walk around.

Administer liquids sparingly, and not at all if medical attention will be available within a short time. If necessary, small amounts of warm water, tea, or coffee may be given to a victim who is conscious. Persons having serious burns are an exception. Burn victims require large amounts of fluids. Water, tea, fruit juices, and sugar water may be given freely to a victim who is conscious, able to swallow, and has no internal injuries. Slightly salted water is also beneficial. **Never give alcohol to a person in shock.**

An injured person may or may not be in pain. The amount of pain felt depends in part on the person's physical condition and the type of injury. Extreme pain, if not relieved, can increase the degree of shock. Make

Student Notes:

the victim as comfortable as possible. Fractures should be immobilized and supported. Immobilization greatly reduces, and sometimes eliminates, pain.

An injured person's body heat must be conserved. Therefore, heat is important in the treatment of shock. Exposure to cold, with resulting loss of body heat, can cause shock to develop or to become worse. You will have to judge the amount of covering to use by considering the weather and the general circumstances of the accident. Often a light covering will be enough to keep the casualty comfortable. Wet clothing should be removed and dry covering provided, even on a hot day. Use blankets or any dry material to conserve body heat. Artificial means of warming (hot water bottles, heated bricks, heated sand) should not ordinarily be used. Artificial heat may cause loss of body fluids (by sweating), and it brings the blood closer to the surface, defeating the body's own efforts to supply blood to the vital organs and to the brain. Also, the warming agent may burn the victim.

REVIEW 4 QUESTIONS

Q1. What is shock?

Q2. List the symptoms of shock.

a.

b.

c.

d.

e.

f.

g.

Q3. True or false. Keep an injured person warm enough for comfort, but do not let the victim become overheated.

Q4. If you suspect a person to be in shock, what is the best position for that person?

SUICIDE

Learning Objective: When you finish this chapter, you will be able to—

- Recognize suicidal tendencies and possible treatment.

Suicide among young adults is a serious and growing problem. Among Navy personnel, approximately 10% of the Navy's nonhostile active-duty deaths are caused by suicide. Among the leading causes of nonhostile deaths in the Navy, suicide ranks third behind accidents and heat-related causes. The most frequent suicide victims in the Navy are enlisted males between the ages of 17 and 24 and in paygrades E-1 to E-6.

Why suicide? There isn't a simple answer as to why people choose to kill themselves. Usually, some emotional trauma is so great they "just want to stop the pain." They feel helpless, hopeless, and worthless. They feel that suicide is the only way out.

CAUSES OF SUICIDE

Most suicides are caused by a combination of events that lead a person to believe that suicide is the only way out. The following are some common causes of suicide:

- The breakup of a close relationship with a loved one or difficulties in interpersonal relationships
- The death of a loved one, spouse, child, parent, sibling, friend, or even a pet
- The loss of social or financial status of the family

Student Notes:

- The compounding and disorienting effects of drugs and/or alcohol

DEPRESSION

Depression is often associated with suicide. In 75% to 80% of all suicides, depression is a contributing factor. Sadness and an occasional "case of the blues" are normal emotions. However, depression isn't a normal emotional state. Depression is a deep sadness that's present almost daily for at least 2 weeks.

WHAT TO DO

If you believe someone you know is suicidal, remember the following:

- Take all threats seriously
- Answer cries for help
- Confront the problem
- Tell the person you care
- Listen actively
- Get professional help
- Don't leave the person alone

REVIEW 5 QUESTIONS

Q1. In the Navy, who is the most frequent suicide victim?

Q2. List the common causes of suicide.

a.

b.

c.

d.

Q3. What condition is often associated with suicide?

Q4. List some actions you should take if someone you know might be suicidal.

a.

b.

c.

d.

e.

f.

g.

BURNS

Learning Objective: When you finish this chapter, you will be able to—

- Recognize the symptoms of, classification of, and first-aid treatment for burns.

The seriousness of a burn depends on two factors—the extent of the burned area and the depth of the burn. Shock can be expected from burns involving 15% or more of the body. Burns involving 20% endanger life. Without adequate treatment, burns of over 30% are usually fatal. The depth of the injury determines whether it is a first-, second-, or third-degree burn.

First-degree burns. First-degree burns are mildest. Symptoms are slight pain, redness, tenderness, and increased temperature of the affected area.

Second-degree burns. Second-degree burns are more serious. The inner skin may be damaged, resulting in blistering, severe pain, some dehydration, and possible shock.

Student Notes:

Third-degree burns. Third-degree burns are worst of all. The skin is destroyed, and possibly also the tissue and muscle beneath it. The skin may be charred, or it may be white and lifeless (from scalds). After the initial injury, pain may be less severe because of destroyed nerve ends. There may be chilling of the body. Some form of shock will result.

Probably the most important aspect is the extent of the burned area. A first-degree burn covering a large area could be more serious than a small third-degree burn. A sunburn, for example, ranging from mild to serious, is easily obtained, particularly if you aren't accustomed to the exposure. If you fall asleep while sunbathing, possible second- or even third-degree burns might occur and could be fatal.

The most effective immediate treatment of burns and of pain is as follows:

1. If the burn area covers **less than 20% of the body**, immerse the burned area in cold water, or apply cold compresses if immersion is impracticable. Cold water not only minimizes pain but also reduces the burning effect in the deeper layers of the skin. Gently pat dry the area with lint-free cloth or gauze.

2. If the burn area covers **more than 20% of the body**, apply sterile, dry bandages. Aspirin is also effective for the relief of pain. Continue treatment until no pain is felt when the burned area is exposed to the air.

Burn victims require large amounts of water, which should be slightly salted. Because of the nature of the injury, most burns are sterile. Therefore, the best treatment for uninfected burns is merely to protect the area by covering it with the cleanest (preferably sterile) dressing available.

Some actions that should **not** be taken when dealing with burns are as follows:

- Never apply ointments to a burn or use petrolatum gauze.
- Don't attempt to break blisters or to remove shreds of tissue or adhered particles of charred clothing.
- Never apply a greasy substance (butter, lard, or Vaseline™), antiseptic preparations, or ointments. These may cause further complications and interfere with later treatment by medical personnel.

REVIEW 6 QUESTIONS

Q1. Define the following types of burns:

a. First-degree burn

b. Second-degree burn

c. Third-degree burn

Q2. If a burn covers less than 20% of a victim's body, you should immerse the burned area in cold water or apply cold compresses. Why should you take these actions?

Q3. If a burn covers more than 20% of a victim's body, what actions should you take?

Q4. When treating burns, you should NEVER take which of the following actions?

a. Apply petrolatum gauze

b. Break blisters

c. Apply butter, lard, or Vaseline™

d. Each of the above

HEAT-RELATED PROBLEMS

Learning Objective: When you finish this chapter, you will be able to—

- Recognize the symptoms of and first-aid treatment for heat-related illnesses.

Student Notes:

Look at figure 14-16. Here, you see a comparison of the symptoms of heatstroke and heat exhaustion. These are dangers you face when working or exposed to conditions that might cause heatstroke or heat exhaustion.

HEATSTROKE

Sunstroke is more accurately called *heatstroke* since it is not necessary for a person to be exposed to the sun for this condition to develop. It is a less common but far more serious condition than heat exhaustion, since heatstroke has a 20% mortality rate. The more important feature of heatstroke is the extremely high body temperature (105°F [41°C] or higher) that accompanies it. In heatstroke, the victim has a breakdown of the sweating mechanism and is unable to eliminate excessive body heat built up while exercising. If the body temperature rises too high, the brain, kidneys, and liver may be permanently damaged.

Signs and symptoms of heatstroke. Sometimes the victim may have preliminary symptoms such as headache, nausea, dizziness, or weakness. Breathing will be deep and rapid at first, later shallow and almost absent. Usually the victim will be flushed, very dry, and very hot. The pupils will be constricted (pinpoint) and the pulse fast and strong.

Treatment of heatstroke. When you provide first aid for heatstroke, remember that this is a true life-and-death emergency. The longer the victim remains overheated, the higher the chances of irreversible body damage or even death. First-aid treatment for heatstroke is designed to reduce body heat and includes the following:

- Reduce body heat immediately by dousing the body with cold water, or applying wet, cold towels to the whole body.
- Move the victim to the coolest possible place and remove as much clothing as possible.
- Maintain an open airway.
- Place the victim on his or her back, with the head and shoulders slightly raised.
- If cold packs are available, place them under the arms, around the neck, at the ankles, and on the groin.

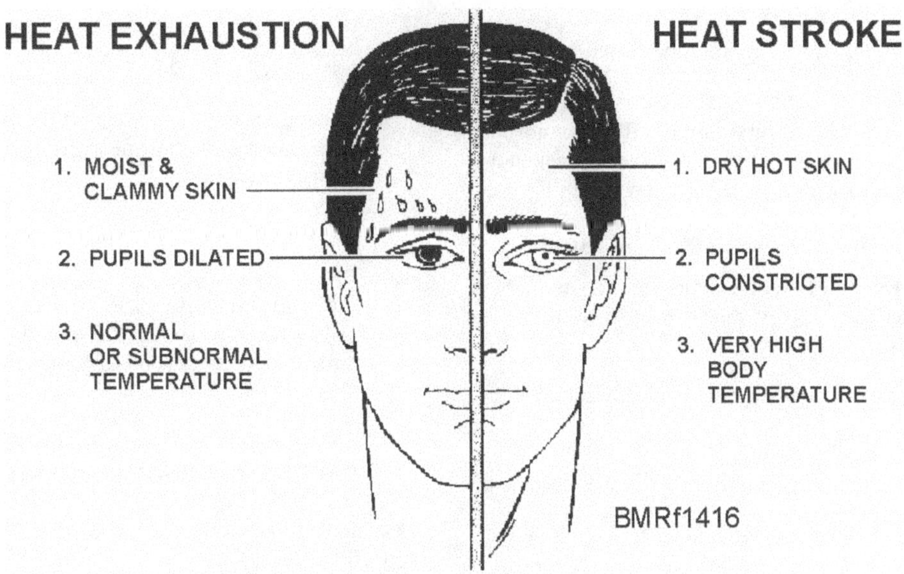

Figure 14-16.—Symptoms of heatstroke and heat exhaustion.

Student Notes:

- Expose the victim to a fan or air-conditioner since drafts will promote cooling.

- Immersing the victim in a cold water bath is also effective.

- Give the victim (if conscious) cool water to drink. Do not give any hot drinks or stimulants.

- Get the victim to a medical facility as soon as possible. Cooling measures must be continued while the victim is being transported.

HEAT EXHAUSTION

Heat exhaustion (heat prostration or heat collapse) is the most common condition caused by working or exercising in hot spaces. Heat exhaustion produces a serious disruption of blood flow to the brain, heart, and lungs. This disruption of blood flow causes the victim to experience weakness, dizziness, headache, loss of appetite, and nausea.

Signs and symptoms of heat exhaustion. Signs and symptoms of heat exhaustion are similar to those of shock: for example—

- The victim will appear ashen gray; the skin cold, moist, and clammy.

- The pupils of the eyes may be dilated (enlarged).

- The vital signs (blood pressure, temperature, pulse, and respiration) usually are normal; however, the victim may have a weak pulse together with rapid and shallow breathing.

- Body temperature may be below normal.

Treatment of heat exhaustion. To treat heat exhaustion victims, you should treat them as if they were in shock.

- Loosen the clothing; apply cool, wet cloths.

- Move the victim to either a cool or an air-conditioned area, and fan the victim.

- Do not allow the person to become chilled.

- If the victim is conscious, administer a solution of 1 teaspoon of salt dissolved in a quart of cool water.

- If the victim vomits, don't give any more fluids.

- Transport the victim to a medical facility as soon as possible.

REVIEW 7 QUESTIONS

Q1. List the three most important signs of heatstroke.

a.

b.

c.

Q2. List the three most important signs of heat exhaustion.

a.

b.

c.

Q3. What is the most important action when treating someone who is showing signs of heatstroke or heat exhaustion?

Q4. True or false. In case of heatstroke/heat exhaustion, you should transport the victim to a medical facility as soon as possible.

FRACTURES, SPRAINS, AND STRAINS

Learning Objectives: When you finish this chapter, you will be able to—

Student Notes:

- Recognize the classification of, symptoms of, and first-aid treatment for fractures.
- Recall the first-aid treatment for strains and sprains.

Simply put, a fracture is a broken bone. The severity of the injury depends on the part of the body affected, the type of fracture, and the amount of tissue damaged.

FRACTURES

In this section, you will learn about fractures—how they're classified and the first-aid you would give the victim. Additional information is given on how to transport victims.

Classification

Fractures may be classified in several ways. However, they are generally classified as are either closed or open. A closed fracture is one in which the skin remains intact. An open fracture is one in which the bone protrudes from the skin. These fractures are shown in figure 14-17.

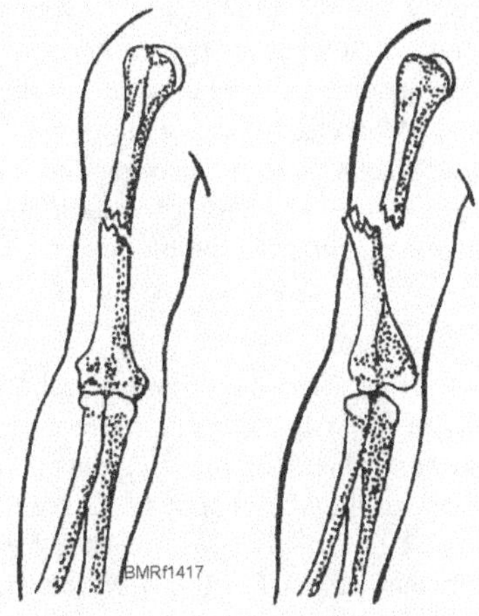

Figure 14-17.—Types of fractures.

Symptoms

You can't always tell that a fracture has occurred. However, if the victim has been involved in some form of violence, you may suspect that one or more bones have been broken. The victim may even have heard the bone snap. Some symptoms of a fracture are as follows:

- Pain and tenderness
- Inability to use the part
- Creaking or cracking
- Motion at points other than joints
- Swelling
- Deformity
- Discoloration of skin

Treatment

If you are required to give first aid to a person who has suffered a fracture, you should follow these general rules:

- If there is any possibility that a fracture has been sustained, treat the injury as a fracture.
- Get medical aid at the first possible opportunity. All fractures require medical treatment.
- Don't move the victim until splints have been applied to the injured parts, unless the victim's life is in danger.
- Treat for shock.
- Don't attempt to locate a fracture by grating the ends of the bone together.
- Don't attempt to set a broken bone.
- When a long bone in the arm or leg is fractured, the limb should be carefully straightened so that splints can be applied. Pulling gently with your hands in the long axis of the limb is permissible, and it may be all that is necessary to get the limb back into position.
- Apply splints. Emergency splinting may be placed over clothing if the victim will be seen very soon by a medical officer or if the victim will be transported for a short distance. Otherwise, it's best to remove just enough clothing so you can apply well-padded splints

Student Notes:

directly to the injured part. If you decide to remove clothing over the injured part, cut the clothing or rip it along the seams. In any case, **be careful!** Rough handling of the victim may turn a closed fracture into an open fracture. That could increase the severity of shock and cause extensive damage to the blood vessels, nerves, muscles, and other tissues around the broken bone.

If the fracture is open, you must treat the wound before you can deal with the fracture. Bleeding from the wound may be serious. Most bleeding can be stopped by direct pressure on the wound or by finger pressure at the appropriate point. If, after your best efforts, these methods are not successful, use a constricting band; then treat the fracture.

Use of Splints

An essential part of the first-aid treatment is immobilizing the injured part with splints so that the sharp ends of broken bones won't move around and cause further damage to nerves, blood vessels, or vital organs. Splints are also used to immobilize severely injured joints or muscles and to prevent the enlargement of extensive wounds. Before you can use a splint, you need to have a general understanding of the use of splints.

In an emergency, almost any firm object or material can be used as a splint. Such things as umbrellas, canes, swords, rifles, tent pegs, laths, sticks, oars, paddles, spars, wire, leather, boards, pillows, heavy clothing, corrugated cardboard, and folded newspapers can be used as splints. A fractured leg may sometimes be splinted by fastening it securely to the uninjured leg.

Splints, whether ready-made or improvised, must meet the following requirements:

- Be light in weight, but still be strong and fairly rigid.
- Be long enough to reach the joints above and below the fracture.
- Be wide enough so the bandages used to hold them in place won't pinch the injured part.
- Be well padded on the sides that touch the body. If they're not properly padded, they won't fit well and won't adequately immobilize the injured part.
- To improvise the padding for a splint, use articles of clothing, bandages, cotton, blankets, or any other soft material.
- If the victim is wearing heavy clothes, apply the splint on the outside, allowing the clothing to serve as at least part of the required padding.

Although splints should be applied snugly, **never** apply them tight enough to interfere with the circulation of the blood. When applying splints to an arm or a leg, try to leave the fingers or toes exposed. If the tips of the fingers or toes become blue or cold, you will know that the splints or bandages are too tight. You should examine a splinted part approximately every half-hour, and loosen the fastenings if circulation appears to be cut off. Remember that any injured part is likely to swell, and splints or bandages that are all right when applied may be too tight later.

Figure 14-18 shows a method of immobilizing the leg of a person with a broken kneecap. To secure the limb to the splint, belts, neckerchiefs, rope, or any suitable material may be used. If possible, tie the limb at two places above and two places below the break.

Leave the treatment of other types of fractures, such as jaw, ribs, and spine, to medical personnel. **Never try to move a person who might have a fractured spine or neck.** Moving such a person could cause permanent paralysis. Don't attempt to reset bones.

SPRAINS AND STRAINS

A person with a sprain or a strain might have some of the same symptoms as a person who has a fracture. The information contained in this section will help you

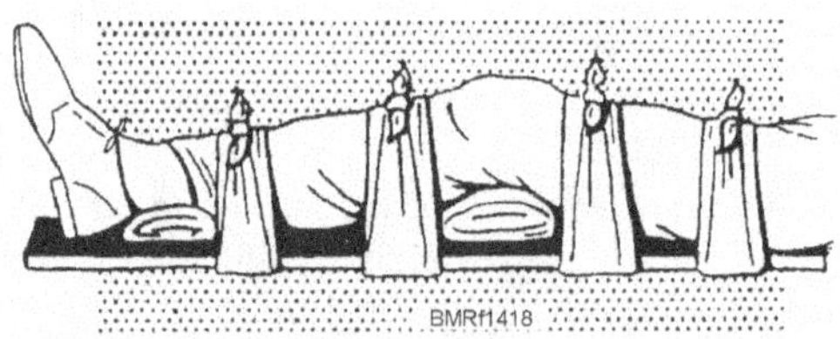

Figure 14-18.—Splinting.

Student Notes:

know what to do if a there is a possibility a shipmate has sustained a strain or a sprain.

Sprains

A sprain is an injury to the ligaments and soft tissues that support a joint. A sprain is caused by the violent wrenching or twisting of the joint beyond its normal limits of movement. Any joint may be sprained; however, sprains of the ankle, wrist, knee, and finger are most common.

SYMPTOMS.—Symptoms of sprains include pain or pressure at the joint, pain upon movement, swelling and tenderness, possible loss of movement, and discoloration.

TREATMENT.—Treat all sprains as fractures until ruled out by X-rays. To treat a sprain, you should take the following actions:

- Application of cold packs for the first 24 to 48 hours.
- Elevation and rest of the affected area.
- Application of a snug, smooth, figure-eight bandage to control swelling and to immobilize (keep from moving) the affected area. (**NOTE**: Check bandaged areas regularly for swelling that might cause circulation problems and loosen bandages if necessary.)
- After the swelling stops (24 to 48 hours), apply moist heat for short periods (15 to 30 minutes).

CAUTION

Do not apply heat until 24 hours after the last cold pack.

After applying first aid, make sure the victim has a follow-up examination by a medical officer. This exam includes X-rays to rule out fractures.

Strains

A strain is an injury caused by the forcible over stretching or tearing of a muscle or tendon. A strain may be caused by lifting excessively heavy loads, sudden or violent movements, or any other action that pulls the muscles beyond their normal limits.

SYMPTOMS.—Symptoms of strains include pain, lameness or stiffness, moderate swelling at the place of the injury, discoloration caused by blood escaping from injured blood vessels into the tissues, possible loss of power, and a distinct gap felt at the site of the injury.

TREATMENT.—To treat a strain, you should take the following actions:

- Elevate the affected area.
- Apply cold packs for 24 to 48 hours.
- After the swelling stops, apply mild heat to increase circulation and aid in healing.

NOTE

Do not apply heat until 24 hours after the last cold pack.

The victim should be evacuated to a medical facility where X-rays can be taken to rule out the possibility of a fracture.

Student Notes:

REVIEW 8 QUESTIONS

Q1. Label the following fractures.

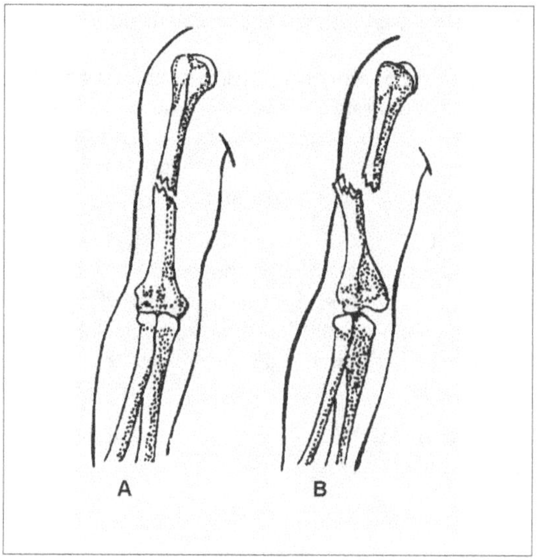

Q2. List the symptoms of a fractured leg or arm.

a.

b.

c.

d.

e.

f.

g.

Q3. Briefly describe how to give first aid to someone with a fractured leg or arm.

a.

b.

c.

d.

e.

f.

g.

Q4. List the types of fractures that should be treated by medical personnel.

a.

b.

c.

Q5. What is the reason that you should never move a person who might have a fractured spine or neck?

Q6. List the symptoms a victim might have with a sprained or strained leg.

a.

b.

c.

d.

e.

Student Notes:

f.

g.

Q7. Describe the first aid that should be given to a victim suspected of having a sprained or strained leg.

RESCUE PROCEDURES

Learning Objective: When you finish this chapter, you will be able to—

- Recall the procedures to rescue a person.

There are many ways to move victims. The method used depends on several factors—where the victim is located and where the victim is to be taken, assistance available, equipment on hand, and so forth. If available, litters or stretchers should be used.

In you don't have any help, there are several methods you can use to move a victim alone. One method is simply to pick up and carry the victim in your arms, but it can be quite a task if the victim weighs more than you. If a blanket is handy, the victim can be placed upon it and dragged. Two other means are the fireman's carry (fig. 14-19) and the tied-hands crawl (fig. 14-20).

FIREMAN'S CARRY

One of the easiest ways to carry an unconscious person is by the fireman's lift, also called the *fireman's carry* (fig. 14-19).

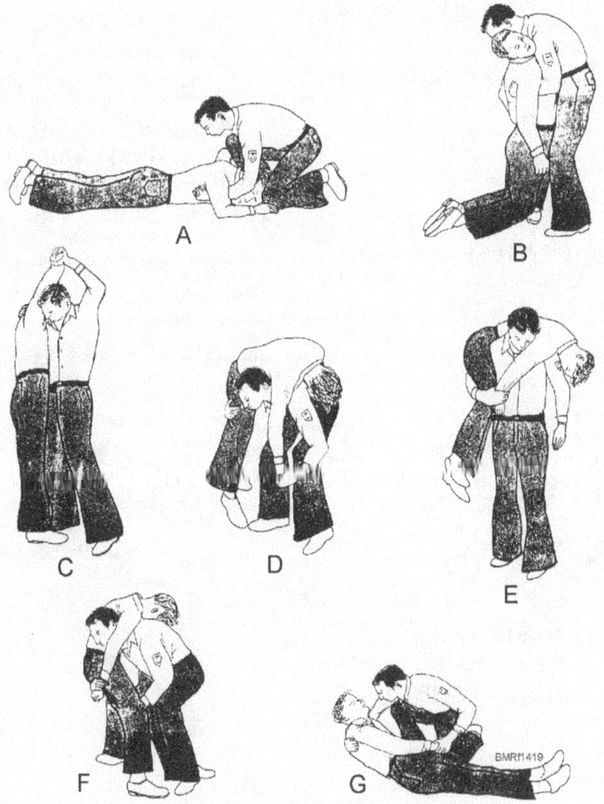

Figure 14-19.—Fireman's carry.

Student Notes:

1. Place the victim face down, as shown in figure 14-19, view A. Kneel on one knee at the head, facing the victim. Pass your hands under the armpits.
2. Raise the victim, as shown in figure 14-19, view B. Take a better hold across the back.
3. Raise the victim to a standing position and stick your right leg between the victim's legs, as shown in figure 14-19, view C. Grasp the victim's right wrist in your left hand and swing the arm around the back of your neck and down your left shoulder.
4. Stoop quickly and pull the victim across your shoulders and, at the same time, put your right arm between the victim's legs, as shown in figure 14-19, view D.
5. Grasp the victim's right wrist with your right hand and straighten up, as shown in figure 14-19, view E.

The procedure for lowering the victim to the deck is shown in figure 14-19, views F and G.

TIED-HANDS CRAWL

The tied-hands crawl shown in figure 14-20 may be used to drag an unconscious person for a short distance; it is particularly useful when you must crawl underneath a low structure.

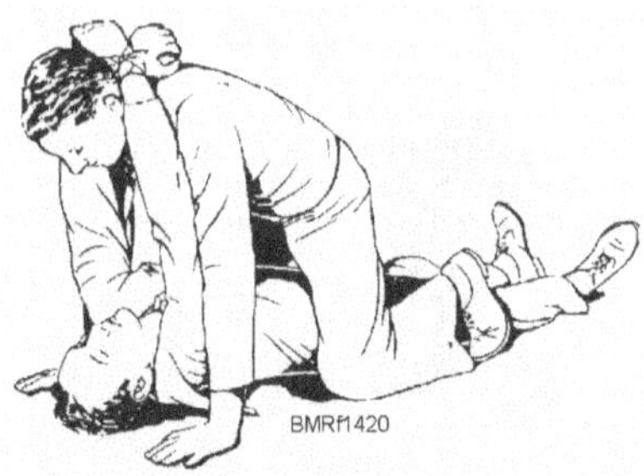

Figure 14-20.—Tied-hands crawl.

RESCUE FROM ELECTRICAL CONTACT

Rescuing a person who has received an electric shock is likely to be difficult and dangerous. Use extreme caution or the rescuer may also be electrocuted.

Don't touch the victim's body, the wire, or any other object that may be conducting electricity.

Some procedures you might use to rescue a person who's received an electric shock are as follows:

- Look for the switch first of all, and if you find it, turn off the current immediately. Don't waste too much time hunting for the switch; however, every second is important.
- If you cannot find the switch, you should try to remove the wire from the victim with a dry broom handle, branch, pole, oar, board, or similar nonconducting object (fig. 14-21).
- It may be possible to use dry rope or dry clothing to pull the wire away from the victim.
- You can also break the contact by cutting the wire with a wooden-handled axe, but that is extremely dangerous because the cut ends of the wire are likely to curl and lash back at you before you have time to get out of the way.

When you are trying to break an electrical contact, always stand on some nonconducting material, such as a dry board, newspaper, or clothing.

Administer artificial ventilation immediately after freeing the person from the wire if the electric shock caused breathing to stop. Check the victim's pulse, since electric shock may also cause the heart to stop. If

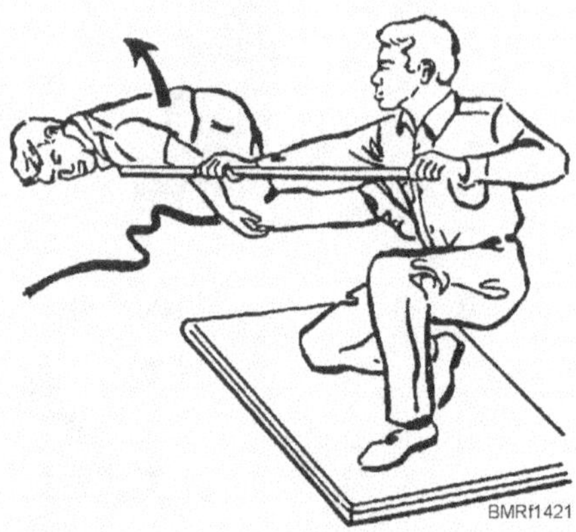

Figure 14-21.—Pushing a victim away from a power line.

Student Notes:

you do not feel a pulse, immediately administer CPR. Get the victim to a medical facility as soon as possible.

TRANSPORTATION PROCEDURES

Learning Objective: When you finish this chapter, you will be able to—

- Recall the procedures to transport a person.

So far, you've learned about the emergency methods used to get an injured person out of danger and into a position where first aid can be administered. As you have learned, these emergency rescue procedures often involve substantial risk to the victim and should be used only when clearly necessary.

Once you've rescued the victim from the immediate danger, **slow down!** Handle and transport the victim with care, being careful about the injuries that have been sustained. In the excitement and confusion that almost always accompany a mishap, you are likely to feel rushed, as though you must do everything rapidly. This is a reasonable way to feel. Speed is essential in treating many injuries and in getting the casualty to a medical officer or hospital. However, it's **not** reasonable to let yourself feel so hurried that you handle the victim roughly or carelessly or transport the victim in a way that will make the injuries worse.

GENERAL PRECAUTIONS

The basic precautions to observe when transporting an injured person are summarized as follows:

- Give necessary first aid **before** attempting to transport the victim if possible. Be sure all injuries have been located. Treat serious breathing problems, bleeding, and shock in that order. Immobilize all fractures, sprains, and dislocations. Do whatever you can to reduce the victim's pain and to make the victim as comfortable as possible under the circumstances.

- Use a regular stretcher if one is available. If you must use an improvised stretcher, be sure it is strong enough. Also, be sure that you have enough personnel to carry the stretcher so that you won't run any risk of dropping the victim.

- Whenever possible, bring the stretcher to the victim instead of carrying the victim to the stretcher.

- Fasten the victim to the stretcher to prevent slipping, sliding, or falling off. Tie the victim's feet together, unless the injuries make it impracticable.

- Use blankets, garments, or other material to pad the stretcher and to protect the victim from exposure.

- As a general rule, an injured person should be lying down, face up, while being moved. However, in some instances the type or location of the injury will necessitate the use of another position. If the victim has a chest wound, raising the head and shoulders may give greater comfort, and ease any breathing difficulties the victim may have. A person who has a broken bone should be moved very carefully so that the injury will not be made worse. If the victim has received a severe injury to the head, the victim should be kept lying on the side or on the back with the head turned to one side to prevent choking on saliva, blood, or vomit while being transported. In all cases, it is important to place the victim in a position that prevents further injuries.

- The stretcher should be carried in such a way that the victim will be moved feet first, so that the rear stretcher bearer can continually watch the victim for signs of breathing difficulty.

- If you must use a motor vehicle to transport a seriously injured person, the best means is an ambulance. If no ambulance is available, a truck or station wagon makes a fairly good substitute. If it is necessary to use a passenger car to transport a seriously injured person, the victim should be put in a place that requires the least amount of bending, twisting, or turning.

- Don't turn the victim over to anyone without giving a complete account of the situation. Be sure the person taking over knows what caused the injury and what first-aid treatment has been given. If a constricting band has been applied, make sure that is known to the victim.

Student Notes:

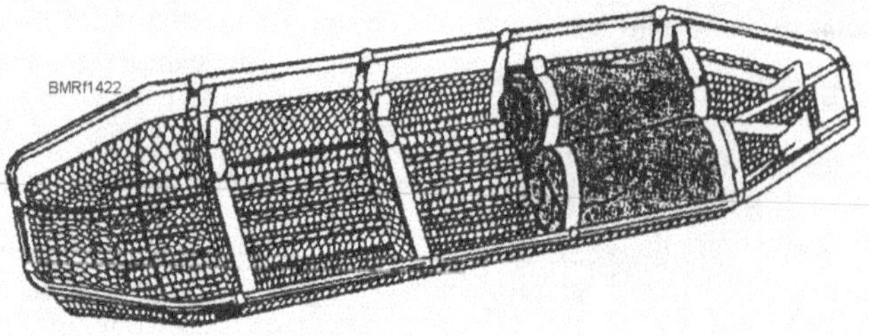

Figure 14-22.—Stokes stretcher.

STOKES STRETCHER

The Navy service litter most commonly used for transporting sick or injured persons is called the *Stokes stretcher* (fig. 14-22). The Stokes stretcher is a wire basket supported by iron or aluminum rods. It's adaptable to a variety of uses, since the victim can be held securely in place, even if the stretcher is tipped or turned. The Stokes stretcher is particularly valuable for transferring injured persons to and from boats. It is also used for direct ship-to-ship transfer of injured persons.

NEIL ROBERTSON STRETCHER

The Neil Robertson stretcher is designed for removing an injured person from engine-room spaces, holds, and other compartments where access hatches are too small to permit the use of regular stretchers.

The Neil Robertson stretcher is made of semirigid canvas. When firmly wrapped around the victim mummy-fashion, it gives sufficient support so the victim may be lifted vertically (fig. 14-23). To keep the injured person from swaying against bulkheads and hatchways while being lifted, tie a guideline to the victim's ankles.

Stretchers of this type can be made on board ship and kept in appropriate places ready for use. If a Neil Robertson stretcher is not available when needed, a piece of heavy canvas, wrapped firmly around the victim, will serve somewhat the same purpose.

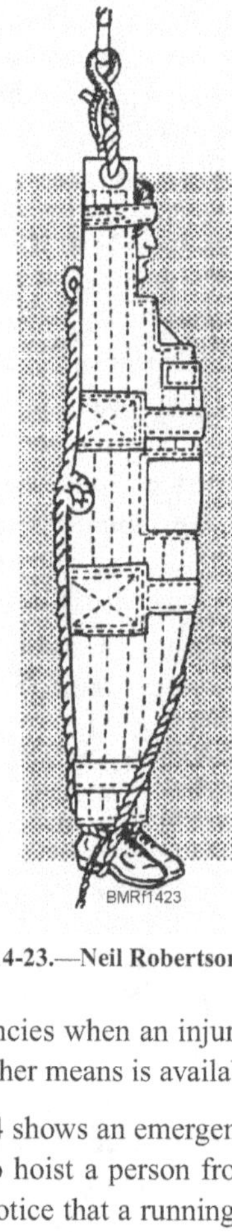

Figure 14-23.—Neil Robertson stretcher.

EMERGENCY RESCUE LINES

An emergency rescue line can be made from any strong fiber line. These lines should be used only in extreme emergencies when an injured person must be moved and no other means is available.

Figure 14-24 shows an emergency rescue line that could be used to hoist a person from a void or small compartment. Notice that a running bowline is passed around the body, just below the hips, and a half hitch just under the arms. Again, a guideline is tied to the victim's ankles.

Student Notes:

Figure 14-24.—Emergency rescue line.

REVIEW 9 QUESTIONS

Q1. What is one of the easiest ways to carry an unconscious person?

Q2. Describe the precautions you should take when rescuing a person who has received an electric shock.

Q3. How should you carry a stretcher?

Q4. What type of stretcher is used to transport an injured person from engine-room spaces?

Q5. When are emergency rescue lines used?

PERSONAL HYGIENE

Learning Objectives: When you finish this chapter, you will be able to—

- Identify the purpose for personal hygiene.
- Recognize the consequences of not following a personal hygiene program.

Because of the close living quarters in the Navy, particularly aboard ship, personal hygiene is very important. Developing good personal hygiene habits is essential for the good health of the individual and for the protection of the entire crew. For the same reasons, sanitary conditions aboard ship must be maintained at all times. Clean spaces are a necessity. Dirt breeds disease. When spaces are kept clean and orderly, the general well-being of the crew improves and morale increases. No one wants to live or work in a filthy environment. In the Navy and at home, everyone should make it a habit to keep living and working spaces as clean as possible. Maintaining a clean, healthy environment reduces the chances of illness.

Negligence in reporting to the medical officer any matter that affects one's health is inexcusable. It can lead to a more serious illness. Don't ignore minor injuries. An untreated cut or scratch can lead to infection, loss of a limb, and even death. If you can't report for treatment right away, wash the injury with soap and clean water. A large wound should not be washed; cover it with a clean dressing until it can be attended to by medical personnel.

Some practices you can take to be healthy include the following:

Showering. Shower and change underwear daily. After showering, dry thoroughly, particularly your feet to prevent fungus development. Wear shower shoes when taking a shower to avoid contracting athlete's foot.

Shoes and socks. Wear properly fitted shoes and socks. The inner dimensions of the shoe should be about 1/4 inch longer and wider than the foot. Improperly fitted socks and socks with holes can cause blisters. Change your socks daily.

Student Notes:

Toenails and feet. Cut your nails straight across to prevent ingrown toenails. If corns or other foot ailments develop, have them treated at once.

Fingernails. Keep fingernails trimmed and clean.

Hair. Keep your hair neatly trimmed and wash it often.

Bunk linen. Change it at least weekly.

Exercise and sleep. Daily exercise improves bodily functions, increasing muscle tone and physical endurance. Even aboard small ships, it's possible to exercise in some manner. Get as much sleep as watch and work conditions permit.

Diet. Navy food is good and wholesome. It provides a well-balanced diet. Don't be a finicky eater, even though you don't like some foods. Learn to eat a variety of foods; try to avoid putting more on your tray than you care to eat.

ORAL HYGIENE

Many dental disorders begin with the buildup of bacterial plaque that remains undisturbed around the teeth. The purpose of personal oral hygiene is to remove this plaque buildup. Plaque can be removed by proper tooth brushing and flossing techniques.

There are three common dental conditions that are caused by poor dental hygiene:

1. Tooth decay
2. Reddening of the gums
3. Gum and bone disease

Any of these can cause the loss of a tooth; but with proper oral hygiene, these conditions can be controlled or prevented.

Tooth decay can be reduced by cutting down on sweets and by brushing properly. For most people, cavities and gum and bone disease occur primarily between the teeth. No toothbrush can effectively cleanse these areas or the areas behind the last tooth in each arch. You must use dental floss to clean such hard-to-reach areas. You should floss at least once a day, either just before or just after brushing. Unwaxed dental floss should be used in most cases.

Dental cleansing devices, oral irrigators, and commercial mouthwashes are aids to oral hygiene. They may be used in addition to—but not in place of—tooth brushing and flossing. If these devices are electrically powered, they must be safety checked by electrical safety personnel before use.

NOTE

Oral irrigation may be harmful for individuals with cardiovascular problems.

In addition to all of these procedures, you should also have a dental checkup every 6 months or at least once a year. Your dental technician or dentist can show you the proper way to brush and floss your teeth.

SEXUALLY TRANSMITTED DISEASES

Sexually transmitted diseases (STDs) are illness caused by organisms that are transmitted through sexual intercourse or by forms of other intimate body contact with an infected person. The disease germs that cause syphilis and gonorrhea are very fragile and can live for only short periods of time outside the body. Venereal disease is not spread from inanimate objects such as toilet seats, drinking glasses, bed linens, or clothes.

Syphilis and gonorrhea are the two most common sexually transmitted diseases in the United States. Syphilis has had the worst reputation, but it is gonorrhea that is out of control.

Syphilis

Syphilis can attack any tissue or organ of the body and is especially damaging to the brain, spinal cord, blood vessels, and heart.

A painless sore, called a *chancre*, is the first sign of syphilis. The sore usually appears on or around the sex organs about 9 to 90 days after contact with an infected person. The chancre will heal within a few weeks, even without treatment.

Other signs of syphilis that may develop either before or after the chancre goes away are a rash that may cover any part of the body; white, glistening spots in the mouth; and fever, sore throat, and headaches. The rash

Student Notes:

and other signs may not appear or may be so slight as to be unnoticed.

After these signs disappear, the germs may stay hidden for 10 to 20 years. If untreated, the disease causes mental illness, blindness, heart disease, or even death.

Syphilis is not inherited, but a pregnant woman with the disease can give it to her unborn child. These babies are born with congenital syphilis. A baby with congenital syphilis may be born dead or deformed. Congenital syphilis can be prevented if it is detected and treated in time.

The signs of syphilis may resemble many other diseases, or the signs may be slight and be unnoticed. The disease can be detected by a blood test for syphilis.

Gonorrhea

If you have gonorrhea and don't get treatment, you may become sterile. Gonorrhea can damage the sperm ducts in males and the fallopian tubes in females. In men and women, gonorrhea may result in crippling arthritis, meningitis, or heart disease.

The signs of gonorrhea in males usually appear 3 to 5 days after sexual contact with an infected partner. Most men have a pus discharge from the sex organ and a painful, burning sensation during urination. Women rarely have painful symptoms until gonorrhea has seriously damaged their reproductive system. There may be some vaginal discharge or burning during urination, but women will usually have no symptoms and will not know that they have gonorrhea until a sexual partner has been infected.

If you have syphilis or gonorrhea, a cure is as near as your medical department. But early treatment is important. These diseases can be cured even in people who have had the disease for a long time, but the damage to the reproductive organs may be irreversible.

NOTE

Self-treatment or pills from a friend are extremely dangerous.

Genital Herpes Infection

Genital herpes is an increasingly common viral infection that produces recurrent, painful genital sores similar to cold sores that occur around the mouth. At this time, there is no known cure for genital herpes; the infected person may have recurrences of lesions throughout life. Individuals should avoid sexual intercourse when the sores are present because the herpes virus is infectious in this phase of the disease.

Acquired Immune Deficiency Syndrome

The Acquired Immune Deficiency Syndrome (AIDS) was first reported in the United States in mid 1981. AIDS is a serious illness and a public health problem. It's the number one priority of the U.S. Public Health Service.

AIDS is a serious condition characterized by a defect in natural immunity (defense) against disease. People who have AIDS are vulnerable to serious illnesses that aren't a threat to anyone whose immune system is functioning normally. These illnesses are referred to as "opportunistic" infections or diseases.

Investigators have discovered the virus that causes AIDS. The virus is called either *human immune virus* (*HIV*); *human T-lymphotropic virus, type III* (*HTLV-3*); *lymphadenopathy associated virus* (*LAV*); or *AIDS-related virus* (*ARV*). Most people infected with the AIDS virus have no symptoms and feel well. Some develop symptoms that may include tiredness; fever; loss of appetite and weight; diarrhea; night sweats; and swollen glands (lymph nodes), usually in the neck, armpits, or groin. Anyone with these symptoms should see a doctor if the symptoms continue for more than 2 weeks

AIDS is spread by sexual contact, needle sharing, or less commonly, through blood or its components. The risk of getting AIDS is increased by having multiple sexual partners, either homosexual or heterosexual, and sharing needles with people who use illicit drugs. The occurrence of the AIDS in hemophilia patients and persons receiving transfusions provides evidence of transmission through blood. It may be transmitted from infected mother to infant before, during, or shortly after birth.

Student Notes:

Prevention

Using a condom during sex offers some protection. Birth control pills offer no protection against STDs. If you had the disease once and have been successfully treated, that does not grant you immunity against contracting an STD again.

If you have been diagnosed as having an STD and are receiving treatment at the present time, don't attempt to hide the name(s) of your sexual partners. The chances are that one of them infected you or have been infected by you. They deserve the benefit of treatment too. The health department will contact the persons named and treat them. These steps, which are done confidentially, can help in stopping an outbreak of a sexually transmitted disease.

REVIEW 10 QUESTION

Q1. List some of the reasons why personal hygiene is important.

a.

b.

c.

d.

Q2. List the three most common dental conditions caused by poor dental hygiene.

a.

b.

c.

Q3. What methods should you use to avoid dental problems?

a.

b.

c.

Q4. What are the two most common sexually transmitted diseases?

a.

b.

Q5. How is the Acquired Immune Deficiency Syndrome (AIDS) spread?

a.

b.

c.

SUMMARY

In this chapter, you have learned some of the basic steps and procedures required when administering first aid. You may never have the need to use these procedures, but if the situation should arise, by following the procedures outlined, and with additional training, you may be in a position to render what could be life-saving assistance. You also learned the recommended ways of transporting injured personnel so they can receive proper medical attention.

Personal hygiene is an important part of living closely together. A shipmate not overly concerned with keeping himself or herself clean and squared away could affect your physical well-being, but could also affect the morale of a great number of crew members. Keeping yourself clean and squared away will benefit you and the people you come into contact with on a daily basis.

Another topic covered here is sexually transmitted diseases. Being attracted to a member of the opposite sex is a natural reaction. Be aware of the possibility that if you engage in multiple sexual relations, you could become infected with one of the sexually transmitted diseases discussed in this chapter. Being responsible in your sexual relations and using approved protective measures will go a long way toward protecting yourself.

Student Notes:

REVIEW 1 ANSWERS

A1. The primary purpose of first aid is **to safe lives, prevent further injury, and limit infection**.

A2. The primary tasks to take when you administer first aid are to—

a. **maintain breathing,**

b. **stop bleeding and maintain circulation**, and

c. **prevent or treat shock.**

A3. The general first-aid rule for—

a. **shock is to place the victim on his/her back with the head slightly lower than the feet**

b. **broken bones is to keep the person still until you immobilize the injured part**

c. **transport of injured persons is on the litter with the litter carried feet first**

REVIEW 2 ANSWERS

A1. The first-aid treatment for respiratory failure is **artificial ventilation**.

A2. Artificial ventilation should be administered **only when natural breathing has stopped. NEVER give artificial ventilation to a person who is still breathing**.

A3. The three types of artificial ventilation are—

a. **Mouth to mouth**

b. **Mouth to nose**

c. **Back pressure/arm lift**

A4. Cardiac arrest is **the complete stoppage of heart function**.

A5. To be effective, CPR **must be started within 4 minutes** of the onset of cardiac arrest.

A6. When you use the one-rescuer technique of CPR, the ratio of compressions to ventilations is **15 compressions to 2 ventilations**.

A7. When you use the two-rescuer technique of CPR, the ratio of compressions to ventilations is **5 compressions to 1 ventilation**.

A8. The symptoms of airway blockage are—

a. **Inability of the victim to speak**

b. **Exaggerated breathing efforts**

c. **Skin turning blue**

A9. The four methods you can use to clear a person's airway are—

a. **Standing abdominal thrust**

b. **Reclining abdominal thrust**

c. **Standing chest thrust**

d. **Reclining chest thrust**

REVIEW 3 ANSWERS

A1. The three types of blood vessels the body uses to circulate blood are—

a. **Arteries—large vessels that carry blood away from the heart**

b. **Veins—large vessels that carry blood back to the heart**

c. **Capillaries—a connecting network of smaller vessels between the arteries and the veins**

A2. Hemorrhage is severe enough to endanger life **when arteries or veins are cut**

A3. A loss of **2 pints of blood is usually enough to cause shock.**

A4. Blood carried by—

a. **Capillaries is brick red**

b. **Veins is dark red**

c. **Arteries is bright red**

A5. The only way to stop serious bleeding is the **application of pressure**.

A6. A constricting band is **a pad, a band, and a device for tightening the band so that the blood vessels will be compressed. Only use a constricting band when hemorrhage can't be controlled any other way. Constricting bands are used above the wound. They aren't used for wounds on the head, neck, or body**.

A7. When a constricting band or a battle dressing has been applied, **only medical personnel should release/remove it**.

A8. A battle dressing is **a combination compress and bandage, in which a sterile gauze pad is fastened to a gauze, muslin, or adhesive bandage**.

A9. When applying a battle dressing, you should make sure that the **dressing covers the entire wound**.

REVIEW 4 ANSWERS

A1. Shock is **a condition where the blood circulation is seriously disturbed**.

A2. The symptoms of shock in a person are—

a. **Weak and rapid pulse**

b. **Shallow, rapid, and irregular breathing**

c. **Lower temperature—the arms, face, and legs feel cold to the touch**

d. **Sweating**

e. **Pale skin color; however, in some cases, it may be bluish or reddish**

f. **Dilated (enlarged) pupils**

g. **Thirst and an feeling of weakness, faintness, or dizziness**

A3. **True**, you should keep an injured person warm enough to be comfortable, but not warm enough to become overheated.

A4. If you suspect that a person is in shock, **you should keep the person lying flat on his/her back with the feet slightly elevated (raised) so that the position encourages the blood to flow back to the brain**.

REVIEW 5 ANSWERS

A1. In the Navy, the most frequent suicide victim is **an enlisted male between 17 and 24 years old and in paygrades E-1 through E-6**.

A2. The most common causes of suicide are—

a. **Breakup of a close relationship**

b. **Death of a loved one**

c. **Loss of social or financial status**

d. **Effects of drugs and/or alcohol**

A3. **Depression** is often associated with suicide.

A4. Some actions you can take if you believe someone is suicidal are—

a. **Take all threats seriously**

b. **Confront the problem**

c. **Answer cries for help**

d. **Let the person know you care**

e. **Listen**

f. **Get professional help**

g. **Don't leave the person alone**

REVIEW 6 ANSWERS

A1. Burns are defined as follows:

a. First-degree burn—**Mildest burn. Slight redness, tenderness, and increased temperature of the burned area**.

b. Second-degree burn—**More serious than first-degree burn. Inner skin may be damaged, blistering, severe pain, some dehydration, and possible shock**.

c. Third-degree burn—**Most serious burn. Skin is destroyed and possibly tissue and muscle beneath it. Skin may be charred or white and lifeless (from scalds). Some form of shock will result**.

A2. By immersing the burned area in cold water or by applying cold compresses, you **minimize pain and reduce the burning effect in deeper layers of the skin**.

A3. If a burn covers more than 20% of a victim's body, you should **apply sterile, dry bandages**.

A4. When treating burns you should **NEVER apply petrolatum gauze, break blisters or apply butter, lard, or Vaseline™**.

REVIEW 7 ANSWERS

A1. The three most important signs of heatstroke are—

a. **Dry, hot skin**

b. **Constricted pupils**

c. **Very high body temperature (usually above 105°F)**

A2. The three most important signs of heat exhaustion are—

a. **Moist, clammy skin**

b. **Dilated pupils**

c. **Normal or subnormal temperature**

A3. The aim of first-aid treatment for heatstroke or heat exhaustion is **to reduce body temperature**.

A4. **True**, in case of heatstroke/heat exhaustion, you should transport the victim to a medical facility as soon as possible.

REVIEW 8 ANSWERS

A1. Fractures are—

a. **Closed fracture**

b. **Open fracture**

A2. The symptoms of a fractured leg or arm include—

a. **Pain and tenderness**

b. **Discoloration of the skin**

c. **Creaking or cracking**

d. **Inability to use the part**

e. **Motion at points other than joints**

f. **Swelling**

g. **Deformity**

A3. To give first aid to someone with a fractured leg or arm, you should—

a. **Get medical aid as soon as possible**

b. **Don't move the victim until splints have been applied, unless the victim's life is in danger**

c. **Treat for shock**

d. **Don't try to find a fracture by grating the ends of the bone together**

e. **Don't try to set a broken bone**

f. **If a long bone in the leg is fractured, carefully straighten the leg so it can be immobilized**

g. **Apply splints**

A4. The types of fractures that should be treated by medical personnel are—

a. **Jaw**

b. **Ribs**

c. **Spine**

A5. You should never move a person who might have a fractured spine or neck because **moving that person might cause permanent paralysis.**

A6. The symptoms a victim might have with a sprained or strained leg include—

a. **Pain, lameness, stiffness, or pressure**

b. **Pain on movement**

c. **Swelling and tenderness**

d. **Discoloration**

e. **With a strain, a distinct gap at the site of the injury**

A7. The first aid that should be given to a victim suspected of having a sprained or strained leg includes **treating all sprains as fractures until ruled out by X-rays.**

REVIEW 9 ANSWERS

A1. One of the easiest ways to carry an unconscious person is to use the **fireman's lift/carry**.

A2. When rescuing a person who has received an electric shock, you should **not touch the victim's body, wire, or any other object that may conduct electricity**.

A3. You should carry a stretcher **with the victim's feet first so the rear stretcher bearer can see the victim for signs of breathing difficulty**.

A4. To transport an injured person from engine-room spaces, a **Neil Robertson stretcher is usually used**.

A5. Emergency rescue lines are used **when an injured person must be transported and no other means is available**.

REVIEW 10 ANSWERS

A1. Personal hygiene is important for the following reasons:

a. **Close living quarters**

b. **Well-being of the crew**

c. **Reduced chance of illness**

d. **Morale increase**

A2. The three most common dental conditions caused by poor dental hygiene are—

a. **Tooth decay**

b. **Reddening of the gums**

c. **Gum and bone disease**

A3. To avoid dental problems, you should—

a. **Brush your teeth**

b. **Floss your teeth**

c. **Have dental checkups every 6 months**

A4. The two most common sexually transmitted diseases are—

a. **Syphilis**

b. **Gonorrhea**

A5. AIDS is spread through—

a. **Sexual contact**

b. **Needle sharing by drug users**

c. **Transfusions**

CHAPTER 15

SURVIVAL

Without a decisive naval force we can do nothing definitive, and with it, everything honorable and glorious.

—George Washington

As you learned in earlier chapters, being a professional Sailor is dangerous. These dangers aren't limited to just your job in the Navy. In times of conflict, your ship or shore station may be in contact with an enemy force or ship. Regardless of your rate, rating, or duty station, you may need to stay alive in the water until you can reach land or be rescued. You must have the knowledge required to live in the field with limited equipment (survival) and to avoid the enemy (evasion). If captured, you also have the responsibility to flee from the enemy (escape) if possible.

This chapter contains information on the principles and techniques of sea survival, evasion, land survival, and escape that have been used successfully worldwide. The information given here is by no means all-inclusive, but should serve to help you if the need arises.

SURVIVAL AT SEA

Learning Objectives: When you finish this chapter, you will be able to—

- Recall the methods and procedures for abandoning ship.
- Identify the techniques for swimming through oil, flames, and debris.
- Recognize the techniques for using clothing and buoyant objects to stay afloat.
- Recognize the procedures used to care for and use personal floatation devices and the use of lifeboats and associated survival gear.
- Recall the characteristics of, use of, and adjustment to CO_2 inflatable and inherently buoyant life preservers.
- Identify the responsibilities and authority of the senior person in a survival situation.

Survival at sea depends on your knowledge, self-control, training, and equipment. The time to learn as much as possible about survival and rescue at sea is **before** you abandon ship, not after you find yourself in the water. The information for survival at sea is general in nature and applies to all Navy ratings.

ABANDONING SHIP

Having to abandon ship isn't pleasant. Your "home" is gone along with most of your possessions and possibly some of your shipmates. You don't know how long you must wait for rescue. However, with the proper knowledge and training, frightening aspects can be greatly reduced. Don't panic and don't give up hope. Remember, the Navy knows you're missing and is searching for you. Also, remember that thousands of persons have survived ships sinking in both wartime and peacetime.

If time permits, the crew will abandon the ship in a planned and orderly manner. In the prepare-to-abandon-ship stage, all personnel go topside and muster at their abandon ship stations, don life jackets, and rig lines and ladders over the side. Bearing and distance to the nearest land, sea and wind conditions, and water temperature are passed over the 1MC (ship's general announcing system). When the order to abandon ship is given, all boats are lowered and lifeboats are released. The crew members then go over the side and board them as quickly as possible.

Know Escape Routes

Many survivors have reported that their shipmates were lost because they were unable to get topside before the ship sank. In many of these cases, the compartments in which personnel were trapped were not cut off—the individuals only thought they were.

Once on board a particular ship, most Sailors learn the easiest ways from their berthing compartments to their stations and automatically use these routes day after day. The habit of using the same hatches and ladders day after day becomes so strong that a person finds it difficult to use other routes. This habit is especially true of persons whose stations are in the lower part of the ship. However, a hit from a torpedo or bomb or a collision with another ship may flood the compartments normally used or knock out a ladder. Often, some measure to control flooding taken by the damage control party closes off the normal method of travel.

The only answer to this situation is to know your ship. Small ships don't present much of a problem because they have only a few routes you can follow. However, large ships are another matter. Aboard an aircraft carrier or cruiser, learning all the passageways, doors, and ladders takes a long time. During leisure time, learn escape routes from various below-deck sections to the weather decks. Ask the individuals who work in those sections the best way to get topside; then follow that route. The time to experiment is before an emergency occurs, not during one.

Going Over the Side

As in everything else, there is a right way and a wrong way to abandon ship. Whenever possible, go over the side fully clothed. Shoes and clothing may hinder you while swimming; but in lifeboats, a covering of any kind offers protection against the effects of sun and salt water. In a cold climate, wear a watch cap to keep your head warm. Take along a pair of gloves and extra clothes if you can. Even in tropical waters you may feel cool at night because you can do little to keep warm.

Normally, you should leave from whichever side of the ship is lower in the water; but, if the propellers are turning, leave from the bow. Leave by the windward side whenever possible. Leaving from the lee side might protect you from a stiff wind, but the same wind causes the ship to drift down on you, often faster than you can swim. Also, if oil is on the water, you can clear the slick sooner by swimming into the wind.

Never dive, and do not jump unless you have to. Use a ladder, cargo net, line, or fire hose. If you must jump, do so feet first, legs together, and body erect. (First, check the water so you will not land on debris or on other personnel.) Except when jumping into flames, be sure your life preserver is fastened securely, including the leg straps. If you are wearing a vest-type preserver, place one hand firmly on the opposite shoulder to keep the preserver from riding up sharply when you hit the water (in a long drop, the force of impact might hurt your chin or neck). Hold your nose with your other hand. If you are wearing an inflatable preserver, inflate it after you have entered the water.

In the Water

Once you are in the water, your immediate concern is to clear the ship as quickly as possible. Before you rest, you should try to be 150 to 200 yards away from the ship. When the ship goes down, it may create a strong whirlpool effect, which might draw you down with the ship if you are too close. Another advantage of distance is that you will be safer if an explosion occurs.

After you are safely away from the ship, conserve your energy. Don't splash about or shout unnecessarily. If any danger of underwater explosions exists, float or swim on your back with your head and chest as far out of the water as possible. Help your shipmates all you can, and try to stay in groups (fig. 15-1). Get on a lifeboat, of course, as soon as you can. In the meantime, grab anything floatable that comes by, or just relax in the water. Above all, **remain calm!**

SWIMMING AND FLOATING.—Check the chart shown below. It tells you the requirements you must meet to qualify as a third class, second class, and first class swimmer.

Meeting the requirements for swimmer third class won't help you if you have to swim ½ mile to a lifeboat. You can see that by qualifying for swimmer second class, you'd have a better chance to survive. Better yet, qualifying for swimmer first class gives you the best chance for survival.

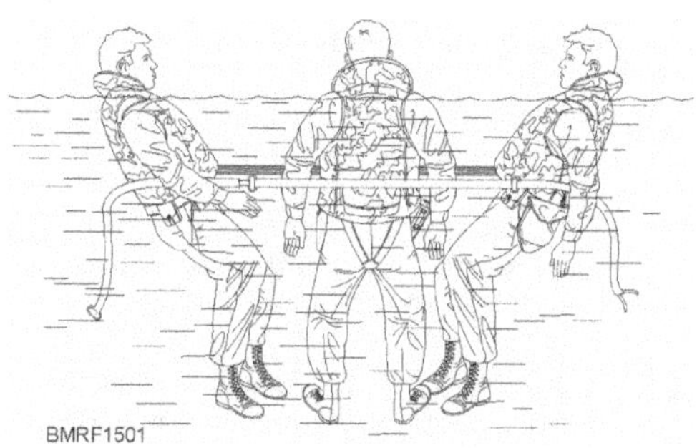

Figure 15-1.—Joining life preservers.

Student Notes:

After abandoning ship, you may have to swim fast, slow, on the water, or under the water. You may have to put on or take off clothes; carry or search for objects; float for hours; or in shark-infested waters, lie still and keep your arms and legs from dangling. There is a lot you might have to do. You can get ready by practicing all the strokes you know.

THIRD CLASS SWIMMER

1. Enter the water feet first from a height of 5 feet
2. Remain afloat for 5 minutes
3. Swim 50 yards

SECOND CLASS SWIMMER

1. Jump from a height of 10 feet
2. Remain afloat for 10 minutes
3. Swim 100 yards, using three survival strokes for at least 25 yards each:
   - Breast stroke
   - Side stroke
   - Elementary back stroke

FIRST CLASS SWIMMER

1. Swim 220 yards
2. Enter the water feet first and immediately swim 25 yards underwater (you may surface for air twice at 25-foot intervals)
3. Remove your trousers or slacks in the water and inflate them
4. Tow another person 25 yards, using the following methods:
   - Cross-chest carry
   - Extended reach (recommended for struggling victims)
   - Grabbing the victims hair from behind and use side stroke (recommended for towing unconscious victims)

Almost all the Navy's shore installations have swimming facilities for your use. Here, you can practice swimming. You should practice various strokes and extend your swimming range. Then, you will feel more confident that you can stay afloat and swim to a distant lifeboat or floating object.

SWIMMING THROUGH FLAMES.—Flame-covered water is a terrifying sight. However, you don't need to be afraid of jumping into flames. If you follow the procedures listed here, you will clear the burning area safely (fig. 15-2).

1. Don't wear an inherently buoyant life preserver (if you have one on, get rid of it).

BMRF1502

Figure 15-2.—Swimming through flames.

2. If you're wearing a CO_2 preserver, keep it on but don't inflate it.
3. Discard your shoes because they will hinder your underwater swimming.
4. Take a deep breath when you jump from the ship and cover your nose and mouth with one hand and your eyes with the other.

Student Notes:

5. Swim as far underwater as possible.
6. When you must come up for air, extend your arms above your head, then pull them back in a wide sweep to force the upper part of your body above the surface.
7. When you surface, use your hands and arms to make wide sweeping movements across the surface to splash the water and drive away the flames.

NOTE

As you pop up above the surface, try to turn your back to the wind before you take a breath.

8. Submerge again feet first, and repeat the procedure until you're clear of the burning oil.

When going into oil that isn't burning, save your preserver to use as a raft. Keep your face above the surface. Keeping your head above the surface helps keep oil from getting into your eyes and mouth.

AIDS FOR STAYING AFLOAT.—If you're in the water without a life jacket, don't become frightened that you can't stay afloat—you can. Several articles of clothing, including the white hat, provide some flotation when used properly. The most useful article is your trousers or slacks, which you can inflate to serve as water wings.

1. To remove your trousers, lean forward in the water and slowly slip them down over your hips and legs. Don't let go of them—they may sink. To inflate your trousers—
2. Zip them; then float them on the surface with the fly or front turned down.
3. Tie a knot in each leg as close to the cuff as possible.
4. Work the garment around on the surface until the legs are over your shoulders and the knots are behind you, leaving the crotch in front of you.
5. Grasp the waist of the trousers with one hand on each side; then extend your arms straight upward, kicking your feet to get your body as high out of the water as you can.
6. When this position is reached, pull the trousers downward smartly on the surface, trapping a pocket of air in each leg.
7. Then gather the waist under the water and hold in one hand (fig. 15-3). Keep the trousers legs wet by splashing water on them to reduce the loss of the trapped air.

You may use mattress covers, sea bags, laundry bags, and pillowcases in a similar manner. A large amount of debris, such as pieces of wood, empty shell boxes, powder cans, and so forth, is usually present. You can use this debris to stay afloat.

SURVIVAL EQUIPMENT

The two basic categories of flotation devices are life preservers and lifeboats. Each is vital to the survival of a ship's crew if the ship sinks. Other than the lifeboat, the life preserver (commonly called a *life jacket*) is the most important piece of abandon ship equipment.

The inherently buoyant (vest-type) preserver is designed so that, if adjusted properly, it supports you and keeps your head out of the water even if you are unconscious. With a life preserver on, you can stay afloat for many days. Without a life preserver, you have little chance of surviving in the water for any great length of time.

The lifeboat presents the greatest chance of survival because it contains food and water, provides shelter from the elements, and contains equipment that greatly

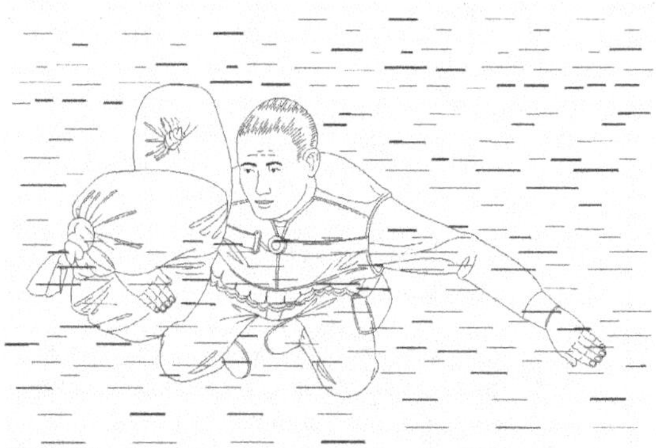

BMRF1503

Figure 15-3.—Using inflated trousers/slacks for support.

Student Notes:

enhance your chances for survival.

During wartime, each person aboard ship is issued a life preserver. Wear it or keep it handy at all times. During peacetime, life preservers are stowed in ready-use lockers. Know where your preserver is stowed, how to put it on, and how to release and inflate the lifeboat.

Life Preservers

The Navy uses two types of life preservers—the inherently buoyant and the inflatable types. The inherently buoyant type has several designs. The vest type is the most widely used.

INHERENTLY BUOYANT TYPE.—The inherently buoyant vest type of life preserver (fig. 15-4) uses fibrous glass pads to provide buoyancy. The pads are sealed in plastic waterproof bags placed in an outer

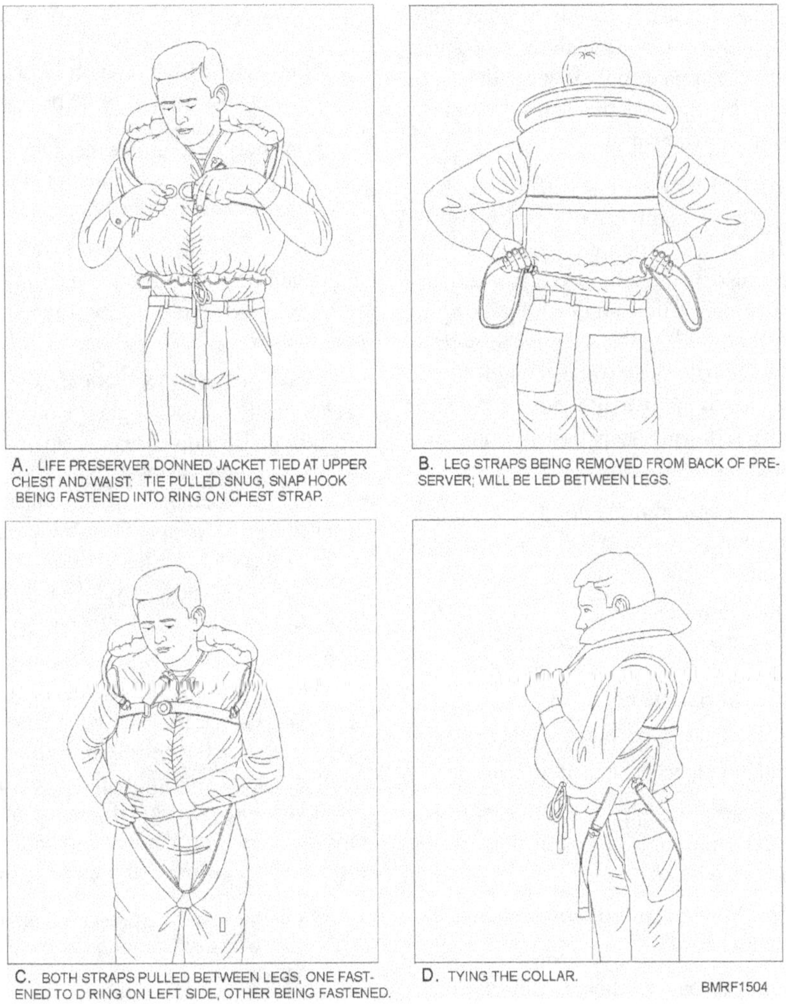

Figure 15-4.—Adjusting the inherently buoyant vest-type life preserver.

Student Notes:

cover or envelope. The preserver has cloth tapes to pull tight for a close fit. Leg straps prevent it from riding up while you are in the water. A body strap across the chest helps give a snug fit and provides a hold for lifting you out of the water. You can also use the strap to attach yourself to a life raft or to other persons in the water.

Put on the vest type of life preserver over your clothing. Tie the upper tapes to make it fit comfortably, and pull the tape at the waist fairly tight to keep the preserver from sliding up in the water. Then adjust the chest strap and fasten the snap hook into the ring. Pull the leg straps as tight as possible without producing discomfort. Tie the collar tapes tightly under the chin. The collar holds the head upright and helps prevent an unconscious person from drowning.

INFLATABLE TYPE.—The inflatable life preserver (fig. 15-5) is made of lightweight, neoprene-coated nylon. It's carried in a pouch container held around your waist on a web belt. You blow up the inflatable preserver either by mouth or by using a carbon dioxide (CO_2) cylinder. It's equipped with a lifting harness, a waist belt, and a wooden toggle and a line for attaching yourself to a life raft or another survivor. Take the following steps when using the inflatable life preserver:

1. Pull the pouch around to the front, remove the preserver from its pouch, and slip it over your head.
2. Grasp the lanyard attached to the CO_2 cylinder and jerk downward. If you need more buoyancy, the life preserver can be orally inflated by taking the following steps:
   a. Turn down the knurled ring at the base of the oral inflation tube as far as it will go.
   b. Depress the mouthpiece by force of the mouth, and blow into the tube as if you were blowing up a balloon.
   c. Release the mouthpiece when inhaling to prevent escape of the air.
3. When the preserver is inflated, lock the oral valve by turning the knurled ring against the mouthpiece.

NOTE

Always wait until you have entered the water to inflate this type of life preserver.

The automatically inflatable work-type life preserver provides you maximum lifesaving protection. At the same time, it doesn't interfere with the jobs you do, such as working over the side, performing underway replenishment (UNREP) duties, working as part of a boat crew, or manning selected battle stations. The automatically inflatable work-type life preserver will—

- Inflate the life preserver if you go into the water in an unconscious or helpless state.
- Allow you to inflate the auto inflatable preserver orally by the auto function device or by using a combination of the two.

The auto-function device uses a water-degradable paper to release a spring that causes two CO_2 cylinders to be punctured and inflate the preserver.

PIN-ON LIGHTS.—Small watertight flashlights or chemically activated light sticks have been developed for use with life preservers to help rescuers see a person in the water more easily at night. The flashlight consists of a one-cell battery case to which is permanently attached a heavy metal safety pin for fastening the light to the preserver. The lens is dome-shaped, providing 360° visibility from above. The chemically activated light sticks are activated by a chemical reaction in the stick.

Wear these lights whenever you use the life preserver. Check the battery at least once a week to see that it works. Replace the battery at least every 6 months. Check the light stick each time you use the preserver, and replace it if you see any indication that the stick has been damaged or used. Remember the following tips when using these lights:

- On the vest-type preserver, pin the light near the top of your shoulder so that the lens points upward.
- When pinning the light on the vest-type preserver, take care not to pierce the waterproof

Student Notes:

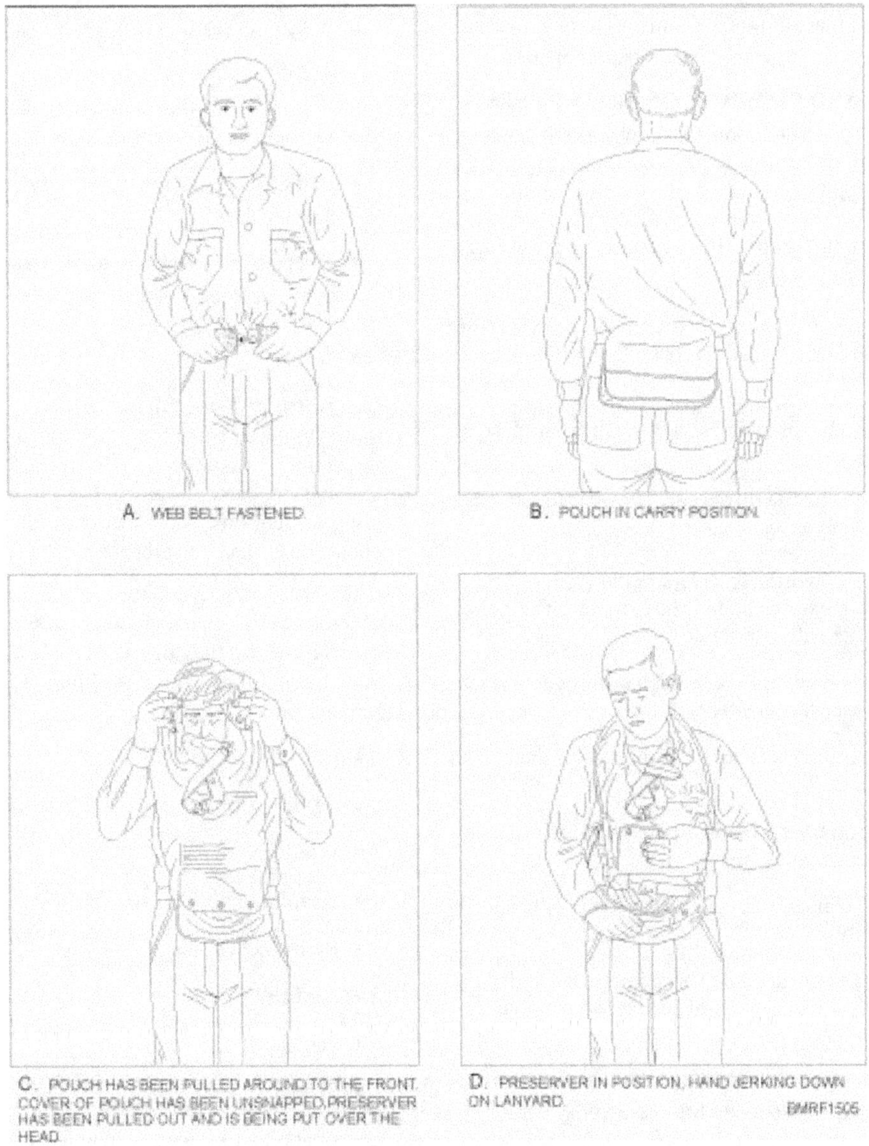

Figure 15-5.—Inflatable life preserver.

covering in which the fibrous glass pads are wrapped.

- Attach the light to the inflatable preserver to the tab provided for this purpose.

Some ships may issue strobe lights. These lights have a brighter intensity. The battery screws in and is water-resistant.

Some commands are issuing chemical lights as life vest pin-on lights. The light used for a pin-on light has a green color when the chemical is activated. You activate the chemical light by squeezing the lens, which crushes

Student Notes:

an inner vial; that allows the chemicals to mix, causing the wand to glow. Dispose of these lights after one use.

CARE AND STOWAGE OF PRESERVERS.— Some of the rules you should follow when taking care of and stowing your preservers are contained in the following section:

Laundering your life preserver. Inherently buoyant life preservers—

- Launder the outer covers after removing the fibrous glass pads. (**NOTE**: Don't launder the pads.)
- Clean the inflatable types with a mild soap solution only.

Stowing your life preserver. The rules for stowing life preservers include—

- Don't stow life preservers in the vicinity of oil, paint, grease, heat, moisture, or dirt. The nylon material will deteriorate.
- Keep preservers clear of sharp edges, which increase wear and tear.
- Keep preservers away from steam lines and radiators.
- Dry preservers thoroughly before stowing them to prevent mildew.
- Don't tamper with your life preserver or handle it roughly.
- Don't sit or lie on it. This compresses and mats the filler pads and reduces the buoyancy of the preserver.

Inspecting your life preserver. The following rules apply when inspecting life preservers:

- Inspect your inflatable life preserver every time you put it on and at least once every month (when in your custody).
- Inflate it by mouth to locate possible leaks in the air chamber or inflation valve.
- Make sure the piercing pin of the CO_2 valve is in good working order and the cylinder itself has not been punctured.
- Weigh the cylinder on a gram scale to make sure it is fully charged.

Other actions. Other actions you should take with regard to your life preserver include—

- Being able to put the life preserver on and adjust it in the dark.
- Treat it like a friend; someday it might turn out to be the best one you have!

Lifeboats

A warship doesn't have room to carry all the powerboats needed to transport the entire crew. At sea, a powerboat is usually difficult and sometimes impossible to launch rapidly. For these reasons, the Navy has spent time and expense developing efficient lifeboats other than powerboats.

The Navy uses several types of inflatable lifeboats. Each boat has sufficient equipment to support the number of survivors for which the boat was designed to carry. Each boat's gear includes the following equipment:

- Canopy
- Sea anchor
- Lifeline
- Boarding line
- Rain-catcher tube
- Air hand pumps
- Paddles
- Sponges
- Boat repair kit for patching leaks
- Floatable knife

The inflatable lifeboat (fig. 15-6) also carries—

- Desalter kits for turning seawater into freshwater.
- Survival kits containing food rations, sea marker dye, a flashlight, batteries, a signal mirror, a

Student Notes:

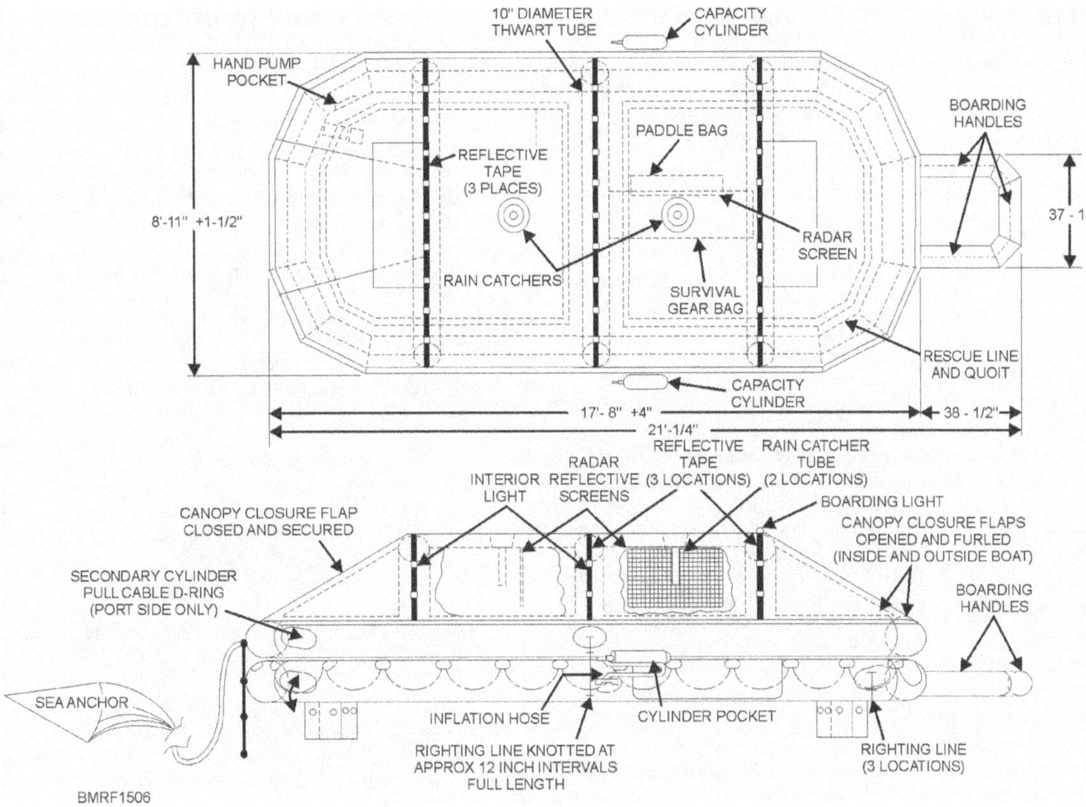

Figure 15-6.—Inflatable lifeboat.

whistle, a first-aid kit, a distress signal kit, and containers of freshwater.

- Survival kits in the large boats are designed to sustain 15 to 20 people for 5 days on regular rations.

SIGNAL EQUIPMENT.—Using signaling equipment in the lifeboat correctly might be the difference between rescue or remaining adrift. The opportunity to attract the attention of friendly aircraft or surface vessels may pass quickly; you must be prepared at all times to use the signaling equipment.

The following chart (next page) describes how to use signaling equipment.

CARE AND USE OF SURVIVAL AND SIGNAL EQUIPMENT.—When using survival and signal equipment, stow it in containers for safekeeping and protection against the elements. Some of the items, such as the mirror and whistle, have a lanyard for wearing around the neck. Keep all items as dry as possible. After using any item, replace it in its container. Protect flashlights and knives from salt spray; otherwise, they will soon become corroded. About the only items that should be left out continuously are the sponges.

EQUIPMENT FOR OBTAINING WATER.—Never discard (throw away) any article that will hold water. When it rains, every container that can possibly hold water is invaluable. A rain-catcher tube attached to the lifeboat canopy will help you fill the containers. Even in a light rain, some water will drain from the canopy down through the tube. After filling all available

Student Notes:

EQUIPMENT	DESCRIPTION	HOW TO USE
Signal mirror	The mirror is an effective device when the sun is shining. Rough water makes focusing the mirror on a rescue ship or aircraft difficult. If the mirror is lost or is unusable, make another one from a piece of shiny metal.	To signal with the mirror— 1. Punch a cross-hole in its center. 2. Hold the mirror about 3 inches in front of your face and sight through the cross at the ship or aircraft. The spot of light shining through the hole onto your face will be seen in the cross-hole. 3. While keeping a sight on the ship or aircraft, adjust the mirror until the spot of light on your face disappears in the hole. The bright spot, seen through the sight, will then be aimed directly at the search ship or aircraft **NOTE** The survival kit contains instructions for using the mirror
Distress signal kit	The signal kit contains 12 (Mk 13 Mod 0) distress signals for day and night use and for providing wind drift information to helicopters rescuing personnel. One end of the signal tube produces an orange smoke for day use; the other end produces a red flare for night use. You can identify the night flare end in the dark by a series of small beadlike projections embossed around it. Each signal will burn for approximately 18 seconds.	To use the signal— 1. Select the proper flare, tear off the sealing tape from around the end of the cylinder, and remove the plastic cap to expose a metal pull ring (fig. 15-7). (Only the night end of the flare has a metal ring; the smoke [day] end does not have the ring.) 2. To ignite the MK 13 signal, grasp the pull ring and flip it over the rim of the signal case, as shown in view A. 3. Press down the overhanging ring with your thumb until the seal snaps, as shown in view B. (If the seal refuses to snap, continue pressing on the ring so that it bends over the rim and against the signal body, as shown in view C). 4. Flip the ring back to the top of the signal and press down, as shown in view D, using the bent pull ring as a lever. 5. After the seal breaks, point the signal away from your face and body and give a sharp yank on the pull ring. 6. Hold the signal at an angle of approximately 45° from the horizontal position with your arm fully extended. The contents are hot, so take care not to drop any of the contents on yourself or the lifeboat. 7. After using one end of the signal, cool it by dipping it in water; then save it until you use the other end. Make sure the distress signal is cool before storing it.
Dye marker	The dye marker shown (fig. 15-8) produces a brilliant yellowish-green fluorescence when it is submerged in water. Under good conditions, the dye will be a good target for only about 1 hour, but it will retain some of its color for up to 4 hours. From an altitude of 3,000 feet, a rescue plane may see the dye marker as far away as 10 miles. The range decreases as the dye spreads or is diluted by the water.	See the front of the dye marker cover.

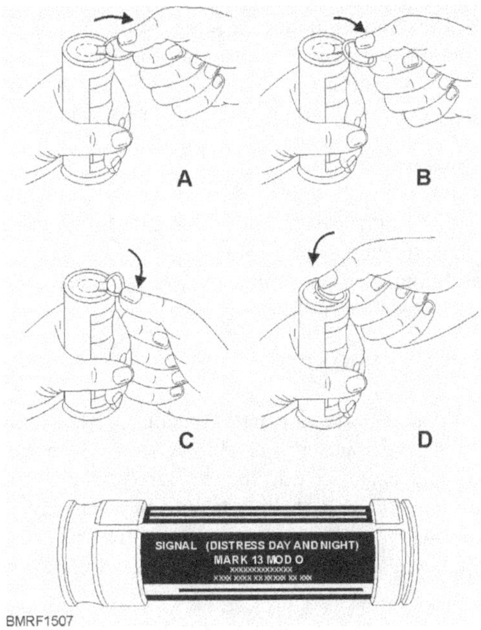

Figure 15-7.—Igniting the MK 13 distress signal.

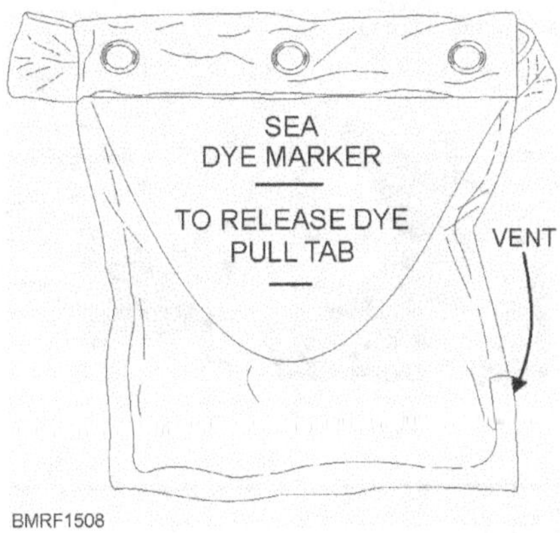

Figure 15-8.—Dye marker.

containers, stow them carefully so that you won't lose any water. Cover all open containers to slow down evaporation; use those you don't have covers for first. During the rain, drink all you can hold.

In polar areas, you can obtain freshwater from old sea ice. Old sea ice is a bluish color, splinters easily, and is nearly free from salt. New ice is milky in color, hard, and salty. You may also obtain freshwater from icebergs, but use caution. As its underwater portion melts, an iceberg gets top heavy and can capsize without warning.

SURVIVAL STEPS

Most of the following survival information applies to persons in lifeboats, but some of this information applies to persons in the water. In trying to survive at sea, you will face thirst, hunger, and exposure whether you are in a lifeboat or in the water. You can endure these conditions, however, if you take the proper steps.

Thirst

The one absolutely essential requirement for survival is drinking water. Without it, death will most likely occur in 8 to 12 days. Normally, you need about 2 quarts of water a day; but because of inactivity and lack of food, you can survive on as little as 6 ounces a day in a lifeboat.

Water is lost from the body by the evaporation of perspiration and through the digestive process. Some actions you can take to reduce water loss include the following:

- Keep your clothes wet during the day (weather permitting, of course), but dry them before sundown.
- Wear the least amount of clothes possible, depending upon your need for protection from the elements.
- If water is scarce, eat sparingly.
- **Never drink seawater or urine.** To do so would only aggravate your thirst and increase body water loss with a subsequent speedup in dehydration.
- Do not drink your entire daily water ration at one time. It is better to drink small amounts three or four times daily.

Hunger

The food rations supplied with each lifeboat are

Student Notes:

specially designed to maintain your physical and mental abilities and aren't thirst-provoking. The ration is based on an allowance of one packet per person per day; but, you should eat only when you feel the greatest need. Don't take any food or water the first 24 hours. Food is much less important for survival than water. With water, a person can survive for 4 weeks or longer without food.

Nearly all forms of sea life are edible. Some fish are poisonous; for example, jellyfish (which you should never eat). Each lifeboat has a fishing kit for catching fish.

All sea birds are edible, and practically the entire bird is useful. In addition to the food and liquid obtained from sea birds, you can fashion fishhooks and lures from the bones and feathers. In cold weather, a bird's skin (with feathers) will protect exposed parts of your body.

Birds sometimes settle on the raft or boat, and survivors have reported instances where birds landed on their shoulders. If birds are shy, try dragging a baited hook through the water or throwing a baited hook into the air.

You can catch gulls, terns, gannets, and albatrosses by dragging a baited hook behind the boat or raft. You can attract them within shooting distance by dragging a bright piece of metal or shell behind the raft. It's possible to catch a bird if it lands within reach. Most birds, however, are shy and will settle on the raft out of reach. In that case, try a bird noose. Make it by tying a loose knot with two pieces of line, as shown in figure 15-9. Bait the center of the loop with fish entrails or similar bait. When the bird settles in the loop to eat the bait, tighten the noose around its feet.

The North Atlantic and the North Pacific have relatively few birds, and these are found mostly along the coasts. You may see many species of birds, often hundreds of miles from land, in southern waters.

Exposure

Exposure presents many dangers. Some dangers include sunburn, hypothermia, frostbite, and immersion foot. Some actions you can take to survive these conditions are as follows:

MAN OVERBOARD

All the information in this section applies mainly to ship disasters when your ship is sunk. Such events normally occur in wartime but rarely in peacetime. However, a mishap that can happen to you at any time, and usually without warning, is to fall overboard. One minute you are walking along the main deck; the next

Figure 15-9.—Bird noose.

CONDITION	ACTION
Cold	You can't survive for any great length of time in cold water without a special exposure suit. In water cooler than 75 °F, you face a serious condition called *hypothermia.* Hypothermia occurs when your body is exposed to subnormal temperatures. To overcome hypothermia, minimize heat loss from your head, neck, sides, and groin. Raise as much of your body as possible out of the water; wear a hat; and assume the fetal position or huddle in close, side-by-side contact with others. Don't move about. Stay calm and encourage others not to panic.

Student Notes:

CONDITION	ACTION
Cold (Continued)	In cold waters, your greatest danger after abandoning ship is the effects of the cold. Wear as much clothing as you possibly can, especially heavy undergarments. Ordinary clothing gives you no protection against cold if you are immersed in water. You must get out of the water as quickly as possible Huddle together for warmth. A huddled group can survive cold that might be fatal to one person alone. Rig wind and spray shields, but don't block the sun's heat. Exercise mildly, if possible, to increase body heat; but never do so to the point of exhaustion. Lifeboats are uncomfortable and cold. In frigid temperatures, you must keep both ends of the inflatable lifeboat closed to keep the temperature comfortable; but this confinement creates other discomforts. Closing the ends reduces ventilation and raises the humidity. Then you must reopen the ends to let out the impure air and to bring in fresh air, which, of course, is cold.
Sunburn	Shoes and clothing are a real protection against sunburn and exposure. Remove clothing only when it is absolutely necessary. If you must remove your clothes while in the water, take off only the heaviest articles. Because your shirt or jumper offers warmth at night as well as protection from the sun during the day, don't remove it. Sunburn is easier to prevent than to treat. Try to remain out of the direct rays of the sun. If you can't avoid direct exposure, keep your hat on and cool your body by wetting your clothing.
Dampness	Although remaining dry on a lifeboat is always difficult, make every effort to keep your clothing dry. Since continuous condensation of moisture causes it to drop like rain, sponge out the boat whenever possible. Cold weather aggravates these uncomfortable conditions.
Frostbite and immersion foot	Frostbite and immersion foot are serious injuries that can happen even when you're wearing enough clothing to stay fairly comfortable. Frostbite usually affects the hands, face, or feet, and it most often occurs on windy, very cold days. Affected parts of the body turn stiff, pale, and numb. To prevent frostbite, keep exposed parts of the body as warm as possible and maintain circulation. If frostbite occurs, treat the affected part immediately by placing it in contact with a warm part of your body. Cover it with your hand or put frozen fingers inside your clothing. Don't rub the affected parts; that could result in damage to frozen tissue. Immersion foot is the swelling of the foot accompanied by numbness and pallor (lack of color) or discoloration. Immersion foot is caused by poor circulation in the legs, particularly when the foot remains wet for several days. To prevent immersion foot, exercise the ankles and toes for a few minutes several times each day. Keep your feet warm, dry, and elevated as much as possible. Unlace your shoes or take them off. If you have no dry socks or wrappings for your feet, put them under the arms or in the lap of a shipmate. Never treat immersion foot by rubbing. As with frostbite, tissue damage may result. Rewarming is the only proper treatment.

Student Notes:

minute you are in the water, swimming for your life.

If you fall overboard and someone hears or sees you (one of the purposes of the lookout watch), you can count on being rescued within a few minutes. Such rescues are made in nearly every instance. However, if no one sees you fall overboard or hears a cry for help, you'll be missed and rescue procedures will then be put into action.

If you fall overboard, the most important thing to do is stay calm. Panic will cause you more harm than almost anything else. If you see any floating debris nearby, hang on to it. Otherwise, remove and inflate your trousers. Remember, you can stay afloat for a long time, even without help, if you use the floating positions. Don't swim after the ship, because you'll only exhaust yourself needlessly, and the ship may waste valuable time searching for you at the point where you fell overboard.

The method used to rescue a person overboard depends on the circumstances at the time. In daylight, with good weather, a helicopter (if available) is normally used. Otherwise, the ship's motor whaleboat is used, or you may be recovered directly over the side of the ship.

Helicopters use three basic devices for recovering a person in the water—

1. Sling. If the sling is used, adjust it so that it is across your back and under your arms with the hoisting cable in front of you.
2. Net. If the net is used, simply sit in it and hold on.
3. Two- or three-pronged seat. If a two- and three-pronged seat is used, sit on the prongs and wrap your arms around the upright portion.

When a motor whaleboat is used for rescue, the boat crew helps you into the boat. Also, a swimmer provides assistance if you are injured or exhausted. Don't try to enter the boat from astern; you may be injured by the propeller.

If neither a helicopter nor a whaleboat can be used for rescue, the ship will maneuver to a position where a swimmer, towing a line, can reach you. After the line is fastened around your body, personnel on deck will haul you in and hoist you aboard.

While awaiting rescue, remain calm. If sharks are in the area, float on your back, using as little arm and leg movement as possible.

To decrease your chances of having to be rescued at all, observe all safety regulations. Don't lean on lifelines. Don't go on deck in bad weather unless you have to. Always wear a life preserver when working in areas where you are in danger of falling overboard. Aboard aircraft carriers, don't walk behind a jet plane turning up its engines because the blast can blow you overboard.

Ships frequently hold man-overboard drills. In spite of precautions, accidents happen. Therefore, when you are at the beach, don't spend all your time sunbathing. Practice swimming and floating. Someday your life may depend on your ability to swim and float.

REVIEW 1 QUESTIONS

Q1. When aboard ship, you should know escape routes for what reason?

Q2. True or False. You should dive into the water to abandon ship.

Q3. What swimming classification gives you the best chance for survival if you have to abandon ship?

Q4. If you have to jump from a ship into burning water, you should—

Q5. Which of the following items can you use to stay afloat?

a. Trousers

b. Sea bag

c. Pieces of wood

d. All of the above

Student Notes:

Q6. List the two types of life preservers used by the Navy.

a.

b.

Q7. When you have custody of your life preserver, how often should you inspect it?

Q8. List the contents of survival kits carried by inflatable lifeboats.

a.

b.

c.

d.

e.

f.

g.

h.

i.

Q9. What is the one essential requirement for survival?

Q10. The food ration carried by lifeboats is based on how many packets per person per day?

Q11. List some of the dangers you might face by exposure.

a.

b.

c.

d.

SURVIVAL ASHORE

Learning Objectives: When you finish this chapter, you will be able to—

- Recall the methods and procedures for survival ashore to include individual survival, group survival, and methods of evasion and escape.
- Identify the responsibilities and authority of the senior person in a survival situation.

Survival is largely a matter of mental outlook, and the will to survive is the deciding factor. The experiences of hundreds of service personnel isolated during World War II and the Korean conflict and Vietnam police action prove that survival is largely a matter of mental outlook. These experiences also prove that the will to survive is the deciding factor in survival. Whether with a group or alone, you will experience emotional problems resulting from fear, despair, loneliness, and boredom. Your will to live will also be taxed by injury and pain, fatigue, hunger, and thirst. Being prepared mentally to overcome all obstacles and accept the worst greatly increases your chances of coming out alive.

INDIVIDUAL SURVIVAL

The shock of being isolated behind the enemy lines, in a desolate area, or in enemy hands can be reduced or even avoided if you remember what each letter in the key word *S-U-R-V-I-V-A-L* stands for.

S ize up the situation

U ndue haste makes waste

R emember where you are

V anquish fear and panic

I mprovise

Student Notes:

V alue living

A ct like the natives

L earn basic skills

S — Size up the situation by considering yourself, the country, and the enemy.

In considering yourself, hope for the best, but be prepared for the worst. Get to a safe, comfortable place as quickly as possible. Once there, look things over, think, and form a plan. Your fear will lessen and your confidence will increase. **Be calm!** Take it easy until you know where you are and where you are going.

Part of your fear may come from being in a strange country; therefore, try to determine where you are by landmarks, compass directions, or by recalling intelligence information passed on to you by your leaders.

In considering the enemy, put yourself in the enemy's shoes. What would you do? Watch the enemy's habits and routines. Base your plan on your observation. Remember, you know where the enemy is; the enemy does not know where you are.

U — Undue haste makes waste.

Don't be too eager to move. That will make you careless and impatient. If you begin to take unnecessary risks, you have a good chance of being captured. Don't lose your temper; doing so may cause you to stop thinking. When something irritating happens, stop, take a deep breath, relax, and start over.

Face the fact that danger does exist. To try to convince yourself otherwise only adds to the danger.

R — Remember where you are.

You may give yourself away because you're used to acting in a certain way. Doing "what comes naturally" could be the tip off that you don't belong there.

V — Vanquish fear and panic.

To feel fear is normal and necessary. It's nature's way of giving you that extra shot of energy just when you need it. Learn to recognize fear for what it is and control it. Look carefully at a situation and determine if your fear is justified. When you investigate, you will usually find many of your fears unreal.

When injured and in pain, you'll have difficulty controlling fear. Pain sometimes turns fear into panic and causes you to act without thinking. Loneliness can also cause panic. It can lead to hopelessness, thoughts of suicide, carelessness, even capture or surrender. Recognizing these signs helps you overcome panic.

I — Improvise.

You can always do something to improve the situation. Figure out what you need, take stock of what you have, and then improvise. Learn to put up with new and unpleasant conditions. Keeping your mind on SURVIVAL will help. Don't be afraid to try strange foods.

V — Value living.

Conserve your health and strength. Illness or injury will greatly reduce your chances of survival and escape. Hunger, cold, and fatigue lower your efficiency and stamina, make you careless, and increase the possibility of capture. Knowing that will make you especially careful because you'll realize your low spirits are the result of your physical condition and not the danger. Remember your goal of getting out alive. Concentrating on the future—on the time when you will return home—will help you value living during your survival situation.

A — Act like the natives.

"At a railroad station, there were German guards," one World War II male escapee related. "I had an urgent need to go to the rest room. The only rest room was an exposed one in front of the station. I felt too embarrassed to relieve myself in front of all passersby. I walked throughout the entire town, occasionally stopping and inquiring if a rest room were available."

This man was detected and captured because he failed to accept the customs of the natives. When you are in a foreign country, accept and adopt native behavior to avoid attracting attention to yourself.

L — Learn basic skills.

The best life insurance is to make sure you learn the techniques and procedures for survival so thoroughly that they become automatic. That increases the chances that you will do the right thing, even in panic. What you know about survival could save your life. Be inquisitive

Student Notes:

and search for additional survival knowledge.

GROUP SURVIVAL

Just as you must make your reactions to survival situations automatic, so must the entire squad, platoon, or other group that you might be a member of or be leading. The best chance for survival belongs to the group that works **together** and has a leader who accepts responsibility for the group. When you are the senior person, accept responsibility for your group by taking steps to lead members to work together. Some actions you can take include the following:

Organize group survival activities. Group survival depends largely upon the organization of its manpower. Organized action by group members who know what to do and when to do it, during ordinary circumstances and during a crisis, prevents panic. Keeping the group informed, devising a plan, and sticking to the plan helps achieve organization.

Assume command and establish a chain of command that includes all members of the group. Good leadership lessens panic, confusion, and disorganization. Make certain each person knows his or her position in the chain of command and is familiar with the duties of every other person, especially your duties as the senior member. Under no circumstances leave leadership of the group to chance acceptance by some member after a situation arises.

Maintain respect for your leadership by using it wisely; be the leader and set the example. Group survival is a test of effective leadership. Watch out for problems that could turn into serious arguments. Keep troublemakers from attracting undue attention, and keep those who may "crack up" from disrupting the group. Prevent carelessness caused by fatigue, hunger, and cold. Know yourself and the members of your group; take responsibility for each person's welfare.

Develop a feeling of mutual dependence within the group by stressing that each person depends on the others for survival. Emphasize that the group will not leave the wounded or injured behind—that each member's responsibility is to make sure the group returns intact. A feeling of mutual dependence fosters high morale and unity. Each member receives support and strength from the others.

Make the decisions no matter what the situation. However, base your decisions on the information and advice of other members of the group—much as admirals make decisions based on input from their staff. Above all else, never appear indecisive.

If situations require you to act immediately, consider the facts and make decisions rapidly. The ability to think on your feet usually determines successful survival.

STRESS OF SURVIVAL

Survival is a state of mind. Your ability to return to your group or to be rescued depends in a great part on your ability to cope with frustrations. You may become frustrated because you find you are unable to accomplish specific tasks. Perhaps you are hungry, cold, lost, injured, or lack the proper equipment. Being able to improvise equipment, care for your physical needs, and provide first aid for your injuries will help you to control your environment, reactions, and emotions. Don't be afraid to experiment and use your imagination. A logical experimental approach is the best way to solve most problems.

Remember the following rules:

1. Almost everything is useful—don't throw away anything.
2. You can be lazier than you would expect, if you just think. The least effort can be the most efficient.
3. Everything you do should be oriented toward rescue.
4. If your surrounding conditions don't suit your needs, do what you can to change them.

SURVIVAL TECHNIQUES

As a member of the armed forces, you always face the chance of being exposed to conditions that can force you into a life-or-death struggle. However, you can remain alive anywhere in the world when you keep your wits. Remember that nature and the elements are neither your friend nor your enemy. By using your wits, you can make them work for you instead of allowing them to work against you.

Student Notes:

Survival depends on you. You must be physically fit and know how to locate or collect water. You must know what plants and animals are available for food, how to find or catch them, how to prepare them, and how to recognize those which will harm you. The more you know about the conditions peculiar to the region you are in, including the plant and animal life, the better are your chances for survival.

Water

Without water your chances of living are slight, and all the food in the area means little. That is especially true in hot climates where you sweat a lot. Even in cold weather your body needs at least 2 quarts of water each day; a lesser amount reduces your efficiency.

When you can't find surface water, tap through the earth's water table for groundwater (rain or melted snow that has filtered through the ground). Getting to the water table and its supply of generally pure water depends on the contour of the land and the characteristics of the soil.

In the desert or arid regions, watch for water indicators. Some signs of water include—

- Plants covering animal trails and the direction in which certain birds fly. By searching in areas toward which these birds fly, you will probably find water.

- Places that are visibly damp, where animals have scratched, or where flies hover indicates recent surface water. Dig in those spots for water.

Leave your handkerchief out on clear nights to collect dew; then squeeze the water into a container. During a heavy dew, you should be able to collect about a pint an hour.

You may find runoff water above the water table. Runoff water includes streams, stagnant pools, and water in bogs. Consider this water contaminated and dangerous even if it is away from human habitation. Boil or treat this water with water purification tablets before you drink it.

If you are unsuccessful in your search for ground or runoff water or if you don't have time to purify questionable water, a water-yielding plant may be your best bet. You can easily get clear, sweet sap that is pure and chiefly water from many plants. Many plants with fleshy leaves or stems store drinkable water. Try them wherever you find them. Desert plants often have their roots near the surface. Pry these roots out of the ground and cut them into 24- to 36-inch lengths. Remove the bark and suck out the water.

Not all vines yield palatable water, but try any vine you find. Use the following method for tapping a vine. It will work on any species.

1. Cut a deep notch in the vine as high up as you can reach.
2. Then cut the vine off close to the ground and let the water drip into your mouth or a container.
3. When the water ceases to drip, cut another section off the vine.
4. Repeat this procedure until the supply of fluid is exhausted (fig. 15-10).

NOTE

If the liquid is a white sap or very dark in color, it is not drinkable. If the liquid is clear, test it for odor. If it is slightly pink or red in color, that normally indicates the presence of tannic acid, which isn't harmful. If it has no taste, or does not taste bad, it is a good source of water.

Food

It takes little reasoning to recognize that your second requirement is food. That's especially true during a time of survival when you need every ounce of energy and endurance that you can muster.

People have been known to live for more than a month without food; but unless you are in extremely difficult circumstances, you don't need to deprive yourself of something to eat. Used properly, nature can provide you with food. Apply the following rules as soon as you realize you are isolated:

1. Inventory your rations and water. Estimate the length of time you will be on your own.
2. Divide your food—two thirds for the first half of your isolation and one third for the second half.

Student Notes:

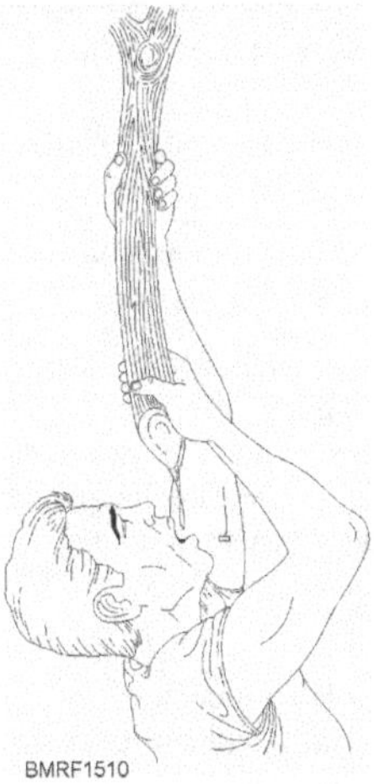

Figure 15-10.—Extracting water from vines.

3. Avoid dry, starchy, and highly flavored foods and meats if you have less than 1 quart of water for each day. Remember eating makes you thirsty. Eat food high in carbohydrates, such as hard candy and fruit bars.
4. Keep strenuous work to a minimum. The less you work, the less food and water you require.
5. Eat regularly if possible—don't nibble. Plan one good meal each day and cook it if you can. Cooking makes food safer, more digestible, and better tasting. Also, the time you spend cooking will give you a rest period in which you can relax.
6. Always be on the lookout for food. With few exceptions, everything you see that walks, crawls, swims, or grows from the soil is edible. Learn to live off the land.

PLANTS.—Experts estimate that about 300,000 classified plants grow on the earth's surface, including many that grow on mountain tops and ocean floors. Of these, 120,000 varieties are edible. Obviously, you won't be able to learn about all of these plants from reading this chapter. But if you know what types of food to look for in the area in which you are stranded, can identify them, and know how to prepare them properly, you should find enough to keep you alive. You may even surprise yourself with a delicious meal.

Eat those plants available in the area to provide you with needed energy while you search for meat. You can depend on them to keep you alive if you're injured, unarmed in enemy territory, or in an area where wildlife is not abundant. Although plant food may not provide a balanced diet, especially in the Arctic where heat-producing qualities of meat are essential, it will sustain you. Many plant foods, like nuts and seeds, will give you enough protein for normal efficiency. In all cases, plants provide energy and calorie-giving carbohydrates.

Most sources of plant foods (fruits, nuts, and berries) have one or more parts that have a lot of food value. For example, certain roots and other underground parts of plants that are rich in starch are excellent sources of food. Some examples are shown on the following page.

ANIMALS.—Foods derived from animals have more food value per pound than those derived from plants. Learning what parts of animals you can eat or use in other ways and learning how to prepare animals for cooking increase your chances of survival.

Methods of Cooking and Preserving Foods

Besides making most foods more tasty and digestible, cooking makes them safer to eat by destroying bacteria, toxins, and harmful elements in the food. Your survival chances increase as your knowledge of field survival skills increases. Survival skills include your ability to improvise and to apply the following principles of cooking and preserving the foods you obtain in the field.

Harmful Plant and Animal Foods

Although you will encounter relatively few poisonous plants and animals, you should learn to recognize and avoid them.

Some places, such as the Arctic and subarctic regions, have less than a dozen plants that are poisonous. These include the water hemlock (fig. 15-16) and the poisonous mushrooms (figs. 15-17 and 15-18).

Student Notes:

FOOD	CHARACTERISTICS
Wild potato	The wild potato is an example of an edible tuber (fig. 15-11). This small plant is found throughout the world, especially in the tropics.
Solomon's seal	Tubers of Solomon's seal (fig. 15-12) grow on small plants found in North America, Europe, Northern Asia, and Jamaica. Boiled or roasted, they taste much like parsnips.
Water chestnut	The water chestnut is a native of Asia, but it has spread to both tropical and temperate areas of the world including North America, Africa, and Australia. It is found as a free-floating plant on rivers, lakes, and ponds in quiet water. The plant covers large areas wherever it grows. It has two kinds of leaves—the submerged leaf, which is long, rootlike, and feathery, and the floating leaf, which forms a rosette on the surface of the water. Beneath the water, the plant bears nuts that are 1 to 2 inches broad with strong spines that give them the appearance of a horned steer (fig. 15-13). You can roast or boil the seed inside the horny structure.
Nut grass	Nut grass is widespread in many parts of the world. Look for it in moist, sandy places along the margins of streams, ponds, and ditches. It occurs in both tropical and temperate climates. The grass differs from true grass because it has a three-angle stem and thick underground tubers that grow ½ to 1 inch in diameter. (See fig. 15-14.) These tubers are sweet and nutty. Boil, peel, and grind them into flour; you can use the flour as a coffee substitute.
Bullrush	Bullrush is a tall plant found in the wet, swampy areas of North America, Africa, Australia, the East Indies, and Malaya. (See fig. 15-15.) You may eat the roots and white stem base cooked or raw.

Figure 15-11.—Wild potato.

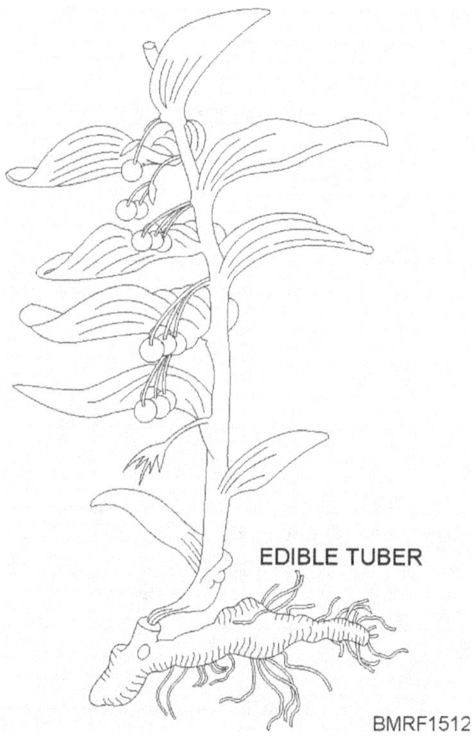

Figure 15-12.—Solomon's seal.

Student Notes:

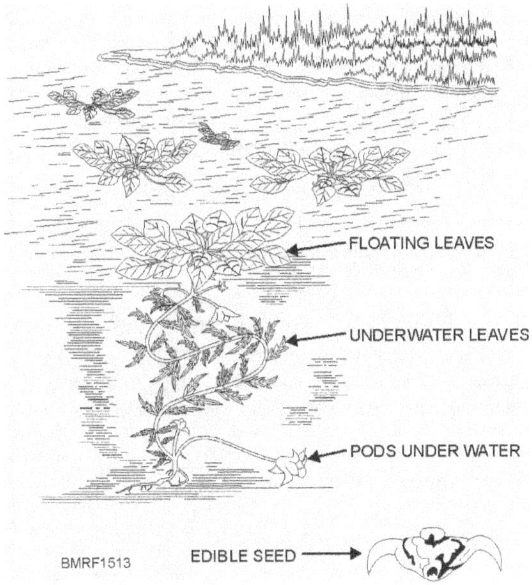

Figure 15-13.—Water chestnut.

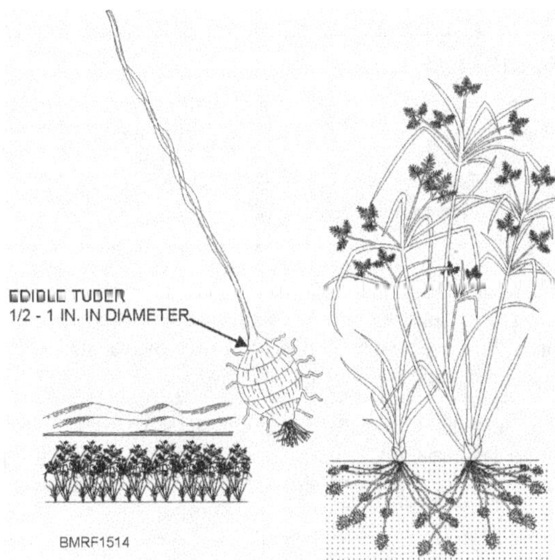

Figure 15-14.—Nut grass.

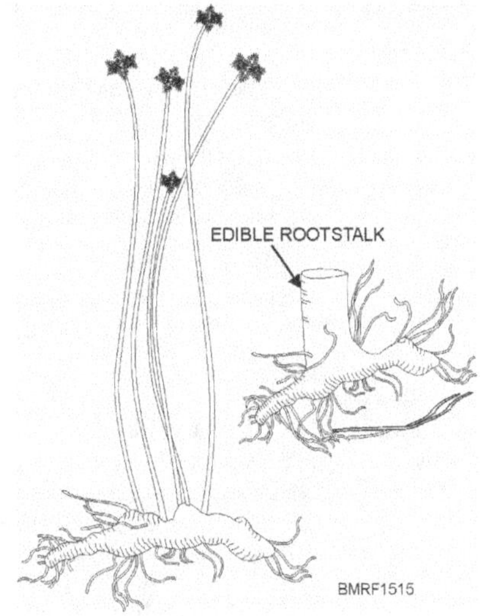

Figure 15-15.—Bullrush.

The tropics have no greater proportion of poisonous plants than the United States. If you're in doubt about whether plants are poisonous or nonpoisonous, observe the habits of vegetable-eating animals, such as birds, rodents, monkeys, baboons, and bears. Usually the foods these animals eat are safe for humans. Cook all plant foods because cooking removes plant poisons (except those in poisonous mushrooms).

NOTE

Avoid eating plants that taste bitter. Also avoid eating untested plants that have milky juices. Don't let the milky juice contact your skin.

You may eat most animals. However, some, like mollusks, may introduce parasites into your body, especially if you eat them uncooked or when they aren't fresh. Crustaceans are almost always edible, but they spoil rapidly and may harbor harmful parasites. Be sure to cook the freshwater variety; eat the saltwater variety raw if you desire.

You have no simple way of telling whether a fish is edible. That depends on the place in which they live, their source of food, or even the season of the year. Often fish that are edible in one area of the world are not in another. At first, eat only small portions of any fish. If you feel no ill effects, you can probably continue to eat the fish safely.

Student Notes:

TYPE OF ANIMAL	PROCEDURE
Birds	Cook most birds with the skin on to retain their food value. After plucking a bird, cut off the neck close to the body and take out the internal organs through the cavity. (**NOTE**: Scalding most birds makes them easier to pluck. Waterfowl are an exception; they are easier to pluck when dry.) Wash out the cavity with fresh, clean water. Save the neck, liver, and heart for stew. Boil scavenger birds, like buzzards and vultures, at least 20 minutes before you cook them to kill parasites. Birds' eggs are among the safest of foods. You can hard boil eggs and carry them for days as reserve food. Save all the feathers you pluck from the birds. You may use them for insulating your shoes or clothing or for bedding.
Fur-bearing animals	Clean and dress the carcass of a fur-bearing animal as soon as possible after death. Any delay will make your job harder. Cut the animal's throat and allow the blood to drain into a container. The boiled blood is a valuable source of food and salt. Save the kidneys, liver, and heart. Use the fat surrounding the intestines. All parts of the animal are edible, including the meaty parts of the skull, such as the brain, eyes, tongue, and flesh.
Shellfish	Crabs, crayfish, shrimp, prawns, and other crustaceans are excellent sources of food. However, crustaceans spoil rapidly so boil them alive immediately after capture. You can steam, boil, or bake shellfish such as clams, oysters, and conchs in the shell. Shellfish make an excellent stew when cooked with greens or tubers.
Other foods	You can easily catch grasshoppers, locusts, large grubs, termites, ants, and other insects to provide nourishment in an emergency.

METHOD	DESCRIPTION
Roasting or broiling	This is a quick way to prepare wild plant foods and tender meats. Roast meat by putting it on a stick and holding it near the embers of your fire. Roasting hardens the outside of the meat and retains the juices.
Baking	Baking is cooking in an oven over steady, moderate heat. The oven maybe a pit under you fire, a closed vessel, or a leaf or clay wrapping. Pit cooking protects food from flies and other pests and reveals no flame at night.
Steaming	You can steam foods that require little cooking, like shellfish. Place your food in a pit filled with heated stones over which leaves are placed. Put more leaves over your food. Then force a stick through the leaves down to the food pocket. Pack a layer of dirt on top of the leaves and around the stick. Remove the stick and pour water to the food through the holes that remains. Steaming is a slow but effective way to cook.
Parching	Parching may be a desirable method of preparing some foods, especially grains and nuts. To parch, place the food in a metal container and heat slowly until it is thoroughly scorched. In the absence of a suitable container, use anything that holds food or water—a heated, flat stone; turtle shells; seashells; leaves; bamboo; or a section of bark.
Drying	Drying preserves food by ridding it of moisture. You can dry plant food and meat by exposing them to wind, sun, air, fire, or any combination of these. To produce jerky, cut meat into 1/4-inch strips and place it across grates; allow it to dry in either the wind or smoke until brittle.

Student Notes:

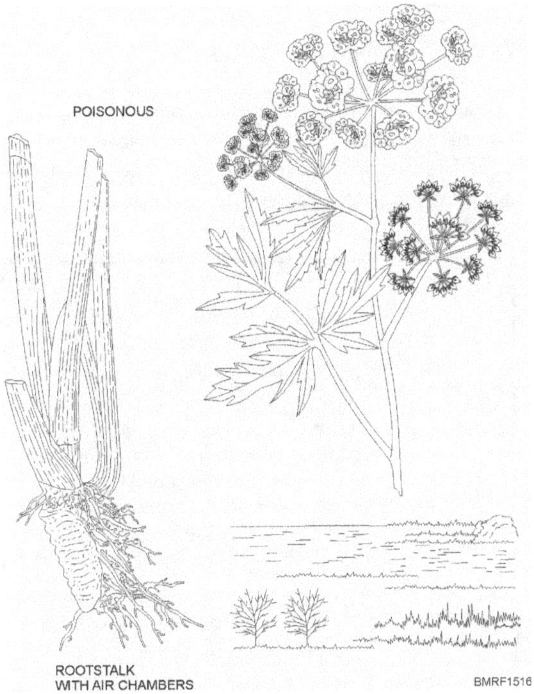

Figure 15-16.—Water hemlock.

Figure 15-17.—Fly agaric.

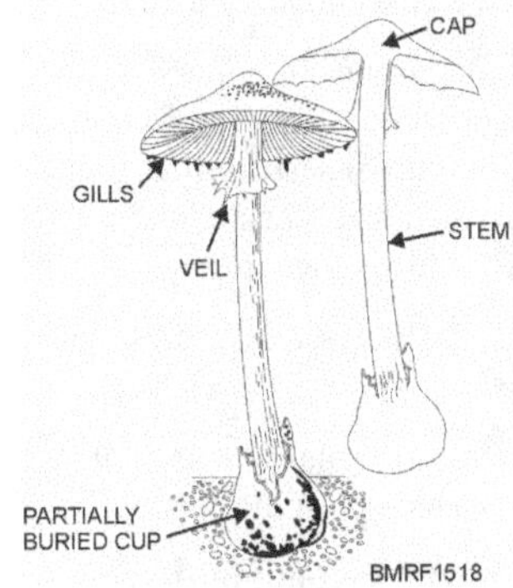

Figure 15-18.—Death angel.

EVASION

According to the Code of Conduct for Members of the Armed Forces of the United States, it is your duty to evade capture by the enemy. Your job is to **get back** to your unit. Your survival will depend on your ability to apply the techniques of evasion. No other reason is more important for making evasion techniques part of your basic combat skills.

Evasion means traveling through enemy-held territory without being captured. Falling into the hands of the enemy is an event that no military person wants to experience. However, at some point in your career you may find yourself in a situation where capture is a possibility. You need to know a few basic evasion principles to decrease your chances of winding up as a guest of the enemy.

During World War II and the succeeding actions in Korea and Vietnam, many of our soldiers, Sailors, and marines were able to avoid the enemy and safely return to friendly forces. They were successful because they applied some or all of the guidelines presented in the following paragraphs. You need to learn this information so that you know how to evade the enemy. It could mean the difference between freedom or capture; interrogation; and possibly, inhumane treatment by enemy forces.

Obviously, the most important consideration in evasion is knowing where the enemy is located. If you don't know the enemy's location, watch for the

Student Notes:

following signs. They can tell you the enemy's location as well as other valuable information.

1. Signs that groups have passed, such as crushed grass, broken branches, footprints, cigarette butts, or other discarded trash, may reveal their identity and size, their direction of travel, and the time they passed through.
2. Workers in fields may indicate absence of the enemy.
3. Apparently normal activities in villages may indicate absence of the enemy.
4. Less obvious conditions may indicate the presence of the enemy, such as the following:
    a. The absence of workers in fields is an indication that the enemy is near.
    b. The absence of children in a village is an indication that the children have been hidden to protect them from action that may take place.
    c. The absence of young people in a village is an indication that the enemy controls the village.

Some evasion techniques you may find useful are cover, concealment, and camouflage. To keep yourself from being seen, you may have to hide in bushes or lie flat in shallow ditches using brush as a cover or camouflage.

When evading the enemy, remember the following points:

1. Conceal yourself from enemy aircraft and nearby enemy troops.
2. Move quietly; noises carry in fog, fallen snow, heavy foliage, and over rock faces.
3. Maintain personal hygiene to prevent body odor; cover body waste and scraps of food; avoid activities, such as cooking and smoking, that produce smells; such smells can reveal your location.
4. Don't make sudden, rapid movements that can reveal your location.
5. Select routes for movement that avoid exposed areas and don't show your silhouette against the skyline. Don't leave obvious tracks.

Crude Direction-Finding Techniques

How do you determine direction without a compass? Nature can help you or nature can fool you. The two best crude sources of direction are the sun and the stars, but you must know how to use them.

Sun	The sun travels from the eastern sky to the western sky. How can you use the sun to determine an east-west direction? You can use shadows (even on a cloudy day) made by the sun to get an accurate east-west line. On a flat surface, drive a stick 3 or 4 feet high in the ground. Then mark the tip of the stick's shadow with a rock. If you wait awhile and then mark the shadow again, you will see that the line connecting the tips of the shadows inscribes an east-west line on the ground.
Stars	To use the stars, you must have a clear night. You may locate north by finding the North Star (Polaris), the outermost star in the handle of the Little Dipper.

These are very crude direction-finding techniques; you may only use them in the Northern Hemisphere. If your ship or aircraft is going to be operating in the Southern Hemisphere, you should learn the techniques for that area of the world.

Evasion Travel

The route that you select to travel while trying to evade the enemy depends on your situation, the weather conditions, and the nature of the terrain. Whether you select a ridge, stream, valley, coastline, dense forest, or mountain range to follow, be sure it is the safest, rather than the easiest, way. Experience has proved that the most difficult route is frequently the safest.

Travel Tips

Some tips you can use when traveling include the following:

Student Notes:

• Be patient, cautious, and avoid overconfidence. An enemy's approach isn't a cause for panic. Normally, you have a good chance of remaining unobserved.

• Conserve your strength by avoiding exhaustion. When you have to remain in one place for an extended period, exercise moderately to keep fit.

• Generally, avoid eating uncooked food or drinking unboiled water. Select a hiding place to cook the food and boil the water you will use en route to the next evasion objective.

• Hold on to items of personal clothing and equipment; they serve a useful purpose during evasion. Keep some items that will identify you as a military person, such as your dog tags. If you can't positively

Along a ridgeline	Using a route along a ridgeline is usually easier to follow than one through a valley. You can frequently use animal trails on top of ridges to guide your travel. When following a ridge-top trail, stay below the trail and move parallel to it. **Never travel along the top of a ridge.** Doing so makes you an easily identifiable silhouette against the skyline.
Use of a stream	Using a stream as a route is of particular advantage in a strange country. It provides a fairly definite course and might lead to populated areas. It's a potential food and water source and may provide you a means of travel by boat or raft.
Following a coastline	Following a coastline leads you on a long, roundabout route. However, a coastline serves as a good starting point. It is an excellent base line from which to get your bearings and a probable source of food.
In a dense forest	When traveling in a dense forest, you probably won't be able to spot distant landmarks. You can stay on course by lining up two trees forward of your position in your direction of travel. As soon as you pass the first one, line up another beyond the second. You might find it helpful to look back occasionally to check the relative positions of landmarks.
Marking your route	You can mark your route with bent bushes, rocks, or notches placed on the backsides of trees at approximately eye level. Make bush marks by cutting vegetation or bending it so that the under, lighter sides of the leaves are facing upward. These signs are especially conspicuous in dense vegetation, but you should be cautious in using them. By plainly marking your route, you risk discovery.
Trails in your general direction	Follow trails that lead in your general direction; when you come to a fork, follow the path that appears most traveled. If you follow the wrong trail and become lost, stop and try to remember the last time you were sure of where you were. Mark your location and start backtracking. Sooner or later you will discover a recognizable feature with which you can pinpoint your position.
Detouring in rough country	You might have to detour frequently in rough country. To do that, try to follow the method shown in figure 15-19 for estimating distance and average angle of departure for short detours. On your return from the detour, estimate the angle and distance to regain your original line of travel. For greater accuracy, count paces and use a compass. Another method (fig. 15-20) lets you select a prominent landmark ahead and behind your line of travel. On returning from your detour, walk until you are again lined up on the two landmarks; then follow your original course.

Student Notes:

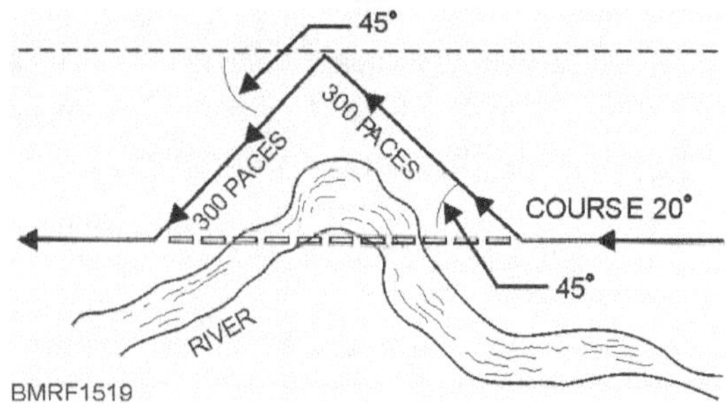

Figure 15-19.—Estimating distance and average angle of departure.

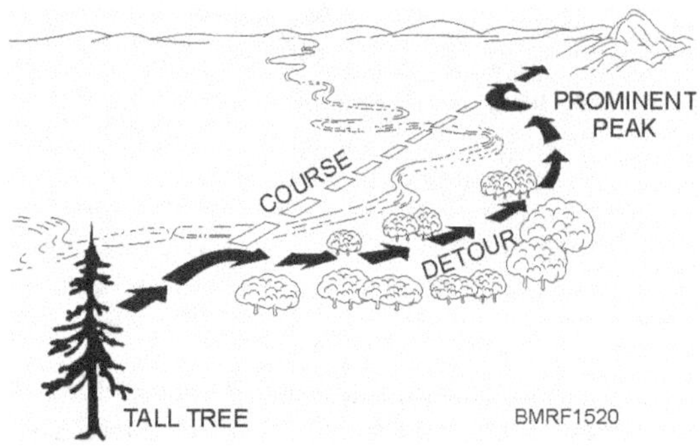

Figure 15-20.—Using prominent landmarks.

identify yourself as a military person, you may be treated as a spy if captured or be refused assistance by escape organizations or friendly natives.

- Don't leave or throw away any articles that, if found, could give the enemy a clear picture of your direction of travel. Bury, or otherwise dispose of, the effects of your campsite.

- Practice supply economy. You may have to use the same jacket or pair of shoes throughout the entire evasion trip, which could cover hundreds of cross-country miles during both winter and summer seasons. Build up your food and water supplies. Carefully ration them so that they will last until you can reach an evasion objective or can replenish them. If you have food but no water, don't eat. Since the digestive processes require water, you will dehydrate faster if you eat.

- Use firearms only in an emergency. Keep them concealed at all times during your evasion unless a situation arises that requires a show of arms.

- Avoid contact with people as long as possible. However, if you can't proceed on your own because of sickness, lack of food, or other reasons, then, and only then, seek out native assistance. Natives who are sympathetic to the allied cause or members of the underground who operate escape lines for the purpose of returning evaders to allied control may offer assistance. Be wary in contacting natives or accepting their help, regardless of what they claim to be.

- If you're fortunate enough to travel through an area where an organized escape line exists, the chances are good that a spotter will seek you out. Spotters for resistance or underground organizations are particularly alert when they have reason to believe allied evaders are in their area but so are enemy police and counterintelligence agents. Persons wearing civilian clothing in enemy-held territory are not necessarily civilians.

Crucial Phase of Evasion

To establish contact with friendly lines or to cross the border to a neutral country is the most crucial point of evasion. All of your patience, planning, and hardships will be in vain if you aren't careful when contacting friendly frontline forces. Many personnel attempting to pass through friendly lines have been killed because they didn't identify themselves properly. Most of these people wouldn't have been shot if they had been cautious and followed proper procedures. The normal tendency is to throw caution to the wind when in sight of friendly forces. You must control this tendency.

Regular patrols or special mission personnel operating behind enemy lines are given the challenge and password of the day as a security measure. Challenges and passwords provide for the identification of the patrol as it approaches a friendly position. In addition, frontline troops are told the time and place where patrols will leave and enter the lines. These conditions exist only if you are able to rejoin your outfit within 24 hours following your separation. After 24 hours, you must follow certain established procedures and hope the frontline troops will also

Student Notes:

follow them. Usually frontline troops, especially those employed several miles forward of the forward edge of the battle area, shoot first and ask questions later. Contacting these troops is, at the very least, sensitive and a calculated risk. However, in the absence of an opportunity to contact a friendly patrol, contact with frontline troops may be your only alternative. Generally, frontline troops are told to honor the display of a white flag or another white object and to advance the unknown person to be recognized.

Once back in friendly hands, you'll naturally want to talk about your exploits and will undoubtedly receive countless questions from frontline troops. However, that is the time you should remain silent. If you talk at this point, you may endanger the lives of those who helped you. In addition, you may compromise methods other service personnel might use to evade the enemy and get out safely. Give only information of immediate tactical importance to frontline units. Advise the first officer or petty officer contacted that you are returning to duty from missing in action, prisoner of war, or internment status. Then request to be taken to someone authorized to receive evasion and escape information.

These survival techniques are but a few of the ways you can stay alive and live to return to friendly forces. You can gain an in-depth knowledge of survival, evasion, and escape techniques through special training. The Navy provides this special training at survival, evasion, resistance, and escape (SERE) schools located at strategic locations throughout the world.

ESCAPE

If I am captured I will continue to resist by all means available. I will make every effort to escape and aid others to escape. I will accept neither parole nor special favors from the enemy.

—Code of Conduct, Art. III

What happens if you become a prisoner of war (POW)? After all, it is possible. Isolation, fear, and injury all work in favor of the enemy to increase your chances of capture in spite of a determined effort on your part to evade. The surrender of your arms, however, does not mean you forfeit your responsibilities as a member of the American armed forces. **The armed forces Code of Conduct directs that you begin planning your escape the minute you are taken prisoner**.

Escape is tough; making it work is even tougher. It demands courage, cunning, and much planning in seeking ways out, determining what routes to follow, and locating friends. Above all, it demands physical stamina under the worst conditions imaginable. Experience has proven that "model" camps with regular rations and considerate treatment are the exception. But no matter what extremes you encounter as a POW, try to keep yourself physically able and sufficiently equipped to escape as soon as possible.

If captured, try to make your escape early. You may never be in any better physical condition to escape than at the moment you are captured. Prison rations are barely enough to keep you alive; they certainly won't supply you with a reserve of energy. The physical treatment, lack of medical care, and insufficient rations of prison life soon have effects such as physical weakness; night blindness; and loss of coordination, reasoning power, and morale.

There are other reasons for making your escape early after your capture. Friendly artillery fire or air strikes occurring during that time may increase your chances of getting away. The first guards you will have are not as well trained in handling prisoners as those farther back from the front lines. Some of the line guards may even be walking wounded who are distracted by their own condition. In addition, you know something about the terrain where you are captured, and you know the approximate location of friendly units. Several days later and many miles away, you may be in strange territory. An escape from a prison camp is much more difficult and requires more detailed planning. It must be organized and supported as any other military operation. The method you should use to escape depends on your particular situation. The only general rules are to make an early escape and to escape when the enemy's attention is distracted.

Save, Add to, Take Care of (S-A-T)

Since the conditions in various POW camps differ, it is impossible to provide a specific escape or survival plan for each situation. What you need is a guide to help

Student Notes:

CODE OF CONDUCT

ARTICLE I

I am an American, fighting in the forces which guard my country and our way of life. I am prepared to give my life in their defense.

ARTICLE II

I will never surrender of my own free will. If in command I will never surrender the members of my command while they still have the means to resist.

ARTICLE III

If I am captured I will continue to resist by all means available. I will make every effort to escape and aid others to escape. I will accept neither parole nor special favors from the enemy.

ARTICLE IV

If I become a prisoner of war, I will keep faith with my fellow prisoners. I will give no information or take part in any action which might be harmful to my comrades. If I am senior, I will take command. If not, I will obey the lawful orders of those appointed over me and will back them up in every way.

ARTICLE V

When questioned, should I become a prisoner of war, I am required to give name, rank, service number and date of birth. I will evade answering further questions to the utmost of my ability. I will make no oral or written statements disloyal to my country and its allies or harmful to their cause.

ARTICLE VI

I will never forget that I am an American, fighting for freedom, responsible for my actions, and dedicated to the principles which made my country free. I will trust in my God and in the United States of America.

you plan to make the best of what you have. One such guide is to remember the word S-A-T–SAVE, ADD TO, TAKE CARE OF

Maintaining Your Health

Good physical health is essential to survival under any circumstances. It is especially important in a POW camp where living conditions are crowded and food and shelter are lacking. That means you must use every device possible to keep yourself well.

Soap and water provide a basic preventive medicine; so keep clean. If water is scarce, collect rainwater, use dew, or simply rub yourself daily with a cloth or your bare hands. Pay attention to areas on your body that are likely to develop rash and fungus infection–your crotch, your scalp, and between your toes.

Student Notes:

Save	**Save** what you can in a POW camp—clothing, pieces of metal, cloth, paper, string anything! A piece of twine may mean success or failure when the time comes for you to break out. Hide these items under the floor or in a hole in the ground. Since they appear harmless, little or nothing will be done to punish you if they are discovered. **Wear** as few clothes as possible during your imprisonment. SAVE your shoes, underwear, shirts, jacket, and any other items of clothing that will protect you from the elements to wear during your escape. **Save** any nonperishable foods you receive from the Red Cross or your captors. Candy, for example, comes in handy as a quick source of energy when you are traveling. If no candy source is available, SAVE each issue of sugar given you by the enemy. When you get enough, boil it down into hard candy. SAVE it until you build up your supply. Store any canned foods you receive. The enemy might puncture the cans to prevent you from saving them. However, you can recook some food into another form that preserves it. Other foods to hoard against the day of your escape include suet (a hard fat), cooked meat, nuts, and bread. **Save** pieces of metal no matter how insignificant they may seem. Nails and pins can serve as buttons or fasteners. You can use old cans to improvise knives, cups, or food containers. If you are fortunate enough to have a razor blade, guard it. Use it for shaving only. Devise ways of sharpening it; rub it on glass or stone or some other hard surface. A clean shave is a good morale booster. **Save** your strength but keep active. A walk around the compound or a few mild calisthenics will keep your muscles toned. Sleep as much as you can. You will not get much rest on your way back.
Add to	**Use your ingenuity**. Select those items that you cannot get along without and supplement them; for example, your rations. There is more to eat in and around your compound than you think. When you are allowed to roam around the prison campgrounds, look for natural foods native to the area, such as roots, grasses, leaves, barks, and insects. If possible, ADD these foods to your escape cache (supplies). They will keep you alive when the going gets tough. **Supplement** your clothing so that the more durable garments are in good repair when you escape. A block of wood and a piece of cloth make good moccasins; that saves wear on your shoes. Substitute rags for gloves; weave straw into hats. Do not forget to salvage clothing from the dead.
Take care of	Probably the most important part of any plan for survival is the **take-care-of** phase. **Maintain** what you have. You won't receive a reissue of shoes or clothes that you wear out or lose. Also, maintain your health; it is not easy to regain once you lose it. **Put** some of your clothing into your escape cache. Watch the rest for early signs of wear, and repair them with improvised material if needed. Use a needle made from a thorn, nail, or splinter and thread from unraveled cloth to mend a torn pair of trousers. Wood, canvas, or cardboard bound to the soles of your shoes will save them from wear. Even paper will suffice as a reinforcing insole if your shoes do wear through.

Student Notes:

Keeping clean also applies to your clothing. Use soap and water when you can spare it. Hang your clothes in the sun to air if soap and water are not available. Examine the seams of your clothing and the hairy portions on your body frequently for lice and their eggs. Disease-infected lice can kill. A possible way to get laundry service, or even a bath, is to tell your guard that you are infested with lice, whether or not your complaint is true. The prison authorities, fearing that lice on prisoners may cause an outbreak of louse-borne disease among the civilian and guard population, might provide this service.

If you become ill, report your condition to the camp authorities. The chance that you will receive aid is worth the try.

After You Escape

Once you escape, you may have trouble contacting friendly units even when you know where they are. Approach the problem as you would if you were a member of a lost patrol. Time your movements so that you pass through the enemy forward areas at night and arrive between the enemy and friendly units at dawn. A good plan is to find a ditch or shell hole where you have cover from both friendly and enemy fire. Attract the attention of the friendly forces by waving a white cloth, shouting, exposing or laying out a panel, or some other method. In doing so, you alert friendly forces who are prepared to accept any small group that appears willing to regain contact. When you alert friendly forces, they are not as likely to shoot you on sight.

REVIEW 2 QUESTIONS

Q1. Give the meaning of the letters in the key word *S-U-R-V-I-V-A-L.*

S

U

R

V

I

V

A

L

Q2. If in a group, what action(s) makes(s) for the best chance of survival?

Q3. List the sources of drinking water.

a.

b.

c.

d.

Student Notes:

Q4. True or False. Food derived from animals has more food value per pound that food from plants.

Q5. List plants that you should not eat.

a.

b.

c.

Q6. List some techniques that are useful to evade the enemy.

a.

b.

c.

Q7. What does the armed forces Code of Conduct direct you to do?

SUMMARY

You will probably spend the majority of your naval career aboard ship. Hopefully you will never fall or be washed overboard or have to abandon ship.

The U.S. Navy operates in all parts of the world from the tropics to polar regions. Each region has its own special survival problems. You may encounter the extreme cold of the polar regions or the heat and humidity of a tropical jungle. Your survival in these places will depend on your ability to take care of yourself. Knowing how to combat hypothermia or heat exhaustion will greatly increase your chances for survival.

Although it could happen, hopefully you will never find yourself stranded in enemy-held territory. To be captured by an enemy force is one of the worst situations you could face. Being properly prepared to make an escape and return to your unit is not only your duty, but it is what every POW thinks about. Knowing how to make that escape work is even more difficult. Knowing what the local environment has to offer in food and water supplies will help you survive during your escape. Maintaining the proper state of mind will greatly increase your chances of making a successful escape.

REVIEW 1 ANSWERS

A1. When aboard ship, you should know escape routes **so you won't be trapped or cut off in case of an emergency or if you must abandon ship**.

A2. **False**. You should **never** dive into the water to abandon ship. Use a ladder, cargo net, line, or fire hose.

A3. The swimming classification that gives you the best chance for survival if you have to abandon ship is the **First Class Swimmer**.

A4. If you have to jump from a ship into burning water, you should **take a deep breath, cover your nose and mouth with one hand and your eyes with the other, and swim under water as far as possible**.

A5. You can use **trousers/slacks, sea bag, and pieces of wood** to stay afloat.

A6. The two types of life preservers used by the Navy are the—

a. **Inherently buoyant type**

b. **Inflatable type**

A7. When you have custody of your life preserver, you should inspect it **once each month**.

A8. The contents of survival kits carried by inflatable lifeboats include—

a. **Food rations**

b. **Sea marker dye**

c. **Flashlight**

d. **Batteries**

e. **Signal mirror**

f. **Whistle**

g. **First-aid kit**

h. **Distress signal kit**

i. **Containers of fresh water**

A9. The one essential requirement for survival is **drinking water**.

A10. The food ration carried by lifeboats is based on **one packet of food per person per day**.

A11. Some of the dangers you might face by exposure include—

a. **Sunburn**

b. **Hypothermia**

c. **Frostbite**

d. **Immersion foot**

REVIEW 2 ANSWERS

A1. The meaning of the letters in the key word *S-U-R-V-I-V-A-L* are—

S ize up the situation

U ndue haste makes waste

R emember where you are

V anquish fear and panic

I mprovise

V alue living

A ct like the natives

L earn basic skills

A2. In a group, **working together** is the best chance of survival.

A3. Some sources of drinking water include—

a. **Dig to the water table**

b. **Collect dew during the night**

c. **Runoff water**

d. **A water-yielding plant**

A4. **True**. Food derived from animals has more food value per pound than food from plants.

A5. Plants that you should not eat include—

A. **Water hemlock**

B. **Fly agaric**

C. Poisonous mushrooms

A6. Some techniques that are useful to evade the enemy include—

a. **Cover**

b. **Concealment**

c. **Camouflage**

A7. The armed forces Code of Conduct directs you to **make every effort to escape**.

CHAPTER 16

CAREER AND EDUCATION INFORMATION

I came into the Navy feeling I could spend a few years away from home, save some money, see different places, and maybe get some training or education that I could use later. I've had to change my attitude about several things since joining. I didn't realize there are so many different opportunities open to me. I guess I'm only limited to how far I can go by how much effort I'm willing to put forth.

—A letter home

As this letter points out, you have opportunities in the Navy. You can advance, get an education, and have a rewarding career. Since the Navy is an all-volunteer organization, its success is influenced by the personal satisfaction of its personnel. Your desire to serve and your patriotism are two factors that contribute to your job satisfaction.

This chapter doesn't provide a detailed explanation of all the available rights and benefits; but it does introduce you to some of them. Remember, the Navy and the Department of Defense make frequent changes to personnel policies. Therefore, some of the information may have changed by the time you read this manual. You should check with your LPO, division or department career counselor, or the command career counselor for the latest information about any Navy program.

THE NAVY GOAL CARD

Learning Objective: When you finish this chapter, you will be able to—

- Identify the purpose of the Navy Goal Card.

The Navy helps first-term Sailors set and achieve both short-term and long-term goals while in the service. The Goal Card Program is one way to keep the volunteer, high-quality Sailor in the Navy. This program is made up of the Navy Goal Card and the Navy Pocket Goal Card. It reinforces goal setting and goal accomplishment by first-term Sailors.

The Navy Goal Card is a two-page document of rating and advancement career information for each new recruit and first-term Sailor. Some of the topics covered by the Goal Card include the following:

- Advanced training and education for your rating
- Montgomery G.I. Bill benefits and goals
- Voluntary education, including Tuition Assistance and SOCNAV
- Officer programs
- Advancement
- Career milestones
- The Apprenticeship Program
- Job descriptions

The Navy Pocket Goal Card (fig. 16-1 and Appendix V) shows a sample of the trifold form for newly recruited Sailors. Appendix V contains a Navy Pocket Goal Card for your use. Areas of goal setting covered on the Pocket Goal Card include the following:

- Delayed Entry Program (DEP) goals
- Navy Core Values
- Recruit training goals
- The Sailor's Creed
- Fleet goals, personal priorities (including education)
- Space for Sailors to write in their own goals

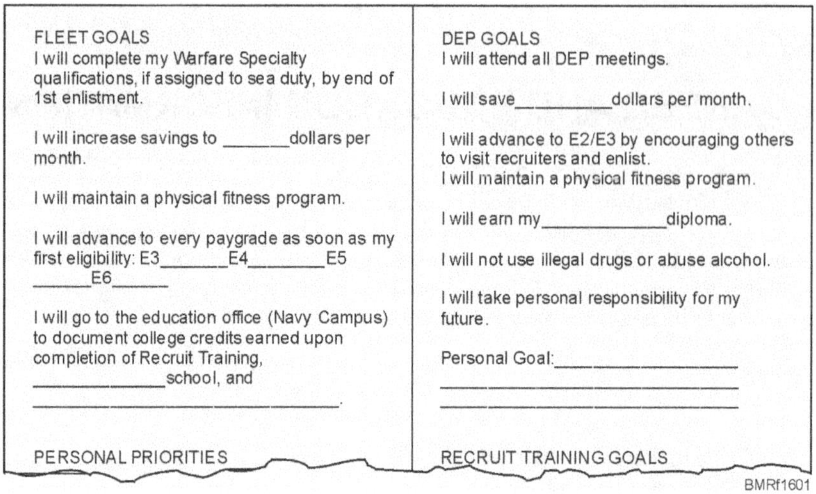

FLEET GOALS
I will complete my Warfare Specialty qualifications, if assigned to sea duty, by end of 1st enlistment.

I will increase savings to ______dollars per month.

I will maintain a physical fitness program.

I will advance to every paygrade as soon as my first eligibility: E3_______E4________E5 ______E6______

I will go to the education office (Navy Campus) to document college credits earned upon completion of Recruit Training, _____________school, and ______________________________.

PERSONAL PRIORITIES

DEP GOALS
I will attend all DEP meetings.

I will save__________dollars per month.

I will advance to E2/E3 by encouraging others to visit recruiters and enlist.
I will maintain a physical fitness program.

I will earn my____________diploma.

I will not use illegal drugs or abuse alcohol.

I will take personal responsibility for my future.

Personal Goal:__________________

RECRUIT TRAINING GOALS

BMRf1601

Figure 16-1.—Navy Pocket Goal Card.

REVIEW 1 QUESTIONS

Q1. What means does the Navy use to help new Sailors set and achieve goals while in the service?

Q2. List some of the areas covered in the Pocket Goal Card.

a.

b.

c.

d.

e.

f.

g.

PROFESSIONAL DEVELOPMENT

Learning Objectives: When you finish this chapter, you will be able to—

- Identify the requirements for professional development.
- Recognize the purpose of the Professional Development Board.

One purpose of the Professional Development Board is to give Sailors a chance for greater responsibility. The board interviews Sailors who want advancement training and who want to attend special programs or programs that need command endorsement (approval). Also, the board advises career Sailors who find it difficult to be selected for advancement or to complete command-required personnel qualification standards (PQS). All recommendations made by the board are forwarded to the CO for approval.

Permanent board members include the—

- Command master chief,
- Command career counselor,
- Personnel officer, and the
- Educational service officer.

Student Notes:

Supplemental board members include the—

- Division officer,
- Division chief, and the
- Division career counselor.

ENLISTED CAREER STRUCTURE

Learning Objectives: When you finish this chapter, you will be able to—

- Recognize the paths of advancement and recall the requirements for advancement of nonrated personnel.
- Identify the eligibility requirements for advancement to E-2 and E-3 and petty officer.
- Recognize selection criteria for advancement and preparation for advancement.
- Identify the career enlistment objectives.

The objective of the enlisted advancement system is to provide qualified petty officers to operate the Navy's ships, squadrons, and shore stations. Advancements, in turn, provide the opportunity for the orderly progression of qualified enlisted personnel to higher levels of responsibility throughout their naval career. Information about the advancement system is contained in BUPERSINST 1430.16. The advancement system offers you increased pay, prestige, and privileges, as well as additional responsibilities and authority.

PATH OF ADVANCEMENT

The enlisted advancement structure is organized into paygrades. E-1 is the lowest enlisted paygrade and E-9 is the highest. The path of advancement from E-1 to E-9, along with the title of each paygrade, is shown in figure 16-2.

The lowest three paygrades (E-1 to E-3) are referred to as apprenticeships and identified as one of the following:

- Seaman apprenticeship (SR, SA, SN)
- Fireman apprenticeship (FR, FA, FN)
- Airman apprenticeship (AR, AA, AN)

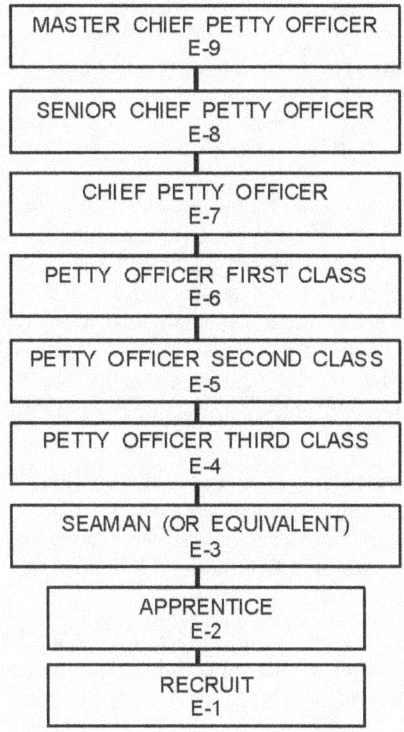

BMRf1602

Figure 16-2.—Path of advancement.

- Constructionman apprenticeship (CR, CA, CN)
- Hospitalman apprenticeship (HR, HA, HN)
- Dentalman apprenticeship (DR, DA, DN)

Petty officers (E-4 to E-9) and designated strikers belong to a rating. Ratings are divided into two categories—general ratings and service ratings.

General Ratings

A general rating is a broad occupational field (a group of jobs) that requires the same general qualifications and includes similar duties. Boatswain's Mate, Quartermaster, and Storekeeper are all examples of a general rating. Each rating has its own rating badge. These rating badges are shown in chapter 10 of this manual.

In some cases, two or more related general ratings will combine at the E-8 or E-9 level to form a new

Student Notes:

general rating. That is called compressing. For example, the two ratings Electrician's Mate and Interior Communications Electrician compress into Electrician's Mate at the E-9 level.

Service Ratings

Some general ratings are subdivided into service ratings to allow for special training or the assignments of personnel who have received special training. Service ratings indicate specialties within a general rating; for example, Aviation Boatswain Mate is a general rating, but Aviation Boatswain Mate is divided into three service ratings: Catapults and Arresting Gear Equipment (ABE), Handling (ABH), and Fuels (ABF).

Service ratings may be established within a general rating at any paygrade and may extend to any other paygrade. For example, a general rating may have service ratings at E-4 and E-5 but not at E-6 through E-9.

Service ratings are not identified by special rating badges. They use the rating badge of the general rating to which they belong.

Designated Strikers

A designated striker is a person in paygrade E-1, E-2, or E-3 who has been designated (appointed or specified) as technically qualified for a particular rating. Personnel in the general apprenticeships (E-1, E-2, and E-3) are identified as strikers for ratings for which they—

- have demonstrated their technical qualifications through on-the-job training (OJT) or
- have received formal school training.

Commanding officers may designate personnel in their commands as strikers if certain qualifications are met. These qualifications are spelled out in the Navy's advancement manual. For more information on the requirements to be a striker in a rating, go to your career counselor or personnel office.

QUALIFICATIONS FOR ADVANCEMENT

Before you are advanced, you must fulfill (meet) the qualifications for the paygrade you wish to be advanced to. You must also fulfill other eligibility requirements, and then you must be selected to be advanced.

How can you find out what is required of you for you to be considered qualified for the next paygrade? The Navy has created standards for every enlisted paygrade and rate. These standards are of two types: Naval Standards (NAVSTDs) and Occupational Standards (OCCSTDs).

The NAVSTDs and OCCSTDs are published in the *Manual of Navy Enlisted Manpower and Personnel Classifications and Occupational Standards,* NAVPERS 18068. This publication should be available at your ESO or personnel office. Parts of this publication are reprinted in booklet form. There are two different types of booklets. One type lists the occupational standards for a particular rating. The other booklet lists the naval standards for all paygrades and the occupational standards for AN, CN, FN, and SN. These booklets are helpful when you are preparing for advancement and are available at your educational services office (ESO).

To help you study and prepare for your advancement examination, refer to the Advancement Handbook (AH) for your rate. Also, information about advancement can be found in the Navy Enlisted Advancement System (NEAS).

Navy Enlisted Advancement System (NEAS)

The NEAS contains general information about the enlisted advancement system, exam study tips, how exams are developed, final multiple computations, and explanation of the tear sheet and the profile form, and other information useful to all Navy enlisted advancement candidates.

Advancement handbooks (AHs) provide the occupational skills for a rating, the knowledge factors that relate to those skills, and references to read to understand the knowledge factors. Also, AHs contain a section titled "Exam Expectations," a narrative that describes how knowledge factors could be tested.

Student Notes:

The NEAS and AHs are available in electronic form at the Naval Education and Training Professional Development and Technology Center (NETPDTC) at www.cnet.navy.mil/netpdtc/nac/download/ah_intro.htm.

Naval Standards

NAVSTDs are military requirements for a paygrade. They apply to all enlisted personnel in the Navy. NAVSTDs are skills and knowledges required for enlisted personnel to be able to perform their duty. They include military requirements and essential qualities of professionalism and pride of service in support of your oath of enlistment. They also include basic skills and knowledges relating to the maintenance of good order and discipline, as well as those that directly contribute to the mission of the Navy. To be qualified for a paygrade, you are responsible for knowing all the naval standards for that paygrade and all the naval standards for all lower paygrades.

Occupational Standards

OCCSTDs are a listing of the things you must be able to do to be considered professionally qualified for a rate. OCCSTDs are the minimum occupational requirements of a particular rate and are separate and different from NAVSTDs. In other words, to be an SN you would have to fulfill the OCCSTDs for SN as well as the NAVSTDs for E-3.

If you wanted to look at the OCCSTDs for a particular rate, you would need to look at the OCCSTDs for that rate and all lower paygrades in the same rating as well as the OCCSTDs for the appropriate apprenticeship. For example, to see all the OCCSTDs for Boatswain's Mate second class (BM2), you would have to look at the OCCSTDs for BM2, BM3, and Seaman (SN).

ELIGIBILITY FOR ADVANCEMENT

In addition to all the naval and occupational standards for a rate, other requirements must be met for you to be eligible for advancement. However, being eligible does not guarantee advancement. To be advanced, you must be selected for advancement.

Eligibility for Advancement to E-2 and E-3

The eligibility requirements for E-2 and E-3 are relatively simple. The requirements are as follows:

- Have a certain amount of time in rate
- Be recommended by your commanding officer
- Complete *Basic Military Requirements*, NAVEDTRA 12018

Additionally, your command may require you to pass a written examination. An examination for E-2 would be prepared by your command. For an E-3, the examination would either be prepared locally or prepared by the Naval Education and Training Professional Development and Technology Center (NETPDTC).

Remember, these are eligibility requirements. Meeting these requirements means you are eligible for advancement; but they don't mean you'll be automatically advanced. Selection for advancement is discussed later in this chapter.

Eligibility for Advancement to Petty Officer

Advancement to petty officer has more eligibility requirements than advancement to E-2 or E-3. The eligibility requirements are as follows:

- Have a certain amount of time in rate
- Complete all personnel advancement requirements (PARs)
- Demonstrate knowledge of material in your mandatory rate training manual
- Be recommended by your commanding officer (CO)

TIME IN RATE.—You must fulfill time-in-rate requirements to be eligible for advancement to petty officer. That means you must have been in your present paygrade for a specific minimum period of time to be eligible for the next paygrade.

Student Notes:

PERSONNEL ADVANCEMENT REQUIREMENTS (PARS).—PARS are skills and abilities that can best be demonstrated (shown) by actual performance. Generally, each PAR contains one or more OCCSTDs on the same or similar subject and is written in on-the-job rating language. PARs aren't competitive; that is, no mark is assigned. Completion indicates that you can perform the tasks.

Completion of PARs is mandatory for advancement. Commands should make sure you complete PARs before you are recommended for advancement. Because of limitations in command equipment, mission, and operations, you may not be able to demonstrate all PARs. In that case, actual demonstration of ability isn't mandatory. However, your being recommended for advancement must be based on the command being satisfied that you have the necessary ability to perform properly at the higher paygrade at the present command and at other commands.

PARs are not designed to replace other qualification programs, such as PQS. However, PAR items that duplicate "sign off" items in other programs can be signed off as PAR items if they have already been signed off under any other program.

PARs and BIBS are available in electronic format from:

- NETPDTC Web Site: Get PARS, www.cnet.navy.mil/netpdtc
- Streamlined Automated Logistics Transmission System (SALTS), www.salts.navy.mil

BIBLIOGRAPHY (BIB) FOR ADVANCEMENT-IN-RATE EXAM STUDY.—The BIB is developed by exam writers (chief petty officers) to help Sailors study for advancement-in-rate examinations. The BIB is a list of references that includes training courses (TRAMANs/NRTCs), instructions, technical manuals, guides, and other publications commonly used in a rating.

BIBs are posted (issued) three times a year and are only available in electronic format. You can find the BIBs at the NETPDTC web site. The E-4/E-5/E-6 BIBs are posted in March and September; and E-7 BIBs are posted in July. The BIBs posted in March are for the exam given the following September; the BIBs posted in September are for the exam given the following March.

TRAINING MANUAL INFORMATION.—Training manuals (TRAMANs) and their associated nonresident training courses (NRTCs) are prepared as self-study packages to help you develop the knowledge required for your rating. You may also use them when preparing to take an advancement examination. The information in some TRAMANs is considered mandatory.

You must complete certain courses (mandatory courses) to meet advancement eligibility requirements. For example, if you're going up for E-3 and didn't attend the Apprenticeship Training Program (ATP), you must complete *Basic Military Requirements (BMR)* and either the *Airman* (*AN*), *Fireman* (*FN*), or *Seaman* (*SN*) (depending on your rate) TRAMANs. If you graduated from the ATP, you have satisfied the requirement for completion of the *AN*, *FN*, or *SN* apprenticeship TRAMAN. However, you still must complete the *BMR*.

Remember, you are responsible for the information in training manuals concerning the rating in which you wish to be advanced and the appropriate apprenticeship and general rate training manuals.

COMMANDING OFFICER'S (CO'S) RECOMMENDATION.—This eligibility requirement is, perhaps, the most important of all. For your CO to recommend you for advancement, he/she must be satisfied that you are fully qualified for advancement.

To a great extent, your CO relies on the recommendations of the people in the chain of command to decide if you are fully qualified for advancement. Your supervisor constantly evaluates your performance to see if you can handle the duties and responsibilities of an advancement.

In addition, your CO can add requirements to the eligibility requirements shown here. These additional requirements should be met for you to receive your CO's recommendation. Check with your supervisor or personnel office to see if your command has local requirements.

ADDITIONAL ELIGIBILITY REQUIREMENTS.—In addition to the eligibility requirements

Student Notes:

already mentioned, some ratings require a specific school and/or a performance test for advancement. Figure 16-3 is a presentation of the general requirements for advancement. To get specific information on advancement to a particular rate, see your ESO or personnel office.

SELECTION FOR ADVANCEMENT

Once you meet all the eligibility requirements, you are considered eligible and qualified for advancement. However, to be advanced, you must be selected for advancement. In all advancements, your commanding officer has the final word—you are always advanced by your commanding officer.

Selection for Advancement to E-2 or E-3

The selection for advancement to E-2 or E-3 is done by your CO. The Navy has no limits on the number of people who can be advanced to E-2 or E-3. Therefore, the CO may select and advance people to E-2 or E-3 as soon as they have met all the eligibility requirements.

Selection for Advancement to Petty Officer

Selection for advancement to petty officer (up to E-6) is done on the basis of a final multiple among those who pass the Navywide advancement examination. The number of persons who may be advanced is limited by the number of vacancies that exist in each rate and rating. Therefore, when the number of those who pass the Navywide advancement examination is greater then the number of vacancies, a final multiple system is used to determine which personnel may be advanced to paygrades E-4, E-5, and E-6.

Three separate categories are taken into consideration when a final multiple is computed. The final multiple score is based on these three things:

1. Merit rating
2. Personnel testing
3. Experience

Merit rating gives people who have shown they are outstanding performers an advantage in promotion. Merit rating is done by averaging your performance marks for the last 3 years.

Personnel testing refers to the Navywide advancement examination. These examinations are prepared and administrated by NETPDTC. Each test consists of 200 multiple-choice questions based upon the occupational standards for the rating and Naval Standards. If you pass this examination but are not selected for advancement, you are considered to have PNA (passed, not advanced) status for the examination. Personnel testing includes your examination score in computing the final multiple score.

You receive credit for your **experience** in the final multiple score. Experience includes longevity—your total active federal military service (TAFMS)—and time in rate (TIR). It also includes certain awards and PNA credits.

To sum it all up, the following factors are considered in your final multiple computation:

- Performance mark average
- Examination score
- Length of service (TAFMS)
- Service in paygrade (TIR)
- Awards
- PNA credit

Your final multiple score is computed by NETPDTC at the time your Navywide advancement examination is scored.

REVIEW 2 QUESTIONS

Q1. List the permanent board members on the Professional Development Board.

a.

b.

c.

d.

Q2 What is the purpose of the Professional Development Board?

Student Notes:

Requirements	E-1 to E-2	E-2 to E-3	E-3 to E-4	E-4 to E-5	E-5 to E-6	E-6 to E-7	E-7 to E-8	E-8 to E-9
Time in rate	9 months	9 mo as E-2	12 mos as E-3	36 mo as E-4	36 mo as E-5	36 mo as E-6	36 mo as E-7	36 mo as E-8
School	RTC (CO may advance up to 20% of company)	None	Class "A" for AME, BU, CE, CM, CTA, CTI, CTM, CTO, CTR, CTT, DT, EA, EO, EW, FT, HM, IS, JO, NM, MT, MU, PR, SW, UT	Naval Justice School for LN3	None	Navy School for AGC, MU	Navy School for MUCS	None
Performance Test	None	None	Specified ratings must complete applicable performance test before taking Navywide advancement examination.				None	None
Nonresident Training Course (NRTC) training manual (TRAMAN)	None	Required for E-3 and all petty officer advancements unless waived because of completion of Navy school. Courses need not be completed but once; i.e., those who complete the 3&2 course for PO3 need not complete the same course again for advancement to PO2.				Nonresident training course recommended (See NAVEDTRA 12061*) *Catalog available in electronic format only		
Examinations	Locally prepared tests	NETPDTC exams or locally prepared test	Navywide advancement examinations required for advancement to E-4 to E-7.				None	None
Selection board	None	None	None	None	None	Navywide CPO or SCPO/MCPO selection		
Obligated service requirement	There is no set amount of obligated service required either to take the Navywide advancement examination or to accept advancement to paygrades E-1 through E-6.					All CPO candidates must have two years remaining to accept appointment to a CPO paygrade.		
Enlisted performance evaluation	As used by CO when approving advancements		Counts toward performance factor credit in advancement final multiple for all E-4 through E-6 candidates.					
CO recommendation	All Navy Advancements require the commanding officer's recommendation for advancement.							
Authorization for advancement	Commanding officer		Naval Education and Training Professional Development and Technology Center (NETPDTC) for advancement to E-4 through E-9 in addition to command approval.					

Figure 16-3.—Requirements for advancement.

Q3. The lowest three paygrades are referred to as—

Q4. What are the two categories ratings divided into?

a.

b.

Q5. What is a designated striker?

Q6. What manual contains a list of Navy standards (NAVSTDs) and occupational standards (OCCSTDs)?

Q7. What's the difference between a NAVSTD and an OCCSTD?

Q8. What three requirements do you need to meet to be eligible to advance from E-2 to E-3?

a.

b.

c.

Q9. List the eligibility requirements to advance to Petty Officer.

a.

b.

c.

d.

Q10. The final multiple score of a rating exam is based on what three things?

a.

b.

c.

TYPES OF DUTY

Learning Objectives: When you finish this chapter, you will be able to—

- Recall the definitions of sea duty, shore duty, and neutral duty.
- Recognize the methods used for enlisted assignments.

You often hear about three types of duty: sea duty, shore duty, and neutral duty. These three designations refer to duty for rotation purposes.

Everyone in the Navy has sea/shore rotation. The amount of time spent on sea duty or shore duty depends on your rate, rating, and individual circumstances. Each rate and rating in the Navy has a designated sea/shore rotation cycle. You can find out what the current sea/shore rotation for your rate and rating is from your supervisor or career counselor.

For example, if your sea/shore rotation is listed as 36/36, that means that you spend 36 months in sea duty billets and 36 months in shore duty billets. In other words when you complete 36 months of sea duty, your next 36 months is shore duty. After 36 months of shore duty, you have 36 months of sea duty. That is your sea/shore rotation.

You might ask, "What is sea duty, and what is shore duty"? There are eight types of duty designations used for sea/shore rotation. Each of these duty types is credited as sea, shore, or neutral duty for rotation purposes.

Student Notes:

1. Shore duty (sea/shore Code 1). Shore duty, Code 1, is performed in CONUS (the 48 contiguous states) land-based activities and long-term schooling programs. (Long term is defined as 18 or more months; school assignments of less than 18 months are considered neutral duty.) Members are not required to be absent from the corporate limits of their duty stations in excess of 99 days.

2. Preferred overseas shore duty (sea/shore Code 6). Preferred overseas shore duty, Code 6, is duty performed in overseas land-based activities that are credited as shore duty for rotational purposes as determined by BUPERS.

3. Sea duty (sea/shore Code 2). Sea duty, Code 2, is duty performed in commissioned vessels or activities home ported/home based in CONUS that operate away from their home port/home base in excess of 150 days per year.

4. Overseas shore duty (sea/shore Code 3). Overseas shore duty, Code 3, is duty performed in overseas land activities that is credited as sea duty for rotational purposes as determined by BUPERS.

5. Nonrotated sea duty (sea/shore Code 4). Nonrotated sea duty, Code 4, is duty performed in commissioned vessels home-ported overseas (outside the contiguous 48 states) or in activities that operate away from their overseas home port/home base in excess of 150 days per year.

6. Neutral duty (sea/shore Code 5). Neutral duty, Code 5, is duty in activities normally designated as shore duty for rotation, but that requires members to be absent 100 to 150 days per year from the corporate limits of their duty station while accomplishing their assigned task. School assignments of less than 18 months are included in this category.

7. Partial sea duty (sea/shore Code 7).Partial sea duty, Code 7, is duty performed in overseas, land-based activities credited as shore duty for rotational purposes, but credited as partial sea duty according to established guidelines.

8. Double sea duty (sea/shore Code 8). Double sea duty, Code 8, is duty performed in commissioned vessels or activities in an active status that operate away from their home port/home base in excess of 50 days a year credited as double sea credited because of the nature of the mission.

ENLISTED DETAILERS AND USE OF THE DUTY PREFERENCE FORM, NAVPERS 1306/63

Learning Objective: When you finish this chapter, you will be able to—

- Identify the entries made on the Duty Preference Form, NAVPERS 1306/63.

Every rate and, in most cases, every paygrade has a senior enlisted person who matches personnel within a particular rate or specialty with the available billets Navywide. This person is referred to as the enlisted detailer. When detailers work to fulfill requisitions (vacant billets), several factors are involved. To assign you to a billet, the enlisted detailer for your rate must match you with a billet you are qualified for and within a certain time frame.

DUTY PREFERENCE FORM, NAVPERS 1306/63

You may sometimes ask yourself "How did I get the job I have now"? Your detailer had a lot to do with it, of course. However, the detailer determines what jobs you are qualified to hold by the information you submitted on your Enlisted Duty Preference Form, NAVPERS 1306/63.

Your detailer has access to your Enlisted Duty Preference Form and a record of your training through a computer terminal. The computer contains a record of the on-the-job and formal training you have received.

Although you may not have total control over your training and qualifications, you are completely responsible for the information the duty preference sheet contains. *You* are also responsible for submitting the form.

Student Notes:

FILLING OUT THE NAVPERS 1306/63

You can get NAVPERS 1306/63 (fig. 16-4) from your division or command career counselor. The form contains instructions for filling it out. If you need help, contact your division or command career counselor.

The information on this form tells your detailer where you would like to be stationed, what type of duty you prefer, and your career intentions. The Remarks section tells the detailer if you or your family has special qualifications that would make a particular duty station advantageous to you, the Navy, or both. The form contains this section because the Navy recognizes that no one can be completely described in encoded, check-block-type symbols. Other information you might want to include in the Remarks sections includes the following:

- If you are volunteering for overseas duty, all community support skills your family has; for example, qualification as a teacher, nurse, dental technician, hairdresser, or secretary
- Any handicap a family member may have, and the areas where treatment or support facilities exist
- If your wife is pregnant, her expected delivery date
- Dates and terms of a reenlistment within 24 hours of reenlisting
- If you are married to another service member, your spouse's full name, military service, social security number, rate, and present duty station

SUBMITTING THE NAVPERS 1306/63

Although you have no guarantee of getting the duty you want, your detailer will try to match your desires with the needs of the Navy. Without a NAVPERS 1306/63 on file, your detailer assumes you don't care where or what duty you are assigned. Unfortunately, a large number of Sailors don't submit any duty preference.

You may want duty in a location, or of a type, that isn't listed on the form. In that case, you will find a detailed listing of duty choices you may request in chapter 25 of the TRANSMAN. Another handy reference, available from your command, is *Homeports and Permanent Duty Stations of Activities of the Operating Forces of the Navy*, OPNAVINST 3111.14. This instruction contains the location of home ports of ships and activities and can help you choose realistic duty preferences. Once you have completed the NAVPERS 1306/63, submit it through your command to BUPERS. BUPERS enters the information into the database detailers use to determine your qualifications. Be sure to keep a copy of the form you submit for your own reference.

WHEN TO SUBMIT NAVPERS 1306/63

You should submit a duty preference form after 6 months at your first duty station. After submitting the first NAVPERS 1306/63, you may submit a new one at any time. Submit a revised form anytime you change duty stations or when important personal data, such as status of dependents or location of household goods, changes. Within 24 hours of a reenlistment, you must submit a new NAVPERS 1306/63 that indicates the date and number of years of reenlistment in the Remarks section.

REVIEW 3 QUESTIONS

Q1. List the three types of duty.

a.

b.

c.

Q2. Overseas shore duty Code 3 is classified as what type of duty?

Student Notes:

ENLISTED DUTY PREFERENCES

PRIVACY ACT STATEMENT The authority to request this information is contained in 5 USC 301 Department Regulations. The principal purpose of the information is to enable you to make known what you want for a future duty assignment. The information will be used to assist officials of the Department of the Navy in determining your future duty assignment. Completion of this form is mandatory, failure to provide required information may result in no consideration of your desires for a future duty assignment.

Social Security Number — Rating/Rate — Date Submitted (YYMMDD) — UIC

COMMAND/DIVISION CAREER COUNSELOR'S INITIALS

Last Name (First five letters in blocks) — First Name — M.I.

Sea — Overseas — Shore

-Rank in order of desirability (1, 2, 3)

-Indicate whether LOCATION (L) or TYPE duty (T) is more important

SEA: 1st Choice / 2nd Choice / 3rd Choice — Location — Code — Duty/Activity type

OVERSEAS: 1st Choice / 2nd Choice / 3rd Choice — Location — Code — Duty/Activity type

SHORE: 1st Choice / 2nd Choice / 3rd Choice — Location — Code — Duty/Activity type

School Preferences: (Write in school or code from TRANSMAN, Chapter 25, e.g., HM1 A8 or 3393)

Volunteer: F- Flight Duty; S- Submarine Duty

Last Deployment: When returned Mo. Yr. — Length Mos. — Where Location Code

Duty willing to extend for: Duty type (Code) — Location (Code)

DEPLOYMENT LOCATION CODES:

7FLT - 7th Fleet	MEST - Mid-East
MED - Mediterranean	INDO - Indian Ocean
CARB - Caribbean	SOAM - South America
NEUR - North Europe	ANT - Antarctic

Career intentions: Codes: Place one of these in block at left

R - Reenlist at EAOS; P - Ext onboard Present Duty Station; E - Extension at EAOS; U - Uncertain; D - Discharge at EAOS; F - Fleet Reserve/Retire

Number of months for desired ext./reen

STATUS — Marriage/Dependents (Ages): M - Married; S - Single; D - Divorced; W - Married, Spouse Working; P - Single Sponsor

Location Code — Primary Dependents M F Ages 0-4, M F Ages 5-11, M F Ages 12+ — Secondary Dependents No. Location Code — Military Spouse SSN (Navy) Branch of Service (Other)

Quarters information: Present Duty Station — Types: H -Home G -Gov't Qtrs A -Apt. M -Mobile Home — Own/Rent (O or R) — Other Area (Type) — Own/Rent (O or R) — Location Code

Household Effects: Pres duty wgt. (in thous. of pounds) — Other loc-wgt — Location Code

REMARKS *(Space between each word)*

Signature

NAVPERS 1306/63 (Rev. 10-86) S/N 0106-LF-013-0917

BMRf1604

Figure 16-4.—Enlisted Duty Preference Form, NAVPERS 1306/63 (front).

Q3. What form do you submit to your detailer to let him/her know what duty station you want?

Q4. List the kind of information found on the Enlisted Duty Preference Form.

a.

b.

c.

d.

ENLISTED EVALUATION REPORT AND COUNSELING RECORD

Learning Objectives: When you finish this chapter, you will be able to—

- Recognize the purpose of the enlisted performance evaluation system.
- Identify the traits to be evaluated.

The Enlisted Evaluation Report and Counseling Record is used to document an individual's qualifications, performance, conduct, and eligibility for increased responsibility. **The Evaluation Report and Counseling Record is the most significant personnel management tool in your service record**. It is used primarily by BUPERS to make advancement-in-rate and assignment decisions. It may also be used for the following purposes:

- To determine eligibility for Good Conduct Medals
- For reenlistment
- To determine the type of discharge
- As a basis for selecting members for advancement
- For continuation of service
- For appointment to commissioned status
- For assignment to special duties
- For special educational programs

The Evaluation Report and Counseling Record is very important. Both the command and you, the individual Sailor, need to pay attention to it.

DEVELOPMENT AND REVIEW

All Sailors need to submit information they believe should be included in their evaluation to their reporting senior. Types of information you may submit include but are not limited to—

- Off-duty educational achievements
- Completed correspondence courses
- Community involvement

Also, you have the right to review your own evaluation before final disposition is made. You need to take an active role in developing and reviewing your evaluation. Your career and your future depend on it.

TRAITS TO BE EVALUATED

The reporting senior compares your performance against others of the same rate and rating as yourself. When you aren't assigned duties of your rate or rating, comparison is made against others of the same paygrade who are performing similar duties. The reporting senior will make a concerted effort to evaluate you objectively in each trait. Each trait is assigned a numerical value and there are meanings as follows:

5.0—Greatly Exceeds Standards

4.0—Above Standards

3.0—Meets Standards

2.0—Progressing

1.0—Below Standards

Some of the traits you may be evaluated on are professional knowledge, quality of work, equal

Student Notes:

opportunity, military bearing/character, personal job accomplishment/initiative, teamwork, and leadership.

Professional Knowledge

In the professional knowledge trait, you are rated on your knowledge and performance of your job-related duties, your application of technical and professional skills, your problem-solving abilities, and your ability to accept instructions and directions.

Quality of Work

In the quality of work trait, you are rated on the extent to which you can be depended on to perform assigned tasks successfully and the quality of the work you performed. You're also rated on how much supervision is required for you to perform an assigned task.

Equal Opportunity

In the equal opportunity trait, you are evaluated on your contribution to command morale, unit cohesiveness, and your support of the Navy's Command Managed Equal Opportunity Program.

Military Bearing/Character

In the military bearing/character trait, you are evaluated on your personal appearance, including physical fitness; wearing of your uniform; and, when appropriate, neatness of your civilian clothing. You are also graded on your knowledge and practice of military courtesies and the way you adhere to the Navy Core Values—Honor, Commitment, and Courage.

Personal Job Accomplishment/Initiative

In the personal job accomplishment/initiative trait, you are evaluated on your ability to act appropriately, independently, and without specific direction, while exercising sound judgement. You're also rated on your ability to plan/prioritize wisely, seek extra responsibility, and willingness to take on the hardest jobs.

Teamwork

In the teamwork trait, you're evaluated on your contributions to team building and your ability to work successfully with subordinates, peers, and superiors. Finally, under this trait, you're rated on your ability to understand team goals.

Leadership

In the leadership trait, you're evaluated on your ability to organize and motivate people, as well as developing in others their ability to accomplish goals. Your ability to delegate, to gain commitment from others, and to challenge and inspire subordinates while maintaining positive and realistic expectations are taken into account.

NOTE

For personnel in paygrades E-1 through E-3, a grade in this trait is not required unless abilities are clearly demonstrated.

SUBMISSION AND DISPOSITION

The Evaluation Report and Counseling Record for E-3 and below is submitted on a biyearly basis or when a person is transferred. In addition, counseling is performed on a biyearly basis to record your progress and make you aware of your performance.

You must sign your Evaluation Report and Counseling Record. Your signature on your Evaluation Report and Counseling Record does not indicate agreement with the evaluation; it indicates you have seen the Evaluation Report and Counseling Record and your rights have been explained. Your signature also indicates you have verified the identification data in the evaluation.

Once signed the Evaluation Report and Counseling Record is sent to BUPERS (counseling documentation is retained at the command and not sent to BUPERS). A copy of your Evaluation Report and Counseling Record is placed in your field service record, a copy is retained by the reporting activity, and you are given a copy.

Student Notes:

REVIEW 4 QUESTIONS

Q1. What is the purpose of the Evaluation Report and Counseling Record?

Q2. What is the numerical grading scale used on the Evaluation Report and Counseling Record?

Q3. List the evaluation traits that are found on the Evaluation Report and Counseling Record.

a.

b.

c.

d.

e.

f.

g.

Q4. Once your Evaluation Report is signed, where is it sent and who gets a copy?

ENLISTED SERVICE RECORD

Learning Objective: When you finish this chapter, you will be able to—

- Identify the components of the Enlisted Service Record, NAVPERS 1070/600, to include the Record of Emergency Data, Navy Occupation/Training and Awards History, Enlisted Performance Record, and the Enlisted Remarks Form.

The Enlisted Service Record, NAVPERS 1070/600, is the official history of a person's Navy career. The information contained in the service record starts when you apply for enlistment and is added to throughout your naval service. The record is the property of the Navy. It must be safeguarded against loss and against access by unauthorized persons. Only those personnel given the authority by the CO make service record entries.

The Enlisted Service Record is a folder that contains various forms concerning your enlisted service. The right-hand side has various forms in a specific order. (**NOTE**: Your service record will contain only the forms that apply to you.) There are 15 different forms altogether. The order in which these forms are filed has led to their being referred to as pages. For example, your enlisted contract is the first, or bottommost, form. It's referred to as a *Page 1*.

Other papers required for safe keeping or record purposes are filed on the left-hand side of the folder. A separator entitled Career Performance Data, NAVPERS 1070/617, divides the left-hand side. Beneath this separator, all your performance evaluations, commendations, and awards correspondence are filed. If you have a previous enlistment, a certified copy of the Enlisted Performance Record from the previous enlistment and copies of any Certificates of Release or Discharge from Active Duty, DD Form 214s, are also filed beneath the separator. All other papers are filed above the separator in chronological order, the latest date on top.

Only three of the forms from the Enlisted Service Record are covered in this chapter. They include the

- Enlisted Qualifications History, NAVPERS 1070/604,
- Dependency Application/Record of Emergency Data, NAVPERS 1070/602W, and the
- Administrative Remarks Form, NAVPERS, 1070/613

The remaining forms are more or less of an administrative nature. Some pages require several sheets during an enlistment; for example, there are usually several Page 13s.

Student Notes:

DEPENDENCY APPLICATION/RECORD OF EMERGENCY DATA, NAVPERS 1070/602W

The Dependency Application/Record of Emergency Data, NAVPERS 1070/602W, Page 2, is a multipurpose form. It is used for both officer and enlisted personnel. Figures 16-5 and 16-6 show the worksheet used to enter information. When the worksheet is complete, PSD personnel enter the information into the computer. This then becomes a computerized record that is entered into your Enlisted Service Record.

The Dependency Application/Record of Emergency Data serves as an application for dependency allowances. This form is normally completed at the recruit training command, or first duty station, for all personnel with dependents. Information on this form provides an immediately accessible, up-to-date record of emergency data for casualty reporting and notification of the next of kin. Therefore, you need to update this part of the form whenever there is any change in family member status, such as marriage, birth, divorce, a change of address, etc.

ENLISTED QUALIFICATIONS HISTORY, NAVPERS 1070/604

The Enlisted Qualifications History, NAVPERS 1070/604, Page 4, is another service record of interest to you and the Navy (figs. 16-7, 16-8, 16-9, 16-10). This form consists of the following 12 parts:

1. Educational Experience Level
2. Classification/ASVAB Testing Qualifications
3. Record of Off-Duty Education/VOC/TECH Training and Non-Required Correspondence Courses
4. Other Training Courses/Instructions Completed
5. Navy Service Schools/Military Training Courses
6. Correspondence Courses Required for Advancement
7. Navy Enlisted Classifications
8. Pers. Adv. Req. (PARS) no longer required.
9. Enlisted Rate/Rating
10. Designator Record
11. Awards
12. Personnel Qualification Standards (PQS)

The information contained in the various parts of NAVPERS 1070/604 is valuable, both to you and to the Navy. The information provides a complete chronological record of the following types of information:

- Navy enlisted classification (NEC) codes
- Designators assigned, changed, or revoked
- Navy service schools attended
- Navy training courses, performance tests, and personnel qualification standards completed
- Maintenance and/or technical qualifications attained
- Advancements, reductions, changes in rate or rating
- General educational development (GED) tests and off-duty courses completed
- Decorations received and good conduct, unit, marksmanship, campaign/service, and other awards received

If you reenlist, transfer to the Fleet Reserve, or enlist in the Naval Reserve at your place of discharge, the Enlisted Classification Record is removed from your closed (old) service record and inserted in your new record.

When you are discharged and do not immediately reenlist, this form is given to you. Upon application for enlistment/reenlistment, this form should be presented to the recruiter along with the discharge certificate.

ADMINISTRATIVE REMARKS FORM NAVPERS 1070/613

When complete, the Administrative Remarks Form, NAVPERS 1070/613, becomes Page 13 of your service record. Page 13 contains miscellaneous entries

Student Notes:

WORKSHEET

DEPENDENCY APPLICATION/RECORD OF EMERGENCY DATA

1. UNIT I.D. | 2. SHIP OR STATION | 3. INITIAL ☐ | 4. CHANGE ☐

5. NAME OF SPOUSE | 6. DATE OF BIRTH OF SPOUSE | 7. RELATIONSHIP

8. PLACE OF MARRIAGE (CITY & STATE OR COUNTRY) | 9. DATE MARRIED | 10. CITIZENSHIP OF SPOUSE

11. ADDRESS OF SPOUSE | 12. DEP

13. NAME OF CHILD OR DEPENDENT | 14. DATE OF BIRTH | 15. RELATIONSHIP

16. ADDRESS (INCLUDE NAME OF CUSTODIAN IF OTHER THAN CLAIMANT) | 17. DEP

18. NAME OF CHILD OR DEPENDENT | 19. DATE OF BIRTH | 20. RELATIONSHIP

21. ADDRESS (INCLUDE NAME OF CUSTODIAN IF OTHER THAN CLAIMANT) | 22. DEP

23. NAME OF CHILD OR DEPENDENT | 24. DATE OF BIRTH | 25. RELATIONSHIP

26. ADDRESS (INCLUDE NAME OF CUSTODIAN IF OTHER THAN CLAIMANT) | 27. DEP

28. NAME OF CHILD OR DEPENDENT | 29. DATE OF BIRTH | 30. RELATIONSHIP

31. ADDRESS (INCLUDE NAME OF CUSTODIAN OF OTHER THAN CLAIMANT) | 32. DEP

33. NAME OF FATHER

34. ADDRESS OF FATHER (SEE SPECIAL INSTRUCTIONS BEFORE COMPLETING BLOCK 35) | 35. DEP

36. NAME OF MOTHER

37. ADDRESS OF MOTHER (SEE SPECIAL INSTRUCTIONS BEFORE COMPLETING BLOCK 38) | 38. DEP

39. WERE YOU PREVIOUSLY MARRIED? ☐ YES ☐ NO | 40. PRIOR MARRIAGE DISSOLVED BY ☐ DEATH ☐ ANNULMENT ☐ DIVORCE | 41. DATE | 42. PLACE (CITY & STATE OR COUNTRY)

43. WAS SPOUSE PREVIOUSLY MARRIED? ☐ YES ☐ NO | 44. PRIOR MARRIAGE DISSOLVED BY ☐ DEATH ☐ ANNULMENT ☐ DIVORCE | 45. DATE | 46. PLACE (CITY & STATE OR COUNTRY)

47. OTHER | 48. ADDRESS | 49. RELATIONSHIP

50. NEXT OF KIN OS SPOUSE (NOT HUSBAND, WIFE OR MINOR CHILD) | 51. ADDRESS | 52. RELATIONSHIP

53. BENEFICIARY(S) FOR UNPAID PAY AND ALLOWANCES | 54. ADDRESS | 55. RELATIONSHIP | 56. %

57. PERSON TO RECEIVE ALLOTMENT IF IN A MISSING STATUS. SUBJECT TO SECNAV DETERMINATION | 58. ADDRESS | 59. %

60. BENEFICIARY(S) FOR GRATUITY PAY (NO SPOUSE OR CHILD SURVIVING) | 61. ADDRESS | 62. RELATIONSHIP | 63. %

64. LIFE INSURANCE DATA (NAME OF CO) (DO NOT INCLUDE SGLI) | 65. ADDRESS | 66. POLICY NUMBER

67. RELIGION | 68. TOTAL NO DEPENDENTS THIS PAGE | 69. EFFECTIVE DATE | 70. RANK/RATE | 71. PAGE | 72. OF PAGES

73. NAME OF APPLICANT/DESIGNATOR (LAST, FIRST, MIDDLE) | 74. SSN | 75. USN ☐ | 76. USNR ☐

NAVPERS 1070/602W (Rev. 7-72) S/N 0106-018-6031

BMRf1605

Figure 16-5.—Dependency Application/Record of Emergency Data (Page 2), NAVPERS 1070/602W (front).

NAVPERS 1070/602W (Rev. 7-72) (BACK)

77. LOCATION OF WILL OR OTHER VALUABLE PAPERS

78. REMARKS

Is beneficiary designation of S. G. L. I. on file? ☐ YES ☐ NO — DATE (If Yes)

NOTE: THIS FORM DOES NOT DESIGNATE OR CHANGE BENEFICIARIES OF GOV'T LIFE INSURANCE.

79. SIGNATURE OF DESIGNATOR	80. SIGNATURE OF APPROVING OFFICER, TITLE, AND DATE

CERTIFICATION OF DESIGNATOR

I have reviewed the data entered on this form and certify that it is correct.
Execute a new NAVPERS 1070/602 if data is not correct.

DATE	SIGNATURE OF DESIGNATOR	DATE	SIGNATURE OF DESIGNATOR

BMRf1606

Figure 16-6.—Dependency Application/Record of Emergency Data (Page 2), NAVPERS 1070/602W (back).

36

E N L I S T E D   Q U A L I F I C A T I O N S   H I S T O R Y

1. EDUCATIONAL EXPERIENCE LEVEL

GED (HS) EQUIVALENT TEST			COLLEGE LEVEL GENERAL EXAMS		PRESENT LEVEL OF EDUCATION					
DATE ISSUED	ISSUING STATE	INIT	DATE PASSED	INIT	12	13	14	15	16	17

2. CLASSIFICATION/ASVAB TESTING QUALIFICATIONS

TEST FORM ID	DATE ADMIN.	AFQT	GS	AR	WK	PC	NO	CS	AS	MK	MC	EI	VE

ASVAB ADMINISTERED BY:

SPECIAL TEST SCORES

NAME	FORM	DATE	SCORE

CLASSIFIER'S SIGNATURE:

3. RECORD OF OFF-DUTY EDUCATION/VOC/TECH TRAINING AND NON-REQUIRED CORRESPONDENCE COURSES

NUMBER/TITLE OF COURSE OR TEST	SCHOOL	DATE COMPLETED	GRADE	INIT	NUMBER/TITLE OF COURSE OR TEST	SCHOOL	DATE COMPLETED	GRADE	INIT

4. OTHER TRAINING COURSES/INSTRUCTIONS COMPLETED

DATE COMPLETED	TYPE OF COURSE AND/OR INSTRUCTION	DURATION	LOCATION	INIT

NAME (Last, First, Middle)	SOCIAL SECURITY NUMBER	BRANCH AND CLASS

NAVPERS 1070/604 (Rev. 7/91)
S/N 0106-LF-012-2500

4

BMRf1607

Figure 16-7.—Enlisted Qualifications History, NAVPERS 1070/604 (front).

36

5. NAVY SERVICE SCHOOLS/MILITARY TRAINING COURSES							
COURSE TITLE/SCHOOL		NEC	DATE ENROLLED/COMPLETED	COURSE TITLE/SCHOOL		NEC	DATE ENROLLED/COMPLETED
COURSE LENGTH	GRADE	MANNER OF COMPLETION ☐GRADUATED ☐DROPPED	INIT	COURSE LENGTH	GRADE	MANNER OF COMPLETION ☐GRADUATED ☐DROPPED	INIT
COURSE TITLE/SCHOOL		NEC	DATE ENROLLED/COMPLETED	COURSE TITLE/SCHOOL		NEC	DATE ENROLLED/COMPLETED
COURSE LENGTH	GRADE	MANNER OF COMPLETION ☐GRADUATED ☐DROPPED	INIT	COURSE LENGTH	GRADE	MANNER OF COMPLETION ☐GRADUATED ☐DROPPED	INIT

6. CORRESPONDENCE COURSES REQUIRED FOR ADVANCEMENT

DESCRIPTION OF COURSE, RATE OR NAVPERS NUMBER	DATE COMPLETED	INIT	DESCRIPTION OF COURSE, RATE OR NAVPERS NUMBER	DATE COMPLETED	INIT

7. NAVY ENLISTED CLASSIFICATIONS

PRIMARY CODE	SECONDARY CODE	DATE	INIT

8. PERSONNEL ADVANCEMENT REQUIREMENTS

DESCRIPTION	DATE COMPLETED	INIT

9. ENLISTED RATE/RATING

RATE	DATE	TIME IN RATE	INIT

10. DESIGNATOR RECORD

DATE	DESIGNATOR	QUAL/REVOCATION	INIT

NAME (Last, First, Middle)	SOCIAL SECURITY NUMBER	BRANCH AND CLASS

NAVPERS 1070/604 (Rev. 7/91) PAGE 2

4 ☐

BMRf1608

Figure 16-8.—Enlisted Qualifications History, NAVPERS 1070/604 (page 2).

36

11. AWARDS				
AWARD NAME	DATE OF AWARD	AUTHORITY	ADV PNTS	INIT

NAME (Last, First, Middle)	SOCIAL SECURITY NUMBER	BRANCH AND CLASS

NAVPERS 1070/604 (Rev. 7/91) PAGE 3

4

BMRf1609

Figure 16-9.—Enlisted Qualifications History, NAVPERS 1070/604 (page 3).

36

12. PERSONNEL QUALIFICATION STANDARDS (PQS)							
PQS TITLE	PQS STATION #	DATE	INIT	PQS TITLE	PQS STATION #	DATE	INIT

NAME (Last, First, Middle)	SOCIAL SECURITY NUMBER	BRANCH AND CLASS

NAVPERS 1070/604 (Rev. 7/91) PAGE 4

4

BMRf1610

Figure 16-10.—Enlisted Qualifications History, NAVPERS 1070/604 (page 4).

of information not recorded elsewhere or of detailed information that may be required in the clarification of entries on other pages of the service record. The original is retained in your service record, and a copy is forwarded to BUPERS.

REVIEW 5 QUESTIONS

Q1. Your service record contains several pages. What form is page 1 of your service record?

Q2. Your evaluations are kept in what part of your service record?

Q3. The Dependency Application/Record of Emergency Data form is what page of your service record?

Q4. How often should you update your Page 2?

Q5. DELETE

Q6. What type of information is recorded on your Page 4?

a.

b.

c.

d.

e.

f.

SIGNATURE AUTHORITY

Learning Objective: When you finish this chapter, you will be able to—

- Identify the purpose of signature authority.

The commanding officer, officer in charge, or other person acting in either position is responsible for signing all command documents. Some documents require the commanding officer's personal signature. Documents that require the CO's personal signature include those that establish policy or deal with aspects of military justice. Other documents that require the CO's signature are those he/she is required by law to sign, such as ships' deck logs.

The CO may delegate (give) signature authority to both military and civilian subordinates. However, this authority is normally limited to their specific area of responsibility. This responsibility may include the work center supervisor signing a PQS requirement or the division chief or officer signing off advancement requirements.

Command personnel authorized to sign command correspondence are normally listed in a unit organization manual or instruction. A signature above the words "By direction" shows that the CO has authorized that person to sign the document.

DIVISIONAL LOGS AND FILES

Learning Objective: When you finish this chapter, you will be able to—

- Identify the procedures used to maintain publications, logs, and files.

There are many logs and files division personnel maintain. Therefore, not all of them are shown here. They may range from a QM3 keeping a list of all required chart corrections, an ENFN maintaining a fuel log for the ship's boats, or an ET2 listing all field changes for the surface search radar. Each division of every ship, squadron, or facility has a certain number of logs and files that must be kept up-to-date so that the command can operate efficiently. Here are a couple of examples:

Student Notes:

- 1,000 flying hours are logged on an F-14 *Tomcat's* engines—these engines should have been replaced after 750 hours.

- The USS *Missouri* (BB-63) fired 400 rounds of 16" projectiles in practice but deployed with only 20 rounds on board.

Both of these situations were avoidable. The division concerned should have kept up-to-date files.

You are aboard a ship under way in the North Atlantic. Think about going on a lookout watch at midnight in December. You arrive for watch at the prescribed time to find no foul weather gear available. The person responsible for maintaining an inventory of special gear didn't do the inventory because he/she didn't think it was that important last June in sunny Florida.

You must remember that besides your division, the entire crew and even the ship itself may depend on how well you maintain your assigned logs and files.

3-M SYSTEMS

Learning Objectives: When you finish this chapter, you will be able to—

- Recognize the concepts of the 3-M Systems.

- Identify the basic procedures used in the 3-M Systems.

Equipment must be cared for. One way to take care of equipment is through preventive maintenance. Preventive maintenance is maintenance done before a problem exists. The Navy has procedures (ways to do things) and schedules for accomplishing (doing) preventive maintenance. These procedures and schedules are part of the Maintenance and Material Management Systems—the 3-M Systems. The objectives of the 3-M Systems are shown below.

- Maintain equipment at maximum operating efficiency

- Reduce equipment downtime

- Reduce the cost of maintenance in both money and man-hours

- Provide data on the expenditures of spare parts, failure rates, man-hours expended, and other information directly related to maintenance

Essentially, the 3-M Systems is used to improve the material readiness of the fleet. The main feature of the 3-M Systems you will be concerned with is the planned maintenance system (PMS).

PMS simplifies maintenance procedures by—

- Defining the maintenance required,

- Scheduling its performance,

- Describing the tools and methods to be used, and

- Providing for the detection and prevention of impending casualties.

Your department head uses PMS to manage, schedule, and control the maintenance of assigned equipment. The components (parts) of the PMS are—

- PMS manual,

- Cycle, quarterly, and weekly maintenance schedules; and

- Maintenance requirements cards (MRCs).

PMS also provides a good foundation for training in equipment operation and maintenance. As you become more familiar with your shipboard duties and are assigned the responsibility for equipment maintenance, PMS will play a big part in your daily activities on the job.

REVIEW 6 QUESTIONS

Q1. What type of documents would require the CO's personal signature?

Q2. Where can you find a list of command personnel that has signature authority to sign command correspondence?

Student Notes:

Q3. What does 3-M stand for?

Q4. What are the objectives of the 3-M system?

a.

b.

c.

d.

PERSONNEL QUALIFICATION STANDARDS (PQS) PROGRAM

Learning Objectives: When you finish this chapter, you will be able to—

- Recognize the purpose of the PQS program.
- Identify the provisions of the PQS program.

The PQS program is a way you can qualify to perform your assigned duties. A personnel qualification standard (PQS) is a written list of knowledges and skills you must have to—

- Qualify for a specific watch station,
- Maintain a specific equipment or system, or
- Perform as a team member within an assigned unit.

Most PQS standards are divided into three sections—Fundamentals, Systems, and Watch Stations.

The 100 Series. The **Fundamentals** section contains the facts, principles, and fundamentals about the subject you are qualifying for.

The 200 Series. The **Systems** section deals with the major working parts of the installation, organization, or equipment the PQS is concerned with.

The 300 Series. The **Watch Stations** section defines the actual duties, assignments, and responsibilities you must perform to obtain your qualification. The Watch Stations section also contains spaces for your supervisor's or qualifying officer's signature once you have proved your abilities.

If you have any questions about PQS in general or a specific PQS, see your supervisor or training petty officer.

TRAINING AND EDUCATION

Learning Objectives: When you finish this chapter, you will be able to—

- Identify the duties of the educational services officer (ESO).
- Recognize the purpose of various types of training to include on-the job training (OJT), general military training (GMT), and various Navy schools.
- Recognize the purpose of distance education.
- Recall the incentives for reenlistment, education, and special duty.

The Navy offers you training and education. If you take advantage of various programs the Navy offers, you can increase your knowledge and skills. By increasing your knowledges and skills, you are more valuable to the Navy, civilian employers, and yourself.

Training and education are closely related. The following are definitions of these terms as used in this chapter:

Training. Training is being taught skills directed to specific tasks. Training is usually based on knowledge you already have. Usually, Navy training refers to those things related to your job or Navy skills.

Education. Education is being taught broad, general, and specific knowledge. This knowledge prepares you for the specific skills you will receive through training. Education refers to schooling not directly related to your naval career. Because of that, education programs are sometimes referred to as off-duty educational opportunities.

Student Notes:

EDUCATIONAL SERVICES OFFICER

The educational services officer (ESO) is your point of contact for all the Navy's training and education programs. The ESO gives all locally administered tests, fills all orders for correspondence courses, and arranges off-duty education. In short, the ESO is responsible for all the training within and for your unit.

PURPOSE OF TRAINING

The purpose of training in the Navy is to support and improve fleet readiness. All training in the Navy is directed toward accomplishing the Navy's mission. Training helps you to do your job better. Remember, training refers to skills directed to specific tasks.

ON-THE-JOB TRAINING

On-the-job training (OJT) takes place during daily operation and maintenance situations. In OJT you learn to perform a task or duty while performing it. For example, when you have a new job or are standing a watch for the first time, someone shows you how to do that job or what is involved in standing the watch. That is OJT. When your supervisor corrects you or shows you a better or faster way to do a job, that is OJT also. OJT is usually informal; but if a group of people are being indoctrinated about a job or watch, OJT may be conducted in a more formal, classroomlike way. OJT is probably the most common form of training in the Navy.

Remember, that even as you work at a familiar job, such as painting, watch standing, boat details, and so on, you are qualifying yourself to be a better Navy member. Do your daily jobs with snap and precision. Your officers and petty officers will recognize your ability and will let you take on jobs of increased responsibility, thereby assisting you in your overall preparation for advancement. Moreover, a job done halfheartedly becomes twice as boring and seems to last twice as long. By trying to do a job faster, more economically, or more neatly, the work becomes more interesting. At the same time you will be training yourself in better attitudes.

During the day, your petty officers will take the opportunity to instruct you in various jobs as they occur. Think about what they tell and show you. Practice as much as you can. Ask questions of experienced personnel so that you understand what you are doing, how and why it should be done, and why the work is important to the Navy and to you. Don't wait for the chief to come along and tell you what to do. Use some initiative, observe what others do, think about what you see, ask questions, and keep learning as you work.

GENERAL MILITARY TRAINING

General military training (GMT) is nonoccupational training that all naval personnel are required to take on a periodic basis. GMT is an important part of the Navy's Leadership Continuum. GMT is an important source of needed *booster shots*. It calls attention to the leadership responsibilities and Navy core values at all levels—both officer and enlisted. At the same time, GMT makes the CO's duty to provide continuing training easier.

In the GMT, you'll get training that has a value-based approach in the following five areas:

1. Healthy lifestyles
2. Interpersonal relationships
3. Pride and professionalism in the Navy
4. Personal and professional growth
5. Risk management

Navy military training (NMT) is a part of GMT. NMT is a combination of formal and informal training, staff leadership, supervision, mentoring, counseling, and positive reinforcement. NMT does this within the framework of a strong military environment. It spans the new Sailors first year in the Navy (after completion of recruit training) and continues developing the Sailor's professional behavior and military knowledge and skills the Sailor needs in military life.

NMT is a shared responsibility. The length of NMT you'll get depends on the amount of time you stay in the training command, often less than 1 year. As you graduate and transfer, you will continue NMT in the fleet.

NAVY SCHOOLS

Navy schools, sometimes referred to as service schools, are divided into several classifications. Each

Student Notes:

class of school has a particular purpose. They usually train you in a specific skill or for a particular job. The classes and their purposes are given in the following paragraphs.

Class "R" Schools

Class "R" schools provide general indoctrination and teach skills and knowledge in basic military subjects. You have already attended a class "R" school—recruit training. Recruit training is considered GMT as well as a class "R" school.

Class "A" Schools

Class "A" schools provide basic technical knowledge and skills required to prepare you for a Navy rating and further specialized training. An example of a class "A" school is Electrician's Mate "A" school.

Class "C" Schools

Class "C" schools provide you with the advanced knowledge, skills, and techniques to perform a particular job in a billet. A Navy enlisted classification (NEC) code may be awarded to identify the skill achieved. An example of a class "C "school would be a school on a particular type of radar system.

Class "F" Schools

Class "F" schools provide team training to officer and enlisted fleet personnel who normally are members of ships' companies. They also provide refresher training, including operator and technical courses of short duration to meet the needs of a fleet or type commander.

Class "P" Schools

Class "P" schools provide undergraduate education and indoctrination and basic training in fundamentals, preliminaries, or principles to midshipmen officer candidates and other newly commissioned officers (except those schools acquired through class "V" programs). The Naval Academy, Naval Reserve Officer Training Corps (NROTC), and Officer Candidate School (OCS) are all class "P" schools.

Class "V" Schools

Class "V" schools provide training in the skills that lead to the designation of naval aviator or naval flight officer.

Obligated Service Requirements for Schools

Normally, you must have a certain amount of obligated service to be eligible to attend a Navy school. The amount of obligated service required depends on the length of the school. Obligated service is counted from the time you start the school until the end of your active obligated service (EAOS) date. You may increase your obligated service to qualify for a school by agreeing to extend your enlistment or reenlist. Your personnel office can give you the obligated service requirement for any particular school.

TRAINING MANUALS AND NONRESIDENT TRAINING COURSES

A training manual (TRAMAN) provides you with basic information about a particular rating. You may also use it to study for advancement examinations. The Naval Education and Training Professional Development and Technology Center (NETPDTC) publishes TRAMANs. Navy schools may use them as texts or references. They may also be used as references for questions in personnel qualification standards (PQS), as texts for correspondence courses, or as self-study manuals. TRAMANs cover the qualifications necessary for advancement by covering the material directly or by directing you to some other reference. TRAMANs include general TRAMANs, such as this text and other military requirements texts, and texts written for a specific rating, such as *Equipment Operator Basics*. Other TRAMANs cover a wide range of subjects, such as basic machines, fluid power, blueprint reading and sketching, and leadership.

The nonresident training course (NRTC) is a self-study, enlisted training course used with a TRAMAN. Generally, the NRTC is locally administered, which means your ESO scores it. TRAMANs and NRTCs are usually printed in one book and referred to as a TRAMAN/NRTC.

The *Catalog of Nonresident Training Courses*, NAVEDTRA 12061, contains a current list of available

Student Notes:

courses. This catalog can be found on the web at http://www.cnet.navy.mil/netpdtc/nac/neas.htm.

OFF-DUTY EDUCATIONAL OPPORTUNITIES

This section covers some of the off-duty educational programs designed to help you in your career and allow you to improve your education. Some programs are Navywide and others are local. Take advantage of as many of the available educational programs as you can. For detailed information on off-duty educational programs, contact your ESO.

Navy Campus

Navy Campus is the name given to in-service voluntary educational programs and the supporting services provided by the Navy to help you with your education. Navy Campus includes all educational activities, from basic education skills preparation to graduate study, that contribute to the general academic and vocational development of naval personnel.

In the following paragraphs, you will learn about some of the educational opportunities available to you through Navy Campus. For further information on those opportunities or to find out about other Navy Campus programs, see your ESO, career counselor, or Navy Campus representative.

BASIC SKILLS PROGRAM.—Many shore stations and some large ships provide tuition-free, on-duty courses to help Sailors improve their skills and military performance. The basic courses offered include subjects such as English, mathematics, and reading.

The Basic Skills Program offers courses to people who need to earn high school diplomas. The Navy pays for all high school completion courses personnel take during off-duty hours. However, the Navy encourages young people to stay in school and graduate before enlisting in the Navy.

PROGRAM FOR AFLOAT COLLEGE EDUCATION (PACE).—The Program for Afloat College Education (PACE) provides undergraduate courses from accredited colleges or universities to shipboard personnel. Civilian instructors teach the courses aboard ship. The Navy fully funds PACE courses; however, students must pay course registration fees and purchase their own books.

Tuition Assistance (TA) Program

The Tuition Assistance (TA) Program provides financial assistance to eligible personnel who attend educational institutions on a voluntary, off-duty basis.

Defense Activity for Nontraditional Education Support (DANTES)

The Defense Activity for Nontraditional Education Support (DANTES) provides support to the voluntary education programs of all the military services. DANTES is not a Navy activity, but is part of the Department of Defense. DANTES administers nontraditional education; that is, education that does not take place in a formal classroom. DANTES provides a wide range of examination and certification programs, operates an independent study support system, and provides other support and developmental activities.

DANTES EXAMINATION PROGRAMS.—DANTES administers and sponsors examination programs at over 560 test centers throughout the world. DANTES offers aptitude and interests tests as well as examinations for college admission, academic credit, professional certification, and high school equivalency.

DANTES INDEPENDENT STUDY PROGRAMS.—DANTES Independent Study Programs let you take correspondence courses from many colleges and universities. Those courses range from high school to graduate level.

OTHER DANTES SERVICES.—DANTES provides many other services besides those just described. To find out more about DANTES, contact your ESO or Navy Campus representative.

Enlisted Education Advancement Program

The Enlisted Education Advancement Program (EEAP) lets career-motivated individuals get an associate of arts/sciences degree in 24 calendar months or less. If you're accepted into the program, you must pay all educational expenses, such as tuition, fees, and books. Upon enrolling in this program, you must

Student Notes:

obligate for 6 years' active duty. While attending college, you may compete for advancement.

REVIEW 7 QUESTIONS

Q1. What are the three sections that PQS is divided into?

a.

b.

c.

Q2. To find a list of Navy TRAMANs, you would refer to the ____________________ on the web at ______________________________.

Q3. What type of training are you receiving when you're learning a skill while working?

Q4. What type of training is recruit training?

Q5. List the different classifications of Navy schools.

a.

b.

c.

d.

e.

f.

Q6. What program does the Navy offer to help Sailors earn a high school diploma or improve their skills and military performance?

Q7. What financial program does the Navy have to help Sailors with their off-duty education?

Q8. What type of services does DANTES provide?

PROGRAMS LEADING TO A NAVAL COMMISSION

Learning Objective: When you finish this chapter, you will be able to—

- Identify the programs that can lead to a Navy commission.

Navy personnel may follow many paths to a Navy commission. Certain enlisted men and women who are outstanding performers may qualify for a commissioning program. This section briefly describes the Navy's basic commissioning programs.

NAVAL ACADEMY

Each year, the Secretary of the Navy may appoint the following to the Naval Academy at Annapolis, Maryland:

- 85 enlisted men and women from the Regular Navy or Regular Marine Corps and
- 85 enlisted men and women from the Naval or Marine Corps Reserve (active or inactive)

Those who are appointed receive a fully subsidized undergraduate education that leads to a commission in the Navy or the Marine Corps.

Students at the Naval Academy are appointed as midshipmen, U.S. Navy. They receive pay equal to about one-half an ensign's basic monthly pay, plus tuition, room, and board. Upon graduation, they are

Student Notes:

awarded a Bachelor of Science degree in one of 18 majors and an ensign's or second lieutenant's gold bars.

NAVAL ACADEMY PREPARATORY SCHOOL

The Naval Academy Preparatory School (NAPS) is located in Newport, Rhode Island, as a part of the Naval Education and Training Center. With up to 300 male and female students, the school offers a balanced academic, military, and physical program patterned after the Naval Academy.

Academically, the school teaches mathematics, chemistry, physics, and English at three levels of difficulty. It also teaches an introductory computer course.

Although not required to gain a Secretary of the Navy appointment to the Academy, attendance at NAPS greatly improves the chances for obtaining one of these appointments.

NROTC SCHOLARSHIP PROGRAM

The Naval Reserve Officer Training Corps (NROTC) Scholarship Program leads to an appointment as a Reserve or Regular officer in the Navy or Marine Corps at the grade of ensign or second lieutenant. If you qualify and are selected for this program, you will receive a scholarship to a college or university with an NROTC unit. You must sign an agreement to spend 6 years in the Navy upon completion of or withdrawal from school. If you are on active duty at the time you sign the agreement, you will be discharged to attend school for a maximum of 40 months. During that time you will receive tuition, books, and fees. Personnel who have entered the program from active duty will also receive a subsistence allowance.

BROADENED OPPORTUNITY FOR OFFICER SELECTION AND TRAINING (BOOST) PROGRAM

If you are interested in the Naval Academy or the NROTC Scholarship Program and qualify in all respects except academically, you may want to apply for the BOOST program. BOOST stands for Broadened Opportunity for Officer Selection and Training. The BOOST program is intended to help people who have been educationally deprived but have demonstrated they have the basic qualities and desires needed to gain a commission.

If you are selected for BOOST, you will receive academic, physical fitness, and general military training as well as counseling. Selection for BOOST does not guarantee your selection for the Naval Academy or the NROTC Scholarship Program, but it certainly increases your opportunities.

ENLISTED COMMISSIONING PROGRAM

If you are interested in a commission and have enough college credit to complete all the requirements for a baccalaureate degree within 2 years, you may qualify for the Enlisted Commissioning Program (ECP). If you are selected for the ECP, you will attend a college of your choice and receive full pay and allowances while you do so. However, you will pay your own educational expenses. Upon graduation, you will attend Officer Candidate School and be commissioned.

You must agree to a 6-year obligation for active enlisted service from the date of enrollment in the Enlisted Commissioning Program. Upon your commission, that obligation is canceled and you assume an obligation of 4 years of commissioned service.

NAVAL RESERVE OFFICER PROGRAMS

Six programs lead to commissions in the U.S. Naval Reserve for enlisted personnel who possess a baccalaureate degree or higher. These programs are as follows:

1. Unrestricted Line Appointment
2. Program Nuclear Propulsion Officer Candidate
3. Program Aviation Officer Candidate School
4. Program Navy Judge Advocate General (JAG) Corps
5. Program Civil Engineer Corps
6. Direct Appointment Program Nuclear Power Instructor and Naval Reactor Engineer Direct Appointment Program

Student Notes:

All candidates attend either the Aviation Officer Candidate School (AOCS) program, Officer Candidate School (OCS), or Officer Indoctrination School (OIS) held in Pensacola, Florida.

If you are interested in any of the Naval Reserve Officer Programs, contact your career counselor. The *Retention Team Manual* contains information on these programs.

CHIEF WARRANT OFFICER PROGRAM

The Chief Warrant Officer Program provides personnel an opportunity to earn a commission as an officer without possessing a college degree. You must be in one of the senior enlisted paygrades to qualify as a chief warrant officer candidate. The specific requirements of the program, which are published each year, are available from your career counselor.

LIMITED DUTY OFFICER PROGRAM

The Limited Duty Officer (LDO) Program is another way in which you can obtain a commission without a college degree. LDOs are commissioned officers who are selected from the senior enlisted paygrades. The requirements for this program, also published each year, are available from your career counselor.

SEAMEN TO ADMIRAL PROGRAM

The Seaman to Admiral program is applicable to enlisted personnel of the Regular Navy and Naval Reserve. Eligible applicants will be considered by a board convened by the Chief of Naval Personnel (CNP). The board will select the best qualified for appointment in the program, within quotas authorized. If selected, you will be appointed a permanent ensign in the U.S. Navy after successful completion of Officer Candidate School (OCS). Following commissioning, officers will be assigned to a warfare community. Upon successful completion of initial sea duty and warfare qualification, officers will be screened for selection to a bachelor's degree program at the Naval Postgraduate School.

REVIEW 8 QUESTIONS

Q1. The Secretary of the Navy can appoint a specific number of enlisted personnel to the Naval Academy. What number of (a) Regular Navy or Regular Marine Corps and (b) Naval or Marine Corps Reserve (active or inactive) can be appointed?

a.

b.

Q2. What is the maximum time allowed to attend college in the NROTC program?

Q3. Who was the BOOST program intended for?

Q4. What is the major requirement to be eligible for the Enlisted Commissioning Program?

Q5. In what two programs can senior enlisted personnel obtain a commission without a requirement for a college degree?

a.

b.

DISCHARGES

Learning Objectives: When you finish this chapter, you will be able to—

- Identify the types of discharges.
- Recognize the effects of the various types of discharges.

If you separate from the Navy before the end of your active obligated service (EAOS) or after 8 years of service or if you reenlist, you will receive a discharge

Student Notes:

from the Navy. If you separate from the Navy at your EAOS but before completing 8 years of service, you will not receive a discharge but will be "separated" from active naval service. The Navy gives five types of discharge. Each type of discharge has a specific meaning and affects you in a way different from any of the others. The type of discharge you receive depends on the reason for your discharge.

REASONS FOR DISCHARGE

You may receive a discharge for many reasons. Under almost all conditions, whenever and however you leave the Navy, you will receive a discharge. Some of the reasons for receiving a discharge are as follows:

- Expiration of enlistment
- Disability, dependency, or hardship
- Fulfillment of service obligation
- Convenience of the government
- Unsuitability

If discharged for any of the above reasons, you will receive an honorable or a general discharge.

TYPES OF DISCHARGE

The five types of discharge are as follows:

1. Honorable
2. General (under honorable conditions)
3. Other than honorable
4. Bad conduct
5. Dishonorable

Some personnel think because a general discharge is given under honorable conditions, it is as good as the honorable discharge itself. However, that assumption is not true. A general discharge indicates satisfactory service but not to the established standard of the Navy.

Honorable Discharge

To receive an honorable discharge, you must have received a rating from good to excellent for your service to the Navy. Even though you only qualify for a general discharge, you may receive an honorable discharge under two circumstances.

1. When you are being separated because of a disability incurred in the line of duty
2. When you receive any awards for gallantry in action, heroism, or other meritorious service

General Discharge

You receive a general discharge when you separate from the service, under honorable conditions, without a sufficiently meritorious military record to deserve an honorable discharge.

Other Than Honorable Discharge

You receive an other than honorable discharge for misconduct or security reasons.

Bad Conduct Discharge

You receive a bad conduct discharge (BCD) when you separate from the service under conditions other than honorable. You receive a bad conduct discharge only by an approved sentence of a general or a special court-martial.

Dishonorable Discharge

You receive a dishonorable discharge (DD) when you separate from the service under dishonorable conditions. You may receive a dishonorable discharge only by a general court-martial and as appropriate for serious offenses calling for dishonorable separation as part of the punishment.

EFFECTS OF THE TYPE OF DISCHARGE

Some people will try to convince you (or themselves) that the type of discharge they receive will make no difference in their civilian lives. Others will tell you that a discharge under less than honorable conditions can be upgraded if they show themselves to have been good citizens for a time. How wrong they are! Although some discharges have been upgraded by the Board for Correction of Naval Records, the percentage

Student Notes:

is small. The Board is not interested in your civilian life, but how you performed while in the Navy.

When you leave the Navy, you want to do so with an honorable discharge. An honorable discharge has many advantages for you throughout your life. Some of the immediate advantages are the entitlements to various veterans' benefits and rights. When you apply for a job or for entry to a school or college, you will find an honorable discharge is advantageous, and, in many instances, an absolute necessity. Most important of all, and vital for your future self-respect and peace of mind, is the knowledge that your service to your country was up to standard.

Receiving an honorable or general discharge makes you eligible for all federal benefits (and they are considerable). Receiving a dishonorable or bad conduct discharge by a general court-martial disqualifies you for any benefits. A bad conduct discharge from a special court-martial even disqualifies you for any military benefits such as transportation home or payment for accrued leave. A bad conduct discharge bars you from receiving civil service employment preference, reemployment rights, or other related benefits. The Veterans' Administration decides your entitlement to veterans' benefits on an individual basis.

Failing to receive an honorable discharge also has consequences of a more personal and far-reaching nature. You bring shame to your family. You will have difficulty explaining your dishonorable or bad conduct discharge to friends who have honorable military service. You will have difficulty getting good jobs and getting accepted into good schools. Everybody knows the Navy does not give bad conduct discharges except for serious or repeated offenses. Thus, you may have a hard time proving that people can trust you as a friend or to do a job.

Receiving an honorable discharge means you can face the world proudly and secure in the knowledge that your years served in the Navy were well spent. On the other hand, receiving a dishonorable or bad conduct discharge means you must admit to wasted years in the Navy. It means you failed in your duty to your country and in meeting the high standards of the Navy.

NAVY GOOD CONDUCT MEDAL

Learning Objective: When you finish this chapter, you will be able to—

- Identify the requirements for the Good Conduct Medal.

You may earn many awards while you are in the Navy. One of the most important of these awards is the Navy Good Conduct Medal. That medal is the highest precedence award among the campaign and service awards.

Make every effort to earn the Navy Good Conduct Medal. Earning that award can affect your promotion. Meeting the requirements for the Navy Good Conduct Medal means you also meet the requirements for reenlistment, overseas duty, certain Navy schools, and Navy commissioning programs.

Your CO can recommend you for a Navy Good Conduct Medal as a reward for 3 years of good conduct.

REVIEW 9 QUESTIONS

Q1. List the five types of discharges the Navy gives.

a.

b.

c.

d.

e.

Q2. When getting out of the Navy with less than 8 years of service, you are _________ from naval service instead of discharged.

Student Notes:

Q3. List some advantages of having an honorable discharge from the Navy.

a.

b.

c.

Q4. DELETE

SUMMARY

This chapter provides a wealth of information useful to you in your continuing professional development.

The Navy's Enlisted Performance Evaluation System is a system that documents a Sailor's qualifications, performance, conduct, and increased responsibilities. It is the Navy's prime personnel management tool.

Many incentive programs are available for Navy personnel. The purpose of incentive programs is to attract an individual to a rewarding, enjoyable Navy career. Other considerations that often persuade an individual to select a Navy career are job security, paid vacations, travel, family protection plans, retirement, and many other incentives. Often, a combination of these factors causes a person to choose a naval career.

Education is a key to professional development and a better understanding of the world in which we live. The Navy gives Sailors the opportunity to meet their career and educational needs. Navy-sponsored programs, as well as civilian schools, are available to all Navy personnel. As a Navy member, you should take part in academic programs to increase your formal educational background and to further develop your potential for a rewarding career in the Navy.

In this chapter we have also discussed a wide variety of programs designed to help you in making decisions that affect your career.

The Navy has several programs that provide professional training and off-duty educational opportunities. As the Navy has long recognized, the more education you get, the more you will benefit your organization and the Navy.

Many programs lead to a commission as a naval officer. Each year, hundreds of enlisted personnel receive a commission through one of these programs. They then continue to make significant contributions to the Navy's mission as an officer.

The Navy gives various types of discharges. Your eligibility for benefits and other programs after separation or retirement depends on the type of discharge you receive. An other than honorable discharge has certain negative social effects.

The many helping resources and programs within the Navy's organization can benefit everyone. Look into these programs. Ask questions and learn all you can about them. They can benefit you in many ways during your naval career.

REVIEW 1 ANSWERS

A1. The Navy uses the **Goal Card Program** to help new Sailors set and achieve goals while in the service.

A2. Some areas covered by the Pocket Goal Card include—

a. **DEP goals**

b. **Navy core values**

c. **Recruit training goals**

d. **The Sailor's Creed**

e. **Fleet goals**

f. **Personal priorities (including education)**

g. **There is also space for Sailors to write their own goals**

Student Notes:

REVIEW 2 ANSWERS

A1. The permanent board members of the Professional Development Board include the—

a. **Command Master Chief**

b. **Command Career Counselor**

c. **Personnel Officer**

d. **Educational Service Officer**

A2. The Professional Development Board **interviews Sailors who want advancement training or who want to attend special programs**.

A3. The three lowest grades are known as **apprenticeships**.

A4. Ratings are divided into—

a. **General**

b. **Service**

A5. A designated striker is a person in **paygrades E-1, E-2, or E-3 who has been designated as technically qualified for a particular rating**.

A6. The ***Manpower and Personnel Classifications and Occupational Standard,* NAVPERS 18068,** contains a list of NAVSTDs and OCCSTDs.

A7. A NAVSTD is **a military requirement that deals with all enlisted personnel**, while an OCCSTD is **an occupational requirement that is rate specific**.

A8. The three requirements you need to meet to be advanced to E-3 are—

a. **Time in rate**

b. **The CO's recommendation**

c. **Complete *Basic Military Requirements,* NAVEDTRA 12018**

A9. To be advanced to petty officer, you must meet the following requirements:

a. **Time in rate**

b. DELETE

c. **Demonstrate knowledge of material in your TRAMAN**

d. **CO's recommendation**

A10. The final multiple score of an advancement exam is based on—

a. **Merit rating**

b. **Personnel testing**

c. **Experience**

REVIEW 3 ANSWERS

A1. The three types of duty are—

a. **Sea**

b. **Shore**

c. **Neutral**

A2. Overseas shore duty Code 3 is classified as **sea duty for rotational purposes**.

A3. To let your detailer know what duty station you want, you should submit **Enlisted Duty Preference Form, NAVPERS 1306/63**.

A4. The kind of information found on the Enlisted Duty Preference Form includes—

a. **Where you want to go**

b. **What type of duty you prefer**

c. **Career intentions**

d. **Family status**

REVIEW 4 ANSWERS

A1. The Evaluation Report and Counseling Record is used to **record your qualifications, conduct, performance, and eligibility for advancement**.

A2. The numerical grading scale used on the Evaluation Report and Counseling Record is similar to the A, B, C, D, F scale used in most high schools. The following scale shows the grading scale used on the Evaluation Report and Counseling Record:

a. **5.0—greatly exceeds standards**

b. **4.0—above standards**

c. **3.0—meets standards**

d. **2.0—progressing**

e. **1.0—below standards**

A3. The evaluation traits that are found on the Evaluation Report and Counseling Record include—

a. **Professional knowledge**

b. **Quality of work**

c. **Equal opportunity**

d. **Military bearing and character**

e. **Personal job accomplishment and initiative**

f. **Teamwork**

g. **Leadership**

A4. After you sign your Evaluation Report, **it is sent to BUPERS and copies go to the field service record, reporting activity, and to you, the service member**.

REVIEW 5 ANSWERS

A1. Page 1 of your service record contains your **enlisted contract**.

A2. Your evaluations are kept in the **left-hand side of your service record, underneath the separator**.

A3. Record of Emergency Data form is **Page 2** of your service record.

A4. You should update Page 2 **anytime you or your family member has a change of address or change in status**.

A5. DELETE

A6. Page 4 contains the following information:

a. **NECs; designators, assigned, changed, or revoked**

b. **Navy schools attended**

c. **Navy training courses completed**

d. **Personal qualifications; technical qualifications**

e. **GED and off-duty courses completed**

f. **Decoration and awards**

REVIEW 6 ANSWERS

A1. Documents that require the CO's personal signature include **documents dealing with law or aspects of military justice and documents that by law are required to have the CO's signature, such as ships' deck log**.

A2. You can find a list of command personnel that has signature authority to sign command correspondence in your **unit's organizational manual or instruction**.

A3. 3-M stands for the **Maintenance and Material Management Systems**.

A4. The objectives of the 3-M system include—

a. **Maintain equipment at maximum operating efficiency**

b. **Reduce equipment downtime**

c. **Reduce cost of maintenance in both money and man-hours**

d. **Provide data directly related to maintenance**

REVIEW 7 ANSWERS

A1. PQS is divided into three sections that include—

a. **Fundamentals**

b. **Systems**

c. **Watch stations**

A2. To find a list of Navy TRAMANs, you would refer to the *Catalog of Nonresident Training Courses* on the web at **http://www.cnet.navy.mil/netpdtc/nac/neas.htm**.

A3. When you're learning a skill while working, you are receiving **on-the-job (OJT) training**.

A4. Recruit training is **General Military Training (GMT) and a class "R" school**.

A5. Different Navy schools include—

a. **Class "R"**

b. **Class "A"**

c. **Class "C"**

d. **Class "F"**

e. **Class "P"**

f. **Class "V"**

A6. The **Basic Skills Program** offers Sailors a chance to earn a high school diploma or improve their skills and military performance.

A7. The **Tuition Assistance Program** is the Navy's financial program that helps Sailors with their off-duty education.

A8. DANTES offers **examinations and certification programs, operates an independent study support system, and provides other support and development activities**.

REVIEW 8 ANSWERS

A1. The Secretary of the Navy can appoint the following enlisted personnel to the Naval Academy:

a. **85 Regular Navy or Regular Marine Corps**

b. **85 Naval or Marine Corps Reserve (active or inactive)**

A2. The **maximum time allowed to attend college on the NROTC program is 40 months**.

A3. The BOOST program was intended for **Sailors who meet all the requirements for the Naval Academy or NROTC program except for academics and people who have been educationally deprived**.

A4. The major requirement to be eligible for the Enlisted Commissioning Program is the **candidate must be able to complete a baccalaureate degree within 2 years**.

A5. The two programs that can lead to an enlisted member being commissioned are—

a. **Chief Warrant Officer**

b. **Limited Duty Officer**

REVIEW 9 ANSWERS

A1. The five types of discharge are—

a. **Honorable**

b. **General**

c. **Other than Honorable**

d. **Bad Conduct**

e. **Dishonorable**

A2. When getting out of the Navy with less than 8 years of service, you are **separated** from naval service instead of discharged.

A3. Some advantages of having an honorable discharge from the Navy include—

a. **Entitlements to various veterans' benefits and rights**

b. **Job preferences**

c. **Entry into a school or college**

A4. DELETE.

CHAPTER 17

FINANCIAL MANAGEMENT AND STRESS MANAGEMENT

The policy of the Navy is "to promote habits of thrift and encourage... conduct of financial affairs in such a manner as to reflect credit upon the naval services." As a Navy sailor it is your responsibility to seek out financial information to avoid any financial problems.

You may wonder why this chapter is titled "Financial Management and Stress Management," or why financial management and stress management are covered in the same chapter. Although there are many causes of stress, one primary cause of stress in families is not having enough money to meet needs. This cause of stress can result in spouse and child abuse, which is **not acceptable behavior**. All commands have a Family Advocacy Program (FAP) to help families undergoing stress.

Many commands provide financial counselors to advise Sailors in financial difficulties. Family service centers or your leading petty officer (LPO) are some examples of who you can seek for financial counseling. The Naval Military Personnel Manual section 62 offers some good advice to all paygrades.

MILITARY PAY SYSTEM

Learning Objectives: When you finish this chapter, you will be able to—

- Identify the various types of military pay, the Leave and Earnings Statement, and the method used to deposit military pay.
- Recognize the responsibilities of making sure that pay and earnings statements are correct.
- Identify liberty and leave and recognize their differences.

The military pay system affects you directly. The amount you receive every payday is determined by the military pay system. Therefore, you should have a basic understanding of the difference between pay and allowances and the different types of pay and allowances. You should also understand a little about allotments and government insurance.

In this section, you will learn about the basics of the military pay system. The pay system is very complex and pay and allowances are subject to change. If you need specific information about your pay, you should consult your disbursing office.

PAY

Pay is money paid to you for services rendered. All pay is taxable as income. The Navy has three types of pay:

1. Basic pay
2. Incentive pay
3. Special pay

You may receive all three types of pay if you are qualified, or you may receive only basic pay.

Navy personnel paychecks are deposited automatically into their checking or savings account via the Direct Deposit System (DDS). To get paid, you must open up a savings or checking account.

Basic Pay

Basic pay is the pay you receive based upon your paygrade and your length of service. All people on active duty in the Navy receive basic pay.

Navy personnel receive longevity (length of service) raises after 2, 3, and 4 years of service. After that, they generally receive a longevity raise for every 2 years of service. Personnel in paygrades E-1 and E-2 don't receive longevity raises. An E-3 doesn't receive longevity raises after 4 years of service. Length of service for pay purposes includes active-duty and inactive Reserve time, former service (if you have a broken-service enlistment), and service in other branches of the U.S. armed forces.

Incentive Pay

Incentive pay is pay you receive for certain types of duty. These types of duty are usually considered hazardous. Therefore, incentive pay is sometimes referred to as *hazardous duty pay*. Duty for which you may receive incentive pay includes aviation duty, submarine duty, parachute duty, flight deck duty, demolition duty, and experimental stress duty.

You receive incentive pay based on the following guidelines:

- You may receive a maximum of two incentive pays if you meet the requirements for more than one.
- You may not receive incentive pay if you receive special pay for diving duty. (Special pay is covered next.)
- You receive the same basic rate of pay for all types of incentive pay with the exception of aviation duty and submarine duty pay, which vary according to your paygrade and longevity.

Special Pay

Special pay is pay for special circumstances, such as reenlistment or a particular type of duty. Duty for which you may receive special pay includes foreign duty, sea duty, medical duty, special assignment duty, hostile fire duty, and diving duty. You may also receive special pay in the form of a selective reenlistment bonus (SRB).

ALLOWANCES

An allowance is money used to reimburse you (pay you back) for expenses necessary for you to perform your job. Because they are reimbursements for expenses, allowances are not taxable as income. You receive allowances for expenses, such as clothing, quarters, and food. You may also receive allowances for various other expenses.

Clothing Allowance

Enlisted members of the Navy, including Naval Reservists on extended active duty, normally receive an initial allowance for uniforms. You may receive a clothing allowance by two methods.

1. You may receive a reimbursement of cash for your purchases of the uniforms and uniform accessories required for your paygrade.
2. You may receive issues of clothing equal to the cash value of your allowance.

Following an initial 6-month active-duty period, you are entitled to receive an annual clothing maintenance allowance. The purpose of the maintenance allowance is to provide you with cash for the purchase of replacement clothing or for the repair of clothing.

Basic Allowance for Subsistence

Entitlement to a basic allowance for subsistence (BAS) depends on your status and the availability of a government mess. Enlisted members are entitled to a daily ration in kind. Each enlisted member receives a daily ration in kind in the form of three meals a day in a government mess. An enlisted member may receive a daily subsistence allowance for each day a government mess is not available or not used.

Normally, entitlement to BAS depends on the conditions at your permanent duty station. If the station doesn't have a government mess, you are entitled to BAS. If the station has a government mess but you are authorized to mess separately, you are entitled to separate rations (RATS SEP). When authorized BAS, you receive the applicable rate for each calendar day of the month for which you don't receive a ration in kind.

If you are authorized to mess separately, are receiving RATS SEP, and your duties prevent you from purchasing certain meals in a government mess, you are entitled to a supplemental BAS.

Basic Allowance for Quarters

The purpose of basic allowance for quarters (BAQ) is to help you pay the cost of suitable living quarters when government quarters are unavailable or not assigned. Entitlement to BAQ depends on your paygrade, whether you have dependents, and whether you and your dependents have been assigned quarters. The receipt of BAQ involves many restrictions and conditions of entitlement.

Student Notes:

BAQ is divided into two basic categories—BAQ for members without dependents and BAQ for members with dependents. The rates payable vary within each category and with each paygrade. To find out whether you are entitled to BAQ and the amounts payable, check with your personnel or disbursing office.

Other Allowances

In addition to the allowances mentioned above, you may receive a family separation allowance (FSA), cost of living allowance (COLA), overseas housing allowance (OHO), variable housing allowance (VHA), or other allowances. Your disbursing or personnel office can provide you with information about the type of allowances, if any, you are entitled to.

Basic Allowance for Housing

Basic allowance for quarters and variable housing allowance are a single allowance called basic allowance for housing(BAH). Your LES will show only the BAH amount.

ALLOTMENTS

Allotments are amounts of money you designate to be withheld from your pay and paid directly to someone else. You may authorize many types of allotments, including the following:

- C (charity drive donation)—allotments to a charity such as the Combined Federal Campaign
- D (dependent)—allotments directly to your dependents
- H (housing)—allotments to a lending institution to pay home-loan payments
- I (insurance)—allotments to a commercial insurance company for life insurance premiums
- S (savings)—allotments directly to an account in your name at a savings institution such as a bank or credit union

For information on making allotments and rules governing their use, see your disbursing office.

OVERPAID

You aren't responsible for calculating your pay, but you are responsible for questioning anything that isn't normal. If you don't question something that isn't normal with your pay, you could be at risk for being charged with larceny. Computerized systems, equal pay periods, and Leave and Earnings Statements (LES) have made budgeting your pay easy. You should be getting the same amount every payday. But computers are only as smart as their operators and the electricity they run on. When you notice a large difference in your pay from last payday and you aren't due for a longevity raise, promotion, or annual pay raise, there may be an error in your pay.

Sailors who haven't reported the difference to their disbursing offices have found themselves held liable for stealing. Even if you do notice and report a questionable payday and nothing changes, you are still liable for the overpayment. Regular disbursing audits balance payments made with those due. Eventually, you'll have to reimburse (give back) that amount, so bank the overage. Look at it this way: You would rush in to your disbursing office and insist on knowing why you were paid too little—right? So—rush in if you're being paid too much, too.

LEAVE AND EARNINGS STATEMENT

Based on the Navy's Joint Uniform Military Pay System (JUMPS), the Navy must provide you a monthly Leave and Earnings Statement (LES). JUMPS is a computerized pay and leave accounting system located at the Defense Finance and Accounting Service, Cleveland, Ohio. The monthly leave and earnings statement provides you with a complete and accurate record of the following:

- Pay
- Allowances
- The type and amount of each allotment requested
- The amount deducted for withholding tax, Social Security, and Servicemen's Group Life Insurance

Student Notes:

Earned and Used Leave

The LES (fig. 17-1) contains all the details you need to keep a personal record of these items. Most of the blocks are self-explanatory. Some of the abbreviations and the use of some of the blocks are explained on the back of the form.

After receiving your LES, check it carefully to verify (make sure) that the information is correct. If it isn't correct or if you have any questions, go to your personnel office or disbursing office.

Leave and Liberty

Leave and liberty consist of the times you are authorized to spend away from work and off duty. Each is a separate category, and the two cannot be combined.

LEAVE.—Leave is an authorized absence similar to vacations in civilian jobs. Basically, you will earn 30 days of leave in each year of active duty. The various terms applied to leave are covered after you learn about the way leave is computed and earned. **Leave is shown on your LES (fig. 17-1) in the row "LEAVE."**

Vacations and short periods of rest from duty provide benefits to morale and motivation that are essential to maintaining maximum effectiveness. The lack of a break from the work environment adversely affects your health, your availability, and your performance.

Normally, you're encouraged to use your entire 30 days of leave each year. Congress has provided compensation for you if military requirements prevented you from using your leave. You should not be required to expend leave immediately before separation simply for the purpose of reducing your leave balance.

LIBERTY.—Liberty is an authorized absence from work or duty for a short period. The Navy grants two types of liberty—regular and special. **Liberty is not shown on your LES.**

Regular liberty is usually granted from the end of one work period to the beginning of the next. That period may be from one day to the next or over a weekend or holiday.

Special liberty is liberty granted outside of regular liberty periods for unusual reasons, such as compensatory time, emergencies, or voting. You may also receive special liberty for special recognition or to allow you to observe major religious events. Special liberty is granted as 3-day or 4-day periods.

Three-day special liberty is a liberty period designed to give a servicemember three full days absence from work or duty. Three-day special liberty usually begins at the end of normal working hours on a given day and ends with the start of normal working hours on the fourth day—for example from Monday evening until Friday morning. When a 3-day special liberty is during regular liberty time, such as a Saturday and Sunday with Monday or Friday a national holiday (special work hours aren't included), the time off is treated as regular liberty.

Four-day liberty is a special liberty period granted by the CO that gives the servicemember four full days absence from work or duty. Usually, special liberty begins at the end of normal working hours on a given day and ends with the start of normal working hours on the fifth day. Four-day special liberty includes at least two consecutive nonwork days—for example, from Wednesday evening until Monday morning.

CONVALESCENT LEAVE.—Convalescent leave is a period of authorized absence given as part of care and treatment prescribed for your recuperation and convalescence. If you have a medical problem that requires a period of recovery but does not require hospitalization, your doctor may prescribe convalescent leave. Convalescent leave is not charged to your earned, advance, or excess leave account; it is computed separately.

REQUESTING LEAVE.—To request either regular or emergency leave, you should use the Leave Request/Authorization, NAVCOMPT Form 3065. When you submit a leave request, forward the completed form through the normal chain of command. Emergency leave requests are hand-carried for approval. When emergency requests need approval after normal working hours, the command duty officer usually approves the request.

Student Notes:

```
Division : FNV6
---------------------------------------------------------------------------------------------------
      DEFENSE  FINANCE   AND  ACCOUNTING  SERVICE  MILITARY  LEAVE  AND  EARNINGS  STATEMENT
---------------------------------------------------------------------------------------------------
    |Name (Last, First, MI)     |SSN      |Grade|PyDate|Yrs| ETS  |Branch|ADSN PPO | Period Covered
 ID |                           |         |  E  |      |   |      | NAVY |         |
---------------------------------------------------------------------------------------------------
          ENTITLEMENTS        |         DEDUCTIONS        |        ALLOTMENTS        |     SUMMARY
---------------------------------------------------------------------------------------------------
   TYPE                 AMOUNT|TYPE                 AMOUNT|TYPE                AMOUNT|+AmtFwd
---------------------------------------------------------------------------------------------------
A|  BASE PAY                  |FEDERAL TAXES              |COMB FED CAMPAIGN    15.00|
B|  BAS                       |FICA-SOC SECURITY          |                          |+Tot Ent
C|  BAH                       |FICA-MEDICARE              |                          |
D|                            |SGLI FOR 200,000           |                          |-Tot Ded
E|                            |AFRH                       |                          |
F|                            |DENTAL                     |                          |-Tot Allt
G|                            |MID-MONTH-PAY              |                          |
H|                            |                           |                          |=Net Amt
I|                            |                           |                          |
J|                            |                           |                          |-Cr Fwd
K|                            |                           |                          |
L|                            |                           |                          |----------------
M|                            |                           |                          |EOM Pay
N|                            |                           |                          |
O|                            |                           |                          |
 |        |---------------------------------------------------------------------------|----------------
 |  Total |                   |                           |                          |
---------------------------------------------------------------------------------------------------
    |Bf Bal|Ernd|Used| CrBal|ETSBal|Lost|Paid|UseLose|FED   |WagePeriod|  Wage YTD|Ms|Ex|  AddlTax|  Tax YTD
LEAVE|     |    |    |      |      |    |    |       |TAXES|          |          |  |  |         |
---------------------------------------------------------------------------------------------------
FICA |WagePeriod|SocWageYTD|  SocTaxYTD|Med Wg YTD|Med Tx YTD|STATE|St|Wg Period|  Wage YTD |Ms|Ex|   Tax YTD
TAXES|          |          |           |          |          |TAXES|FL|         |           |  |  |
---------------------------------------------------------------------------------------------------
PAY  |BAQ Type|BAQ Depn|VHA Zip| Rent Amt |Share|Sta| JFTR|Depns|2d JFTR|BAS Type|  Charity|Tpc|Pacidn
DATA |        |        |       |          |     |   |     |     |       |        |         |   |
---------------------------------------------------------------------------------------------------
 REMARKS:
      YTD ENTITLE                          YTD DEDUCT
               ----------                            ----------

   FOREIGN DUTY PAY WAS CHANGED TO                   |SEA DUTY TIME
   HARDSHIP DUTY PAY EFFECTIVE FEBRUARY 4, 1999.     |BAH BASED ON
   BANK                                              |
                                                     |




                                                          OFFICER IN CHARGE
                                                          PERSONNEL SUPPORT DETACHMENT
                                                          CODE
```

BMRF1701

Figure 17-1.—Leave and earning statement.

REVIEW 1 QUESTIONS

Q1. What's the main difference between pay and allowance?

Q2. List the three types of pay the Navy uses.

a.

b.

c.

Q3. What system is used to deposit Navy personnel paychecks?

Q4. As an E-4, you have served more than 4 years of active-duty service. How often will you receive a longevity raise?

Q5. How often do you receive your clothing maintenance allowance?

Q6. DELETE

Q7. What person is responsible for making sure your paycheck and LES are accurate?

Q8. How many days of leave do you earn per year?

Q9. The CO may grant how many days of special liberty?

PERSONAL FINANCIAL MANAGEMENT

Learning Objective: When you finish this chapter, you will be able to—

- Recognize the procedures for managing personal finances to include money management, use of credit, and indebtedness.

The consumer debt of the United States is the amount Americans borrow for large purchases, such as cars, stereos, appliances, and furniture. The consumer debt also includes revolving credit (which is a type of loan), such as credit cards. This debt keeps spiraling up (getting larger). Repayment of consumer loans slices more then a quarter of every dollar a wage earner takes home. You are probably no exception.

As a young service member, your take-home pay may be less than the national average. You should learn to plan your finances so you can balance your income, savings, and spending.

The following section on personal financial management gives you information you can use. Paying attention to this information will help you manage your money.

MONEY MANAGEMENT

Managing money can be hard to do. You will probably have checking and savings accounts, have allotments, and keep some cash to spend. There are advantages and disadvantages to each of these.

Student Notes:

Checking Account

A checking account usually serves as the safest and the easiest way for you to keep track of your money. A checking account is a financial arrangement with a bank, savings and loan association, or credit union for safeguarding money. It provides a system that allows you to account for your money—both what you've received and what you've spent. Money you *receive* might be your paycheck, while money you *expend* might be a bill you pay.

Some terms that deal with checking accounts are shown below.

Check. A check (fig. 17-2) is a written order telling your bank to withdraw a sum of money from your account to pay another person or business.

Check register. A check register is a booklet used to record transactions involving your checking account.

Deposit ticket or deposit slip. A deposit ticket (fig. 17-3) is a slip of paper used to place money into your account. Deposits can be done either electronically or by you actually going to the bank, filling out a deposit ticket, and handing it to a teller.

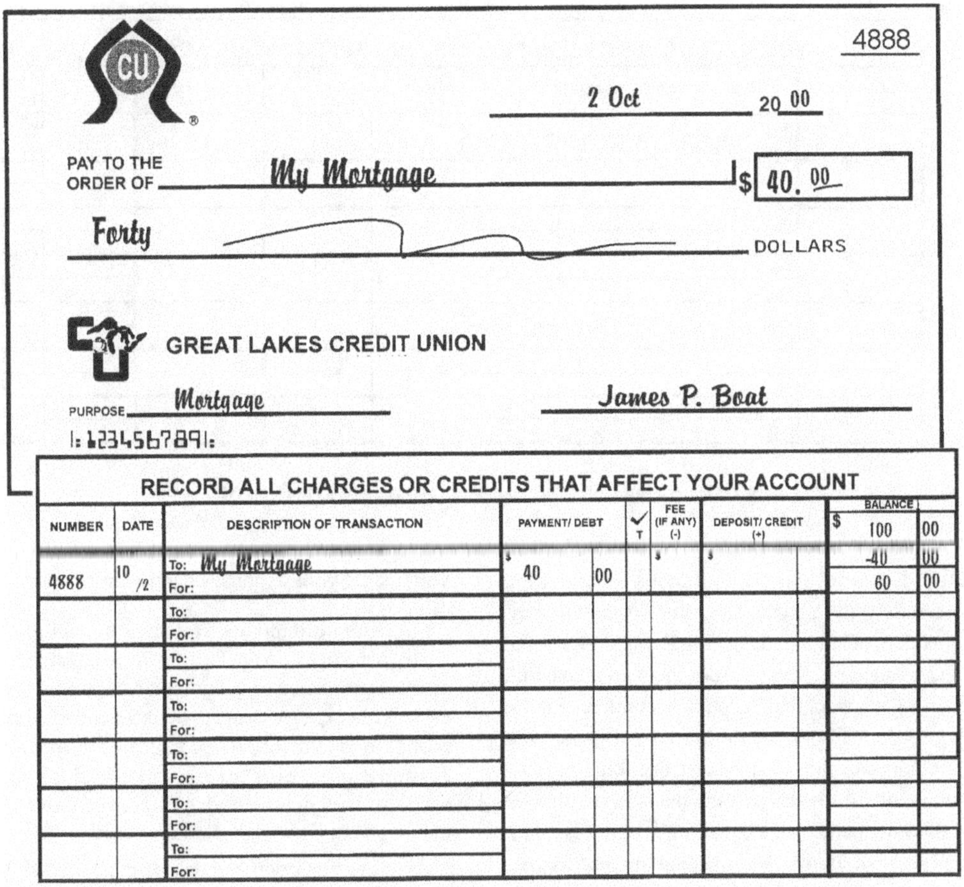

BMRF1702

Figure 17-2.—A check and check register.

Student Notes:

DEPOSIT TICKET

James P. Boat

1763 SHERIDAN DRIVE
YOUR CITY, STATE 12345

DATE 19 Sep 20 00

DEPOSITS MAY NOT BE AVAILABLE FOR IMMEDIATE WITHDRAWAL

SIGN HERE FOR CASH RECEIVED (IF REQUIRED)

DATE			
	CURRENCY		
	COIN		
LIST CHECKS SINGLY			
		100	00
TOTAL FROM OTHER SIDE			
TOTAL		100	00
LESS CASH RECEIVED		0	
NET DEPOSIT		100	00

00-6789/0000

USE OTHER SIDE FOR ADDITIONAL LISTING

BE SURE EACH ITEM IS PROPERLY ENDORSED

DELUXE CHECK PRINTERS

⑆000067894⑆ 123456789⑈

CHECKS AND OTHER ITEMS RECEIVED FOR DEPOSIT SUBJECT TO THE PROVISIONS OF THE UNIFORM COMMERCIAL CODE OR ANY APPLICABLE COLLECTION AGREEMENT.

RECORD ALL CHARGES OR CREDITS THAT AFFECT YOUR ACCOUNT

NUMBER	DATE	DESCRIPTION OF TRANSACTION	PAYMENT/ DEBT	✓ T	FEE (IF ANY) (-)	DEPOSIT/ CREDIT (+)	BALANCE
							$ 15 00
	9/19	To:	$		$	$	+100 00
		For: Deposit				100 00	115 00
		To:					
		For:					
		To:					
		For:					
		To:					
		For:					
		To:					
		For:					
		To:					
		For:					
		To:					
		For:					

BMRF1703

Figure 17-3.—A sample deposit ticket and corresponding check register entry.

BENEFITS OF HAVING A CHECKING ACCOUNT.—One benefit of having a checking account is safety. It is safer to carry checks than money. Another benefit of having a checking account is proof of payment. A canceled check is proof that you paid a bill. Also, having a checking account is convenient. A checking account allows you to receive and spend your money without carrying cash. Also, a checking account lets you pay your bills through the mail, rather than in person. Another benefit of a checking account is that it lets you establish credit. A well-maintained checking account is an asset to establishing and obtaining credit. Finally, a checking account helps you budget your money. Keeping a record of checking activities helps you budget your expenses and income.

As you need money, you *draw* or *transfer* funds by writing a check. You can issue a check payable to another person or to a company to pay bills or to get cash. A checking account provides a canceled check as a receipt of payment. Also, checks are available with carbonless copies of the original check. This easy-to-maintain method can conveniently help you manage your financial affairs.

Before you open a checking account, ask the bank or credit union the questions shown in the following chart.

Student Notes:

1. Is there a minimum balance required?
2. Does the account pay interest?
3. Is there a monthly service fee? Are there other service charges?
4. Is there a limit on how many checks per month I can write?
5. What is the cost to order checks?
6. Are canceled checks returned or photocopied?
7. Is overdraft protection available?

RESPONSIBILITIES OF HAVING A CHECKING ACCOUNT.—You have responsibilities when you have a checking account. You must maintain your check register with exactness to avoid checks being returned for insufficient funds. This is known as *bouncing* a check. For example, if you write out a check and there isn't enough money in your account to cover the check, the check will bounce. The check will usually be sent back to the payee with "Non-sufficient Funds" stamped on it. The bank and the payee will charge you more money because you wrote a bad check. To avoid bouncing a check, always balance your checkbook.

Here are some tips you can use to avoid bouncing a check.

1. Each month, your bank will send a statement of your transactions. Check it for accuracy and balance your checkbook each month (fig. 17-4).
2. Always record transactions in your check register as they occur.
3. Be aware of any service fees and deduct them promptly.

It's unlawful to knowingly write a check when you don't have the necessary funds in your account. In fact, UCMJ, article 123a, prohibits this action. Also, it's a federal offense in civilian courts. Further, writing checks without having sufficient funds can do the following:

- Ruin your credit history
- Destroy your reputation
- Land you in jail (civilian and/or military)

Convenience Cards

Convenience cards are available from your financial institution. These cards make it easier to get money and to make purchases from your bank account. Two types of convenience cards are covered in this section. If you have a convenience card, you will have a personal identification number (PIN). A PIN is a secret access code that you must provide to use your convenience card. Do not tell your PIN to anyone.

WARNING

Do not make purchases that will exceed the balance in your checking account.

One thing to remember, make sure that you update your check register each time you make a transaction using a convenience card. Updating your check register will prevent you from overdrawing your checking account.

Finally, a record of all your convenience card transactions will appear on your monthly bank statement.

AUTOMATIC TELLER MACHINE (ATM) CARDS.—ATM cards are available from your financial institution. ATM cards can be used to make deposits or withdrawals; to make inquiries about account balances; or to move money among your accounts. ATM cards can also be used 24 hours a day, 7 days a week.

CHECK (DEBIT) CARDS.—You can use a debit card instead of writing a check. When used to pay for merchandise or services, the amount is automatically deducted from your checking account. You can use your

Student Notes:

Monthly Bank Statement

24 HOUR TELEPHONE TRANSFER LINE - 123-5678
CUSTOMER SERVICE NUMBER - 567-1234 EXT 296

★★★

DEPOSIT ACCOUNTS

DETAIL CHECKING
REGULAR CHECKING
ACCOUNT:
SOC. SEC.

THIS STATEMENT SHOWS ALL ACCOUNT TRANSACTIONS FROM SEP 14,19?? - THRU OCT 12, 19??

DEPOSITS		CHECKS AND DEDUCTIONS						DAILY BALANCES	
DATE	AMOUNT	NO	DATE	AMOUNT	NO	DATE	AMOUNT	DATE	AMOUNT
9/19	100.00	4882	9/15	32.00				9/15	2533.40
		****						9/16	2503.45
		****						9/19	2603.45
9/28	269.00	4885	9/18	29.95				9/26	2593.45
		4886	9/26	10.00				9/28	2862.45
		****						10/02	2822.45
		4888	10/02	40.00					

**** INDICATES ONE OR MORE MISSING CHECKS

BEGINNING BALANCE 9/14/??	DEPOSITS & CREDITS		CHECKS & DEBITS		ENDING BALANCE 10/12/??
	NO	AMOUNT	NO	AMOUNT	
2565.40	2	369.00	4	111.95	2822.45

ENCLOSURES: 8

Figure 17-4.—Monthly bank statement.

debit card to withdraw funds from your checking or savings account, transfer funds, and check your account balance day or night at ATMs.

SAFETY PRECAUTIONS FOR USING ATMs.—Some precautions you should use when using ATMs are—

- Be alert, Don't use an ATM if the lights aren't operating or you see suspicious activity. At drive-up ATMs, keep car doors locked, other windows closed, and the engine running. If you feel that something is wrong, leave.

- Take someone with you if you must make a transaction at night.

- Keep a low profile. Have your card ready when you approach the ATM. Remember to take your card, cash, and receipt and put them away. Count your money only when you are safely away from the ATM.

Savings Account

One way for you to manage your money is to have a savings account. Savings accounts draw interest (earn money), while checking accounts sometimes do not. A savings account is an excellent way to earn interest and keep from spending money.

Balancing Your Account

Depending on the bank and type of account, your monthly bank statement might include the following:

- Actual or miniphoto copies of your canceled checks.

- A list of your checks. The bank keeps photocopies of your checks on file.

- A listing of your savings account transactions.

The part of your statement dealing with your checking account includes—

- All processed checks,

- Deposits and withdrawals, including those made via convenience cards, and the

- Balance as of the end of your statement.

You use the bank statement to balance your checking account. Compare your statement and register and identify any discrepancies to your accounts.

If you have any questions, the family service center, your command financial specialist, or your LPO can teach you how to balance a checkbook.

Allotments

Allotments provide a good method for you to handle your financial affairs. The following paragraphs describe voluntary and involuntary allotments.

VOLUNTARY ALLOTMENTS.—Voluntary allotments are requested by you. Some of the reasons for making a voluntary allotment are as follows:

- Savings

- Purchase of U.S. saving bonds

- Loan payments

- Life insurance payments

- Mortgage payments

- Pledges to the Combined Federal Campaign payments

- Payment to family members and relatives

INVOLUNTARY ALLOTMENTS.— Involuntary allotments from a Navy member's pay usually mean one thing—financial irresponsibility. Involuntary allotments are usually garnishment of your pay.

Budgeting

Preparing and using a budget is the key to successful money management. A budget is a plan to spend money or a plan of money management. Many Navy members have false images of the meaning of a budget. They often associate budgets with detailed bookkeeping, stacks of paper, ledgers, and so forth. A

Student Notes:

budget gives you records of your income vice your expenses and helps you manage your financial affairs.

If you're married, budgeting involves both you and your spouse. For married couples, handling money matters is a joint effort. With two-income families, money management is a different ball game. The yours-mine-ours approach usually comes up, requiring definite understandings. Certain inherent expenses become greater when both the husband and wife earn wages. Couples also need to have an understanding as to what expenses they will pay from what funds. A written budget, properly prepared and followed, helps couples work out these problems.

In budget preparation you determine income and expenses; examine spending habits; and see what, if anything, you need to correct or improve. To help you improve your spending habits, you need to be familiar with the following terms used in financial management:

Gross income. The total amount of pay before any deductions.

Deductions. The amount of money taken from pay for income taxes, Social Security, Service Group Life Insurance(SGLI), and so forth.

Allotments. The money taken from gross income for savings, checking accounts, family support or to pay debts, such as car payments and debts due the United States.

Net income. The money paid to a member after all deductions and allotments are paid. Also known as *take-home pay.*

Fixed expenses. Expenses that are the same each month.

Flexible or variable expenses. Expenses that are different each month.

Fixed expenses include rent and mortgage payments and time payments for expenses, such as autos, furniture, and insurance. The difference between fixed expenses and net income is optional income. This is the income available for planning purposes, which you can apply to variable or flexible expenses. These expenses include items such as savings, food, utilities, entertainment, clothes, and gifts.

When preparing a budget, plan for savings first. Planning for savings first is important. If you save first, then you can plan your budget and still save money.

Everyone needs a savings program for unforeseen expenses in the future. In addition, using a systematic, planned savings program will help you to achieve set goals. In determining how much to save, have a realistic percentage of your optional income. This percentage could be as little as 5% to 10% or as high as 20% of your optional income.

After savings comes a fixed expense, followed by variable expenses. The U.S. Department of Labor suggest these percentage of take-home-pay for budget preparation:

Fixed Expenses		**Variable Expenses**	
Housing	25%	Food	23%
Transportation	9%	Clothing	11%
		Gifts and contributions	5%
		Savings and unforeseen expenses	22%

These percentages are approximate and will vary from area to area and person to person.

To prepare a personal budget, you should keep close track of your income, expenses, and savings for several months. This information will help you understand your spending habits. It will also help you determine average non-fixed expenses. Understanding your spending habits puts you in a position not only to budget your income but also to correct undesirable spending habits.

Plans for spending extend to many areas and vary according to the person's status and requirements. The basics of spending are to spend money wisely and in as small amounts as possible.

INVESTMENT RULE OF 72

What is the rule of 72? The Rule of 72 gives you an easy method of estimating the number of years it takes for an investment's value to double at a specific interest rate or rate of return. The general formula for the Rule of 72 is as follows:

Student Notes:

$72 = I \times Y$,

where,

I is the interest rate, and

Y is the number of years needed to double your investment.

Divide 72 years by your interest rate to estimate the number of years it will take to double your investment. For example, at a rate of 8%, an investment's value will double in 9 years.

CREDIT

Credit is based largely on trust. The average person in the Navy is trustworthy and expects to receive a fair deal in business and financial dealings. On the other hand, the way people handle their finances is a reliable sign of their general character and trustworthiness.

Usually, when you think of credit, you think of time payment purchases or charge accounts. Actually credit has a much broader scope.

The entire country runs on credit, including industries; banks; and local, state, and federal governments. In fact, if credit were to stop suddenly, the result would be catastrophic. For example, almost no one would be able to buy a home, an automobile, furniture, or a television or stereo set. Without these sales, unemployment would skyrocket. These salaries, not available for the retail market, would in turn adversely affect the sale of other goods. The effect would continue from the highest to the lowest level, and economic chaos would result.

Principles of Credit

Credit literally means buy now, pay later. The system permits you to purchase goods as you need them, but pay for them over a certain period. Credit means you receive a loan of money, and you always pay extra when you borrow money. Credit, if used wisely, ensures a reasonable standard of living. However, you cannot substitute credit for sound financial planning and a systematic savings plan. Additionally, improper use of credit can create a financial nightmare that can adversely affect your job, family life, and mental and physical health.

Cost of Credit

Have you ever rented a motorcycle or sailboat? You always know in advance that it will cost you so much an hour or day. The rent or cost of using the bike or boat has its base on length of use.

The rent paid for using borrowed money or credit is known as *interest*. Sometimes, you may have difficulty figuring interest. Some lenders and businesses quote interest rates plus other charges in a way that hides the actual figures. Then, people don't know the total cost of loans or installment purchases.

When you borrow or buy something *on time*, keep your eyes open for extra charges in addition to the interest charge for the use of the money. Some of these additional charges include credit life insurance, fees for credit investigations, loan-handling fees, and health and accident insurance. Often, the down payment and the monthly payments are the only figures stated.

Ask for the total charges in writing, including early repayment penalties and monthly rates. If you don't receive the amount in writing, you can figure it your self. First, find the total amount you will pay for the loan or the purchase. Then subtract the actual price of the goods from the total cost of the loan. The difference shows the total cost of credit. Taking the time to get the facts pays off.

Credit Rating

Most people find it to their advantage to build a good credit rating. Some people object to buying anything on credit and insist on paying for everything in cash. They save until they have the cash to make a major purchase, and they often do get better buys for cash. However, a good credit rating is like money in the bank. When you have a good credit rating, it means that you pay your bills on time. Navy personnel usually have a good credit reputation and should have no problem getting a loan or credit when needed. A good credit rating can be priceless in an emergency, such as a medical crisis, fire, or death in the family.

You can establish a good credit rating by paying for time purchases according to the purchase agreement. Time purchases include items, such as furniture or cars and items bought on credit card accounts. You can also

Student Notes:

establish credit by repaying a loan from a bank or a credit union according to the loan agreement. Making these payments according to their agreements means that you pay the amount agreed upon by a certain date. You can then use these companies, banks, or credit unions as credit references if you apply for credit at any future time.

Use of Credit by Navy Personnel

The Navy expects all its members to discharge their financial responsibilities in a timely manner. The Navy expects its members to be a credit to themselves and the naval service. Knowing about credit lets you handle your financial affairs better and often saves you money.

If Navy personnel are to use credit wisely, they need to know the cost of credit. They especially need to know how to avoid some of the problems young Navy men and women often have.

Credit plays an important part in the financial world. Use it wisely and carefully, and pay attention to the following principles:

- Use credit for those necessary goods that you can't afford with one or two paychecks.
- Use credit mainly for goods that have a useful life longer than the time needed to pay for them.
- Make as large a down payment as possible. This reduces the total amount spent because of interest charges.
- Know what your income will be. Set a spending limit equal to the smallest paycheck received to be sure of having enough money to meet the payment when due.
- Don't buy another item on credit just because you have finished paying for one.
- Avoid the temptation to use credit for splurging. For example, buying too much on credit at Christmas becomes a shock in January when you receive the bills.
- Check with consumer affairs offices about local credit regulations. For example, some states allow up to 3 days to change your mind on a credit purchase or a loan received.

When using credit, remember the following facts about credit:

- Credit costs money, but many credit plans exist. Some plans are much less expensive than others. When you buy a car or furniture, you shop for the best bargain. Do the same when you shop for the best bargain in credit.
- Consider carefully before borrowing from finance companies. These companies often charge high interest rates on loans.
- The faster you pay off a debt, the less interest charges you'll pay.
- Use credit only for unforeseen emergencies and for higher-cost purchases, such as furniture, cars, or houses.

While buying on credit has advantages, you also need to recognize some of the disadvantages of using credit. The following are some of the problems you may encounter:

- Credit customers may overbuy.
- Credit customers may buy at the wrong time or place.
- Credit prices may be higher than cash prices.
- Credit ties up future income.
- Payments must be made on time.
- Because of the addition of interest charges to the price, the purchase costs more.

REVIEW 2 QUESTIONS

Q1. What is the safest and most convenient way to keep track of the money you spend?

Student Notes:

Q2. You have paid for an item with a check; however, you don't have enough money in your checking account to cover the check. What is the result of this action?

Q3. You are having money taken out of your pay to make loan payments. What type of allotment are you making?

Q4. What's the first thing you should plan for when making out a budget?

Q5. The money charged for using borrowed money or using credit is known as—

Q6. If total charges of a loan or purchases agreement are not listed, what is a simplest way to find the total cost of credit?

Q7. How do you establish a good credit rating?

GOVERNMENT-SUPERVISED LIFE INSURANCE

Learning Objective: When you finish this chapter, you will be able to—

- Recognize the purpose of life insurance.

The government has provided premium-free or low-cost life insurance for service members and veterans since World War I. Since 1919, various insurance programs have been offered as insurance needs have changed over the years.

SERVICEMEN'S GROUP LIFE INSURANCE

Servicemen's Group Life Insurance (SGLI) is a low-cost group insurance program open to active-duty personnel without regard to special qualifications, such as disability. You may secure SGLI only in increments of $10,000, up to a maximum of $200,000. You are automatically issued the $200,000 coverage, unless you choose a lower amount. The cost of SGLI is deducted automatically from your pay.

Unlike some commercial insurance policies, SGLI has no loan, paid-up, or cash-surrender value. In other words, you can't borrow money against this insurance; if you stop payment on the policy or cancel it, you will receive neither paid-up insurance nor cash.

SGLI coverage continues for 120 days after your separation. If you are separated for a disability, coverage may be extended up to 1 year after your separation date.

VETERANS GROUP LIFE INSURANCE

The Veterans Insurance Act of 1974 established a program of post-separation insurance called Veterans Group Life Insurance (VGLI). That act provides for the automatic conversion of SGLI to a 5-year nonrenewable term policy at reasonable rates and with a "no physical exam" advantage. That is, you can have insurance coverage at reasonable rates for 5 years after you separate from the Navy. You can convert the policy at any time during that 5 years to a commercial insurance policy with the same amount of coverage without a physical examination. Like SGLI, the Office of Servicemen's Group Life Insurance (OSGLI) administers the VGLI program, and the Veterans' Administration supervises it.

You can get VGLI coverage in amounts equal to, but not exceeding, the amount of SGLI in force at the time of your separation. This insurance, like SGLI, has no cash, loan, paid-up, or extended insurance value.

REVIEW 3 QUESTIONS

Q1. You can secure SGLI in what increments?

Student Notes:

Q2. What is the maximum amount of coverage for SGLI?

Q3. You have separated from the service. You will be covered by SGLI for up what maximum number of days after your separation?

YOU AND YOUR FAMILY

Learning Objectives: When you complete this chapter, you will be able to—

- Identify types of abuse to include spouse and child abuse.
- Recognize the effect of abuse on self, family, and the Navy.
- Identify procedures to follow to obtain help.

As part of the naval tradition of taking care of our own—it's the responsibility of each Sailor to ensure the safety, health, and well being of his/her family. The military family deals with the challenges posed by the demands of military life and family life. Sometimes, military life creates stress and friction within the family.

WHAT IS ABUSE?

Stress and friction within the family can lead to abuse, either physical or emotional. Navy personnel are expected to show the Navy leadership core values of honor, courage, and commitment. Child and spouse abuse is unacceptable and incompatible with these high standards of professional and personal discipline. The result of abusive behavior by Navy personnel is—

- Destroyed lives.
- A detraction from military performance.
- A negative affect on the efficient functioning and morale of military units.
- A bad reputation and loss of prestige of the military service in the civilian community.

The following information will help you understand what is meant by the term abuse.

Victim. An individual who is abused or whose welfare is harmed or threatened by acts of omission or commission by another individual or individuals.

Emotional abuse. Actions including, but not limited to active, intentional berating, disparaging, or other behavior towards the victim that adversely affects the psychological well-being of the victim.

Spouse abuse. Spouse abuse includes, but is not limited to, assault, battery, threat to injure or kill, or any other act of force, violence, or emotional abuse, or undue physical or psychological trauma, or fear of physical injury. This includes physical injury, sexual assault, intentional destruction of property, psychological abuse, and stalking.

Stalking. Actions of a person performed in a repeatedly harassing manner, including, but not limited to, following another person in a manner to induce, in a reasonable person, fear of sexual battery, bodily injury, or death of that person or that person's immediate family.

Child abuse/neglect. The physical injury, sexual abuse, emotional abuse, deprivation of necessities, or other abuse of a child by a parent, guardian, employee of a residential facility, or any person providing out-of-home care, who is responsible for the child's welfare, under circumstances that indicate the child's welfare is harmed or threatened. The term encompasses both acts and omissions on the part of such a responsible person. This term includes offenders whose relationship is outside the family and includes, but is not limited to, individuals known to the child and living or visiting in the same residence who are unrelated to the victim by blood or marriage, and individuals unknown to the victim. Child abuse/neglect includes the following:

- Physical abuse. In the case of child abuse, physical abuse includes, but is not limited to, acts that result in death or other physical injury that seriously impairs the health or physical well-being of the victim.

Student Notes:

- Sexual abuse. In the case of child abuse, sexual abuse is actions that include, but are not limited to, the employment, use inducement, enticement, or coercion of any child to engage in, or have a child assist any other person to engage in, any sexually explicit conduct or any simulation of such conduct. Actions include, but are not limited to, rape, molestation, prostitution, or other sexual activity between the offender or a third party and a child, when the offender is in a position or a power over the child.

WHAT CAN THE COMMAND AND THE FAMILY DO ABOUT ABUSE?

Child and spouse abuse are serious behavioral, social, and community problems. These problems need a comprehensive, community-based response. The most effective response to family violence occurs when individuals, families, commands, and communities act as a community to keep the victim safe.

The Department of the Navy (DoN) has a Family Advocacy Program (FAP) that addresses child and spouse abuse. It involves the prevention, evaluation, identification, intervention, rehabilitation/behavioral education and counseling, follow-up, and reporting of child and spouse abuse. The Navy uses this program as a tool to assist victims and to reduce the occurrence of child and spouse abuse.

The five primary goals of the DoN FAP are as follows:

1. Victim safety and protection
2. Offender accountability
3. Rehabilitative education and counseling
4. Community accountability
5. Responsibility for a consistent appropriate response

A continuous effort to reduce and eliminate child and spouse abuse is actively pursued at every level of command. Each command has a Family Advocacy Program. The CO at each installation appoints a family advocacy officer (FAO). The CO also ensures that a family advocacy committee (FAC) and a case review committee (CRC) are established. The primary goal of the FAP is prevention of abuse. The FAP establishes education, support, and awareness programs so that families and their command understand the risk factors of child and spouse abuse. Programs emphasize prevention, recognition, prompt notification and reporting, and availability of responsive services.

Early intervention involving cases of spouse or child abuse of any kind is very important. Victims can report incidents of abuse directly to the FAO, family service center, medical treatment facility, Chaplain, or the Ombudsman. The important thing is to report it.

STRESS MANAGEMENT

Learning Objectives: When you finish this chapter, you will be able to—

- Recognize factors that cause stress.
- Identify ways to combat stress.

Everybody experiences stress. It's the body's natural reaction to tension, pressure, and change. Most people think of stressors (or things that cause stress) as negative, such as traffic, a difficult job, or a divorce. However, stressors can be positive experiences. For example, having a baby, bowling a perfect 300 game, or completing a satisfying project. These are all changes that can cause stress.

Your body can't tell the difference between a good and a bad stressor. Both too much stress and too little stress are bad for you, while the right balance keeps you going. Positive, or good stress, can keep you going. It makes life more challenging and less boring.

Too much stress can be bad for you, both physically and mentally. Prolonged, unrelieved stress can lead to accidental injury, serious illness, or inappropriate behavior. For the sake of your health, safety, and happiness, you need to recognize and manage stress before it gets the best of you.

Stress occurs when there is an imbalance between the demands of our lives and the resources we have to deal with those demands. An imbalance may happen when there are changes in our lives. It's not the changes themselves that cause stress but our reaction to those changes or events.

Student Notes:

Reactions to stress vary and can take their toll, both mentally and physically. Common stress symptoms include upset stomach, fatigue, tight neck muscles, irritability, and headaches. Some people react to stress by eating or drinking too much, losing sleep, or smoking cigarettes.

On-the-job pressures, changes in lifestyle, financial difficulties, and family tensions are stressful. All too often, people use alcohol or drugs to control the stress they feel. However, alcohol and drugs can increase both mental and physical stress. Regular use of alcohol and drugs can lead to dependency.

The first step to managing stress is to identify your stressors—what things make you react. Stressors aren't only events that cause you to feel sad, frightened, anxious, or happy. You can cause stress through your thoughts, feelings, and expectations. A key to dealing with the big and little everyday stressors is to cope with stress in a positive way. The following are some ways you can use to cope with stress:

Acceptance. Many of us worry about things that we have no control over. Learn to accept when things are beyond your control.

Attitude. Try to focus on the positive side of situations. By focusing on the positive, you'll find solutions come more easily and your stress level will be reduced.

Perspective. Too often, we worry or become upset about things that never happen. Keep things in perspective.

There are many healthy ways to combat stress. Regular exercise, proper diet, meditation, laughter, relaxation techniques, and involvement with outside activities can positively affect your attitude and enhance your life as well as reduce stress.

REVIEW 4 QUESTIONS

Q1. When service members or their families are a victim of spousal or child abuse, what Navy program was established to help them?

Q2. List some of the ways that the FAP can help a family.

Q3. How does stress occur?

Q4. What's the first step when dealing with stress?

Q5. List some of the ways you can combat stress.

SUMMARY

Being a member of the Navy gives you various responsibilities, including that of your own financial management. Learn to use credit wisely and don't bite off more than you can chew. You can use your leave and earnings statement to help you develop a budget to keep from overextending yourself financially. The Navy takes matters of indebtedness very seriously. Therefore, take advantage of the programs available through the Navy to help you with money problems.

Trying to balance a military life with a family at best can be very challenging. Budgeting and preplanning for periods of long deployment can help lessen the strain. Through the Family Advocacy Program, families can get help in times of family distress.

Stress is like body temperature. If it's too low or too high, you can't survive; but, the right balance can keep you going strong. It makes sense to use stress energy positively, to meet life's challenges, experiences and goals. Stress is not all bad. In fact, positive stress can make life both rich and satisfying.

REVIEW 1 ANSWERS

A1. The main difference between pay and allowance is that **pay is taxable income and allowance is nontaxable income**.

Student Notes:

A2. The three types of pay are—

a. **Basic**

b. **Incentive**

c. **Special**

A3. The Navy uses the **Direct Deposit System (DDS)** to deposit personnel paychecks.

A4. When you have served more than 4 years of active-duty service, you will receive a longevity raise every **2 years**.

A5. You receive your clothing maintenance allowance **once a year**.

A6. DELETE

A7. **You** are responsible for making sure your paycheck and LES are correct.

A8. You earn **30 days a year or 2.5 days** of leave per month.

A9. The CO may grant **3- or 4-day** special liberty periods.

REVIEW 2 ANSWERS

A1. A checking account is the **safest and most convenient way to keep track of the money you spend.**

A2. If you don't have enough money in your checking account to cover a check, you have **bounced a check**. You are usually charged a fee by the bank to process this check and charged a fee by the company you wrote the check to.

A3. When you have money taken out to make loan payments, you have a **voluntary allotment**.

A4. The first thing to do when making out a budget is to **start a savings plan—pay yourself first**!

A5. The money you're charged to use borrowed money is known as **interest**.

A6. The simplest way to find the total cost of credit is to **subtract the actual price of goods from the total amount of the loan**.

A7. You establish good credit by **paying loans or purchase agreements according to your contract and on time**.

REVIEW 3 ANSWERS

A1. SGLI is available in **increments of $10,000 only**.

A2. The maximum amount of coverage under SGLI is **$200,000**.

A3. Normally, you are covered for **a maximum of 120 days after separation from the service**.

REVIEW 4 ANSWERS

A1. The Family Advocacy Program was established **to help service members or their families when they are a victim of spousal or child abuse.**

A2. The FAP can help a family through—

a. **Education programs**

b. **Counseling**

c. **Intervention in cases of abuse**

A3. Stress occurs when there's **an imbalance between the demands of our lives and resources we have to deal with those demands**.

A4. The first step to take when dealing with stress is to **identify your stressors; that is, find out what causes the stress.**

A5. Some of the ways you can combat stress are—

a. **Exercise**

b. **Diet**

c. **Meditation**

d. **Laughter**

e. **Relaxation techniques**

f. **Involvement with outside activities**

CHAPTER 18

SURFACE PRESERVATION

Summer seas and a good ship—life has nothing better.

—Mark Twain

Just about everyone has been involved in cleaning, preserving, and maintaining something. Painting the family home or washing and waxing your car are good examples. What you did was to protect a surface from the effects of weather or exposure, to extend its lifetime, and to improve its appearance.

The U.S. Navy has a far greater problem because all Navy ships operate in a much harsher environment than your home or car. Constant exposure to the sea and saltwater corrosion can quickly turn the exterior of a ship into a mass of rust. Interior spaces have their problems as well. Constant changes in the weather and in the surrounding water temperature cause moisture, humidity, and chemical reactions that affect electrical systems and machinery. To overcome these harsh conditions, the Navy expends a great deal of time, effort, and money applying surface preservatives. These preservatives range from detergent and fresh water to paint and lubricants. How well these preservatives work depends on you.

CLEANING

Learning Objectives: When you finish this chapter, you will be able to—

- Recognize the purpose of cleaning and preserving.
- Identify the cleaning bill.
- Recall the purpose of compartment cleaning, sweepers, cleaning process, field day, and zone inspections.

Maintaining clean conditions aboard ship and ashore is an important job. Cleaning involves practically every member, from the compartment cleaner to the inspecting officer. Navy life requires each of us to have a personal interest in our living and working areas, not only for the sake of appearance but for our health and safety as well.

THE CLEANING BILL

Each area of the ship is divided into various departments for upkeep. The Cleaning, Preservation, and Maintenance Bill describes these areas and outlines the department that is responsible for them. This bill is carefully planned to make sure all interior areas and exterior areas of the ship's hull are assigned to personnel for upkeep and that no areas overlap or are left out. Each division within the department assigns its personnel to the spaces it's responsible for. Division personnel carry out the duties of cleaning, preserving, and maintaining.

COMPARTMENT CLEANING

The term *compartment cleaner* generally applies to persons assigned to clean living or berthing compartments or spaces, such as passageways and heads. If you are assigned compartment cleaner duties, you will be responsible for keeping your spaces clean, preserved, and in good order. Newly assigned personnel are closely supervised to make sure they understand what to clean and how to clean it. Items, such as electrical and mechanical devices, might be unfamiliar to you. These types of items are located in almost every space aboard ship. With this in mind, caution must be observed at all times. Ask your supervisor to point out any hazardous items located in your compartment and observe all special cleaning instructions.

Cleaning gear is stocked in and issued from the first lieutenant's storeroom. Each division is periodically issued cleaning gear and is then responsible for its proper stowage and care. Because cleaning compounds and solvents are often flammable or toxic, or both, they must **never** be left unattended or improperly stowed. You should always read warning labels and follow their directions carefully. Gear, such as brooms and swabs (mops), must be cleaned after each use and placed in their stowage racks. Gear adrift, such as rags, clothing, or personal gear, must be "policed up" immediately. If left adrift, these items are a tripping or fire hazard—or worse, they might clog up dewatering equipment if the space were flooded.

SWEEPERS

"Sweepers" is piped shortly after reveille, before the end of the regular working day, and at other times as scheduled. At these times, all persons assigned as sweepers draw their gear and sweep and swab down their assigned areas. All trash and dirt are picked up in a dustpan and placed in a trash receptacle.

NOTE

If dirt is swept over the side, the wind may blow it back on board or the dirt may stick to the side of the ship. In either case, additional work is necessary to clean the ship.

At this time you should empty all butt kits (make sure no butts are still burning) and trash receptacles as instructed. **Never** dump trash or garbage over the side of the ship without first obtaining permission from the officer of the deck. At times, all trash must be kept in a safe area aboard the ship until it can be properly removed.

CLEANING PROCESS

Dirt, soil, and contamination all describe the same thing—a foreign material on a surface where it is not wanted. Soil includes grease, oil, tarnish, rust, food residue, and stains. Most exposed surfaces that have been soiled may be cleaned with the proper use of cleaning agents.

Detergents are materials that have the ability to remove contamination and soil. There are other ways of cleaning besides using detergents or cleaning compounds. These include purely mechanical processes, such as removing rust from steel by sand-blasting or cleaning decks by sweeping. For many cleaning problems, chipping, sweeping, sanding, or brushing may be needed. However, when detergent compounds are coupled with the mechanical action, a cleaner surface is usually produced with less time and work.

The steps used in most detergent cleaning operations are as follows:

1. Wetting—The soil and the surface of the object being cleaned must be wetted. If the surface is not wetted properly, cleaning results will be poor. Contrary to popular belief, water has very poor wetting properties. Its wetting ability, and therefore its cleaning ability, is improved by adding other materials, such as soap or synthetic detergents. Adding soap or synthetic detergents cause the water to flow into tiny crevices and around small particles of soil.

2. Scrubbing—Dirt is loosened by the mechanical action of rubbing or scrubbing. For example, oil droplets are emulsified; that is, they are coated with a thin film of soap and prevented from recombining, and then they rise to the surface. In a somewhat similar manner, solid particles are suspended in solution.

3. Rinsing—Rinsing is very important. Rinsing removes loosened dirt from the surface along with the cleaning material.

FIELD DAY

Field day is cleaning day. Periodically, a field day is held. All hands "turn to" and thoroughly clean the ship inside and out, usually in preparation for an inspection. Fixtures and areas that sometimes are neglected during regular sweepdowns (overhead cables, piping, corners, spaces behind and under equipment, and so on) are cleaned. Bulkheads, decks, ladders, and all other accessible areas are scrubbed. Knife edges and door gaskets are checked; any paint, oil, or other substances are removed; all brightwork is shined; and clean linen is placed on each bunk. Field days improve the appearance and sanitary condition of the ship, aid in the preservation of the ship by extending paint life, and reduce the dirt intake caused by operating equipment.

Because of weather conditions, there are many days at sea when the ship's topside areas can't be cleaned. At the first opportunity, all topside surfaces are cleaned with freshwater and inspected for signs of rust and corrosion. If such signs are discovered, you should tend to the area immediately. A little work at that time will save you a lot of work later.

Student Notes:

DECK COVERS

Aboard ship, deck coverings get more wear than any other material. Unless deck coverings are properly cared for, costly replacement is required. There are several materials used for covering decks, but only two types are covered here. These are the resilient and the nonslip (nonskid paint) types.

Resilient deck coverings include vinyl tile, vinyl asbestos tile, and linoleum. These deck coverings do not need painting; however, daily sweeping and wiping away spills as soon as possible are required. Resilient deck covering is clamped down (cleaned with a damp swab) frequently, allowed to dry, and then buffed with a buffer. For more thorough cleaning when the deck is unusually dirty, apply a solution of warm water and detergent with a stiff bristle brush or buffer and rinse with clean water to remove residual detergent. Stubborn dirt and black marks left by shoes can be removed by rubbing lightly with a scouring pad, fine steel wool, or a rag moistened with mineral spirits.

After the deck covering is washed and dried, it can be polished (with or without waxing) with a buffer, or it may be given a coat of self-polishing wax and allowed to dry without buffing. Deck coverings can be buffed several times before rewaxing.

No wax should be applied to the deck when the ship is going out to sea or when heavy weather is anticipated. This is an added precaution against slipping, even though the approved floor waxes are designed to be slip resistant.

Nonslip (nonskid paint) deck coverings contain pumice, which provides a better footing. To clean a nonskid painted deck, use a cleaning solution of detergent and dishwashing compound. To make the solution, mix 1 pint of detergent and 5 tablespoons of dishwashing compound. You can mix this compound with freshwater to make 20 gallons of cleaning solution. Apply the solution with a hand scrubber, let it soak for 5 minutes, and then rinse with freshwater. **Don't** wax or paint nonskid deck coverings. Waxing or painting reduces their nonskid properties.

NOTE

If it becomes necessary to spruce up the appearance of a nonskid deck cover, brush it with deck paint diluted with mineral spirits. The diluted paint should be as thin as possible so that the nonskid properties are not affected.

ZONE INSPECTION

Frequent inspections are held to make sure that all spaces, machinery, and equipment are in a satisfactory state of operation, preservation, and cleanliness. One type of inspection, the zone inspection, divides the ship or station into various sections. Each zone is then assigned to an inspection party or team. Usually the CO will head one team, while an officer or chief petty officer will head each of the remaining teams. If you are assigned to present a compartment, you present the space to the inspecting officer by saluting and greeting the inspector in the following manner: "Good morning (afternoon), sir/ma'am; Seaman Apprentice Frost (your rank and name) standing by compartment (name or number), for your inspection, sir/ma'am." You will then stay with the inspecting officer during the inspection of your spaces to answer questions and provide assistance. Such things as stowage cabinets, lockers, and drawers should be unlocked before the inspection for easy access. Usually the inspecting officer will give an overall grade to the space; for example, a grade of outstanding would indicate that no new discrepancies were noted and all previous discrepancies have been corrected. You can be proud of an outstanding grade.

REVIEW 1 QUESTIONS

Q1. The responsibility for cleaning and maintaining certain spaces in the ship is listed in what publication?

Q2. What person is generally assigned to clean living or berthing spaces?

Student Notes:

Q3. You should pick up and put away gear that has been left adrift for what reason?

Q4. True or False. When sweeping exterior decks, you can sweep dirt over the side.

Q5. List the three steps used in most detergent cleaning.

a.

b.

c.

Q6. List two types of deck covers.

a.

b.

CLEANING SOLVENTS

Learning Objectives: When you finish this chapter, you will be able to—

- Identify types of cleaning solvents.
- Recall the precautions to be followed when working with cleaning solvents.

No matter what the job, from paint removal to swabbing the decks, take precautions against carelessly using cleaning solvents. Look at the following example:

> Seaman Joe Frost didn't read the labels on the chlorine-based cleaning material he was using to clean the commode. He decided to clean the drains at the same time and added a granulated drain cleaner to the chlorine-based cleaner. Then he left the head. A few minutes later he heard a loud explosion. The reaction between the chlorine-based cleaner and granulated cleaner caused the explosion. Luckily, no one was hurt, but the head was a mess.

Solvents used in paints, adhesives, rubber and plastic materials, and degreasing solutions are hazardous to your health. Most solvents are toxic and, with a few exceptions, are flammable. Take the appropriate measures to reduce their toxic and flammable effects. In addition, solvents that come in contact with your skin can cause serious skin problems. When using solvents, always observe the following precautions:

- **Make** sure the space in which you are working has adequate ventilation.
- **Wear** protective clothing, goggles, respirators, gloves, and other appropriate equipment.
- **Make** sure accessible fire-fighting equipment is nearby.
- **Take** every precaution to prevent excessive vapors from contaminating the air.
- **Check** the labels on all containers of liquids.
- **Wipe** up spilled solvents immediately.
- **Avoid** contact with your eyes, skin, or clothing.
- **Never** swallow solvents.
- **Avoid** breathing the vapors.
- **Keep** solvent containers tightly closed when you are not using them.
- **Check** containers for leakage.
- **Transfer** solvents from a defective/leaking container to a new container.
- **Make** sure containers are empty before you discard them. You must observe the approved practices for disposal of solvents, cleaners, and their containers.
- **Label** all containers used to store solvents.

Student Notes:

- **Read** and comply with all instructions and precautions on the label.

PRECAUTIONS

Always follow safety precautions when working with solvents. **Never** use solvents in an unventilated space under any circumstances. Special clothing requirements also must be observed when using some solvents. **Always follow safety precautions!** Carelessness on anyone's part could cause a mishap, resulting in injuries or even deaths. By observing safety precautions, you will reduce mishaps and save lives.

Ventilation

When you think of ventilation, you usually think of air conditioning and cooling. However, when working with solvents, the term *ventilation* means providing fresh air and exhaust to the area in which you are working. Make sure the work area is properly ventilated. That includes topside areas of a ship because some topside areas are enclosed on three sides and will not allow proper ventilation.

When applying flammable coatings or using solvents, you **must** provide adequate ventilation, which will help prevent accidental ignition. You may have to use extra fans or local exhaust to ventilate a space, especially in spaces where pockets can develop. A *pocket* is the buildup of vapors and poisonous air in an area, causing an explosion. **Always follow safety precautions and make sure spaces are ventilated properly when solvents are used!** When in doubt, contact your supervisor for additional guidance.

Preventing Excess Vapors

Any type of solvent will give off some type of vapor. These vapors may be toxic or flammable. Always use proper ventilation to prevent a buildup of vapors. As you have learned, some vapors can linger in pockets of spaces; therefore, make sure the complete work area is fully ventilated. Before starting a job, ask a gas free engineer to examine the area for toxic gases and ask for the proper ventilation plan for the space. Be sure to have the space checked frequently for excessive vapors. If vapors are found to be excessive, stop all work immediately and have all personnel clear the area until it is safe to return.

Protective Clothing

When working with solvents, you always face the risk of their contacting your skin through splatters or spills. Some caustic solvents will actually eat the skin off your body. Make sure that you have all the protective clothing needed for the job.

When working with solvents, you **must** wear adequate protective clothing and gloves to prevent skin contact with the solvents and cleaning materials. **Do not** wear jewelry or clothing with cuffs, loose pockets, rips, or loose ties. Observe the following safety precautions when working with solvents:

- Wear chemical splash goggles at all times.
- Wear acid-resistant aprons, face shield with goggles, gloves, and boots when handling acid or caustic cleaners.
- Wear nonskid rubber-soled shoes when working in enclosed spaces or when flammable vapors may be present (spark prevention).
- Never work in an enclosed space without using the buddy system.
- Respiratory protection, with either an organic vapor cartridge or supplied air, should be worn when dispensing, handling, or cleaning using solvents.

Using solvents for cleaning saves time; but, make sure you read all the labels before using the solvent. Many solvents are corrosive and can irritate or cause serious injuries to your eyes, skin, and lungs. **Always check the caution labels before using any solvents!**

Fire-Fighting Equipment Required While Using Solvents

Nothing ruins a CO's day faster than receiving word that the ship is burning. A fire can cause injury and loss of life **and** take a ship off the line for a long time. When working with solvents, you have no room for error. If you're on a work detail that requires the use of solvents

Student Notes:

or solvent-based paint, make sure the proper fire-fighting equipment is located close to the work area. One little spark can set the vapors of some solvents into a roaring fire that can take life and destroy a ship. Proper equipment may include fire extinguishers, charged fire hoses, or foam. You always need to be prepared. An ounce of prevention goes a long way. Ask your supervisor to check the type of fire-fighting equipment you are going to use to see if more equipment or some other type is needed for the job at hand.

While working with people using solvents or solvent-based paint, make sure you know the location of the nearest fire alarm. Also, make sure all the people working know the nearest fire escape route. Always notify damage control central (DCC) when you are using flammable materials.

Wiping Up Solvent Spills

When using solvents, be careful not to spill them on the deck or get them on anything except what you are cleaning. Solvents may cause paint to bubble and peel off surfaces. The corrosive nature of some solvents can damage equipment. When mixed with some tile compositions, solvents can form toxic vapors that can irritate your lungs and make you sick. If you spill solvent, clean it up as soon as possible. If you think the spill has caused some type of damage, contact your supervisor for guidance.

When a spill involves more than 5 gallons of solvent or presents a threat to the ship or the health of the crew, report it immediately to your supervisor, DCC, or the OOD. Each ship has a hazardous material response kit to handle such emergencies. Spilled material and contaminated clothing or rags become hazardous waste and must be treated as hazardous material (HAZMAT). Your supervisor will tell you the proper disposal procedures for your command.

Dangers

Working with solvents is dangerous. Avoid inhaling vapors. Personnel with a history of chronic skin disease, allergies, or asthma should not be permitted to work with paint, solvents, and thinners.

When you handle a solvent, don't let it contact your skin. If a solvent does contact your skin, flush it with clear water as soon as possible. If solvent contacts your skin or eyes, report to the nearest medical facility as soon as possible for treatment.

When working with solvents, wear an approved respirator and protective clothing at all times. If you think that your respirator isn't working properly, request an air line mask. The safety department of your ship usually provides these items.

If you breathe some of the vapors given off by solvents, get to a doctor as soon as possible.

Respirators

The National Institute for Occupational Safety and Health/Mine Safety and Health Administration (NIOSH/MSHA) must approve all respirators and pumps. Users must be medically qualified and fit-tested before wearing a respirator. The following text describes the air-purifying respirators and air-supplied or self-contained breathing apparatus (SCBA) approved for use by the Navy:

- Air-purifying respirators use a filter, a chemical cartridge, or a combination of the two to remove air contamination. Filters capture particles of dust or metal fumes. The cartridges may contain a chemical or carbon to absorb vapors or gases. A combination of filter and cartridge is used for a combination of hazards, such as spray painting. The filter captures the spray mist and the cartridge absorbs the paint vapors, protecting the wearer.

- Air-supplied or self-contained breathing apparatus (SCBA) provides fresh air when the vapor or gas concentration is too high or the area lacks oxygen. Air-supplied or SCBA is required for all internal shipboard spray painting operations. Air for supplied air masks is provided by certified breathing air compressors or breathing air pumps.

The use of proper equipment may save your life and the lives of your shipmates. If you are in doubt about the type of equipment to use, be sure to check with your supervisor.

Student Notes:

Keep Solvent Containers Tightly Closed

Most fires in paint and solvent storage areas are caused by a buildup of vapors. Usually, vapors escape from containers that are not closed tightly. It only takes a small spark to ignite these vapors. Since vapors can quickly displace the oxygen in a storage space, there may not be enough oxygen left to sustain life.

All containers **must** be tightly closed when not in use. Besides the danger of vapors accumulating, air can cause a chemical breakdown of some solvents. After a short time, the solvent may evaporate or decay to the point that it can't be used.

Check Containers for Leakage

The Navy uses many types of corrosive materials that can eat through a container. You must make sure this doesn't happen. How can you do that? **Don't accept containers until they are inspected!** If you are inspecting the containers, check all the seams carefully for leaks or cracks. Check the sides of the containers for dents. If a container is dented, that means the side of the container may have been weakened and will eventually leak. Don't sign for material in damaged containers. If you are unsure of the condition of a container, ask your supervisor to inspect it.

If you discover a leaking container while inspecting your storage areas, find the name of the material or solvent on the label of the container. Then immediately inform your supervisor of the problem. The material or solvent in the container may be caustic and highly flammable. You and your supervisor should inspect the damaged container and the surrounding area. Then the material or solvent should be transferred to another container using a standard Navy transfer pump. This container must be equal to or surpass the storage requirements of the damaged container.

Ensure Containers Are Empty Before Discarding

One of the most dangerous practices is to discard (throw away) a container partially filled with a solvent or some type of caustic or flammable substance. For example, several years ago a container of highly flammable liquid was discarded into a dumpster in Charleston, South Carolina. The dumpster was taken to the trash disposal area where the driver, thinking that the dumpster only contained burnable trash, dumped the contents of the dumpster into a small fire. As the flammable liquid drained from the container, it ran into the fire and created an explosion that was heard for miles around. The dumpster and the truck were destroyed, and the driver was killed instantly.

Solvent containers are considered hazardous waste, so you **must** dispose of them according to local hazmat regulations. When at sea, **never** throw solvent containers over the side; they contaminate the seas. Stow containers in a disposal storage area until you reach your next port of call and then have them disposed of in the proper manner.

Label All Containers

On board naval ships, paints and solvents are stored in a storeroom designated for flammable liquids. The storeroom should be neat, clean, cool, and dry. Make sure a label appears on the door of the space to show the space contains flammable liquids. Store paints or solvents in tightly sealed cans or containers. Mark the container with the name, formula number, solvent composition, Navy hazard identification label or Department of Transportation hazard identification label, and manufacture date of the paint or solvent it contains.

Inspect the contents of any paint or solvent container more than 2 years old. If the container is unfit, properly dispose of it. If you're not sure whether the paint or solvent is usable (particularly large quantities), send samples to the nearest laboratory for testing.

Working in Closed compartments

Tests are performed with a combustible gas indicator (explosimeter), toxic gas detectors, and an oxygen indicator. Personnel who test a space are required to wear an oxygen-breathing apparatus (OBA) or air-line mask. If the atmosphere is found unsafe, the space is thoroughly ventilated and provided with adequate forced fresh air circulation. Only after the space has been retested and pronounced safe to enter can personnel without an OBA or air-line mask enter it. After personnel (other than testing personnel) have

Student Notes:

entered a declared safe space, periodic tests are made to determine that it is still safe. Upon the detection of an unsafe condition, an order must be given for all personnel to evacuate the space.

Because a space cannot be guaranteed to remain safe, you should be aware of the symptoms of bad air. Symptoms of bad air include the following:

- Labored breathing
- Excessive fatigue from slight exertion
- Headache
- Dizziness

If you feel any of these symptoms, warn others and get to fresh air immediately.

A more dangerous situation exists if a compartment has no, or very little, oxygen. If this happens, a person can lose consciousness almost immediately without warning. If such an incident occurs while you are in an area, **do not** enter the space without wearing an OBA or air-line mask; otherwise, you will become a casualty. Always summon (call for) help before making a rescue attempt. Also, have a person stationed at the entrance to maintain communications while watching to see that you are not overcome.

TYPES OF SOLVENTS

As you have already learned, the Navy uses many types of solvents for many cleaning assignments. You also know that many of these solvents are highly toxic and some are highly flammable. Take special care when using many of these solvents; make sure you store them in cool, dry areas. Material Safety Data Sheets (MSDSs) list the storage requirements for solvents. You should refer to the MSDS for solvents you are using.

Most cleaning solvents contain toxic substances. These substances can cause injuries if they are inhaled, absorbed by the skin, or ingested. All toxic materials must be handled carefully to prevent injury. Many of them have additional hazards, such as flammability. The following paragraphs contain information about general categories of toxic cleaning solvents. If you have any questions about the solvent you are going to use, check the Maintenance Requirement Cards (MRCs) for the task or ask your supervisor.

The three types of solvents covered in this section are chlorinated cleaning solvents, organic cleaning solvents, and fluorocarbon refrigerants and solvents.

Chlorinated Cleaning Solvents

Chlorinated cleaning solvents can be highly toxic if used improperly. They may be irritating to skin and toxic if ingested. In confined spaces, in spaces with inadequate ventilation, or when the vapor concentration is increased by heating, toxic vapors may cause damage to the lungs, eyes, and nervous system. Solvents decompose at high temperatures and produce gases more toxic than the solvents themselves. Solvents react with alkalies, oxidizers, and powdered metals to produce toxic gases.

Common types of chlorinated cleaning solvents are trichloroethane (inhibited methyl chloroform), trichloroethylene, tetrachloroethane, and tetrachloroethylene (perchloroethylene, dry-cleaning solvent). Because of the extreme dangers involved, the Navy severely restricts the use of these solvents.

You should observe the following precautions when working with chlorinated cleaning solvents:

- Never stow chlorinated cleaning solvents near heat sources or open flames.
- Don't allow them to come in contact with hot surfaces.
- Make sure stowage areas are well ventilated and monitored regularly by the gas free engineer. Don't stow these solvents near incompatible materials. (**NOTE**: Incompatible materials include strong alkalies, such as sodium hydroxide; oxidizers, such as calcium hypochlorite and sodium nitrate; or powdered metals, such as aluminum.)

When handling chlorinated cleaning solvents, wear the following personal protective equipment (PPE):

- Neoprene gloves

Student Notes:

- Safety goggles that will protect against splashes, or a face shield
- A chemical cartridge respirator for protection against small amounts of organic vapors or for protection for a short duration; or an air-line respirator (or some other type of supplied-air respirator) if use is extensive or in a confined space
- Coveralls

Make sure work areas in which you use chlorinated cleaning solvents have proper ventilation. For enclosed spaces, an air change every 3 minutes is recommended. Consult the gas free engineer to determine if the ventilation is adequate.

Organic Cleaning Solvents

Organic cleaning solvents include the following:

- Toluene
- Xylene
- Some alcohols
- Acetone
- Methyl ethyl ketone
- Ethyl acetate
- Dry-cleaning solvent
- Kerosene
- Petroleum
- Ether
- Turpentine
- Morpholine and other related compounds

These compounds are highly flammable and highly to moderately toxic. Some are also corrosive. Inhalation of concentrated vapors may cause dizziness, nausea, or vomiting.

Organic cleaning solvents should be stowed as follows:

- Stow organic cleaning solvents in a flammable liquid storeroom, ready service storeroom, or a flammable locker.
- Keep them away heat, open flames, or spark-producing devices.
- Stow them away from oxidizers, such as calcium hypochlorite, sodium nitrate, and hydrogen peroxide.

When handling organic cleaning solvents, wear the following PPE:

- Neoprene gloves
- Safety splash goggles
- Protective coveralls (recommended)

In addition, if vapors accumulate over 100 parts per million (ppm), wear an OBA and notify the gas free engineer.

Fluorocarbon Refrigerants and Solvents

Fluorocarbon refrigerants and fluorocarbon solvents, such as trichlorotrifluoroethane (Freon 113, Freon TF, Genetron 113, R-113), are commonly found aboard ship. They are used in food storage compartments and air-conditioned spaces and as solvents in engineering spaces.

Fluorocarbon vapors have the following characteristics:

- They are colorless and almost odorless.
- They cannot be detected without special instruments.
- They are nonflammable and nonexplosive; however, exposure to flames or hot surfaces will cause these compounds to generate hydrogen chloride, hydrogen fluoride, and other poisonous gases.
- They aren't irritating, but contact may cause frostbite.

Operations involving trichlorotrifluoroethane (Freon 113) are considered hazardous. An industrial hygienist or a gas free engineer must evaluate and approve these operations to ensure the work process meets safety requirements.

Student Notes:

When exposed to the atmosphere, fluorocarbon vapors will accumulate in low spaces unless local ventilation is provided. Since these vapors are heavier than air, they can displace oxygen. Inhaling vapors at high concentrations (4,500 ppm or greater) will cause dizziness or narcosis. If fluorocarbon vapors displace oxygen, suffocation occurs.

When handling fluorocarbon refrigerants and solvents, wear the following PPE:

- Rubber gloves
- Safety splash goggles
- Protective clothing

In addition, if vapors accumulate over 1,000 ppm, wear an OBA or air-line respirator; and notify the gas free engineer.

REVIEW 2 QUESTIONS

Q1. When you are working with solvents, what does the term *ventilation* mean?

Q2. List the protective equipment you should wear when handling acid or caustic cleaners.

a.

b.

c.

d.

Q3. How must the material and rags used to clean up a solvent spill be treated?

Q4. List two types of respirators used by Sailors when handling solvents.

a.

b.

Q5. What causes most fires in paint and solvent storage areas?

Q6. List the symptoms personnel might have when working in a compartment having bad air?

a.

b.

c.

d.

PAINTING AND PRESERVATION

Learning Objectives: When you finish this chapter, you will be able to—

- Identify the equipment and procedures used for surface preparation.
- Identify types of paint and recognize their use.
- Recall fixtures, devices, and surfaces that should not be painted.
- Recall painting safety precautions.
- Recall methods used when painting to include care of brushes and rollers.

The Navy uses from 25 to 30 million gallons of all types of paint a year. Roughly 20 million gallons are used for preservation, some of which you will apply. Paintbrush purchases also run into millions of dollars. It is no exaggeration to state that millions of man-hours a year are expended in cleaning, chipping, and painting.

To paint a ship's exterior with one coat takes 20 gallons on a tugboat, 50 gallons on a submarine, and as much as 950 gallons on a carrier. The average basic

Student Notes:

paint requirements for preservation of a destroyer every 60 to 70 days are 270 gallons. All of this is a way of saying the Navy uses a lot of paint. The more attention you pay to the basic instructions, the less paint you will have to use.

The Navy uses paint primarily to preserve surfaces. It seals the pores of steel and other materials, prevents decay, and arrests (stops) rust and corrosion. Paint also serves several other purposes. It is valuable as an aid to cleanliness and sanitation because of its antiseptic properties and because it provides a smooth, washable surface. Paint is also used to reflect, absorb, or redistribute light. For example, light-colored paint on a ship's interior distributes natural and artificial light to its best advantage.

Learning to paint properly requires the selection of suitable paints for the surfaces to be covered, the proper preparation of the surfaces before painting, and the correct methods of applying paint. Though the selection of suitable paints won't concern you now, you should know how to prepare the surface and how to apply paint with a brush and roller. Improper surface preparation and paint application, in that order, are the greatest reasons for paint failure.

PREPARING THE SURFACE

For paint to stick to a surface, all salt, dirt, oil, grease, rust, and loose paint must be removed completely, and the surface must be thoroughly dry. Salt and most dirt can be removed with soap or detergent and freshwater. Firmly imbedded dirt may require scrubbing with scouring powder. When scrubbing won't remove oil and grease, they may be removed with paint thinner or other approved solvents. After scrubbing or scouring, always rinse the surface with freshwater.

Equipment and Procedures

The removal of rust, scale, and loose paint requires the use of hand tools or power tools, paint and varnish removers. Hand tools are usually used to clean small areas. Power tools are used to clean larger areas and for completely cleaning decks, bulkheads, and overheads covered with too many coats of paint. Paint and varnish removers are used to remove paint from wood.

HAND TOOLS.—The most commonly used hand tools are sandpaper, wire brushes, and hand scrapers.

Sandpaper.—Use sandpaper to clean corners and to feather paint. (**NOTE**: To feather paint, you taper the edges of chipped areas down to the cleaned surface so that no rough edges remain.) Paint will bond best to a clean surface that has been lightly sanded.

Sandpaper is graded from 12 to 600, which corresponds to the size of the abrasive grit on its surface. For example, the coarsest sandpaper is 12 grit and the finest is 600 grit. Very fine emery (a natural abrasive) paper is sometimes used to polish unpainted steel surfaces. However, **never** use abrasives, such as sandpaper, on unpainted galvanized metal (brass, copper, nickel, or aluminum) surfaces.

Hand Wire Brush.—A hand wire brush is a handy tool for light work on rust or on light coats of paint. You can also use hand wire brushes for brushing weld spots and cleaning pitted surfaces.

Hand Scraper.—Hand scrapers are made of tool steel. The most common type is L-shaped, with each end tapered to a cutting edge like a wood chisel. Hand scrapers are useful for removing rust and paint from small areas and from plating less than 1/4 of an inch thick when it's impractical or impossible to use power tools.

Chipping (Scaling) Hammer.—Occasionally, it's necessary to use a chipping or scaling hammer. However, take care to use only enough force to remove the paint. Too much force dents the metal, resulting in high and low areas. In painting, the paint naturally is thinner on the high areas. Therefore, if you leave high and low spots, rust will form on the high spots and, in time, spread under the good paint.

PORTABLE POWER TOOLS.—The most useful power tool is the portable grinder (fig. 18-1). Portable grinders are usually equipped with a grinding wheel that may be replaced by either the rotary wheel wire brush or the rotary cup wire brush. Light-duty brushes are made of crimped wire. Use them to remove light rust. Heavy-duty brushes are made by the twisting of several wires into tufts. Use them to remove deeply imbedded rust.

Student Notes:

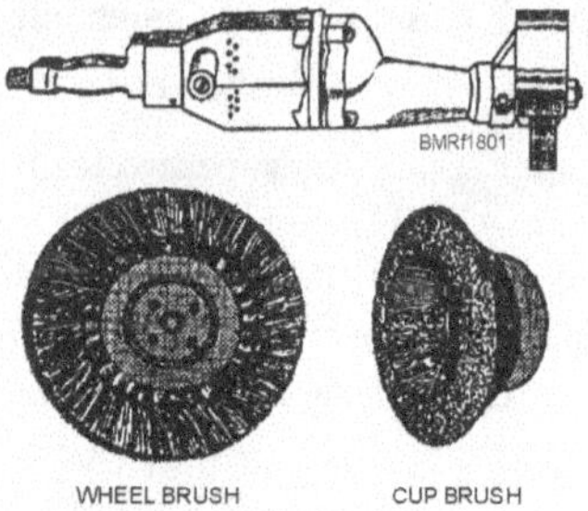

Figure 18-1.—Portable grinder and wire brush.

Scaling is done by using either tool shown in figure 18-2. A chisel is used with the pneumatic hammer and must be held so that the chisel strikes the surface at approximately a 45° angle. As with the hand scaling hammer, take care that you don't dent the surface. The rotary scaling and chipping tool shown in figure 18-2 (called a *deck crawler*) has a bundle of cutters or chippers mounted on either side. As it is pushed along the surface to be scaled, the rotating cutters do the work. This tool is particularly helpful on large deck areas.

The electric disk sander is another useful tool for preparing surfaces. However, it must be used with care. If too much pressure is applied or it is allowed to rest in one place too long, it will quickly cut into the surface, particularly wood and aluminum surfaces.

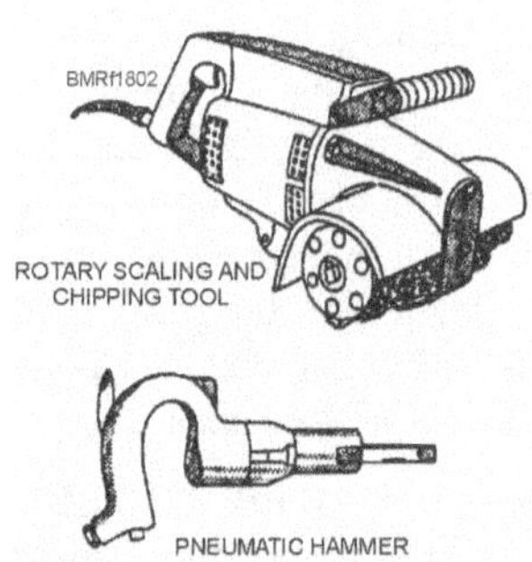

Figure 18-2.—Power scaling tools.

POWER TOOL SAFETY PRECAUTIONS.— You must be trained and qualified before you operate portable power tools. You **must** observe the following safety precautions when working with electrical and pneumatic (air) tools:

- Wear eye and ear protection while chipping, grinding, sanding, or wire brushing. If dust is excessive, also wear a respirator. Do not wear jewelry or loose fitting clothing.
- Do not use defective tools. If you have any doubt about the condition of any tool, show it to your supervisor, who will have its condition determined.
- Make certain that electrical power tools are grounded properly. Every portable electrical power tool must be provided with a ground lead that connects the tool casing to the ship's structure and an up-to-date electrical safety tag.
- Give your full attention to your job.
- Give electricity the respect it is due—115 volts can and does kill.
- Do not operate power tools in areas where flammable vapors, gases, liquids, or exposed explosives are present.
- Do not allow power cords and air hoses to kink or come in contact with oil, grease, hot surfaces, or sharp objects.
- Do not lay power cords and air hoses over ladders, steps, scaffolds, or walkways in such a manner as to cause a trip hazard.
- Do not use compressed air to clean clothing being worn or to blow dust off the body.

PAINT

Paint consists of four essential ingredients:

- Pigment
- Vehicle (known as the *base*)
- Drier
- Thinner

Pigment provides the coloring, rust prevention (in primers), and the lasting quality of the paint. The most common pigments are made of metals, such as lead, zinc, or titanium.

Student Notes:

The vehicle is the liquid portion in a paint. It wets the surface being painted, penetrates into the pores, and ensures adhesion. Until recently, the base of most paints was oil, such as linseed oil, but few paints today contain oils. Some have vehicles of processed oils in combination with synthetic resins; others have vinyl chlorinated bases that are quick drying.

To add to the drying properties of paint, certain metallic compounds, called **driers**, are added to the paint. When mixed with oil, they act as conveyers of oxygen, which they take from the air and add to the oil, speeding up the drying process.

Thinners are used for thinning the paint to the proper degree for spraying, brushing, or rolling. They also increase the penetration of the paint into the surface and cut down the gloss. Too much thinner affects the durability of the paint. The most common type of thinner is made of mineral spirits, but the proper type to use depends on the paint base. **Never use diesel oil or kerosene to thin paint**.

Types of Paint

Paints are of many different kinds, and the Navy constantly works and experiments to improve them. As a result, you are provided the best paints available for the type of surface to be covered. Most Navy paints are named according to color and/or use, such as exterior gray deck and pretreatment coating (primers).

PRIMERS.—Primers are base coats of paint that stick firmly to bare woods and metals, providing a smooth surface for finishing coats. They also serve to seal the pores, and those applied on steel are rust inhibitors as well.

A minimum of two coats of primer should always be used after the surface is cleaned down to the bare metal. A third coat should be added at all outside corners and edges. At least 8 hours of drying time should be allowed between primer coats.

SYNTHETIC PAINTS.—Synthetic resin coatings, such as epoxies, urethanes, and inorganic zinc, are used for areas subject to severe service or exposure, such as bilges, tanks, and decks. The base coating is mixed with a converter (hardener) to cure or harden the paint film.

EXTERIOR PAINTS.—Vertical surfaces above the upper limit of the boot topping (waterline area, painted black) are given two coats of haze gray. Horizontal surfaces are painted with exterior deck gray (darker than haze gray) except the underside of deck overhangs, which are painted white.

A nonskid deck paint is used on main walkways, flight decks, and hangar decks. It contains a small amount of pumice, which helps to give a better footing. Top-hamper areas subject to discoloration from smoke and stack gases and the tops of stacks are painted black.

INTERIOR PAINTS.—Depending on the use of individual compartments, several colors are authorized or prescribed for interior bulkheads, decks, and overheads.

The choice of colors for berthing, messing, and recreation spaces usually is left to the individual ship. All other shipboard spaces are painted the color prescribed by the Naval Sea Systems Command. Deck paint colors, for example, are dark green in the wardroom and officers' quarters, dark red in machinery spaces, and light gray in enlisted personnel living spaces.

Some common bulkhead colors are green for offices, radio rooms, the pilothouse, and medical spaces; gray for the flag plot, the combat information center, and the sonar control room; and white for storerooms and sanitary and commissary spaces. Overhead colors are either the same as the bulkhead or white.

Student Notes:

REVIEW 3 QUESTIONS

Q1. List the most common hand tools used to remove paint and rust from small areas.

a.

b.

c.

Q2. How should you prepare chipped edges of paint to make ready for painting?

Q3. List the two main reasons for a bad paint job.

a.

b.

Q4. List the four main essential ingredients in paint.

a.

b.

c.

d.

Q5. How many coats of primer should be applied to bare metal?

WHAT NOT TO PAINT

Never paint the following items:

- Start-stop mechanisms of electrical safety devices and control switchboards on machinery elevators
- Bell pulls, sheaves, annunciator chains, and other mechanical communications devices
- Composition metal water ends of pumps
- Condenser heads and outside surfaces of condensers made of composition metal
- Sprinkler piping within magazines
- Exposed composition metal parts of any machinery
- Glands, stems, yokes, toggle gear, and all machined external parts of the valves
- Heat exchange surfaces of heating or cooling equipment
- Identification plates
- Joint faces of gaskets and packing surfaces
- Lubricating gear, such as oil holes, oil or grease cups, grease fittings, lubricators, and surfaces in contact with lubricating oil
- Lubricating oil reservoirs
- Machined metal surfaces (working surfaces) of reciprocating engines or pumps
- Metal lagging
- Rods, gears, universal joints, and couplings of valve operating gear
- Rubber elements of isolation mounts
- Ground plates
- Springs
- Strainers
- Threaded parts
- Zincs
- Working surfaces
- Hose and applicator nozzles
- Knife edges; rubber gaskets; dogs; drop bolts; wedges; and operating gear of watertight doors, hatches, and scuttles

Student Notes:

- Electrical contact points and insulators
- The original enamel, lacquer, or crackle finish on all radio, electrical, and sound equipment, unless existing damage makes refinishing essential
- Decorative plastic, such as tabletops

SURFACES TO PAINT

The Navy uses a variety of metal, metal compounds, and synthetic materials to build a ship or boat. Each type of surface requires special preparation and special primers and paint to extend its life cycle. In this section, you will learn about various surfaces and the procedures needed to maintain them properly.

Aluminum Surfaces

Aboard ship, aluminum surfaces are a special problem. If they're not treated properly, corrosion results. Corrosion is greater when dissimilar metals (for example, aluminum and steel) are in contact with each other **and** are exposed to seawater. Seawater is an electrolyte (an electrical conductor). As such, the seawater causes an electrical current to flow between the steel and aluminum surfaces, resulting in galvanic corrosion of the aluminum. The first sign of aluminum corrosion is a white, powdery residue in the area where the two dissimilar metals make contact. Later, the aluminum surface is pitted and scarred. Finally, there is a complete deterioration of the aluminum area. Holes in aluminum plate enlarge and screws, bolts, or rivets pull out, or they may even disintegrate.

Before joining aluminum to another metal, give each surface a pretreatment formula and two coats of primer formula.

NOTE

Never use red lead as a primer on aluminum.

If the joint is exposed to the weather, use insulation tape between the two surfaces, and fill the joint with caulking compound. When aluminum is joined to wood, give the wood one coat of phenolic varnish. Replace any missing fasteners (screws, bolts, rivets, and so on) with items of the original type. (**NOTE**: Replacements of stainless or galvanized steel may be used.) When painted, the best way to prepare the aluminum surface for repainting is to use hand scrapers, hand and power wire brushes, or fine grit sandpaper. Be careful if you use a power sander to prepare the aluminum surface for repainting.

NOTE

Never use scaling hammers on aluminum.

Steel Surfaces

When painting a steel surface, preparation of the surface is important. Steel surfaces **must** be completely free of rust, loose paint, dirt, scale, oil, grease, salt deposits, and moisture before they are painted. Old paint in good condition is an excellent base for repainting. Smooth, thoroughly clean, and dry the surface before applying new paint.

In touch-up painting (when only small areas or spots need repainting), remove old paint to the edges of the spot or area until an area of completely intact paint is reached. (**NOTE**: This area must be free of rust or blisters underneath the paint.) Feather the edges of the remaining paint.

When completely reworking an old painted surface, take the old paint down to the bare metal. Then apply a primer before painting. Never leave a base metal surface exposed overnight. Always put on a primer coat before you secure for the day.

Fillers

Fill holes, dents, and cracks in all surfaces and open-grained woods before they are finished. Putty, wood fillers, and even sawdust mixed with glue can be used on wood. Use epoxy fillers on steel and aluminum surfaces. The method you use varies with the type of filler. Therefore, follow the instructions carefully. Allow all fillers to dry and then sand them smooth before you apply the first finishing coat.

Paint and Varnish Removers

Paint and varnish removers are most often used on wood surfaces. However, you can use paint and varnish

Student Notes:

on metal surfaces that are too thin to be chipped or wire brushed. The three types of removers generally used are flammable, nonflammable, and water-base alkali. They are hazardous materials, and you must strictly observe safety precautions when you use them. Use these removers only in well-ventilated spaces. Don't use the alkali type on aluminum or zinc because of its corrosive properties.

The procedures you follow when using paint and varnish removers are the same regardless of type. Wet the surface with a smooth coat of the remover and let it soak thoroughly until the paint or varnish is loosened. Then lift the paint off with a hand scraper. After the surface is cleaned, wet it again with the remover and wipe it off with a rag. Finally, wash the surface thoroughly with paint thinner or soap and water. The final rinse gets rid of any wax left by the remover and any acids that may have worked into the grain of the wood.

Paint and Varnish Remover Safety Precautions

The following safety precautions should be observed when you use paint and varnish removers:

- Never use paint and varnish removers around an open flame. Some types are highly flammable.
- Do not use removers in confined spaces because their dangerous anesthetic or toxic properties can kill or cause injury if you are exposed to them for long periods.
- Do not use paint or varnish removers if you have an open cut or sore on your hand **unless** you wear rubber gloves.
- Do not let the remover touch your skin; watch out particularly for your face, eyes, and mouth. If paint or varnish remover should come in contact with the skin, wash it off immediately with cold water; seek medical attention as soon as possible if it gets into your eyes or mouth.
- Never use turpentine or mineral spirits as hand cleaners because they are absorbed through the skin pores. Gasoline also is dangerous and must never be used. To clean paint or varnish remover from your hands, use soap and water only.

PAINTING SAFETY PRECAUTIONS

Painting can be dangerous if one is careless. Many paints are highly flammable, others are poisonous, and some are both flammable and poisonous. To increase your chances of remaining alive and healthy, observe the following precautions:

- Keep paint off your skin as much as possible. Wash your hands, arms, and face with soap and warm water before eating. Do not put your fingers, food, or cigarettes in your mouth if they are contaminated with paint.
- Be sure you have adequate ventilation, and wear an approved paint/spray respirator whenever there is reason to believe the ventilation is inadequate in the place you are painting. At the first sign of dizziness, leave the space and get to fresh air.
- Do not smoke, use an open flame, or use spark-producing tools in the vicinity of painting operations.
- Use only explosion-proof lights near painting operations.
- Do not wear nylon, orlon, or plastic clothing or covering. These materials generate static electricity, which may spark and ignite paint vapors.
- Do not carry matches or cigarette lighters or wear steel buckles or metal shoe plates. Too often one forgets and strikes a match or lights a cigarette lighter in areas filled with explosive vapors. Also, steel buttons, buckles, and tabs can strike sparks that are invisible to your eyes but are capable of igniting paint vapors.
- When pouring solvents, make sure the containers are touching each other to prevent sparks.
- Never paint during electrical storms.
- Keep food and drink away from areas being painted.

Student Notes:

- Do not use gasoline, turpentine, mineral spirits, or other solvents to remove paint from the skin, as the skin will absorb them.
- Follow the instructions of your supervisor carefully.

PAINT ISSUE

Before paint is issued, several events must occur.

1. The division petty officer inspects the area to make sure all preparations have been made. The petty officer will check for the following:

   - Are all items not to be painted properly identified or masked?
   - Are all safety precautions understood and properly observed?
   - Is the surface ready to be painted?

2. Having checked out these items, you must fill out the paint request; and describe the area to be painted, including the paint color, type, and approximately how much paint is needed. Completing the paint request chit reduces the waste of materials and time spent redoing a paint job. Your division officer may also inspect the area to be painted before signing the paint chit.

3. The next step is the approval of the request by the first lieutenant, who regulates the issue of paint.

As you can see, sometimes getting ready to paint takes longer than the actual painting. Remember, if you spill paint (oil, grease, and so on), you are responsible for cleaning it up. At the end of working hours, return all paint and brushes to the paint locker. Store the paint in its proper container, and clean all brushes and rollers.

PAINTING

Three means of applying paint are used in the Navy—brush, roller, and spray. The majority of Sailors don't use paint sprayers; therefore, they aren't covered in this section. However, you will learn about using brushes and rollers to apply paint. Everyone in the Navy should be familiar with these items.

Paint Application by Brush

Smooth and even painting depends as much on good brushwork as on good paint. There is a brush for almost every purpose. You should use the proper brush and keep it in the best condition.

The two most useful brushes are the flat brush and the sash tool brush. These brushes and some others commonly used aboard ship are shown in figure 18-3. With a flat brush, you can paint almost anything aboard ship. Flat brushes are wide and thick. They carry a large quantity of paint and provide a maximum of brushing action. Sash brushes are handy for painting small items, for cutting in at corners, and for hard-to-get-at spaces. The fitch brush also is useful for small surfaces. The painter's dusting brush is used for cleaning surfaces.

The following are hints to help you use a paintbrush properly:

- Grip the brush firmly, but lightly as shown in figure 18-4. Don't put your fingers on the bristles below the metal band (ferrule). The grip shown permits easy wrist and arm motion. To hold it otherwise restricts your movements and causes undue fatigue.

- When using a flat brush, don't paint with the narrow edge. This practice wears down the corners and spoils the shape and efficiency of the brush. When using an oval brush, don't revolve it too much or it soon wears to a pointed shape and becomes useless. Do not poke oversized brushes into corners and around moldings. Such a practice bends the bristles, eventually ruining a good brush. Use a smaller brush that fits into such odd spots.

- Dip the brush into the paint, but not over halfway up the bristles. Remove the excess paint by patting the brush on the inside of the pot. (Avoid overfilling the

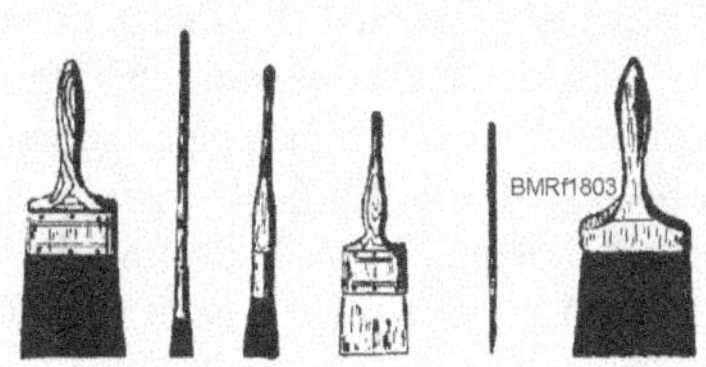

Figure 18-3.—Types of brushes.

Student Notes:

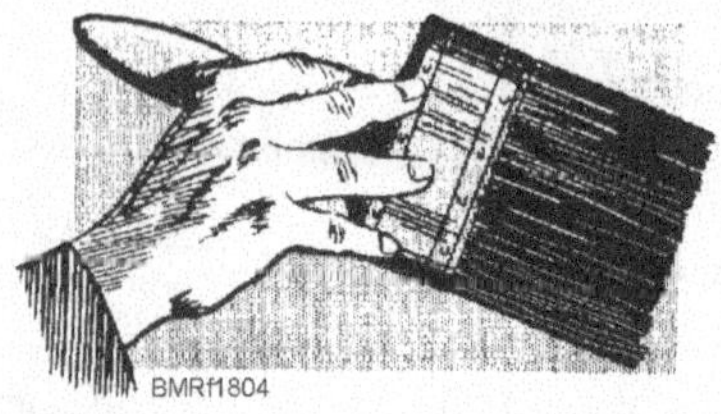

Figure 18-4.—Correct way to hold a brush.

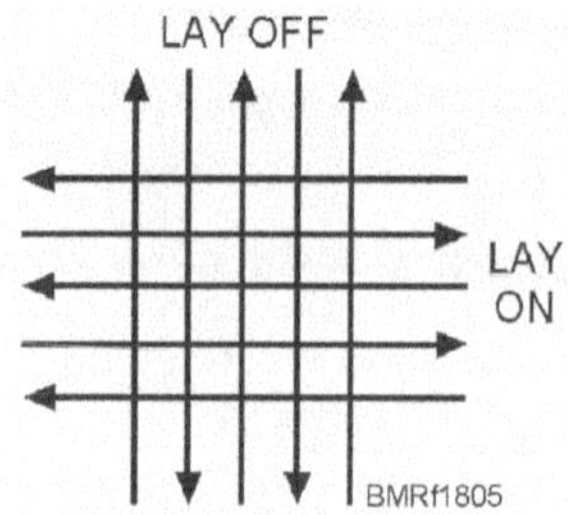

Figure 18-5.—Laying on and laying off.

brush; otherwise, paint will drip on the deck or other surfaces and run down the handle.)

- Hold the brush at right angles to the surface being painted, with the ends of the bristles just touching the surface. Lift the brush clear off the surface when starting the return stroke. If the brush is not held correctly and is not lifted, the painted surface will be uneven, showing laps and spots and a daubed appearance. Also, a brush that is held at any angle other than a right angle will soon wear away at the ends.

For complete and even coverage, follow the Navy method and first lay on, and then lay off. "Laying on," means applying the paint first in long strokes in one direction. "Laying off," means crossing your first strokes. The proper method is shown in figure 18-5. By using the recommended Navy method and crossing your strokes, you can distribute the paint evenly and completely with a minimum amount of paint being used.

Always paint the overhead first, working from the corner that is farthest from the entrance of the compartment. By painting the overhead first, you can wipe drippings off the bulkhead without smearing the bulkhead paint.

When overhead surfaces are being painted, sections should normally be painted in a fore-and-aft direction; beams, in an athwartship direction. But where sections of the overhead contain many pipes running parallel with the beams, it is often difficult to lay off the paint in a fore-and-aft direction. In such situations, better results are obtained by laying off the paint parallel with the beams.

To avoid brush marks when finishing up an area you have painted, use strokes directed toward the last section finished, gradually lifting the brush near the end of the stroke while the brush still is in motion. Every time the brush touches the painted surface at the start of a stroke, it leaves a mark. For this reason, never finish a section by brushing toward the unpainted area. Instead, always end up by brushing back toward the area already painted.

When painting pipes, stanchions, narrow straps, beams, and angles, lay the paint on diagonally, as shown in figure 18-6. Lay off along the long dimension.

Always carry a rag for wiping up dripped or smeared paint. Carefully remove loose bristles sticking to the painted surface.

Cutting In

After you master the art of using a paintbrush properly, learn to cut in. Cutting in is a simple procedure that you can learn in a short time.

Suppose you have to cut in the angle between an overhead and a bulkhead, as shown in figure 18-7. Start at one corner. Hold your brush at an angle of about 76° to 80° from the bulkhead and about 10° from the overhead. Draw your brush along in fairly long, smooth strokes. This is one job where working slowly does not produce better results. The slower you stroke, the wavier your line will be.

Use of Rollers

The type of paint roller (fig. 18-8) used in the Navy is equipped with a replaceable cylinder of soft fabric over a solvent-resistant paper core. It rotates on the shaft of a corrosion-resistant steel frame.

Student Notes:

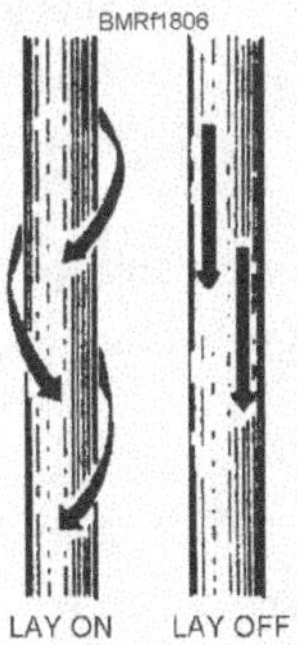

Figure 18-6.—Painting pipes and stanchions.

Figure 18-7.—Cutting in.

Large areas, such as ships' decks and sides (free of rivets, bolts, cable, pipes, and so on), can be covered with paint quickly by the roller method. The paint should be laid on and laid off the same way as when brushes are used. Apply a moderate amount of pressure to the roller to make sure the paint is worked into the surface. If pressure is not applied, the paint doesn't stick and soon peels off. When the paint roller is properly used, it will apply a more even coat and use less paint than with a brush.

CARE OF BRUSHES AND ROLLERS

Unfortunately, too many good paintbrushes and rollers are ruined because painters have little or no idea how to care for them, or they are too lazy to clean them.

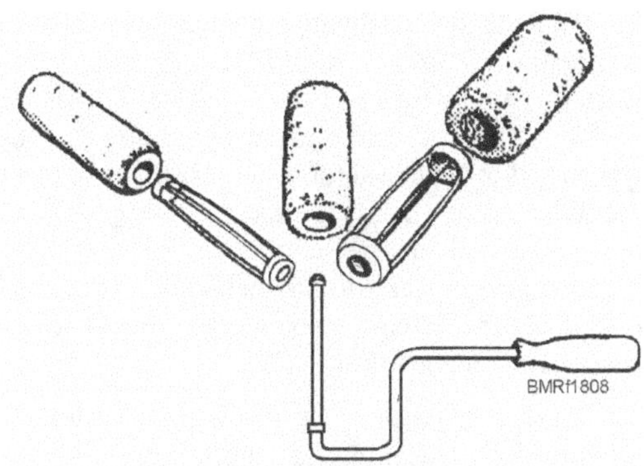

Figure 18-8.—Parts of a paint roller.

To avoid ruining paintbrushes and rollers, pay attention to the following hints. Treat applicators as though you paid for them yourself, and replace them when they no longer are usable.

- **Do not let a brush stand on its bristles in a pot of paint for more than a few minutes**. The weight of the brush bends the bristles, making it almost impossible to do a good job.

- **Never allow paint to dry on a brush**. If you intend to leave a paint-filled brush for long periods, as over the noon hour, fold wax paper or other heavy paper around the bristles and ferrule in such a way that air is kept away from the bristles. Twist the paper around the handle and secure it with rope yarn or sail twine. Cover your pot of paint, and place both it and the brush in a safe place. Before starting to paint again, stir the paint thoroughly with a paddle—not the brush.

- **At the end of the day, clean as much paint from the brush as possible by wiping it across the edge of the paint pot or mixing paddle.** Then turn in your paint and brush to the paint locker.

Ordinarily, the person or persons working in the paint locker will clean and stow the brushes turned in. Occasionally, though, they require help; and you may be detailed to the job. If so, follow instructions carefully; and do a thorough job of cleaning the brushes.

Student Notes:

Paint lockers usually have containers with divided compartments for stowing different types of brushes (that is, paint, varnish, shellac, and so on) for short periods of time. These containers normally have tight covers and are equipped for hanging brushes so that the entire length of the bristles and the lower part of the ferrule are covered by the solvent or cleaner oil kept in the container. Brushes are suspended so that the bristles don't touch the bottom, preventing them from becoming permanently misshapen.

Brushes to be used the following day should be cleaned in the proper cleaner and placed in the proper compartment of the container. Those not to be used again soon should be cleaned, washed in soap or detergent and water, and hung to dry. After drying, they should be wrapped in heavy paper and stowed flat. Do not leave a brush soaking in water. Water causes the bristles to separate into bunches, flare, and become bushy.

The proper cleaners for paint applicators are shown below:

PAINT/FINISH	SOLVENT/CLEANER
Natural and synthetic oil-base paints and varnishes; chlorinated alkyd paints	Turpentine or mineral spirits
Latex emulsion paints	Water
Chlorinated rubber paints	Synthetic enamel thinner xylene
Shellac	Alcohol (denatured)
Lacquer	Lacquer thinner

Paint rollers are cleaned in a different fashion. After use, the fabric cylinder is stripped from the frame, washed in the cleaner recommended for the paint used, washed in soap and water, rinsed thoroughly in fresh water, and replaced on the frame to dry. Combing the pile of the fabric while it is damp prevents matting.

REVIEW 4 QUESTIONS

Q1. What is the first sign of aluminum corrosion?

Q2. True or False. Old paint in good condition makes an excellent base for repainting.

Q3. For painting small areas and cutting into corners, what type of paintbrush is best?

Q4. What method of painting does the Navy use to completely and evenly cover an area?

SUMMARY

We live close together aboard ship. The daily routine of cleaning the berthing compartment and head areas is not only beneficial for our own welfare but for our shipmates as well. It also makes those long cruises easier if we take the time to make our living spaces as pleasant as possible. The occasional zone inspection will help in keeping all our spaces up to speed. Looking for problems that exist, or ones that could arise in the future, will benefit us all.

We also discussed some of the more important aspects of surface preservation. Most of our ships serve for over 20 years, and in the case of carriers, over 30 years. That is testimony to how well the Navy cares for its ships. This care would not be possible without personnel having the proper equipment and materials, being properly trained in the correct application of these materials, and taking pride in doing a good job. Anyone can paint, but taking that extra step to ensure the assigned job is completed with the best possible results is the difference in a job that really looks sharp and one that just gets by.

REVIEW 1 ANSWERS

A1. To find the responsibilities for cleaning and maintaining spaces, you should refer to the **Cleaning, Preservation, and Maintenance Bill**.

A2. The **compartment cleaner** is responsible for cleaning living and berthing spaces

Student Notes:

A3. You should pick up and stow gear that has been left adrift. This **reduces tripping and fire hazards and keeps dewatering equipment from clogging.**

A4. **False**. You should not sweep dirt and debris over the side.

A5. The three steps used in most detergent cleaning are

a. **Wetting**

b. **Scrubbing**

c. **Rinsing**

A6. The two types of deck covers are—

a. **Resilient**

b. **Nonslip**

REVIEW 2 ANSWERS

A1. When working with solvents, the term *ventilation* means **fresh air moving in and through the space with proper exhaust.**

A2. When handling acid or caustic cleaners, you should wear the following protective equipment:

a. **Acid-resistant apron**

b. **Face shield with goggles**

c. **Gloves**

d. **Boots**

A3. Treat material and rags used to clean up a solvent spill as **HAZMAT material**.

A4. The two types of respirators used by Sailors when handling solvents—

a. **Air-purifying**

b. **Air-supplied**

A5. Most fires in paint and solvent storage areas are caused by **vapor buildup**.

A6. Personnel who work in a compartment having bad air might have one or all of the following symptoms.

a. **Dizziness**

b. **Headache**

c. **Labored breathing**

d. **Excessive fatigue**

REVIEW 3 ANSWERS

A1. The most common hand tools used to remove paint and rust from small areas are—

a. **Sandpaper**

b. **Wire brush**

c. **Hand scraper**

A2. To prepare chipped edges of paint for painting, you should **feather the edge of chipped paint with sandpaper**.

A3. The two main reasons for a bad paint job are—

a. **Improper surface preparation**

b. **Improper paint application**

A4. The four main essential ingredients in paint are—

a. **Pigment**

b. **Vehicle**

c. **Drier**

d. **Thinner**

A5. **At a minimum, two coats of primer** should be applied to bare metal.

REVIEW 4 ANSWERS

A1. The first sign of aluminum corrosion is a **white, powdery residue**.

A2. **True,** old paint in good condition makes an excellent base for repainting.

A3. When painting small areas and cutting into corners, you should use the **sash tool brush**.

A4. To completely and evenly cover an area, you should use the Navy **laying on and the laying off method**.

CHAPTER 19

SAFETY AND HAZARDOUS MATERIALS

I wish to have no connection with any ship that does not sail fast for I intend to go in harms way.

—John Paul Jones

Naval warships are inherently dangerous. Crowded living conditions, confined working spaces, and long hours, often at night, are just a few reasons why you must use caution at all times. Some evolutions, such as underway replenishment, conducting flight operations, testing weapons systems, or just a change in weather conditions, greatly increase the dangers of being at sea. All Navy ships have a comprehensive shipboard safety program. This program was developed over many years to make life at sea safe. This program is designed to follow established procedures in conducting the day-to-day business aboard ship, and it places special emphasis on observing certain precautions.

The safety program stresses constant awareness of the hazards of being at sea. The word *mishap* is often used in referring to an incident that *just happened.* **Mishaps don't just happen; they are caused**. Most mishaps could have been prevented if the individuals involved had followed established procedures and safety precautions.

Most of the precautions discussed in this chapter are from a shipboard viewpoint, but many of them also apply ashore. Don't depend on memory to remember safety precautions. Almost every task you perform has safety precautions that must be followed. **Get the operator's manual, planned maintenance system (PMS) card, or technical manual and read these precautions**. If you don't understand them or can't find them, ask your supervisor for help. The few minutes you take to read and understand these safety precautions will make your job safer. Don't be one of the casualties reported during a mishap. It's better to be safe than hurt or possibly worse—**dead**!

PERSONAL RESPONSIBILITY

Learning Objective: When you finish this chapter, you will be able to—

- Recognize that safety is a personal responsibility.

Your personal responsibilities for safety are as follows:

- Observe all safety precautions related to your work or duties.

- Report unsafe conditions. Do not walk around a ladder well with missing safety chains and forget it. Report it! If you use a piece of equipment that is damaged, report it!

- Warn others of hazards. If you see someone knowingly, or unknowingly, placing themselves or others in danger, say something. If that particular person will not listen, tell your supervisor.

- Protective equipment and clothing is issued to you for a purpose—use them.

- Wear eye and/or full-face protection. It's hard to explain to the chief that you had to go to sick bay to get something removed from your eye when you were given a full-face shield before you started working.

- Report all injuries or illnesses. If you should become injured or feel sick, tell your supervisor. A little scratch could become infected or your illness could be a sign of something more serious. A little time having the corpsman check you now is better than being in the hospital later.

- Remain alert. Look for any possibilities of danger. Be safety conscious.

- Don't rush into a job. Look at what you are supposed to do. Is the equipment you have suited to the job? Check the safety precautions for the equipment you were issued. Is the equipment in good condition?

A shipboard environment introduces factors affecting safety that are not found ashore. Danger exists in every naval operation and aboard every naval vessel. Going to sea involves working with powerful machinery; high-speed equipment; high-temperature, high-pressure steam; volatile fuels and propellants;

heavy lifts; high explosives; stepped-up electrical voltages; and the unpredictable forces of wind and waves.

Underway refueling, multiship exercises, storms, and other situations require personnel at sea to be constantly alert. A mishap (there's that word again) at sea can involve all hands in a matter of seconds. Therefore, you must be continually alert to hazardous conditions. If you observe unsafe practices or conditions, report them to your supervisors.

REVIEW 1 QUESTIONS

Q1. List some of the safety precautions that could save you and your shipmate's life.

a.

b.

c.

d.

e.

Q2. What are some shipboard environments that are dangerous?

a.

b.

c.

d.

e.

f.

g.

h.

SAFETY PRECAUTIONS AND HAZARDS TO SAFETY

Learning Objectives: When you finish this chapter, you will be able to—

- Recall the purpose and use of Material Safety Data Sheets (MSDS).
- Recognize safety precautions when you are embarked in a small boat.
- Recognize the purpose of safety precautions when working around various equipment and working in spaces to include the following: steam; lifelines, ladders, and scaffolding; heavy weight and moving equipment; personnel aloft or over the side; antennas; flammable liquids, paints, and solvents; weapons, ammunition, and explosives; electrical and electronic equipment; compressed gases; fiber glass and asbestos; power tools; cutting and welding operations; liquids under pressure; rotating machinery; marine sanitation systems; high noise levels; lifting objects; shipyards and docks; aircraft and flight deck operations; when involved in sporting and recreational events; and operating motor vehicles.

The safety precautions and hazards discussed are of a general nature only. Following them will help you to avoid injury to yourself and others and to prevent loss of or damage to equipment.

MATERIAL SAFETY DATA SHEET (MSDS)

Material Safety Data Sheets (MSDS) are technical bulletins that contain information about hazardous material (figs. 19-1 and 19-2). Manufacturers create MSDSs based on their testing and research of their products. **By law, manufacturers must provide the data to hazardous material users**. They tell users how to use, store, and dispose of hazardous material. According to OPNAVINST 5100.19, all hands are required to follow these guidelines. MSDSs must be in English and contain at least the following information about the material:

- Identity

Student Notes:

BMR1901

DOD Hazardous Materials Information System
DOD 6050.5 5-LR
AS OF MAY 1992
For U. S. Government Use Only

FSC: 6505
NIIN: 001538480
Manufacturer's CAGE: 1BF21
Part No. Indicator: A
Part Number/Trade Name: HYDROGEN PEROXIDE SOLUTION 3%

General Information

Item Name: HYDROGEN PEROXIDE TOPICAL SOLUTION, USP
Manufacturer's Name: MALLINCRODT INC
Manufacturer's Street:
Manufacturer's P.O. Box: M
Manufacturer's City: PARIS
Manufacturer's State: KY
Manufacturer's Country: US
Manufacturer's Zip Code: 40361-0315
Manufacturer's Emerg Ph #: 314-982-5000
Manufacturer's Info Ph #: 314-982-5000
Distributor/Vender # 1: RUGER CHEMICAL CO. INC.
Distributor/Vender # 1 Cage: 64220
Distributor/Vender # 2:
Distributor/Vender # 2 Cage:
Distributor/Vender # 3:
Distributor/Vender # 3 Cage:
Distributor/Vender # 4:
Distributor/Vender # 4 Cage:
Safety Data Action Code:
Safety Focal Point: D
Record No. For Safety Entry: 001
Tot Safety Entries This Stk#: 006
Status: SE
Date MSDS Prepared: 21JAN86
Safety Data Review Date: 05DEC90
Supply Item Manager: KX
MSDS Preparer's Name:
Preparer's Company:
Preparer's St Or P.O. Box:
Preparer's City:
Preparer's State:
Preparer's Zip Code:
Other MSDS Number:
MSDS Serial Number: BJFCN
Specification Number:
Spec Type, Grade, Class:
Hazard Characteristic Code: P2
Unit Of Issue: BT
Unit Of Issue Container Qty: 1 PT
Type Of Container: BOTTLE
Net Unit Weight: 1.16 LBS
NRC/State License Number:
Net Explosive Weight:
Net Propellant Weight-Ammo:
Coast Guard Ammunition Code:

Ingredients/Identity Information

Proprietary: NO
Ingredient: HYDROGEN PEROXIDE, SOLUTION 3%
Ingredient: Sequence Number: 01
Percent: UNKNOWN
Ingredient Action Code:
Ingredient Focal Point: D
NIOSH (RTECS) Number: 1005359HP
CAS Number: UNKNOWN
OSHA PEL: UNKNOWN
ACGIH TLV: UNKNOWN
Other Recommended Limit: NONE SPECIFIED

Physical/Chemical Characteristics

Appearance and Odor:; CLEAR, COLORLRSS SOLUTION, ODORLESS.
Boiling Point: 212F,100C
Melting Point: 32.0F,0.0C
Vapor Pressure (MM Hg/70 F): N/K
Vapor Density (Air=1): N/K
Specific Gravity: N/K
Decomposition Temperature: UNKNOWN
Evaporation Rate and Ref: N/K
Solubility In Water: INFINITELY
Percent Volatiles By Volume: N/R
Viscosity:
pH: N/R
Radioactivity:
Form (Radioactive Matl):
Magnetism (Milligauss):
Corrosion Rate (IPY): UNKNOWN
Autoignition Temperature:

Fire and Explosion Hazard Data

Flash Point: N/K
Flash Point Method: N/K
Lower Explosive Limit: N/K
Upper Explosive Limit: N/K
Extinguishing Media: MAY USE WATER SPRAY TO PUT OUT SURROUNDING FIRE & COOL EXPLOSED CONTAINERS. WATER SPRAY WILL REDUCE FUME & IRRITANT GAS.
Special Fire Fighting Proc: USE NIOSH APPROVED SCBA WITH FULL POSITIVE PRESSURE FACE PIECE & CLOTHING. NOT COMBUSTIBLE, BUT LIQUID IS OXIDIZER & MAY IGNITE/W REDUCING AGENTS/COMBUSTIBLES.
Unusual Fire and Expl Hazards: EXTREMELY VIOLENT COMBUSTION IF CONTACT WITH OXIDIZING SUBSTANCES. DRIED CONCENTRATED HYDROGEN PEROXIDE/COMBUST. MATERIAL ON CLOTHES MAY CAUSE FIRE OR EXPLOSION.

Reactivity Data

Stability: NO
Cond To Avoid (Stability): HEAT. SPONTANEOUS COMBUSTION MAY OCCUR ON STANDING IN CONTACT WITH READILY FLAMMABLE MATERIALS.
Materials to Avoid: REDUCING AGENTS, ORGANIC MATERIALS, DIRT, ALKALIES, RUST & MANY METALS.

Figure 19-1.—Material Safety Data Sheet (front).

BMRf1902

Hazardous Decomp Products: DECOMPOSES TO WATER & OXYGEN WITH RAPID HEAT RELEASE. USE VENTED CONTAINERS. SOLUTION CAN DECOMPOSE RAPIDLY UPON HEATING
Hazardous Poly Occur: NO
Conditions To Avoid (Poly): NOT APPLICABLE

Health Hazard Data

LD50-LC50 Mixture: LD50 UNKNOWN
Route Of Entry–Inhalation: YES
Route of Entry–Skin: YES
Route of Entry–Ingestion: YES
Health Haz Acute and Chronic: ACUTE-EYES: IRRITATION. SKIN: INTACT SKIN, NONE. CONTACT ON BURN/OPEN SKIN MAY CAUSE IRRITATION. INHALE: UPON HEATING, MAY CAUSE IRRITATION TO MUCOUS MEMBRANES OF NOSE & THROAT. INGEST: IRRITATION TO MOUTH, THROAT & ABDOMEN. CHRONIC-NONE.
Carcinogenicity–NTP: NO
Carcinogenicity–IARC: NO
Carcinogenicity–OSHA: NO
Explanation Carcinogenicity: THIS PRODUCT IS NOT LISTED BY IARC, NTP, OR OSHA AS A CARCINOGEN.
Signs/Symptoms Of Overexp: EYE: REDNESS & PAIN. SKIN: STINGING PAIN. INHALE: IRRITATION TO MUCOUS MEMBRANES OF NOSE & THROAT. INGEST: BLISTERING TO MOUTH, THROAT & ABDOMEN. ABDOMINAL PAIN, VOMITING & DIARRHEA.
Med Cond Aggravated By Exp: NONE
Emergency/First Aid Proc: FIRST AID-INHALATION: REMOVE TO FRESH AIR. SEE DOCTOR IF NEEDED. EYES: WASH WITH PLENTY OF WATER FOR 15 MINUTES. SEE DOCTOR. SKIN: WASH WITH SOAP & WATER. IF IRRITATION PERSISTS, GET MEDICAL ADVICE. INGEST: GIVE SEVERAL GLASSES OF WATER TO DRINK TO DILUTE. GET MEDICAL ADVICE.

Precautions for Safe Handling and Use

Steps If Matl Released/Spill: VENT SPILL AREA. MAY REQUIRE PROTECTIVE CLOTHING. ABSORB SPILL WITH DRY ABSORBENT OR DILUTE WITH LARGE AMOUNTS OF WATER AND HANDLE AS NONHAZARDOUS WASTE. CONTAINERIZE UNSTABLE MATERIAL FOR DISPOSAL IN AN APPROVED WASTE FACILITY.
Neutralizing Agent: NOT APPLICABLE.
Waste Disposal Method: DISPOSE OF IN ACCORDANCE WITH FEDERAL, STATE AND LOCAL LAWS.
Precautions-Handling/Storing: STORAGE-STORE IN A COOL, WELL-VENTILATED DARK AREA. ISOLATE FROM INCOMPATIBLE SUBSTANCES. PROTECT FROM PHYSICAL DAMAGE.
Other Precautions: NONE

Control Measures

Respiratory Protection: NONE
Ventilation: DILUTION VENTILATION IS SATISFACTORY. HOWEVER IF WORKER FEELS DISCOMFORT, LOCAL EXHAUST SYSTEM SHOULD BE USED.
Protect Gloves: RUBBER
Eye Protection: CHEMICAL SAFETY GOGGLES/FULL FACE SHIELD
Other Protective Equipment: EYEWASH STATION & QUICK-DRENCH FACILITY.
Work Hygienic Practices: OBSERVE GOOD PERSONAL HYGIENE PRACTICES AND RECOMMENDED PROCEDURES.
Suppl. Safety & Health Data: NONE

Transportation Data

Transportation Action Code:
Transportation Focal Point: D
Trans Data Review Date: 90339
DOT PSN Code: ZZZ
DOT Proper Shipping Name: NOT REGULATED FOR THIS MODE OF TRANSPORTATION
DOT Class: N/R
DOT Label: N/R
Limited Quantity: NO
DOT Mode Indicator:
Identification Number: N/R
Reportable Qty.–Trans File: NO
DOT/DoD Exemption Number:
IMO PSN Code: ZZZ
IMO Proper Shipping Name: NOT REGULATED FOR THIS MODE OF TRANSPORTATION
IMO Regulations Page Number: N/R
IMO UN Number: N/R
IMO UN Class: N/R
IMO Subsidiary Risk Label: N/R
IATA PSN Code: ZZZ
IATA UN ID Number: N/R
IATA Proper Shipping Name: NOT REGULATED BY THIS MODE OF TRANSPORTATION
IATA UN Class: N/R
IATA Subsidiary Risk Class: N/R
IATA Label: N/R
AFR 71-4 PSN Code: ZZZ
AFR 71-4 Prop. Shipping Name: NOT REGULATED FOR THIS MODE OF TRANSPORTATION
AFR 71-4 Class: N/R
AFR 71-4 Label: N/R
AFR 71-4 ID Number: N/R
AF MMAC Code:
Tech Entry NOS Shipping Name:
Additional Trans Data:

Disposal Data

Disposal Data Action Code:
Disposal Data Focal Point:
Disposal Data Review Date:
Rec # For This Disp Entry:
Tot Disp Entries This Stock#:
Landfill Ban Item:
Disposal Supplemental Data:
1st EPA Haz Wst Code UnUsed:
1st EPA Haz Wst Name UnUsed:
1st EPA Haz Wst Char UnUsed:
1st EPA Acute Hazard UnUsed:
2nd EPA Haz Wst Code UnUsed:
2nd EPA Haz Wst Name UnUsed:
2nd EPA Haz Wst Char UnUsed:
2nd EPA Acute Hazard UnUsed:
3rd EPA Haz Wst Code UnUsed:
3rd EPA Haz Wst Name UnUsed:
3rd EPA Haz Wst Char UnUsed:

Figure 19-2.—Material Safety Data Sheet (back).

- Hazardous ingredients
- Physical and chemical characteristics
- Physical hazards
- Reactivity
- Health hazards
- Precautions for safe handling and use
- Control measures
- Routes of entry into the body
- Emergency and first-aid procedures for exposure
- Date of preparation of the MSDS or last change
- Name, address, and phone number of a responsible party who can provide additional information on the hazardous material and appropriate emergency procedure

Manufacturers may use any format or arrangements of this information, but every MSDS must include all the items. Every hazardous material user must be trained on the precautions associated with that material. MSDSs must be available upon request to any user. If you have a question, check with your command's hazardous material/hazardous waste coordinator.

REVIEW 2 QUESTIONS

Q1. Manufacturers provide data to people who use hazardous materials. What publication contains information on using, storing, and disposing of hazardous materials?

Q2. What instruction dictates that all hands are to follow Material Safety Data Sheets guidelines?

BOAT SAFETY

The major concern of Navy personnel aboard small boats is safety—for passengers and crew members. This section covers safety precautions to follow aboard small boats. Every Sailor should be thoroughly familiar with boat safety precautions. When you are on or boarding a boat, observe the following precautions:

- Obey all orders from the coxswain.
- Embark in a quiet, orderly manner and move as far forward as possible. Once embarked, stay in place.
- Keep all parts of your body in the boat; do not sit on gunwales.
- Don't engage in horseplay.
- Never distract the attention of crew members from their duties.
- Don't sit on life jackets; this will compress the filler and reduce buoyancy.
- When told to do so, don your life jacket immediately.
- Don't smoke in a boat.
- If told to embark or disembark, do so without argument. During heavy weather, boat loads will be reduced.
- If a boat swamps or capsizes, do not panic. Fear can spread quickly from person to person. A terrified person drowns easily. Stay with the boat or huddle with other passengers. A large group can be found much easier than individual swimmers.

DECK SAFETY

Weather decks of ships at sea are extremely hazardous places, particularly aboard small ships. The ship may be level one minute and take a sharp roll the next. At any moment, a large wave could submerge the main deck to a depth of several feet, or a wave could come unexpectedly over the fantail.

Vigilance (alertness) is always a necessity aboard ship. In foul weather, you must be even more alert. If your duties don't require your presence on the main deck, don't go there. Use interior passageways or superstructure decks for moving about the ship. When

Student Notes:

you must be on the main deck in foul weather, wear your life jacket. You must always wear an inherently buoyant life jacket whenever you are handling lines or are otherwise involved in underway replenishment or transfer operations.

A ship's deck has many tripping hazards, such as cleats, bitts, and pad eyes, as well as larger obstacles, such as boat davits and winches. Learn their locations so that if you must go on deck at night, you will have a better chance of avoiding these hazards.

Don't sit or lean on lifelines. When the sea is unusually rough, a safety line may be rigged on the main deck. When you are moving along the deck, you should stay inboard of, and hold on to, the safety line.

The flight decks of aircraft carriers are particularly hazardous areas. **Beware of propellers and jet blast**! Often, propellers are invisible because of the speed at which they rotate. They can act just like a meat slicer; so you need to use extreme care when walking or working near propeller-driven aircraft.

Jet planes present other hazards—a person can be sucked into the jet's intake, be burned, or be blown overboard (or against an object) by its exhaust. Keep off the flight deck if you don't work there. Because of minimum lighting requirements, nighttime is especially hazardous on the flight deck. When working on the flight deck, always wear your ear protectors when jet engines are running. One other caution—Smoking is prohibited on the flight and hangar decks and in all fuel and ammunition-handling spaces.

In general, the same rules apply to ships with operating helicopters. Only authorized personnel are permitted in the landing area during helicopter operations. Those personnel must wear proper protective clothing and equipment. During vertical replenishment operations, keep out from under loads and stay clear of the unloading area until the helicopter has departed. Keep the landing area free from loose debris or "foreign object damage" (FOD) that may be blown about by the downwash from the rotor blades or sucked up by jet intakes.

During flight quarters, the flight deck of an aircraft carrier is a dangerous place. This deck, combined with the hangar deck, magazines, and shops, provides the equivalent operating facilities of a large airfield. However, the hazards associated with aircraft operations are focused into a relatively small area. Therefore, personnel are exposed to a greater potential of danger.

REVIEW 3 QUESTIONS

Q1. List four boat safety precautions that every Sailor should know.

a.

b.

c.

d.

Q2. If a boat swamps, what usually causes a loss of personnel?

Q3. Why should you learn the location of cleats, bitts, and pad eyes on a ship's deck?

Q4. What are two hazards found on flight decks of aircraft carriers?

a.

b.

LIFELINES, LADDERS, AND SCAFFOLDING

Lifelines, as used here, refer to lines erected around the edges of weather decks. They are safety barriers to prevent personnel from falling or being washed over the side. Never sit, lean, or stand on any lifeline—if the ship takes a sudden roll while you are leaning against a lifeline, you could fall overboard.

Never remove lifelines without permission from the proper authority. When removing a lifeline,

Student Notes:

immediately rig a temporary line. Don't hang or secure any weight on a lifeline.

When working near a ladder, Sailors have the bad habit of placing paint cans, buckets, or tools on the steps to minimize bending over. This practice could cause a mishap. Because water will cause a ladder to become very slippery, you should be especially careful on rainy days. Paint drippings are equally dangerous for the same reason.

Never unship (take down) a ladder without permission. Rope off all open hatches and gangways leading to unshipped ladders.

The smooth deck of a ship does not provide a good hold for scaffolding. The base of scaffolding must be properly braced and lashed down to prevent it from sliding. The use of makeshift scaffolds is prohibited. Scaffolds must be erected only when needed to do a job and dismantled as soon as the job is completed.

You should not work on a scaffold in high winds or when the scaffold is covered with ice or snow. Never throw or drop objects from a scaffold; use handlines for raising or lowering objects. Do not paint scaffolds, because the paint might conceal defects. Use lifelines and safety belts when working on a boatswain's chair or on unguarded scaffolds above a height of 10 feet.

HANDLING CARGO

Serious, sometimes long-lasting injuries can result from improperly handling heavy objects and from the failure to observe basic safety precautions. By observing the following precautions, you can prevent injury to yourself or to others and prevent damage to cargo and equipment:

- When lifting a heavy or bulky object, crouch close to the load with feet solidly placed and slightly spread. Get a good grip on the object and lift with your arm and leg muscles, keeping your back as nearly vertical as possible. If the load is bulky or heavy, don't feel embarrassed to ask for help.

- Don't throw articles from elevated places; lower them by a line or carry them.

- Wear appropriate safety clothing and equipment, such as safety shoes, a hard hat, gloves, and a life preserver, for the job at hand. Remove rings, wristwatches, and bracelets when handling cargo.

- Stow hatch covers and strongbacks in such a manner that they won't interfere with traffic or be knocked into the hatch or over the side.

- When steadying loads, don't stand between the load and a fixed object. Don't stand under a suspended load. Never ride loads. Use the nonworking side of a ship for fore-and-aft travel.

- Never stand in the bight of a line. Keep clear of lines under a strain. A line (particularly nylon) can part with a whiplike snapback, which can cause severe bruises, broken bones, amputations, or even death.

- Don't engage in horseplay.

- When going up or down a ramp with a hand truck, keep the load below you. Thus you pull the load up and push it down.

WORKING ALOFT OR OVER THE SIDE

Before any work may be done aloft, permission must be obtained from the OOD. Before granting permission, the OOD makes sure that all power on appropriate radio and radar antennas is secured and that controls associated with the antennas are tagged "SECURED. PERSONNEL ALOFT." The OOD also notifies the engineer officer where the personnel will be working so that the necessary precautions can be taken to prevent operations such as the lifting of boiler safety valves or the blowing of tubes. After the work has been completed, a report is made to the OOD, who, in turn, will notify the appropriate officers.

When you are working aloft, wear a standard Navy-approved safety harness with a safety line attached. Radio and radar transmissions, even from another ship, can induce a charge in guy wires, stays, ladders, and other metal fittings. If you touch one, you may receive a shock. The shock itself may not be dangerous, but a natural reaction when shocked is to jerk away. Without a safety harness you could easily fall.

Student Notes:

Secure all tools and equipment with lanyards to prevent dropping them and injuring personnel below. Burning and welding or the presence of any open flame isn't permitted on a stage or boatswain's chair unless the suspension ropes and bridle are made of steel. Always check equipment for weakened or broken fittings before going aloft.

When working over the side, you must wear a standard Navy safety harness with a safety line attached and tended by someone on deck. You must also wear an inherently buoyant life jacket with a hole in the back. The hole in the life jacket will allow you to wear a safety harness. The line should be only long enough to permit freedom of movement.

Wear a life jacket when you work at underway replenishment stations, when you are in a lifeboat at sea, when you work on weather decks during heavy weather, or whenever you are directed to do so. While the ship is under way, you must be given permission by the CO to work over the side.

ANTENNAS

Personnel aren't permitted to go aloft in the vicinity of energized antennas. The voltages set up in a ship's structure or section of rigging by electromagnetic radiation (EMR) can shock or burn you. When deck force personnel or others work on rigging, they must be familiar with the hazards that exist and know the precautions to be observed. Safety harnesses are used when working aloft to guard against falls.

The previously mentioned precautions should be observed also when other antennas in the immediate vicinity are energized by electronic transmitters, unless it is definitely known that no danger exists. Other antennas may be interpreted to mean any antennas on board another ship moored alongside or across a pier or at a nearby shore station.

Personnel aloft are in danger from falls caused by radar or other antennas that rotate or swing through horizontal or vertical arcs. Motor switches controlling the motion of radar antennas should be locked open and tagged before you go aloft to work in the vicinity of such antennas.

REVIEW 4 QUESTIONS

Q1. Handling cargo improperly can result in injury and death. What precautions should you follow in the following cases?

a. Working over the side

b. Lifting heavy objects

c. Steadying a load

Q2. What person grants permission for any work done aloft?

Q3. What precautions are taken before permission is given for personnel to work aloft?

Q4. Describe the purpose of lifelines.

Q5. Describe the equipment you should wear when working over the side.

STEAM

Most accidents involving steam occur in engine rooms and firerooms. However, steam lines run throughout a ship; therefore, observe proper precautions at all times. Some practices can be applied to almost any situation regardless of the type of equipment, the steam pressure, or any other job-related condition.

Live steam is often invisible and it is always dangerous. If you are not familiar with a system or have not been trained for the task at hand, do not attempt the job.

Always drain lines before removing valves or otherwise opening the system. Close all associated

Student Notes:

valves to isolate the system to be opened, and tag these valves to ensure they remain shut while you are working on the equipment. Wear proper protective clothing. Do not try to take shortcuts and do not skylark. Carelessness has been a factor in nearly all reported mishaps involving steam. Observe all appropriate precautions.

CLOSED COMPARTMENTS AND UNVENTED SPACES

Never enter a closed space until it is certified safe by a gas free engineer.

Closed compartments may contain hidden dangers, both to yourself and to the ship as a whole. One possibility is toxic or explosive gases. After these spaces are opened, your gas free engineer will make sure that it s safe for you to work there.

If the ship's been damaged, other dangers may exist. The manhole access cover to a damaged tank or compartment might be all that's preventing flooding. Additionally, water entering a closed compartment pressurizes the air already there. Don't try to open a pressurized compartment or void without venting the pressure first. If you don't vent the pressure first, the hatch cover/access will fly open violently, possibly injuring you or a shipmate. Check with your supervisor for help in learning to recognize these and other hazards.

Consider all compartments dangerous if they've been closed for any length of time. If the bulkheads, deck, or overhead are rusted, they have absorbed oxygen from the air. This means there may not be enough oxygen left for you to breathe. If the compartment was painted before it was closed, the hardening paint has absorbed oxygen and given off carbon monoxide. Carbon monoxide is particularly dangerous because it gives no warning. If you're working in a compartment that's been closed and you notice a sudden feeling of weakness, drowsiness, or a slight headache, **call for help and get to fresh air**.

In storage compartments, several toxic gases may be generated by mildewing or rotting foodstuffs or by materials such as cloth, leather, and wood. Mildewing and rotting are speeded up when the space is warm and humid, such as when a ship is cruising in the Tropics or when an area has been flooded as a result of damage or accident.

Carbon dioxide is frequently found in refrigerator spaces, even though the spaces are undamaged and the foodstuffs are still good. This condition results from lack of ventilation and the fact that foods slowly absorb oxygen and give off carbon dioxide. If personnel stay in such spaces longer than a few minutes at a time, they may be overcome and eventually suffocate.

Sulfur oxides are acrid, corrosive, poisonous gases produced when fuels containing sulfur are burned. For example, aboard ship the primary producer of sulfur oxides is fuel oil, which contains sulfur as an impurity.

Government agencies and industries have sought to reduce sulfur oxide emissions in three ways:

1. Switching to low-sulfur fuels (those with less than 1% sulfur).
2. Removing sulfur from fuels entirely.
3. Removing sulfur oxides from combustion gases.

To reduce the sulfur oxide problem on ships, the Navy developed a fuel oil called *Navy distillate fuel.*

Sulfur oxides produce an offensive odor and can cause eye and lung irritation. Tanks that have held petroleum products and compartments in which oil, gasoline, solvents, and organic products that have been spilled will contain the vapors of these products.

Tanks that have held petroleum products, and compartments in which oil, gasoline, solvents, and organic products have been spilt, will contain the vapors of these products.

You must assume that any closed space, double bottom, tank, cofferdam, pontoon, or void contains gases that can poison or suffocate you or can explode. (**NOTE**: Never enter any such space until it has been thoroughly ventilated and checked by a gas free engineer to make sure there is no danger of poisoning, suffocating, or igniting flammable gases.) Before entering a closed space, make sure that it's been ventilated for 24 hours. Also, the gas free engineer must certify the safety of the space and recertify it every 8 hours while personnel are working in the space. Always have a person stationed at the entrance to maintain

Student Notes:

communications and to watch to see that you are not overcome.

Symptoms of bad air include the following:

- Labored breathing
- Excessive fatigue from slight exertion
- Headaches
- Dizziness

If you feel any of these symptoms, warn others and get to fresh air immediately.

A more dangerous situation occurs if there is very little or no oxygen in a compartment. In this case, a person can lose consciousness almost immediately and without warning. If this happens and you're tending the person, don't enter the space without wearing an oxygen breathing apparatus (OBA). If you do, you'll become a casualty yourself. Always summon (call for) help before making a rescue attempt.

Another hazard of working in closed compartments or connected spaces is the use of internal combustion engines in these spaces. For example, if a P-100 pump for fire fighting or dewatering is used in a closed compartment, the engine used to drive the pump takes in the air through the carburetor and exhausts poisonous carbon monoxide. If you need to use an internal combustion engine in a closed space, make sure the exhaust is carried (vented) to the open atmosphere.

REVIEW 5 QUESTIONS

Q1. Where do most accidents involving steam occur?

Q2. Describe the reason why you should never enter a closed space until its certified by the gas free engineer.

Q3. List the symptoms caused by bad air.

a.

b.

c.

d.

FLAMMABLES

Rules for preventing fuel fires were presented in chapter 13 of this manual. Our discussion here will include fire hazards and toxic hazards of flammable materials and applicable safety precautions.

The vapors of petroleum products cause anesthetic effects when inhaled. Breathing air where petroleum vapors have a concentration of only 0.1 percent by volume can result in the inability to walk straight after only 4 minutes. Longer exposure or greater concentration may cause unconsciousness or death. When lead is added to the fuel, toxicity is increased. The lead may be inhaled or it may be absorbed through the skin. Proper ventilation, therefore, must be provided at all times when personnel are working in fuel tanks. An air-line respirator is recommended when personnel enter such spaces.

Symptoms of exposure to toxic vapors are headache, nausea, and dizziness. If you are working in a space that formerly held oil, gasoline, or other fuels and you experience these symptoms, get to fresh air at once. Recovery is usually prompt in fresh air; but if you are overcome by the vapors, you may require immediate medical attention. First-aid measures are to prevent the victim from becoming chilled and to administer artificial ventilation if breathing has stopped.

All fuel spills must be wiped up immediately to prevent the spread of vapors to a possible ignition source. Never use gasoline for cleaning purposes, and avoid getting gasoline on the skin. Repeated contact causes drying, chapping, and cracking and may cause infection.

Student Notes:

OPEN FLAME AND NAKED LIGHT NEAR FUELS

The use of open flame, naked lights, or any apparatus that is likely to cause a spark is not permitted in spaces or areas where fuel is exposed or during fueling. The term *open flame* includes all forms of fuel or gas lanterns, lighted candles, matches, cigarette lighters, and so on. The term *naked lights* includes any unprotected electrical lighting device. Permanently installed electrical apparatus necessary for maintenance of power or services in the ship could produce sparks.

PAINTS

Paints, varnishes, lacquers, cleaners, solvents, or other finishing materials contain flammable solvents that can ignite at comparatively low temperatures and, therefore, present a fire hazard. They also give off toxic vapors that are harmful when inhaled. When using paints and finishing materials, you should observe the following precautions:

- Do not smoke or use an open flame in areas where paint, varnishes, lacquers, and solvents are mixed or applied.
- Wipe up spilled paint or solvents immediately to reduce fire and vapor hazards.
- Place rags or other items used for cleaning up paint in a separate container with a closed top.
- Take care to prevent paint products from coming in contact with the eyes and skin.
- Wear goggles when chipping and cleaning surfaces to be painted.
- Wear gloves and a filter respirator when mixing paint and when painting.

SOLVENTS

Solvents used in paints, adhesives, rubber and plastic materials, and degreasing solutions are hazardous to your health. Most solvents are toxic and, with a few exceptions, are flammable. Appropriate measures must be taken to reduce their toxic and flammable effects. In addition, exposure of the skin to solvents can cause serious skin problems. Therefore, you should observe the following precautions when using solvents:

- Use adequate ventilation.
- Wear protective clothing, goggles, gloves, and other appropriate safeguards.
- Have readily accessible fire-fighting equipment nearby.
- Take every precaution to prevent excessive vapors from contaminating the air.
- Check all liquids before using. If in doubt of any cleaning fluids, consult the officer in charge.
- Wipe up spilled solvents immediately.
- Avoid contact with your eyes, skin, or clothing. Do not take solvents internally, and avoid breathing solvent vapors.
- Keep solvent containers tightly closed when not in use.
- Check containers for leakage; if a container is defective, transfer the solvent to a new container.
- Be sure containers are empty before they are discarded. Observe approved practices for disposal of solvents and cleaners and their containers.
- Label all containers in which solvents are to be stored.
- Store solvents in an appropriate solvent storage locker.

REVIEW 6 QUESTIONS

Q1. Define the following terms.

a. Open flame—

b. Naked lights—

Student Notes:

Q2. When storing solvents, what actions should you take?

a.

b.

WEAPONS AND EXPLOSIVES

You should observe the following general precautions when handling any type of weapon:

- Consider every weapon loaded until you examine it and find it otherwise.
- Never point a weapon at anyone you do not intend to shoot or in a direction where an accidental discharge may do harm.
- Place a cartridge in the chamber only when you intend to fire the weapon.
- Whenever you handle a weapon, think about what you are doing. Accidents seldom "just happen." They frequently are caused by persons ignorant of safety precautions. All too often they are caused by carelessness.
- Make sure the ammunition is suited to the type of weapon you intend to fire.

Ammunition is stowed aboard ship in specially constructed compartments called magazines, which are located as far as possible from firerooms and engine rooms. Each magazine is equipped with a sprinkler system, and many are equipped with a quick-flooding system for use in an emergency to prevent explosion of the magazine. Lighting is accomplished with vaportight fixtures. Naked lights, matches, or other flame-producing apparatus must never be taken into a magazine. Heel plates or other spark-producing materials are also forbidden. Magazines must be kept scrupulously clean and dry at all times. Particular attention must be paid to ensure that no oily rags, waste, or other materials that may cause spontaneous combustion are stored in magazines.

Extreme care must be exercised when handling ammunition. Remember, the purpose of ammunition is to cause destruction. Be sure the destruction is to the enemy and not to your own ship. Figure 19-3 shows the tragic results of careless handling of ammunition. A ship was lost and over 150 persons were killed or injured.

An important part of ammunition handling is identifying the type of ammunition. Projectiles of 3-inch and greater diameter are color-coded to indicate the projectile type and the kind of bursting charge they contain. Armor-piercing, antiaircraft, illuminating, and chemical projectiles are identified by their own distinctive color markings. Whenever you are handling ammunition, keep projectiles of the same type (same color) together.

A few additional rules are given here for handling ammunition. These rules are general in nature and are not all-encompassing, but they apply to all types of ammunitions.

- Loading or unloading ammunition is not a contest. Racing against other handling parties only increases the possibility of a mishap.
- Be careful not to dent cartridge cases. Dented casings may jam in the bore. Some thin-cased explosives are known to have detonated when their casings were dented.
- Avoid obliterating (blotting out, blurring, etc.) identification marks.
- Grommets are used to protect the rotating bands of projectiles; don't lose the grommets.
- Don't smoke in magazines or in the vicinity of explosives-handling and explosives-loading operations.
- Unless you are involved, keep clear of ammunition-handling operations.
- Never tamper with explosive devices.
- Don't store drill charges in magazines with live ammunition.

All pyrotechnic materials are kept in special stowage spaces, usually located on topside decks. Any pyrotechnic material that shows signs of damage to its safety device is considered unserviceable and must be segregated for prompt disposal. Extreme caution must

Student Notes:

Figure 19-3.—A result of carelessness.

be taken to prevent accidental ignition of loose pyrotechnics made ready for disposal, because damaged material can be ignited by rough handling.

REVIEW 7 QUESTIONS

Q1. When handling a weapon, you need to think about what you're doing for what reason?

Q2. Projectiles that have a 3-inch or greater diameter are color-coded. What information is shown by the color code?

ELECTRICAL AND ELECTRONIC EQUIPMENT

All electrical and electronic equipment is hazardous; therefore, strictly observe all safety precautions. Most people treat high-voltage equipment carefully, but they tend to treat the common 115-volt equipment lightly. Yet, 115-volt equipment is the cause of more deaths than any other voltage. Cases of fatal shock have been recorded from the use of equipment such as portable grinders and drills, fans, movie projectors, and coffee makers. In most cases, death would have been avoided if proper grounding instructions had been observed. The precautions that follow must be observed by personnel working on or near other types of equipment:

- Most electronic equipment has a metal grounding strap connecting the equipment to the ship's hull. The straps keep the equipment's frame and the ship's hull at the same electrical potential. Never paint, loosen, disconnect, or

Student Notes:

otherwise tamper with the straps without proper authority.

- Never replace or pull a fuse. Only authorized personnel are allowed to do such work.

- Motors and generators often have openings in their casings. Avoid dropping tools or other objects into the openings. Some machinery and electrical circuits generate magnetic fields, so be alert; don't let magnetic tools you are holding be drawn to such equipment.

- Electrical and electronic equipment and power cables are identified by nameplates, tags, or other markings. Never paint over such identification markings.

- Don't hang items on, or secure lines to, any power cable, antennas, wave guide, or other electrical or electronic equipment.

- Don't use personal electrical equipment aboard ship without the approval of the engineer officer.

COMPRESSED GASES

Compressed gases includes air, oxygen, acetylene, carbon dioxide, and other gaseous or gas-forming compounds held under pressure in steel bottles, cylinders, or tanks. In general, three types of hazards are connected with compressed gases as follows:

1. Cylinders containing compressed gas are usually round and long. They are made of heavy steel. Unless secured to a structure, they can roll, tip over, or bang around. If not secured properly, they can roll around and cause damage by bumping into a person or an object.

2. The cylinders contain gas under pressure—often under very high pressure. A cracked cylinder can fly apart. Air or gas from a valve or hose connected to a cylinder can blow dirt into your eyes; or the hose can whip around and strike you, causing an injury. If you drop or mishandle an oxygen cylinder so that its valve breaks off, you may see the heavy steel bottle take off like a rocket—causing injury and damage.

3. The cylinders may contain gases that are poisonous, flammable, or explosive, and often all three. Acetylene cylinders are common aboard ship. If you ignite acetylene, it will blaze with intense heat; if it's mixed with air and a spark gets to it, it will explode. In fact, an acetylene cylinder can explode if it is overheated and then given a sudden blow. If oxygen comes into contact with oil or grease, you can be sure you will have a fire. CO_2 used in fire extinguishers is particularly dangerous; you will suffocate in a room filled with it. Also, CO_2 is extremely cold when it is discharged. It may cause painful blisters if it comes in contact with your skin.

You must handle, work with, and work around compressed gas cylinders with care and caution. The cylinders are heavy and can easily be tipped.

In general, weather-deck stowage will be provided for flammable and explosive gases. However, in specific cases, the approval of below-deck stowage depends on the particular type, mission, and arrangement of the ship. In such cases, these approved locations are shown on the ship's plans.

Compressed gases aboard all ships, except cargo ships, should be stowed in compartments designed for stowage of gas cylinders. In such cases, the following precautions must be observed:

- Take the necessary steps to prevent the maximum temperature of the stowage compartment from exceeding 130°F.

- When provisions are made for mechanical ventilation, operate this ventilation according to the damage control classification assigned.

- The classification for closure of this system is ZEBRA (Z), CIRCLE WILLIAM [(W)], and WILLIAM (W).

- In compartments designated for stowing flammable or explosive gases, the installation of portable electric wiring and equipment isn't permitted.

Student Notes:

- Flammable materials, especially grease and oil, must be kept out of the stowage space used for oxygen cylinders.

- Each cylinder must be securely fastened in the vertical position (valve end up) by using such means as metal collars. On cargo ships fitted especially for cylinder transport, other arrangements are approved.

- Oxygen and chlorine must be stowed in compartments separate from flammable gases. Inert or nonflammable gases may be stowed in compartments designated for compressed gas stowage.

- Compartments containing compressed gases must be ventilated for 15 minutes before entry if the ventilation has been secured; a suitable sign to this effect should be posted on the outside of the access door.

When compressed gas is stowed on the weather deck, the following additional precautions must be observed:

- Oxygen and chlorine cylinders must not be in close proximity to fuel-gas cylinders.

- Cylinders containing compressed gases should be stowed so that they will be protected. During the winter, cylinder valves must be protected against the accumulation of snow and ice. Warm water (not hot) should be used to thaw ice accumulations in cylinder valve caps and outlets. During the summer, cylinders must be screened from the direct rays of the sun. Every effort should be taken to prevent corrosion of threaded connections of cylinders in stowage for extended periods of time. The use of grease or flammable corrosion inhibitors on oxygen cylinders is not permitted.

- The stowage area should be as remote as practical from navigating, fire control, and gun stations.

ASBESTOS

Asbestos is a fibrous material that is incombustible (doesn't burn), possesses high tensile strength, has good thermal and electrical insulating properties, and has moderate to good chemical resistance. Because of these characteristics, the Navy has had many uses for asbestos. Asbestos was used as the primary insulation and lagging material for high-temperature machinery, boilers, and piping on board ships. Other applications included floor tile, tile underlayment (especially decks above engineering spaces), rope and pressed gaskets, brake and clutch facings, and expansion joints.

When intact and not disturbed, asbestos doesn't normally present a hazard. Problems arise when repair work causes the generation of asbestos dust. Inhaling asbestos fibers present in the dust may lead to various forms of asbestos-related diseases. Most symptoms of asbestos-related diseases do not show up until 10 to 45 years after exposure. Since the total removal of all asbestos materials on board Navy ships is not feasible, the Navy has instituted a program to control the use and replacement of asbestos with nonasbestos substitutes.

Only specially trained and medically qualified personnel are authorized to remove asbestos. When asbestos material is being handled, complex safety requirements and precautions are used. **Never enter a space that has been designated as an asbestos hazard area unless specifically told to do so**. For more detailed instructions on the hazards and control of asbestos, refer to *Navy Safety Precautions for Forces Afloat*, OPNAV 5100.19.

FIBERGLASS

Reinforced plastic materials are currently being used by the Navy in—

- Boat hulls,

- Submarines,

- Minesweeping equipment,

- Protective coverings for wood and steel, and

- Many other types of equipment and materials.

Reinforced plastic is made of glass fibers, resin, and chemicals, which gives it the name *fiberglass*. The resin and activating chemicals bond the glass fibers together, producing a very tough and rugged material. Polyester or epoxy resins are used to make fiberglass.

Student Notes:

Fiberglass isn't totally safe to work with. Certain safety precautions must be observed when working with or around fiberglass. If fiberglass is cut or ground, a fine dust is produced. This dust is abrasive and can irritate your skin and eyes. Use a filter mask respirator when working in this type of atmosphere.

The chemicals used in making fiberglass and fiber glass patches are very flammable and toxic. Provide adequate ventilation to remove the fumes and dust particles. Most important, never smoke in areas where fiberglass work is being carried out.

REVIEW 8 QUESTIONS

Q1. What is the cause of many fatal shocks received from drills and fans?

Q2. List the three types of hazards associated with compressed gases.

a.

b.

c.

Q3. True or false. Oxygen and chlorine are stowed in compartments separate from flammable gases.

Q4. Why should you use a filter mask respirator when working with fiberglass?

POWER TOOLS

During your career in the Navy, you may be required to use a variety of power tools. Whether these tools are electrical, pneumatic, or hydraulic, the same common sense safety precautions apply to all of them.

Before you use a portable electric tool for the first time, have it inspected and approved by the ship's electrical department for safety. If it has a current ship's inspection mark, visually examine the attached cable for any cracks, breaks, exposed conductors, or a damaged plug. If any defects are found, turn the tool in to the ship's electrical shop for repair. Before plugging an electric tool into a receptacle, make sure the tool is turned off. When using portable electrical tools, wear safety glasses or goggles if the job involves danger from flying objects, such as paint or metal chips. You should also wear ear protection devices if the tool has a "Produces Hazardous Noise" label on it.

Metal-cased portable electric tools must have a three-pronged plug on the power cord. If an extension cord is used, it must be the three-pronged type with a three-pronged plug at one end and a three-pronged receptacle at the other end. When using an extension cord with an electric tool, you must first plug the tool into the extension cord and then the extension cord into the receptacle. When you are finished with the electric tool, switch it off, unplug the extension cord from the receptacle, and then unplug the tool.

Portable tools should be kept clean and in good repair. Arcing portable tools are not to be used in areas where flammable vapors, gases, liquids, or exposed explosives are kept.

CUTTING AND WELDING OPERATIONS

The convenience of arc and gas welding and cutting allows the performance of repair jobs in almost any location. Failure to use proper safety precautions during welding or cutting operations presents a serious fire hazard. **Only properly trained personnel should operate gas welding or cutting equipment**. Because cutting and welding operations are continuously being performed throughout the ship, you may be called upon to stand a fire watch and must be familiar with the safety precautions of such operations. The following are some basic precautions to be taken during welding or cutting operations:

- The gases used in welding and cutting are explosive. When one of these gases is mixed with air, the mixture will burst into flames if a spark or flame is brought near it.

Student Notes:

• Remove all combustible materials, flammable or explosive, from the area where welding or cutting is to be done.

• When welding or cutting a bulkhead, deck, floor, or other structure, you should check both sides of the structure to ensure that no materials near the structure will be damaged or will become a possible fire hazard.

• Post fire watches on both sides of a deck or bulkhead before welding or cutting operations can be started. Personnel assigned fire watches should be thoroughly familiar with fire watch responsibilities and outfitted with the proper safety gear, such as gloves, proper eye protection (particularly when arc welding), and safety shoes. To make sure no fire hazards exist, personnel assigned to the duties of a fire watch must remain at the location at least 30 minutes after the job is completed.

• Keep approved fire-extinguishing equipment near welding and cutting operations. Usually, a CO_2 extinguisher is adequate. If the space is small or if the access is only a small opening, CO_2 is not the extinguishing agent to use. CO_2 could fill the small space, and the small opening would not allow for breathable air to enter. The small entry or exit may also hinder any rescue attempts should you be overcome by suffocation. If CO_2 is not recommended, the use of water spray from a fog nozzle is preferred. In the event the fire is caused by electricity, secure power before using the water spray.

• Welding or cutting operations aren't permitted in or on the outer surfaces of a compartment or tank that contains or has contained a flammable or explosive substance, unless applicable safety precautions are observed.

ROTATING MACHINERY

The safe operation of rotating machinery and tools requires the operator to be thoroughly knowledgeable in the equipment operation. It also requires strict adherence to established operating procedures. The operators should be familiar with the safety precautions for their own particular machinery. However, when operating rotating machinery, the following general safety precautions should be observed:

• Never place any part of your body into moving machinery. Never attempt to ride machinery that is not designed for human conveyance.

• Never wear jewelry, neckties, or loose-fitting clothing.

• Wear proper protective clothing and equipment suited to the operation being performed (hearing protection; eye, hand, and foot protection; dust and paint respirators; and so on).

• Before attempting to perform repairs or preventive maintenance on any equipment, ensure that it is de-energized and/or depressurized and properly tagged out of service before beginning to work.

• When working in the vicinity of electrical equipment or electrical cables, be alert to the presence of dangerous voltages and avoid striking such equipment with tools of any kind. Should such damage inadvertently occur, report it immediately to the ship's electrical officer.

• Don't use compressed air to clean parts of your body or clothing or to perform general space cleanup. Compressed air may be used to clean machinery parts that have been properly disassembled provided that the supply air pressure does not exceed 30 psi and a safety shield tip is used.

• Reinstall shaft guards, coupling guards, deck plates, handrails, flange shields, and other protective devices removed as interference immediately after removal of machinery, piping, valves, or other system components during maintenance to prevent injury to personnel.

• Inspect and/or test, according to scheduled PMS and other type commander (TYCOM) requirements, all installed safety devices, alarms, and sensors. Assign a high priority to repair of defective safety devices.

• Cleanliness of machinery and its spaces profoundly affects the safety of personnel and equipment. Correct oil leaks at their source. Wipe up spills of any kind immediately, and dispose of the

Student Notes:

wiping rags immediately or store them in firesafe containers. Avoid trip hazards by maintaining proper stowage. Do not allow fire hazards to accumulate.

REVIEW 9 QUESTIONS

Q1. Before beginning work to repair a piece of equipment, you should take what action?

Q2. What personnel are authorized to operate gas welding or cutting equipment?

Q3. When working around rotating machinery, what types of clothing/equipment should you wear?

LIQUIDS UNDER PRESSURE

Any liquid in a system that has been pressurized is to be considered dangerous until the pressure has been removed. For example, the ship's fire-main system uses salt water that has been pressurized to make the water available throughout the ship. The pressurized water in the system is not dangerous, but the misuse of it is. Therefore, you should observe the following safety practices when using the fire-main system or any other system that may have pressurized liquid in it:

- Never connect or disconnect a hose from the system until the pressure has been removed. This can be done by shutting off the valve on the fire-main system.
- Never use ruptured or worn hoses with any system that has pressure in it.
- Never point a charged (pressurized) fire hose at anyone.
- Spray paints, butane fluids, lacquers, and other aerosol products contain liquids under pressure. Be extremely careful with these containers. Don't use these containers near a flame, throw them in a fire, or puncture the containers.

ACIDS, ALKALIES, AND OTHER CHEMICALS

Acids and alkalies are used in the Navy in the form of pure compounds and mixtures. Acids and alkalies are hazardous because they're corrosive (cause chemical burns) when they come in direct contact with the skin, eyes, or other body tissues. They can cause breathing difficulties or injure respiratory organs if too much of the acid mist is inhaled. The acids and alkalies can also cause dangerous chemical reactions if not handled properly.

When handling acids, alkalies, or other chemicals, you should observe the following precautions:

- Wear chemically resistant rubber or plastic gloves.
- Wear chemically resistant rubber or plastic goggles. You may need to wear a plastic face shield in addition to the goggles.
- Wear chemically resistant rubber boots or overshoes with resistant soles. Wear trousers outside of the boots.
- Wear a rubber or plastic apron.
- Wear a respirator when indicated for the chemical you are working with.

Persons who have been exposed to acids or alkalies should seek medical attention immediately.

MARINE SANITATION SYSTEMS

Sewage is a mixture of all liquid domestic wastes, especially human body wastes (fecal matter and urine). Sewage contains large numbers of microorganisms, some of which are disease bearing. Bacteria and viruses enter the human body through the mouth, nose, open sores, and so on. Therefore, you must observe the following basic precautions when working in sewage-handling areas.

Student Notes:

- Never take food or drink of any nature into sewage-handling areas.
- Never work on sewage-handling equipment if you have open cuts or sores.
- Maintain cleanliness of equipment at all times.
- Wash down any spilled sewage immediately (before it dries) with water and a good quality nonscented disinfectant. Don't use liquid soaps or scented disinfectants because they may temporarily disguise inadequate cleanup procedures.
- Always follow personal hygiene routines after working in a sewage-handling area or after being in contact with sewage-handling equipment.

NOTE

Notify the medical department and the damage control assistant (DCA) on the status of any holding or other marine sanitation device (MSD) whenever the ship is threatened by hostilities, fire, flooding, or conditions that could turn the MSD into a biological hazard to the ship's crew. Each ship should have developed plans to eliminate or control the biological hazards from these occurrences.

WARNING

Do not smoke in the vicinity of the sewage-handling equipment. Fuel leaks or spills can occur in the incinerator area where temperatures may exceed the flash point of the fuels used. Methane and hydrogen sulfide may be emitted by any tank or tank leaks. These gases are also flammable and under some conditions are explosive.

REVIEW 10 QUESTIONS

Q1. List the safety precautions to follow when working with systems having pressurized liquids.

a.

b.

c.

d.

Q2. If you've been exposed to acids or alkalis, what action should you take?

Q3. Why shouldn't you smoke near sewage-handling equipment?

HIGH NOISE LEVELS

Continuous exposure to noise at a high level can cause temporary or permanent hearing loss. Electrical/electronic equipment, portable power tools, machinery, and weapons are a prime source of loud noise.

The Navy has different types of hearing protection for use in subduing noise, such as earplugs (regular and disposable), headband earplugs, and the circumaural muff. If the noise is too loud, you may need to wear the earplugs and the circumaural muff for double protection.

RECREATION AND SPORTS

Participation in recreational activities is responsible for many injuries to personnel. Practically all sports involve some type of hazard. The principles of attack and retreat in body-contact sports arouse emotions that can lead to hazardous circumstances. When participating in sports, you should be familiar with and observe protective measures, rules, regulations, procedures, and applicable safety precautions.

When you engage in recreational activities, observe the following precautions:

Student Notes:

- Don't engage in recreational activities unless you are physically able to do so without harm.
- Wear necessary and prescribed protective equipment and clothing.
- Avoid overexertion and excessive fatigue. Such conditions can lead to injuries.
- Don't engage in an activity if you have an old injury that may be aggravated by additional activity.
- Warm-up properly before engaging in any vigorous sport.
- Avoid horseplay. This is a common cause of accidental injuries.
- Obtain medical attention immediately if you are injured, feel faint, become dizzy, or ill.
- Alcohol and sports do not mix. Drinking while participating in sports increases your chances of injuries.
- Don't try to play a new game or practice new athletic skills unless you are under the direction of a qualified instructor. Don't take unnecessary chances.
- Always keep a safe distance from sporting equipment being thrown, such as the discus and bats.

Remember, when engaged in a recreational activity, you are responsible for protecting yourself from injury. Therefore, you must observe all rules and safety measures.

MOTOR VEHICLES

You may be assigned duties as a driver. As a driver, you are responsible for the safe operation of the vehicle while it is assigned to you and for the safety of the passengers and cargo. **You (as a passenger or operator) are required to wear seat belts**. You are to make daily inspections of the vehicle assigned to you. If the vehicle is found to be unsafe, you aren't permitted or required to operate that vehicle until it has been repaired. You must obey all local traffic laws and ordinances while operating a motor vehicle on and off duty.

Except under extreme emergencies when no relief is available, you should only drive for short periods of time. If you must drive for a long period of time or if you become fatigued (tired) or drowsy when driving, pull off the road and stop for a few minutes to rest. Never operate a vehicle if you have been drinking alcoholic beverages, if you are taking medication that will make you drowsy, or if you are sick or physically unfit to drive.

LIFTING

Lifting is so much a part of our everyday jobs that we don't think about it, and most of the time we do it wrong. Results of improper lifting may be a painful hernia, a strained or pulled muscle, or a disk lesion. For the sake of your back, you should observe the following rules and precautions for lifting:

- **Don't lift an object if it is too heavy or too clumsy for good balance**. Get help, or use mechanical aids such as a dolly or hand truck.

- **Keep the load close to the center of your body**. The farther the load is from the small of your back, the greater the strain. That is the reason a heavy compact load is easier to lift than a bulky, lighter load—you just cannot get the bulky object close to you. The best way to handle a compact load is to squat down close to the load with one foot alongside it and the other foot behind it. With the feet comfortably spread, you will have better stability with the rear foot in the position for the upward thrust of the lift.

- **Pull the load toward you; then lift it gradually**. Avoid quick and jerky motions. Push up on your legs (fig. 19-4) while keeping your back straight. A straight back keeps the spine, back muscles, and other organs of the body in correct alignment. Tucking in your chin helps to align the spine. No matter what size the load, get as close to it as you can; then get a good grip by using the full palm and extending your fingers and hands around the object. Remember that your fingers

Student Notes:

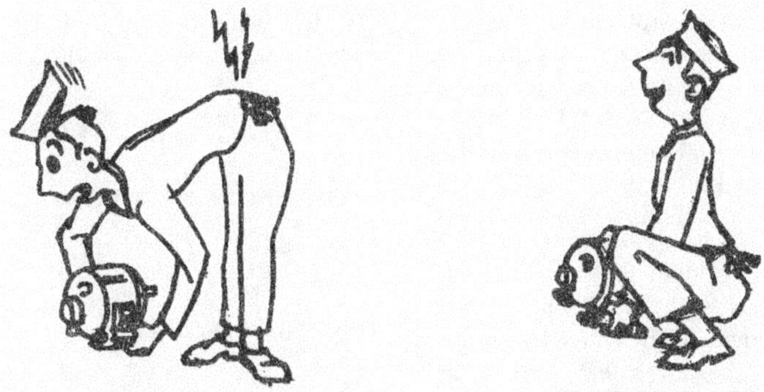

Figure 19-4.—Lift using your legs, not your back.

have very little power and need the strength of the entire hand. Keep your arms and elbows tucked in to the side of your body to help keep the body weight centered. Avoid twisting your body during the lift or while moving the load; change directions by moving your feet. Twisting your body during a lift is one of the most common causes of back injury.

- Be sure you have a clear vision over the load you are carrying.
- Don't change your grip while carrying the load.
- Face the spot where you intend to set the object; bend your knees, keeping your back as straight as possible and the weight of the object close to your body.
- Always allow enough room for the load to prevent injury to your toes and fingers.
- When you are placing a load on a table or bench, set it on the edge and push it forward with your arms and body.

Remember, if the load is too heavy or too awkward for you to move alone, get help! Remember, also lift with your legs, not with your back!

SHIPYARDS AND DRY DOCKS

Sooner or later every ship in the Navy will enter a shipyard or dry dock, usually during a predetermined scheduled overhaul. At times, ships go into shipyards or dry docks between overhauls for necessary repairs.

Shipyards and dry docks are dangerous places to work. So much work is scheduled, normally in a limited time frame, that safety is sometimes sacrificed for expediency (speed doing the work). During these times, look at what's taking place around you. Notice things like missing lifelines on deck (it's a long way to the bottom of a dry dock) and hatch or manhole covers removed without warning barriers erected. Working inside previously sealed compartments, voids, or tanks can be extremely dangerous if the proper safety precautions are not followed.

Often a lot of different evolutions are going on in a confined space. Welding or cutting operations could and often are conducted in the same small space as heavy equipment removal and chipping and painting.

Sometimes all lighting in a compartment or passageway may be removed for various reasons. That presents several safety concerns. You may trip on equipment or tools someone has left behind or bang

Student Notes:

your head on wire runs or ventilation ducts hanging down where they shouldn't be.

Fire hazards are always a problem in shipyards. Often, there is a large amount of equipment removal or repairs that require welding or cutting, the repainting of spaces, or opening fuel tanks and voids. Therefore, the need to make sure that all flammable material is removed from the ship everyday is significantly increased.

You won't be able to stop all shipyard accidents; but, by following the prescribed safety precautions, you can make the shipyard environment a lot safer place to work. Every ship has a shipyard safety doctrine and conducts safety training before entering a shipyard. If you pay attention at safety lectures and read the safety doctrine, your ship's stay in the dry dock will be much safer.

AIRCRAFT STRESS AREAS

Flight decks and hangars are dangerous, and the danger to personnel goes beyond the possibility of crashes. Engine exhaust tailpipes, engine-starting units, liquid oxygen (LOX) bottles, and connectors are all capable of causing severe injury. Engine-starting equipment (known as *huffers*) generate high temperatures that could severely burn personnel. If not sufficiently separate from the area where fuel tanks, ammunition, or other hazardous materials are being handled, they could cause fires or explosions. Jet engines also generate very high temperatures. Before attempting any type of repair or service work on these engines, make sure they have had enough time to cool down to avoid any possibility of burns.

Any area in which LOX is being used requires extra safety precautions. LOX in liquid form flows like water, but it also boils into gaseous oxygen at -297°F and is capable of immediately freezing any object it contacts. When LOX expands as a gas and is confined and allowed to warm, it exerts extremely high pressures (up to 12,000 psi), causing it to be very dangerous. Always keep clothing and tools free of oil and dirt. Never smoke or have any spark or flame-producing materials near an area where LOX is being handled. A spark or flame in this oxygen-rich atmosphere could be extremely dangerous with violent results. If your skin comes into contact with LOX, get medical attention immediately. Once again, safety procedures and precautions must be followed when you work with aircraft and equipment. By following these procedures and precautions, you significantly reduce your chances of getting hurt.

REVIEW 11 QUESTIONS

Q1. List the three types of hearing protection.

a.

b.

c.

Q2. List the three major precautions that you should follow when lifting heavy loads.

a.

b.

c.

Q3. What precaution is taken before a ship enters a shipyard for dry dock work?

Q4. LOX is dangerous and requires that you follow extra safety precautions. List two reasons why you should be careful when handling LOX.

a.

b.

HEAT STRESS PROGRAM

Heat stress is a combination of air temperature, thermal radiation, humidity, airflow, and workload that places stress on the body. The Navy's Heat Stress Program evaluates and monitors heat stress conditions

Student Notes:

to establish safe work schedules in heat stress environments.

Aboard ship, heat stress conditions can occur in almost any space. The causes of heat stress conditions are steam and water leaks, ventilation system deficiencies, missing or deteriorated insulation, and weather conditions of high heat and humidity. Prolonged exposure to heat stress conditions can cause heat exhaustion or heatstroke. These injuries occur when the body temperature continues to increase. The first signs are—

- Increased body temperature causing fatigue
- Severe headache
- Nausea
- Reduced physical and mental performance

If not immediately and properly treated, these injuries can be life threatening.

The best way to control heat stress hazards is to follow recommended work practices and procedures. Every ship in the Navy has a heat stress monitoring program. This program is designed to assist personnel that may be required to work in a heat stress environment by limiting the time they spend in a high heat stress situation. Personnel required to work in a heat stress environment receive training at regular intervals. Heat stress not only affects personnel that work below decks or in confined spaces but also personnel that work topside. Read your command's heat stress instruction; it may help you work smarter and safer.

COLD WEATHER

The Navy conducts operations in areas where weather is often a problem. You have already learned about the possible problems that you may meet in the Tropics. Now, you will learn about the problems you may face when the Navy operates in severe cold weather areas. These problem areas range from the Antarctic to the northern regions of the Pacific or Atlantic Oceans.

Your major health risk when working in these areas is hypothermia. Hypothermia results when the temperature of the body reaches subnormal levels. First aid for hypothermia, like that for heat stress, must be immediate. Other safety factors involved with operations in colder regions include ice accumulation on ships' decks and superstructures or when outside bulkheads or fittings become so cold that, when touched with bare skin, the skin sticks to these objects.

To protect yourself from hypothermia if you're working topside or go topside as part of your duties, you need to wear clothing designed to maintain body heat. You need to limit the amount of time you're exposed to such conditions. If you work topside and start to lose feeling on any part of your body, get inside and warm up. **Safety is paramount**!

REVIEW 12 QUESTIONS

Q1. Heat exhaustion and heat stroke are life threatening. List the signs of heat exhaustion and heat stroke.

a.

b.

c.

d.

Q2. What is the major health threat of cold weather?

GENERAL PRECAUTIONS

The precautions that follow are general, all-around safety practices that don't fit into any particular category. Some apply to several situations. Failure to observe any one of these practices could result in a serious mishap.

- Use tools that fit the work being done. Screwdrivers aren't meant to be used as punches.

Student Notes:

- If you are issued protective gear, wear it when performing work for which the gear was designed.
- Never overload electrical outlets.
- Keep file drawers closed when they are not in use. Avoid making files top-heavy and be sure drawer stops are operative.
- Don't hang extension cords where somebody can be snagged by them. Extension cords can become a trip hazard also. When using an extension cord, make certain it won't be cut by a closing hatch or door or by any other means while it is lying on the deck.
- Keep all tools in good condition.
- Don't watch a welder's arc if you aren't wearing dark goggles.
- Report defective equipment.
- When you open a hatch, always secure it open with the equipment provided.
- Secure all loose articles when heavy weather is expected.
- Take heed of all warning signs: HIGH VOLTAGE, STACK GAS, RF HAZARD, and so on.
- Never smoke in NO SMOKING areas, when the smoking lamp is out, when painting, or when handling ammunition or flammables.
- Follow good housekeeping practices at all times. Don't allow loose gear to accumulate where it might present a tripping hazard.
- Learn and follow all safety precautions for the job you are doing.

EQUIPMENT TAG-OUT PROCEDURES

Learning Objective: When you finish this chapter, you will be able to—

- Recall the purpose and procedures of the Navy Tag-out System.
- Identify and interpret HAZMAT labels.
- Identify the purpose of hazardous materials labels, signs, and symbols.

Post DANGER tags, CAUTION tags, and instrument OUT-OF-COMMISSION or OUT-OF-CALIBRATION labels following the authorized procedures. Those tags and labels help ensure the safety of personnel and prevent improper operation of equipment. Don't remove or break posted safety tags without proper authorization.

Practically every day, you are involved with tagging out a piece of equipment. You tag out a switch or a motor to secure the equipment to perform planned maintenance.

Why do we have tag-out procedures? We have them because our ships are complex and personnel can get hurt because of improper equipment operation. For example,

> A submarine was moored to a pier, where it was waiting for the local diving team to do some repair work on the hull. Since it was Saturday, only the duty officer, the duty chief, and the duty section were aboard. The diving tag-out had been written out and hung, and the diving team was waiting for the senior diver to get to the ship and check the tag out. Once that was done, the divers entered the water. The duty chief decided to check the tag outs in the torpedo room. The chief found the DANGER tags for the torpedo high-pressure air ejection system properly hung in place and second-checked—just like they were supposed to be. But, when the duty chief checked the position of the air valves, they were in the open instead of shut position. With the valves in the open position, the divers were subject to a blast of air of 5,000 pounds from the torpedo tubes. The duty chief immediately ordered the divers out of the water.
>
> What happened? The persons who hung the tags hadn't changed the position of the air

Student Notes:

valves. That time, no one was hurt. The persons who had hung the tags were reprimanded, and the repairs were completed.

During your career, you will probably tag out some type of equipment. Be alert, do the job right, and you shouldn't have any problem. Take your time and do the tag out right the first time. Don't let anyone or anything distract you while you're hanging a tag or second-checking one. If you're not sure of a tag-out procedure, get a copy of your command's tag-out bill. The *Standard Organization and Regulations of the U.S. Navy*, OPNAVINST 3120.32, govern the Navy's equipment tag-out bill.

PURPOSE OF EQUIPMENT TAG-OUT BILL

An equipment tag-out bill has three purposes—

1. To provide a procedure for personnel to use to prevent the improper operation of a component, piece of equipment, system, or portion of a system that is isolated or in an abnormal condition.

2. To provide a procedure for personnel to use in operating an instrument that is unreliable or not in a normal operating condition. (**NOTE**: This procedure is like the tag-out procedure, except that it requires the use of labels instead of tags to indicate instrument status.)

3. To provide separate procedures for personnel to use when accomplishing certain planned maintenance (PMS) actions. These procedures apply only to non-nuclear surface ships and craft and non-nuclear, non-propulsion areas of nuclear surface ships. PMS tag-out procedures aren't authorized aboard submarines, submarine tenders, submarine rescue vessels, in propulsion areas of nuclear surface ships, or within submarine support facilities.

All U.S. Navy ships and repair activities must use standardized tag-out procedures.

ORGANIZATION

The CO or officer in charge heads the tag-out bill organization. Department heads are responsible for making sure personnel in their departments understand and follow bill procedures.

When a repair activity performs repairs on a ship, the ship is responsible for and controls the tag-out system for the equipment being repaired. The repair activity is responsible for complying with (following) tag-out bill procedures.

Commanding Officer

The CO or officer in charge is responsible for the safety of the entire command. The CO must make sure that all concerned persons know and comply with the applicable safety precautions and procedures of the tag-out system.

Officer of the Deck (OOD)

The OOD may be the OOD or the ship's duty officer, depending on the ship's condition. The OOD keeps track of the systems being tagged out and the condition of readiness of the ship.

Departmental Duty Officer (DDO)

The departmental duty officer (DDO) is designated (named) on the approved watch bill or in the plan of the day. The DDO is responsible for knowing the material condition of a department and the state of the readiness at all times. This officer must know what systems are tagged out for periodic maintenance or for repairs requiring long downtime.

Engineering Officer of the Watch (EOOW)

The engineering officer of the watch (EOOW) keeps up with the status of the engineering plant at all times and whether a tag-out bill affects the readiness of the plant. Depending on the engineering plant conditions, the engineering duty officer may serve as the EOOW. The EOOW informs the proper persons of the status and readiness of the plant and when it will be repaired and returned to normal status.

Authorizing Officer

The authorizing officer signs the final authorization placing a system or piece of equipment off line for

Student Notes:

repairs or maintenance. The authorizing officer has the authority to sign tags and labels and the authority to cause tags and labels to be issued or cleared. The authorizing officer is always the officer responsible for supervising the tag-out log. The CO designates authorizing officers by billet or watch station.

Repair Activity Representative

If a tag out has been requested by a repair activity, a representative (shop supervisor or equivalent) signs the tag-out record sheet. This person's signature indicates repair activity satisfaction with completeness of the tag out. The repair activity representative should check and sign each tag that has been hung as he or she makes sure each system is completely isolated. Only after taking that safety precaution should the representative sign the tag-out record. When verified, the tags alert personnel that the repair activity must approve removal of the tags. The repair activity representative approves removal of the tags by signing a tag-out sheet stating that the work is completed and no more work is to be done on the system(s).

Person Attaching the Tag

The person who attaches the tag (along with the person who second-checks the tag) can make or break the tag-out system. The person hanging the tag actually shuts a valve or secures a switch that takes a piece of equipment off line for repairs or maintenance. When you secure a switch or shut a valve, you hang the danger or caution tag securely so that it won't fall off, then you sign it. By signing the danger or caution tag, you verify that you have secured the items that need to be secured and that they are secured.

Person Checking Tag

As you know, the person checking the tag is an important person in the tag-out procedure. The process of checking a tag or label is called *second-checking*. The second-checker examines the tag or label to make sure it corresponds to the equipment that is supposed to be secured and checks the position of the switch or valve. If no mistakes are found, the second-checker signs the tag or label. The signature tells everyone concerned that "all is okay" with the tag or label and that the equipment is secured. If the second-checker finds something wrong, he or she notifies the first person (person attaching the tag) and the authorizing officer that something's wrong. The person who tags a system and the second-checker have a big responsibility—the lives of their shipmates as well as their own rely on how well they do their jobs.

TAGS, LABELS, AND LOGS

The various tags, labels, and logs used in the tag-out system have a definite purpose. The tags and labels indicate the equipment is out of order or unable to perform its normal functions. These tags are red and yellow, and both are used as warning tags.

- A red tag means a certain DANGER exists if the valve or equipment lineup is changed.
- A CAUTION tag is yellow and usually has a set of instructions printed on it about the operation of the equipment.
- Two labels are associated with the tag-out system—the OUT-OF-COMMISSION(red) and OUT-OF-CALIBRATION (orange) labels.

The tags, labels, and logs used in the tag-out system help to ensure personnel safety. Let's look at how you use each of them.

Caution Tag

Use a yellow CAUTION tag, NAVSHIPS 9890/5 (fig. 19-5), as a precautionary measure to provide temporary special instructions or to show personnel that they must use extra caution in operating equipment. In the instructions, state the specific reason for the tag. Don't use phrases such as "Do not operate without EOOW permission." Personnel don't operate equipment on systems without permission from the responsible supervisor. Don't use a CAUTION tag if personnel or equipment can be endangered while performing evolutions using normal procedures. Use a DANGER tag in these circumstances.

Danger Tag

Attach a red DANGER tag, NAVSHIPS 9890/8 (fig. 19-6), to prohibit operation of equipment that could jeopardize the safety of personnel or endanger equipment, systems, or components. Never operate or

Student Notes:

remove equipment tagged with DANGER tags. Operating a piece of equipment tagged out because of an electric short could cause an injury or death. It could also cause damage to equipment that could stop a ship from operating.

Out-of-Calibration Label

Many gauges and devices are used to monitor how equipment is operating. When regularly monitored, these gauges or devices tell us when something is wrong with the equipment. Check all monitoring devices periodically to ensure they are measuring accurately. Attach orange OUT-OF-CALIBRATION labels, NAVSEA 9210/6 (fig. 19-7), to identify instruments that give inaccurate measurements because they are out of calibration. This label means you must use the instrument only with extreme caution, if at all. When using an out-of-calibration label, mark the label with the magnitude sign (6 or 4) and units of the required correction or the word *overdue*.

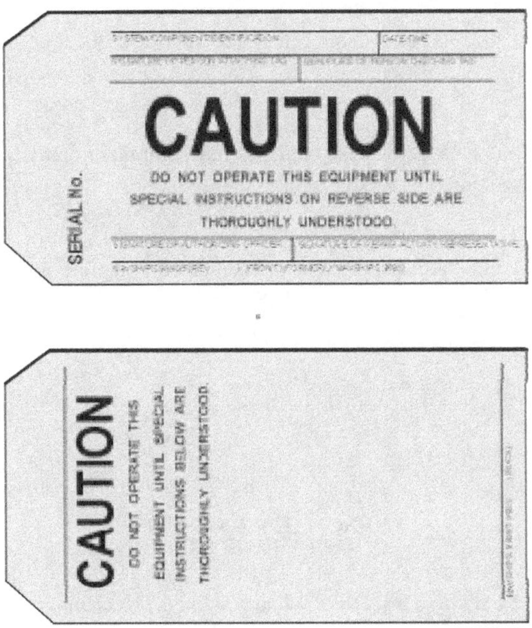

BMRf1904

Figure 19-5.—Caution tag.

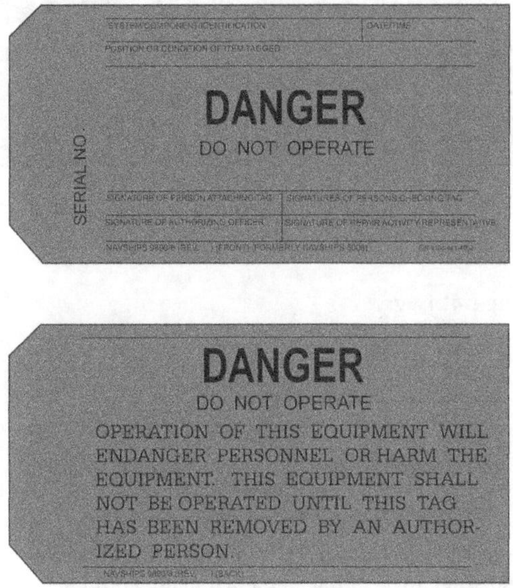

BMRf1905

Figure 19-6.—Danger tag.

Out-of-commission Label

Use red OUT-OF-COMMISSION labels, NAVSHIPS 9890/7 (fig. 19-8), to identify instruments that give incorrect measurements because they are defective or isolated from the system. This label shows that you cannot rely upon the instrument or use it properly until it has been repaired and recalibrated or reconnected to the system.

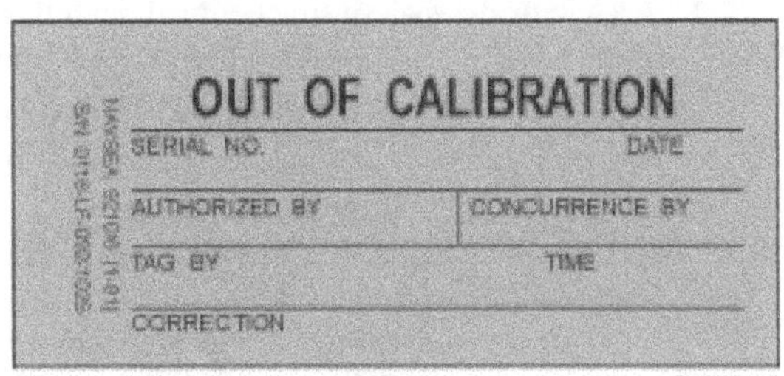

BMRf1906

Figure 19-7.—Out-of-calibration label.

Student Notes:

Figure 19-8.—Out-of-commission label.

Tag-out Logs

Tag-out logs are used to control the entire tag-out procedure. The number of tag-out logs required depends on ship size. For example, a minesweeper may only require one tag-out log for the whole ship, while a major surface combatant may require a separate log for each department. Individual force commanders specify the number of logs various ship classes must maintain and what areas of the ship must maintain them.

On ships maintaining more than one tag-out log, authorizing officers must exchange information on tag-out actions. When a tag-out affects other authorizing officers, the initiating party obtains verbal permission from those officers to tag out the system or equipment in question before the tag out is authorized. Examples of systems that may require such coordination are ship service electrical distribution, hydraulics, air, ventilation, and air-conditioning chill water systems.

The tag-out log is a record of authorization of each effective tag-out action. It contains the following documents:

1. A copy of the main instruction and any other amplifying directives for administering the system. These documents are kept in the front of the log.
2. A DANGER/CAUTION tag-out index and record of audits (index/audit record). The index/audit record provides a sequential list of all tag outs and ensures serial numbers are sequentially issued. They are used in audits of the log as well to provide a ready reference of existing tag outs. The cognizant department head may remove the index pages with all tag outs listed as cleared.
3. Cleared DANGER/CAUTION tag-out record sheets that have been cleared and completed. These sheets are kept in the log until received and removed by the cognizant department head.

Tags in a common system (for example, ship's radar or a fire-control system) are logged on one DANGER/CAUTION tag-out record sheet. Subsequent sheets on the same system are kept together.

REVIEW 13 QUESTIONS

Q1. List three purposes of the tag-out bill.

a.

b.

c.

Q2. What person can make or break the tag-out system?

Q3. A DANGER tag identifies equipment that is in what condition?

Q4. What documents are contained in tag-out logs?

a.

b.

c.

PERSONAL PROTECTIVE EQUIPMENT

Learning Objective: When you finish this chapter, you will be able to—

Student Notes:

- Recall the procedures for use and maintenance of personal protective equipment.

Personal protective devices do nothing to reduce or eliminate hazards. They merely establish a "last line of defense." Some devices that are not worn properly or that are subjected to improper maintenance may not work as designed. For this reason, proper equipment selection, maintenance, personnel training, and mandatory enforcement of equipment use are key elements in the use of personal protective equipment.

You should know what equipment to wear, when to wear it, and how to wear it. You should also know how to take care of the equipment. If you take care of the protective devices, they will take care of you.

The following paragraphs describe some of the protective equipment available to personnel and the procedures to follow in upkeeping this equipment:

- Eye protection includes such articles as personal eyeglasses, common-use goggles, and common-use face shields. These articles should be kept clean and disinfected. Personal eyeglasses are the responsibility of the owner/wearer. Eye protection should be stored where it will be protected from dust, moisture, and the weight of other objects placed directly on it. The best container is probably the box it was packaged in by the manufacturer.

- Respiratory protection, such as respirators, should be assigned to you for your exclusive use, if practical. Respirators should be cleaned and disinfected regularly. While cleaning, you should check for wear or deterioration. This type of protection should be stored in a container that will protect it from dust, moisture, and the weight of objects placed on top of it.

- Hearing protection includes articles such as circumaural protection and earplugs. Earplugs should be washed often (with the exception of the disposable plugs, of course). The circumaural protective devices should have the ear pads cleaned and disinfected periodically. Most small earplugs come in a small container especially made for them. The circumaural device can be hung from the headband.

- Foot protection includes steel-toed boots or shoes, which should fit properly. When they wear out, replace them.

- Head protection includes helmets and hats that are worn to protect the head from falling or flying objects and low overheads. Check these periodically for worn headbands or cracks in the shell.

- Electrical protective devices include rubber gloves, rubber mats, rubber hoods, rubber sleeves, and rubber blankets. Keep these items clean and free of moisture. Check these periodically for cracks or holes in the rubber material. When storing the gloves, return them to the box they came in and do not stack anything on them that would crush them.

PROCEDURES FOR REPORTING SAFETY HAZARDS/VIOLATIONS

Learning Objective: When you finish this chapter, you will be able to—

- Recall the procedures for reporting safety hazards and violations.

The first part of this chapter explained your personal responsibilities. If you detect a safety hazard, you are required to report this hazard to your immediate supervisor. The supervisor will then have the hazard corrected or seek assistance from the ship's safety officer on ways to correct it. *Navy Safety Precautions for Forces Afloat*, OPNAVINST 5100.19, contains the information on Navy safety.

REVIEW 14 QUESTIONS

Q1. List the personal protection equipment you should use in each of the following categories.

a. Head protection

b. Electrical protective devices

c. Eye protection

d. Respiratory protection

Student Notes:

Q2. If you see a safety hazard, whom should you notify?

SUMMARY

Throughout your Navy career you will continually hear the phrase "Think safety!" and rightfully so. As said at the beginning of this chapter, our profession is inherently dangerous. We can make our place of work considerably safer simply by paying attention to what goes on in our work space on a daily basis.

We have covered a wide variety of safety factors in this chapter. How to properly and safely embark and disembark a liberty boat was discussed. You learned how to use cleaning supplies and equipment properly to keep your berthing compartment shipshape. The proper use of paint and utensils to keep your ship looking good was also covered. How to use the tag-out system to repair or replace equipment, systems, or components to avoid hazards to personnel or equipment was stressed. Numerous evolutions conducted aboard ship on a daily basis would be safer if people would take a few minutes to observe what is going on. Hopefully, observing the safety precautions associated with doing a particular task will reduce mishaps.

Every job in the Navy has a set of safety guidelines. In their haste to get the job done, people sometimes cut corners. They do not realize that just around the corner lies an overlooked or disregarded safety precaution waiting to get us. Paying attention to what goes on around you and your shipmates and observing the proper safety precautions will reduce the number of mishaps considerably. Think safety!

REVIEW 1 ANSWERS

A1. Some safety precautions that could save lives include—

a. **Observe all safety precautions**

b. **Report unsafe conditions**

c. **Warn your shipmates of hazards**

d. **Wear protective clothing and equipment**

e. **Stay alert**

A2. Being aboard ship is dangerous. Some dangerous shipboard environments you may work in or work around involve—

a. **Powerful machinery**

b. **High-speed equipment**

c. **High-temperature, high-pressure steam**

d. **Volatile fuels and propellants**

e. **Heavy lifts**

f. **High explosives**

g. **Electrical voltages**

h. **Wind and waves**

REVIEW 2 ANSWERS

A1. The publication that contains information on using, storing, and disposing of hazardous materials is the **Material Safety Data Sheets (MSDS)**.

A2. According to **OPNAVINST 5100.19**, you should follow MSDS guidelines when handling hazardous materials.

REVIEW 3 ANSWERS

A1. The boat safety precautions that every Sailor should know include—

a. **Obey all orders from the coxswain.**

b. **Embark in a quiet, orderly manner and move as far forward as possible. Once embarked, stay in place**.

Student Notes:

c. **Keep all parts of your body in the boat; do not sit on gunwales.**

d. **Don't engage in horseplay.**

e. **Never distract the attention of crew members from their duties.**

f. **Don't sit on life jackets; this will mat the filler and reduce buoyancy.**

g. **When told to do so, don your life jacket immediately.**

h. **Don't smoke in a boat.**

i. **If told to embark or disembark, do so without argument. During heavy weather, boat loads will be reduced.**

A2. If a boat swamps, **don't panic**! Panic is easily spread from person to person causing people to lose their lives.

A3. You should learn the location of cleats, bitts, and pad eyes on a ship's deck because they're **tripping hazards**; if you know where hazards are located, you stand a better chance of avoiding the hazard.

A4. Two hazards found on flight decks of aircraft carriers are—

a. **Propellers**

b. **Jet engines**

REVIEW 4 ANSWERS

A1. Handling cargo improperly can result in injury and death. In the following cases you should take the indicated precautions.

a. When working with line, **never stand in the bight of a line. Keep clear of lines under strain because a line under strain can break with a whiplike snap that can cause severe bruising, broken bones, amputations, or death.**

b. When lifting heavy objects, **crouch close to the load with feet spread. Grip the object and lift with your arm and leg muscles (not your back). If the load is too heavy for one person to lift, ask for help.**

c. When steadying a load, **use the nonworking side of a ship for fore-and-aft travel. Don't stand between the load and a fixed object; don't stand under a suspended load; and never ride loads.**

A2. The **OOD** grants permission for any work done aloft.

A3. Before permission is given for personnel to work aloft, the following precautions are taken:

a. **Power is secured on radio and radar antennas and controls associated with antennas are tagged.**

b. **The engineer officer is notified to prevent operations such as lifting boiler safety valves or blowing tubes.**

A4. Lifelines are **safety barriers to prevent personnel from falling or being washed over the side.**

A5. When working over the side, you should wear the following equipment:

a. **Standard Navy safety harness with safety line attached and tended by someone on deck**

b. **An inherently buoyant life jacket with a hole in the back, allowing you to wear a safety harness**

REVIEW 5 ANSWERS

A1. Most accidents involving steam happen in **engine rooms and firerooms.**

A2. You should never enter a closed space until it's certified by the gas free engineer **because closed compartments contain unexpected dangers, including pressures, toxic gases, carbon monoxide, carbon dioxide, and possibly no oxygen.**

A3. The symptoms caused by bad air include—

a. **Labored breathing**

b. **Excessive fatigue**

c. **Headache**

d. **Dizziness**

REVIEW 6 ANSWERS

A1. Open flame and naked lights are defined as follows:

a. The term ***open flame*** **includes all forms of fuel or gas lanterns, lighted candles, matches, cigarette lighters, and so on.**

b. The term ***naked lights*** **includes any unprotected electrical lighting device.**

A2. You should take the following actions when storing solvents:

a. **Label all containers used to store solvents**

b. **Store solvents in appropriate lockers**

REVIEW 7 ANSWERS

A1. When handling a weapon, you need to think about what you're doing **because accidents don't "just happen;" they're caused. In fact, they're often caused by personnel who don't follow safety precautions or who are careless.**

A2. Projectiles that have a 3-inch or greater diameter are color-coded **to show the projectile type and the kind of bursting charge that they contain.**

REVIEW 8 ANSWERS

A1. **Treating common 115-volt equipment lightly** is the cause of many fatal shocks received from drills and fans.

A2. Three types of hazards associated with compressed gases are—

a. **Cylinders not secured**

b. **Cylinders under high pressure**

c. **Cylinders containing poisonous, flammable, or explosive material**

A3. **True**, oxygen and chlorine are stowed in compartments separate from flammable gases.

A4. You should use a filter mask respirator when working with fiberglass **because fiberglass dust is abrasive and an irritant to skin and eyes.**

REVIEW 9 ANSWERS

A1. Before beginning work to repair a piece of equipment, you should **make sure that the equipment is de-energized and/or depressurized and tagged out of service.**

A2. **Only properly trained** personnel should operate gas welding or cutting equipment.

A3. When working around rotating machinery, **you should remove jewelry and watches and you shouldn't wear loose fitting clothing; wear protecting clothing and equipment, such as hearing protection, eye, hand, and foot protection, dust and paint respirators, and so on.**

REVIEW 10 ANSWERS

A1. The safety precautions to follow when working with systems having pressurized liquids include—

a. **Never connect or disconnect a hose from the system until the pressure has been removed.**

b. **Never point a charged (pressurized) fire hose at anyone.**

c. **Never use ruptured or worn hoses.**

d. **Don't use spray paints, butane fluids, lacquers, and other aerosol products near a flame; don't throw them into a fire; and don't puncture the container.**

A2. If you've been exposed to acids or alkalis, **you should immediately seek medical attention.**

A3. You shouldn't smoke near sewage-handling equipment for the following reasons:

a. **Fuel leaks or spills can occur in the incinerator area where temperatures may exceed the flash point of the fuels used.**

b. **Methane and hydrogen sulfide may be emitted by any tank or tank leaks. These gases are also flammable and under some conditions are explosive.**

REVIEW 11 ANSWERS

A1. The three types of hearing protection are the—

a. **Headband,**

b. **Earplugs**, and the

c. **Circumaural muff**.

A2. List the three major precautions you should follow when lifting heavy loads.

a. **Don't lift an object if it is too heavy or too clumsy for good balance**.

b. **Keep the load close to the center of your body**.

c. **Pull the load toward you; then lift it gradually**.

A3. Before entering a shipyard for dry dock work, **every ship has a shipyard safety doctrine and conducts safety training before entering a shipyard**.

A4. LOX is dangerous to handle because—

a. **It freezes immediately on contact**.

b. **As a gas, it exerts extremely high pressure**.

REVIEW 12 ANSWERS

A1. The symptoms of heat exhaustion and heat stroke include—

a. **Increased body temperature**

b. **Severe headache**

c. **Nausea**

d. **Reduced mental and physical performance**

A2. The major health threat of cold weather is **hypothermia**.

REVIEW 13 ANSWERS

A1. Purposes of the tag-out bill include—

a. **To provide personnel a way to prevent the improper operation of a component, piece of equipment, system, or a part of a system that's isolated or in an abnormal condition.**

b. **To give personnel a way to operate an instrument that's unreliable or not in a normal operating condition.**

c. **To give personal a way to accomplish certain planned maintenance system (PMS) procedures**.

A2. The tag-out system is made or broken by the **person attaching the tag**.

A3. A DANGER tag identifies equipment **whose operation is prohibited because its use could jeopardize the safety of personnel or endanger equipment**.

A4. Tag-out logs contain—

a. **A copy of the main instruction and any other amplifying directives for administering the system**.

b. **A DANGER/CAUTION tag-out index and record of audits (index/audit record)**.

c. **Cleared DANGER/CAUTION tag-out record sheets that have been cleared and completed**.

REVIEW 14 ANSWERS

A1. Personal protection equipment you should use in each of the following categories:

a. Head protection	**Helmets and hats**
b. Electrical protective devices	**Rubber gloves, rubber mats, rubber hoods, rubber sleeves, and rubber blankets**
c. Eye protection	**Personal eyeglasses, common-use goggles, and common-use face shields**
d. Respiratory protection	**Respirators**

A2. If you see a safety hazard, you should **notify your immediate supervisor**

CHAPTER 20

SEA POWER

Control of the seas means security. Control of the seas means peace. Control of the seas can mean victory. The United States must control the sea if it is to protect our security.

—John F. Kennedy

The United States is in a position of world leadership. Maintaining that position is a never-ending task that becomes harder with each crucial world situation. The Navy has a vital role in protecting world freedom. We can only maintain this freedom through a Navy that has total dedication to that end. You are an important link in our Navy's commitment to freedom.

In the Navy, we, like our forefathers, must make many sacrifices to maintain our goals. That often means being away from our homes for long periods, standing long watches, or doing arduous work. The result is fulfilling the goal of keeping the world free.

As you study for advancement to petty officer, you should begin to realize your importance to the overall mission of the Navy. Advancement will be just one of the rewards you will receive for dedication and sacrifice.

UNITED STATES SEA POWER

Learning Objectives: When you finish this chapter, you will be able to—

- Recognize the importance of sea power in relation to today's world.
- Identify the operational components of the U.S. Navy sea power.

Sea power as a concept means more than military power at sea. Sea power describes a nation's ability to protect its political, economic, and military interests through control of the sea. The principal parts of sea power are naval power, ocean science, ocean industry, and ocean commerce.

Sea power encompasses commercial rivalries in peacetime, diplomatic maneuvering and the clash of fleets in wartime. The concept of sea power has been valid whether the fleets were wooden men-of-war or mighty battleships. It remains sound today, although technology has caused ship-to-ship battles to become part of history instead of part of contemporary tactics. Captain Alfred Thayer Mahan, USN, was the first person to use the term *sea power.* He used it in his principal work, *The Influence of Sea Power Upon History, 1660-1783,* published in 1890. Mahan proposed that there were six conditions required for a nation to have sea power:

1. An advantageous geographical position
2. Serviceable coastlines, abundant natural resources, and a favorable climate
3. Extent of territory
4. A population large enough to defend its territory
5. A society with an aptitude for the sea and commercial enterprise
6. A government with the influence to dominate the sea.

In the decades immediately following the Civil War, the primary role of the U.S. Navy was as coastal defender and commerce raider. The United States did not exercise sea power, but believed in the concept of national isolation. In effect, the nation stressed naval expansion within its own country. By 1890, however, the nation began naval expansion toward other countries; its concept of national isolation began to ebb.

Those groups in the Navy and in the government who believed in sea power endorsed Mahan's doctrine. They based their endorsement on the belief that history provides clues to achieving maritime supremacy. Mahan's concept, therefore, became the intellectual force behind the United States' development of its Navy into a sea power.

During World War II the emerging effects of aircraft, aircraft carriers, and radar meant we fought fewer battles with ships within sight of each other. In modern naval tactics, we employ gunfire for protection against aircraft and missiles or for bombarding shore targets. If aimed at ships, the targets will most likely be small, fast, patrol craft. These crafts deliver missile or torpedo attacks in coastal waters.

Sea power today includes many aspects of the naval strength of a nation that did not exist in the last century. Sea power now encompasses maritime industry and marine sciences. These industries and sciences add to our national economy by exploring new resources for food, freshwater, minerals, and even living space.

Figure 20-1 shows a Carrier Task Group, one concept of sea power today. Sea power is a unique resource that nations can use in the oceans. We use it to reach political, economic, and military goals in times of peace and war.

The seas are our lifeline for survival. In addition to being a barrier between nations and a broad highway for ships, the seas are an important source of food, minerals, and metals. We use oceangoing craft to get to these riches. The development of these craft has resulted in the need to provide for their protection.

A well-established theory for the economic advantage of a nation is to produce goods and services and exchange them with other nations. Throughout history, nations that have traded this way and conducted a strong foreign trade have prospered and grown in economic and political strength. Those that have failed in commerce have also failed as world powers. Throughout history, no country has ever become a world power without a strong foreign trade. All countries generally have raw materials, but they often have limited quantities. Countries then trade with each other to get needed materials. Modern nations with highly complex economies need more raw materials from other countries. We can often obtain many manufactured goods cheaper from other countries than we can produce them locally. As a matter of economic reality, most nations must trade or decline in strength.

Until recently, Americans believed that our raw materials would last forever and that we could live without help from any other nation. With our population growth and the advanced technology of the United States, this concept has changed. Today we rely heavily on trade with our world neighbors for raw materials. We need that kind of trade to keep our economy strong and our work force employed.

Figure 20-1.—U. S. naval sea power.

Student Notes:

The United States is not as independent as people think. We must import most of our raw materials. Actually, we import no fewer than 77 resources to maintain our present economy. As an example, we import 85 percent of the manganese we need to make steel. We use columbite to make nuclear reactors, stainless steel, rockets, and missiles; we import 90 percent of it. We also import bauxite (used to refine aluminum) and chromite (used to strengthen steel). More than 90 percent of the tin we need in this country we import. At one time, the U.S. consumed more than one-third of the entire world's supply of oil. However, through conservation efforts we have reduced that oil consumption. Half of the free-world mineral production goes into the industrial needs of the United States. Of all our needed minerals, only about 11 are found within our borders; the U.S. is a raw-material-deficient nation. The United States could not possibly produce enough aircraft to move all the goods that now travel by water. Our economy depends on waterborne commerce.

The United States, like all nations of the world, acknowledges freedom of the seas under international law. When fighting wars, nations do whatever is in their power to prevent the enemy from using the seas. They aim to cut commercial shipping lanes to prevent the enemy from receiving critical raw materials for the war effort. Throughout history, the great nations have been those which controlled the seas. From the ancient times of Persia to the World War II days of Japan, loss of sea power has caused many nations to fail.

Before World War I, we were a quiet nation and stayed mostly to ourselves. When we were drawn into World War I, we became the most industrialized nation in the world. Our economy slowed down after the war; when World War II started, we once more became highly industrialized. We have remained that way ever since. Our defense depends on a highly productive industrial system. We must keep the sea-lanes open so that the supply of essential raw material continues to flow in our direction. Halting the flow would be a great blow to the safety and economy of the United States. In the wars of this nation, we have managed to maintain a constant supply of raw materials. But, to keep our troops supplied, we have had to ship over 97 percent of our products overseas.

You should realize the importance of the United States' ability to maintain control of the seas for the use of the free world. To protect our national security and sustain our economy, our nation must continue to take the following actions:

- Import raw materials from throughout the world, convert them into manufactured goods, and export them to the world marketplaces by ocean shipping.
- Keep the sea-lanes open and secure in times of peace and tension, and deny them to the enemy in times of war.

Many areas of sea power are covered in the remainder of this chapter. Keep in mind that no matter where your station is, your job plays an important role in our nation's sea power. Your job helps keep us all free and secure.

REVIEW 1 QUESTIONS

Q1. What is sea power?

Q2. List the principal operational components of our nation's sea power.

a.

b.

c.

d.

Q3. According to Alfred Mahan, there are six conditions required for a nation to have sea power. List these conditions.

a.

Student Notes:

b.

c.

d.

e.

f.

Q4. In today's world, what aspects of naval strength exist that didn't exist in the 19th century.

Q5. As this century closes, no nation is totally independent. To protect ourselves and to keep our economy going, this country must take the following actions:

a.

b.

THE U.S. NAVY'S RESPONSIBILITY IN SEA POWER

Learning Objectives: When you finish this chapter, you will be able to—

- Identify the missions and functions of the U.S. Navy in wartime and peacetime.
- Identify the functions of the U.S. Navy to include strategic nuclear deterrence and security of sea-lanes communications.

At this point in your Navy career, if you haven't done so already, you may soon find yourself asking several questions. Why are we spread out far and wide from our shores? Why do we have a Navy? What is the purpose of this deployment? If you look at the goals of our nation, you will see what our mission is. First, you should understand why we need a strong Navy to support our national objectives. Some of these reasons are as follows:

- Two of our states are outside the continental United States (Hawaii and Alaska).
- Four U.S. territories lie overseas (Puerto Rico, the Virgin Islands, Guam, and the Northern Marianas).
- Two of our allies (Canada and Mexico) border the United States; the rest of our allies, some 42 of them, are overseas.
- NATO countries and Japan, our principal allies, are highly dependent on U.S. support and imports, the bulk of which comes to them by sea.
- Ninety-nine percent of all U.S. overseas trade is transported by sea lines of communications (world trade routes).
- The U.S. industrial output depends on continued shipments of raw materials and energy-producing resources from overseas.
- Our ability to control the seas is essential in the deterrence of a general war and aggression against any nation or area vital to our interest.

Now, let's look at the primary functions of the Navy. The Navy and the Marine Corps organize, train, and equip Navy and Marine Corps forces to conduct prompt and sustained combat operations at sea. These operations involve sea-based aircraft and land-based naval air components. These forces have five primary tasks:

1. They must seek and destroy enemy naval forces
2. Suppress enemy sea commerce gain
3. Maintain general naval supremacy
4. Control vital sea areas
5. Protect vital sea lines of communications

The Navy's business is to clear the way for the operating forces to accomplish their task, whatever it is. The Navy must drive the enemy's fighting forces off the high seas, out of the air, and across the seas. The Navy

Student Notes:

must block the enemy's sea-lanes and sink its merchant ships and transports.

In recent years, we have exercised control of sea-lanes in the Middle East. During the 1987-1989 "tanker wars" in the Persian Gulf (fig. 20-2), the U.S. Navy protected merchant ships and oil tankers flying the U.S. flag. In 1990, we conducted a naval blockade of Iraq to enforce United Nations sanctions following Iraq's invasion of Kuwait.

The Navy also provides forces for joint amphibious operations. It trains all forces assigned to these operations in amphibious warfare as directed by the Joint Chiefs of Staff. It also conducts naval reconnaissance, antisubmarine warfare, mine laying and controlled mine-field operations, and protects shipping. Operation Desert Shield/Storm is a typical example. The Navy joins with the other services in defending the United States against air attack.

As you can see, the Navy's mission is very complex. As a result of that complexity, the United States is undertaking a massive modernization of Navy ships, aircraft, and weapons in three forms. The first involves the speedup of research and development to find new weapons. The second entails the laying up of old ships to save operating and overhauling costs and the shifting of that money into new construction. The third consists of the "hi-low balanced mix" concept. That concept involves the purchase of a few highly effective aircraft and ships, such as nuclear propulsion aircraft carriers (CVNs) and submarines (SSBNs). At the same time, we are developing new classes of low-cost ships, such as guided-missile frigates and sea-control ships.

Our nuclear-age world has resulted in a nuclear-age Navy. Although the Navy uses nuclear weapons and guided missiles as its primary destructive weapons, it still maintains, and is improving, conventional weapons. Such weapons enable the Navy and Marines to rapidly deploy and to apply the necessary force to fight a limited war.

The Navy leads the way in scientific projects. In the area of navigation, Navy ships can navigate on and under the oceans for days at a time. They no longer rely on traditional sources such as landmarks and stars to fix

BMR2002

Figure 20-2.—U. S. ships blowing up an oil platform in the Persian Gulf.

Student Notes:

their position. The Navy continues to improve its propulsion systems. The Navy's continued improvements in propulsion systems allow Trident submarines to operate undetected beneath the oceans. The newer, faster, and quieter fast-attack submarines prowl the oceans at will. These ships have added a new dimension to the world of undersea warfare. We have made great strides in underwater acoustics, oceanography, and other scientific fields.

Throughout history, the shores of the enemy and the range of our ship's guns have limited the Navy's radius of action. Now with the development of long-range aircraft and ballistic missiles, the Navy's radius of action spans the world.

In the past, when ships sailed in a task force, they traveled together in formation. However, that tactic increased the number of losses during an attack. Today, ships are dispersed over a wide area, which increases their chances of survival in the event of a nuclear attack.

Although the tactics of our fleets have changed, the meaning of sea power and the need for sea power have remained constant. The Navy will always seek positive change, using weapons dictated by the times and situation, to protect our nation from enemy invasion. America's sea power will play a vital part in tomorrow's world and will have a great influence on peace.

Our nation and the other countries of our world rely on the U.S. Navy to guard their liberties. We must continue to guard these liberties as an instrument of peace, not as an instrument of terror or offensive threat. We must join with other free nations in promoting freedom throughout the world.

THE U.S. NAVY'S MISSION

Today, the Navy, together with the Army and the Air Force, is a member of the National Military Establishment. Their mission is to be prepared to conduct prompt and sustained combat operations in support of the national interest. As part of the National Military Establishment, the U.S. Navy's mission is to assure continued maritime superiority for the United States. The National Security Act, passed by Congress in 1947, instituted the National Military Establishment. The aim of the National Military Establishment is the coordination of the security of the United States under the Secretary of Defense.

You have an important part to play in the mission of the Navy. Your responsibility grows as you advance in rate. Before you start to take on that responsibility, you should be familiar with certain terms so that you can fully understand the mission of the Navy. They are national strategy, national interests, and national objectives, as stated in Naval Warfare Publication 1 (NWP-1). Naval Warfare Publication 3 (NWP-3) defines naval strategy. Those publications outline our commitment to the security of the United States.

National Strategy

National strategy is that broad course of action designed to achieve national objectives in support of national interests. To satisfy that objective, the defense forces must have the capability to deter aggression and to prevent coercion. They must also have enough influence to shape world events in favor of U.S. interests. The United States maintains its defense forces to preserve its physical security and protect its political independence.

National Interests

National interests are conditions that are to the advantage of our nation to pursue or protect. These conditions frequently are of a continuing nature. They range from the ultimate interest—national survival—to specific regional interests. Collectively, those interests determine the importance of a particular region to the security of the United States.

National Objectives

National objectives are specific goals our nation seeks to advance, support, or protect. We primarily have political, economic, and security objectives.

Naval Strategy

Naval strategy is our nation's use of naval forces (including naval aviation and Marine Corps forces) to achieve its naval objectives. National strategy determines our naval objectives. Our overall naval strategy objective is control of the seas and the denial of

Student Notes:

an enemy's use of those seas important to our operations.

The Navy's job goes hand in hand with the national interest and the objectives of the rest of the U.S. armed forces. Title 10 of the U.S. Code states that the Navy is to be prepared to conduct prompt and sustained combat operations in support of the national interest. That means we must assure continued maritime superiority for the United States. We must be able to totally defeat any threat to the continued free use of the high seas by the United States. Therefore, we must maintain the ability to destroy hostile aircraft, surface ships, and submarines that threaten our seaborne forces and those of our allies. The national strategy determines the Navy's mission. We carry out that mission in joint coordination with the other armed forces and in combined planning with U.S. allies. In carrying out that mission, the Navy has two major functions—**sea control** and **power projection**.

THE FUNCTIONS OF THE U.S. NAVY

Sea control, total control of the seas for the free movement of all, is the first function of the U.S. Navy. It means control of set air, surface, and subsurface areas, when and where needed. Sea control is crucial to national strategy. It allows us to use the oceans as barriers for defense and as avenues to extend our influences overseas.

Power projection is the second function of the Navy. It is the ability to use sea power throughout the world in the timely and precise manner needed to accomplish a given goal. This covers a wide area. We accomplish power projection by using a broad spectrum of offensive naval operations. These operations include the tactical employment of carrier-based aircraft and the use of amphibious forces and naval gunfire support forces. They also include the strategic nuclear response by the fleet ballistic missile forces.

The functions of sea control and power projection are closely related. Depending on the type of force we are to use, we need some degree of sea control in the sea areas from which we are to project power. The United States developed the naval forces' capability to project power largely as one means of achieving or supporting control of the seas.

To carry out the functions of sea control and power projection in support of its mission, the U.S. Navy has three functions.

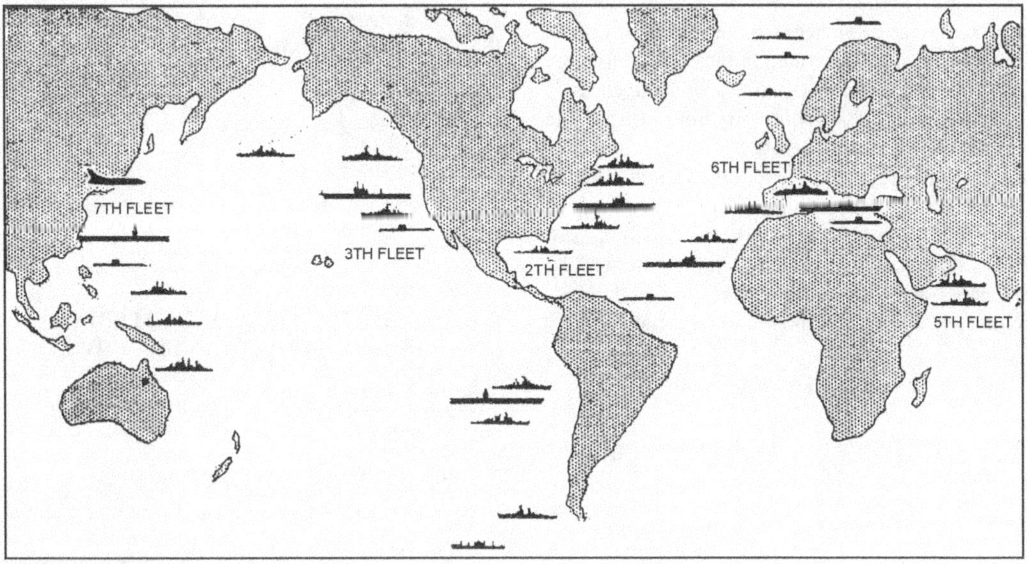

Figure 20-3.—U.S. naval presence throughout the world.

Student Notes:

1. Strategic nuclear deterrence
2. A strong naval presence
3. Security of the sea lines of communications

Strategic Nuclear Deterrence

The effectiveness of the submarine-launched ballistic missile provides the strongest deterrent in our strategic nuclear forces. Thus that deterrent is a stabilizing factor in the strategic nuclear balance.

Naval Presence

To achieve naval presence, the Navy deploys operationally ready naval forces to various overseas locations throughout the world (fig. 20-3). From these locations, our forces can combat hostile forces and support forward-positioned U.S. ground and air forces as well as U.S. allies.

Security of the Sea Lines of Communications

The success of a forward military strategy depends upon the Navy's ability to keep the sea lines of communications open. These lines are between the United States and its forward deployed forces, its allies, and those areas of the world essential for imports. The most vulnerable areas of these sea lines are those closest to potential hostile bases and farthest from friendly territory. Land-based air and patrol combatant craft aid in the protection of shipping in those areas. The protection of the most vulnerable sea areas requires that U.S. Navy forces be present in enough strength to defeat hostile air, surface, and submarine threats. One of the most demanding requirements upon the capabilities of U.S. naval forces is overseas deployment. The deployments place great demands upon both Navy personnel and our multipurpose combatant ships.

REVIEW 2 QUESTIONS

Q1. List the primary tasks of the Navy's operating forces.

a.

b.

c.

d.

e.

Q2. List three of the ways that the Navy uses to modernize its arsenal.

a.

b.

c.

Q3. As determined by national strategy, what are the missions of the U.S. Navy?

a.

b.

Q4. Navy missions are determined by national strategy. List some of the ways the Navy carries out their missions.

a.

b.

c.

THE U.S. MERCHANT MARINE RESPONSIBILITY IN SEA POWER

Learning Objective: When you finish this chapter, you will be able to—

- Identify the missions and functions of the U.S. Merchant Marine in wartime and peacetime.

Our Navy evolved from the American merchant marine. Practically every Navy member of the

Student Notes:

American Revolution was an experienced merchant mariner. The merchant marines were volunteers at that time, as you are today. When it first came into being, the U.S. Navy converted merchant ships into fighting ships by adding cannons to the decks. Through determination and the skills these merchant mariners had learned on the high seas, we won a great war. Congress authorized the first six frigates of the Continental Navy on 27 March 1794. Ex-merchant mariners commanded and manned these frigates. Until World War II, the officers and personnel trained in the merchant marine formed the most important manpower reserve for the Navy.

With the threat of World War II in Europe and Asia, Congress enacted the Merchant Marine Act of 1936. That act provided for a strong merchant marine to service the fleet as a naval auxiliary during times of war and national emergency.

When World War II started, merchant ships were scarce. Since the United States needed to get ships quickly to supply the war effort, we seized the ships of the enemy in our ports. We also took possession of ships from foreign private operators in both domestic and foreign trade. We bought foreign ships and redoubled our U.S. shipbuilding efforts.

Within a year and a half after we entered the war in 1941, shipyards produced ships faster than the enemy could sink them. By mass-producing ships for the war effort, the Kaiser Shipbuilding Company produced a ship a day. Most shipyards built liberty ships that made only one trip to the war zone. If ships did come back, the Navy loaded them and sent them out again. Shipyards also mass-produced larger and faster ships—victory ships and tankers. Many of them were still in service 20 years later. We produced more than 6,000 merchant ships during World War II and somehow found and trained the crews to sail them.

The Army and Navy used many merchant ships as auxiliaries. We used them as hospital ships, repair ships, airplane carriers, and for other special uses. We devised and used new methods of loading and replenishment. Every inch of the ship's cargo holds and topside areas was loaded for increased carrying capacity.

The U.S. merchant marine plays an important part in the sea power of this country. Besides importing essential raw materials for defense of the free world, the merchant marine transports Army and Air Force personnel during times of war or national emergency. It also transports large amounts of equipment, ammunition, fuel, and other supplies that must follow our forces. In previous wars, we moved most of our troops to the war zone by ship. Although we airlifted most of our forces to the war zone during the Vietnam conflict, the merchant marine transported about 97 percent of needed supplies. We must supply about 5 tons of supplies to take care of each person at the front during war. Getting those vital supplies to the right place is a major task. The experience gained from two World Wars and the Korean and Vietnam conflicts taught us how important the merchant marine is.

PEACETIME MISSION

The merchant marine today consists of all commercial oceangoing vessels flying the U.S. flag. Although the U.S. merchant marine is not part of the armed forces, it serves with them in wartime. It is subject to unified control under the Maritime Administration during times of war. The merchant marine includes all waterborne transportation—combination cargo-passenger ships, tankers, dry-cargo vessels, river barges, and harbor tugs. We have restricted our discussion of the merchant marine in this chapter to oceangoing ships of 1,000 gross tons and over. Ships of that group include the liner fleet (ships operating on regular schedules). They also include ships contracted to carry cargo to all areas of the world and ships in domestic and foreign trade. The term *merchant marine* refers to all these ships and their crews.

WARTIME MISSION

In a war, the mission of the U.S. merchant marine includes the following:

- Transport essential materials and cargo needed for the U.S. economy and needed to aid in supplying the economic needs of overseas allies
- Resupply American and allied military forces overseas
- Provide underway replenishment for wet or dry cargo and other direct services to Navy ships at sea

Student Notes:

- Increase combatant naval forces by being armed to carry out convoy, antiaircraft, or antisubmarine duties

In wartime or a national emergency short of war, our government can get much-needed ships to perform merchant marine tasks from several sources. These sources include merchant ships flying the U.S. flag or a foreign flag, the National Defense Reserve Fleet, and the Military Sealift Command (MSC).

REVIEW 3 QUESTIONS

Q1. Describe the peacetime mission of the U.S. Merchant Marines.

Q2. List the wartime mission of the U.S. Merchant Marines.

a.

b.

c.

d.

THE U.S. COAST GUARD RESPONSIBILITY IN SEA POWER

Learning Objective: When you finish this chapter, you will be able to—

- Identify the missions and functions of the U.S. Coast Guard in wartime and peacetime.

The multimission nature of the Coast Guard makes it unique among the armed services of the United States. It has an operational peacetime role and is the only U.S. military service outside the Department of Defense.

The Coast Guard is the nation's oldest continuous seagoing service. It was set up in 1790 as the United States Revenue Marine (later renamed the Revenue Cutter Service). The United States Revenue Marine was an arm of the Treasury Department, under then Secretary Alexander Hamilton. The Revenue Marine was primarily a law enforcement agency. Its responsibility was to collect custom duties from ships entering United States waters.

Although the original role of the service was law enforcement, revenue cutters took part in almost every conflict involving the United States. These involvements showed the military readiness of the service.

In the mid-1800s, Congress set up the U.S. Lifesaving Service, consisting of stations scattered along U.S. coasts. Shortly after the turn of the century, the Lifesaving Service and the Revenue Cutter Service merged to form the U.S. Coast Guard. That merger provided the Coast Guard with its traditional image—the *lifesavers.*

In 1939, the Coast Guard joined the Lighthouse Service and assumed responsibility for setting up and maintaining aids to navigation in U.S. waters. That responsibility has grown to such an extent that today the Coast Guard maintains nearly 50,000 navigational aids, including worldwide electronic navigation systems.

PEACETIME MISSION

The modern-day mission of the Coast Guard is an interesting mixture of duties, including the following:

- Enforcement of maritime laws and treaties
- Search and rescue operations
- Enforcement of U.S. drug and contraband laws
- Installation and maintenance of aids to navigation
- Icebreaking operations that keep commercial vessel traffic moving in domestic waters and support scientific research in the Arctic and Antarctica

As the primary maritime law enforcement agency of the United States, the Coast Guard enforces the following maritime regulatory laws:

Student Notes:

- Safety regulations for all U.S. commercial vessels, offshore structures, and recreational boating

- Port safety and security, including ports, harbors, and their approaches

- The movement of vessels in ports and waterways during crisis situations

- Marine environmental protection to prevent and contain spills of oil and other hazardous substances

Finally, because the Coast Guard is a military service—one that has ships, planes, and boats—it also has a military readiness mission. The Coast Guard works closely with the Navy, undergoes regular refresher training for its major cutters, and participates in joint operational exercises.

The Coast Guard by itself is among the largest navies in the world, ranking 9th or 10th based on the number of armed vessels. Figure 20-4 shows a 378-foot Coast Guard cutter. The Coast Guard gives significantly to the nation's sea power.

The Coast Guard has continued to grow and shoulder additional responsibilities. In the last 30 years, it has gained responsibilities for polar and domestic icebreaking, cleanup and protection of the marine environment, and recreational boating safety.

WARTIME MISSION

With the start of World War II, the Coast Guard assumed the responsibilities of in-port safety and security and commercial vessel safety. In 1967, the Coast Guard became part of the newly formed Department of Transportation.

In wartime the U.S. Coast Guard has always served with pride. Today, during a wartime condition, the U.S. Coast Guard operates directly under the Chief of Naval Operations. It still has the same mission as it did during World War II, plus added roles. The Coast Guard assumes convoy duties as well as antisubmarine warfare missions. Its cutters are well suited for convoy duties as they have a long cruising range and room for armament. The air search and rescue section of the Coast Guard flies rescue missions. It also flies reconnaissance and antisubmarine aircraft. The Coast Guard's mission in wartime will strain its limited assets.

BMRf2004

Figure 20-4.—U.S. Coast Guard—an element of sea power.

Student Notes:

REVIEW 4 QUESTIONS

Q1. List the peacetime missions of the U.S. Coast Guard.

a.

b.

c.

d.

e.

Q2. List the wartime missions of the U.S. Coast Guard.

a.

b.

c.

U.S. MILITARY SEALIFT COMMAND (MSC) RESPONSIBILITY IN SEA POWER

Learning Objective: When you finish this chapter, you will be able to—

- Identify the missions and functions of the U.S Military Sealift Command (MSC) in wartime and peacetime.

In 1949, the United States set up the Military Sealift Command (MSC) by combining the sealift missions of the Naval and Army Transport Services. (The MSC was originally called the Military Sea Transportation Service.) Today, the MSC is an operating agency within the Department of Defense.

MSC ships fall into two general classes—the nucleus fleet and privately owned ships under charter by MSC (fig. 20-5). The nucleus fleet consists of government-owned ships and chartered tankers. All of these ships have the title United States Naval Ships (USNS). Most nucleus fleet ships have crews of civilian mariners who have civil service status. They enjoy the normal benefits of federal employees, but their pay and work rules stem from those of the commercial maritime industry. Private contractors with union crews operate some ships of the nucleus fleet (tankers). The bulk of the nucleus fleet consists of special project ships such as research vessels and those involved in direct support of the Navy fleet.

MSC transports dry and liquid cargo primarily aboard chartered ships and tankers of the nucleus fleet. MSC contracts most of these ships as voyage charters but occasionally contracts them as time charters. Voyage charters contract ships to carry specific cargo to a certain destination. Time charters contract for the use of an entire ship for months or years. All chartered ships are operated by their owners and manned with union seamen. This segment of the MSC fleet varies in size depending on the command's current requirements.

The ships of the Military Sealift Command fleet go where and when needed to support our armed forces. On any given day some ships may be operating in both polar regions or sailing to and from Alaskan military bases. At the same time other ships may be delivering cargo for military units in Europe and the Far East. In peacetime and wartime, the MSC fleet is ready to respond immediately if needed to support national, military, economic, and diplomatic policies.

PEACETIME MISSION

In peacetime the Military Sealift Command relies heavily on the U.S. merchant marine. The MSC ships nearly 25 percent of all military cargo on privately owned U.S. flagships and other merchant marine vessels. The small size of the MSC-controlled fleet requires the MSC to add to its available sealift forces during United States involvement in armed conflict.

WARTIME MISSION

During peacetime, the MSC supports the fleet by supplying fuel and supplies. During wartime, MSC ships used in moving troops and supplies to the war zone bear arms for protection. Besides moving troops to

Student Notes:

Figure 20-5.—Civilian-operated MSC oiler refueling an LPH.

the front, these ships provide underway replenishment to allow Navy ships to stay on station. They carry Navy personnel to handle areas such as weapons and communications to allow the civilian crew to continue its normal work. The MSC ships travel alone or in convoys, but they go wherever the fleet goes during a war. They move vital supplies at the front as well as at sea.

REVIEW 5 QUESTIONS

Q1. What is the peacetime mission of the Military Sealift Command?

Q2. What is the wartime mission of the Military Sealift Command?

SUMMARY

Sea power is a nation's ability to use the oceans for its political, economic, and military interests to achieve its national objectives. Nations exercise sea power in times of peace and war.

Today, the United States depends on other nations for many goods and commodities needed to keep the economy strong and to keep people working.

The U.S. merchant marine, Military Sealift Command, U.S. Coast Guard, and the U.S. Navy make up the essential ingredients for U.S. sea power. Together they support the United States in its national strategy, interests, and goals. The mission of the Navy is to be prepared to conduct prompt and sustained combat operations. To accomplish its mission, the Navy must perform two main functions—sea control and power projection. Sea control is the basic function of the Navy. Power projection is the ability of the Navy to project military power from the sea worldwide.

To carry out these two functions in support of its mission, the Navy has three main functions: strategic nuclear deterrence, naval presence, and security of the sea lines of communications.

A balanced sea power is the essential ingredient of our national strategy. It is not limited to any one course of action and can meet any type of aggression from the most primitive to the most sophisticated. Today the very survival of our country and of our way of life depends on sea power.

REVIEW 1 ANSWERS

A1. Sea power is **a nation's ability to protect its political, economic, and military interests by controlling the seas.**

A2. The principal operational components of our nation's sea power are—

a. **Naval power**

b. **Ocean science**

c. **Ocean industry**

d. **Ocean commerce**

A3. The six conditions required for a nation to have sea power according to Mahan are—

a. **An advantageous geographical position**

b. **Serviceable coastlines, abundant natural resources, and a favorable climate**

c. **Extent of territory**

d. **A population large enough to defend its territory**

e. **A society with an aptitude for the sea and commercial enterprise**

f. **A government with the influence to dominate the sea**

A4. In today's world, sea power includes **maritime industry and marine sciences.** Maritime industry and science add to our national economy by exploring new resources for food, fresh water, minerals, and new living spaces.

Student Notes:

A5. To protect ourselves and to keep our economy going, this country must—

a. **Import raw materials, convert them into manufactured goods, and transport them to marketplaces throughout the world via shipping**

b. **Keep sea-lanes open and safe in times of peace and tension, and deny sea-lanes to the enemy in times of war**

REVIEW 2 ANSWERS

A1. The primary tasks of the U.S. Navy's operating forces are to—

a. **Seek out and destroy enemy naval forces**

b. **Suppress enemy sea commerce gains**

c. **Maintain general naval supremacy**

d. **Control vital sea areas**

e. **Protect vital sea lines of communication**

A2. The Navy is modernizing its arsenal by—

a **Researching and developing new weapons**

b. **Laying up old ships to save the cost of operating and overhauling so money can be shifted to constructing modern ships**

c. **Purchasing highly effective aircraft and ships, such as nuclear propulsion aircraft carriers (CVNs) and ballistic submarines (SSBNs), and at the same time, developing new classes of cost-effective ships**

A3. The missions of the Navy determined by our national strategy are—

a. **Sea control**

b. **Power projection**

A4. Navy missions, as determined by national strategy, are carried out by—

a. **Maintaining a ready and capable submarine-launched variety of ballistic missiles**

b. **Deploying operationally ready naval forces to various overseas locations throughout the world**

c. **Maintaining an open and secure sea line of communication between the U.S. and its forward deployed forces allies and areas of the world essential for imports**

REVIEW 3 ANSWERS

A1. In peacetime, the U.S. Merchant Marines **transport essential materials to and from the United States for the defense of the free world.**

A2. In wartime, the mission of the U.S. Merchant Marines is to—

a. **Resupply American and allied military forces overseas**

b. **Provide wet and dry replenishments and other direct services to ships underway**

c. **Increase combatant naval forces by being armed to carry out convoy antiaircraft, and antisubmarine duties**

d. **Transport essential materials and cargo needed for the U.S. economy and the economy of allies overseas**

REVIEW 4 ANSWERS

A1. The peacetime mission of the Coast Guard includes—

a. **Enforcing maritime laws and treaties**

b. **Conducting search and rescue operations**

c. **Enforcing U.S. drug and contraband laws**

d. **Installing and maintaining aids to navigation**

e. **Icebreaking operations that keep commercial vessel traffic moving in domestic waters and support scientific research in the Artic and Antartic**

A2. The wartime mission of the U.S. Coast Guard includes—

a. **Maintaining in-port safety and security**

b. **Maintaining commercial vessel safety**

c. **Assuming convoy duties as well as antisubmarine warfare duties**

REVIEW 5 ANSWERS

A1. The peacetime mission of the Military Sealift Command is to **support the mission-ready ships at sea by providing fuel and other essential supplies.**

A2. The wartime mission of the Military Sealift Command is to—

a. **Move troops, equipment, and other supplies**

b. **Provide replenishment to ships on station and under-way**

CHAPTER 21

LEADERSHIP AND SUPERVISION

To lead, you must first be able to follow; for without followers, there can be no leader.

—Navy Saying

Today's Navy operates with fewer people and resources than before. Therefore, good leadership is more important than ever. You may think that because you are nonrated, leadership doesn't apply to you. You're wrong! Learn as much as you can about leadership. Your leadership skills will have a strong impact on your Navy career and your personal life. It doesn't matter whether you're an apprentice, a chief petty officer, a division officer, or a commanding officer; you will assume responsibility and exercise authority within the chain of command. As you advance to higher rates, you'll assume more authority and responsibility as a leader. Now is the time for you to learn about leadership.

BASIC PRINCIPLES OF LEADERSHIP AND FOLLOWERSHIP

Learning Objectives: When you finish this chapter, you will be able to—

- Recognize the purpose of followership and leadership.
- Identify the fundamentals of leadership, including core values.

The Navy defines leadership as *the art of influencing people to progress towards the accomplishment of a specific goal.* Leadership occurs when one person influences other people to work toward a definite goal.

Leadership is based on personal example, good management practices, and moral responsibility. Every person in the Navy must set an example of military ideals and give personal attention and supervision to personnel below them in the chain of command.

You can determine your leadership ability by—

- Examining your conduct
- Reviewing your duties and responsibilities
- Determining how well you're performing

If you don't measure up to Navy standards, take steps to raise your performance level as well as the performance of the personnel who work for you.

ELEMENTS OF LEADERSHIP

You've heard the expression "leaders are born, not made" or "that person's a born leader." Forget these phrases; no one is a "born leader." Many people are "natural" leaders because of their strong, magnetic personality or because of their natural ability to learn rapidly (fast). However, such people are the exception, not the rule. Because leaders aren't "born," they must be "made" (trained). There are three elements that make an effective Navy leader:

1. Moral principles
2. Personal example
3. Administrative ability

Moral Principles

Moral principles include honesty, integrity, and loyalty. These principles of human conduct provide direction, solidity, and consistency to leadership.

The key to leadership is the emphasis you place on personal moral responsibility. You show personal moral responsibility by being honest and loyal. Your shipmates see those traits as your moral character. And a strong moral character influences others in a positive manner.

Personal Example

Leading by personal examples goes along with moral responsibility. Effective leaders have many different leadership traits, such as know-how, sincerity, and courage. Which trait is the most important is a matter of opinion. However, if you show weakness in any trait a worker thinks is important, you lose that person's respect.

Respect isn't automatically given to a leader because of authority. You have to earn respect and confidence of personnel working for you by setting a good example. Lead your workers; don't drive them.

Administrative Ability

Administrative ability is more than maintaining logs, records, and other paper work. *Administrative ability* is another term for good management practices. Good management practices include the ability to organize, manage, and work with people. Learn to apply a personal touch in dealing with your workers. Always remember, everyone wants to be treated as an individual who has worth. Emphasize each person's importance in getting a job done.

Giving Orders

When you're the leader of a group, part of your job is to give orders. Give orders that are simple, clear, and complete; and make sure that everyone understands what's to be done.

A good order makes the following facts clear:

- What's to be done.
- When to do it.

Then, as circumstances require or permit, you may add the following information:

- How to do it.
- Why it must be done.

How you give an order is important. The way you speak is important. Speak in a tone that shows you mean business. When you act as though you expect the job to be done well, it usually will be. With experience and when you closely follow the rules for giving an order, you'll develop an effective technique for giving orders.

Praise and Reprimand

Learn when to praise and when to reprimand. Your workers do better work when they know that you appreciate their efforts. Tell them you appreciate their work; that's the only way they'll know. When a person does more than required, show your approval. If possible, show your approval in front of the other personnel.

At times, you'll have to reprimand. You probably don't like to do that, but warning and reprimanding are part of your responsibility as a leader. Remember, the purpose of a reprimand is to teach, not to embarrass. Therefore, give reprimands in private. Always be sure of your facts—the person may have a reason for the behavior that led to the reprimand. Tell the person what was wrong and why it was wrong. Then explain how the person can improve.

Remember to do the following:

- Praise in public.
- Reprimand in private.

Promoting Morale

Morale means different things to different people. If you ask your shipmates about their morale, you'll get different answers. For example, a person who's just been promoted will tell you morale is high. However, a person who's just been restricted will tell you morale is low.

Keeping morale high helps accomplish the Navy's mission. The Navy realizes the need for high morale; therefore, several ongoing programs are conducted to meet the need. These programs include moral and spiritual guidance, educational opportunities, and personal affairs counseling. Encourage your shipmates to take advantage of these programs.

Organized recreation programs, such as ball games, organizational parties, picnics, and sightseeing tours, contribute to good morale. They bring members of the organization together. Let your people know about all of your organization's recreational programs and activities. Showing interest in your peoples welfare and morale helps keep morale high.

Student Notes:

PRIDE.—Many Navy units have an outstanding reputation for their professional ability and their ability to get the job done. Other units can't seem to do anything right. What makes the difference? The answer is simple—the outstanding outfit has *esprit de corps*. The members of the unit have pride in self, Navy, and their country!

Help your unit be a winner. Show your pride in self, Navy, and country. Wear your uniform proudly. Compliment personnel working for you on their sharp appearance and good work. By doing this, you help your unit become an efficient, tightly knit crew.

KEEP PERSONNEL INFORMED.—You can boost morale and promote *esprit de corps* and pride by keeping your personnel informed. Everyone likes to know what's going on. When will the ship get underway? What's the workload for tomorrow? When will the squadron deploy? This is the type of day-to-day information you can pass on to your personnel. Let them know about upcoming drills. Explain the reasons for the drills. Letting people know what to expect promotes good morale.

INTEGRITY.—Always be honest with yourself, your shipmates, and your superiors. Make promises only when you can keep them and only when you intend to keep them. Keeping promises earns you respect from your shipmates, and you must have their respect to be an effective leader.

FOLLOWERSHIP

Everyone in the Navy is in a position of followership. No matter how high you go in the chain of command, you still report to someone higher. Even the President, as Commander in Chief, reports to the people of the United States. To be a good leader, you must know how to be a good follower. Always carry out your orders promptly, to the best of your ability, and as cheerfully as possible. Show your workers that even if an order is disagreeable or causes personal inconvenience, you still must carry it out. Loyalty, both up and down the chain of command, is essential to effective leadership.

Commands and Orders

A good follower obeys all orders received from personnel higher in the chain of command. The Navy has two kinds of obedience—immediate and reasoned.

COMMAND.—Immediate obedience is an automatic response to a command. You must follow a command immediately and exactly as given without asking questions. For example, if you receive an order to make a turn while steering your ship, you do so immediately. If you didn't respond at once, you could endanger the ship.

ORDER.—Reasoned obedience is the proper response to an order. An order lets you ask questions if you don't understand. You can use your own judgment in carrying out an order. For example, if your leading petty officer (LPO) tells you to paint your living space, you decide the number of brush strokes to use. Reasoned obedience lets you obey an order while learning from your experience in carrying it out.

Followership Qualities

To be a good follower, try to develop the following qualities:

Loyalty—Always be loyal to the personnel above you in the chain of command, whether or not you agree with them.

Initiative—Do what must be done without waiting to be told. Showing initiative demonstrates your ability to be a leader.

Dependability—Be dependable. The person in charge must have help in carrying out the mission. The leader must be able to depend on the followers to get the job done. Dependable followers increase the efficiency of the leader and the command.

CONTINUOUS IMPROVEMENT PROGRAM

Learning Objective: When you finish this chapter, you will be able to—

- Identify the fundamental concept of the Continuous Improvement Program.

Student Notes:

The primary goal of the Continuous Improvement Program is to increase productivity and produce better quality through leadership. The most important part of this program is the process, or, how the job gets done.

You might ask, "Who is the most familiar with the job?"

The answer should be, "The person doing the job."

Often, the way the job gets done is complicated or just doesn't work. In most workplaces, it's almost impossible for workers to get management to change the way the job is done. Under the Continuous Improvement Program, supervisors make sure that job improvement suggestions are heard and, if practical, made part of the way the job is done.

REVIEW 1 QUESTIONS

Q1. List the elements of a good Navy leader.

a.

b.

c.

Q2. List the principles of conduct that give direction, solidity, and consistency to leadership.

a.

b.

c.

Q3. What is another term used to describe administrative ability?

Q4. True or false. To be a good leader, you must know how to be a good follower.

Q5. List three followership qualities.

a.

b.

c.

Q6. Describe the purpose of a reprimand.

Q7. List the ways you can help build morale.

a.

b.

c.

Q8. What is the primary goal of the Continuous Improvement Program?

SUMMARY

In this chapter, you were introduced to the concepts of leadership and supervision. You learned that in order to be an effective leader, you first need to be a good follower.

Also in this chapter, the idea behind the Continuous Improvement Program was presented. Through this program, valuable suggestions about the work place can be acted on.

Student Notes:

REVIEW 1 ANSWERS

A1. The elements of a good Navy leader are—

a. **Moral principles**

b. **Personal example**

c. **Administrative ability**

A2. The principles of conduct that give direction, solidity, and consistency to leadership include—

a. **Honesty**

b. **Integrity**

c. **Loyalty**

A3. Another term for administrative ability is ***good management practices***.

A4. **True**, to be a good leader, you must know how to follow orders.

A5. The three followership qualities are—

a. **Loyalty**

b. **Initiative**

c. **Dependability**

A6. The purpose of a reprimand is **to teach, not to embarrass; therefore, give reprimands in private**.

A7. You can help build morale through—

a. **Pride—being proud of what your personnel have accomplished**

b. **Integrity—being honest with yourself**

c. **Keeping personnel informed—making sure your personnel know what is happening**

A8. The primary goal of the Continuous Improvement Program is **to increase productivity and produce better quality through leadership**.

CHAPTER 22

SECURITY REQUIREMENTS AND INTERNATIONAL AGREEMENTS

There is no way of estimating how many battles have been lost, how many ships have been sunk, or how many lives have been sacrificed because someone intentionally or unintentionally betrayed a military secret.

—Author unknown

Security is the safeguarding of classified information in the interest of national security. The safety of the United States in general and naval operations in particular depends on protecting classified material.

SECURITY

Learning Objectives: When you finish this chapter, you will be able to—

- Recognize the basic security policies, requirements, and procedures for handling classified material and information to include security classification and protection.
- Recall the procedures and principles involved in applying for personnel clearances.
- Identify the various classified material markings.
- Recognize the purpose of downgrading and declassifying classified material.
- Recall the procedures used to transmit classified material.
- Identify the basic security requirements concerning classified information and material and their security levels.
- Identify the types of equipment/material covered by automated data processing (ADP) security.
- Identify the terms used to describe the compromise of classified material.
- Recognize the procedures used to report a suspected compromise or a security violation.
- Identify the basic personal censorship requirements concerning classified information and material.
- Identify the procedures for reporting subversive activities on station or in a leave or liberty status.
- Identify when and where terrorism can occur.
- Identify the most common forms of terrorism.
- Recognize the terms *bomb threat* and *bomb incident*.
- Recall the procedures to follow when a bomb threat is received.

Security involves more than safeguarding classified printed information, such as photographs, blueprints, manuals, and charts. Security also includes safeguarding communications, such as mail, visual signals, radio transmissions, ship movements, or telephones. It includes anything that affects the security of our government in domestic and foreign affairs. It involves protection against sabotage, subversion, or any other illegal acts designed to weaken or destroy the United States. It's important for you to understand what classified information is and how to safeguard it.

Student Notes:

SECURITY CLASSIFICATION LEVELS

All information or material considered vital to the safety of the United States is given a security classification level. Each security classification level indicates (tells) the amount of protection the information and material requires to safeguard it against unauthorized disclosure. There are only three security classification levels—Top Secret, Secret, and Confidential.

The Secretary of the Navy (SECNAV) or his/her designees have the authority to originally classify information as Top Secret, Secret, or Confidential. The SECNAV's designees are listed in the *Department of the Navy Personnel Security Program,* SECNAVINST 5510.30A and *Department of the Navy (DON) Information Security Program (ISP) Regulation,* SECNAVINST 5510.36.

Top Secret

Top Secret is the classification level applied to information whose unauthorized disclosure could reasonably be expected to cause **exceptionally grave damage** to the national security. Some examples of information that could cause grave damage to national security include—

- Armed hostilities against the United States or its allies
- A disruption of foreign relations vitally affecting the national security
- The compromise of vital national defense plans
- The disclosure of complex cryptographic and communications intelligence systems
- The disclosure of sensitive intelligence operations
- The disclosure of significant scientific or technological developments vital to national security

Secret

Secret is the classification level applied to information whose unauthorized disclosure could reasonably be expected to cause **serious damage** to the national security. Some examples of information that could cause serious damage to national security include information that could—

- Disrupt foreign relations significantly affecting the nation's security
- Significantly impair a program or policy directly related to the national security
- Disclose significant military plans or intelligence operations
- Compromise significant scientific or technological developments relating to national security

Confidential

Confidential is the classification level applied to information whose unauthorized disclosure could reasonably be expected to cause **damage to** the national security. Some examples of information that could cause damage to national security include information that could—

- Indicate ground, air, and naval forces (such as force levels and force dispositions)
- Reveal performance characteristics, such as design, test, and production data of U.S. munitions and weapons systems

Controlled Unclassified Information

Controlled unclassified information is defined and governed by laws, international agreements, and regulations that address the identification, marking, protection, handling, transmission, transportation, and destruction of controlled unclassified information. Controlled unclassified information includes—

- For Official Use Only (FOUO) information under the Freedom of Information Act (FOIA)
- Department of State (DOS) Sensitive But Unclassified (SBU) information
- DOD and DOE Unclassified Controlled Nuclear Information (UCNI)

Student Notes:

- Drug Enforcement Administration (DEA) Sensitive Information
- Sensitive Information as defined by the Computer Security Act of 1987
- Unclassified information in technical documents requiring distribution statements and unclassified NNPI

SECURITY CLEARANCES

Sailors in many Navy ratings require some access to classified information. The commanding officer (CO) determines your need for a security clearance. The CO bases your need for a security clearance on your assignment at his/her command or potential assignment on transfer. To apply for a security clearance, you must be a U.S. citizen. There is a security investigation made on each Sailor needing a clearance. This investigation determines the Sailor's potential to protect information during the course of his/her duties.

Security clearances are granted to Sailors when their conduct and behavior are such that they can be entrusted with classified information or they can be assigned to sensitive duties. These are Sailors who—

- are loyal to the United States,
- comply with laws,
- have demonstrated dependability in accepting and discharging responsibilities,
- demonstrate good social adjustment and emotional stability, and
- have the ability to exercise sound judgment in meeting adversity.

To receive and keep a security clearance, you must have and maintain a good record. Your commanding officer can suspend a clearance if you don't maintain a good record. According to *Department of the Navy Personnel Security Program*, SECNAVINST 5510.30A, your command must report any of the following to the DON Central Adjudication Facility (CAF) (the DON CAF grants or revokes clearances):

- Involvement in activities or association with people who unlawfully practice or advocate overthrow or alteration of the United States government by unconstitutional means
- Foreign influence concerns or close personal association with foreign nationals or countries
- Foreign citizenship (dual citizenship) or foreign monetary interests
- Bad conduct, such as excessive drinking, gambling, promiscuity, or illegal or improper drug use/involvement
- Conduct involving questionable judgment, untrustworthiness, unreliability or unwillingness to comply with rules and regulations, or unwillingness to cooperate with security processing
- Unexplained affluence or excessive indebtedness
- Apparent mental, emotional, or personality disorder(s)
- Criminal conduct
- Noncompliance with security requirements
- Engagement in outside activities that could cause a conflict of interest
- Misuse of information technology systems
- General inaptitude
- General disciplinary causes—habitual or accumulated discrepancy causes

A security clearance is granted on your need to know and your meeting the standards for the level of clearance required. To get a security clearance, you must undergo a background investigation by an approved federal government agency. The higher the level of security clearance required, the more thorough the investigation. During the investigation, you are asked questions about your military, civilian, and personal conduct. You must answer the background questions completely and correctly.

Student Notes:

Just because you have a clearance doesn't automatically mean you have access to classified information. Having a clearance means you may be granted access if your duties require access to the information. This is called the *need to know*.

Security clearances and access to classified information are based on a *need to know*. Only Sailors who have a real need to know are cleared for access to the appropriate classified material. The command that has the classified material determines who has the need to know.

If you're cleared to work with classified material, censor what you say by keeping what you know to yourself. The following guidelines will help you safeguard classified material:

- Never reveal (talk about) classified information just to show your shipmates how smart you are or to act important. If they don't need to know the information to carry out their duties, don't tell them.
- Don't talk about classified information to unauthorized persons, including family, friends, shipmates, and especially strangers. Classified information can be unintentionally revealed to unauthorized persons in many ways.
- Interest in your own job is natural and desirable, but it must not lead you to reveal classified information to unauthorized persons. Never add to a news story that's incomplete, no matter how much you may know. If you do, you may make public what the Navy has tried to keep secret.

The SECNAV has designated the Department of the Navy Central Adjudication Facility (DON CAF) as the single clearance granting authority for the Department of the Navy. The DON CAF issues final security clearances for civilian and military personnel at the request of DON commands and activities once it has determined that granting the clearance is clearly consistent with the interests of national security. Once issued, a security clearance remains valid provided the Sailor continues compliance with personnel security standards and has no subsequent break in service exceeding 24 months.

SECURITY AREAS

Classified information is always protected at the level of control appropriate with its assigned security classification level. This policy encompasses all classified information, regardless of media.

Personnel who work with classified information, work with it only in a secure facility. They use an accredited automated information system (AIS) under conditions that prevent unauthorized persons from gaining access to the material. If you have classified material in your possession, you are responsible for protecting that information. Lock classified material in an appropriate security container or facility when you're not using it or when it's not under your direct control.

If you work with classified material, you must follow procedures so unauthorized persons do not gain access to the classified information. In a facility that contains classified material, access is restricted and movement is controlled so personnel without a need to know do not have access to classified material. **All personnel must comply with the *need-to-know* policy**.

If you are using classified material, you can't remove it from the designated office or working area except to perform official duties and under conditions providing the protection required by SECNAVINST 5510.36.

Don't discuss classified material with any person that doesn't have a need to know.

STORING CLASSIFIED MATERIAL

The General Service Agency (GSA) sets and publishes minimum standards, specifications, and supply schedules for containers, vault doors, modular vaults, alarm systems, and associated security devices suitable for the storage and destruction of classified information.

When classified information isn't under the personal control or observation of a cleared person, it's guarded or stored in a locked GSA-approved security container or vault, modular vault, or secure room. For

Student Notes:

information about storage requirements, refer to SECNAVINST 1550.36.

MARKING CLASSIFIED MATERIAL

Classified material is marked so that personnel know the classified nature of the material, to make sure the material receives the degree of protection required, and to help extract, paraphrase, downgrade, and declassify the material.

All classified material is marked so you know the following information about the material:

- The level of classification
- The part(s) that contain(s) or reveal(s) classified information
- How long the material is to remain classified
- Additional measures needed to protect the material

Overall Markings

Material is marked so the security markings are easy to see and recognize. Classified documents are marked on their face and back cover and top and bottom center to show the highest overall classification level of the information they contain. (**NOTE**: Titles of classified documents are usually unclassified.) On documents, the classification level is marked or stamped in capital letters larger than the type used in the text to alert anyone handling the document that it is classified. Material is marked as follows:

AUTOMATED INFORMATION SYSTEM (AIS).—Removable AIS (fig. 22-1) storage media and devices used with AIS and word processors are marked using the appropriate SF label to indicate the highest overall classification level of information contained in the storage media.

PHOTOGRAPHS, SLIDES, AND TRANSPARENCIES.—The face of a classified photograph is marked with its highest overall classification level and associated markings. If this is not possible, these markings are placed on the back of the photograph. These markings are stamped or permanently affixed by pressure tape, labels, or other similar means.

Slides or transparencies (fig. 22-2) are marked with their highest overall classification level and association markings on the image area, border, holder, or frame. Groups of slides or transparencies used and stored together as a set are marked with their highest overall classification level and associated markings. Associated markings "Classified by," "Reason," "Derived from," and "Declassify on" are marked on the image area of the cover slide or transparency only.

MOTION PICTURE FILMS, VIDEOTAPES, AND CONTAINERS.—Classified motion picture films (fig. 22-3), videotapes, and their titles are prominently marked with the highest overall classification level and associated markings of the information they contain. The markings are visible when projected at the beginning and end of the production. Classified films, videotapes, and their containers are marked in the same manner.

SOUND RECORDINGS AND CONTAINERS.—Classified sound recordings (fig. 22-4) have an audible statement at the beginning and end of each recording. This statement identifies the highest overall classification level and associated markings of the recorded information. Containers of classified reels, cassettes, videotapes, and motion picture films are prominently marked with the highest overall classification level and associated markings of the information contained.

ROLLED OR FOLDED DOCUMENTS.—Rolled or folded blueprints, maps, charts, or other large items are clearly marked to show their highest overall classification level (fig. 22-5).

Portion Markings

Each portion such as the title, section, part, paragraph, or subparagraph of a classified document is marked to show its classification level. By doing this, a document is marked so you **know** what part or parts contain or reveal protected information. The classification level of a part of a document is shown by a classification symbol—TS for Top Secret, S for Secret, C for Confidential, and U for unclassified. The symbol

Student Notes:

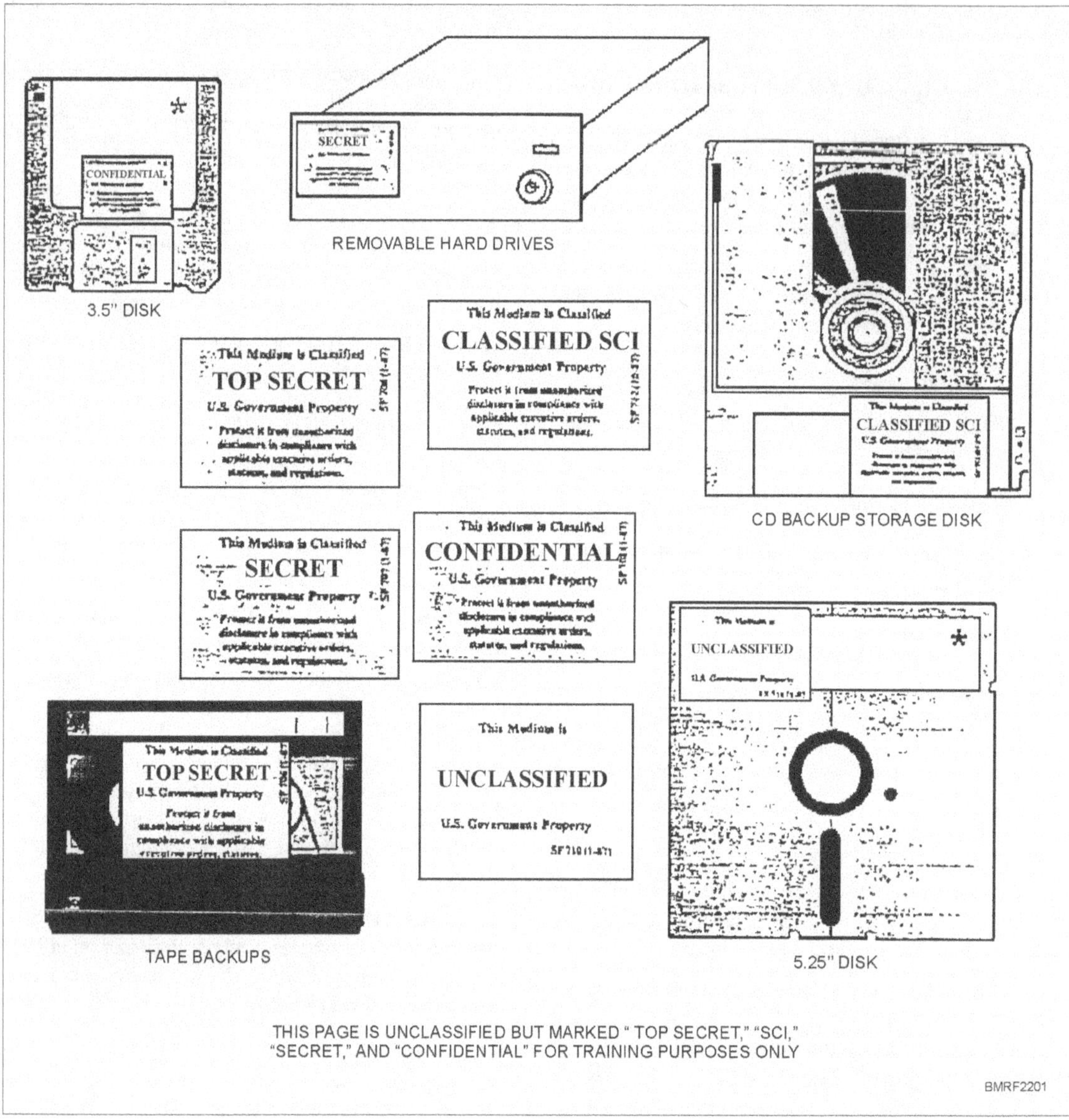

Figure 22-1.—AIS storage media.

is placed in parentheses immediately following the part letter or numbers. If there aren't any part letters or numbers, place the abbreviation immediately before the beginning of the portion.

1. (U) This introductory sentence is Unclassified.

A. (C) This subparagraph is Confidential.

(1) (S) This subparagraph is Secret.

Examples of portion markings are shown in figure 22-6.

Student Notes:

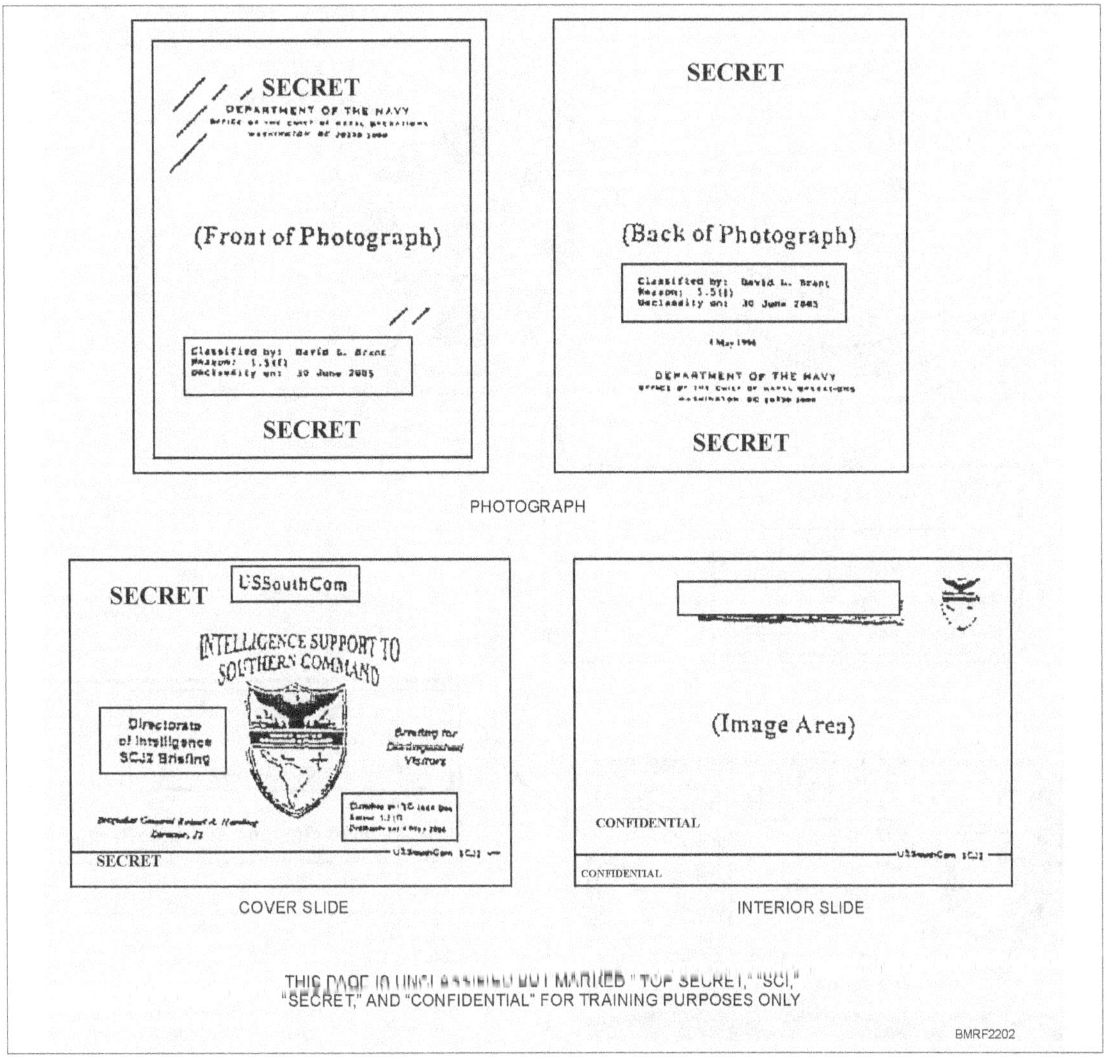

Figure 22-2.—Photographs, slides, and transparencies.

Marking Messages

Messages are marked in a manner similar to documents. They are marked with the highest overall classification level of the information contained in the message. Classified messages are marked to indicate the following:

- The nature of the classification—original or derivative
- The source of classification
- Downgrading instructions (if applicable)
- Declassification instructions (if applicable)

Student Notes:

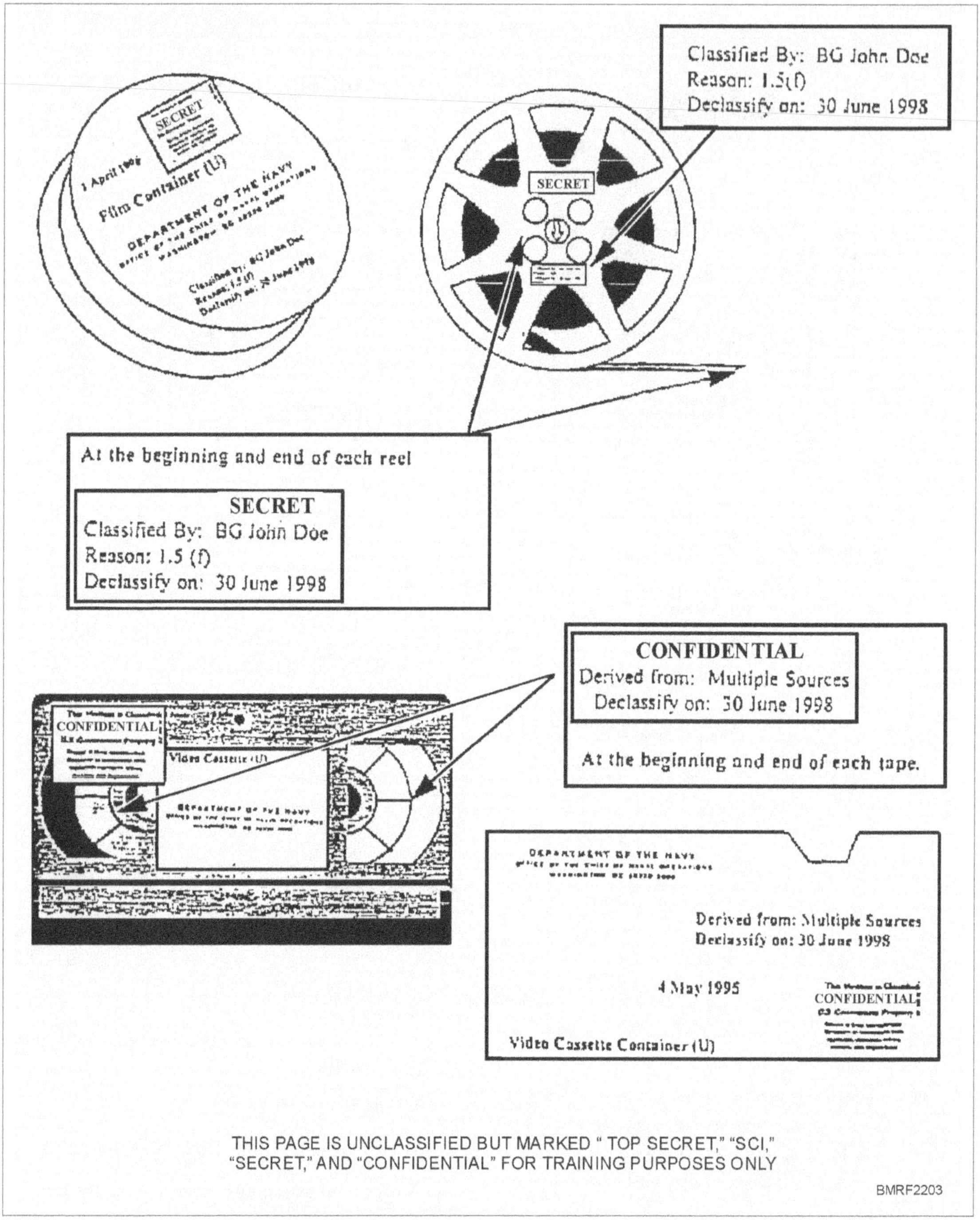

Figure 22-3.—Motion picture films, videotapes, and containers.

For more information on marking classified messages, refer to SECNAVINST 5510.36.

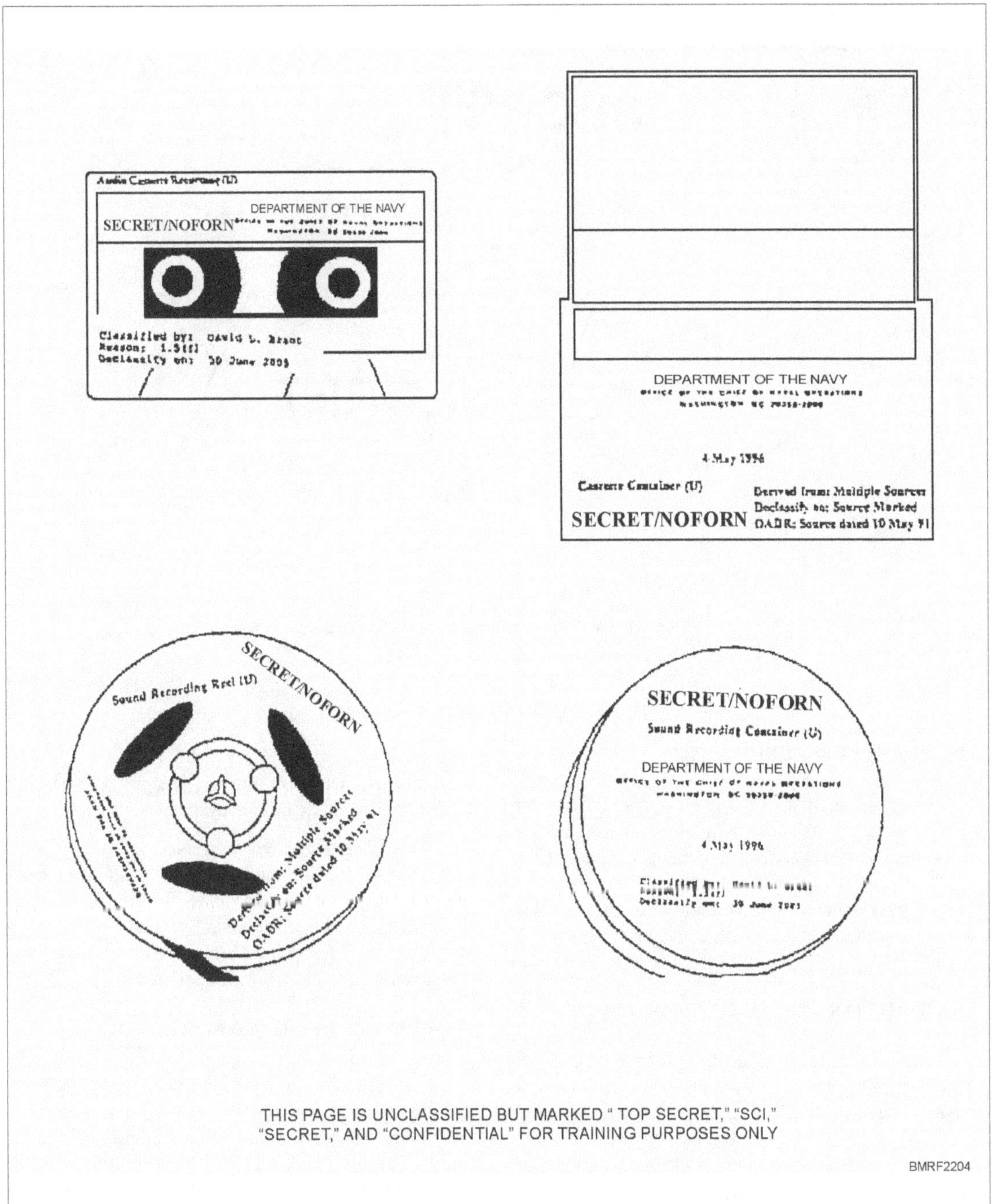

Figure 22-4.—Sound recordings and containers.

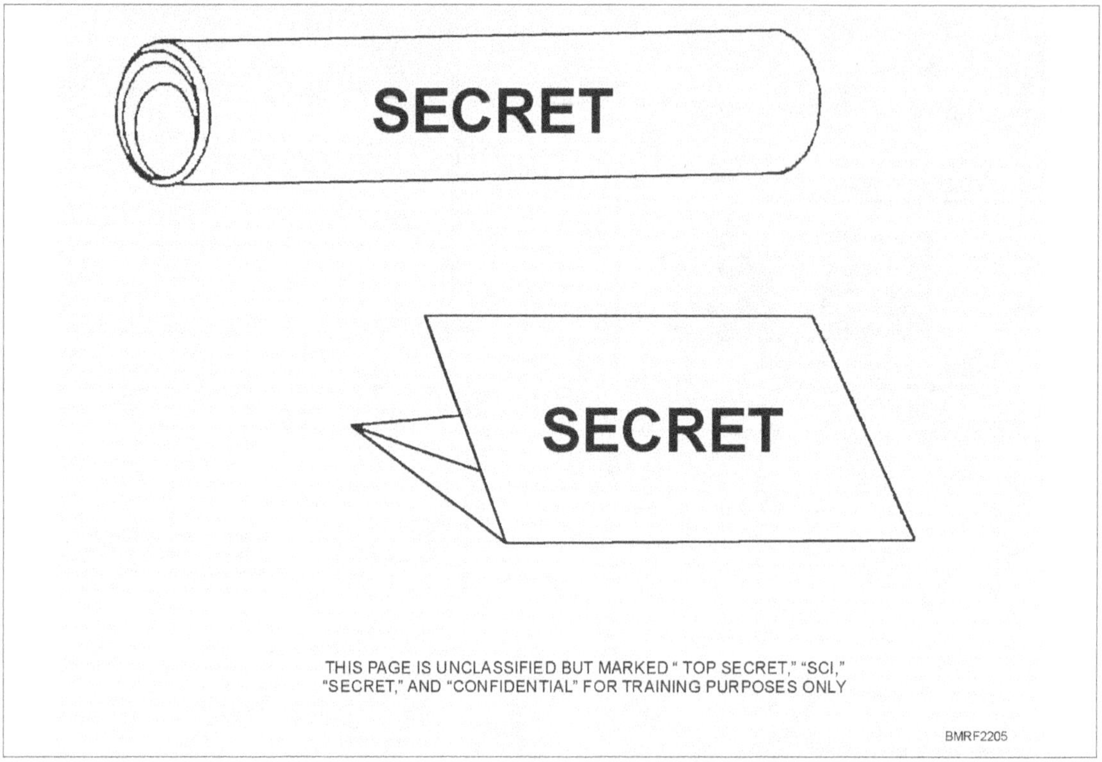

Figure 22-5.—Rolled or folded documents.

Miscellaneous Classified Material

Materials such as rejected copies, typewriter ribbons, carbons, and other similar items used during the production of a classified document are handled in a way that protects the material. Destroy such material when you no longer need it. You don't need to mark this material as classified unless it's necessary to ensure its protection.

TRANSMITTING CLASSIFIED MATERIAL

The rules for transmitting classified material can be found in the *Department of the Navy (DoN) Information Security Program*, SECNAVINST 5510.36. According to SECNAVINST 5510.36, commanding officers must make sure that only appropriately cleared personnel or carriers transmit, transport, escort, or hand-carry classified information. Unless a specific kind of transmission or transportation is restricted, the means selected should minimize the risk of a loss or compromise while permitting the use of the most cost-effective mode of conveyance.

Classified telephone conversations are permitted only over secure communication circuits. These circuits must be approved for the classification level of the information being discussed. Every attempt must be made to make sure that the classified information is not compromised to unauthorized personnel.

COPYING CLASSIFIED MATERIAL

U.S. classified information can be reproduced only to the extent required by operational necessity. However, the agency that originates the information may restrict reproduction of the material, or reproduction of the information may be restricted because of applicable statutes or directives.

Student Notes:

SECNAVINST 5510.36

SECRET

OCT 16 1999

DEPARTMENT OF THE NAVY
OFFICE OF THE CHIEF OF NAVAL OPERATIONS
Washington, DC 20350-2000

IN REPLY REFER TO

5510
Ser NO9N2/9C1234556
(Date)

SECRET—CONFIDENTIAL upon removal of enclosure (2)

From: Chief of Naval Operations
To: Director, Special Programs Office

Subj: CLASSIFIED LETTER OF TRANSMITTAL, TRANSMITTING A CLASSIFIED ENCLOSURE (U)

Encl: (1) CNO ltr 5510 Ser NO9N2/7U123445 of 12 Oct 96
(2) CNO ltr 5510 Ser No9N2/7S12345 of 28 Sep 96

1. (U) A classified letter of transmittal shall be marked as any other classified document with all applicable associated markings.

2. (C) This classified letter of transmittal contains Confidential information and has a Secret enclosure, therefore, its highest overall classification level is Secret, but Confidential when the Secret enclosure is removed. Instructions to this effect are annotated on the face of the letter of transmittal, top let corner, as shown.

3. (U) The declassification instructions, bottom left, reflect the disposition of the Confidential information contained in the classified letter of transmittal after the classified enclosure is removed.

John Boat
By direction

Derived from: OPNAVINST 5513.11B, enclosure (7)
Declassify on: Completion of test or 1 Jan 00

THIS PAGE IS UNCLASSIFIED BUT MARKED "TOP SECRET," "SCI," "SECRET," AND "CONFIDENTIAL" FOR TRAINING PURPOSES ONLY

SECRET

BMRf2206

Figure 22-6.—Portion markings.

DESTROYING CLASSIFIED MATERIALS

Classified material is destroyed in accordance with procedures contained in SECNAVINST 5510.36. Burn bags are used to store classified information awaiting destruction at a central destruction facility.

AUTOMATED DATA PROCESSING (ADP) SECURITY

Automated data processing (ADP) is a Navywide responsibility. It encompasses security aspects that contribute to the protection of the total ADP activity, office information system, or network. ADP security involves the following elements:

- Physical
- Administrative/operating procedures
- Hardware
- Software
- Data

Your command will have an automated data processing security officer (ADPSO) who reports to the CO on matters that concern the protection of electronically generated data. The ADPSO is responsible for the physical security of each computer workstation. The protection of each workstation involves physical security, physical access control, data file protection, and natural disaster protection. Seek out your ADPSO and make sure your workstation complies with Navy and command regulations for the protection of classified material.

Levels of ADP Security

Data processed electronically have three levels of security: Level I, Level II, and Level III. If your command processes Level I and/or Level II data, it must provide a specific degree of protection. The following chart defines the three levels of data:

LEVEL	MEANING
Level I	Classified data
Level II	Classified; requires special protection, such as For Official Use Only and data covered by the Privacy Act of 1974
Level III	All other unclassified data

Marking Removable Classified Automated Information System (AIS)

Pages or portions removed from AIS printouts (fig. 22-7) for separate use or maintenance are marked as individual documents. They are marked with the highest overall classification level and include all the required associated markings for all pages or portions that are removed.

Software used to produce classified material is programmed so that each classified file stored by the system is marked with the highest overall classification level and all associated markings. Also, the outside of AIS media storing classified files is programmed in a readily usable format with the highest overall classification level including all applicable warning notices and intelligence markings. AIS media that contains classified files not programmed in a readily accessible format are marked on the outside with the highest overall classification level and all applicable associated markings (normally a sticker or tag) or have marked documentation kept with the media.

The computer system and its associated peripherals require controlling and safeguarding at all times. This includes the disks, diskettes, disk drives, monitors, printer ribbons, and generated hard copy. Security procedures for electronic data is found in the *Department of the Navy ADP Security Manual,* OPNAVINST 5239.1.

Marking Disks

As a general rule, the two types of electronic media are the working copy media and finished media. Working copy media is temporary information. It stays in your work area and under the control of your activity. After creating a working copy, retain it for 180 days before destruction. Finished media is permanent information. It can be released to other commands and activities. Finished media contains information that doesn't change or is pertinent for more than 180 days.

Student Notes:

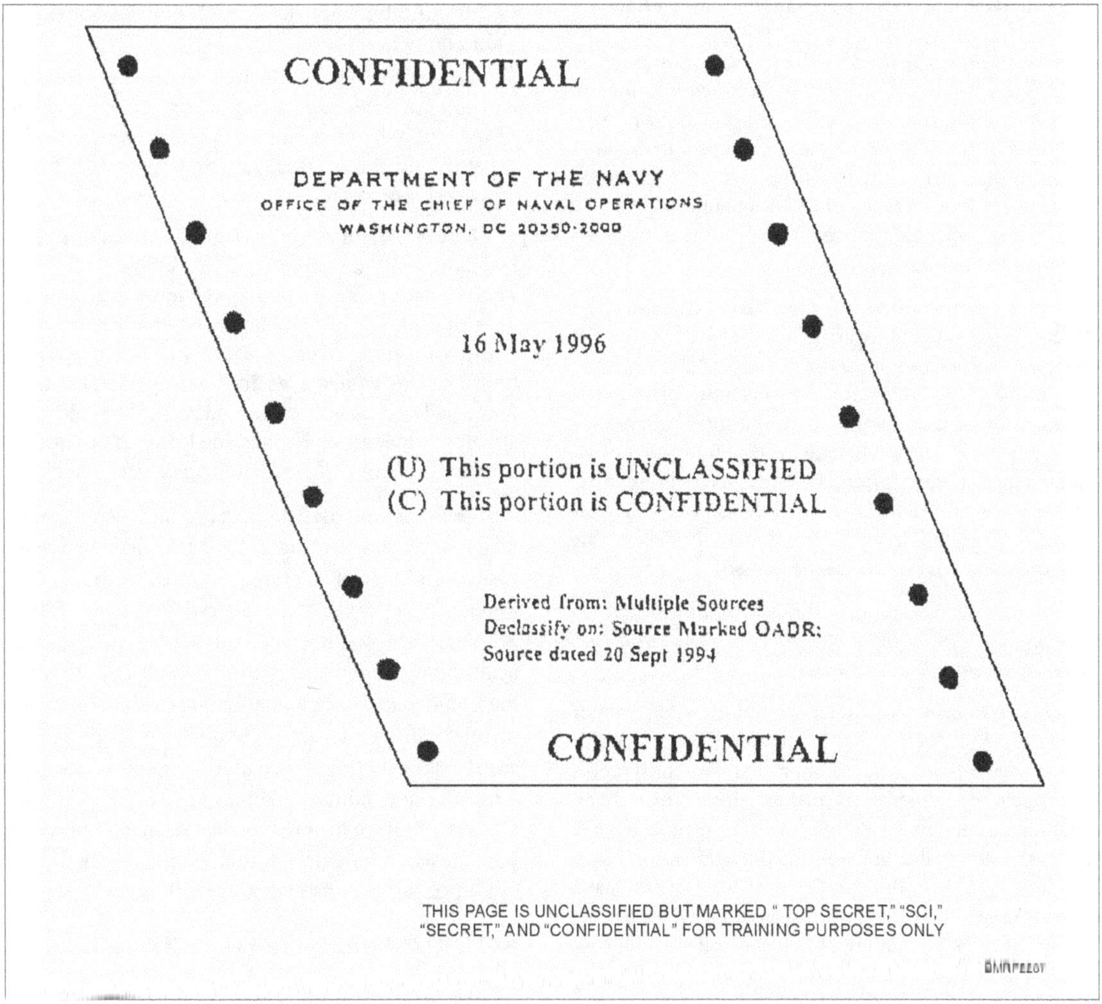

Figure 22-7.—Automated information system printout markings.

Electronic media is dated and the classification marked when it's created. Disks classified as Secret or Top Secret are assigned a sequential identification number so they can be tracked. Electronic media is controlled just like other classified material. Electronic media is protected according to the highest classification ever recorded on the disk.

Disks (see fig. 22-1) are marked with stick-on labels that identify the overall security classification and permanently assigned identification numbers.

The ADP security program protects ADP activities, office information systems, and networks. The management of the ADP security system is continuously monitored and reviewed for effectiveness. The *ADP Security Manual,* OPNAVINST 5239.1, contains a complete description of ADP security policies and procedures.

Student Notes:

COMPROMISE OF CLASSIFIED MATTER

According to SECNAVINST 5510.36, compromise is *An unauthorized disclosure of classified information to one or more persons who do not possess a current valid security clearance.* This means that material is compromised if someone loses, steals, captures, salvages, or sees the material without being cleared. The material is also compromised if a person who has seen the material defects.

The compromise of classified information threatens our national security. How much of a threat the compromise is depends on the nature and classification of the compromised material. If you know that material is compromised or subject to compromise, report the facts to your superiors right away. If you find classified documents where they don't belong, such as lying in the street or on a beach, turn the documents in to your superior or to the nearest military activity. While this doesn't seem possible, it has happened!

A security violation is defined as *any failure to comply with the regulations for the protection and security of classified material.*

If you find an unattended open or unlocked safe or container in which classified material is stowed, a security violation has been committed. You must report the discovery immediately to the senior duty officer. Then, guard the material until the duty officer arrives. After inspecting the material, the duty officer will lock the safe. If it's believed that the material is or may have been compromised, the duty officer will have the person responsible for the material make a detailed inventory.

PERSONAL CENSORSHIP

One form of classified material that can't be physically safeguarded is the information you carry around in your head. You are the only person who can prevent its disclosure. Be constantly on guard to prevent revealing classified information—either by talking or by writing.

A World War II slogan that's still effective is "Loose lips sink ships." Loose talk, even to a person who has the same knowledge you have, may be overheard by unauthorized persons. All of us like to talk about our ships, our jobs, and our travels. However, when we do, we should be sure we don't discuss classified information in our conversations.

Loose talk in public places can be especially damaging. Intelligence agents are trained to collect bits of seemingly harmless information. Putting all the bits together might produce a comprehensive file of classified information.

Never discuss classified information over telephones, as they constitute one of the least secure systems of communication. Telephones are subject to wiretapping—both physically and electronically. Long-distance circuits use microwave radio transmission, which is easily intercepted. The use of homemade or unauthorized codes, double-talk, or an attempt to talk around a classified subject provides no protection against trained intelligence personnel.

The methods used by foreign intelligence agents take many forms. An agent could be male or female, young or old, or of any national origin or background. Foreign agents exist in our everyday lives as ordinary people. They could blackmail you or make threats against you or members of your family. They may take the friendly approach and offer you friendship, money, or other things of value. They may even promise to assist your relatives living in a foreign country. They may offer any number of things in return for classified material or bits of information that seem unimportant to you. Always remember that people who deal in espionage are experts in dealing with people.

REPORTING SUBVERSIVE ACTIVITIES

Whether you have access to classified material or not, you must report to your commanding officer, through your chain of command, anyone you suspect is involved with espionage, sabotage, or is compromising classified material. If a stranger approaches you asking inappropriate questions when you are on leave or liberty status and you cannot contact your chain of command, report this information to the nearest military activity.

Being security conscious and following security standards and requirements is a big responsibility. However, maintaining proper security can be accomplished if you realize that security really is a personal concern.

Student Notes:

TERRORISM

Terrorism is the unlawful use or threatened use of force or violence against individuals or property. Terrorists intend to coerce (force) or intimidate governments or societies. Terrorism is used for political, religious, or ideological purposes. Acts of terrorism directed against naval personnel, activities, or installations can destroy critical facilities and injure or kill personnel. Terrorism can delay mission accomplishment and cause damage through adverse publicity and public perception (the way people see the action) of incident handling and results.

Terrorists use many methods of operation, which may include bombings, ambush, armed attack, sabotage, or taking hostages. The two most publicized terrorist methods are bombings and taking hostages. The terrorist method generally used toward military forces is bombing. However, at times, naval or military personnel have been taken hostage as a result of an aircraft highjacking or of highjacking personnel using some other means of transportation. Military personnel, and particularly naval personnel, are often stationed in or visit foreign countries. Some of these countries have significant levels of terrorist activity.

Indications and warnings of terrorist activity against naval installations or personnel are normally received from U.S. security authorities or through the security agencies of host countries. These warnings usually come in the form of threat conditions (THREATCONS). Threat conditions range from THREATCON ALPHA (the lowest degree of readiness) to THREATCON DELTA (the highest degree of readiness). Each threat condition contains several measures that must be adopted before that degree of readiness is fully set. When stationed in or visiting foreign countries, you will receive a brief concerning the threat condition in force at that time.

When visiting foreign countries, you must be constantly aware of what is going on around you. The actions of terrorist groups are rarely advertised. Terrorists normally choose places of business that have a high volume of target personnel present (such as nightclubs, restaurants, airports, and shopping centers). Be more careful at night, when the cover of darkness helps the terrorist hide his or her activities. Be alert and notice anything out of the ordinary and report it to the proper authorities. You could identify a possible terrorist operation.

Although terrorist attacks within the United States aren't as common as in other countries, they have happened. The same levels of awareness that you practice when visiting foreign countries are necessary here as well. Being alert when you are on or around military installations could mean the difference between the success or failure of a terrorist operation, not to mention the lives of your shipmates.

BOMB THREATS

When detonated or ignited, a bomb can injure or kill personnel and damage material. Bombs are classified as explosive or incendiary. An *explosive bomb* causes damage by fragmentation, heat, and blast. The heat produced often causes a secondary incendiary effect. An *incendiary bomb* generates fire-producing heat without substantial explosion when ignited. Bombing occurs when an explosive bomb detonates or an incendiary bomb ignites.

A bomb threat may happen anytime or anywhere. It can be made by a terrorist group or a disgruntled employee. Many bomb threats are unfounded (not real). False bomb threats make people complacent (at ease). Don't assume a bomb threat is a hoax (not real) until you're sure. **Safety is the major concern**!

Bomb threat. A bomb threat is a message delivered by telephone or letter. A bomb may be delivered through the mail as a letter or a suspicious package. A bomb threat may or may not contain the following information:

- The bomb's location
- The time for detonation/ignition
- An ultimatum related to the detonation/ignition or concealment of the bomb

Bomb incident. A bomb incident is the detonation/ignition of a bomb, discovery of a bomb, or receipt of a bomb threat.

Student Notes:

There are a few things you can do to reduce vulnerability of your ship or station to a bomb threat/incident. You can—

- Strictly comply with your command's procedures for personnel identification and access control procedures to department/division spaces,
- Be suspicious of all articles whose origin is unknown or obviously "out of place" within the space,
- Maintain tight control of locks and keys,
- Lock all rooms/spaces when not in use or manned by authorized personnel, and
- Immediately report suspicious personnel and their actions.

Each telephone at your command should have a copy of the Telephonic Threat Complaint, OPNAV Form 5527/8 (fig. 22-8). When a bomb threat is received by telephone, the person receiving the call should take the following actions:

- Try to keep the caller on the line and obtain as much information as possible. Complete the Telephonic Threat Complaint form while the caller is on the line or immediately thereafter.
- Record in writing the exact words of the caller.
- Try to identify the location of the bomb, the type of device, what it looks like, and the expected time of detonation.
- Attempt to determine the sex, approximate age, and attitude of the caller.
- Note any background sounds that may provide clues to the caller's location.
- Note any accent or peculiarity in speech that may help identify the person.

REVIEW 1 QUESTIONS

Q1. List the security classifications.

a.

b.

c.

Q2. What does FOUO stand for?

Q3. Who is authorized to initiate a request for a security clearance and background investigation?

Q4. A background investigation is required for what levels of security clearances?

Q5. What does a letter in parentheses, such as (S), after a publication title tell you about the publication?

Q6. How are classified material such as videotapes, cassettes, and computer disks marked?

Q7. A publication contains Confidential material, except for one paragraph that contains Top Secret material. How is this publication marked?

Student Notes:

DEPARTMENT OF THE NAVY

TELEPHONIC THREAT COMPLAINT

IF BOMB THREAT, ASK THE CALLER

- WHEN IS THE BOMB TO GO OFF?
- WHERE IS THE BOMB TO GO OFF?
- WHAT KIND OF BOMB IS IT?
- WHAT DOES THE BOMB LOOK LIKE?
- WHERE ARE YOU CALLING FROM?

1. COMMAND

a. Name & Address	b. Phone No

2. COMPLAINANT

a. Name

3. PERSON RECEIVING CALL

a. Name	b. Date & Place of Birth
c. Command Name & Address	d. Phone Number *(Work)* *(Home)*

4. TELEPHONE CALL RECEIVED ON

a. Phone Number *(Included area code)*	b. Location

c. Phone Number listed in *("x" all that apply)*

☐ Command Directory ☐ Base Directory ☐ Local Directory

☐ Unlisted ☐ Other *(list)*

5. DETAILS OF CALL

a. Date	b. Day of Week	c. Time

6. CONTEXT OF CONVERSATION

a. Recipient

b. Caller

c. Recipient

d. Caller

e. Recipient

f. Caller

7. BACKGROUND NOISES *(Describe street sounds, voices, music, etc.. If more space is needed, continue on reverse.)*

8. INFORMATION ABOUT CALLER/VOICE CHARACTERISTICS

a. Sex	b. Age	c. Race	d. Accent
e. Educational Level		f. Attitude *(Calm, Nervous, Serious)*	
g. Other			

9. WERE THERE ANY WITNESSES TO THE CALL? ☐ No ☐ Yes *(List name)*

10. DO YOU HAVE ANY SUSPICION AS TO THE IDENTITY OF THE CALLER? ☐ No ☐ Yes *(List name)*

11. NOTIFICATION OF AUTHORITY *("x" all notified)*

☐ CO ☐ XO ☐ OOD ☐ Security ☐ NISRA ☐ Telephone Company ☐ EOD ☐ Fire Dept

OPNAV 5527/8 (12-82) S/N 0107-LF-055-2740 BMRf2208

Figure 22-8.—Telephonic Threat Complaint, OPNAV Form 5527/8.

Q8. What type of area is used to keep classified material?

Q9. What type of material is safeguarded through ADP Security?

Q10. You are making your rounds as a roving security patrol and discover that the door to the radio room is unlocked and the room unattended. What action should you take?

Q11. The least secure system of communication should never be used to discuss classified material. What is the least secure communications means and why should it never be used to discuss classified material?

Q12. You are on leave away from your command. You meet someone who starts asking questions about your command and its mission. What should you do?

Q13. What are the two most publicized methods of terrorism?

a.

b.

Q14. Where is the likely spot for a terrorist bombing to occur?

Q15. What form is used to record bomb threats received over the phone?

Q16. If you receive a bomb threat over the phone, what should you do?

INTERNATIONAL AGREEMENTS

Learning Objectives: When you finish this chapter, you will be able to—

- Identify the purpose of international agreements.
- Recall the general provisions of the Status of Forces Agreement, the Geneva Convention concerning treatment and rights of prisoners of war, and the Law of Armed Conflict.

Many agreements are made between the government of the United States and governments of other countries. Some of the agreements that directly affect you are discussed in this chapter. These international agreements are the Status of Forces Agreement (SOFA), the Geneva Convention, and the Law of Armed Conflict.

During your tour of duty in the Navy, you will have the opportunity to visit other countries. You may visit as a member of a ship's company, or you may be assigned to a duty station overseas. In either case, remember that you are a guest of the country you are visiting. A small percentage of people feel because they are members of the U.S. Navy, local laws don't apply to them. **That is not true**. If you are on leave or liberty in a foreign country, you must obey the laws of that country.

STATUS OF FORCES AGREEMENT

It is the policy of the Department of Defense (DOD) to protect your rights as much as possible if you are subjected to criminal trial by foreign courts. To do that, the United States has entered into an agreement with several of our allied countries. That agreement is called the *Status of Forces Agreement* (*SOFA*). The SOFA says, in part, that the host country will give up some of

Student Notes:

its jurisdiction to the visiting country in some criminal and civil cases. **The main purpose of the SOFA is to clearly define the status of military personnel of one country stationed in the territory of another**. Some of the topics covered by the Status of Forces Agreement are as follows:

- Freedom of troop movement within the host country
- Passport requirements
- Criminal jurisdiction
- Taxes
- Imposition of customs duties
- Regulations covering driver's licenses

These are just a few of the items covered by the SOFA. (Provisions of the SOFA vary from country to country.) Remember, when you are overseas, YOU are the foreigner. Many customs of the host country may seem strange to you, but you must follow them as well as the local laws. You should receive a briefing on the Status of Forces Agreement that pertains to the country you are visiting. If you have any questions concerning the SOFA while you are in a foreign country, consult your division officer.

GENEVA CONVENTION

Prisoners of war (POWs) have certain rights and are required to observe certain rules, as established by the Geneva Convention of 1949. The Geneva Convention prescribes the following rights of POWs:

- To be treated humanely at all times
- To be protected against insults and public curiosity
- To have decent housing, nourishing food, and adequate clothing
- To be permitted to communicate with their families
- To be given medical care
- To be allowed to worship
- To be allowed to excerise and participate in sports and intellectual pastimes

The Geneva Convention prohibits punishment for refusing to answer questions other than your name, date of birth, rate, and social security number.

A prisoner must salute enemy officers and may be required to perform work if such work is not related to military operations. POWs are subject to the laws, regulations, and orders of the armed forces of the captors and may be punished for violating them. The Geneva Convention recognizes the prisoner's right to try to escape by limiting punishment for such attempts to disciplinary action only, which may consist of 2 hours extra duty daily, loss of half a month's pay (earned as a prisoner), stoppage of any extra privileges, and confinement. A prisoner may not be punished more severely for repeated escape attempts. Prisoners of war are prohibited from renouncing any of the rights to which they are entitled under the Geneva Convention.

Most countries of the world follow the articles of the Geneva Convention. North Vietnam agreed to the convention in 1957 but violated most of its provisions. In 1965, Hanoi violated the convention by announcing the execution of three American POWs in retaliation for the legal execution of Viet Cong terrorists. The Communists also paraded handcuffed Americans through the streets of Hanoi where the people subjected them to ridicule and humiliation. The Geneva Convention expressly forbids such actions. Evidence also indicates that Iraq violated some articles of the convention during the Persian Gulf crisis.

If you have contact with enemy prisoners of war, treat them according to the articles of the Geneva Convention, just as you would expect to be treated by them. If you should become a POW, you should conduct yourself according to the Code of Conduct as well as the Geneva Convention.

LAW OF ARMED CONFLICT

Every nation calls upon its military personnel to defend its national interests by going to war. Our country believes those people involved in armed conflict during war are entitled to fundamental human

Student Notes:

rights regardless of their conduct or beliefs. Because of this belief, our nation has adopted the Law of Armed Conflict to govern the conduct of its military forces engaged in fighting.

Because naval operations frequently involve fighting between major units, you don't need a detailed knowledge of the Law of Armed Conflict. However, you need a basic knowledge of it since even in large-scale naval operations some people may violate the Law of Armed Conflict.

Small-scale operations require a more detailed knowledge of the Law of Armed Conflict by the naval personnel involved. You will receive this detailed knowledge if the need arises.

As a member of a military force, you are allowed during periods of hostilities to attack and even kill the lawful combatants of your enemy. Generally speaking, the term *lawful combatants* means members of the military force and civilian personnel engaged in hostilities.

Just as the Law of Armed Conflict permits certain hostile actions, it limits the way you may conduct these actions. It provides for the protection of certain targets in a war zone to safeguard people and property not directly involved with military activity. For example, it expressly forbids attacking or firing on nonmilitary targets not being used by the enemy for military purposes. The use of illegal techniques and tactics, such as rape, pillage, and plunder, is also prohibited. Unlawful techniques and tactics can backfire on the user because often they are dangerous in themselves. They are also likely to enrage the enemy, causing the enemy to fight harder or respond by using illegal methods, such as killing POWs. Personnel who violate the Law of Armed Conflict will find themselves in serious trouble, including the possibility of trial by court-martial upon return to the United States.

The fundamental terms of the Law of Armed Conflict are as follows:

- Fight only enemy combatants.
- Destroy no more than your mission requires.
- Do not attack enemy soldiers, sailors, airmen, or marines that surrender. Disarm them and turn them over to your superior.
- Never torture or kill prisoners of war and other detainees.
- Collect and care for wounded, sick, or shipwrecked survivors, whether friend or enemy, on land or at sea.
- Protect medical personnel and chaplains, medical and religious facilities, and medical transportation of the enemy. Treat them with respect and do not attack them.
- Treat all civilians humanely and respect their property. Do not attack them.
- Do your best to prevent any violation of these fundamental rules. Report any violations to the appropriate authority promptly.
- Do not violate these rules; an order to do so is illegal.

Discipline yourself to obey these rules during combat. Disobedience of the Law of Armed Conflict dishonors your nation, the Navy, and you. Far from weakening the enemy's will to fight, such disobedience strengthens it. Disobedience of the Law of Armed Conflict is also a crime punishable under the *Uniform Code of Military Justice.*

REVIEW 2 QUESTIONS

Q1. What is the main purpose of the SOFA?

Q2. What document dictates the treatment of POWs?

Q3. What is the purpose of the Law of Armed Conflict?

Student Notes:

SUMMARY

Security of classified material is serious business. Potential enemies are always looking for a chance to gain access to our most guarded secrets. Just one day of failing to safeguard classified material could result in the compromise of extremely sensitive material. The security of classified material not only rests with the personnel that have access to it on a daily basis, but also includes every member of a command. We all have a duty to ensure that only the people requiring access to classified material are allowed to see or use it. The same is true of how we discuss our daily routine. Even if you don't have access to classified material on a daily basis, you could possibly have knowledge of certain exercises or deployment times that world be of benefit to potential enemies. Think carefully before you start talking about upcoming events. Every person in the room is not cleared to have this type of information. Putting pieces of information together to determine what is happening is easy for foreign agents. The same is true when talking on the telephone. Very few phones aboard ship and almost none in the civilian community are secure. Electronic eavesdropping is another way foreign agents collect intelligence data. Be careful of what you say; someone other than the person you called could be listening.

Terrorist activity, particularly when you are visiting a foreign country, should always be of concern. While you should not let it interfere with your enjoyment of visiting a foreign country, you must always be alert to what is going on around you. By taking an extra few minutes to survey your surroundings, you could identify a potentially hazardous situation

The international agreements discussed were designed to protect members of the armed forces. The Status of Forces Agreement protects you when you are stationed in or visiting foreign countries. The Geneva Convention affords you protection if you become a POW. The Law of Armed Conflict protects you in the event of a war. The articles and rules of these agreements will only protect you if you conduct yourself according to U.S. and international law. You have a duty to conduct yourself in a manner that will not bring discredit upon your country, your service, or yourself.

REVIEW 1 ANSWERS

A1. The three levels of security are—

a. **Top Secret**

b. **Secret**

c. **Confidential**

A2. FOUO means **For Official Use Only**.

A3. **Commanding officers** are authorized to initiate a request for a security clearance and background investigation.

A4. A background investigation is required for **Top Secret** and **Secret** clearances.

A5. A letter in parentheses, such as (S), after a publication title tells you **the classification of that publication.**

A6. Classified material, such as videotapes, cassettes, and computer disks, are **marked by tags, stickers, decals, and so on.**

A7. Publications carry the security marking of the highest level of material contained in the publication; therefore, **this publication is marked Top Secret**.

A8. **Security areas** are used to keep classified material.

A9. ADP security is used to safeguard **data processing equipment (computers) including hardware, software, administrative and operating procedures, communications, and personnel and spaces**.

A10. If you find an unattended room with an open and unlocked security container, you should **contact the senior duty officer to report a security violation.** Then, **stand guard over the space until the duty officer arrives**.

A11. The least secure communications means is the **telephone**. **Never use telephones to discuss classified material because they can be physically and electronically wiretapped.**

Student Notes:

A12. If you meet someone who starts asking questions about your command and its mission, you should **report the incident to the nearest military activity**.

A13. The two most publicized forms of terrorism are—

a. **Taking hostages**

b. **Bombing**

A14. Terrorists are likely to bomb **places of business that serve a high volume of people such as airports, nightclubs, and restaurants**.

A15. To report a bomb threat made over the telephone, use **Telephonic Threat Complaint, OPNAV Form 5527/8**.

A16. If you receive a bomb threat over the phone, you should—

a. **Keep the caller on the line and get as much information as possible.**

b. **Record in writing the caller's conversation**.

c. **Ask caller where's the bomb, what type of bomb, time of detonation, and what it looks like**.

d. **Try to determine sex, age, attitude of caller, and accents or speech impediments; try to remember background noises**.

REVIEW 2 ANSWERS

A1. The main purpose of the SOFA is to **define the status of military personnel of one country stationed in a territory of another**.

A2. The treatment of POWs is covered by the **Geneva Convention**.

A3. The purpose of the Law of Armed Conflict is **to govern the conduct of military personnel engaged in fighting**.

APPENDIX I

GLOSSARY

When you start a new job, you're usually faced with the task of learning the vocabulary of that job. The Navy has a language all of its own. One way to learn the vocabulary of the job is to look up terms in a glossary.

This glossary contains many terms used by the Navy. It's not all-inclusive; that is, not all terms are here. If you want to find out more about Navy terms, refer to *Naval Terminology*, Naval Warfare Publication 3 (Revision E).

AA—Antiaircraft.

ABAFT—Further aft, as "*Abaft* the beam."

ABEAM—On a relative bearing of 90° (*abeam* to starboard) or 270° (*abeam* to port).

ABOARD—In or on a ship. Extended to use ashore, as *aboard* a naval station.

ABREAST—Same as abeam.

ACCOMMODATION LADDER—A ladder suspended over and inclining down the side of a ship to let people board the ship from boats.

ADRIFT—Loose from moorings and out of control. Applied to anything that is lost, out of hand, or left lying about.

AFT—Toward the stern. Not as specific as abaft.

AFTER—That furthest aft, as *after* fireroom.

AFTERNOON WATCH—The 1200 to 1600 watch.

AGROUND—When any part of a ship or boat is resting on the bottom. A ship runs *aground* or goes *aground.*

AHOY—A hail or demand for attention, as "Boat *ahoy.*"

ALEE—In the direction toward which the wind is blowing; downwind.

ALIVE—Lively, energetic.

ALL FAST—Tied or lashed down as necessary.

ALL HANDS—The entire ship's company.

ALOFT—Generally speaking, any area above the highest deck.

ALONGSIDE—By the side of the pier or ship.

AMIDSHIPS—An indefinite area midway between the bow and stern. Rudder *amidships* means that the rudder is in line with the ship's centerline.

ANCHOR—(1) Any device used to make a floating body fast to the bottom. (2) The act of so making fast. (3) The act of securing or fixing the lower end of a guy or stay or the lower end of a shore.

ANCHORAGE—An area designated to be used by ships for anchoring.

ANCHOR BALL—A black circular shape hoisted to indicate that the ship is anchored.

ANCHOR BUOY—A small float secured to the anchor by a light line to mark the position of the anchor.

ANCHOR CABLE—The line, wire, or chain that attaches a vessel to its anchor.

ANCHOR WATCH—A group of persons available to the OOD during the night for such duties as heaving in or paying out the cable.

ARMAMENT—The weapons of a ship.

ARMORED DECK—A deck, below the main deck, that provides added protection to vital spaces.

ASTERN—Directly behind a ship.

ATHWART—Across; at right angles to.

AUXILIARY—(1) Extra, or secondary, as *auxiliary* engine. (2) A vessel whose mission is to supply or support the combatant forces.

AVAST—Stop, as "*Avast* heaving."

AYE AYE—Reply to a command or order, meaning "I understand and will obey."

BACK—(1) To go backwards. (2) Act of the wind in changing direction counterclockwise.

BACKSTAY—Piece of standing rigging leading aft.

BAIL—(1) To rid a boat of water by dipping it out. (2) A rigid member supporting two end points, as the *bail* (handle) of a bucket or the support for an accommodation ladder.

BALLAST—Weight (solid or liquid) loaded into a ship to increase stability.

BAR—A long, narrow shoal across a harbor entrance.

BARGE—(1) A blunt-ended, flat-bottomed, waterborne craft, usually nonself-propelled, used to haul supplies or garbage. (2) A type of motorboat assigned for the personal use of a flag officer.

BATTEN—(1) A long strip of steel wedged against the edges of tarpaulins on a hatch to make the hatch watertight. (2) Removable wood or steel members used in ship's holds to keep cargo from shifting.

BATTEN DOWN—The act of applying battens to a hatch. Extended to mean the closing of any watertight fixture.

BATTLE LANTERN—A battery-powered lantern for emergency use.

BEAM—(1) The extreme breadth (width) of a vessel. (2) A transverse frame supporting a deck.

BEAR—The act of locating a particular point, or bearing, as "The lighthouse *bears* 45°."

BEAR A HAND—(1) Provide assistance, as "*Bear a hand* with rigging this stage." (2) Expedite, as "*Bear a hand* with readiness for sea reports."

BEARING—The direction of an object from an observer, measured in degrees clockwise from a reference point. See MAGNETIC BEARING, RELATIVE BEARING, and TRUE BEARING.

BECKET—(1) An eye for securing one end of a line to a block. (2) A rope eye on a cargo net. (3) Shortened form of becket bend.

BECKET BEND—A knot used to tie two lines together.

BELAY—(1) To secure a line to a fixed point. (2) Order to disregard a previous order or to stop an action, as "*Belay* the last order," or "*Belay* the small talk."

BELOW—Downward, beneath, or beyond something, as to lay *below*; *below* the flight deck; *below* the horizon.

BEND—To join two lines together; the type of knot so used.

BERTH—(1) A bunk. (2) A duty assignment. (3) Mooring space assigned to a ship.

BIGHT—The middle part of a line or a loop in a line.

BILGE—(1) Bottom of the hull near the keel. (2) To fail an examination. (3) *Bilge* water is foul water, so to apply the term to something implies that it is worthless.

BILLET—Place or duty to which one is assigned.

BINNACLE—Stand containing a magnetic compass.

BINNACLE LIST—List of persons excused from duty because of illness.

BITT—Cylindrical upright fixture to which mooring or towing lines are secured aboard ship.

BITTER END—The free end of a line.

BLOCK—A frame containing a pulley, called a *sheave*, around which a line (known as a *fall*) is attached.

BLOCK AND TACKLE—See PURCHASE.

BOARD—(1) The act of going aboard a vessel. (2) A group of persons meeting for a specific purpose, as an investigation board.

BOAT—A small craft capable of being carried aboard a ship.

BOAT BOOM—A spar rigged out from the side of an anchored or moored ship to which boats are tied when not in use.

BOAT FALLS—Tackle used to hoist and lower a boat in davits.

BOATHOOK—A staff having a hook at one end. Used for fending a boat off, hooking a line, and so forth.

BOATSWAIN'S CHAIR—A seat attached to a gantline for hoisting a person aloft.

BOATSWAIN'S LOCKER—A compartment, usually forward, where line and other equipment used by the deck force are stowed.

BOLLARD—A strong, cylindrical upright fixture on a pier to which a ship's mooring lines are secured.

BOOM—A spar used for hoisting loads; usually movable.

BOOT TOPPING—Black paint applied to a ship's sides along the waterline.

BOW—The forward end of a ship or boat.

BOW HOOK—Member of a boat's crew whose station is forward.

BREAK OFF—To walk away with a line or run a line in; let go, return to the point from which the line is being hauled; take a new hold, and walk away again.

BREAK OUT—To bring out supplies or equipment from a storage space.

BREAST LINE—Mooring line leading from the ship to the pier at right angles to the ship.

BRIDGE—Area in the superstructure from which a ship is operated. *See* CONN.

BRIDLE—A span of rope, chain, or wire with both ends secured and the strain taken on the midpart.

BRIG—Naval term for jail.

BROACH TO—To get crosswise (without power) to the direction of wave travel; particularly dangerous near a beach.

BROAD—Wide, as *broad* in the beam.

BROAD ON THE BOW—Halfway between dead ahead and abeam.

BROAD ON THE QUARTER—Halfway between abeam and astern.

BROADSIDE—(1) The act of firing all main battery guns to one side at once. (2) Sidewise, as "The current carried the ship *broadside* toward the beach." *Broadside to* is to have the side toward something, as "The ship hit the pier *broadside to*."

BROW—Navy term for gangplank. Used as a crosswalk from one ship to another and from a ship to a pier.

BULKHEAD—A vertical partition in a ship; never called a wall.

BULKHEADING—Complaining or grumbling with the intention of being overheard by seniors.

BULWARK—Solid barrier along the edges of the weather deck that serves as a protection against the weather.

BUOY—An anchored float used as an aid to navigation or to mark the location of an object.

CABIN—Living compartment of a ship's commanding officer.

CABLE—A line, wire, or chain that connects a ship to its anchor.

CAISSON—Gate at the end of a drydock that keeps out the water.

CALL—(1) The boatswain's pipe. (2) A signal sounded on the boatswain's pipe.

CAMEL—Large float or rectangular structure used as a fender between a ship and the pier.

CAN BUOY—A navigational buoy, cylindrical in shape, that marks the port side of a channel from seaward; odd-numbered and painted green.

CANOPY—A cover fitted over part of a boat.

CAPSTAN—The part of a vertical shaft windlass around which a working line is passed; used for heaving in anchors and hawsers.

CARRICK BEND—A knot used for joining two lines. The single carrick bend isn't often used because it jams tight; instead, a double carrick bend is used, particularly for bending towing hawsers together.

CARRY AWAY—To break loose, as "The rough seas *carried away* the lifelines."

CAULK—The act of stuffing the seams between wooden planking with oakum for watertightness.

CHAFING GEAR—Material used to protect lines from excessive wear.

CHAIN LOCKER—Spaces where anchor chain is stowed.

CHAIN MARKINGS—A series of turns of wire and stripes of paint on certain links of each anchor chain. They show the scope or amount of chain that has run out.

CHAINS—Area (a platform on large ships) where the leadsman stands when taking soundings with the hand lead.

CHART—Nautical counterpart of a road map, showing land configuration, water depths, and aids to navigation.

CHECK—(1) To slow or ease; to *check* a line is to pay out just enough line to prevent its parting when under a strain. (2) To investigate or examine something.

CHEEK—One of the sides of a block.

CHOCK—Deck fitting through which mooring lines are led.

CHOW—Feed.

CHRONOMETER—An accurate clock used in navigation.

CLAMP DOWN—To sprinkle the deck with water and dry it with a swab.

CLEAT—A metal casting with two projecting arms to which a line is belayed.

COAMING—Bulwark around a hatch opening.

COFFERDAM—A void between compartments or tanks of a ship for purposes of insulation.

COIL—To lay down a line in circular turns piled loosely on top of one another.

COLLISION BULKHEAD—A bulkhead, stronger than normal, located forward to control flooding in the event of a head-on collision.

COLORS—(1) The national ensign. (2) The ceremony of raising and lowering the ensign.

COMBATANT SHIP—A ship whose primary mission is combat.

COMPANIONWAY—Deck opening giving access to a ladder (includes the ladder).

COMPARTMENT—Interior space (room) in a ship.

COMPLETE DECK—Any deck that extends the length of a ship from side to side.

CONN—Station, usually on the bridge, from which a ship is controlled; the act of controlling the ship's movements.

COURSE—A ship's desired direction of travel, not to be confused with heading, which is the direction in which the bow is pointed at any given instant.

COVER—(1) To protect. (2) A shelter. (3) Headgear, and the act of donning same.

COXSWAIN—Enlisted person in charge of a boat.

DARKEN SHIP—To turn off all external lights and close all openings through which lights could be seen from outside the ship.

DAVITS—A crane or mechanical arms that project over the side of a ship and are used to lower or hoist a boat in or out of the water.

DEAD AHEAD—Directly ahead; a relative bearing of 000°. *Dead astern* is 180° relative.

DEAD IN THE WATER—A ship that has stopped and has no way on, or no movement through the water.

DECK—Horizontal planking or plating that divides a ship into layers.

DECK SEAMANSHIP—The upkeep and operation of all deck equipment.

DEEP SIX—To throw something overboard.

DIP—The act of lowering a flag partway down the staff as a salute to, or in reply to a salute from, another ship.

DISTANCE LINE—A line stretched between two ships engaged in replenishment or transfer operations under way. The line is marked at 20-foot intervals to help the conning officer in maintaining station.

DIVISION—(1) A main subdivision of a ship's crew (1st, E, G, and so forth). (2) An organization made up of two or more ships of the same type.

DOCK—Commonly refers to any pier or wharf; but, strictly speaking, it refers only to the space alongside a pier or in drydock.

DOG—(1) A lever or bolt and thumbscrews used for securing a watertight door. (2) The act of dividing a 4-hour watch into 2-hour watches.

DOG DOWN—To set the dogs on a watertight door.

DOG WATCH—The 1600 to 1800 and 1800 to 2000 watches.

DOLPHIN—(1) A cluster of piles at the end of a pier. (2) A porpoise.

DOUBLE UP—To double mooring lines for extra strength.

DRAFT—The vertical distance from the keel to the waterline.

DRAFT MARKS—The figures fastened to the stem and stern, the center of which indicates the draft of the ship.

DRIFT—The speed at which a ship is pushed off course by wind and current.

DROUGUE—*See* SEA ANCHOR.

DRYDOCK—A dock from which the water may be removed for the purpose of inspecting or working on a ship's bottom; it may be either floating or built into the shore.

EASE—To relax, to slack.

EASE HER—Reduce the amount of rudder the ship is carrying.

EBB, EBB TIDE, ON THE EBB—A falling tide.

EIGHT O'CLOCK REPORTS—Reports received shortly before 2000 by the executive officer from the heads of departments.

ENGINE-ORDER TELEGRAPH—Electro-mechanical device that transmits orders to the engine room concerning the speed of the engines.

ENSIGN—(1) The national flag. (2) The lowest grade of commissioned officer.

EYES—The most forward part of the forecastle.

FAIRLEAD—A device, usually a block, for leading a line around a corner.

FAIRWAY—Thoroughfare for a ship.

FALL—A line, wire, or chain rove on a purchase.

FANTAIL—The after end of the main deck.

FATHOM—Unit of measurement equal to a depth of 6 feet.

FENDER—A cushioning device hung over the side of a ship to prevent contact between the ship and the pier or another ship.

FID—A long, tapered, wooden tool used to open the strands of a line for splicing.

FIELD DAY—A day devoted to general cleaning, usually in preparation for an inspection.

FIREMAIN—Piping system to which fire hydrants are connected.

FIRST WATCH—The 2000 to 2400 watch. Also called the evening watch.

FIRST CALL—A routine call sounded as a warning signal for roll call formations and many other ceremonies; also sounded 5 minutes before morning and evening colors.

FISHHOOK—A broken end of wire protruding from a wire rope.

FLAG OFFICER—An officer of the rank of rear admiral or higher.

FLAGSTAFF—Vertical staff at the stern to which the ensign is hoisted when moored or at anchor.

FLAT—Partial deck (often a grating) to provide walking and working surfaces; used extensively in engineering spaces.

FLEET—An organization of ships, aircraft, marine forces, and shore-based fleet activities, all under one commander, for the purpose of conducting major operations.

FLOOD—(1) To fill a space with water. (2) A rising tide.

FOC'SLE—See FORECASTLE.

FOGY—(Pronounced fo-gee.) A longevity pay increase.

FORE—Forward.

FORE AND AFT—The entire length of a ship, as in "Sweep down *fore and aft*."

FORECASTLE—(Pronounced fok-sul.) Forward section of the main deck, generally extending from the stem aft to just abaft the anchor windlass.

FOREMAST—First mast aft from the bow.

FORENOON WATCH—The 0800 to 1200 watch.

FOUL—(1) Entangled, as "The lines are *foul* of each other." (2) Stormy.

FOUNDER—To sink because of being overwhelmed by the sea.

FRAME—The athwartship strength member of a ship's hull.

FRAPPING LINES—Lines passed around boat falls to steady the boat when hoisting or lowering.

FREEBOARD—Vertical distance from waterline to weather deck.

GAFF—A light spar set at an angle from the upper part of a mast from which the ensign is flown when a ship is under way.

GALLEY—Space where food is prepared. Never called a kitchen.

GANGWAY—(1) The opening in a bulwark or lifeline to provide access to a brow or an accommodation ladder. (2) Given as an order it means "Clear the way."

GANTLINE—Line used for hoisting and lowering a boatswain's chair.

GENERAL ALARM—A sound signal of a pulsating ringing tone used only on board ship for calling all hands to general quarters.

GENERAL QUARTERS (GQ)—The condition of full readiness for battle.

GIG—Boat assigned for the commanding officer's personal use.

GIRDER—A longitudinal supporting a deck.

GRANNY KNOT—A bungled square knot.

GRAPNEL—A small, four-armed anchor used to recover objects in the water.

GRIPE—Device for securing a boat at its davits or in a cradle.

GROUND TACKLE—Equipment used in anchoring or mooring with anchors.

GUNWALE—(Pronounced gunnel.) The upper edge of the sides of a ship.

GUY—A line used to steady a spar or boom.

HALF DECK—A partial deck below the main deck.

HALYARD—A light line used to hoist a flag or pennant.

HAND—A ship's crew member.

HANDSOMELY—Slowly and carefully.

HARD OVER—Condition of a rudder that has been turned to the maximum possible rudder angle.

HASHMARK—(Service stripe.) A red, blue, or gold diagonal stripe across the left sleeve of an enlisted person's jumper or coat; each stripe indicates 4 years service.

HATCH—A square or rectangular access in a deck.

HAUL—To pull in or heave on a line by hand.

HAUL OFF—Changing a vessel's course to keep clear of another vessel.

HAWSEPIPE—Opening through which the anchor cable runs from the deck out through the side of the ship.

HAWSER—Any heavy wire or line used for towing or mooring.

HEAD—(1) The upper end of a lower mast boom. (2) Compartment containing toilet facilities. (3) Ship's bow."

HEADING—The direction toward which the ship is pointing at any instant.

HEAVE—To throw.

HEAVE AROUND—(1) Th act of hauling in a line, usually by means of a capstan or winch. (2) General term for "Get to work."

HEAVE IN—Take in line or cable.

HEAVE OUT AND TRICE UP—Announcement given at reveille to persons sleeping in hammocks. It means "Get up and lash up your hammocks." This term now applies to ships equipped with bunks.

HEAVE TO—Stopping or reducing headway of a vessel just enough to maintain steerageway.

HEAVING LINE—A line with a weight at one end that is heaved across an intervening space for the purpose of passing over a heavier line.

HELM—Mechanical device used to turn the rudder; usually a wheel aboard ship; a lever in boats.

HELMSMAN—Person who steers the ship by turning the helm.

HIGHLINE—The line stretched between the ships under way on which a trolley block travels back and forth for transfer of material and personnel.

HITCH—(1) Used to bend a line to or around a ring or cylindrical object. (2) Common term for an enlistment.

HOLD—Large cargo stowage space aboard ship.

HOLDING BULKHEAD—The innermost of a series of bulkheads that form the tanks and voids of the torpedo protection.

HOLIDAY—Space on a painted surface that the painter neglected to cover.

HOOK—Familiar term for the anchor.

HORN—One of the projections of a cleat.

HOUSE—The act of two-blocking (pulling up tight) an anchor in its hawsepipe.

HULL—The shell, or plating, of a ship from keel to gunwhale.

HULL DOWN—Refers to a ship that is so far over the horizon that only its superstructure or top hamper is visible.

INBOARD—Toward the centerline.

INHAUL LINE—Line used to haul the trolley back to the delivering ship during highline transfers.

INLET—A narrow strip of sea extending into the land.

INSHORE—Close to the shore.

IRISH PENNANT—Loose, untidy end of line left adrift. Also called a deadman or cow's tail.

ISLAND—Superstructure on the starboard side of the flight deck of an aircraft carrier.

JACK—Starred blue flag (representing the union of the ensign) flown at the jackstaff of a commissioned ship not under way.

JACKSTAFF—Vertical spar at the stem to which the jack is hoisted.

JACKSTAY—Any horizontal line or wire for the support of articles (such as seabags).

JACOB'S LADDER—A portable rope or wire ladder.

JETTY—A structure built out from shore to influence water currents or to protect a harbor or pier.

JUMP SHIP—The act of deserting ship.

JURY RIG—Any makeshift device or apparatus.

KAPOK—Material used to stuff life jackets and other lifesaving apparatus.

KEDGE—(1) A small anchor. (2) The act of moving a ship by hauling it ahead by heaving in on a line to a laid-out anchor.

KEEL—The lowermost longitudinal strength member from which the frames and plating rise.

KEEL BLOCK—One of a series of blocks along a drydock bed; used to support the keel of a vessel in drydock.

KEELSON—That part of a boat's keel that is inside the boat.

KING POST—One of a pair of short, strong uprights used to support the cargo booms of cargo vessels.

KING SPOKE—Spoke on the steering wheel that's upright when the rudder is amidships; usually distinctively marked, as with a Turk's head.

KNOCK OFF—Quit working.

KNOT—(1) A unit of measurement of speed equal to 1 nautical mile (6,080 feet) per hour. (2) A collective term for hitches and bends.

LADDER—A shipboard flight of steps.

LANDING CRAFT—Vessels especially designed for landing troops and equipment directly on a beach.

LANDING SHIP—A large seagoing ship designed for landing large numbers of personnel and/or heavy equipment directly on a beach.

LANYARD—(1) Any short line used as a handle or as a means for operating some piece of equipment. (2) A line used to attach an article to the person, such as a pistol *lanyard.*

LASH—To secure an object by turns of line, wire, or chain.

LASHING—Line, wire, or chain used to lash an article.

LASH-UP—An uncomplimentary term applied to a rig, device, or system meaning it's in disorder. For example, "What a *lash-up* they have there."

LAUNCH—(1) To float a vessel off the ways in a building yard. (2) A power boat, usually over 30 feet long.

LAY—(1) To go to a specific place, such as "*Lay* aloft." (2) To put something down, as to *lay* tile. (3) The direction of a twist of the strands in a line or wire.

LEAD LINE—A narrow block of lead weighing from 7 to 14 pounds attached to a marked line. Used by leadsman to determine depth of water.

LEADSMAN—Person who uses the lead line.

LEE—An area sheltered from the wind; downwind.

LEE HELMSMAN—A spare helmsman who usually operates the annunciator.

LEE SHORE—A shore that is leeward of the ship.

LEEWARD—(Pronounced loo-urd.) Side of the ship opposite to the direction the wind is blowing from.

LIBERTY—Permission to be absent from a ship or station for a short time.

LIE OFF—To heave to at some distance from shore.

LIFE BUOY—A buoyant ring or some other floating device, except a life jacket or life belt, designed to support a person in the water.

LIFE JACKET—A buoyant jacket designed to support a person in the water; a life belt fits only around the waist.

LIFELINE—(1) In general, the lines erected around the edges of weather decks, specifically, the topmost line. From top to bottom, the lines are named lifeline, housing line, and foot rope.

LIGHTEN SHIP—To make a ship lighter by removing weight.

LIGHT SHIP—The act of dispensing with blackout precautions.

LINE—Any rope that isn't wire rope.

LINNER BOTTOM—The inside bottom in a system of double bottoms.

LOG—(1) A ship's speedometer. (2) The act of a ship in making a certain speed, as "The ship *logged* 20 knots." (3) Book or ledger in which data or events that occurred during a watch is recorded.

LOOK ALIVE—Admonishment meaning "be alert" or "move faster."

LOOKOUT—Person stationed topside as a formal watch who reports all objects sighted and sounds heard to the OOD.

LOOM—The glow seen in the sky from a light that's below the horizon.

LUBBER'S LINE—Line engraved on the inside of a compass bowl, representing the ship's head, by which the ship's course is steered.

LUCKY BAG—Locker, under the charge of the master-at-arms, used to stow gear found adrift and deserters' effects.

MAGAZINE—Compartment used for stowage of ammunition.

MAGNETIC BEARING—The direction of the object measured on a magnetic compass.

MAIN DECK—The uppermost complete deck.

MAINMAST—Second mast aft from the bow.

MAN—To assume a station, as to *man* a gun.

MAN-O-WAR—*See* COMBATANT SHIP.

MARLINE—Two-strand, left-laid, tarred hemp.

MARLINSPIKE—Tapered steel tool used to open the strands of wire for splicing.

MARLINESPIKE SEAMANSHIP—The art of caring for and handling all types of line and wire.

MASTER-AT-ARMS—A member of a ship's police department.

MASTHEAD LIGHT—A 20-point, white running light located in the fore part of the ship. May or may not be on the foremast.

MATE—A shipmate; another Sailor.

MEET HER—Slow the swing of a ship by putting on opposite rudder.

MESS—(1) Meal. (2) Place where meals are eaten, as *mess* hall. (3) A group of personnel who take meals together, as the officers' *mess*.

MESSENGER—(1) A line used to haul another heavier line across an intervening space. (2) One who delivers messages.

MIDWATCH—The watch that begins at 0000 and ends at 0400.

MIND YOUR RUDDER—An order to the helmsman to steer the proper course.

MONKEY FIST—A complicated knot worked into the end of a heaving line to provide weight.

MOOR—(1) To anchor, using two anchors. (2) To make fast to a mooring buoy. (3) To make fast to a pier or another ship.

MOORING BUOY—A large, anchored float a ship may moor to.

MORNING WATCH—The 0400 to 0800 watch.

MOTOR WHALEBOAT—A double-ended power-boat.

MUSTER—(1) A roll call. (2) The act of assembling for a roll call.

NEST—(1) Two or more boats stowed one within the other. (2) Two or more ships moored alongside each other.

NOTHING TO THE RIGHT (LEFT)—Order given to the helmsman not to allow the ship to come to right (left) of the course because of some danger lying on that side of the course.

NUN BUOY—A navigational buoy, conical in shape, that marks the starboard side of a channel from seaward. Even numbered and painted red.

OAKUM—Tarred hemp fiber used to caulk seams in wooden decks and boats.

OOD—Officer of the deck.

OFFSHORE—Some distance off the shore, as contrasted to inshore.

ON THE BEACH—Ashore; also applied to a Sailor who is assigned to shore duty or is unemployed, retired, or otherwise detached from sea duty.

OUTBOARD—Away from the centerline.

OVERBOARD—Over the side.

OVERHAND KNOT—Simplest of all knots; made by passing one end of a line once around its standing part.

OVERHAUL—(1) To repair or recondition. (2) To overtake another vessel.

OVERHEAD—The underside of a deck forming the ceiling of the compartment below. Never called a ceiling.

PAINTER—Line used to make a boat fast by its bow. When used under way, the *painter* causes the boat to swing out from the side of the loop.

PARCEL—The act of wrapping a line with narrow canvas strips to provide waterproofing or to build up a symmetrical shape for further covering.

PARTY—A group having a common temporary assignment or purpose, as a working *party*, a line-handling *party*, or a liberty *party*.

PASSAGEWAY—A corridor used for interior horizontal movement aboard ship.

PAY—Monthly salary.

PAY OUT—To feed out, or lengthen, a line.

PELORUS—A gyrocompass repeater used to take bearings.

PIER—A structure extending from land out into the water to provide a mooring for vessels.

PIER HEAD—Seaward end of a pier.

PIGSTICK—Small staff from which the commission pennant is flown.

PILOTHOUSE—Enclosure on the bridge housing the main steering controls.

PILOTING—Branch of the science of navigation in which positions are determined by reference to visible objects on the surface or by soundings.

PIPE—The act of sounding a particular call on the boatswain's pipe.

PITCH—Vertical rise and fall of a ship's bow caused by head or following seas.

PLAIN WHIPPING—A whipping made without using a palm and needle.

POLLIWOG—A person who has never crossed the equator.

PORT—To the left of the centerline when facing forward.

PROTECTIVE DECK—*See* ARMORED DECK.

PROW—That part of the stem (bow) above the waterline.

PURCHASE—A machine that's a combination of one or more blocks rove with a line or wire. When rove with chain, called a chain fall.

PYROTECHNICS—Ammunition containing chemicals that produce smoke or a brilliant light when burning; used for signaling or for illumination.

QUARTER—Area between dead astern and either beam.

QUARTERDECK—Deck area designated by the commanding officer as the place to carry out official functions; the station of the OOD in port.

QUARTERMASTER—An enlisted assistant to the navigator.

QUARTERS—(1) Stations for shipboard evolutions, as general *quarter*, fire *quarters*, *quarters* for muster. (2) Living spaces.

QUAY—(Pronounced key.) A solid structure along a bank used for loading and off-loading vessels.

RADAR—A device that uses reflected radio waves to detect objects.

RANGE—(1) The distance of an object from an observer. (2) An aid to navigation consisting of two objects in line. (3) A water area designated for a particular purpose, as a gunnery *range*.

RAT GUARD—A hinged metal disk that can be secured to a mooring line to prevent rats from using the line to gain access to the ship.

RAT-TAILED STOPPER—A braided tapering line used on boat falls, mooring lines, and so forth.

REDUCER—Fitting applied to a fire hydrant to permit the attachment of a hose of smaller diameter than the hydrant outlet.

REEF—An underwater ledge rising abruptly from the floor of the ocean.

REEVE—To thread a line through a pulley.

RELATIVE BEARING—The angle between the ship's head and the object.

RELIEF—Person assigned to assume the duties of another.

RELIEVE—(1) To take the place of another. (2) To ease the strain on a line.

RIDE—A ship at anchor *rides* to its anchor as it swings on the chain attached to the anchor.

RIDING LIGHT—Light required to be shown by a vessel at anchor.

RIG—To set up any device or equipment, as *rig* a stage over the side.

RIGGING—Lines that support a ship's masts are called standing rigging; those used to hoist or otherwise move equipment are called running rigging.

RISER—A pipe leading from the firemain to fireplugs on upper deck levels.

ROLLER CHOCK—A mooring chock that contains a roller for reducing friction.

ROPE—General reference to both fiber and wire rope. Fiber rope usually is referred to as line; wire rope is called rope, wire rope, or just wire.

ROPE YARN SUNDAY—Free time given during a workday (usually an afternoon) to allow personnel to take care of personal business.

RUDDER—Device attached to a ship's stern that controls the ship's direction of travel.

RUNNER— A purchase containing one single-sheave movable block.

RUNNING BOWLINE—A slipknot made by tying a small bowline around a line's own standing part.

RUNNING LIGHTS—Navigational lights required to be shown at night by a vessel under way.

SACK—Bunk.

SCUPPER—The waterway along the gunwales.

SCUTTLE—(1) Round, watertight opening in a hatch. (2) The act of deliberately sinking a vessel.

SCUTTLEBUTT—(1) Originally a ship's water barrel (called a butt), which was tapped (scuttled) by the insertion of a spigot from which the crew drew their drinking water; now applied to any drinking fountain. (2) In the old days the scuttlebutt was a place for personnel to exchange views and news when they gathered to draw their water; hence the term *scuttlebutt* is applied to any rumor.

SEA—(1) The ocean in general. (2) The individual undulations (rolls) of the surface are called waves, but as a whole they are referred to as *seas*. Also, a ship takes a big *sea*, not a wave, over the bow.

SEA ANCHOR—A device streamed from the bow of a vessel for the purpose of holding end-on to the sea.

SEAMANSHIP—(1) The art or skill of handling a vessel. (2) Skill in the use of deck equipment, boat handling, and the care and use of line and wire.

SEAWORTHY—A vessel capable of withstanding normal heavy weather.

SECOND DECK—First complete deck below the main deck.

SECURE—(1) To make fast, as to *secure* a line to a cleat. (2) To cease, as to *secure* from fire drill.

SERVICE FORCE—The organization providing logistic support to the combatant forces.

SET—The direction toward which a ship is pushed by the effects of wind and current. See DRIFT.

SETUP—To tighten up, with particular reference to dogs and turnbuckles.

SHAKE A LEG—An admonishment to move faster.

SHAKEDOWN—The training of a new crew to develop efficiency in operating a ship.

SHEAVE—Pulley in a block around which the fall (line) runs.

SHEER STRAKE—The uppermost strake in a ship's side plating.

SHEET BEND—Same as a becket bend.

SHELL—A vessel's hull plating from the keel to the main deck; also called skin.

SHELLBACK—A person who has crossed the equator.

SHIFT—(1) The act of the wind in changing direction. (2) The act of moving a rudder with angle on it to the same angle on the opposite side.

SHIFT COLORS—To change the arrangement of the colors on getting under way or coming to moorings.

SHIP—(1) Any large vessel capable of extended independent operation. (2) To take on water unintentionally.

SHIPOVER—To reenlist in the Navy.

SHIPSHAPE—Neat, clean, taut, in fine shape.

SHOAL—Similar to a reef, but more gradual in its rise from the floor of the ocean.

SHORE—(1) The land in general, but usually refers to that part adjacent to the water. (2) A timber used in damage control to brace bulkheads and decks.

SHROUD—A line or wire that provides athwartship support for a mast.

SICK BAY—Shipboard space used as a hospital.

SIDE BOY—One of a group of seamen who form two ranks at the gangway as part of the ceremonies conducted for visiting officials.

SIDE LIGHT—One of the required running lights. The starboard *side light* is green and the port *side light* is red.

SIDE PORT—A watertight opening in a ship's side that is used as a doorway.

SIGHT—(1) To see for the first time, as to sight a ship on the horizon. (2) A celestial observation.

SKYLARK—To engage in irresponsible horseplay.

SLACK—(1) To allow a line to run out. (2) A *slack* ship is one that has little or no discipline.

SLIP—(1) To free a ship of its anchor by disconnecting the cable or by allowing its bitter end to run out. (2) A narrow space between two piers, or the space between two rows of piles that guide a ferryboat into its berth.

SMALL CRAFT—Any less-than-ship-sized vessel.

SMALL STORES—Personal needs for Sailors, such as articles of clothing.

SMART—Snappy, seamanlike, shipshape.

SNAKING—Netting stretched between the gunwales and footrope (see LIFELINE) to prevent objects from going over the side.

SNUB—The act of suddenly checking a line that is running out under a strain.

SOPA—Abbreviation for senior officer present afloat.

SOUND—(1) To determine the depth of water. (2) The act of a whale or similar creature in diving deep. (3) A body of water between the mainland and a large coastal island.

SPANNER—A wrench used for tightening couplings on a fire hose.

SPAR—A along cylindrical member of wood or metal, tapered at the ends; usually attached to a mast for use as a boom or for the attachment of equipment such as signal halyards. See BOAT BOOM; YARDARM.

SPAR BUOY—A buoy shaped like a spar. Usually indicates special areas, such as a quarantine anchorage (yellow) or normal anchorage (white), but may be used to indicate a channel (painted red or green, as appropriate).

SPECIAL SEA DETAIL—Personnel aboard ship assigned special duties connected with leaving and entering port.

SPLICE—The act of intertwining strands of lines or wires to join them together or to make an eye; the joint so made.

SPRING—A mooring line that leads forward (or aft) at an angle from ship to pier. Its purpose is to check the fore-and-aft movement of the ship.

SPRING LAY—Wire rope in which each strand consists partly of wire and partly of tarred hemp or similar fiber.

SQUADRON—Two or more divisions of ships or aircraft.

SQUARE AWAY—Put in proper order; make things shipshape.

SQUARE KNOT—Simple knot used for bending two lines together or for bending a line to itself.

STACK—Shipboard chimney.

STANCHIONS—Vertical posts used for supporting decks; smaller, similar posts used for supporting lifelines, awnings, and so forth.

STAND BY—To "prepare for" or "make ready to."

STANDING LIGHTS—Red night-lights throughout the interior of a ship.

STANDING PART—The main part of a line, as distinguished from its ends.

STARBOARD—Direction to the right of the center line as one faces forward.

STATEROOM—A living compartment for an officer or for a small number of officers.

STATION—(1) An individual's place of duty. (2) Position of a ship in formation. (3) Location of persons and equipment having a specific purpose, as a gun control *station* . (4) Order to assume a post of duty, as "*Station* the special sea and anchor detail."

STAY—Any piece of standing rigging, except a shroud, providing support only.

STEADY (STEADY SO) (STEADY AS YOU GO) (STEADY AS SHE GOES)—Order to the helmsman to steer the ship on the course it is heading at the time the order is given.

STEM—The forward vertical extension of the keel.

STERN—The aftermost part of a vessel.

STERN HOOK—Member of a boat's crew whose station is aft.

STERN LIGHT—White navigation light that can be seen only from astern to 6 points on either quarter (total of 12 points, or 135°).

STERNPOST—The after vertical extension of the keel.

STERN SHEETS—The after passenger space in an open boat.

STOP—A short line attached to the edge of an awning, boat cover, and so forth; used to lash the cover to a support.

STOW—To store or pack articles or cargo in a space.

STRAKE—Fore-and-aft strip of plating in the shell or in a deck.

STRAND—(1) One of the main subdivisions of a line or wire. (2) The act of a vessel in going aground.

STRINGER—(1) A longitudinal frame providing strength to a ship's sides. (2) A long timber between piles at the edge of a pier.

STRUCTURAL BULKHEAD—Transverse-strength bulkhead that forms a watertight boundary.

SUPERSTRUCTURE—The ship's structure above the main deck, exclusive of the top hamper.

SWAB—The same as, but never referred to as a mop.

SWAMP—The filling of an open boat with water taken over the side.

TACKLE—See PURCHASE.

TAFFRAIL—The rail around the stern of a ship or boat.

TARPAULIN—Canvas used as a cover.

TAUT—Under tension. A ship noted for its high state of discipline and efficiency is known as a *taut* ship.

TENDER—(1) One who serves as a precautionary standby, as the line *tender* for a diver. (2) An auxiliary vessel that acts as a support ship for other ships, as a destroyer *tender*.

THREEFOLD PURCHASE—A tackle containing two three-sheave blocks.

THWART—Plank set athwartships just below the gunwales in an open boat; acts as a seat and provides support to the sides.

TOPSIDE—Generally refers to weather decks.

TRANSVERSE FRAME—Structural member that extends outward from the keel and upward to the main deck.

TRICE UP—To secure bunks by hauling them up and hanging them off (securing them) on their chains.

TRUE BEARING—The angular difference between lines drawn from the observer to true north and to the object.

TRUNK—The uppermost tip of a mast.

TURNBUCKLE—Device for setting up a tension, as in a lifeline, by turning a buckle into which two eyebolts are threaded.

TURN OF THE BILGE—Where the side meets the bottom.

TURN IN—(1) Retire to bed. (2) Return articles to the issue room.

TURN OUT—(1) Get out of bed. (2) Order out a working party or other groups, as to *turn out* the guard.

TURN TO—Start working.

UP ALL LATE BUNKS—An order to personnel entitled to sleep after reveille to get up.

UPPER DECK—The first deck above the main deck.

VEER—(1) To allow a line, wire, or chain to run out by its own weight. (2) To swerve. (3) Act of the wind in changing direction clockwise.

VOID—An empty tank.

WAIST—The amidships section of the main deck.

WAKE—Trail left by a vessel, or other object, moving through the water.

WARDROOM—Officers' messing compartment.

WATCH—(1) One of the periods (usually 4 hours) into which a day is divided. (2) A particular duty, as lifebuoy *watch*. (3) The act of a buoy or other marker in indicating the position of a sunken object.

WATERTIGHT INTEGRITY—The degree of quality of watertightness.

WAY—(1) Horizontal motion of a floating body. (2) Launching track in a shipbuilding yard.

WEATHER DECK—Any deck exposed to the elements.

WET DOCK—A basin formed by the construction of barriers with gates in a harbor of great tidal ranges to prevent ships from being stranded during low tides. Ships enter the basin at high tide, the gates are closed, and the water is retained in the basin when the tide ebbs.

WHARF—Similar to a quay, but constructed in the fashion of a pier.

WHIPPING—Binding on the end of a line or wire to prevent unraveling.

WILDCAT—That portion of a windlass that engages the links of the anchor chain so that the anchor can be heaved in.

WINDWARD—Toward the direction from which the wind is blowing.

YARD—Spar set athwartships across the upper part of a mast.

YARDARM—The port or starboard half of the horizontal crosspiece of the mast that is either the port or starboard yardarm.

YAW—The act of a vessel when its heading is thrown wide of its course by a force from astern, such as a heavy following sea.

APPENDIX II

BASIC READING LIST

Beach, Edward L., *Run Silent, Run Deep*

Beach, Edward L.,*The United States Navy: A Two Hundred Year Tradition*

Clancy, Tom, *Hunt for Red October*

Clancy, Tom, *Red Storm Rising*

Coontz, Steven, *Flight of the Intruder*

Hawking, Stephen M., *Brief History of Time: From the Big Bang to Black Holes*

Leaman, John F., *Command of the Sea: A Personal Story*

Manchester, William, *American Ceasar: Douglas McArthur*

Marson, Samuel E, *Two Ocean War*

McKenna, Richard, *The Sand Pebbles*

Mitchner, James, *The Source*

Petters, Thomas J., *In Search of Excellence*

Remarque, Erich M, *All Quiet on the Western Front*

Smith, Hendrick, *The Russians*

Stocksdale, James B. and Sybile, *In Love and War*

Wolfe, Tom, *The Right Stuff*

Woulk, Herman, *War and Remembrance*

Woulk, Herman, *Winds of War*

Zumwalt, Elmo, *On Watch*

APPENDIX III

SHIP'S CLASSIFICATION

This appendix contains the types of Navy ships by class. It also identifies the abbreviation for each of the ships within the class.

UNITED STATES NAVAL SHIPS BY CLASS			
Aircraft Carriers		**Mine Warfare Ships**	
Aircraft carrier	CV	Mine countermeasures ship	MCM
Aircraft carrier (Nuclear)	CVN	Minehunter coastal	MHC
Surface Combatant		Minesweeper ocean	MSO
Guide missile cruiser	CG	Minesweeping boats/drones	MSB/MSD
Guided missile cruiser (nuclear)	CGN	**Amphibious Warfare Craft**	
Destroyer	DD	Landing craft, air cushion	LCAC
Guided missile destroyer	DDG	Landing craft, mechanized	LCM
Frigate	FF	Landing craft, Personnel, Large	LCPL
Guide missile frigate	FFG	Landing craft, Utility	LCPL
Patrol Combatants		Landing craft, Vehicle, Personnel	LSSC
Patrol combatant missile (hydrofoil)	PHN	Light Seal Support Craft	LSSC
Submarines		Amphibious Warping Tug	LWT
Ballistic missile submarine (nuclear)	SSBN	Medium Seal Support Craft	MSSC
Attack submarine (nuclear)	SSN	Swimmer Delivery Vehicle	SDV
Auxiliary submarine	SSAG	Side Loading Warping Tug	SLWY
Amphibious Warfare Ships		Special Warfare Craft, Light	SWCL
Amphibious command ship	LLC	Special Warfare Craft, Medium	SWCM
Amphibious assault ship (multipurpose)	LHA/LHD	**Patrol Craft**	
Amphibious cargo ship	LKA	Mini-Armored Troop Carrier	ATC
Amphibious transport dock	LPD	Patrol Boat (Coastal)	PB(C)
Amphibious assault ship (helicopter)	LPH	River Patrol Boat	PBR
Dock landing ship	LSD	Patrol Craft	PC
Logistic support vessel (Army)	LSV	Patrol Craft (Fast)	PCF
Tank landing ship	LST	Patrol Craft (Coastal)	PCC

UNITED STATES NAVAL SHIPS BY CLASS-continued			
Auxiliary Ships		**Service craft**	
Ammunition Ship	AE	Small Auxiliary Floating Dry Dock (NSP)	AFDL
Combat Store Ship	AFS	Medium Auxiliary Floating Dry Dock (NSP)	AFDM
Miscellaneous	AG	Barracks Craft (NSP)	APL
Deep Submergence Support Ship	AGDS	Auxiliary Repair Dry Dock (NSP)	ARD
Hydrofoil Research Ship	AGEH	Medium Auxiliary Repair Dry Dock (NSP)	ARDM
Miscellaneous Command Ship	AGF	Deep Submergence Rescue Vehicle	DSRV
Missile Range Instrumentation Ship	AGM	Deep Submergence Vehicle	DSV
Oceanographic Research Ship	AGOR	Harbor Security Boats	HSB
Ocean Surveillance Ship	AGOS	Submersible Research Vehicle	NR
Surveying Ship	AGS	Open Lighter (NSP)	YC
Auxiliary Research Submarine	AGSS	Car Float (NSP)	YCF
Hospital Ship	AH	Aircraft Transportation Lighter (NSP)	YCV
Cargo Ship	AK	Floating Crane (NSP)	YD
Vehicle Cargo Ship	AKR	Diving Tender (NSP)	YDT
Auxiliary Lighter	ALS	Covered Lighter	YF
Oiler	AO	Ferry Boat or launch	YFB
Fast Combat Support Ship	AOE	Yard Floating Dry Dock (NSP)	YFD
Gasoline Tanker	AO	Covered Lighter (NSP)	YFN
Replenishment Oiler	AOR	Large Cover Lighter (NSP)	YFNB
Transport Oiler	AOT	Dry Dock Companion Craft (NSP)	YFND
Transport	AP	Lighter (Special purpose) (NSP)	YFNX
Self-Propelled Barracks Ship	APB	Floating Power Barge (NSP)	YFP
Cable Repairing Ship	ARC	Refrigerated Covered Lighter	YFR
Salvage Ship	ARS	Refrigerated Covered Lighter (NSP)	YFRN
Submarine Rescue Ship	ASR	Harbor Utility Craft	YFU
Auxiliary Ocean Tug	ATA	Garbage Lighter	YG
Fleet Ocean Tug	ATF	Garbage Lighter (NSP)	YGN
Salvage and Rescue Ship	ATS	Salvage Lift Craft, Heavy (NSP)	YHLC
Guided Missile Ship	AVM	Drege	YM
Large Auxiliary Floating Dry Dock (NSP)	AFBD	Grate Craft (NSP)	YNG

UNITED STATES NAVAL SHIPS BY CLASS-continued			
Service craft (Continued)		Floating Dry Dock Workshop (Hull) (NSP)	YRDH
Fuel Oil Barge	YO	Floating Dry Dock Workshop (Machine) (NSP)	YRDM
Gasoline Barge	YOG	Radiological Repair Barge (NSP)	YRR
Gasoline Barge (NSP)	YOGN	Seaplane Wrecking Derrick	YSD
Fuel Oil Barge (NSP)	YON	Sludge Removal Barge (NSP)	YSR
Oil Storage Barge (NSP)	YOS	Large Harbor Tug	YTB
Patrol Craft	YP	Small Harbor Tug	YTL
Floating Pile Driver (NSP)	YPD	Medium Harbor Tug	YTM
Floating Workshop (NSP)	YR	Water Barge	YW
Repair and Berthing Barge (NSP)	YRB	Water Barge (NSP)	YWN
Repair, Berthing and Messing Barge (NSP)	YRBM		

*NSP—Non self-propelled.

Letter prefixes to classification symbols may be added for further identification.

PREFIX	MEANING
E	Prototype ship in an experimental or development status
T	Assigned to Military Sealift Command
F	Being built for a foreign government
X	Often added to existing classifications to indicate a new class whose characteristics have not been defined
N	Denotes nuclear propulsion when used as last letter of ship symbols

APPENDIX IV

NAVY GOAL CARD

<table>
<tr>
<td>FLEET GOALS
I will complete my Warfare Specialty qualifications, if assigned to sea duty, by end of 1st enlistment.
I will increase savings to ______ dollars per month.
I will maintain a physical fitness program.
I will advance to every paygrade as soon as my first eligibility: E3 ______ E4 ______ E5 ______ E6 ______
I will go the education office (Navy Campus) to document college credits earned upon completion of Recruit Training, ______________ school, and ________________________________.</td>
<td>DEP GOALS
I will attend all DEP meetings.
I will save __________ dollars per month.
I will advance to E2/E3 by encouraging others to visit recruiters and enlist.
I will maintain a physical fitness program.
I will earn my _____________ diploma.
I will not use illegal drugs or abuse alcohol.
I will take personal responsibility for my future.
Personal Goal:__________________
____
________________________________</td>
</tr>
<tr>
<td>PERSONAL PRIORITIES
Education/Training
Discipline
Advancement
Physical Fitness
Saving Money

Habits: Exercise often, avoid alcohol abuse, never use drugs, eat right, avoid smoking, study to earn ____ college credits every year.

Personal Goals:______________________
____
________________________________</td>
<td>RECRUIT TRAINING GOALS
I will report to Recruit Training on ________________________________.

I will complete Recruit Training in 9 weeks.
I will save ________ dollars per month.
I will pass the Navy's physical fitness Requirements.
I will honor the Navy Core Values.
Personal Goal: (Something else you want to accomplish: ______________________
____
____
________________________________</td>
</tr>
<tr>
<td>NAME: ____________________________

Successful Sailors have found that setting goals helps them achieve rewarding careers. Here are just a few examples of goals that will help you on your path to success.

WELCOME ABOARD!

NAVY CORE VALUES

HONOR, COURAGE, COMMITMENT</td>
<td>SAILOR'S CREED

I am a United States Sailor. I will support and Defend the Constitution of the United States of America and I will Obey the orders of those appointed over me. I represent the fighting spirit of the Navy and those who have gone before me to defend Freedom and Democracy around the world. I proudly serve my country's Navy combat team with honor, courage, and commitment. I am committed to excellence and fair treatment to all.</td>
</tr>
</table>

APPENDIX V

REFERENCES

The numbers in parentheses after the reference refer to chapter numbers. The reference was used to develop those chapters. For example, the first reference was used to develop chapter 17, "Financial Management and Stress Management."

Aircrew Survival Equipmentman 1&C, NAVEDTRA 10330, Naval Education and Training Program Management System Support Activity, 1985. (Chapter 15)

Alcohol and Drug Abuse Prevention and Control, OPNAVINST 5350.4B, Office of the Chief of Naval Operations, Washington, DC, 1990. (Chapters 1, 14, and 17)

All Hands, October 1999, Naval Media Center, Publishing Division, Naval Station Anacostia, Washington , DC. (Chapters 1 through 22)

American Naval Fighting Ships, Vol V, Office of the Chief of Naval Operations, Naval History Division, Washington, DC, 1970. (Chapter 8)

Basic Allowances for Housing (BAH), Basic Allowances for Subsistance (BAS) and other FY98 Enacted Pay Legislation, Administrative Message R3119222 Dec 97. (Chapter 17)

Boatswain's Mate, NAVEDTRA 12100, Naval Education and Training Professional Development and Technology Center, Pensacola, FL, 1996. (Chapters 3, 4, 7 and 8)

Career Reenlistment Objectives (CREO), OPNAVINST 1160.4G, Office of the Chief of Naval Operations, Washington, DC, 1987. (Chapter 16)

Catalog of Nonresident Training Courses, NAVEDTRA 12061, Naval Education and Training Professional Development and Technology Center, Pensacola, FL, Jan 99. (Chapter 16)

Correspondence Manual, SECNAVINST 5216.5D, Office of the Secretary of the Navy, Washington, DC, 1996. (Chapter 4)

Department of the Navy Alcohol Abuse Prevention and Deglamorization Campaign, www.chingo.navy.mil/navypalib/people/alcohol/r_spirit/alnav011.txt. (Chapters 1, 14, and 17)

Department of the Navy (DON) Information Security Program (ISP) Regulation, SECNAVINST 5510.36, Department of the Navy , Office of the Secretary, Washington DC, 1999. (Chapter 22)

Department of the Navy Personnel Security Program, SECNAVINST 5510.30A, Department of the Navy, Office of the Secretary, Washington DC, 1999. (Chapter 22)

Department of the Navy Policy on Hazing, SECNAVINST 1610.2, Department of the Navy, Office of the Secretary, Washington DC, 1997. (Chapter 1)

Drugs in the Navy, www.chinfo.navy.mil/navpalib/people/drugs/drug0821.html, 19 Oct 1998. (Chapters 1, 14, 16, and 17)

Enlisted Transfer Manual, NAVPERS 15909-D, Department of the Navy, Naval Military Personnel Command, Washington, DC, 1988. (Chapter 16)

Environmental and Natural Resources Protection Manual, OPNAVINST 5090.1A, Office of the Chief of Naval Operations, Washington, DC, 1990. (Chapter 1)

Family Advocacy Program, SECNAVINST 1752.3A, Department of the Navy, SECNAV, 1000 Navy Pentagon, Washington DC, 11 Sep 1995. (Chapter 1, 14, and 17)

Family Advocacy Program, OPNAVINST 1752.2A, Department of the Navy, Office of the Chief of Naval Operations, Washington DC 20350-2000, 17 Jul 1996. (Chapter 1, 14, and 17)

Gunner's Mate, NAVEDTRA 12443, Naval Education and Training Professional Development and Technology Center, Pensacola, FL, 1996. (Chapter 3, 11)

High Year Policy Update, NAVADMIN 107/96, R 021315Z May 96, CNO Washington DC. (Chapter 16)

Hospital Corpsman 3 & 2, NAVEDTRA 10669-C, Naval Education and Training Professional Development and Technology Center, Pensacola, FL, 1995. (Chapters 14 and 15)

Joint Combat Camera Center, dodimagery.afis.osd.mil/, (Chapter 8)

Law of Armed Conflict (law of War) Program to Insure Compliance by the Naval Establishment, SECNAVINST 3300.1A, Department of the Navy, Washington, DC, 1988. (Chapter 2 and 15)

Life Support Devices, Oxygen Breathing Device (A-4), Air Line Mask, Emergency Escape Breathing Device (EEBD), NAVEDTRA 465-09,00-86, Chief of Naval Education and Training. (Chapters 12 and 13)

List of U.S. Navy Ships, www.chinfo.navy.mil/havpalib/ships/lists/shipalfa.html. (Chapter 8)

Lookout Training Handbook, NAVEDTRA 12968, Naval Education Training Professional Development and Technology Center, 1991. (Chapters 3 and 4)

Manual for Courts-Martial, United States, 1984, Office of the Secretary of the Navy, Washington DC, 1990. (Chapters 2 and 16)

Military Cash Awards Program, OPNAVINST 1650.8C, Office of the Chief of Naval Operations, Washington, DC, 1988. (Chapters 1 and 16)

Military Requirements for Petty Officer Second Class, NAVEDTRA 12045, Naval Education and Training Program Management Support Activity, Pensacola, FL, 1991. (Chapter 16)

NSTM, Personnel Protection Equipment, S9086-CL-STM-010/CH 077R3, Commander, Naval Sea Systems Command, Washington, DC, March 98. (Chapters 12 and 13)

NSTM, Surface Ship Firefighting, S9086-S3-STM-010/CH555V1R5, Commander, Naval Sea Systems Command, Washington, DC, July 98. (Chapters 12 and 13)

NSTM, Damage Control, "Practical Damage Control," S9086-CN-STM-020/CH-079V24R1, Commander, Naval Sea Systems Command, Washington, DC, Aug 98. (Chapters 12 and 13)

NARR/REF A DoD Directive 1327.5 (Change 3), Concerning Leave and Liberty (Oct 95). (Chapter 17)

NARR/REF B NAVPERS 15560C (MILPERSMAN), Oct 9. (Chapter 17)

Naval Orientation, NAVEDTRA 12966, Naval Education and Training Program Management Support Activity, Pensacola, FL, 1991. (Chapters 1, 6, 9, 10, and 20)

Navy Fact File, 9th ed., Office of Information, Department of the Navy, Washington, DC, 1989. (Chapter 1, 5, 8, 11, 17, and 20)

Navy Occupational Safety and Health Program Manual for Forces Afloat, OPNAVINST 5100.19B, Department of the Navy, Office of the Chief of Naval Operations, Washington, DC, 1989. (Chapters 1, 18, and 19)

Personal Finance 101: Rule of 72, www.datalife.com/mil/pages/examples/RULE_72.HTM. (Chapter 17)

Personal Financial Management Education, Training and Counseling Program, OPNAVINST 1740.5, Department of the Navy, Office of the Chief of Naval Operations, Washington, DC, 6 Nov 1990. (Chapter 17)

Petty Officer Quality Control Program, NAVADMIN 253/96, R 231406Z Oct 96. (Chapter 16)

Physical Security and Loss Prevention, OPNAVINST 5530.14B, Department of the Navy, Office of the Chief of Naval Operations, Washington, DC, 21 Dec 1988. (Chapter 22)

Preventing Abuse and Revictimization: Guidelines, Prevention and Victim Assistance, OPNAVINST 1752.2A, Office of the Chief of Naval Operations, Washington, DC. (Chapter 17)

Public Affairs Manual, SECNAVINST 5720.441, Office of the Secretary of the Navy Washington, DC, 1987. (Chapter 1, 8, 16)

Quartermaster, NAVEDTRA 12120, Naval Education and Training Professional Development and Technology Center, Pensacola, FL, 1995. (Chapters 3, 4, and 7)

Reenlistment Quality Control Program, OPNAVINST 1160.5C, Office of the Chief of Naval Operations, Washington, DC, 1993. (Chapters 1 and 16)

Rescue and Salvage Ships—ARS, www.chinfo.mil/navpilb/factfile/ships/ ship-ars.html. (Chapter 8)

Retention Team Manual, NAVPERS 15878G, Office of the Chief of Naval Operations, Washington, DC, 1984. (Chapter 16)

Rights and Responsibilities Pregnancy and Family Care Policies, Recruit Training Command A-950-0001, Recruit Training Command, Great Lakes, IL, 1998. (Chapter 1 and 14)

Seabee Combat Handbook, Volume 1, NAVEDTRA 12003, Naval Education and Training Professional Development and Technology Center, Pensacola, FL, 1993. (Chapter 15)

Seaman, NAVEDTRA 12016, Naval Education and Training Professional Development and Technology Center, Pensacola, FL, 1993. (Chapters 2, 3, and 7)

Secretary Aspin Releases Family Status Study Findings, No. 601-93, www.chinfo.navy.mil/ navypalib/people/families/dodfam.txt. (Chapter 17)

Shipboard Hazardous Materials/Hazardous Waste Management Plan, NAVSEA 59593-A7-PLN-010, Naval Sea Systems Command, Department of the Navy, Washington, DC, 1983. (Chapters 18 and 19)

Standard Organization and Regulations of the U.S. Navy, OPNAVINST 3120.32B, Office of the Chief of Naval Operations, Washington, DC, 1986. (Chapters 2, 5 and 6)

Surface Ship Survivability, NWP 62-1 (Rev. C), Department of the Navy, Office of the Chief of Naval Operations, Washington, DC, 1989. (Chapter 15)

The Right Spirit: Alcohol Abuse Prevention and Deglamorization Campaign for the Navy, www.chingo.navy.mil/navypalib/people/alcohol/r_spirit/navop08.txt. (Chapter 1)

Uniform Code of Military Justice, 1995 edition, (Chapters 2, 15, 16)

United States Navy Chemical, Biological, and Radiological Defense Handbook for Training, S-5080-AA-HBK-010, Naval Sea Systems Command, Washington, DC, 1985. (Chapter 13)

U.S. Navy CBR-Defense/U.S. Marine Corps NBC Defense Handbook, OPNAVINST P-86-1-95, Chief of Naval Operations, Surface Ship Survivability Office N86D, Washington DC, Apr 95. (Chapter 13)

U.S. Navy Regulations, 1990, Office of the Secretary of the Navy, Washington, DC, 1990. (Chapters 1 and 2)

U.S. Navy Uniform Regulations, NAVPERS 12966, Bureau of Naval Personnel, Washington, DC, 1987. (Chapter 10)

INDEX

D

E

N

R

S

T

U

Assignment Questions

<u>Information</u>: The text pages that you are to study are provided at the beginning of the assignment questions.

ASSIGNMENT 1

Textbook Assignment: Chapter 1 "*Policies and Programs*" and chapter 2 "*Military Conduct and Justice*."

1. Pollution can result when which of the following types of agents is/are introduced into the air, water, or soil?
   1. Biological
   2. Chemical
   3. Physical
   4. All of the above

2. Which of the following operations causes the most pollution?
   1. Industrial
   2. Municipal
   3. Transportation
   4. All of the above

3. Which of the following materials is the primary municipal pollutant?
   1. Raw or inadequately treated sewage
   2. Radioactive waste
   3. Petroleum products
   4. Acids

4. Which of the following modes of transportation creates most air pollutants?
   1. Trains
   2. Waterborne vessels
   3. Motor vehicles
   4. Aircraft

5. Which, if any, of the following effects of pollution is the most serious?
   1. Psychological
   2. Biological
   3. Physical
   4. None of the above

6. Steel erodes faster than normal when exposed to which of the following air pollutants?
   1. Pesticides
   2. Herbicides
   3. Zinc oxides
   4. Sulfur oxides

7. Most pesticides fall into which of the following categories?
   1. Selective
   2. Nonselective
   3. Preselective
   4. Control selective

8. What is the primary pollution concern of Navy personnel?
   1. Noise pollution
   2. Shore command wastes
   3. Shipboard wastes

9. Which of the following terms identifies abatement?
   1. Maintaining
   2. Raising
   3. Lowering
   4. Containing

10. Under the Clean Air Act, what government body has the primary responsibility for assuring air quality?
    1. Federal government
    2. Each state
    3. Local municipalities
    4. Department of Transportation

11. Virtually all Navy ships have some type of sanitation device installed. Which of the following types of systems retains sewage on board for discharge ashore or in waters where discharging is allowed?
    1. Direct discharge
    2. Positive flow
    3. Marine sanitation
    4. Collection, holding, and transfer

12. When operating sanitation devices in foreign waters, Navy ships comply with which of the following requirements?
    1. Status of Forces Agreement
    2. Coast Guard instructions
    3. NAVFAC guidelines
    4. All of the above

13. Vessels may not discharge unpulped trash within what minimum distance from the U.S. coastline?

    1. 20 nm
    2. 25 nm
    3. 30 nm
    4. 35 nm

14. Submarines may discharge negatively buoyant compacted trash not less than 12 nm from the U.S. coastline only if the water depth is greater than how many fathoms?

    1. 1,000
    2. 500
    3. 100
    4. 50

15. The Navy's land management program involves which of the following efforts?

    1. Migratory bird management
    2. Production and sale of forest products
    3. Soil and water conservation
    4. Use of off-road vehicles

16. From what part of the world does the U.S. import most of its crude oil?

    1. Southeast Asia
    2. Central Europe
    3. South America
    4. Middle East

17. What program provides information and support for Navy personnel who are guests in foreign lands?

    1. Navy Sponsor Program
    2. Overseas Duty Support Program
    3. Navy Assistance Program
    4. Navy Relocation Program

18. The Military Cash Awards Program (MILCAP) provides monetary recognition of up to what maximum amount?

    1. $ 5,000
    2. $10,000
    3. $20,000
    4. $25,000

19. The Navy's Health and Physical Readiness Program promotes health and fitness at the command level. As part of this program, naval personnel are required to undergo fitness testing at what interval?

    1. Biennially
    2. Annually
    3. Semiannually
    4. Quarterly

20. "Informing the public and members of the naval service about Navy operations and programs" is the mission of

    1. BUPERS
    2. CNO
    3. PAO
    4. SECNAV

IN ANSWERING QUESTIONS 21 THROUGH 23, SELECT THE TERM USED TO DEFINE THE QUESTION.

21. Extravagant, careless, or needless expenditure of government resources.

    1. Fraud
    2. Waste
    3. Abuse
    4. Mismanagement

22. Intentional misleading or deceitful conduct that deprives the government of its resources or rights.

    1. Fraud
    2. Waste
    3. Abuse
    4. Mismanagement

23. Intentional wrongful or improper use of government resources.

    1. Fraud
    2. Waste
    3. Abuse
    4. Mismanagement

24. You can report fraud, waste, abuse, and mismanagement to which of the following offices?

    1. The Navy hotline
    2. The chain of command
    3. The Naval Criminal Investigative Service
    4. All of the above

25. To maintain public confidence in its integrity, all naval personnel must comply with the Standards of Conduct and Professional Ethics.

    1. True
    2. False

26. If you disclose information about a person to unauthorized personnel, you could be fined up to what maximum amount?

    1. $5,000
    2. $3,000
    3. $2,000
    4. $1,000

27. The guidance and policy for making sure that equal opportunity works rests with what office?

    1. Command master chief
    2. Commanding officer
    3. Chief of Naval Operations
    4. Secretary of the Navy

28. Which of the following persons is responsible for making equal opportunity a reality with a command?

    1. Commanding officer
    2. Executive officer
    3. Operations officer
    4. Command master chief

29. Your performance evaluation does not reflect your attitude toward and your conduct in support of the Navy's equal opportunity program.

    1. True
    2. False

30. If a Sailor takes part in insensitive practices, he/she receives counseling on treating people equally. If counseling isn't effective, what action, if any, may take place?

    1. Administrative action only
    2. Disciplinary action only
    3. Administrative or disciplinary action
    4. None

31. On what basis should supervisors assign duties such as food service and compartment cleaning?

    1. Skills and abilities
    2. Seniority
    3. A fair, rotational basis
    4. Time in service

32. The Department of the Navy sets the requirements for advancement for paygrades E-1 through E-9. Which of the following is the determining factor in advancement?

    1. A vacancy
    2. Having a high multiple
    3. Passing the advancement-in-rate exam
    4. All of the above

33. Navy personnel are prohibited from taking part in a civil rights demonstration under which of the following circumstances?

    1. When the demonstration occurs during duty hours
    2. While they are in uniform
    3. When the demonstration occurs on a military reservation
    4. Each of the above

34. If you cannot resolve a complaint among the personnel involved, you can attach a written complaint to a special request chit and forward it through the chain of command. You must do this within 5 days?

    1. True
    2. False

35. Which of the following personnel can be victims of sexual harassment?

    1. Men only
    2. Women only
    3. Both 1 and 2 above

36. Which of the following phrases describes sexual harassment?

    1. Unwelcome sexual advances
    2. Requests for sexual favors
    3. Verbal or physical conduct that is sexual in nature
    4. Each of the above

37. Someone in a command position makes sexual advances towards you, making it impossible to do your job. You are being sexually harassed.

    1. Truc
    2. False

38. Which of the following is a criteria for a person's behavior to be considered sexual harassment?

    1. Unwelcome
    2. Sexual in nature
    3. Occur or impact your work
    4. Each of the above

A. Red
B. Yellow
C. Green

Figure A

IN ANSWERING QUESTIONS 39 AND 40, REFER TO FIGURE A AND SELECT THE ZONE USED TO DESCRIBE THE QUESTION.

39. Sexually explicit pictures.

    1. A
    2. C
    3. B

40. Suggestive posters, calendars, and off-color jokes.

    1. A
    2. C
    3. B

41. What person has the responsibility of appointing the command ombudsman?

    1. Commanding officer
    2. Executive officer
    3. Division officer
    4. Command master chief

42. What person determines the content and priorities of the command ombudsman program?

    1. Commanding officer
    2. Executive officer
    3. Division officer
    4. Command master chief

43. Which of the following statements is a purpose of the Reenlistment Quality Control Program?

    1. To provide a personnel management program to control rating manning
    2. To issue reenlistment criteria
    3. To establish standardized professional growth points
    4. All of the above

44. All first-term Sailors in paygrades E-1 through E-6 requesting reenlistment must be approved for reenlistment through what program?

    1. CREO
    2. ENCORE
    3. HYT
    4. EEO

45. What person directs and supervises the Navy's voting program?

    1. Chairman, Joint Chiefs of Staff
    2. Chief of Naval Operations
    3. Chief of Naval Personnel
    4. Chief of Naval Information

46. Which of the following are sources that set forth the basic disciplinary laws for the U.S. Navy?

    1. *U.S. Navy Regulations*
    2. *Standard Organization and Regulations of the U.S. Navy*
    3. *Uniform Code of Military Justice* (*UCMJ*)
    4. Each of the above

47. Which of the following characteristics are traits of a good Sailor?

    1. Puts the good of the ship and the Navy above personal likes and dislikes
    2. Obeys the rules of military courtesy and etiquette
    3. Demonstrates loyalty, self-control, honesty, and truthfulness
    4. All of the above

48. In what year was the Code of Conduct first prescribed?

    1. 1965
    2. 1955
    3. 1945
    4. 1935

49. The Code of Conduct was adopted to provide guidance for service personnel in which of the following circumstances?

1. When stationed on foreign soil
2. When traveling at home and abroad
3. When facing the enemy as prisoners of war
4. All of the above

50. In what year was Executive Order 12633 issued amending the Code of Conduct to use neutral-gender language?

1. 1987
2. 1988
3. 1989
4. 1990

51. How many articles make up the Code of Conduct?

1. Two
2. Four
3. Six
4. Eight

52. When, if ever, may you voluntarily surrender to the enemy?

1. If alone and completely isolated from friendly troops
2. If no longer able to inflict casualties on the enemy
3. If able to detain the enemy and let others escape capture
4. Never

53. Who may be assigned shore patrol duties?

1. Officers only
2. Petty officers only
3. Officers and petty officers
4. All Navy personnel

54. In areas where different armed services are located, the military police from each service may be combined to form one unit. What term identifies this unit?

1. Armed Forces Police Department
2. Armed Forces Police Detachment
3. Armed Forces Police Service
4. Armed Forces Police Group

55. Aboard ship, the master-at-arms (MAA) force is headed by the chief master-at-arms (CMAA). The CMAA works directly for which of the following officers?

1. Weapons officer
2. Security officer
3. Executive officer
4. Administrative officer

56. Discipline training develops which of the following personal traits?

1. Character
2. Efficiency
3. Self-control
4. All of the above

57. Discipline is important to the Navy for which of the following reasons?

1. To instill fear of punishment
2. To decrease command responsibility
3. To provide punishment for wrongdoers
4. To enable personnel to function as a unit with a high degree of efficiency

58. Punishment is administered in the Navy for which of the following reasons?

1. To serve as an object lesson to the wrongdoer and others
2. To pacify those who have suffered a wrong
3. To correct a wrong
4. To avenge a wrong

59. What chapter of the *United States Navy Regulations* describes the rights and responsibilities of all Navy members?

1. 12
2. 11
3. 10
4. 9

60. What person is responsible for making sure that the *Navy Regs* conforms to the current needs of the Department of the Navy?

1. The Secretary of the Navy
2. The Judge Advocate General
3. The Chief of Naval Operations
4. The Commandant of the Marine Corps

61. *Navy Regs* and changes to it are issued by the Secretary of the Navy after what person approves them?

   1. The President
   2. The Vice President
   3. The Attorney General
   4. The Chairman of the Joint Chiefs of Staff

62. Failure to obey any regulation subjects the offender to charges under what *UCMJ* article?

   1. 91
   2. 92
   3. 93
   4. 94

63. What article of the *Navy Regs* lists the publications that must be made available upon request by any active-duty person?

   1. 1020
   2. 1010
   3. 0917
   4. 0818

64. What article of the *Navy Regs* gives officers the authority necessary to perform their duties?

   1. 1021
   2. 1023
   3. 1025
   4. 1033

65. *Navy Regs*, article 1033, Authority in a Boat, provides which of the following officers the authority and responsibility over all persons embarked?

   1. The senior line officer eligible for command at sea
   2. The junior line officer eligible for command at sea
   3. The senior staff officer
   4. The junior staff officer

66. Which of the following *Navy Regulations* articles outlines the authority of a sentry?

   1. 1037
   2. 1038
   3. 1052
   4. 1053

67. You may not be ordered to active duty without the permission of which of the following persons?

   1. Commandant of the Marine Corps
   2. Commandant of the Coast Guard
   3. Chief of Naval Operations
   4. Chief of Naval Personnel

68. *Navy Regulations*, article 1104, Treatment and Release of Prisoners, prohibits cruel and/or unusual treatment. According to this article, prisoners must be checked on at what minimum interval?

   1. 10 hours
   2. 8 hours
   3. 6 hours
   4. 4 hours

69. During a Saturday duty day, one of your shipmates asks you to change watches with him/her. You agree but fail to get permission from proper authority. Under what article of *Navy Regs* could you be charged?

   1. 1138
   2. 1134
   3. 1133
   4. 1129

70. Sexual harassment is offensive and illegal. Under what article of *Navy Regs* may a person be charged with sexual harassment?

   1. 1166
   2. 1164
   3. 1162
   4. 1160

ASSIGNMENT 2

Textbook Assignment: Chapter 3 "*Watch Standing*" and chapter 4 "*Communications*."

---

1. A ship maintains a watch for which of the following reasons?

   1. Communications
   2. Security
   3. Safety
   4. All of the above

2. A ship's plan for action is contained in what type of bill?

   1. Battle bill
   2. Admin bill
   3. Organization bill
   4. Watch, quarter, and station bill

3. Qualified personnel are assigned to stations by which of the following persons?

   1. Division officer and division chief
   2. Leading petty officer
   3. Leading chief petty officer
   4. Executive officer

A. CONDITION I

B. CONDITION II

C. CONDITION III

Figure A

IN ANSWERING QUESTIONS 4 AND 5, REFER TO FIGURE A AND SELECT THE CONDITION USED TO DEFINE THE QUESTION.

4. General quarters—all battle stations are manned.

   1. A
   2. B
   3. C

5. Normal wartime cruising watch—4 hours on, 8 hours off.

   1. A
   2. B
   3. C

6. If you are scheduled to stand the second dog watch, you should report at which of the following times?

   1. 1745
   2. 1750
   3. 1755
   4. 1800

7. If you are told to report to your duty station at 0745 (24-hour clock), you should arrive at what time?

   1. 6:45 am
   2. 7:45 am
   3. 6:45 pm
   4. 7:45 pm

8. What watch are you standing between 2000 and 2400 hours?

   1. Midwatch
   2. Forenoon watch
   3. First dog watch
   4. Evening watch

9. Watches are split into port and starboard for what reason?

   1. For convenience
   2. For security
   3. To rotate personnel
   4. To allow extra liberty

10. What type of watch do most Sailors stand?

    1. Phone
    2. Security
    3. Admin
    4. Division

11. Which of the following is a type of a security watch?

    1. Sentry duty
    2. Barracks watch
    3. Fire watch
    4. Each of the above

12. Which of the following is a key assignment for officers in the watch organization?

    1. CDO
    2. OOD
    3. JOOD
    4. Each of the above

13. Which of the following is a duty of the QMOW?

    1. To maintain the ship's deck log
    2. To make sure all bells are correctly answered
    3. To stand watch in the bridge and deliver messages
    4. To line up and operate the steering engines

14. What person makes sure all deck watch stations are manned with qualified personnel and all watch standers from previous watches are relieved?

    1. BMOW
    2. QMOW
    3. JOOW
    4. JOOD

15. Where is the fog lookout watch usually stood?

    1. Helm
    2. Aftermast
    3. In the bow where approaching ships can be heard
    4. CIC

16. For what reason does the fog lookout watch normally consist of two Sailors?

    1. In case there is a man overboard
    2. To allow the lookout to work without having his/her hearing impaired by wearing sound-powered phones
    3. The two-man security rule
    4. To verify visual and sound contact

17. What type of watch is set when positive steering control must be maintained?

    1. Helmsman
    2. Lee helmsman
    3. After steering
    4. QMOW

18. What is the purpose of the security watch?

    1. To minimize damage to equipment
    2. To control contact with the CIC
    3. To increase the physical security of the ship

19. Which of the following is a duty of security watches and patrols?

    1. To be alert for fire hazards
    2. To check the security of weapons magazines
    3. To inspect damage control closures
    4. Each of the above

20. You are a member of a security patrol, and you detect a fire hazard that affects the safety of the ship. What action should you take?

    1. Note it on the security log
    2. Inform your LCPO
    3. Investigate it
    4. Report it to the OOD immediately

21. What is the purpose of a shipboard fire watch?

    1. To immediately extinguish fires caused by welding or burning operations
    2. To make sure the welder strikes the welding surface
    3. To relay messages from the work site
    4. To make sure there is a controlled burn of material at the work site

22. When standing a barracks security watch, you have which of the following responsibilities?

    1. Knowing and carrying out provisions of the fire bill
    2. Knowing and carrying out provisions of the emergency bill
    3. Knowing barracks regulations
    4. All of the above

23. When standing a barracks security watch, which of the following is the first action to take if there is a fire?

    1. Report the fire
    2. Spread the alarm
    3. Close doors and windows
    4. Fight the fire, if possible, if you have the proper equipment

24. Sentries are governed by what two types of orders?

    1. Understood and general
    2. Special and verbal
    3. General and special
    4. General and verbal

25. You are required to know the general orders of a sentry. How many general orders are there?

   1. 11
   2. 14
   3. 16
   4. 18

26. When aboard ship, you should refer to what publication for the procedures used to relieve an armed watch?

   1. SOP
   2. FOD
   3. Watch bill
   4. Battle bill

27. Which of the following is a precaution to follow when standing an armed watch with a pistol?

   1. When relieved, unload the pistol in a safe area
   2. Don't surrender the pistol to an unauthorized person
   3. Keep the pistol (which is loaded with one round in the chamber) in its holster unless you have to use it
   4. Each of the above

28. Under which of the following conditions can deadly force be used?

   1. To prevent the escape of a murderer
   2. To prevent sabotage
   3. To protect your life
   4. Each of the above

29. Why is a lookout posted?

   1. To prevent blind spots caused by metal objects
   2. To search for objects radar can't detect
   3. To detect objects low in the water
   4. To search for air attacks

30. The peacetime lookout organization has how many Sailors in each watch station?

   1. One
   2. Two
   3. Three
   4. Four

31. Which of the following is/are types of bearings?

   1. Relative only
   2. True only
   3. Magnetic only
   4. Relative, true, and magnetic

32. Which of the following is Navy phraseology for reporting a bearing of 038°?

   1. O, three, eight
   2. O, three, ate
   3. Zero, tree, ate
   4. Zero, tree, eight

33. Lookouts report what type of bearing?

   1. Magnetic
   2. Relative
   3. True

34. Which of the following statements describes a target angle?

   1. The magnetic north pole is used as the reference point
   2. True north is used as the reference point
   3. An object in the sky
   4. The relative bearing of your ship from another ship

35. A position angle can never be more than what number of degrees?

   1. 0°
   2. 45°
   3. 90°
   4. 180°

36. How are position angles reported?

   1. Three digits, spoken digit by digit
   2. Two digits, spoken digit by digit
   3. Three digits, spoken as a whole
   4. Two digits, spoken as a whole

37. How should you report objects that are low in the water?

   1. By feet above the surface
   2. By the object's approximate distance
   3. In feet from the ship
   4. From the object to the horizon

38. Ranges are reported in what unit of measurement?

   1. Feet
   2. Yards
   3. Rods
   4. Miles

39. When using binoculars, what adjustments should you make?

    1. One for focus
    2. Two for focus and one for proper distance between the lenses
    3. One for proper distance between the lenses
    4. Two for eyepiece and lens

40. When should you use binoculars?

    1. In foggy and rainy conditions
    2. When identifying objects at night
    3. When scanning sectors in the daytime
    4. Both 2 and 3 above

41. How long does it take for you to reach your best night vision?

    1. 10 minutes
    2. 15 minutes
    3. 25 minutes
    4. 30 minutes

42. What is meant by the term *dark adaptation*?

    1. The improvement of vision in dim light
    2. The inability to see in bright light
    3. The red light requirement
    4. Shadows that can't be seen clearly

43. When should you use "off-center vision"?

    1. Below decks
    2. When wearing glasses
    3. When it's dark
    4. In broad daylight

44. What information is contained in an initial report?

    1. The object only
    2. The object's bearing from the ship only
    3. The object and its bearing from the ship
    4. What the object might be

45. To report serial number 23NCI16 over the sound-powered telephone circuit, you would report the serial number in what way?

    1. Too, three, november, charlie, india, wun, six
    2. Too, tree, november, charlie, india, wun, six
    3. Two, tree, november, charlie, india, wun, six
    4. Two, tree, november, charlie, india, wun, sics

46. It's important for you to remember that the mouthpiece and earpiece of sound-powered telephones are interchangeable for which of the following reasons?

    1. Two people can talk at once
    2. They can be interchanged if a piece breaks
    3. Undesirable noises can be fed into the system
    4. Both 2 and 3 above

47. The headset of sound-powered telephones is picked up as a unit for which of the following reasons?

    1. To make sure you have all the parts
    2. To avoid breaking them
    3. Both 1 and 2 above
    4 In case the earpiece is missing

48. When using the mouthpiece of a sound-powered phone set to report contacts, how far from your mouth should you position the mouthpiece?

    1. 1/2 to 1 inch
    2. 1 to 2 inches
    3. 2 to 3 inches
    4. 3 to 4 inches

49. Why should you unplug a phone's headset when it's not in use?

    1. To keep the user costs down
    2. Earpieces will pick up noise and transmit it over the circuit
    3. Carbon will build up at the connectors
    4. Calls from other circuits won't go through

50. Aboard ship, there are how many categories of shipboard sound-powered phone circuits?

    1. One
    2. Two
    3. Three
    4. Four

51. What category of shipboard telephone circuits is designed to maintain vital communications and are preceded by the letter *X*?

    1. Primary system
    2. Auxiliary system
    3. Supplementary system
    4. Command circuit

52. Which of the following sound-powered phone circuits is used as the CO's battle circuit?

   1. JA
   2. JC
   3. JL
   4. 1JV

53. To keep the meaning of a message intact when standing duty as a telephone talker, what action should you take?

   1. Speak loudly
   2. Repeat the message word for word
   3. Paraphrase what you hear
   4. Speak rapidly to transmit the message quickly

54. Which of the following statements is a rule for circuit discipline?

   1. Transmit only official messages
   2. Keep the button in the OFF position when not transmitting
   3. Use only standard words and phrases
   4. All of the above

55. Which of the following elements is included when taking a message?

   1. Name of caller
   2. Message
   3. Time and date
   4. Each of the above

56. The IVCS has which of the following components?

   1. Terminals
   2. Accessories
   3. ICSCs
   4. All of the above

57. Within the IVCS, what is the purpose of the ICSCs?

   1. To perform switching actions
   2. To keep lines clear
   3. To give multi-access to lines
   4. To ensure automatic cutoff for security purposes

58. Which of the following shipboard announcing systems is called the general announcing system?

   1. 1MC
   2. 2MC
   3. 3MC
   4. 4MC

59. Which of the following shipboard announcing systems is used for intership communications?

   1. 5MC
   2. 6MC
   3. 7MC
   4. 8MC

60. Which of the following shipboard announcing systems is used for hangar deck damage control?

   1. 39MC
   2. 51MC
   3. 53MC
   4. 58MC

61. Which of the following persons is authorized to pass calls over the 1MC?

   1. OOD
   2. XO
   3. CO
   4. Each of the above

62. What is the purpose of the 20MC announcing system?

   1. Radio room announcing system
   2. Flag officer's command announcing system
   3. Combat information announcing system
   4. Captain's command announcing system

63. Which of the following types of flags and pennants is/are used by the Navy?

   1. Substitute flags
   2. Numeral pennants
   3. International alphabet flags
   4. All of the above

64. Aboard ship, a man overboard is indicated by what emergency/warning flag?

   1. Code Alfa
   2. Oscar
   3. November Charlie
   4. Bravo

65. What administrative flag is used to recall all personnel to the ship?

   1. Hotel
   2. Juliett
   3. Romeo
   4. Papa

66. What administrative flag is flown in port to indicate the ship has ready duty?

   1. Hotel
   2. India
   3. Romeo
   4. Quebec

67. When under way, the national ensign is normally flown from what location?

   1. The gaff
   2. The aftermast
   3. The flagstaff
   4. The jackstaff

68. Which of the following statements defines the term "colors"?

   1. Colors give recognition of codes
   2. Colors consist of our national ensign along with the union jack
   3. Colors are lights on the flagstaff
   4. Colors are the flags of foreign ships

69. When a naval ship is in port or at anchor, the union jack is flown from what location?

   1. The gaff
   2. The jackstaff
   3. The aftermast
   4. The flagstaff

70. The U.S. Navy flag is flown in which of the following situations?

   1. At official ceremonies or official public gatherings when the Navy is officially a participant
   2 In parades
   3. In official Navy occasions
   4. Each of the above

71. Which of the following flags are half-masted at the death of the CO?

   1. National ensign
   2. Union jack
   3. Commission pennant
   4. Each of the above

72. On small ships, personnel from what watch are responsible for hoisting and hauling down absentee pennants?

   1. Security watch
   2. Quarterdeck watch
   3. Roving watch
   4. DC central watch

73. On large ships, what person is responsible for making sure that special flags or pennants are displayed to indicate changing events aboard ship?

   1. Boatswain's mate
   2. Quarterdeck watch
   3. Duty signalman
   4. Topside watch

74. Where is a list of special flags and pennants normally posted as a ready reference for watch standers?

   1. Combat information center (CIC)
   2. After deck
   3. Quarterdeck area
   4. Half deck

75. An officer in command entitled to a personal flag is embarked in a boat on an official mission. Where should the pennant be flown?

   1. Amid ship
   2. In the bow
   3. In the stern
   4. Yardarm, port

ASSIGNMENT 3

Textbook Assignment: Chapter 5 "*Naval History.*."

---

1. What date commemorates the birthday of the United States Navy?

   1. 5 Sep 1774
   2. 13 Oct 1775
   3. 4 Jul 1776
   4. 14 Feb 1778

2. The Second Continental Congress approved the purchase of how many vessels?

   1. Eight
   2. Six
   3. Four
   4. Two

3. Which of the following were naval vessels in the early 19th century?

   1. Frigates
   2. Sloops of war
   3. Ships of the line
   4. All of the above

4. What category of ship carried the largest number of guns?

   1. Ships of the line
   2. Sloops of war
   3. Schooners
   4. Frigates

5. What type of ships did privateers typically sail?

   1. Ships of the line
   2. Sloops of war
   3. Schooners
   4. Frigates

6. What ship was the first warfare submarine?

   1. *Turtle*
   2. *Hornet*
   3. *Alfred*
   4. *Wasp*

7. Which of the following ships has the distinction of being the U.S. Navy's first flagship?

   1. *Providence*
   2. *Hornet*
   3. *Alfred*
   4. *Wasp*

8. What country was the first to recognize the "Stars and Stripes"?

   1. Germany
   2. France
   3. Spain
   4. Portugal

9. John Paul Jones is often referred to as the "father of our highest naval traditions" because of the example he set as an officer during the Revolutionary War. He is also famous because of which of the following accomplishments?

   1. His appointment as the first U.S. Navy admiral
   2. His selection as the first commander in chief
   3. His victory over the HMS *Serapis*
   4. His capture of the HMS *Nancy*

10. At various times during the Revolutionary War, the U.S. Navy had 56 vessels. What was the peak number of vessels that were operating at any one time?

    1. 45
    2. 32
    3. 27
    4. 15

11. Approximately how many ships did the British loose to privateers?

    1. 1,000
    2. 1,500
    3. 2,000
    4. 2,500

12. What is the oldest U.S. Navy ship still in commission?

    1. *Lexington*
    2. *Constitution*
    3. *Constellation*
    4. *Bonhomme Richard*

13. Who was president when the U.S. Navy Department was established?

    1. George Washington
    2. Thomas Jefferson
    3. James Madison
    4. John Adams

14. When did the expression "Millions for defense, but not one cent for tribute" originate?

    1. During the Revolutionary War
    2. During the "Quasi" War
    3. During the War of 1812
    4. During the Barbary States War

15. Who led the naval forces into Tripoli Harbor and destroyed the captured US frigate USS *Philadelphia*?

    1. Stephen Decatur
    2. James Lawrence
    3. Thomas Truxtun
    4. Edward Preple

16. The War of 1812 was caused, in part, by the efforts to accomplish which of the following goals?

    1. Establishing a naval base in the Mediterranean
    2. Paying ransom payments to the Barbary States
    3. Stopping forced service of American seamen in the British navy
    4. Forcing France to establish trade relations with the United States

17. During the War of 1812, what ship earned the nickname "Old Ironsides"?

    1. *Chesapeake*
    2. *Constitution*
    3. *Constellation*
    4. *Enterprise*

18. On which of the following Great Lakes did Captain Oliver Hazard Perry defeat a British squadron, cutting British supply lines?

    1. Lake Superior
    2. Lake Michigan
    3. Lake Huron
    4. Lake Erie

19. What ship was one of the first ships-of-the line?

    1. *Constitution*
    2. *Enterprise*
    3. *Philadelphia*
    4. *North Carolina*

20. The first half of the 19th century saw a development that was to change navies all over the world. What was that development?

    1. Task forces
    2. Steam power
    3. Steel hulls
    4. Practical submarines

21. In 1843, what invention incorporated in the USS *Princeton* paved the way for progress in the development of propulsion systems?

    1. The screw propeller
    2. The diesel engine
    3. The coal-fired boiler
    4. The stern paddle wheel

22. In 1854, Commodore Perry signed a treaty that opened up what market to American trade?

    1. China
    2. Japan
    3. Russia
    4. India

23. Although neither side could claim victory, the battle between the USS *Monitor* and the *Virginia* (*Merrimack*) was important for which of the following reasons?

   1. Steam engines were used in battle for the first time
   2. The Dahlgren gun was used
   3. The battle began the era of the ironclads
   4. The Union and Confederate navies fought each other

24. The first true submarine attack was conducted against what Union ship?

   1. USS *New Ironsides*
   2. USS *Housatonic*
   3. USS *Hunley*
   4. USS *Custis*

25. During what Civil War battle was the order "Damn the torpedoes! Full speed ahead!" given?

   1. Vicksburg
   2. Mobile Bay
   3. New Orleans Orleans
   4. Kings Bay

26. What person defined sea power, showed the importance of knowing naval needs, and advocated a large, powerful Navy?

   1. Commodore Perry
   2. Admiral Farragut
   3. Andre Foote
   4. Alfred T. Mahan

27. What ship has been labeled as the first modern cruiser in the U.S. Fleet?

   1. USS *Boston*
   2. USS *Atlanta*
   3. USS *Newark*
   4. USS *Chicago*

28. "Remember the Maine," referring to the USS *Maine*, was the battle cry for which of the following wars?

   1. The Quasi War
   2. The Civil War
   3. The Spanish-American War
   4. World War I

29. In what year did the Navy accept its first operational submarine?

   1. 1895
   2. 1898
   3. 1900
   4. 1902

30. Construction of our first destroyer began in what year?

   1. 1895
   2. 1899
   3. 1902
   4. 1905

31. What ship was considered our first "first-class" battleship?

   1. USS *Indiana*
   2. USS *New York*
   3. USS *Texas*
   4. USS *California*

32. Who was the Navy's first aviator?

   1. Lt. Ellyson
   2. Lt. Towers
   3. Lt. Corry
   4. CAPT Chambers

33. Destroyers were first used effectively for antisubmarine warfare during what war?

   1. Civil War
   2. Spanish-American War
   3. World War I
   4. World War II

34. In what war did women first serve as members of the Navy?

   1. Civil War
   2. Spanish-American War
   3. World War I
   4. World War II

35. In what capacity did women first serve as members of the Navy?

   1. Nurse
   2. Yeoman
   3. Radio operator

36. What was the first aircraft carrier designed from the keel up?

   1. USS *Ranger*
   2. USS *Hornet*
   3. USS *Yorktown*
   4. USS *Enterprise*

37. What was the first naval battle of World War II in which two opposing fleets didn't see each other during combat?

   1. The Battle of Midway
   2. The Battle of Okinawa
   3. The Battle of Guadalcanal
   4. The Battle of the Coral Sea

38. What was the decisive battle of World War II that became the turning point of the war in the Pacific?

   1. The Battle of Midway
   2. The Battle of Okinawa
   3. The Battle of Guadalcanal
   4. The Battle of the Coral Sea

39. During World War II, the Japanese loss/losses of what island(s) heralded the end of the war in the Pacific?

   1. Philippines
   2. Solomons
   3. Guadalcanal
   4. Iwo Jima

40. During World War II, the Navy was heavily involved in which of the following Atlantic (European) actions?

   1. The invasion of Normandy
   2. The capture of Navaronne
   3. The Battle of Britain
   4. The fall of Berlin

41. Which of the following were types of ships built during World War II?

   1. Net tenders
   2. Mine sweepers
   3. Repair ships
   4. All of the above

42. Which of the following combat systems came into full use during World War II?

   1. Radar
   2. Sonar
   3. Both 1 and 2 above
   4. SATNAV

43. During World War II, WAVES were eligible for how many ratings?

   1. 28
   2. 30
   3. 34
   4. 40

44. In what year was the Women's Armed Services Integration Act passed?

   1. 1942
   2. 1945
   3. 1948
   4. 1951

45. The first extensive use of jet aircraft and helicopters occurred during what war?

   1. World War I
   2. World War II
   3. The Korean Conflict
   4. The Vietnam Police Action

46. The first U.S. Navy nuclear-powered vessel was what type of ship?

   1. Carrier
   2. Submarine
   3. Merchant ship
   4. Guided-missile cruiser

47. In what year did the USS *Nautilus* make its history-making transpolar voyage?

   1. 1952
   2. 1955
   3. 1958
   4. 1961

48. In what year were the first nuclear-powered surface ships launched?

   1. 1952
   2. 1955
   3. 1958
   4. 1961

49. In what year was the first American satellite placed in orbit?

1. 1952
2. 1955
3. 1958
4. 1961

50. America's first suborbital flight was made by what Navy officer?

1. Commander Conrad
2. Commander Gordan
3. Commander Shepard Jr
4. Commander Kerwin

51. Which of the following warfare tactics was used during the Vietnam Police Action?

1. Gunfire support
2. Riverine operations
3. Coastal interdiction
4. Each of the above

52. Which of the following ships was the world's first nuclear-powered carrier?

1. USS *Nimitz*
2. USS *Carl Vinson*
3. USS *Enterprise*
4. USS *Abraham Lincoln*

53. What moon mission was completely manned by Navy personnel?

1. *Apollo* 5
2. *Apollo* 7
3. *Apollo* 11
4. *Apollo* 12

54. In what year was the *Alvin*, a deep diving vehicle, tested at 6,000-foot depths?

1. 1961
2. 1965
3. 1969
4. 1971

55. In what year was the first nuclear-powered, deep-submergence research and ocean-engineering vehicle launched?

1. 1961
2. 1965
3. 1969
4. 1971

56. Which of the following is/are principle development(s) of the Trident system?

1. A nuclear-powered fleet ballistic missile submarine
2. A strategic weapons system
3. An integrated logistics support system
4. All of the above

57. Which of the following are the most recent additions to the surface fleet?

1. Ticonderoga-class cruisers
2. Arleigh Burke-class destroyers
3. Both 1 and 2 above
4. LHAs

58. The Navy helped move approximately how many pounds of equipment and supplies during Dessert Shield/Desert Storm?

1. 12.4 billion tons
2. 15.8 billion tons
3. 18.3 billion tons
4. 21.6 billion tons

ASSIGNMENT 4

Textbook Assignment: Chapter 6 "*Naval Organization*" and chapter 7 "*Basic Seamanship*."

1. Which of the following is NOT a DoD military department?
   1. Army
   2. Coast Guard
   3. Navy
   4. Air Force

2. By law, what person heads the Department of the Navy (DoN)?
   1. Secretary of Defense
   2. Joint Chief of Staff
   3. Secretary of the Navy

3. Title 10 of the U.S. Code states that which of the following actions is/are part of the Navy's mission?
   1. Oversee construction, outfitting, and repair of naval ships, equipment, and facilities
   2. Station troops in forward positions
   3. Commands U.S. forces in CONUS
   4. Commander and chief of all sea commands

4. What are the three principal components of the DoN?
   1. The Navy Department executive offices, the operating forces including the Marine Corps, and the Shore Establishment
   2. The Navy Department executive offices, the operating forces excluding the Marine Corps, and the Shore Establishment
   3. The Navy Department excluding the executive offices, the operating forces excluding the Marine Corps, and the Shore Establishment
   4. The Navy Department excluding the executive offices, the operating forces including the Marine Corps, and the fleet commands

5. The operating forces are under the command of the
   1. Secretary of Defense
   2. Secretary of the Navy
   3. Chief of Naval Operations
   4. Chief of Naval Personnel

6. What is the purpose of the Shore Establishment?
   1. A last line of defense
   2. To provide support to the operating forces
   3. To provide a supply line
   4. To support the front line

7. Aboard ship, what publication contains information about the ship's organization?
   1. *Standard Organization and Regulations of the U.S. Navy* only
   2. *Shipboard Organization and Regulations Manual* only
   3. *Standard Organization and Regulations of the U.S. Navy and Shipboard Organization and Regulations Manual*
   4. *Uniform Code of Military Justice*

8. A ready source of information about the duties, responsibilities, and authority of personnel assigned to a ship is stated in which of the following documents?
   1. *United States Navy Regulations*
   2. Watch, Quarter, and Station Bill
   3. *Standard Organization and Regulations Manual*
   4. *Ship's Organization and Regulations Manual*

9. What are the two elements of a ship's organization?
   1. Battle organization and damage control organization
   2. Battle organization and administrative organization
   3. Administrative organization and training organization
   4. Administrative organization and damage control organization

10. Each ship is organized into what minimum number of departments?
    1. Five
    2. Two
    3. Three
    4. Four

11. Which of the following is a responsibility of the operations department?

    1. Piloting the ship
    2. Forecasting weather
    3. Conducting and analyzing intelligence information
    4. Both 2 and 3 above

12. The damage control assistant is a member of what department on a ship?

    1. Deck
    2. Supply
    3. Operations
    4. Engineering

13. If a ship doesn't have a deck department, what department is responsible for inspection and maintenance of survival equipment?

    1. Supply
    2. Weapons
    3. Navigation
    4. Engineering

14. Which of the following officers is ultimately responsible for the safe navigation of the ship?

    1. Navigator
    2. Operations officer
    3. Executive officer
    4. Commanding officer

15. When a ship is abandoned, custom and regulation require which of the following actions by the commanding officer?

    1. To be the first person to leave the ship
    2. To be the last person to leave the ship
    3. To exert every effort to destroy the ship before it sinks
    4. To inform all personnel that they are on their own

16. What is the function of the command master chief?

    1. To take charge of and be responsible for the training of enlisted personnel
    2. To assign enlisted personnel to their duties according to their qualification
    3. To relieve the commanding officer of the responsibility for the welfare and morale of enlisted personnel
    4. To transmit ideas and recommendations directly to the commanding officer

17. Of the following duties, which is NOT one of the executive officer's?

    1. Assignment of personnel
    2. Coordination of ship's drills
    3. Assignment of punishment to offenders
    4. Coordination of policing and inspection of the ship

18. If the executive officer becomes incapacitated, what person normally takes over his/her duties?

    1. The next senior line officer assigned to the ship
    2. An officer appointed by the ship's captain
    3. The next senior staff officer on board
    4. The first lieutenant

19. For what reason do commanding officers and executive officers usually have separate battle stations aboard ship?

    1. To decrease the likelihood of their being disabled at the same time
    2. To maintain a high degree of control over personnel
    3. To provide maximum coordination of operations throughout the ship
    4. To divide the areas of responsibility between the executive officer and the commanding officer

20. The department head is responsible for which of the following functions within a department?

    1. General condition of equipment
    2. Administrative matters
    3. Operational readiness of the department
    4. All of the above

21. The division officer has the responsibility of carrying out which of the following duties?

    1. Making frequent inspections of division spaces, equipment, personnel, and supplies
    2. Maintaining copies of division orders and bills and displaying them conspicuously
    3. Training division personnel and preparing them for battle
    4. Each of the above

22. Most of the jobs that are done by the XO's assistants aboard ship are the responsibility of what department in an aircraft squadron?
    1. Administrative department
    2. Maintenance department
    3. Operations department
    4. Safety department

23. Which of the following is a responsibility of the operations department of an aircraft squadron?
    1. Overall maintenance of the ship's aircraft
    2. Operational readiness and tactical efficiency
    3. Squadron safety program
    4. All of the above

24. Which of the following is the definition of the term *job accountability*?
    1. Taking command under duress
    2. Accepting credit for your job
    3. Answering to seniors in the chain of command for the way you do your job
    4. Answering only for personal mistakes

25. What is meant by *effective communications* in the chain of command?
    1. The ability to speak clearly
    2. The use of proper terminology
    3. The proper use of reports, messages, and other types of correspondence
    4. The action of seniors informing juniors about matters that affect the juniors, and the action of juniors informing seniors of existing problems

26. You need help in solving a work-related problem. Which of the following personnel should contact first?
    1. Your supervisor
    2. Your department head
    3. Your division officer
    4. Your executive officer

A. DECK SEAMANSHIP
B. BOAT SEAMANSHIP
C. MARLINESPIKE SEAMANSHIP

Figure A

IN ANSWERING QUESTIONS 27 THROUGH 30, REFER TO FIGURE A AND SELECT THE TERM USED TO DEFINE THE QUESTION.

27. General boat handling.
    1. A
    2. B
    3. C

28. The general work on the ship's deck and the equipment used.
    1. A
    2. B
    3. C

29. Anchoring, mooring, cargo handling, and towing are examples of this type of seamanship.
    1. A
    2. B
    3. C

30. Care and use of line.
    1. A
    2. B
    3. C

31 It is important for you to know shipboard equipment terminology for which of the following reasons?
    1. Equipment changes all the time
    2. You will have to inventory the equipment once each month
    3. You will probably assist the deck force in various seamanship evolutions
    4. There are different names for the same equipment

32. What is ground tackle?

1. Equipment bolted to the deck
2. Equipment used to anchor and moor with anchors
3. Equipment electrically connected to ground
4. Equipment used to refuel the ship

33. Which of the following is/are the most commonly used anchors aboard Navy ships?

1. Lightweight
2. Stockless
3. Both 1 and 2 above
4. Locking pin

34. How long is a standard shot of anchor chain?

1. 15 fathoms
2. 20 fathoms
3. 25 fathoms
4. 30 fathoms

35. What device is used to secure shots of anchor chain together?

1. Link pins
2. Bending shackles
3. Detachable links
4. Securing shackles

36. What types of anchor windlasses are used for lifting the ship's anchor?

1. Vertical shaft type only
2. Horizontal shaft type only
3. Vertical shaft and horizontal shaft types
4. Lateral shaft type

37. What device engages the chain links when hauling anchors on board ship?

1. Wildcat
2. Capstan
3. Gypsy heads
4. Bending shackles

38. Which of the following platforms is/are used in the construction of an accommodation ladder?

1. Middle platform
2. Upper platform
3. Lower platform
4. Both 2 and 3 above

39. What is the Navy term for gangplank?

1. Brow
2. Ramp
3. Platform
4. Accommodation ladder

40. What lines are used to prevent the ship from drifting forward or aft?

1. The bowline and the forward spring lines
2. The stern line and after spring lines
3. The forward and after spring lines
4. The bow and stern lines

41. What means are used to protect the sides of a ship when it is alongside a pier?

1. Doubled lines
2. Camels only
3. Fenders only
4. Camels and fenders

42. What is the main purpose for deck fittings aboard ship?

1. To secure mooring lines
2. To connect electrical power
3. To replace stanchions
4. To secure the anchor

43. Which of the following is NOT a deck fitting found aboard ships?

1. Bitts
2. Cleats
3. Bollards
4. Pad eyes

44. Which of the following is the purpose of boat booms when ships are at anchor or moored to a buoy?

1. To raise and lower supplies
2. To moor their boats well clear of the side
3. Both 1 and 2 above
4. To raise and lower personnel

45. Which of the following is/are types of boats used by the Navy?

1. Service craft
2. Combatant craft
3. Boats in general
4. All of the above

46. A boat is defined as a non-commissioned waterborne vessel that isn't designated as a service craft. According to this definition, which of the following are types of boats?

   1. Personnel boats
   2. Motor whaleboats
   3. Utility boats
   4. All of the above

47. Which of the following is a type of service craft?

   1. Riverine craft
   2. Patrol craft
   3. Ship's boats
   4. Harbor tugs

48. Which of the following is a type of combatant craft?

   1. Patrol craft
   2. Ship's boats
   3. Ferryboats

A. AFTER END
B. AMIDSHIP
C. FORWARD END
D. OUTBOARD
E. STARBOARD
F. PORT

Figure B

IN ANSWERING QUESTIONS 49 THROUGH 53, REFER TO FIGURE B AND SELECT THE TERM DESCRIBED BY THE QUESTION.

49. When facing forward of the boat, your right-hand side is in this direction.

   1. C
   2. D
   3. E
   4. F

50. The stern of the boat.

   1. A
   2. B
   3. C
   4. D

51. The area furthermost from the boat's centerline.

   1. B
   2. C
   3. D
   4. E

52. When facing forward of the boat, your left-hand side is facing this direction.

   1. A
   2. B
   3. E
   4. F

53. The bow of the boat.

   1. B
   2. C
   3. D
   4. F

54. Nylon line is about how many times stronger than manila line of the same size?

   1. 1 1/2
   2. 2 1/2
   3. 3 1/2
   4. 4 1/2

54. How is line termed *small stuff* identified?

   1. By the length of the line
   2. By the number of threads in the line
   3. By the number of strands in the line
   4. By the number of cables twisted together

56. Under safe working conditions, nylon line will stretch what maximum fraction of its length?

   1. 1/4
   2. 1/3
   3. 1/2
   4. 2/3

57. Nylon line will stretch what maximum percentage of its length before it will break?

1. 20%
2. 33%
3. 50%
4. 66%

58. A wire rope designated as 5 by 12 has (a) what number of strands and (b) what number of wires per strand?

1. (a) 5 (b) 12
2. (a) 12 (b) 12
3. (a) 12 (b) 5
4. (a) 5 (b) 5

59. The most secure line whipping is made with which of the following pieces of equipment?

1. Small needle and palm
2. Wire cutters
3. Hammer
4. Pliers

A. KNOTS
B. HITCHES
C. BENDS

Figure C

IN ANSWERING QUESTIONS 60 AND 61, REFER TO FIGURE C AND SELECT THE TERM DESCRIBED BY THE QUESTION.

60. Used to bend a line to or around an object.

1. A
2. B
3. C

61. Used to form eyes or to secure a cord or line around an object.

1. A
2. B
3. C

62. The square knot is also known as a

1. granny knot
2. seaman's knot
3. reef knot
4. top knot

63. The bowline can be used for which of the following purposes?

1. To form an eye
2. To bend two lines together
3. To secure a line to a pad eye
4. Each of the above

64. The main value of the becket bend is that it can be used to bend together two lines of different sizes.

1. True
2. False

65. If there is a great strain on a line, what type of bend should be used?

1. Becket bend
2. Double becket bend
3. Bowline
4. Double bowline

66. What type of hitch will hold as long as there's a strain on it?

1. Two half hitches
2. Two underhanded loops
3. Round and turn and two half hitches
4. Clove hitch

67. Which of the following actions would you perform to "coil down" a line?

1. Lay line in successive circles with the bitter end in the center
2. Lay line in circles, one on top of the other
3. Lay line in long, flat bights
4. Lay line out in full

68. Which of the following actions would you take to "flemish down" a line?

1. Lay line in successive circles with the bitter end in the center
2. Lay line in circles, one on top of the other
3. Lay line in long, flat bights
4. Lay line out in full

69. When making an eye splice, you should unlay what number of line strands?

1. 2 to 4
2. 4 to 6
3. 6 to 8
4. 8 to 10

70. Which of the following procedures is used to prevent the strands of synthetic line from frazzling after a splice has been made?

    1. They are whipped
    2. They are melted together
    3. They are cut off even with the standing part
    4. Each of the above

71. Which of the following is the purpose of using a short splice?

    1. To temporarily join two lines together
    2. To permanently join two lines together
    3. To form an eye
    4. Each of the above

ASSIGNMENT 5

Textbook Assignment: Chapter 8 "*Ship/Aircraft Characteristics*" and chapter 9 "*Customs and Courtesies*."

---

1. Which of the following structural components is the backbone of a ship?

    1. Stringer
    2. Prow
    3. Stem
    4. Keel

2. Which of the following structural components divides the interior of a ship into compartments?

    1. Longitudinals
    2. Bulkheads
    3. Strakes
    4. Gunwales

3. Which of the following structural components form the ship's hull?

    1. Longitudinals
    2. Bulkheads
    3. Strakes
    4. Gunwales

4. The vertical distance from the bottom of the keel to the waterline of the ship is identified by what nautical term?

    1. Freeboard
    2. Strake
    3. Draft
    4. Void

5. Which of the following structural components support decks?

    1. Athwartships deck beams
    2. Fore-and-aft deck girders
    3. Stanchions
    4. All of the above

6. The freeing ports that let water run off during heavy weather are identified by which of the following terms?

    1. Companionways
    2. Bulwarks
    3. Scuppers
    4. Flats

7. Which of the following terms defines the first complete deck below the main deck?

    1. First deck
    2. Second deck
    3. Third deck
    4. Fourth deck

8. The device that bears up tight on wedges and holds watertight doors closed is identified by which of the following terms?

    1. Dogs
    2. Scuttle
    3. Coamings
    4. Belaying pins

9. Which of the following terms defines the horizontal openings for access through decks?

    1. Hatches
    2. Doors
    3. Manholes
    4. Scuttles

10. Which of the following terms defines the solid part of a ship above the main deck?

    1. Superstructure
    2. Upper deck
    3. Forecastle

11. Which of the following is a type of mast?

    1. Mizzenmast
    2. Mainmast
    3. Foremast
    4. All of the above

12. What is the purpose of running rigging?

    1. For stays and shroud support
    2. To support stacks
    3. To hoist, lower, or control booms or boats
    4. To support the mast

13. Commissioned ships of the U.S. Navy fly a commission pennant that is secured to what point?

    1. The forecastle
    2. Aft of the fantail
    3. To a pigstick and hoisted to a truck
    4. Level adjacent to the bridge

14. What term identifies the port and starboard halves of a yard?

    1. Yardarms
    2. Pigstick
    3. Gaff
    4. Peak

15. The national ensign is flown from what part of a ship when it is anchored or moored?

    1. Jackstaff
    2. Flagstaff
    3. Pigstick
    4. Peak

16. What is the additional ship control space used by the squadron commander or admiral called?

    1. Signal bridge
    2. Main control
    3. Flag bridge
    4. Bridge wind

17. In what part of a ship is main control normally located?

    1. Chart hours
    2. Secondary conn
    3. Combat information center
    4. Boiler or machinery spaces

18. Ships of the U.S. Navy are divided into how many categories?

    1. One
    2. Two
    3. Three
    4. Four

19. How many types of ships are included in the warship category?

    1. Five
    2. Six
    3. Seven
    4. Eight

20. What type of ship is the center of a modern naval task force or task group?

    1. Aircraft carrier
    2. Destroyer
    3. Cruiser
    4. Submarine

21. Approximately how many aircraft are embarked on Nimitz class CVs?

    1. 70
    2. 75
    3. 80
    4. 85

22. What class of cruiser is designated as battle force capable?

    1. Ticonderoga
    2. Spruance
    3. Arleigh Burke

23. Which of the following is a principle mission of a destroyer?

    1. Operate offensively against submarines and surface ships
    2. Operate defensively against submarines and surface ships
    3. Both 1 and 2 above
    4. Operate short-range attack against all aircraft

24. What class destroyer represents a return to all-steel construction?

    1. Kidd class
    2. Spruance class
    3. Arleigh Burke class

25. Which of the following is the mission of frigates?

    1. Protective screens
    2. Open ocean escort and patrol
    3. Defensive operations against surface ships
    4. Offensive operations against subsurface ships

26. What class of submarines has the quietest operation?

    1. Sturgeon
    2. Ohio
    3. Seawolf

27. The Ohio class ballistic submarine has how many Trident missile tubes?

 1. 16
 2. 20
 3. 24
 4. 26

28. The LHA carries what means of defense against surface and air attack?

 1. 5-inch guns only
 2. Mk 38 machine guns only
 3. 5-inch guns and Mk 38 machine guns

29. How many troops can be embarked in, transported by, and landed by the Wasp class LHDs?

 1. 1,500
 2. 2,000
 3. 2,500
 4. 3,000

30. What is the purpose of dock landing ships?

 1. To transport amphibious craft only
 2. To transport vehicles only
 3. To transport troops only
 4. To transport a variety of amphibious craft and vehicles with embarked crews and troops

31. Which of the following means is/are used by Avenger class MCMs to find, classify, and destroy moored and bottom mines?

 1. Sonar and video systems
 2. Cable cutters
 3. A mine-detonating device
 4. Each of the above

IN ANSWERING QUESTIONS 32 THROUGH 34, SELECT THE TYPE OF AUXILIARY SHIP DEFINED BY THE QUESTION.

32. An ammunition supply ship.

 1. AOE
 2. ASR
 3. AE
 4. AO

33. Supply dry and refrigerated stores.

 1. AOE
 2. ASR
 3. AE
 4. AO

34. Combat support ship

 1. AOE
 2. ASR
 3. AE
 4. AO

35. Which of the following terms applies to the transfer of fuel, munitions, supplies, and personnel from one vessel to another while ships are under way?

 1. Vertical replenishment
 2. Horizontal replenishment
 3. Replenishment at sea
 4. Replenishment while under way

36. A separation of what approximate distance is maintained between the replenishment ship and the ship it's replenishing?

 1. 50 feet
 2. 75 feet
 3. 100 feet
 4. 125 feet

37. The AOE is designed to operate at what approximate distance between itself and the ship it's replenishing?

 1. 150 feet
 2. 175 feet
 3. 200 feet
 4. 225 feet

38. Most fleet tugs are operated by which of the following organizations?

 1. U.S. Navy
 2. U.S. Coast Guard
 3. Army Corps of Engineers
 4. Military Sealift Command

39. Combatant craft usually operate in what areas?

 1. In open waters
 2. In coastal waters
 3. In intercontinental waters
 4. In the deep sea

40. Support craft designations start with what letter?

 1. S
 2. T
 3. Y
 4. Z

41. Fixed-wing aircraft are divided into how many basic parts?

    1. One
    2. Two
    3. Three
    4. Four

42. What is the primary lifting device of an aircraft?

    1. Tail
    2. Wings
    3. Ailerons
    4. Fuselage

43. What are the three main parts of a helicopter?

    1. Tail, rotors, and empennage
    2. Tail, rotors, and fuselage
    3. Main rotor, fuselage, and tail rotor
    4. Main rotor, empennage, and tail rotor

44. Attack class planes are used in which of the following roles?

    1. Nuclear strikes
    2. Ground support
    3. Low-level bombing
    4. Each of the above

45. What class of aircraft is generally used to gain air superiority?

    1. Attack
    2. Fighter
    3. Patrol
    4. Warning

46. The E-2C *Hawkeye* belongs to what class of aircraft?

    1. Patrol
    2. Warning
    3. Antisubmarine
    4. Fighter

47. The S-3 *Viking* belongs to what class of aircraft?

    1. Patrol
    2. Warning
    3. Antisubmarine
    4. Fighter

48. What helicopter is designated for ASW use?

    1. Ch-46 *Sea Knight*
    2. SH-2 *Seasprite*
    3. SH-60B *Seahawk*

49. What helicopter operates and tows mine countermeasures devices?

    1. CH-46 *Sea Knight*
    2. SH-60B *Seahawk*
    3. CH-53D *Sea Stallion*

ASSIGNMENT 6

Textbook Assignment: Chapter 9 "*Customs and Courtesies*" and chapter 10 "*Uniforms and Formations*."

---

1. Which of the following characteristics define a custom?
   1. An act that is continued consistently over a long period of time
   2. A well-defined and uniformly followed act
   3. A generally accepted act not opposed to a statute, lawful regulation, or order
   4. All of the above

2. Of the following salutes, which one is the most common?
   1. Gun
   2. Hand
   3. Rifle
   4. Ruffles and flourishes

3. When in uniform, Navy personnel are required to salute when which of the following situations occurs?
   1. Meeting officers
   2. Hearing the national anthem
   3. Approaching the national ensign
   4. Each of the above

4. In a normal situation, how many paces from the person being saluted should the hand salute be rendered?
   1. Two
   2. Four
   3. Six
   4. Eight

5. You may salute with your left hand when which of the following situations occurs?
   1. When in civilian dress
   2. When in uniform but uncovered
   3. When in complete uniform and your right hand is injured
   4. Each of the above

6. Salutes are rendered to all officers of the U.S. and foreign armed services. Officers belonging to which of the following organizations are also entitled to salutes?
   1. Local police departments
   2. National Oceanic and Atmospheric Administration
   3. Public Health Service
   4. Both 2 and 3 above

7. When going aboard ship that's flying the national ensign, you must stop on the upper platform on the accommodation ladder or the shipboard end of the brow and take which of the following actions first?
   1. Face the national ensign and salute
   2. Face the brow of the ship and salute
   3. Face the officer of the deck and salute
   4. Face the petty officer of the watch and salute

8. While standing a sentry box, you are approached by an officer. What type of rifle salute should you render?
   1. Present arms
   2. At order arms
   3. At shoulder arms

9. You are not required to salute in which of the following situations?
   1. When standing and talking with an officer and a senior officer approaches
   2. When guarding prisoners and an officer passes within saluting distance
   3. When standing at a bus stop and a car passes carrying officers
   4. When walking and passing an officer going in the same direction

10. The term *honors* is defined by which of the following statements?

  1. Formal acts performed on public occasions
  2. Hand salutes rendered to high-ranking officials
  3. Forms of recognition and respect from one person to another
  4. Salutes rendered by a ship, unit, post, station, or an individual to high-ranking individuals, other ships, or nations

11. Passing honors for ships are exchanged when ships pass within what distance?

  1. 200 yards
  2. 400 yards
  3. 600 yards
  4. 800 yards

12. Passing honors for boats are exchanged when boats pass within what distance?

  1. 200 yards
  2. 400 yards
  3. 600 yards
  4. 800 yards

13. Your ship is about to render honors to another ship passing close aboard to starboard. In what order are the appropriate whistle signals given?

  1. One blast, one blast, two blasts, three blasts
  2. One blast, one blast, three blasts, one blast
  3. Two blasts, two blasts, two blasts, three blasts
  4. Two blasts, two blasts, three blasts, three blasts

14. A crew is paraded at quarters on which of the following occasions?

  1. When the ship is entering a U.S. port for an operational visit
  2. When the ship is returning from an extended deployment
  3. When the ship is entering home port from a local operation
  4. Each of the above

15. Gun salutes are normally fired at what time interval?

  1. 5 seconds
  2. 10 seconds
  3. 15 seconds
  4. 20 seconds

16. On which of the following special occasions is a 21-gun salute fired at 1-minute intervals?

  1. Memorial Day
  2. President's Day
  3. Independence Day
  4. Each of the above

17. Which of the following phrases defines the term *ceremony*?

  1. A regular, expected action
  2. A way of acting
  3. A formal act performed on a public occasion
  4. Each of the above

18. Aboard ship, how many minutes before morning and evening colors is the PREP pennant hoisted?

  1. 1 minute
  2. 3 minutes
  3. 5 minutes
  4. 7 minutes

19. On Navy ships not under way, where is the union jack displayed?

  1. The highest possible point
  2. The flagstaff on the stern
  3. The jack staff on the bow
  4. The gaff

20. A commissioning or command pennant is half-masted only under which, if any, of the following conditions?

  1. When passing Washington's tomb between sunrise and sunset
  2. When passing the Arizona Memorial
  3. When the commanding officer or unit commander dies
  4. None of the above

21. Which of the following naval customs is observed in the U.S. Navy by ships that are under way?

    1. Only the union jack is flown
    2. The national ensign is flown day and night
    3. Morning and evening colors are held each day
    4. Both the national ensign and the union jack are flown

22. The national ensign is hoisted and lowered in which of the following ways?

    1. Hoisted ceremoniously, lowered ceremoniously
    2. Hoisted ceremoniously, lowered smartly
    3. Hoisted smartly, lowered smartly
    4. Hoisted smartly, lowered ceremoniously

23. If you are in uniform and covered, how do you render honors when the national anthem is played indoors but the flag is not displayed?

    1. Face the music and uncover
    2. Stand at attention while facing the music
    3. Hand salute at attention while facing the music
    4. Face the music and hold your hat next to your left shoulder

24. When the national anthem is being played, Sailors in a boat must adhere to which of the following rules?

    1. All persons remain seated or standing and salute
    2. Only the coxswain salutes; all others remain seated but uncovered
    3. All persons standing salute; all others remain seated at attention
    4. Only the boat officer (or, if absent, the coxswain) salutes; all others remain seated at attention

25. What march does the Navy band play to honor the President of the United States?

    1. "Hail, Columbia!"
    2. "Admiral's March"
    3. "Hail to the Chief"
    4. "Stars and Stripes Forever"

26. Upon entering an area where Christian divine services are being held, you, as messenger of the watch, should take which of the following actions?

    1. Uncover only
    2. Remove you duty belt only
    3. Remove you duty belt and uncover
    4. Request permission from the chaplain to enter

27. An enlisted person and two officers are about to board a boat. Which of the following procedures should the enlisted person follow in entering the boat?

    1. Board first and sit aft
    2. Make way for the officers to board, then board and sit in the stern of the boat
    3. Make way for the officers to board, then board and sit in the bow of the boat
    4. Board first and sit forward, leaving room aft for the officers

28. The neckerchief is made from which of the following materials?

    1. Black silk
    2. Black acetate
    3. Both 1 and 2 above
    4. Black cotton

29. What kind of knot is used to tie a neckerchief?

    1. Granny knot
    2. Square knot
    3. Sheep shank
    4. Bowline

30. What material is used to make government-issue dress blue jumpers and trousers?

    1. Navy twill
    2. Wool serge
    3. Nylon
    4. Rayon

31. What material is used to make government-issue dress white jumpers and trousers?

    1. Navy twill
    2. Wool serge
    3. Nylon
    4. Rayon

32. Which of the following uniforms is/are considered working uniforms?

    1. Dungarees
    2. Winter blues
    3. Both 1 and 2 above
    4. Navy twill

33. When large medals are prescribed for wear with the dress blue uniform, the uniform is known as

    1. mess dress
    2. full dress
    3. field dress
    4. service dress

34. Which of the following uniforms is/are considered the working uniforms for female Sailors?

    1. Belted blue slacks and blue winter shirt
    2. Belted blue skirt and blue winter shirt
    3. Dungarees
    4. Each of the above

35. Enlisted Sailors, E-6 and below, are authorized to wear command or Navy ball caps with what type of uniform?

    1. Dungaree
    2. Navy twill
    3. Wool serge

36. Where can you find information on what is the prescribed uniform of the day?

    1. Plan of the Day (POD) only
    2. Plan of the Week (POW) only
    3. POD or POW
    4. Pass down log

37. Division officers are required to inspect the uniforms of nonrated personnel at regular intervals for what reason?

    1. As a part of PQS
    2. To justify clothing allowance
    3. To see if enlisted personnel know what uniforms are prescribed
    4. To make sure that each person has the prescribed uniform

38. Which of the following information is marked on uniforms?

    1. Name
    2. Social security number
    3. Both 1 and 2 above
    4. Rank

39. What is the largest size stencil authorized for marking clothing?

    1. 1/2 inch
    2. 1 inch
    3. 1 1/2 inches
    4. 2 inches

40. The transfer or exchange of enlisted personnel uniform items must be authorized by which of the following persons?

    1. Chief master-at-arms
    2. Division officer
    3. Executive officer
    4. Commanding officer

41. Military personnel may wear authorized military uniform articles of clothing with civilian clothing including shoes, gloves and the knit watch cap.

    1. True
    2. False

42. Occupational groups are identified by what means?

    1. A group mark, which is worn on all uniforms
    2. A rate mark, which is worn on the left sleeve of jumpers and white summer shirts
    3. A group mark, which is worn on the left sleeve of jumpers and white summer shirts
    4. A rate mark, which is worn on all uniforms

43. What is a striker mark?

    1. A specialty designator for seamen only
    2. A rating badge for emergencies
    3. A specialty mark of a particular rating, worn by personnel in paygrades E-1, E-2, and E-3 who have qualified for the rating
    4. A specific date for the air community

44. How many years of service must a person complete to be eligible to wear one service stripe (hash mark)?

    1. 5
    2. 2
    3. 3
    4. 4

45. How many years of continuous good conduct are required before a person becomes eligible to wear gold service stripes?

1. 12
2. 10
3. 8
4. 6

46. How can you determine whether an officer is a line officer or a staff corps officer?

1. By title on the name tag
2. A star is worn on the sleeve or shoulder board of the line officer
3. By the collar devices
4. A designator stripe for the rank

47. What insignia is worn by officers and enlisted personnel who have qualified in all phases of surface warfare?

1. Surface warfare
2. Submarine warfare
3. Aviation warfare
4. Special warfare

48. What insignia is worn by personnel who have qualified to serve in submarines?

1. Surface warfare
2. Submarine warfare
3. Aviation warfare
4. Special warfare

49 What insignia is worn by personnel qualified to serve in flight?

1. Surface warfare
2. Submarine warfare
3. Aviation warfare
4. Special warfare

50. There are how many broad categories of awards?

1. Four
2. Five
3. Six
4. Seven

51. In what year was the Purple Heart founded by President Washington?

1. 1776
2. 1780
3. 1782
4. 1786

52. Which of the following is an example of a campaign or a service award?

1. Medical Scientific Societies
2. Silver Life Saving Medal
3. Good Conduct Medal
4. Navy "E"

53. What is the maximum number of ribbons that may be worn in each row?

1. Five
2. Two
3. Three
4. Four

54. What means of identification must you carry with you at all times?

1. Driver's license
2. Armed forces identification card
3. Liberty card
4. Copy of present set of orders

55. Under which of the following circumstances may you surrender (give up) your ID card?

1. To show a change in rank
2. To correct an error
3. To effect a name change
4. Each of the above

56. Which of the following is the purpose of the armed forces ID card under article 17 of the Geneva Convention?

1. As a means of identification and casualty reporting
2. As a means of grave registration for members who die in a combat zone
3. Both 1 and 2 above
4. As a means of identification for POWs

57. While on active duty, you must wear your ID tags under which of the following conditions?

1. In time of war
2. When engaged in flight operations
3. When prescribed by the CNO
4. All of the above

58. What type of information is found on ID tags?

1. Last name, first name, and middle initial
2. SSN, blood type, and Rh factor
3. Religious preference
4. All of the above

59. According to grooming standards for men, how many rings per hand may be worn ?

    1. One
    2. Two
    3. Three
    4. As many as they wish

60. According to grooming standards for women, what is the maximum length of fingernails (as measured from the tip of the finger)?

    1. 1 inch
    2. 1/2 inch
    3. 3/4 inch
    4. 1/4 inch

61. Enlisted women, E-6 and below, are authorized to wear ball-type earrings of what (a) size and (b) what color?

    1. (a) 6mm (b) gold
    2. (a) 6mm (b) silver
    3. (a) 5mm (b) gold
    4. (a) 5mm (b) silver

62. When personnel are in ranks, the chest of one person and the back of the person ahead should be what distance apart?

    1. 20 inches
    2. 30 inches
    3. 40 inches
    4. 50 inches

63. A pace consists of a full step of what length for (a) men and (b) women?

    1. (a) 18 inches (b) 16 inches
    2. (a) 24 inches (b) 18 inches
    3. (a) 30 inches (b) 24 inches
    4. (a) 36 inches (b) 30 inches

64. Which of the following drill positions is the basic military position?

    1. Rest
    2. At ease
    3. Attention
    4. Parade rest

65. Talk is permitted when you are in which of the following formation positions?

    1. Rest
    2. Parade rest
    3. At ease

66. How many movements are used to perform the ABOUT FACE command?

    1. Five
    2. Two
    3. Three
    4. Four

67. When executing the command FALL IN, the squad forms in line on which of the following persons?

    1. Instructor
    2. Squad leader
    3. Standard bearer
    4. Company commander

68. Personnel in formation align themselves with which of the following persons?

    1. Guide
    2. Leader
    3. Each other
    4. Formation director

69. After the command DRESS RIGHT DRESS, a formation returns to the attention position on which of the following commands?

    1. Cover
    2. Extend
    3. Attention
    4. Ready, on the word FRONT

70. When the command CLOSE RANKS is given to members in formation, the fourth rank moves how many paces forward?

    1. One
    2. Two
    3. Three
    4. Four

ASSIGNMENT 7

Textbook Assignment: Chapter 11 "*Small Arms.*"

---

1. Every firearm used by Navy personnel has some type of safety device built in.
   1. True
   2. False
2. Which of the following is the prime cause of accidental shootings?
   1. Faulty gun
   2. Faulty ammunition
   3. Alcohol or drug use by the user
   4. Negligence or carelessness of the user
3. Which of the following are acceptable ear protective devices?
   1. Insert type
   2. Circumaural type
   3. Both 1 and 2 above
   4. Cotton type
4. The M14 rifle is best described by which of the following groups of characteristics?
   1. Medium weight, recoil-operated, magazine-fed, and fully automatic
   2. Medium weight, gas-operated, clip-fed, and capable of semiautomatic and fully automatic fire
   3. Lightweight, recoil-operated, clip-fed, and semiautomatic
   4. Lightweight, gas-operated, magazine-fed, and capable of semiautomatic or automatic fire
5. Which of the following types of ammunition is used with the M14 rifle?
   1. 7.62-mm NATO cartridge
   2. 20-round magazine
   3. M76 grenade launcher
   4. Each of the above
6. What is the maximum range of the M14 rifle?
   1. 2,500 yards
   2. 3,250 yards
   3. 4,075 yards
   4. 5,250 yards
7. The M16A1 and M16A2 rifles are best described by which of the following groups of characteristics?
   1. Clip-fed, recoil-operated weapons
   2. Magazine-fed, recoil-operated shoulder weapons
   3. Magazine-fed, gas-operated shoulder weapons
   4. Clip-fed, gas-operated weapons
8. For what size cartridge is the M16A1 rifle chambered?
   1. .38 caliber
   2. .45 caliber
   3. 5.56 mm
   4. 7.62 mm
9. What is the maximum magazine capacity of the M16A1 rifle?
   1. 15 rounds
   2. 25 rounds
   3. 30 rounds
   4. 35 rounds
10. What is the muzzle velocity of the M16A1 and M16A2 rifles?
    1. 2,500 feet per second
    2. 3,000 feet per second
    3. 3,500 feet per second
    4. 4,000 feet per second
11. What means is used to adjust the rear sights of the M16A2 rifle?
    1. A windage drum
    2. A windage knob and an elevation knob
    3. A clip lever marked range
    4. A slide adjust to windage
12. What is the first step to take when handling any weapon?
    1. Point the selector lever toward SAFE
    2. Remove the magazine
    3. Visual check of the chamber
    4. Lock the bolt open

13. Why should the selector be on SAFE during assembly and disassembly?
    1. To prevent damage to the automatic sear
    2. To prevent damage to the firing pin
    3. To prevent the barrel from releasing
    4. To prevent the rear slide from disengaging
14. When the selection lever is in the burst position, the M16A1 rifle fires in what way?
    1. In short bursts of two rounds
    2. In short bursts of three rounds
    3. Four rounds each time the trigger is pulled
    4. Six rounds each time the trigger is pulled
15. The M16A1 rifle is clear when which of the following conditions are met?
    1. No case or round is in the chamber and the magazine is out
    2. The bolt carrier is to the rear
    3. The selector lever is on the SAFE position
    4. All of the above
16. When cleaning the barrel bore and chamber of the M16A1 rifle, you should not reverse the brush while in the bore for what reason?
    1. The barrel slide will be damaged
    2. The bore may jam
    3. The trigger pin will need to be replaced
    4. The bore cleaner will not work
17. What parts of the barrel bore and chamber should you lubricate after you've finished cleaning them?
    1. The locking lugs
    2. The extractor ejector
    3. The lugs in the barrel extension
    4. The magazine springs
18. When cleaning ammunition magazines, you need to make sure the magazine is dry for what reason?
    1. The magazine and ammunition can corrode and become dangerous to use
    2. The spring action will tighten up
    3. The ammunition will jam
    4. The magazine won't load
19. The .38-caliber revolver is best described by which of the following groups of characteristics?
    1. Cylinder-loading, single- or double-action, manually operated hand gun
    2. Semiautomatic, recoil-operated, magazine-fed hand gun
    3. Semiautomatic, cylinder-loading double-action hand gun
    4. Cylinder-loading, recoil-operated, manually operated hand gun
20. The .45-caliber service pistol is best described by which of the following groups of characteristics?
    1. Cylinder-loading, single- or double-action, manually operated hand gun
    2. Semiautomatic, recoil-operated, magazine-fed hand gun
    3. Semiautomatic, cylinder-loading double-action hand gun
    4. Cylinder-loading, recoil-operated, manually operated hand gun
21. The .45-caliber revolver has what maximum range and maximum effective range, respectively?
    1. 1,800 yards, 75 yards
    2. 1,500 yards, 50 yards
    3. 1,450 yards, 45 yards
    4. 1,250 yards, 30 yards
22. The magazine capacity of the 9mm service pistol can hold how many rounds in the magazine?
    1. 5 rounds
    2. 10 rounds
    3. 15 rounds
    4. 20 rounds
23. The 9mm service pistol is best described by which of the following groups of characteristics?
    1. Semiautomatic, recoil-operated, magazine-fed hand gun
    2. Semiautomatic, magazine-fed, recoil-operated, double-action pistol
    3. Semiautomatic, cylinder-loading double-action hand gun
    4. Semiautomatic, magazine-fed, single-action pistol

24. Which of the following are safety features incorporated in the 9mm service pistol?

    1. Ambidextrous safety
    2. Firing pin block
    3. Half cock notch
    4. All of the above

25. What safety feature of the 9mm pistol prevents accidental discharge?

    1. Firing pin block
    2. Half cock notch
    3. Muzzle pressure
    4. Rear trigger guard

26. The advantage of shotguns over pistols is that sight alignment is not as critical.

    1. True
    2. False

27. What maximum number of rounds of 12-gauge, 2 3/4-inch ammunition will the magazine of the M870 shotgun hold?

    1. One
    2. Two
    3. Three
    4. Four

QUESTIONS 28 THROUGH 33 REFER TO RIFLE FIRING TECHNIQUES.

28. When firing a rifle, what factors make up the sight picture?

    1. Rear sight
    2. Correct aiming point only
    3. Sight alignment only
    4. Correct aiming point and sight alignment

29. What is meant by the term "eye relief"?

    1. The different distance from the rear sight of your aiming eye, depending on your firing position
    2. The distance of your eye from the peep sight in any particular firing position
    3. The height of the rear sight
    4. The height of the front sight

30. What is meant by the terms "spot weld" or "anchor"?

    1. The distance of your eye from the peep sight in any particular firing position
    2. Holding your rifle in the exact same position to make sure your eye stays the same distance from the peep hole
    3. The distance of your eye from the peep sight, depending on your firing position

31. What is the focus for the eye?

    1. The front sight
    2. The rear sight
    3. The aiming point
    4. The sight picture

32. What is the correct aiming point on a type "A" target?

    1. 3 o'clock
    2. 6 o'clock
    3. 9 o'clock

33. What elements do you need to align to get a correct sight picture?

    1. The rear sight only
    2. The front sight only
    3. The bull's eye only
    4. The rear sight, front sight, and bull's eye

34. How many standard firing positions are taught in the Navy?

    1. Two
    2. Three
    3. Four
    4. Five

35. Which of the following firing positions is most useful when you are constantly firing and moving?

    1. Sitting
    2. Standing
    3. Kneeling

36. Which of the following firing positions is most useful when you are on level ground or firing uphill?

    1. Sitting
    2. Standing
    3. Kneeling

37. Which of the following firing positions is most useful when you are firing downhill?

    1. Sitting
    2. Standing
    3. Kneeling

38. What is the key to trigger control?

    1. Squeeze the trigger smoothly, gradually, and evenly straight to the rear
    2. Placing the finger at the very tip of the trigger
    3. Constant pressure on the trigger
    4. Squeeze the trigger quickly and evenly straight to the rear

39. Remembering which of the following acronyms will help you remember correct shooting techniques?

    1. AIM
    2. RELAX
    3. BRASS
    4. SLACK

40. You should take a breath, let out a little air, and then hold your breath until you fire your rifle. However, if you haven't fired within 10 seconds, what should you do?

    1. Breath normally and continue squeezing the trigger
    2. Take another breath and start the aiming procedure over
    3. Let out more air and firmly jerk the trigger
    4. Continue holding your breath and start the aiming procedure over

QUESTIONS 41 THROUGH 43 REFER TO FIRING TECHNIQUES FOR THE 9mm SERVICE PISTOL.

41. What is a correct sight picture?

    1. Correct sight alignment and correct aiming point
    2. Off center to the target
    3. Bottom of the picture
    4. Top of the picture

42. The pistol can be accurately fired from how many positions?

    1. Two
    2. Three
    3. Four
    4. Five

43. Missing the target is most often caused by

    1. sight misalignment
    2. bent barrel
    3. improper trigger squeeze
    4. bad ammunition

ASSIGNMENT 8

Textbook Assignment: Chapter 12 "*Damage Control.*"

1. The damage control organization is divided into how many parts?

   1. One
   2. Two
   3. Three
   4. Four

QUESTIONS 2 THROUGH 13 REFER TO THE ADMINISTRATIVE ORGANIZATION OF DAMAGE CONTROL.

2. What person coordinates the efforts of repair parties to control damage?

   1. The damage control assistant
   2. The operations officer
   3. The executive officer
   4. The weapons officer

3. Which of the following are duties of the administrative organization of damage control?

   1. Records and schedules for maintenance
   2. Written doctrine and procedures relating to damage control
   3. Ship's bills
   4. All of the above

4. Which of the following are duties of the executive officer?

   1. Ship's survivability training
   2. Readiness to manage casualties
   3. Control and recover from damage
   4. Each of the above

5. What officer is designated as the ship's damage control officer?

   1. The commanding officer
   2. The executive officer
   3. The engineer officer
   4. The operations officer

6. The DCA is the primary assistant to the damage control officer. As such, the DCA has which of the following responsibilities?

   1. Training the ship's DC personnel
   2. Administration of the ship's DC organization
   3. Maintain records of DC personnel PQS accomplishment for all hands
   4. All of the above

7. What requirements must a petty officer have to be designated as the damage control petty officer (DCPO)?

   1. Be a PO2 or above
   2. Complete the PQS
   3. Complete the fire-fighting school
   4. Be designated by the LCPO

8. Normally, the job of DCPO is held for what period of time?

   1. 12 months
   2. 9 months
   3. 3 months
   4. 6 months

9. Which of the following is/are responsibilities of the DCPO?

   1. Assist in the instruction of division personnel in damage control
   2. Prepare and maintain damage control checkoff lists for their spaces
   3. Make required reports
   4. All of the above

10. What person is responsible for determining the safe entry of personnel into closed or poorly ventilated spaces?

    1. The gas free engineer
    2. The fire marshal
    3. The DCA
    4. The XO

11. What person is designated to aid the DCA train personnel and to prevent and fight fires?

    1. The gas free engineer
    2. The fire marshal
    3. The DCPO
    4. The LCPO

12. What action is taken if the fire marshal finds hazards that relate to poor housekeeping during a daily inspection?

    1. Record and advise the responsible division
    2. Submit discrepancy report to DCA with copies to XO and department head
    3. Have the division LCPO schedule corrective action
    4. Report the hazard to the safety department

13. The fire marshal has which of the following responsibilities?

    1. Submitting reports citing hazards and recommendations for their correction
    2. Setting up a fire watch team before regular overhauls
    3. Both 1 and 2 above
    4. Inspecting the engineering department

QUESTIONS 14 THROUGH 17 REFER TO THE BATTLE ORGANIZATION OF DAMAGE CONTROL.

14. What is the purpose of the ship's damage control battle organization?

    1. To stand at ready for the battle to be over
    2. To restore the ship to as near normal operation as possible
    3. To comfort injured crew members
    4. When directed, to take charge of the weapons

15. In the damage control battle organization, what person is responsible for controlling damage; fighting fires; CBR countermeasures; and control of stability, list, and trim?

    1. The DCPO
    2. The DCA
    3. The XO
    4. The CO

16. What are the primary damage control battle organization unit(s)?

    1. Repair parties
    2. Navigation crew
    3. Deck crews
    4. Engineering personnel

17. What is the nerve center of the directing force for directing the damage control organization?

    1. DCC or CCS
    2. Bridge
    3. Aft steering
    4. DC locker 5

IN ANSWERING QUESTIONS 18 THROUGH 21, SELECT THE REPAIR PARTY THAT IS IDENTIFIED BY THE FUNCTION USED AS THE QUESTION.

18. Main deck repair.

    1. Repair 1
    2. Repair 3
    3. Repair 5
    4. Repair 7

19. Propulsion repair.

    1. Repair 1
    2. Repair 3
    3. Repair 5
    4. Repair 7

20. Ordnance.

    1. Repair 2
    2. Repair 4
    3. Repair 6
    4. Repair 8

21. Electronic casualty control.

    1. Repair 2
    2. Repair 4
    3. Repair 6
    4. Repair 8

22. Each repair party should be capable of performing which of the following functions?

    1. Rigging casualty power
    2. Controlling flooding
    3. Extinguishing all types of fires
    4. Each of the above

23. When in port, the ship has which of the following duty section components available to respond to any type of casualty?

  1. In port emergency teams
  2. Salvage teams
  3. Duty DC watch
  4. Deck department

24. The rescue and assistance detail must have which of the following qualifications?

  1. Be qualified as an emergency team member
  2. Be qualified in first aid
  3. Both 1 and 2 above
  4. Have passed the PRT

25. General quarters is an all hands evolution—it is the highest state of readiness of the ship.

  1. True
  2. False

26. Which of the following statements describes a correct GQ route to follow?

  1. Forward in the passageways and down ladders on the starboard side
  2. Aft in the passageways and down ladders on the port side
  3. Forward in the passageways and up ladders on the port side
  4. Aft in the passageways and up ladders on the starboard side

27. Which of the following is an emergency damage control communications system?

  1. 2JZ
  2. 6JZ
  3. X40J
  4. X24J

28. What system signals override microphone control stations to notify the ship's crew of imminent danger?

  1. Alarms for collision, chemical attack, general, and flight crash
  2. General announcing system
  3. DC controls
  4. Bridge alarms

A. COLLISION
B. CHEMICAL ATTACK
C. GENERAL
D. FLIGHT CRASH

Figure A

IN ANSWERING QUESTIONS 29 THROUGH 31, REFER TO FIGURE A AND SELECT THE TERM USED TO DEFINE THE QUESTION.

29. This alarm is sounded by the OOD or PreFly notifying ship's company of a pending or actual flight deck emergency.

  1. A
  2. B
  3. C
  4. D

30. This alarm is sounded when there is a possibility that the ship will be struck by another waterborne unit.

  1. A
  2. B
  3. C
  4. D

31. When this alarm is sounded, all hands report to their preassigned stations and set material condition ZEBRA.

  1. A
  2. B
  3. C
  4. D

32. Which of the following means of communications is used when all other methods have failed?

  1. Messengers
  2. Sound-powered telephones
  3. Morse Code
  4. Bullhorn

33. All Navy ships have how many material conditions of readiness?

  1. One
  2. Two
  3. Three
  4. Four

34. What material condition provides the least degree of watertight integrity?

    1. ZEBRA
    2. YOKE
    3. XRAY

35. What material condition sets the highest degree of watertight integrity?

    1. ZEBRA
    2. YOKE
    3. XRAY

36. Which of the following fittings are closed when condition ZEBRA is set?

    1. DOG Z fittings
    2. Circle X fittings
    3. Y fittings
    4. All of the above

37. What means, if any, is used by repair parties to find damage control fittings and closures in each compartment?

    1. Master alfa list
    2. Compartment checkoff lists
    3. DC compartment checks
    4. None

38. The Damage Control Closure Log is maintained in which of the following locations?

    1. DCC
    2. Quarterdeck
    3. Both 1 and 2 above
    4. Engineering spaces

39. Which of the following logs is a list of all DC-related fittings that don't work properly?

    1. Damage Control Closure Log
    2. Fire marshal pass down log
    3. Inoperative Fittings and Closures Log
    4. Bridge log

40. What fittings are secured when the ship is set for "darken ship"?

    1. WILLIAM
    2. Circle WILLIAM
    3. DOG ZEBRA
    4. Circle ZEBRA

41. The emergency escape breathing device (EEBD) supplies breathable air for what maximum period of time?

    1. 10 minutes
    2. 15 minutes
    3. 20 minutes
    4. 25 minutes

42. With training, you should be able to activate the EEBD within what maximum period of time?

    1. 10 seconds
    2. 20 seconds
    3. 30 seconds
    4. 40 seconds

43. Which of the following breathing devices should NOT be worn for fire-fighting purposes?

    1. OBA
    2. SEED
    3. SCBA

44. Which of the following is the primary fire fighting tool for respiratory protection?

    1. EEBD
    2. SEED
    3. OBA
    4. SCBA

45. From the time it is activated, each cylinder used in the self-contained breathing apparatus (SCBA) will last approximately what length of time?

    1. 15 minutes
    2. 30 minutes
    3. 45 minutes
    4. 60 minutes

46. Which of the following substances must be present to start a fire?

    1. Oxygen
    2. Heat
    3. Fuel
    4. All of the above

47. In the fire tetrahedron, how many components are necessary for combustion?
    1. One
    2. Two
    3. Three
    4. Four

48. Flammable materials give off vapors. What is the lowest temperature that these vapors burn when a spark is applied?
    1. Ambient temperature
    2. Room temperature
    3. Flash point
    4. Ignition point

49. What term is used to describe the lowest temperature at which spontaneous combustion occurs?
    1. Flash point
    2. Ignition point
    3. Ambient temperature
    4. Room temperature

50. Heat from fire can be transferred by how many methods?
    1. One
    2. Two
    3. Three
    4. Four

51. What method of heat transfer occurs when heat moves from one body to another by direct contact?
    1. Conduction
    2. Convection
    3. Radiation
    4. Reflection

52. What method of heat transfer occurs through the motion of smoke, hot air, and heated gases?
    1. Conduction
    2. Convection
    3. Radiation
    4. Reflection

53. What type of heat transfer occurs when heat moves in all direction unless blocked?
    1. Conduction
    2. Convection
    3. Radiation
    4. Reflection

54. Which of the following agents should be used to extinguish a class B fire?
    1. Water
    2. AFFF
    3. PKP
    4. Both 2 and 3 above

55. Which of the following agents should be used to extinguish class A or D fires?
    1. Water
    2. AFFF
    3. PKP
    4. Both 2 and 3 above

ASSIGNMENT 9

Textbook Assignment: Chapter 13 "*Chemical, Biological, and Radiological (CBR) Defense.*"

---

1. Which of the following are weapons of mass destruction?
   1. Chemical weapons
   2. Biological agents
   3. Nuclear weapons
   4. All of the above

2. Which of the following warfare agents are used to kill or disable personnel by affecting their blood, nerves, lungs, or stomach?
   1. CW
   2. BW
   3. Nuclear
   4. All of the above

3. Generally, antipersonnel agents are divided into how many types?
   1. Five
   2. Two
   3. Three
   4. Four

4. Cramps, breathing difficulty, nausea, headache, convulsions, and contractions of the pupils are all symptoms of which of the following types of contamination?
   1. Tear agent
   2. Nerve agent
   3. Choking agent
   4. Blister agent

5. Burns from exposure to mustard vapor will be more serious in which of the following body areas?
   1. Neck
   2. Groin
   3. Armpits
   4. All of the above

6. A person exposed to a blood agent may experience respiratory paralysis within what length of time?
   1. Seconds
   2. Minutes
   3. Hours

7. Atropine and 2-PAM C1 oxime are used to counteract the effects and relieve the symptoms of which of the following agents?
   1. Nerve
   2. Blood
   3. Blister
   4. Choking

8. What type of agents are used to produce temporary misery and harassment?
   1. Blister
   2. Choking
   3. Riot control
   4. Nerve

9. Which of following agents can be used in BW operations?
   1. Living organisms
   2. Toxins
   3. Microtoxins
   4. All of the above

10. Which of the following diseases can be spread as part of a BW attack?
    1. Cholera
    2. Anthrax
    3. Both 1 and 2 above
    4. Mumps

11. In its early stage, which of the following is/are symptoms of BW attack?
    1. Fever
    2. Inflammation
    3. Malaise
    4. Each of the above

12. Nuclear weapons have the capability of destroying areas in which of the following ways?
    1. Blast
    2. Shock
    3. Nuclear radiation
    4. All of the above

13. Nuclear explosions are divided into how many classes?
    1. Five
    2. Two
    3. Three
    4. Four
14. What would be the most effective type of nuclear strike to use against a battle group at sea?
    1. High altitude blast
    2. Air blast
    3. Surface blast
    4. Subsurface burst
15. What type of nuclear strike would be used to destroy satellites and interrupt communications systems through the effects of EMP?
    1. High altitude blast
    2. Air blast
    3. Surface blast
    4. Subsurface burst
16. In what type of nuclear strike would the shock wave near ground zero be greater than the blast wave?
    1. High altitude blast
    2. Air blast
    3. Surface blast
    4. Subsurface burst
17. What are the effects of nuclear weapons?
    1. Blast waves only
    2. Incendiary only
    3. Radiation only
    4. Blast waves, incendiary, and radiation
18. If a nuclear blast at night causes you to experience flash blindness, you can expect your vision to recover in what length of time?
    1. 15 minutes
    2. 2 hours
    3. 3 hours
    4. 45 minutes
19. What type of radiation hazard must enter the body through ingestion or cuts to cause bodily harm?
    1. Alpha particles
    2. Beta particles
    3. Gamma rays
    4. Neutrons
20. What is TREE?
    1. The absorption of EMP by electrical conductors
    2. Interference of passive sonar systems
    3. The affect of gamma or neutron radiation on shipboard electronic systems
    4. Interference of radio transmission through ion fields
21. A survey team is made up of what minimum number of personnel?
    1. Five
    2. Two
    3. Three
    4. Four
22. What person is in charge of a survey team?
    1. Monitor
    2. Recorder
    3. Messenger
    4. Exec
23. Areas contaminated by CW, BW, or nuclear agents are identified by markers having what shape?
    1. Triangular
    2. Hexagonal
    3. Circular
    4. Square
24. Dose rate is expressed in roentgens, which are gamma ray measurements only.
    1. True
    2. False
25. What does the acronym *radiac* stand for?
    1. Radiation decontamination and control
    2. Radiological activity detection and computation
    3. Radioactivity defense, identification, and instrument calibration
    4. Radioactivity detection, indication, and computation
26. The nonself-reading, high-range casualty dosimeter measures what maximum amount of gamma radiation?
    1. 5 roentgens
    2. 200 roentgens
    3. 600 roentgens
    4. 200 milliroentgens

27. What kit is used to check areas suspected to have been contaminated by CW agents?

    1. M248A2 kit
    2. M256A1 kit
    3. M258A1 kit
    4. M262A2 kit

28. Which of the following actions should personnel topside take if an airburst occurs?

    1. Close their eyes
    2. Drop to the deck
    3. Cover as much exposed skin as possible
    4. All of the above

29. Which of the following pieces of protective equipment is/are the most important in protecting you against CBR agents?

    1. Coveralls
    2. Protective mask
    3. Both 1 and 2 above
    4. Steel-toed shoes

30. Protective masks serve how many functions?

    1. One
    2. Two
    3. Three
    4. Four

31. You should be able to don and adjust your protective mask in what maximum amount of time?

    1. 10 seconds
    2. 20 seconds
    3. 30 seconds
    4. 40 seconds

32. The MCU-2/P protective mask has how many voice emitters?

    1. One
    2. Two
    3. Three
    4. Four

33. The chemical protective overgarment consists of how many parts?

    1. One
    2. Two
    3. Three
    4. Four

34. The chemical protective overgarment can be used for protection against radiological contamination.

    1. True
    2. False

35. Which of the following statements best describes the purpose of the mission oriented protective posture (MOPP)?

    1. Provides a means to establish levels of readiness
    2. Provides a method for identifying agents
    3. Provides a means to prevent contaminants from entering the ship
    4. Provides a method for cleansing the ship of CBR agents

36. At what MOPP level would all protective equipment be worn with the hood up and secured?

    1. 1
    2. 2
    3. 3
    4. 4

37. How many levels of decontamination are there?

    1. One
    2. Two
    3. Three
    4. Four

38. A decontamination team usually consists of how many people?

    1. Five
    2. Two
    3. Six
    4. Four

39. What is the most effective way to decontaminate biological agents?

    1. Burning
    2. Using dry heat
    3. Using steam under pressure
    4. Using a chemical disinfectant

40. Aboard ship, a decontamination station has how many parts?

    1. One
    2. Two
    3. Three
    4. Four

41. Showering will destroy nuclear and biological agents.

    1. True
    2. False

42. The Collective Protection System (CPS) consists of how many protection zones?

    1. One
    2. Two
    3. Three
    4. Four

43. Which of the following CPS levels provides the maximum operational protection envelope?

    1. Level I
    2. Level II
    3. Level III
    4. Level IV

ASSIGNMENT 10

Textbook Assignment: Chapter 14 "*First Aid and Health*" and Chapter 15 "*Survival.*"

---

1. First aid has which of the following objectives?
   1. To save lives
   2. To limit infection
   3. To prevent further injury
   4. Each of the above

2. In administering first aid, you are responsible for performing which of the following tasks?
   1. Stop bleeding
   2. Maintain breathing
   3. Prevent or treat for shock
   4. All of the above

3. Under which, if any, of the following circumstances should you touch an open wound with your fingers?
   1. To replace bulging abdominal organs
   2. To remove a protruding foreign object
   3. Only when absolutely necessary to stop severe bleeding
   4. None of the above

4. A person who has stopped breathing is considered dead.
   1. True
   2. False

5. What is the purpose of artificial ventilation?
   1. To restore the function of the heart
   2. To provide a method of air exchange
   3. To clear an upper air passage obstruction
   4. To clear a lower air passage obstruction

6. When using the mouth-to-mouth technique for administering artificial ventilation, how often should you force air into the victim's lungs?
   1. Once every 3 seconds
   2. Once every 4 seconds
   3. Once every 5 seconds
   4. Once every 6 seconds

7. The mouth-to-nose technique for administering artificial ventilation is effective on which of the following victims?
   1. The victim who is breathing very slowly
   2. The victim who is very young
   3. The victim who has extensive facial injuries
   4. Both 2 and 3 above

8. When using the back pressure/arm lift technique for administering artificial ventilation, you should repeat the cycle how many times per minute?
   1. 10 to 12
   2. 8 to 10
   3. 6 to 8
   4. 4 to 6

9. Cardiopulmonary resuscitation (CPR) should be started within how many minutes of the onset of cardiac arrest?
   1. 6
   2. 5
   3. 3
   4. 4

10. When administering CPR, you should place your hands on what area of the victim's chest?
    1. On the upper part of the sternum
    2. About 1 inch below the sternum
    3. Above the tip of the sternum
    4. On the tip of the sternum

11. When using the one-rescuer CPR technique, you should administer how many compressions per minute?
    1. 60 to 80
    2. 40 to 60
    3. 20 to 40
    4. 10 to 20

12. When using the one-rescuer CPR technique, you should give how many ventilations after each set of compressions?

    1. One
    2. Two
    3. Three
    4. Four

13. When using the two-rescuer CPR technique, you should use what ratio of compressions to ventilations?

    1. 1 to 5
    2. 5 to 1
    3. 10 to 4
    4. 4 to 10

14. Which of the following is one of the most reliable indications of a blocked airway in a conscious person?

    1. Inability to speak
    2. Cherry red skin color
    3. Profuse sweating of the face
    4. Partially digested food in the mouth

15. You are assisting a person who is choking. What is the first action you should take?

    1. Apply the standing chest thrust to the victim
    2. Apply the standing abdominal thrust to the victim
    3. Clear the victim's mouth of any food or foreign objects
    4. Sharply slap the victim on the back between the shoulder blades

16. What minimum amount of blood loss usually causes a person to go into shock?

    1. 1 pint
    2. 2 pints
    3. 3 pints
    4. 4 pints

17. How is arterial bleeding from a cut near the surface of the skin indicated?

    1. Spurting dark red blood
    2. Steady flow of dark red blood
    3. Steady flow of bright red blood
    4. Gushing spurts of bright red blood

18. To control bleeding, which of the following methods should you try first?

    1. Direct pressure
    2. A tourniquet
    3. A battle dressing
    4. Pressure points

19. What is meant by the pressure points in the human body?

    1. A place where the artery is protected on all sides by bone or muscle
    2. A place where the main artery is close to the skin surface and over a bone
    3. A point where an artery crosses between the heart and the wound
    4. A point where an artery crosses a joint

20. If the use of a battle dressing is required, who should loosen/remove it?

    1. The on-scene leader
    2. The repair locker leader
    3. A person qualified in first aid
    4. Medical personnel

21. Shock will never be serious enough to cause death.

    1. True
    2. False

22. At which of the following times should you start treatment for shock?

    1. As soon as possible after an injury occurs
    2. Only when symptoms indicate severe shock
    3. Only after other injuries have been treated
    4. As soon as unconsciousness occurs

23. What is the basic position for treating shock?

    1. Putting the head and feet at the same level
    2. Putting the head lower than the feet
    3. Putting the feet lower than the head

24. Which of the following Navy personnel are the most frequent victims of suicide?

    1. Males between the ages of 25 to 37 in paygrades E-6 and E-8
    2. Males between the ages of 17 to 24 in paygrades E-1 to E-6
    3. Females between the ages of 25 to 37 in paygrades E-6 to E-8
    4. Females between the ages of 17 to 24 in paygrades E-1 to E-6

25. Which of the following are actions to take if you believe someone you know is thinking about suicide?

    1. Take all threats seriously
    2. Don't leave the person alone
    3. Get professional help
    4. All of the above

26. Which of the following burns is considered the most serious?

    1. First degree
    2. Second degree
    3. Third degree

27. A closed fracture is one where the skin is intact and an open fracture is one where the skin is broken.

    1. True
    2. False

28. When choosing a material to use as a splint, you should choose material that has which of the following characteristics?

    1. Light weight
    2. Fairly rigid
    3. Strong
    4. All of the above

29. Which of the following is/are symptoms of a broken bone?

    1. Swelling
    2. Deformity
    3. Inability to use the part
    4. Each of the above

30. Which of the following is/are symptoms of a sprain or a strain?

    1. Swelling
    2. Inability to use the part
    3. Each of the above

31. What is one of the easiest ways to carry an unconscious person?

    1. Arm carry
    2. Fireman's carry
    3. Tied-hands crawl
    4. Lift and drag

32. In compartments with access hatches that are too small to permit the use of regular stretchers, you would remove an injured person using what type of stretcher?

    1. Neil Robertson
    2. Gaylord
    3. Stokes

33. Aboard ship, keeping yourself and your spaces clean and orderly has which of the following advantages?

    1. Improves morale
    2. Contributes to the well-being of the crew
    3. Both 1 and 2 above

34. What common dental condition(s) can be prevented by making sure you develop the habit of good oral hygiene?

    1. Tooth decay
    2. Gum and bone disease
    3. Reddening of the gums
    4. All of the above

35. Sexually transmitted diseases may be spread through the use of inanimate objects, such as toilet seats, bed linens, or drinking glasses.

    1. True
    2. False

36. If left untreated, syphilis may cause which of the following conditions?

    1. Heart disease
    2. Mental illness
    3. Blindness
    4. All of the above

37. Sterility is the result of leaving which of the following sexually transmitted diseases untreated?

    1. Syphilis
    2. Gonorrhea
    3. Acquired Immune Deficiency Syndrome
    4. Herpes

38. Use of condoms offers some protection from Acquired Immune Deficiency Syndrome.

    1. True
    2. False

39. If time permits during abandon-ship preparation, a message announced over the 1MC will give which of the following information?
    1. Water temperature
    2. Sea and wind conditions
    3. Bearing and distance to the nearest land
    4. All of the above

40. If you have to go over the side and the ships' propellers are turning, you should leave from what point on the ship?
    1. The windward side, if possible
    2. The lee side, if possible
    3. From the bow
    4. From the stern

41. Personnel have the greatest chance for survival in the water if they meet which of the following swimmer requirements?
    1. First class
    2. Second class
    3. Third class

42. If you have to swim through flames, which of the following is a procedure to follow?
    1. Use your life preserver as a raft
    2. Keep your face above the surface of the water as much as possible
    3. Both 1 and 2 above
    4. Swim underwater until you are clear of the oil

43. If you must abandon ship into oily water that is not burning, which of the following precautions should you take?
    1. Use your life preserver as a raft
    2. Keep your face above the surface of the water as much as possible
    3. Both 1 and 2 above
    4. Swim underwater until you are clear of the oil

44. You can use which of the following items to help you stay afloat?
    1. Seabags
    2. Pillow cases
    3. Mattress covers
    4. All of the above

45. The Navy uses a maximum of how many types of life preservers?
    1. One
    2. Two
    3. Three
    4. Four

46. Which of the following benefits does the collar on the vest-type life preserver provide?
    1. Additional insulation against chill in cold water
    2. Additional buoyancy to keep the head upright
    3. A place to store survival equipment
    4. A place to attach a retrieving line

47. The wooden toggle and line of an inflatable life preserver are used to
    1. permit easy removal of the preserver
    2. make the preserver fit snugly around your body
    3. attach yourself to a life raft or another person
    4. provide a means for retrieving you out of the water

48. When using a pin-on, battery-operated light on a life preserver, you should replace the battery at what minimum interval?
    1. 18 months
    2. 12 months
    3. 3 months
    4. 6 months

49. You may launder the fibrous glass pads in addition to the outer cover of the inherently buoyant preserver.
    1. True
    2. False

50. In addition to inspecting your inflatable life preserver each time you wear it, you should also inspect it for air leaks at what minimum interval?
    1. Daily
    2. Weekly
    3. Monthly
    4. Quarterly

51. The survival kits in large lifeboats are designed to sustain 15 to 20 people on regular rations for what maximum number of days?

   1. 20
   2. 15
   3. 10
   4. 5

52. You can identify the red flare end of a Mk 13 Mod 0 distress signal kit in the dark by which of the following indicators?

   1. A metal pull ring
   2. Beadlike projections
   3. The absence of beadlike projections
   4. The absence of a metal pull ring

53. Under good weather conditions, the dye marker will retain some color for what maximum length of time?

   1. 1 hour
   2. 2 hours
   3. 3 hours
   4. 4 hours

54. In a lifeboat, what piece of survival equipment is provided to assist you in filling containers with freshwater?

   1. Rain catcher tube
   2. Rain cistern
   3. Rain bucket
   4. Funnel

55. In a lifeboat, continuous exposure to the elements will not harm which of the following pieces of survival or signal equipment?

   1. Sponges
   2. Knives
   3. Flashlights
   4. Signal mirrors

56. In a lifeboat, you may survive on as little as how much water a day?

   1. 10 ounces
   2. 8 ounces
   3. 6 ounces

57 People are known to live for 4 weeks or longer in a survival situation if a sufficient amount of water is available.

   1. True
   2. False

58. Which of the following forms of sea life or birds should NEVER be eaten?

   1. Sharks
   2. Jellyfish
   3. Seabirds
   4. Sea turtles

59. Which of the following sea birds is/are edible?

   1. Albatrosses
   2. Gannets
   3. Terns
   4. All of the above

60. At what minimum water temperature are you at risk for a serious condition called *hypothermia*?

   1. 75°F
   2. 80°F
   3. 85°F
   4. 95°F

61. What means should you use to treat frostbitten hands and fingers?

   1. Rub them
   2. Exercise them
   3. Place them in cold water
   4. Place them in contact with a warm part of your body

62. Assume that you have just fallen overboard. What is the most important survival technique for you to remember?

   1. Remain calm and try to stay afloat
   2. Swim after the ship and call for help
   3. Remove your shoes and other heavy clothing
   4. Keep moving your arms and feet for protection from sharks

63. Helicopters use a maximum of how many basic devices for recovering personnel in the water?

   1. One
   2. Two
   3. Three
   4. Four

64. If you fall overboard and sharks are in the area, you should take which of the following actions?

    1. Swim away from the area
    2. Assume the jellyfish position and try to remain motionless
    3. Float on your back and use as little arm and leg movement as possible
    4. Tread water and make wide sweeping movements with your arms to splash water

65. In a group survival situation, good leadership will lessen the effects of which of the following emotional states?

    1. Panic
    2. Confusion
    3. Disorganization
    4. All of the above

66. At least how many quarts of water are required each day to maintain your efficiency?

    1. 1
    2. 2
    3. 3
    4. 4

67. Liquids obtained from vines are undrinkable if they have which of the following characteristics?

    1. White sap
    2. Very dark in color
    3. Both 1 and 2 above
    4. Slightly pink color

68. In a survival situation with less than 1 quart of water per day, you should avoid eating what type of food?

    1. High-carbohydrate
    2. Highly flavored
    3. Excessively sweetened
    4. High-protein

69. Under survival conditions, you would obtain the most food value from which of the following sources?

    1. Nuts
    2. Tubers
    3. Insects
    4. Animal flesh

70. To kill any parasites scavenger birds such as buzzards and vultures might carry, you should boil the birds for what minimum length of time?

    1. 5 minutes
    2. 10 minutes
    3. 15 minutes
    4. 20 minutes

71. When selecting a route for evasion travel, you should always choose the easiest route.

    1. True
    2. False

72. Which of the following sources is best for determining directions under survival conditions?

    1. The position of the stars and the sun
    2. The growth of moss on trees and rocks
    3. The direction of movement of birds and animals
    4. The direction of water flow in streams and rivers

73. During evasion, if you can no longer proceed on your own because of illness, which of the following actions should you take?

    1. Seek help from friendly natives
    2. Display a white flag or other white object
    3. Surrender to enemy troops by walking toward them with raised arms
    4. Select a hiding place and stay there until you are well enough to travel

74. After evading the enemy and returning to an area with friendly forces, you should take which of the following actions?

    1. Fire your weapon three times and give your name
    2. Arouse their attention by shouting at them
    3. Display a white flag or other white object
    4. Try to get through their lines at night

75. You should give friendly frontline troops which, if any, of the following information about your evasion experiences?

    1. All information they request
    2. Immediate tactical information
    3. A description of the methods you used during evasion
    4. None of the above

ASSIGNMENT 11

Textbook Assignment: Chapter 16 "*Career and Education Information.*"

---

1. Which of the following information is contained in the Navy Goal Card?
   1. Navy Core Values
   2. Rating and advancement career information for each new recruit
   3. The Sailor's Creed
   4. All of the above

2. In fleet and shore stations, who is responsible for maintaining the two-page Goal Card?
   1. First term Sailors
   2. Second term Sailors
   3. Newly recruited Sailors

3. Which of the following goal-setting areas is included in the Pocket Goal Card?
   1. Navy core values and recruit training goals
   2. The Sailor's Creed and personal priorities
   3. DEP goals and fleet goals
   4. All of the above

4. Which of the following is a purpose of the Professional Development Board?
   1. To advise Sailors on the necessity of completing PQS
   2. To give Sailors a chance for greater responsibility
   3. Both 1 and 2 above

5. Which of the following personnel make up the Professional Development Board?
   1. Command master chief
   2. Command career counselor
   3. Educational service officer
   4. All of the above

6. What is the objective of the enlisted advancement system?
   1. To keep Sailors from stagnating
   2. To provide qualified petty officers to operate the Navy's ships, squadrons, and shore stations
   3. To train Sailors for fleet commands
   4. To provide equal opportunity to lower enlisted

7. The enlisted rating structure provides paths of advancement for personnel in paygrades E-1 through what maximum paygrade?
   1. E-9
   2. E-7
   3. E-6
   4. E-4

8. Which of the following is an apprenticeship designation?
   1. AT
   2. EM
   3. FN
   4. MM

9. Ratings are divided into how many categories?
   1. One
   2. Two
   3. Three
   4. Four

10. What is a general rating?
    1. An identification of special skills not related to any occupational field
    2. An identification of general skills within a broad occupational field
    3. An occupational field having different qualifications and duties
    4. A broad occupational field requiring the same general qualifications and includes similar duties

11. Which of the following statements defines a designated striker?
    1. Any Sailor in paygrade E-1 through E-3
    2. A Sailor in paygrade E-1 through E-3 who is technically qualified for a specific rating
    3. A Sailor who wants to become qualified for a specific rating
    4. Each of the above

12. What is the difference, if any, between naval standards and occupational standards?

    1. Naval standards only affect paygrades E-1 through E-3, while occupational standards affect all paygrades
    2. Naval standards are the technical standards for a particular paygrade, and occupational standards are the military requirements for a particular paygrade
    3. Naval standards are the military requirements for a particular paygrade, and occupational standards are technical standards for a particular paygrade
    4. None

13. To be eligible for advancement to E-2 or E-3, you must meet which of the following requirements?

    1. Have a certain time in rate
    2. Be recommended by your CO
    3. Complete *Basic Military Requirements*, NAVEDTRA 12018
    4. All of the above

14. What is the purpose for the *Bibliography for Advancement-in-Rate*?

    1. To train Sailors for advancement
    2. To help Sailors study for advancement-in-rate exams
    3. To show what publications are current
    4. To give answers to the tests

15. For you to become eligible for advancement, what person must recommend you?

    1. CO
    2. XO
    3. Division officer
    4. Division LCPO

16. You are an E-2 and are eligible for advancement. What person advances you?

    1. Secretary of the Navy
    2. Chief of Naval Personnel
    3. Commanding officer
    4. Executive officer

17. What factor limits the number of Sailors who can be advanced to petty officer?

    1. The number of vacancies that exist on board the ship
    2. The number of vacancies that exist in each rate and rating
    3. The ability of the Sailors to pass the advancement exam
    4. Each of the above

18. Which of the following are categories of the final multiple that determines which personnel are promoted to paygrades E-4 through E-6?

    1. Merit rating
    2. Personnel testing
    3. Experience
    4. All of the above

19. Which of the following factors are considered in the final multiple computation?

    1. Performance mark average and examination score
    2. Length of service and service in paygrade
    3. Awards and PNA credit
    4. All of the above

20. In the Navy, there are how many types of duty?

    1. One
    2. Two
    3. Three
    4. Four

21. Which of the following statements describes sea duty?

    1. Duty performed in commissioned vessels or activities home ported/home based in CONUS that operate away from the home port/home base in excess of 150 days per year
    2. Duty performed in overseas land-based activities that are credited as sea duty for rotational purposes
    3. Duty in activities normally designated as shore duty but that require members to be absent 100 to 150 days year

22. What form should you submit to indicate your duty preference?

    1. Special request
    2. Personnel requisition
    3. NAVPERS 1306/63
    4. NAVPERS 1170

23. You have just arrived at your first duty station. You should submit a duty preference form after what period of time?

    1. 1 month
    2. 3 months
    3. 6 months
    4. 12 months

24. What is the most significant personnel management tool in your service record?

    1. List of Navy schools
    2. The Evaluation Report and Counseling Record
    3. The Enlisted Duty Preference Form
    4. Commands attached

25. Which of the following is the **main** purpose of the Evaluation Report and Counseling Record?

    1. For continuation of service
    2. For assignment to special duties
    3. For BUPERS to use when making advancement-in-rate assignment decisions
    4. All of the above

26. What is meant if you are assigned a 3.0 on an evaluation?

    1. Your performance exceeds standards
    2. Your performance is above standard
    3. Your performance meets standards
    4. Your performance is progressing

27. How many traits are evaluated on the Evaluation Report and Counseling Record?

    1. Three
    2. Five
    3. Seven
    4. Nine

28. An evaluation for which of the following traits is NOT required for paygrades E-1 through E-3?

    1. Leadership
    2. Teamwork
    3. Personal job accomplishment
    4. Military bearing

29. At what interval are Evaluation Report and Counseling Records for E-3 and below submitted?

    1. Yearly only
    2. Yearly and when transferred
    3. Biyearly only
    4. Biyearly and when transferred

30. Once you have signed your Evaluation Report and Counseling Record, it is sent to what agency?

    1. CINCPAC/LANT FLT
    2. CINC
    3. BUPERS
    4. CNO

31. How many different forms are contained in the Enlisted Service Record?

    1. 7
    2. 9
    3. 13
    4. 15

32. Which of the following data is contained in Page 2 of your Enlisted Service Record?

    1. An application for dependency allowances
    2. An up-to-date record of emergency data
    3. Both 1 and 2 above
    4. Your civilian education

33. What page of your Enlisted Service Record contains information on your occupational training and awards you've received?

    1. Page 1
    2. Page 2
    3. Page 4

34. Where would you find information about your civilian education before you entered the Navy?

    1. Page 1
    2. Page 2
    3. Page 4

35. What type of documents require the CO's signature?

    1. Those that establish policy
    2. Those that deal with aspects of military justice
    3. Those required by law
    4. Each of the above

36. Which of the following is/are objective(s) of the 3-M Systems?
    1. To maintain equipment at maximum operating efficiency
    2. To reduce equipment downtime
    3. To reduce cost of maintenance in money and man-hours
    4. Each of the above

37. Which of the following is a requirement for qualifying for a particular watch station?
    1. Completing a PQS
    2. Completing a PAR
    3. Passing an advancement-in-rate exam
    4. Each of the above

38. What section of the PQS standards deals with the major working parts of an installation?
    1. Fundamentals
    2. Systems
    3. Watch Stations

39. What person/office is your point of contact for all the Navy's training and education programs?
    1. LCPO
    2. ESO
    3. XO
    4. CO

40. What type of training do you receive during daily operation and maintenance situations?
    1. OJT
    2. GMT
    3. NMT
    4. "A" school

41. What training is an important part of the Navy's leadership continuum?
    1. OJT
    2. GMT
    3. "A" school
    4. "C" school

42. What Navy school provides you with advanced skills and knowledge for a particular job or billet?
    1. Class "A" school
    2. Class "C" school
    3. Class "F" school
    4. Class "R" school

43. What publication contains the list of current training manuals?
    1. *Naval Occupational Standards List*
    2. *Catalog of Nonresident Training Courses*, NAVEDTRA 12061
    3. *Military Rate Training Guide*

44. What activity provides support to the voluntary education programs of all the military services?
    1. Tuition assistance
    2. Navy Campus
    3. DANTES
    4. EEAP

45. What program allows an enlisted person to complete a baccalaureate degree within 2 years while receiving full pay and allowances and be commissioned upon graduation?
    1. Enlisted Commissioning Program
    2. Naval Reserve Officer Program
    3. NROTC Scholarship Program
    4. BOOST Program

46. There are how many types of discharge?
    1. Five
    2. Four
    3. Three
    4. Two

47. If you meet the requirements for the Navy Good Conduct Medal, you also meet the requirements for
    1. reenlistment only
    2. overseas duty only
    3. commissioning programs only
    4. reenlistment, overseas duty, and commissioning programs

ASSIGNMENT 12

Textbook Assignment: Chapter 17 "*Financial Management and Stress Management.*"

---

1. How many types of pay may you receive?
   1. One
   2. Two
   3. Three
   4. Four

2. As a Sailor, what action must you take in order to get paid?
   1. Open a savings account only
   2. Open a checking account only
   3. Open a savings or a checking account

IN ANSWERING QUESTIONS 3 AND 4, SELECT THE TERM USED TO DEFINE THE QUESTION.

3. Pay you get for certain types of duty that are usually considered hazardous.
   1. Basic pay
   2. Incentive pay
   3. Special pay

4. The pay you get that's based on your paygrade and length of service.
   1. Basic pay
   2. Incentive pay
   3. Special pay

5. You are getting a selective reenlistment bonus. What type of pay are you receiving?
   1. Incentive pay
   2. Basic pay
   3. Special pay

6. What is an allowance?
   1. Money used to reimburse you for expenses necessary for you to do your job
   2. Money used to pay you for expenses unnecessary for you to do your job
   3. Money paid for services rendered
   4. Money paid for longevity

7. You are entitled to an annual clothing maintenance allowance after you have been on active duty for what length of time?
   1. 12 months
   2. 6 months
   3. 3 months
   4. 9 months

8. Which of the following offices can provide you information about the types of allowances to which you're entitled?
   1. Education services office
   2. Disbursing
   3. Personnel
   4. Both 2 and 3 above

9. How is your housing allowance shown on the leave and earnings statement (LES)?
   1. BAQ only
   2. VHA only
   3. BAQ and VHA
   4. BAH

10. An allotment is money you have withheld from your pay and paid directly to someone else. There are how many categories of authorized allotments?
    1. Six
    2. Five
    3. Three
    4. Four

11. What office should you notify if you think that you're being overpaid?
    1. Division
    2. Disbursing
    3. Admin
    4. ESO

12. By looking at your LES, you can find the amount of allowances you have earned.
    1. True
    2. False

13. Leave and liberty are times you're authorized to spend away from work and off duty. They are combined on the LES.

    1. True
    2. False

14. You earn a certain number of leave days each year you serve on active duty. What is the maximum number of days of leave you can earn in a year?

    1. 10
    2. 20
    3. 30
    4. 40

15. Regular liberty is usually granted as a 4-day period.

    1. True
    2. False

16. Under certain circumstances, what is the maximum number of days special liberty a CO can grant?

    1. 1 day
    2. 2 days
    3. 3 days
    4. 4 days

17. Which of the following types of leave is NOT charged to your earned, annual, or excess leave account?

    1. Authorized regular leave
    2. Convalescent leave
    3. Sick leave
    4. Recovery leave

18. What form should you use to request either regular or emergency leave?

    1. NAVCOMPT Form 3065
    2. NAVCOMPT Form 3180
    3. BUPERS Form 3065
    4. BUPERS Form 3180

19. The safest and most convenient way for you to keep track of your money is to open a checking account.

    1. True
    2. False

20. Which of the following is one way you can avoid bouncing a check?

    1. Only use debit cards
    2. Balance your checkbook
    3. Always pay cash
    4. Get a second job

21. Which of the following is/are types of voluntary allotments?

    1. Life insurance payments
    2. Mortgage payments
    3. Payment to family members
    4. All of the above

22. Which of the following is/are types of involuntary allotments?

    1. CFC pledges
    2. Purchase of U.S. savings bonds
    3. Garnishment of pay
    4. All of the above

23. Which of the following is the key to money management?

    1. Using a budget
    2. Using an ATM
    3. Using a checking account
    4. Each of the above

IN ANSWERING QUESTIONS 24 THROUGH 27, SELECT THE TERM USED TO DEFINE THE QUESTION.

24. The amount of money taken from pay for income taxes, Social Security, SGLI, and so forth.

    1. Allotments
    2. Deductions
    3. Fixed expenses
    4. Net income

25. The money taken from gross income to pay debts to the United States.

    1. Allotments
    2. Gross income
    3. Fixed expenses
    4. Net income

26. The money paid to a member after all deductions and allotments are paid.

    1. Deductions
    2. Fixed expenses
    3. Gross income
    4. Net income

27. Expenses that are the same each month.
    1. Allotments
    2. Deductions
    3. Fixed
    4. Net income

28. Of the following expenses, which one is a fixed expense?
    1. Clothes
    2. Rent
    3. Savings
    4. Food

29. You are planning a budget. What is the first thing for which you should plan?
    1. Clothes
    2. Rent
    3. Savings
    4. Food

30. According to the U.S. Department of Labor, approximately what percentage of your income should be budgeted for housing costs?
    1. 15%
    2. 20%
    3. 25%
    4. 30%

31. Credit is buying now and paying later at no extra cost.
    1. True
    2. False

32. What method, if any, can you use to find the total amount you will pay for a loan?
    1. Add the price of the purchase to the total amount of the loan
    2. Subtract the price of the purchase from the total amount you will pay for the loan
    3. None

33. Good credit is priceless for which of the following reasons?
    1. Buying a house
    2. In emergencies
    3. Making big purchases

34. Which of the following are principles of using credit?
    1. Don't use credit for splurging
    2. Make as large a down payment as possible
    3. Use credit to purchase goods that will last for a long time
    4. Each of the above

35. What is the maximum life insurance coverage under the Serviceman's Group Life Insurance (SGLI) program?
    1. $100,000
    2. $150,000
    3. $200,000
    4. $250,000

36. Who is responsible for the safety, health, and well-being of your family?
    1. Yourself
    2. The Navy
    3. Your spouse
    4. The government

37. What is the result of abusive behavior of Navy personnel?
    1. Destroyed lives
    2. Negative morale of the military unit
    3. Bad reputation of the military in the civilian community
    4. All of the above

38. What program, if any, was established to help families in distress?
    1. Case Review Committee (CRC)
    2. Family Advocacy Program (FAP)
    3. Family Advocacy Committee (FAC)
    4. None

39. Victims of spouse or child abuse can report incidents directly to which of the following persons/activities?
    1. FAO
    2. FSC
    3. Medical treatment center
    4. All of the above

40. Stress happens when there is an imbalance between the demands of our lives and the means we have to deal with those demands.
    1. True
    2. False

41. What are the three means we can use to deal with stress?
    1. Acceptance, attitude, and perspective
    2. Attitude, avoidance, and perspective
    3. Acceptance, avoidance, and perspective
    4. Acceptance, avoidance, and rejection

ASSIGNMENT 13

Textbook Assignment: Chapter 18 "*Surface Preservation.*"

---

1. Your department is responsible for cleaning all the areas listed on which of the following documents?

    1. The compartment checkoff list
    2. The Watch, Quarter, and Station Bill
    3. The Maintenance and Material Management System
    4. The Cleaning, Preservation, and Maintenance Bill

2. Cleaning gear is stocked and issued from what area?

    1. The supply department
    2. The first lieutenant's storeroom
    3. The aft deck storeroom
    4. The common gear locker

3. When using cleaning agents, such as detergents, you should take which of the following steps to ensure good cleaning results?

    1. Wetting and rinsing only
    2. Wetting, scrubbing, and rinsing
    3. Scrubbing and drying only
    4. Scrubbing, wetting, and drying

4. Which of the following is the definition of a *field day*?

    1. The day before an important personal inspection
    2. A period set aside for the maintenance of personal clothing
    3. A day designated by the captain for participation of all hands in organized sports
    4. A period when all hands thoroughly clean the ship inside and out

5. Field days accomplish which of the following functions?

    1. Reduce the dirt intake caused by operating equipment
    2. Aid in the preservation of the ship by extending paint life
    3. Improve the appearance and sanitary condition of the ship
    4. All of the above

6. Vinyl deck coverings should be given what care frequently?

    1. Scrubbed, waxed, and buffed with an electric buffing machine
    2. Clamped down, dried, and buffed with a buffer
    3. Buffed with scouring pads and mineral spirits
    4. Swabbed with detergent and waxed with self-polishing wax

7. What material is added to nonslip deck coverings to provide better footing?

    1. Pieces of sandpaper
    2. Small pebbles
    3. Pumice

8. What type of inspection is taking place when the ship or station is divided into sections?

    1. Cruise inspection
    2. Zone inspection
    3. Captain's inspection
    4. Shakedown inspection

9. Solvents should NEVER be used in unventilated spaces under any circumstances.

    1. True
    2. False

10. When using solvents, proper ventilation must be provided in which of the following areas?

    1. In voids only
    2. In exterior spaces only
    3. In interior spaces only
    4. Each of the above

11. Which of the following precautions will reduce the possibility of vapor buildup in an area?

    1. Wearing protective clothing, goggles, and gloves
    2. Keeping oxygen and first-aid equipment nearby
    3. Using extra fans for ventilation
    4. All of the above

12. When working with solvents in an enclosed space, you must take which of the following precautions?

    1. Wear an OBA at all times
    2. Always use the buddy system
    3. Ensure that the installed CO system is operational
    4. Have damage control personnel present to assist with spills

13. When using solvents, you must make sure that your personnel know the nearest escape route in case of fire and the location of the

    1. head
    2. nearest fire alarm
    3. roving patrol
    4. gas free engineer

14. Before starting a job that involves working with solvents, you should take which of the following precautions?

    1. Secure the roving patrol
    2. Have the area checked by the gas free engineer
    3. Obtain the ventilation plan for the space involved
    4. Both 2 and 3 above

15. If you are told by the gas free engineer that harmful vapors have increased to unsafe levels in the area in which your detail is working, you should take which of the following actions?

    1. Stop work immediately and clear the area until it is safe to return
    2. Notify damage control central and await guidance
    3. Evacuate the area after the second warning
    4. Have each member of the detail don an OBA

16. Solvents spilled on some types of tile may cause what type of problem?

    1. Radiation poisoning
    2. Lung irritation
    3. Skin disease

17. A spill must be reported if it presents a threat to the ship, the health of the crew, or involves more than what amount of solvent?

    1. 1 quart
    2. 1 gallon
    3. 5 quarts
    4. 5 gallons

18. A Sailor with a history of which of the following medical problems should not be permitted to work with paint, solvents, and thinners?

    1. Asthma
    2. Allergies
    3. Both 1 and 2 above
    4. Diabetes

19. If a solvent makes contact with a person's skin, the skin should be immediately flushed with which of the following liquids?

    1. Clear water
    2. Hydrogen peroxide
    3. Sodium chloride solution
    4. Sodium hydroxide solution

20. If someone has breathed vapors from solvents, which of the following actions should you take?

    1. Relieve the Sailor from the work detail
    2. Get the Sailor to a doctor as soon as possible
    3. Place the Sailor on report for unsafe work habits
    4. Make the Sailor wear an OBA for the remainder of the work detail

21. With reference to a self-contained breathing apparatus (SCBA), which of the following statements is correct?

    1. It contains a cartridge that may contain a chemical or carbon
    2. It filters out spray mist and absorbs vapors
    3. Both 1 and 2 above
    4. It is used in areas that lack oxygen

22. In addition to the possibility of fire, which of the following safety hazards may result if paint and solvent containers are not kept tightly closed?

    1. The area may become oxygen rich, causing light-headedness
    2. The paints and solvents may evaporate, causing loss of inventory
    3. The fumes and vapors may react with the deck tile, causing it to become slippery
    4. The oxygen in the area may be displaced, causing a shortage that will not sustain life

23. While inspecting containers of corrosive material being loaded aboard ship, you notice a dented can. What action should you take?

    1. Refuse to accept the damaged container
    2. Sign for the shipment but note the condition of the can on the bill of lading
    3. Sign for the material, but store the damaged container in a locker designed for flammable materials
    4. Circle the dent on the can with an international orange marker and use the can first

24. If you discover a leaking solvent container while inspecting your storage area, which of the following actions should you take?

    1. Check for the type of solvent by reading the contents label
    2. Immediately inform your supervisor
    3. Both 1 and 2 above
    4. Post a hazardous materials warning outside the area and notify the officer of the deck

25. When transferring solvents from one container to another, which of the following types of equipment should you use?

    1. An electric transfer pump using 110 volts ac or less
    2. An electric transfer pump using 28 volts dc or less
    3. A battery-operated transfer pump
    4. A standard Navy transfer pump

26. When at sea, which of the following methods is approved for handling and disposing of empty solvent containers?

    1. Retain all solvent containers onboard to facilitate transfer in the event of damage to a stored container
    2. Stow the containers in a disposal storage area until your next port of call; then dispose of them properly
    3. Puncture and discard the containers over the fantail once outside the 12-mile limit
    4. Puncture and discard the containers over the fantail once outside the 3-mile limit

27. You are inspecting stored paint and solvents. If you're unsure whether a large batch of paint is suitable for use, you should take what action?

    1. Return paint over 2 years old to the manufacturer
    2. Forward a sample to the nearest Navy testing laboratory
    3. Paint a test area and observe the results
    4. Use the paint

28. If you have which of the following symptoms, the compartment you're working in might have bad air?

1. Headache
2. Dizziness
3. Labored breathing
4. All of the above

29. While working in a closed space, a Sailor has lost consciousness. You should immediately enter the space to help your shipmate.

1. True
2. False

30. To find the storage requirements for solvents, you should refer to what document?

1. PMS
2. MRC
3. MSDS

31. Which of the following protective equipment should be worn when working with chlorinated cleaning solvents, organic cleaning solvents, and fluorocarbon refrigerants and solvents?

1. Neoprene gloves
2. Rubber gloves
3. Safety splash goggles
4. All of the above

32. Painting the exterior of an aircraft carrier requires about how many gallons of paint?

1. 950
2. 1,000
3. 1,250
4. 1,500

33. Paint consists of a total of how many essential ingredients?

1. One
2. Two
3. Three
4. Four

34. Paint pigment provides the coloring, rust prevention, and lasting quality of paint. Pigment is made from which of the following metals?

1. Lead
2. Zinc
3. Titanium
4. All of the above

35. What ingredient is the most common one used to make thinners?

1. Denatured alcohol
2. Mineral spirits
3. Linseed oil
4. Water

36. Which of the following petroleum products must NEVER be used to thin paint?

1. Kerosene
2. Diesel oil
3. Both 1 and 2 above
4. Mineral spirits

37. What minimum number of coats of primer should be used on a surface cleaned to bare metal?

1. One
2. Two
3. Three
4. Four

38. What is the minimum amount of drying time required between primer coats?

1. 24 hours
2. 12 hours
3. 8 hours
4. 4 hours

39. What color paint is used to paint the underside of deck overhangs?

1. White
2. Black
3. Haze gray
4. Deck gray

40. The deck of the machinery spaces is painted what color?

1. Haze gray
2. Deck gray
3. Dark red
4. Dark green

41. Feathering the paint edges of chipped or scraped areas should be done with which of the following tools?

1. Scraper
2. Sandpaper
3. Chipping hammer
4. Hand wire brush

42. What tool should you use to remove deeply embedded rust?

1. A power-operated heavy-duty wire brush
2. An electric disk sander
3. A rotary chipping tool
4. A pneumatic hammer

43. You would use a rotary scaling and chipping tool on which of the following jobs?

1. To chip a bulkhead
2. To chip up old tile
3. To chip a large deck area

44. The most important safety precaution to follow when using portable tools is to make sure they are properly grounded.

1. True
2. False

45. What is the first sign of galvanic corrosion on aluminum?

1. Disintegration of the rivets or bolts holding the aluminum
2. Loose rivets, screws, or bolts holding the aluminum
3. Pitting and scaling of the surface
4. Appearance of a white powdery residue

46. Which of the following tools should be used to chip painted aluminum surfaces?

1. Hand scraper
2. Wire brush
3. Sandpaper
4. Each of the above

47. If you have a cut on your hand, what precaution, if any, should you take when using paint remover?

1. Use a simple dressing
2. Put a bandage over the cut
3. Wear gloves
4. None

48. How many methods does the Navy use to apply paint?

1. One
2. Two
3. Three
4. Four

49. What are the two most useful brushes to use when painting?

1. Flat brush and fitch brush
2. Flat brush and sash tool brush
3. Sash tool brush and painter's dusting brush
4. Sash tool brush and fitch brush

50. What is meant by the terms laying on and laying off?

1. Double coat, going from left to right
2. Applying strokes first in long strokes in one direction and then crossing your first strokes
3. Spraying 10 inches away and then misting

51. What type of remover should you use to remove natural oil-based paints?

1. Turpentine
2. Water
3. Alcohol
4 Xylene

ASSIGNMENT 14

Textbook Assignment: Chapter 19 *"Safety and Hazardous Materials."*

---

1. When performing maintenance or upkeep on equipment or machinery, you would find applicable safety precautions in which of the following publications?

    1. Planned maintenance system (PMS) cards
    2. Operator's manuals
    3. Technical manuals
    4. Each of the above

2. Which of the following safety actions should you take?

    1. Observe all safety precautions
    2. Report unsafe conditions
    3. Warn others of hazards
    4. Each of the above

3. What type of information is contained in Material Safety Data Sheets?

    1. Supply codes
    2. Information about hazardous material
    3. Transfer dates
    4. The age of the materials

4. Which of the following is the major concern of Navy personnel aboard small boats?

    1. Crew safety
    2. Passenger safety
    3. Both 1 and 2 above
    4. Other vessels

5. When handling lines or taking part in underway replenishment, you must always wear what article of safety equipment?

    1. Inherently buoyant life jacket
    2. Inflatable life jacket
    3. Safety glasses
    4. Ear plugs

6. Which of the following hazards is/are associated with jet aircraft on flight decks?

    1. Being blown overboard
    2. Being burned by jet exhaust
    3. Being sucked into jet intakes
    4. Each of the above

7. Besides fuel and ammunition handling spaces, smoking is prohibited in which of the following areas?

    1. Flight deck
    2. Hangar deck
    3. Both 1 and 2 above
    4. Crew lounge

8. Lifelines are used for which of the following purposes?

    1. To prevent personnel from being washed overboard
    2. To provide a place to hang heavy weights
    3. To give personnel on deck a place to sit
    4. To provide a neater appearance

9. You should not paint scaffolding for which of the following reasons?

    1. Paint conceals defects
    2. Paint makes scaffolds slick
    3. Paint is too hard to keep clean
    4. Paint makes scaffolds too heavy

10. Which of the following safety items should you use when handling cargo?

    1. Safety shoes
    2. Hard hat
    3. Gloves
    4. All of the above

11. When using a hand truck to move loads on a ramp, you should move the hand truck in what way?

    1. Push the load up, pull the load down
    2. Pull the load up, push the load down
    3. Pull the load both up and down
    4. Push the load both up and down

12. When working aloft, you can receive a shock from which of the following pieces of gear?

    1. Ladders
    2. Guy wires
    3. Metal fittings
    4. All of the above

13. When working over the side, you must wear which of the following pieces of safety equipment?

    1. Inherently buoyant life jacket
    2. Inflatable life jacket
    3. Deck shoes
    4. Gloves

14. Accidents involving steam usually occur in what working spaces?

    1. Engine rooms only
    2. Firerooms only
    3. Engine rooms and firerooms
    4. Galley

15. A person overcome by carbon monoxide has which of the following symptoms?

    1. Sudden feeling of weakness
    2. Headache
    3. Drowsiness
    4. All of the above

16. If you are in a closed compartment and think you're being affected by carbon monoxide, you should take which of the following actions?

    1. Call for help
    2. Get to fresh air
    3. Both 1 and 2 above
    4. Stay where you are and relax

17. What person is authorized to certify that a closed space is safe to enter?

    1. Work center supervisor
    2. Gas free engineer
    3. Division officer
    4. Division chief

18. When using an internal combustion engine in a closed space for de-watering or fire fighting, you should take which of the following actions to ensure personal safety?

    1. Make sure the engine is clean
    2. Make sure the engine is fully fueled
    3. Make sure the exhaust is carried to the open atmosphere
    4. Make sure the exhaust is confined to one section of the compartment

19. During fueling operations, the word is passed "the use of open flame devices is prohibited." Which of the following devices is/are considered an open flame?

    1. Matches
    2. Lighted candles
    3. Cigarette lighters
    4. All of the above

20. Projectile-type ammunition that is 3 inches or greater in diameter can be identified by what method?

    1. Shape
    2. Color code
    3. Storage container
    4. Label

21. You should NEVER take which of the following items into a magazine?

    1. Naked lights
    2. Matches
    3. Both 1 and 2 above

22. Aboard ship, where are pyrotechnic materials usually stored?

    1. In interior passageways
    2. In machinery spaces
    3. In stowage spaces on topside decks
    4. In magazines

23. What person must approve the use of personal electrical equipment before you can use it aboard ship?

    1. The division officer
    2. The engineer officer
    3. The department head
    4. The division chief

24. Compartments used to store compressed gas cylinders should not be allowed to rise above what maximum temperature?

    1. 130°F
    2. 135°F
    3. 140°F
    4. 145°F

25. On noncargo ships, in what position should compressed gas cylinders be stored?

    1. Vertically, valve up
    2. Vertically, valve down
    3. Horizontally, valve up
    4. Horizontally, valve down

26. Compartments that contain compressed gases are ventilated for what length of time if ventilation has been secured?

    1. 5 minutes
    2. 10 minutes
    3. 15 minutes
    4. 20 minutes

27. Oxygen and chlorine cylinders may be stored in close proximity (near) to fuel or gas cylinders.

    1. True
    2. False

28. Only trained and medically qualified personnel are authorized to remove asbestos.

    1. True
    2. False

29. A tool is classified as power-driven if it has which of the following power sources?

    1. Pneumatic
    2. Hydraulic
    3. Electrical
    4. Each of the above

30. Personnel assigned to a fire watch during a welding operation must remain at their location for what minimum length of time after the job is completed?

    1. 10 minutes
    2. 20 minutes
    3. 30 minutes
    4. 40 minutes

31. When operating rotating machinery, you should never wear which of the following items?

    1. Jewelry
    2. Neckties
    3. Loose-fitting clothes
    4. All of the above

32. Compressed air can be used to clean disassembled machinery parts provided the pressure doesn't exceed how many pounds per square inch (psi)?

    1. 30 psi
    2. 45 psi
    3. 60 psi
    4. 75 psi

33. Sewage wastes contain bacteria and viruses. They can enter your body through which of the following means?

    1. Your nose
    2. Your mouth
    3. Open sores
    4. All of the above

34. You should not use liquid soaps or scented disinfectants to clean up spilled sewage for what reason?

    1. They cause too many suds
    2. They have poor cleaning characteristics
    3. They may temporarily disguise inadequate clean-up procedures

35. You shouldn't smoke around sewage-handling equipment for what reason?

    1. Germs found in the sewage can be inhaled
    2. Smoke adds to the odor
    3. Gases found around equipment and given off by sewage are explosive

36. Continuous exposure to high-level noises could cause which of the following kinds of hearing loss?

    1. Temporary
    2. Permanent
    3. Both 1 and 2 above

37. When working in machinery rooms and repair shops, you may be required to wear double-hearing protection.

    1. True
    2. False

38. When driving or riding in a Navy vehicle, you are required to wear seat belts.

    1. True
    2. False

39. Which of the following precautions should you follow when lifting heavy objects?

    1. Keep the load close to the center of your body
    2. Pull the load toward you; then lift gradually
    3. If too heavy to lift alone, get help
    4. All of the above

40. Heat stress is caused by which of the following factors?

    1. Workload
    2. Humidity
    3. Air temperature
    4. All of the above

41. Prolonged exposure to heat stress conditions causes which of the following medical emergencies?

    1. Heat stroke
    2. Heat exhaustion
    3. Both 1 and 2 above
    4. Euphoria

42. What is the major health risk to personnel who are exposed to severe cold weather?

    1. Snow blindness
    2. Hypothermia
    3. Sunburn
    4. Flu

43. Which of the following documents standardizes tag-out procedures aboard ship?

    1. NAVSHIPS 9890/3
    2. NAVSHIPS 9890/5
    3. OPNAVINST 3120.32
    4. OPNAVINST 4450.2

44. Under the tag-out procedures, what person has the authority to place a system off line for repairs or maintenance?

    1. Authorizing officer
    2. Repair activity rep
    3. Person attaching the tag
    4. Person checking the tag

45. How many different tags are authorized for use in identifying defective instruments or pieces of equipment?

    1. One
    2. Two
    3. Three
    4. Four

46. What color is used to identify a danger tag?

    1. Yellow
    2. Green
    3. Red

47. What color identifies a caution tag?

    1. Yellow
    2. Green
    3. Red

48. What means are used to control an entire tag-out procedure?

    1. DC fitting closure tag
    2. Tag-out logs
    3. Engineer boiler log
    4. First lieutenant's deck log

49. Which of the following publications contain information on Navy safety?

    1. OPNAVINST 4450.2
    2. OPNAVINST 5100.19
    3. NAVPERS 4450.2
    4. NAVPERS 5100.19

ASSIGNMENT 15

Textbook Assignment: Chapter 20 "*Sea Power*," Chapter 21 "*Leadership and Supervision*," and Chapter 22 "*Security Requirements and International Agreements*."

1. Sea power is a nation's ability to protect which of the following interests?
   1. Political
   2. Economic
   3. Military
   4. All of the above

2. What are the principle parts of sea power?
   1. Naval power, ocean science, ocean industry, and ocean commerce
   2. Ocean science, ocean industry, ocean commerce, and ASW warfare
   3. Ocean industry, ocean commerce, ocean science, and nuclear propulsion aircraft carriers
   4. Naval power, ocean industry, ocean commerce, and ballistic missiles

3. In peacetime, what does sea power encompass?
   1. Clash of fleets
   2. Commercial rivalries
   3. Diplomatic maneuvering
   4. Both 2 and 3 above

4. What person coined the phrase "sea power"?
   1. Secretary Alexander Hamilton
   2. John Paul Jones
   3. Admiral David Farragut
   4. Captain Alfred Thayer Mahan

5. Which of the following is a requirement for a nation to have sea power?
   1. Serviceable coastlines
   2. Favorable climate
   3. Abundant natural resources
   4. Each of the above

6. Immediately after the Civil War, the primary role of the U.S. Navy was to defend the coast and as a commerce raider.
   1. True
   2. False

7. During World War II, fewer battles were fought between ships within sight of each other for which of the following reasons?
   1. There were fewer ships in sea battles
   2. Submarines were usually used to fight battles
   3. The effects of aircraft, aircraft carriers, and radar began to emerge
   4. The convoy system kept enemy ships away from allied battle groups

8. Today, sea power involves which of the following industries?
   1. Marine science
   2. Maritime industry
   3. Both 1 and 2 above

9. The seas are our lifeline for survival. Which of the following factors make this a true statement?
   1. A barrier between nations
   2. A broad highway for ships
   3. A source for food, minerals, and metals
   4. All of the above

10. An economic advantage for a nation is to produce goods and services and to exchange them with other nations. Those that have failed in commerce have also failed as world powers.
    1. True
    2. False

11. What is the determining factor in the United States that changed our point of view about raw materials?
    1. Population growth and advanced technology
    2. Interrelationships between countries
    3. Growing isolationist policy
    4. Increased loss of farmland

12. What amount of minerals does the United States produces?

    1. 4 minerals
    2. 11 minerals
    3. 33 minerals
    4. 48 minerals

13. The United States acknowledges freedom of the seas under what law or treaty?

    1. Federal law
    2. Treaty of Versailles
    3. International law
    4. Treaty of Zurich

14. Which of the following actions must our country take to protect our national security and sustain our economy?

    1. Import raw materials, manufacture goods, and export goods to world marketplace
    2. Keep the sea lanes open
    3. Both 1 and 2 above

15. What states are outside the continental United States (CONUS)?

    1. New Mexico and Alaska
    2. Hawaii and Alaska
    3. New Mexico and Guam
    4. Alaska and Puerto Rico

16. There are a total of how many overseas U.S. territories?

    1. Two
    2. Three
    3. Four
    4. Five

17. The primary functions of the Navy and Marine Corps forces is to seek and destroy enemy naval forces, suppress enemy sea commerce gain, maintain general naval supremacy, control vital sea areas, and protect sea lines of communication.

    1. True
    2. False

18. When did the "tanker wars" occur in the Persian Gulf?

    1. 1986 to 1988
    2. 1987 to 1989
    3. 1988 to 1990
    4. 1989 to 1991

19. Operation Desert Shield/Desert Storm is an example of what type of exercise?

    1. Army operation only
    2. Air Force operation only
    3. Joint amphibious operations

20. What is meant by the term "hi-low balanced mix"?

    1. Speeding up research and development of new weapons
    2. Purchasing highly effective aircraft and ships and developing new classes of low-cost ships
    3. Laying up of old ships to save money
    4. Training personnel in high and low technology areas

A. NAVAL STRATEGY
B. NATIONAL INTERESTS
C. NATIONAL STRATEGY
D. NATIONAL OBJECTIVES

Figure A

IN ANSWERING QUESTIONS 21 THROUGH 23 SELECT THE TERM FROM FIGURE A THAT DESCRIBES THE QUESTION.

21. A broad course of action designed to achieve national objectives

    1. A
    2. B
    3. C
    4. D

22. Conditions that are to the advantage of our nation to pursue or protect.

    1. A
    2. B
    3. C
    4. D

23. Use of naval forces to achieve naval objectives.

    1. A
    2. B
    3. C
    4. D

24. Which of the following objectives would fall under the term "national objective"?

    1. Political
    2. Security
    3. Economic
    4. Each of the above

25. Sea control and power projection are the Navy's mission in support of

    1. naval strategy
    2. national interests
    3. national strategy
    4. naval objectives

26. Which of the following functions allows the Navy to control the sea and project power?

    1. Strategic nuclear deterrence
    2. Strong naval presence
    3. Security of the sea lines of communications
    4. Each of the above

27. On what date did Congress authorize the first six frigates of the Continental Navy?

    1. 27 Mar 1794
    2. 4 Feb 1776
    3. 20 Aug 1775
    4. 19 Jul 1773

28. In what year did Congress enact the Merchant Marine Act?

    1. 1916
    2. 1926
    3. 1936
    4. 1946

29. During World War II, the U.S. built and manned more than how many merchant ships?

    1. 3,000
    2. 4,500
    3. 6,000
    4. 7,500

30. In wartime, the Merchant Marine is responsible for which of the following missions?

    1. Transporting essential materials and cargo
    2. Resupplying allied military forces overseas
    3. Providing underway replenishment to Navy ships at sea
    4. All of the above

31. In peacetime, which of the following military services is not controlled by the Department of Defense?

    1. U.S. Navy
    2. U.S. Marine Corps
    3. U.S. Coast Guard
    4. U.S. Naval Reserve

32. The U.S. Coast Guard was established as the United States Revenue Marine in what year?

    1. 1760
    2. 1776
    3. 1785
    4. 1790

33. What is the traditional image of the U.S. Coast Guard?

    1. Watchful
    2. Vigilant
    3. Lifesaver
    4. Benefactor

34. Which of the following are modern-day Coast Guard duties?

    1. Enforcement of maritime laws and treaties
    2. Search and rescue operations
    3. Enforcement of drug and contraband laws
    4. Each of the above

35. During wartime, the Coast Guard operates directly under the

    1. Chief of Naval Operations
    2. Secretary of the Navy
    3. Joint Chiefs of Staff
    4. Secretary of Defense

36. Which of the following organizations was established by combining the sealift missions of the Naval and Army Transport services?

    1. Merchant Sea Transportation Service
    2. Department of Transportation
    3. Military Sealift Command
    4. U.S. Coast Guard

37. Military Sealift Command ships use which of the following titles?

    1. United States Naval Ships (USNS)
    2. United States Charter Ships (USCS)
    3. United States Coast Guard Ships (USCGS)
    4. United States Transportation Service Ships (USTSS)

38. In peacetime, the Military Sealift Command ships nearly what percentage of all military cargo on privately owned U.S. flagships and other merchant marine vessels?

1. 14%
2. 25%
3. 35%
4. 45%

39. What are the essential ingredients for U.S. sea power?

1. Merchant Marine, Military Sealift Command, Coast Guard, and the Navy
2. Navy, Marine Corps, Coast Guard, and the Military Sealift command
3. Merchant Marine, Coast Guard, Military Sealift Command, and the Marine Corps
4. Navy, Marine Corps, Coast Guard, and the Merchant Marine

40. Which of the following qualities should Navy leadership exhibit?

1. Administrative ability
2. Moral principals
3. Personal example
4. Each of the above

41. When followed, which of the following moral principals provides direction and consistency to leadership?

1. Integrity
2. Loyalty
3. Honesty
4. All of the above

42. To make sure an order to a job will get it done, orders need to be given so they can be followed. Orders should be given in what way?

1. Simple only
2. Clear only
3. Simple, clear, and complete
4. Complex

43. To be a good leader, you need to carry out your orders in which of the following ways?

1. Promptly
2. Cheerfully
3. To the best of your ability
4. Each of the above

44. Immediate obedience is an automatic response to a command.

1. True
2. False

45. Reasoned obedience lets you obey an order while learning from your experience while carrying it out.

1. True
2. False

46. What is the primary goal of the Continuous Improvement Program?

1. Increased productivity only
2. Produce better quality through leadership only
3. Increase productivity and produce better quality through leadership

47. How many security classifications does the Navy use to identify classified material?

1. One
2. Two
3. Three
4. Four

48. Which of the following security classifications is used for information or material that requires the highest degree of protection?

1. Top Secret
2. Secret
3. Confidential
4. For Official Use Only

49. Having a security clearance automatically grants you access to classified material.

1. True
2. False

50. To get a security clearance, you must be a United States citizen.

1. True
2. False

51. Which of the following infractions will cause a Sailor's CO to report that infraction to DON CAF?

1. Criminal conduct
2. General inaptitude
3. Noncompliance with security requirements
4. All of the above

52. Classified material is assigned a security classification for which of the following reasons?

    1. To ensure personnel are aware of the classified nature of the material
    2. To ensure the material receives the degree of protection required
    3. To assist in extracting, paraphrasing, downgrading, and declassifying actions
    4. All of the above

53. If a publication contains unclassified, FOUO, Confidential, Secret, and Top Secret information, what security classification is assigned?

    1. Top Secret
    2. Secret
    3. Confidential
    4. For Official Use Only

54. If you need to find the rules for transmitting classified material, you should refer to what SECNAV instruction?

    1. 5510.36
    2. 5510.30A
    3. 5510.3
    4. 5510.3A

55. Classified information is not transmitted over the telephone except when authorized on approved, secure communications circuits.

    1. True
    2. False

56. Which of the following is a concern of ADP security?

    1. Hardware
    2. Software
    3. Admin procedures
    4. All of the above

57. What term defines classified material that is lost, stolen, captured, salvaged, or seen by unauthorized personnel?

    1. Secure
    2. Abandoned
    3. Compromised

58. What type of communications is one of the least secure communications system?

    1. Registered U.S. mail
    2. Telephone
    3. U.S. mail
    4. Courier Service

59. What action, if any, should you take if you suspect someone you know is compromising classified material?

    1. Confront the individual
    2. Report it to the command security officer
    3. Report it to your CO through the chain of command
    4. None

60. Terrorists try to force governments or societies to take certain actions for political, religious, or ideological purposes.

    1. True
    2. False

61. The greatest publicity is given to which of the following terrorism methods?

    1. Taking hostages
    2. Bombing
    3. Both 1 and 2 above
    4. Sabotage

62. Which of the following threat conditions affords the highest degree of readiness?

    1. ALPHA
    2. BRAVO
    3. CHARLIE
    4. DELTA

63. The Status of Forces Agreement covers which of the following topics?

    1. Taxes
    2. Criminal jurisdiction
    3. Passport requirements
    4. All of the above

64. In what year did the Geneva Convention establish certain rights for prisoners of war?

    1. 1948
    2. 1949
    3. 1950
    4. 1951

65. The Law of Armed Conflict prohibits which of the following techniques or tactics?

1. Rape
2. Pillage
3. Plunder
4. All of the above

66. The Geneva Convention recognizes a prisoner's right to try to escape. Which of the following disciplinary actions may be taken when a prisoner is caught in an escape attempt?

1. Stoppage of extra privileges
2. Confinement
3. Both 1 and 2 above
4. Torture

www.ingramcontent.com/pod-product-compliance
Lightning Source LLC
LaVergne TN
LVHW082003200926
848481LV00010B/2793